ULTRAFAST LASER PROCESSING

Edited by
Koji Sugioka
Ya Cheng

ULTRAFAST LASER PROCESSING

From Micro- to Nanoscale

Pan Stanford Publishing

Published by

Pan Stanford Publishing Pte. Ltd.
Penthouse Level, Suntec Tower 3
8 Temasek Boulevard
Singapore 038988

Email: editorial@panstanford.com
Web: www.panstanford.com

British Library Cataloguing-in-Publication Data
A catalogue record for this book is available from the British Library.

Ultrafast Laser Processing: From Micro- to Nanoscale

ISBN 978-981-4267-33-5 (Hardcover)
ISBN 978-981-4303-69-9 (eBook)

Printed in the USA

Contents

Preface

The rapid development of ultrafast lasers (i.e., picosecond and femtosecond lasers) over the past few decades has opened up new avenues for materials processing that exploit the many advantages that such lasers have over conventional pulsed lasers (i.e., nanosecond lasers). The extremely short pulses of ultrafast lasers reduce the size of the heat-affected zone (HAZ) in processed regions, resulting in high-quality microfabrication of both soft materials and hard or brittle materials. The negligible HAZ can also give nanoscale spatial resolutions in fabrication. Another important advantage of ultrafast lasers is that they can generate extremely high peak powers. Focusing the laser beam produces sufficiently high peak intensities to induce efficient multiphoton absorption, even in transparent materials such as glass. This permits surface microstructuring and micromachining of transparent materials. Furthermore, by shifting the position of a tightly focused laser beam with a moderate pulse energy in a transparent material, multiphoton absorption can be confined to a region near the focus position, allowing internal modification and microfabrication of transparent materials. Ultrafast lasers have become common tools for micro- and nanoprocessing, and they are currently widely used in both fundamental research and practical applications. In fact, ultrafast lasers were used in over 60% of the studies presented at the 11th International Symposium on Laser Precision Microfabrication (LPM 2011, June 7–10, 2011, Takamatsu, Japan), one of the biggest and most important international conferences in the field of laser micro- and nanoprocessing.

Despite the rapid growth in this field, there are only a limited number of books that review ultrafast laser processing. We, thus, decided to edit a book that covers a broad range from fundamentals to scientific and industrial applications to provide comprehensive information on ultrafast laser processing. This book consists of 12 chapters that cover relevant topics in ultrafast laser processing, which have been reviewed by internationally recognized experts in the field. It includes an overview of ultrafast laser processing (Chapter 1), ultrafast laser systems and optics for materials processing (Chapter 2), fundamental mechanisms in the interaction of ultrafast laser beams with matter (Chapter 3), beam-shaping techniques for micro- and nanoprocessing (Chapter 4), surface patterning, drilling, cutting, micro- and nanostructuring, and nanoablation (Chapters 5–7), ultrafast laser-induced phenomena in transparent materials and internal modification (Chapter 8), applications of internal modification and microfabrication for fabricating three-dimensional (3D) photonic devices and biochips (Chapters 9 and 10), two-photon photopolymerization and 3D lithography and their applications (Chapter 11), and industrial applications (Chapter 12). Each chapter clearly presents the background, state-of-the-art techniques, and future prospects of the topic.

We believe that this book provides a realistic and comprehensive review of ultrafast laser processing and will be beneficial for students and young scientists who either are considering to work or have just started working in this area, as well as for researchers and engineers who are already working in this field.

Finally yet importantly, we would like to thank all the chapter authors for their great effort and wonderful work in writing informative chapters.

Koji Sugioka
Ya Cheng
May 2013

Chapter 1

Overview of Ultrafast Laser Processing

Koji Sugioka[a] **and Ya Cheng**[b]

[a] *Laser Technology Laboratory, RIKEN, 2-1 Hirosawa, Wako, Saitama 351-0198, Japan*

[b] *State Key Laboratory of High Field Laser Physics, Shanghai Institute of Optics and Fine Mechanics, Chinese Academy of Sciences, P.O. Box 800-211, Shanghai 201800, People's Republic of China*

ksugioka@riken.jp, ya.cheng@siom.ac.cn

Ultrafast lasers (i.e., picosecond and femtosecond lasers) have excellent characteristics for materials processing. Thus, ultrafast lasers are currently widely used for both fundamental research and practical applications. This chapter describes the features of ultrafast laser processing and then briefly reviews various processing techniques, including surface micromachining, micro- and nanostructuring, nanoablation, two-photon photopolymerization, internal modification of transparent materials, and biomedical and industrial applications.

Ultrafast Laser Processing: From Micro- to Nanoscale

Edited by Koji Sugioka and Ya Cheng

ISBN 978-981-4267-33-5 (Hardcover), 978-981-4303-69-9 (eBook)

www.panstanford.com

1.1 Introduction

Ultrafast lasers are defined as lasers that emit light pulses shorter than a few tens of picoseconds (ps) and they thus include femtosecond (fs) and ps lasers. Srinivasan *et al.* [1] and Küper and Stuke [2] pioneered materials processing using ultrafast lasers in 1987. They demonstrated clean ablation of polymethylmethacrylate (PMMA) with little formation of a heat-affected zone (HAZ) by using fs UV excimer lasers. They found that the ablation threshold was significantly lower than that for nanosecond (ns) laser ablation. Furthermore, it has been shown that ultrafast lasers can cleanly ablate even transparent materials such as NaCl and polytetrafluoroethylene (PTFE) by multiphoton absorption because they generate extremely high peak intensities [3, 4]. These experiments had a great impact and consequently research in this field expanded rapidly in the 1990s. Ultrafast lasers are currently becoming very common tools for laser materials processing both in fundamental investigations and for various applications. Such rapid evolution is due to the unique characteristics of ultrafast laser processing as well as the development of high-performance laser systems.

One important feature of ultrafast laser processing is that it reduces heat diffusion to surrounding regions of the processed area [5], resulting in high-quality microfabrication of soft materials such as biological tissues [6] and hard or brittle materials such as semiconductors and insulators [7] without HAZ formation. This suppression of heat diffusion to the surroundings also improves the spatial resolution of nanoscale processing [8]. Additionally, ultrafast laser irradiation at intensities near the ablation threshold forms nanoripple structures on various materials with periodicities much shorter than the wavelength [9–12]. Another important aspect of ultrafast laser processing is that, as already mentioned, nonlinear absorption (i.e., multiphoton absorption) can induce strong absorption even in materials that are transparent to the ultrafast laser beam [3, 4]. Multiphoton absorption permits not only surface modification but also three-dimensional (3D) internal microfabrication of transparent materials such as glass and polymers [13–16]. Additionally, multiphoton absorption has a spatial resolution that exceeds the diffraction limit due to its nonlinearity [17].

In addition, the rapid improvement in the performances of ultrafast laser systems has contributed significantly to the expansion of research on ultrafast laser processing. Femtosecond UV excimer lasers were used in the early stages of research on fs laser processing in the 1980s. The generation of chirped-pulse amplification (CPA) in Ti:sapphire regenerative amplifiers [18], which emit energetic fs pulses without inducing damage or undesirable nonlinear effects in the amplification medium, opened up new avenues for fundamental research on ultrafast laser processing in the 1990s. A robust, stable, and very compact fiber chirped pulse amplifier (FCPA) developed in the 2000s [19] increased research into applications. More recently, a rare-earth-doped laser medium was used to realize a compact and high-power ultrafast laser system by diode pumping, although the pulses were much broader than pulses generated by Ti: sapphire systems [20, 21]. Picosecond lasers based on this system that are suitable for industrial applications are now commercially available from some companies.

This chapter first describes the characteristics of ultrafast laser processing. It then reviews various processing techniques, including surface micromachining, surface micro- and nano-structuring, nanoablation, two-photon photopolymerization, internal modification of transparent materials for fabricating photonic devices and biomicrochips, and biomedical and industrial applications. Finally, a summary and outlook of ultrafast laser processing are given.

1.2 Characteristics of Ultrafast Laser Processing

1.2.1 Nonthermal Process

In contrast to ablation by ns and longer pulses in which thermal processes dominate, ablation by ultrafast laser pulses is by nonthermal processes that enable high-precision material processing to be realized. This characteristic is attributed to rapid energy deposition in the material. It takes a few hundred fs to a few ps for the electron distribution to reach thermal equilibrium after ultrafast laser irradiation [22, 23]. On the other hand, the energy transfer time from the electron subsystem to the lattice, which induces thermalization, is of the order of 1–100 ps, depending on the electron–phonon coupling strength of the material; this time is much longer than the time for the electrons to reach thermal equilibrium [24, 25]. Thus, an ultrafast laser can efficiently cause electron heating and generate a hot electron gas, which is far from equilibrium with the lattice. Consequently, only a very small fraction of the laser pulse energy is transmitted as heat; this results in nonthermal processing that enables high-precision microprocessing to be realized.

1.2.2 Suppression of Heat-Affected Zone

Even though ultrafast laser pulses mainly induce nonthermal processes (see Section 1.2.1), they may still generate heat. However, ultrafast laser pulses suppress the formation of a HAZ due to their extremely short pulse widths of several tens of fs to a few ps. This permits high-quality microfabrication, even for metals with high thermal conductivities.

Here, we compare the thermal diffusion lengths in a metal for irradiation by fs and ns pulses. When the pulse width of the laser is shorter than the electron–phonon coupling time in laser–matter interaction, most of the laser energy is absorbed by electrons and is transferred to the lattice very rapidly without wasting energy through thermal diffusion [5]. Therefore, thermal diffusion to the surrounding area of the laser-irradiated region can almost be ignored. For most metals, the electron–phonon coupling time is of the order of picoseconds [26], which is considerably longer than the pulse widths of ultrafast lasers. In this regime, the thermal diffusion length l_d when the metal is heated to near its melting point T_{im} by ultrafast laser irradiation is given by

$$l_d = \left[\frac{128}{\pi}\right]^{1/8}\left[\frac{DC_i}{T_{im}\gamma^2 C'_e}\right]^{1/4}, \tag{1.1}$$

where D is the heat conductivity, C_i is the lattice heat capacity, C'_e is given by $C'_e = C_e/T_e$ (where C_e is the electron heat capacity and T_e is the electron temperature), and γ is the electron–phonon coupling constant [27]. For example, when copper is heated to its melting point of T_{im} = 1356 K by ultrafast laser pulses, l_d is calculated to be 329 nm [28].

On the other hand, when the laser pulse width τ is much longer than the electron–phonon coupling time, l_d is approximately given by

$$l_d = \sqrt{\kappa\tau} \ . \tag{1.2}$$

Here, κ is the thermal diffusivity. For copper, l_d is estimated to be 1.5 μm for τ = 10 ns, which is a typical pulse width for a conventional nanosecond laser (e.g., an excimer laser). Thus, using an ultrafast laser can clearly reduce the thermal diffusion length, which means that it can reduce the HAZ at the processed region, resulting in higher-quality processing. However, thermal diffusion cannot be ignored when ultrafast lasers are used at high repetition rates due to heat accumulation (see Section 1.2.6).

1.2.3 Absence of Plasma Shielding

In laser ablation, an ablation plasma (plasma plume) is generated several hundred ps after the commencement of laser pulse irradiation [29]. Consequently, this plume shields the latter part of the laser pulse in ns laser ablation so that some of the laser pulse energy is not introduced into the material and is thus wasted. In contrast, pulse irradiation finishes before a plume develops when using ultrafast lasers; thus, the entire pulse energy can be absorbed by the material, enabling efficient processing.

1.2.4 Multiphoton Absorption

Ultrafast lasers can induce strong absorption even in materials that are transparent to the laser frequency due to nonlinear multiphoton absorption [3]. Figure 1.1 shows single and multiphoton absorption processes that induce electron excitation. Conventional absorption involves linear single-photon absorption. When light whose photon energy exceeds the bandgap of the material is incident on the material, it is absorbed by the material and an electron is excited from the valence band to the conduction band by a single photon. Light with a photon energy smaller than the bandgap cannot excite electrons, so that no absorption occurs in the stationary state. However, when an extremely high-density of photons (i.e., extremely high intensity light) is incident on the material, an electron can be excited by multiple photons even when the photon energy is smaller than the bandgap. This phenomenon is known as multiphoton absorption. An ultrafast laser can easily induce multiphoton absorption since it generates extremely high peak powers due to the ultrashort pulses. Thus, ultrafast lasers can induce strong absorption even in transparent materials, thereby enabling high-quality microprocessing of transparent materials such as glass.

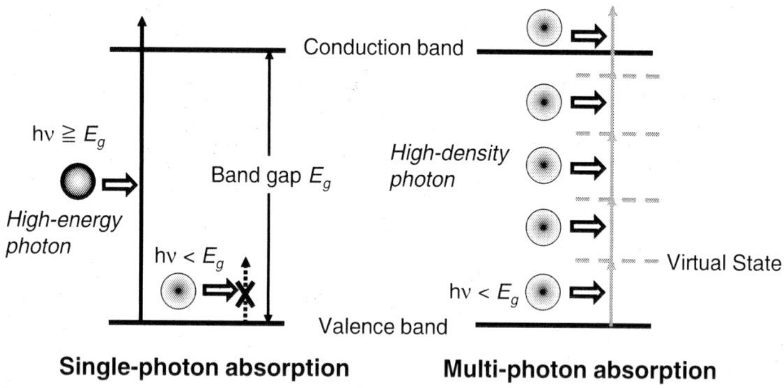

Figure 1.1 Electron excitation in a material by single and multiphoton absorption.

1.2.5 Internal Modification

Multiphoton absorption is a nonlinear process that can be efficiently induced only at intensities above a critical value, which depends on both the material and the pulse width. When an ultrafast laser beam with a sufficiently high pulse energy is focused by a lens into a transparent material (see Fig. 1.2), absorption is confined to a region

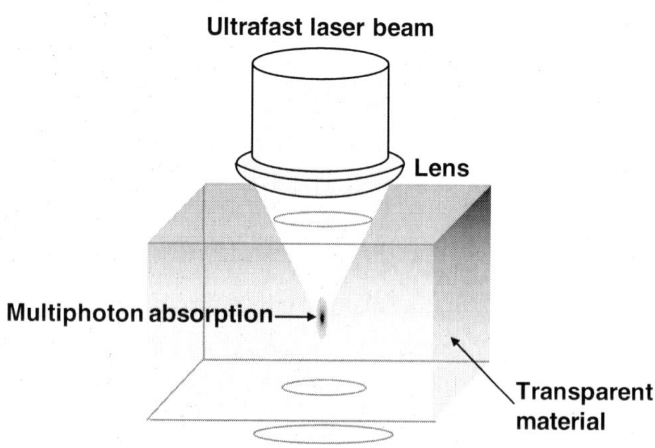

Figure 1.2 Schematic diagram of internal modification of a transparent material by multiphoton absorption using an ultrafast laser.

near the focal point inside the material where the laser intensity exceeds the critical value. Thus, internal modification of transparent materials can be performed using ultrafast laser pulses [13, 14]. Internal modification is widely used to write 3D optical waveguides and to fabricate micro-optical components and microfluidic channels buried inside glass (see Section 1.3.5).

1.2.6 Carrier Excitation in Dielectrics

Ultrafast laser irradiation induces the following electron excitation and relaxation processes in dielectrics such as glass [30]. Electrons are initially excited from the valence band to the conduction band by multiphoton absorption of the ultrafast laser light (multiphoton ionization), as described in Section 1.2.4, or by tunneling ionization when Keldysh parameter γ, which is determined by the laser electric field strength, wavelength, and the ionization potential of material, is much smaller than 1 ($\gamma \ll 1$). The excited electrons can absorb several laser photons sequentially so that they are excited to higher energy states where free carrier absorption is efficient. Otherwise, for sufficiently high laser intensities, the excited electrons are accelerated by the intense electric field of the ultrafast laser beam and collide with surrounding atoms, generating secondary electrons (avalanche ionization). Some of the generated free electrons relax to localize the energy stored in electron–hole pairs, which form self-trapped excitons (STEs). This relaxation often starts within 1 ps after the completion of laser irradiation. Some STEs relax to form permanent defects within a few hundred ps. Glass heating also occurs a few tens of ps after laser irradiation and the irradiated area returns to room temperature after several tens of μs, causing modification or damage. When ultrafast laser pulses were focused above the critical intensity in glass, melting was observed due to heating [31, 32]. This melting has been used to weld glass substrates [33–35]. Additionally, operating the laser at repetition rates higher than a few hundred kHz results in significant heat accumulation in a controllable manner, which is beneficial for writing low-loss optical waveguides with circular cross-sections [31, 32].

1.2.7 Spatial Resolution of Ultrafast Laser Processing

As discussed in Section 1.2.2, ultrafast laser pulses suppress heat diffusion to the surrounding regions of the processed area, which improves the spatial resolutions for fabrication. When a 10 ns laser pulse is irradiated onto Cu with a spot size the same as the laser wavelength (typically several hundred nm to 1 μm), the processed region is larger than the spot size since the thermal diffusion length is 1.5 μm. In contrast, since thermal diffusion can be almost ignored for ultrafast laser irradiation, the processed region is unlikely to extend beyond the spot size.

Using nonlinear multiphoton absorption can further enhance the spatial resolution as compared with the single-photon absorption at the same wavelength. Ideally, the intensity of an ultrafast laser beam will have a Gaussian spatial profile, as indicated by the thick dashed line in Fig. 1.3. For single-photon absorption, the spatial distribution of the laser energy absorbed by the material corresponds to this beam profile. However, for multiphoton absorption, the distribution of the absorbed energy will become narrower as the order (n) of the multiphoton absorption increases since the effective absorption coefficient for n-photon absorption is proportional to the nth power of the laser intensity. Therefore, the effective beam size ω for n-photon absorption is expressed by

$$\omega = \frac{\omega_0}{\sqrt{n}}, \tag{1.3}$$

where ω_0 is the actual spot size of the focused laser beam. Figure 1.3 shows the spatial distributions of the laser energy absorbed by

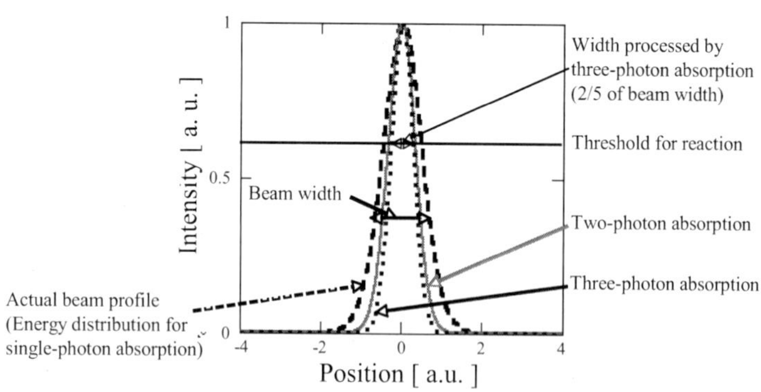

Figure 1.3 Actual beam profile (thick dashed line) and spatial distributions of laser energy absorbed by transparent materials by two- (solid line) and three-photon (thin dashed line) absorption. The solid horizontal line indicates the reaction threshold.

a transparent material for two- (solid line) and three-photon (thin dashed line) absorption. Based on Eq. (1.3), a spatial resolution much smaller than the wavelength is expected for multiphoton absorption. In addition, when there is a threshold laser intensity, above which a reaction occurs on absorption, the fabrication resolution can be further improved by adjusting the laser intensity. For example, if the laser energy is adjusted to match the straight solid line in Fig. 1.3 with the threshold intensity for the reaction, the fabrication resolution can be reduced to $(2/5)\times\ \omega_0$. Thus, nonlinear multiphoton absorption can realize a sub-diffraction-limited resolution [17].

1.3 Ultrafast Laser Materials Processing

1.3.1 Surface Micromachining

As mentioned in Sections 1.2.1 and 1.2.2, ultrafast lasers perform nonthermal ablation and thus reduce HAZ formation. Figures 1.4(a) and (b) show scanning electron microscopy (SEM) images of holes drilled in 100 μm-thick steel foils by ablation using laser pulses with widths of 200 fs and 3.3 ns, respectively [36]. Femtosecond laser ablation produces an ablated hole with a sharp edge and a steep wall and exhibits little indication of HAZ. In contrast, ns laser ablation produces large molten regions around the ablated hole.

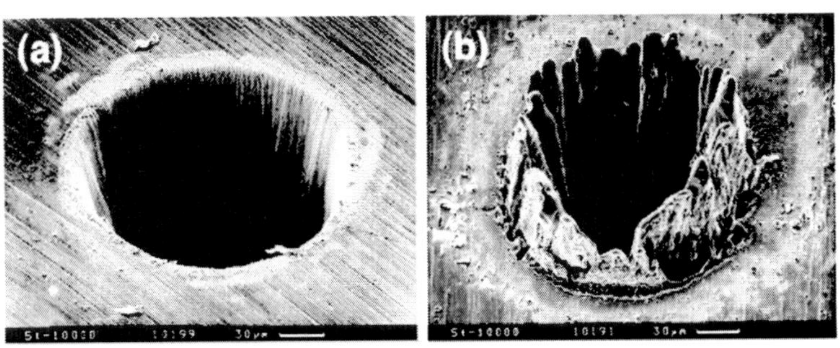

Figure 1.4 Holes drilled in 100 μm-thick steel foils by ablation using laser pulses with the following parameters: (a) pulse width: 200 fs, pulse energy: 120 μJ, fluence: 0.5 J/cm^2, wavelength: 780 nm; and (b) pulse width: 3.3 ns, pulse energy: 1 mJ, fluence: 4.2 J/cm^2, wavelength: 780 nm. The scale bars represent 30 μm. Courtesy of A. Ostendorf.

Ultrafast lasers can perform high-quality micromachining of even brittle materials such as glass. Figures 1.5(a) and (b), respectively, show SEM images of surface micromachining and cutting of glass materials by fs laser ablation. Both images reveal that clean ablation with sharp edges and without cracks was achieved.

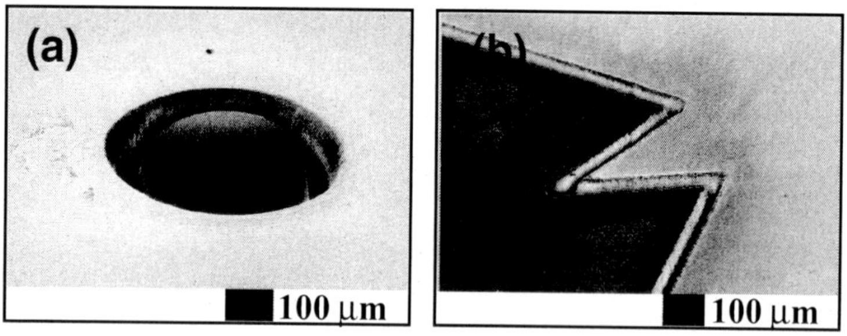

Figure 1.5 SEM images of (a) surface micromachining and (b) cutting of glass materials by fs laser ablation. Courtesy of M. Gower.

One application of ultrafast laser micromachining is to drill inkjet nozzles of specific shapes. Panasonic US used picosecond lasers for the first time to mass produce funnel-shaped inkjet nozzles in stainless steel [37]. Direct writing by a ps laser with a PC-controlled scanning mirror enabled such holes to be fabricated reproducibly, precisely, and rapidly.

Another application is the fabrication of coronary stents, which are used in minimally invasive treatment of arteriosclerosis as an alternative to bypass operations. Stents are typically fabricated from stainless steel or shape memory alloys. It is necessary to chemical post-process these materials to realize the required properties. In addition, they have some medical problems such as the risk of restenosis and limited biocompatibility. To achieve better biocompatibility, Mg-based alloys or special biopolymers should be used. However, these materials also suffer from problems in stent production; specifically, they require a poorly established post-processing technique and they react strongly to thermal loads. Ultrafast laser micromachining can overcome these problems since it is a nonthermal process and produces little debris. Figure 1.6 shows a fabricated prototype medical stent made from a bioresorbable polymer by fs laser ablation with no post-processing [38]. Meanwhile, the use of tantalum would ensure better X-ray visibility. Both soft materials and high melting point metals such as tantalum can be machined (see Fig. 5.20 in Chapter 5). Thus, femtosecond laser micromachining is suitable for processing these functional materials.

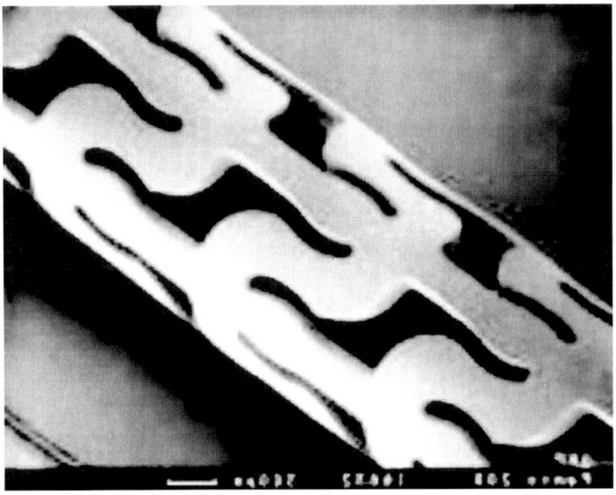

Figure 1.6 Prototype medical stent made from a bioresorbable polymer. Courtesy of A. Ostendorf.

Surface micromachining and patterning by ultrafast lasers are considered in more detail in Chapter 5.

1.3.2 Surface Micro- and Nanostructuring

Surface texturing of various materials is a crucial research topic in ultrafast laser processing. Various micro- and nano-scale textures can be formed by controlling the processing parameters such as the beam intensity, the spatial and temporal beam profiles, the wavelength, the polarization, and the processing environment (ambient gas or liquid).

The most well-known textured structure is nanoripples, which are formed by ultrafast laser irradiation with a fluence near the ablation threshold. It has long been known that periodic grating structures (i.e., laser-induced periodic surface structures: LIPSS) are formed by irradiating materials with ns or longer pulses [39]. Such structures are thought to be caused by the incident laser light interfering with the reflected (scattered) light. Consequently, the ripples are generally oriented perpendicular to the incident polarization. The spacing of the fabricated structure is given by $\lambda/n(1 \pm \sin\theta)$, where λ is the laser wavelength, θ is the laser incident angle, and n is the refractive index of the material [40, 41]. Therefore, the spacing is always of the order of the wavelength or greater. In contrast, periodic grating structures formed by fs laser irradiation have a spacing much smaller than the laser wavelength [42]. Nanoscale periodic grating structures (nanoripples) can be formed not only on the surfaces of various materials, including metals [43], ceramics [11, 44], semiconductors [45], and insulators [9, 46], but also inside transparent materials such as glass [9, 47]. Figure 1.7 shows the laser fluence dependence of the spacing of nanoripples formed on Cu by irradiation by 100 fs laser pulses with a wavelength of 800 nm [48]. The grating structures formed on the Cu surface are oriented perpendicular to the laser polarization irrespective of the laser fluence. For fluences near the low ablation threshold (~0.04 J/cm^2), grating structures have a spacing of 300 nm, which is much narrower than the laser wavelength of 800 nm. The spacing increases as the laser fluence is increased up to 2 J/cm^2, but it remains narrower than the wavelength. The surface nanotexturing produced by fs laser irradiation can be applied to reduce the friction between moving compo-

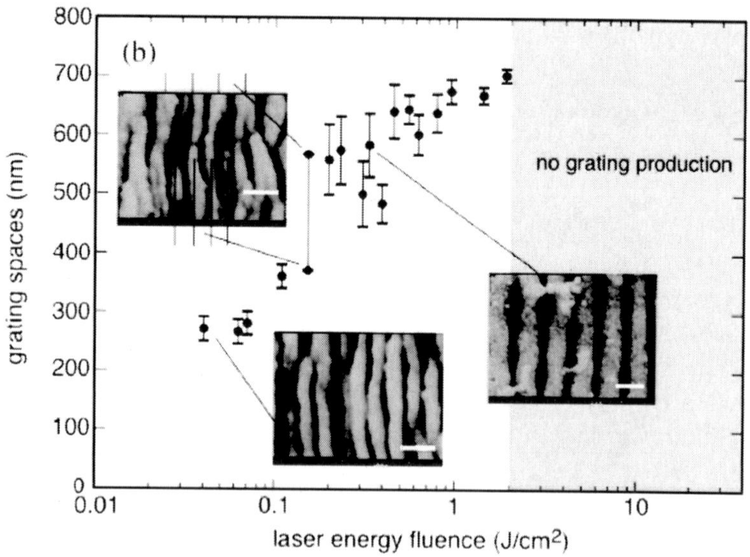

Figure 1.7 Laser fluence dependence of spacing of nanoripples formed on Cu by irradiating 100 fs laser pulses with wavelength of 800 nm (scale bars: 660 nm). Courtesy of M. Hashida.

nents, reduce the adhesive force of micro- and nano-components, and increase the adhesion of thin films and medical implants, etc. Other interesting and useful textures formed by ultrafast laser irradiation are regular arrays of conical microstructures, which can be produced on Si by irradiating it in a halogen atmosphere (e.g., SF_6 or Cl_2) with hundreds of fs laser pulses [49, 50]. The produced structures strongly reduce incident light reflection, giving rise to so-called black silicon, and greatly increase absorption even in the infrared (IR) region [51, 52]. This technique is effective for enhancing the efficiency of photovoltaic solar cells [53]. Meanwhile, coating structured surfaces with a layer of silane molecules produces superhydrophobic surfaces due to the lotus effect, which can be used for producing self-cleaning products [50, 54]. Similar structures can be created in other materials [55].

Chapter 6 describes surface micro- and nano-structuring using ultrafast lasers in more detail.

1.3.3 Nanoablation

The excellent characteristic of ultrafast laser processing of reducing the HAZ enables nanoablation to be performed with subwavelength resolution or smaller [56, 57]. Figure 1.8 shows nanohole arrays formed on a GaN surface by fs laser ablation based on two-photon absorption. The second harmonic of a near-IR fs laser (λ = 387 nm; pulse width: 150 fs) was focused on a single-crystal GaN substrate by an objective lens with a numerical aperture (NA) of 0.9. The ablated samples were post-treated in a hydrochloric acid (HCl) solution to remove residual debris. Ablation craters with diameters as small as 200 nm were formed. The diameters are significantly smaller than the laser wavelength due to the nonlinearity of the reaction in multiphoton absorption (see Fig. 1.3) in addition to the negligible HAZ.

Furthermore, new nanoscale structures (e.g., nanobumps, nanobelts, and nanomeshes) have been formed on Au thin films deposited on silica glass by multibeam interference of light from an fs laser [8, 58, 59]. In this scheme, it is essential to suppress heat diffusion to surrounding areas from the processed region. A demagnification optical system with a transmission beam splitter was used to split and correlate fs laser beams to realize multibeam interference. By carefully adjusting the processing parameters such as the laser flu-

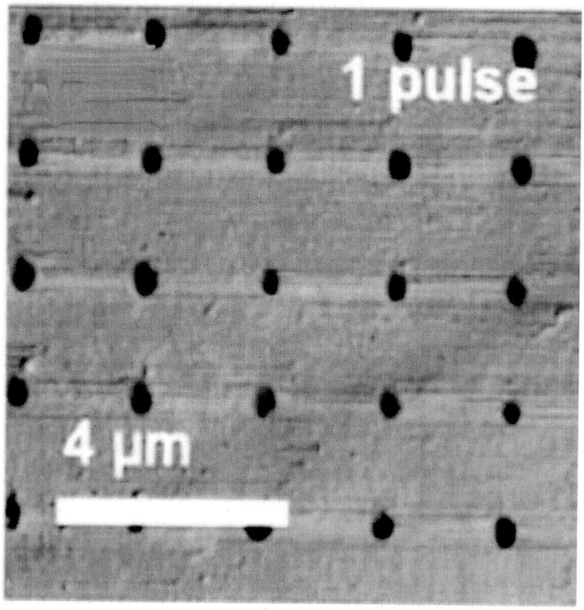

Figure 1.8 2D nanohole arrays on a GaN surface fabricated by fs laser ablation. Wavelength: 387 nm; pulse width: 150 fs; pulse energy: 10 nJ; NA: 0.9.

ence, the interference pattern periods, the Au film thickness, and the substrate material, unique nanostructures could be formed (see Fig. 1.9). Irradiation of a 10 nm-thick gold thin film on a sapphire (0001) substrate by a single pulse from a four-beam-interfering fs laser (130 fs, 790 nm) generated nano water drops (Figs. 1.9(a) and (b)) [60]. This interference pattern had a period of 1.3 µm. The structure was approximately 800 nm high and the bead and neck radii were 75 and 18 nm, respectively. Using a wider-period interference pattern (3.9 µm) and a silicon (100) substrate, which has a higher thermal conductivity than sapphire and is opaque to the laser wavelength, resulted in the formation of nano water crowns (Figs. 1.9(c) and (d)) [61]. On the other hand, for a period of 1.7 µm and a sapphire (0001) substrate, nanowhiskers form on the bump with a sharp top and a curvature radius smaller than 15 nm (Figs. 1.9(e) and (f)) [62]. The spiky structures shown in Fig. 1.9 could generate field enhancements and are promising for use in nanotechnology applications. Despite ultrafast laser processing being usually regarded as a non-thermal technique in which thermal effects are minimized, thermal processes (e.g., melting, flow, bending, and inflation) are responsible for generating these unique nanostructures.

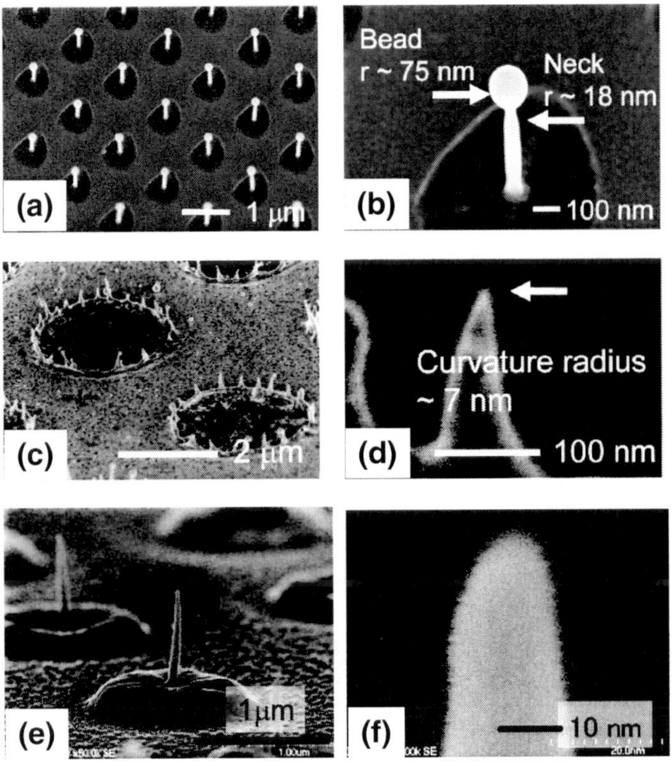

Figure 1.9 Nanostructures generated on gold thin films. (a) Array of nano water drops and (b) magnified image of a nano water drop on a 10 nm-thick film on a sapphire (0001) substrate. (c) Array of nanocrowns and (d) a spike of a nanocrown on a 50 nm-thick thin film on a silicon (100) substrate. (e) Nanowhisker and (f) a magnified image of a nanowhisker on a 50 nm-thick film on a sapphire (0001) substrate. Courtesy of Y. Nakata.

Another approach for nanofabrication is to develop novel irradiation methods that can overcome the diffraction limit of a focused laser beam by using optical near fields. Combining an fs laser beam with nanotips in scanning-probe microscopes such as scanning near-field optical microscopes (SNOM), scanning tunneling microscopes (STM), and atomic force microscopes (AFM) permits patterning with a nanoscale resolution [63, 64]. Recently, nanoablation using metal and dielectric nanospheres has been extensively investigated [65, 66]. The formation mechanism for subwavelength structures is ablation by the enhanced near field near the nanospheres. This method can be used to fabricate nanoholes on various material surfaces, including semiconductors, metals, and dielectrics. Chapter 7 describes nanoablation by optical near fields enhanced by nanochips and nanospheres in more detail.

1.3.4 Two-Photon Photopolymerization

Rapid prototyping of 3D structures by laser stereolithography (i.e., 3D photopolymerization) is commercially used for design verification, fabricating working models, functional and performance testing, fabricating direct injection molding dies, and fabricating medical models [67]. In this process, a focused UV laser beam (typically from a He–Cd laser) is scanned in the plane of the first layer on an elevator stage in photocurable epoxy resin. The epoxy resin is initially liquid, but it solidifies on laser irradiation due to single-photon absorption. The elevator stage is then shifted down and the focused laser beam is scanned in the plane of the second layer. By repeating this process layer by layer, a 3D structure is fabricated.

If a near-IR fs laser is used as the light source for stereolithography, 3D structures can be formed directly without shifting the elevator stage due to two-photon absorption (see Fig. 1.10). Stereolithography performed using a near-IR fs laser is termed two-photon photopolymerization (TPP). An important characteristic of TPP is sub-wavelength spatial resolution (see Fig. 1.3). Kawata *et al.* fabricated the smallest bull in the world by TPP, as acknowledged by the Guinness Book of Records [17]. The width spatial resolution was 120 nm in this case.

A solid resist can replace epoxy resin for direct fabrication of 3D microstructures. Perry's group has fabricated complex 3D microstructures and topologically complex structures using an organic photoactive material (bis-donor phenylene vinylene) by two-photon absorption using an fs laser [68–70]. The fabricated microstructures can be filled or coated with other materials. Unlike TPP, this process can perform not only bottom-up fabrication but also top-down fabrication by selecting a suitable photoresist (negative- or positive resist). For example, a positive tone resist is suitable for fabricating a hollow microchannel embedded in a material.

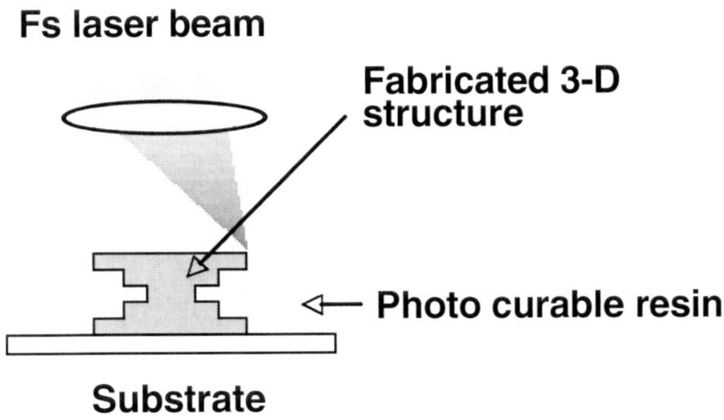

Figure 1.10 Schematic illustration of process for directly fabricating 3D microstructures by two-photon photopolymerization using a near-IR fs laser and a photocurable resin.

Two-photon photopolymerization and photolithography using fs lasers are currently being extensively investigated for fabricating photonic crystals [68, 71, 72], micro- and nano-systems [73–75], and lab-on-a-chip (LOC) devices [76, 77] and for medical and tissue engineering [78, 79]. Figure 1.11 shows examples of 3D microstructures fabricated by TPP for medical and tissue engineering. This technique typically has a resolution of about 100 nm; its resolution can be further improved to 25 nm by very carefully adjusting the laser power and the scanning speed [80].

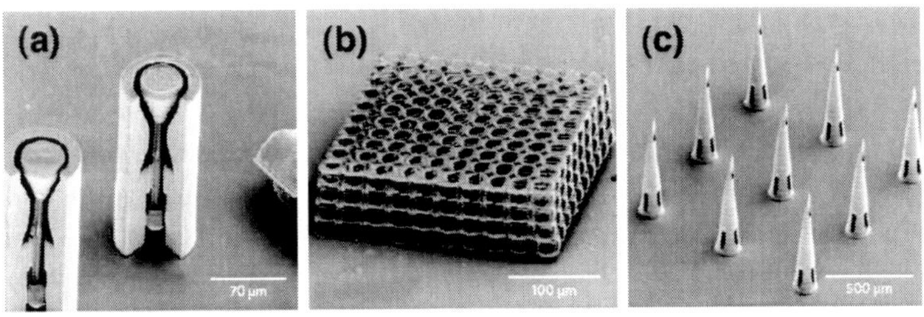

Figure 1.11 Three-dimensional microstructures fabricated by TPP. (a) Microvalves designed to prevent reversal of blood flow in human veins. Only part of the valve cover was constructed to enable the interior to be visualized. (b) Scanning electron microscopy image of high-porosity tissue engineering scaffold. (c) Test microneedle array for transdermal drug delivery. Courtesy of B. Chichkov.

Electrically conductive 3D metal microstructures can be fabricated by using aqueous solutions containing metal ions (e.g., aqueous solutions of silver nitrate ($AgNO_3$) for silver deposition and tetrachloroauric acid ($HAuCl_4$) for gold deposition) instead of a photocurable resin [81]. This process is based on two-photon-induced metal ion reduction in the aqueous solution. Such 3D metal microstructures are used for fabricating microelectromechanical systems (MEMS), LOC, metamaterial devices, etc.

Chapter 11 explains in more detail 3D microstructure fabrication by TPP and its application.

1.3.5 Internal Modification of Transparent Materials

Ultrafast lasers can be used to internally modify glass, as described in Section 1.2.5. In 1996, Davis *et al.* caused a permanent change of the refractive index, which was applied to optical waveguide writing inside glass by an ultrafast laser [13]. Currently, many researchers are investigating writing of optical waveguides embedded in various glasses, including fused silica [31, 82], borosilicate glass [31, 83], and chalcogenide glass [84, 85]. Written optical waveguides have core diameters (whose refractive index is increased by 10^{-4} to 10^{-2}) ranging from 2 to 25 μm depending on the writing parameters used. An optical waveguide written in fused silica was estimated to have a propagation loss as low as ~0.2 dB/cm at a wavelength of 633 nm [86]. Optical waveguides can also be written in transparent polymers [15, 16]. Optical waveguides written in the UV-transparent polymer, CYTOP, can propagate UV light with propagation losses of 0.77 and 0.91 dB/cm at wavelengths of 355 and 266 nm, respectively [16].

Refractive index modification was used to fabricate 3D optical microdevices such as optical couplers and splitters [87], volume Bragg gratings [88], diffractive lenses [89], and lasers [90, 91]. As one example, optical waveguides written in active materials (e.g., erbium–ytterbium-doped phosphate glass) by an ultrafast laser function as compact and efficient single-longitudinal-mode lasers at a wavelength of 1.5 μm and they have a maximum output power of up to 55 mW [90]. By encoding refractive index-modulated volume gratings in the written optical waveguide using a single interfered fs laser pulse, a distributed feedback laser was realized in lithium fluoride [91]. Chapter 9 describes fabrication of 3D photonic devices by internal modification using ultrafast lasers in more detail.

Multiphoton absorption by ultrafast lasers can space-selectively induce other interesting phenomena besides refractive index modification, including valence state change of doped ions [92], elemental redistribution in materials [93], atom precipitation [94], and particle crystallization [95] (see Chapter 8).

Multiphoton absorption by ultrafast lasers can also modify the chemical properties of laser-irradiated regions inside materials. Ultrafast laser direct writing followed by chemical wet etching in dilute HF acid produced 3D hollow microstructures in fused silica, including 3D microfluidic channels with diameters as small as 10 μm and with any angle between the channels and a high aspect ratio [96]. Lithium aluminosilicate glass doped with trace amounts of silver and cerium, which is a photosensitive glass, is a more attractive material for this application as it permits high efficiency, high throughput processing and much smoother etched surfaces, although thermal treatment is required before wet etching [97, 98]. This technique can be applied to fabricate microoptics based on hollow structures inside the glass, such as micromirrors [99] and microlenses [100], as well as fabricating 3D microfluidics. Furthermore, fabricated 3D microstructures can be easily integrated with other optical elements such as optical waveguides and optical filters written by the ultrafast lasers, enabling optofluidic and LOC devices to be realized for chemical and biological applications [101–103]. Figure 1.12(a) shows a schematic illustration of 3D microfluidics integrated with optical waveguides in photosensitive glass fabricated by fs laser direct writing; it can be used for quantitatively analyzing CO_2 concentration in water. For analysis, the microfluidic channel is filled with liquid samples containing bromothymol blue (BTB) solution and white light is then coupled to a facet of waveguide I. White light transmitted by waveguide I propagates across the microfluidic channel, which is filled with a liquid sample, and is then coupled into optical waveguide II. The detector head of the spectrometer is placed at the end facet of waveguide II to measure the spectrum of white light transmitted by waveguide II. The green line in Fig. 1.12(b) indicates the absorption spectrum of pure water containing BTB solution. Carbonic water samples with different CO_2 concentrations produce different absorption spectra due to the different pH. The optical absorption spectrum of pure water containing BTB solution exhibits a high absorbance at a wavelength of about 620 nm. This

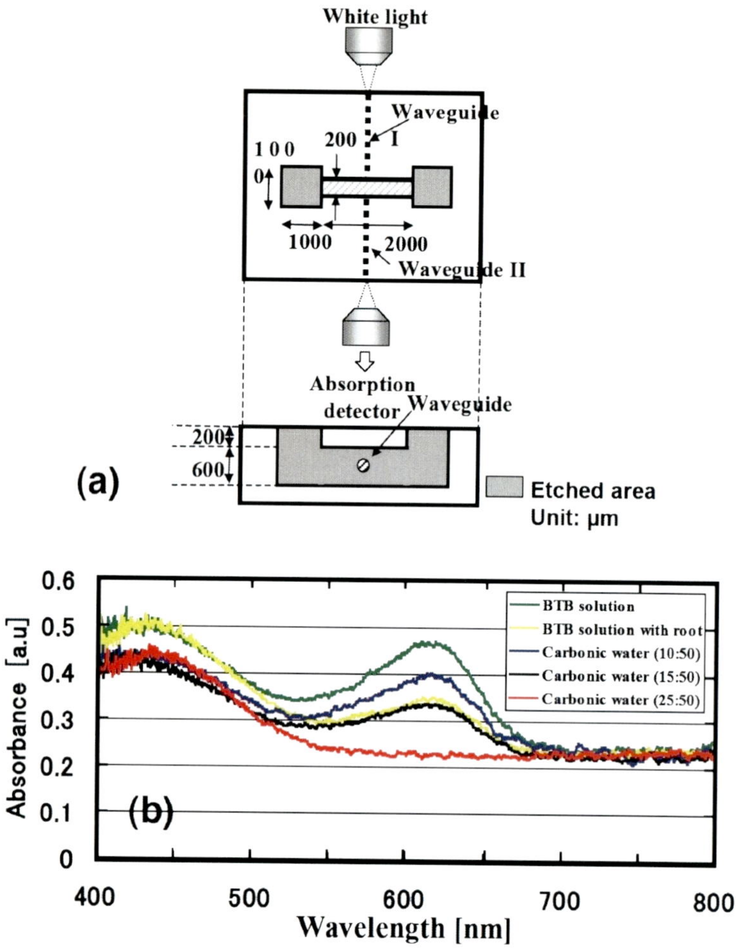

Figure 1.12 (a) Schematic illustration of 3D microfluidics integrated with optical waveguides in photosensitive glass by fs laser direct writing for quantitative analysis of CO_2 concentration in water. (b) Optical absorption spectra of water containing BTB solution (green line), with a seedling root (yellow line), and with carbonic water with different CO_2 concentrations (blue line [10 mL CO_2:50 mL H_2O], black line [15 mL CO_2:50 mL H_2O], and red line [25 mL CO_2:50 mL H_2O]).

peak decreases slightly with the reduction in pH that occurs with decreasing CO_2 concentration (10 mL CO_2:50 mL H_2O) and it decreases further with the introduction of more concentrated carbonic water (15 mL CO_2:50 mL H_2O). The peak disappears entirely for carbonic water with the highest CO_2 concentration (25 mL CO_2:50 mL H_2O). The yellow line in Fig. 1.12(b) shows the absorption spectrum of pure water containing BTB solution with a vegetable seedling root; it is almost the same as that for carbonic water with 15 mL CO_2:50 mL H_2O. This indicates that the CO_2 concentration generated by the root's respiration is comparable to that of carbonic water with 15 mL CO_2:50 mL H_2O. This experiment demonstrates that a CO_2 concentration gradient is necessary to induce the gliding movement of *Phormidium*, a microorganism that promotes vegetable growth. Chapter 10 describes the fabrication of optofluidics and biochips in glass by ultrafast lasers in more detail.

1.3.6 Biomedical Applications

Near-IR fs lasers are promising tools for imaging and processing cells and tissues since they can limit the interaction region to a tiny 3D volume [104]. Focused fs laser beam can perform nanosurgery, including dissection [105], axotomy [106], and transfection [107] of tissues, cells, and intracellular structures. Enhancement of the optical near field using metal nanoparticles by fs lasers is described in Section 1.3.3; it is also a promising nanosurgery technique [108].

When an fs laser beam is focused in a liquid, shock waves and cavitation bubbles are generated [109]. This fs laser-induced phenomenon has been applied to polymer crystallization [110], patterning of protein cubes and cells [111], injection of nanoparticles into cells [112], local stimulation of cultured cells [113], and isolation of a single cultured cell [114].

One practical application of fs lasers in biomedicine is laser *in situ* keratomileusis (LASIK), which is refractive surgery used to correct myopia, hyperopia, and astigmatism performed using a laser [115]. Figure 1.13 shows an SEM image of a primate cornea in which fs laser irradiation has created a flap and a lens body by making cuts inside the cornea. A system equipped with an fs laser for creating corneal flaps (Femto-LASIK) is now commercially available; it is more precise and predictable than conventional mechanical microkeratome [116].

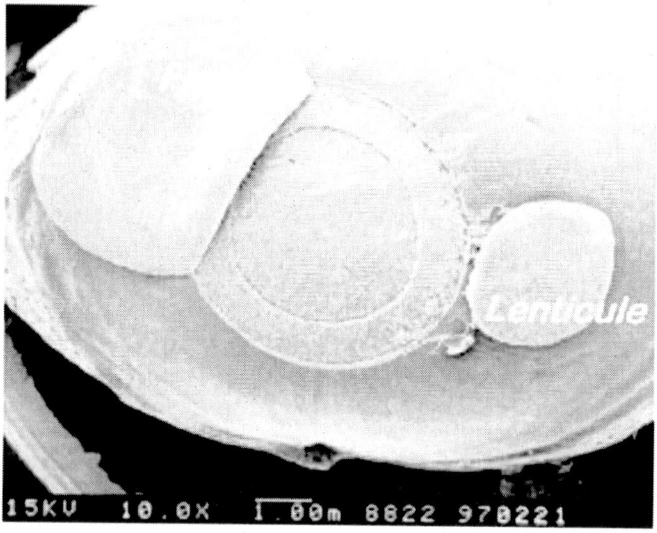

Figure 1.13 SEM image of primate cornea in which fs laser irradiation has formed a flap and a lens body by making cuts inside the cornea. Courtesy of F. Dausinger.

1.3.7 Industrial and Commercial Applications

One of the first applications of ultrafast laser processing in the field of semiconductor manufacturing was the repair of chromium-on-fused-silica photomasks by ablation for deep UV photolithography. 100 fs pulses with a wavelength of 266 nm realized a spatial resolution of better than 100 nm [117]. Ultrafast laser based photomask repair tools were employed at IBM's mask-making facility [37].

The first application of ultrafast lasers to mass production processes was drilling inkjet nozzles with specific shapes by ps laser ablation (see Section 1.3.1) [37]. Currently, ps lasers are superior to fs lasers for industrial applications due to their higher powers, efficiencies, reliability, and stability.

About 10 years ago, ultrafast laser processing showed potential to satisfy many requirements in automotive industry, including miniaturization, high precision, high quality, applicability to diverse materials, variety of variants, smaller lots, and cost effectiveness [118]. Since 2007, ps laser trimming has been used to produce exhaust gas sensors at the Bosch plant in Bamberg, Germany (see Fig. 1.14). They have a special ceramic layer and can measure the exhaust gas properties faster and more precisely than conventional sensors. This enables emissions to be reduced by optimizing combustion control. Moreover, since 2009, ps lasers have been used to fabricate diesel engine injectors in which drain grooves are microstructured around a high-pressure passage by ps laser pulses (see Fig. 12.4 in Chapter 12). The ps-laser structured drainage grooves allow a tight system even at pressures of up to 2000 bar. As a result, diesel injection systems have become more reliable, powerful, and environmental friendly.

Figure 1.14 Cross section of exhaust gas sensors produced by ps laser trimming.

Picosecond lasers are also promising tools for scribing and patterning in manufacturing photovoltaic solar cells based on copper indium gallium (di)selenide (CIGS) or CIS [119]. For patterning, it is necessary to isolate the back contact (Mo) between cells (P1), selectively remove the CIGS layer from the top of the back contact layer (P2), and selectively remove the absorber and front contact (transparent conductive oxide: TCO) layers (P3). Conventional ns lasers are widely used for P1 processes, whereas P2 and P3 processes are typically processed using mechanical scribing tools. This is due to the structure of CIGS photovoltaic solar cells. Specifically, when fabricating CIGS photovoltaic solar cells, a metallic back contact is first deposited on the glass substrate and forms a broadband reflective layer at the rear surface of the module. The absorber and front contact TCO layers are then deposited on top and hence these layers must be processed from the front to prevent damaging the metallic back contact. Picosecond laser processing can replace mechanical scribing for the P2 and P3 processes since it causes little thermal damage of the underlying layers. However, its adoption is currently limited as the process is immature and is more expensive than conventional ns laser processing.

Chapter 12 describes in more detail industrial and commercial applications of ultrafast laser processing.

1.4 Summary and Outlook

Ultrafast laser processing has opened new doors for materials processing since the first experiments in 1987 because it offers great advantages over conventional laser processing using ns or longer pulse lasers. These advantages include nonthermal processing, suppression of HAZ, absence of plasma shielding, processing of transparent materials by multiphoton absorption, internal modification of transparent materials, and nanofabrication ability. Ultrafast lasers are currently becoming common tools for both fundamental research and practical applications. Ultrafast lasers can perform high-quality, high-precision surface micromachining of various materials, including metals, ceramics, soft materials (e.g., polymers and biotissues), and even brittle materials (e.g., glasses). Nano- and micro-structured surfaces formed by fs laser irradiation have unique and useful properties in terms of friction, adhesion, optical absorption, hydrophobicity, etc. Additionally, subwavelength nanoscale ablation is possible and new nanoscale structures such as nanobumps, nanobelts, nanomeshes, nano-water drops, nano water crowns, and nanowhiskers can be generated by optimizing the laser irradiation conditions and the structures of target materials. Spatial resolutions of 100 nm are typically realized when using TPP to fabricate photonic crystals, micro- and nano-systems, LOC devices; a resolution of better than 25 nm has been achieved. Internal modification of transparent materials is only possible using ultrafast lasers; it is widely applied to fabricate 3D optical microdevices and optofluidic and LOC devices. Microwelding of glass substrates is another recent topic in internal modification by ultrafast lasers.

Modifying the spatial and temporal profiles of ultrafast laser pulses is very attractive for micro- and nanoprocessing and it is becoming one of the hottest topics in this field. Spatial manipulation is effective for realizing high-throughput, high-energy-efficiency laser processing [120–122]. Since temporal manipulation of ultrafast laser pulses provides the possibility of controlling the transient free-electron density [123], it has been used for realizing high-quality ablation and nanoscale ablation of transparent materials [123–125]. Chapter 4 describes manipulation of ultrafast laser pulses in more detail.

Although not described in this book, ultrafast lasers have the potential to synthesize metastable materials since they can generate extremely high temperatures and pressures in materials. Femtosecond laser driven shock has been used to create the high-pressure ε phase of iron [126]. Microexplosions triggered by fs laser pulses generated pressures greater than 380 GPa inside sapphire, resulting in the creation of bcc-Al with a crystallite size of $\sim 18 \pm 2$ nm [127, 128]. The unique characteristics of ultrafast laser processing offer the possibility of developing new processes that are peculiar to ultrafast lasers in the future.

Ultrafast lasers, especially ps lasers, are now suitable to be used in industrial and practical applications. Photomask repairs with a nanoscale resolution and inkjet nozzle drilling have already been demonstrated. Furthermore, ultrafast lasers have great potential to be used in automotive manufacturing and electronic industries. In addition to scribing and patterning of photovoltaic solar cells, patterning in active matrix organic light emitting diode (AMOLED) displays and fabricating through Si via (TSV) for 3D assembly of Si IC are possible applications in the electronic industry. For medical applications of fs lasers, the Femto-LASIK system for forming corneal flaps is now commercially available.

The performance of fs lasers is currently not sufficiently good for industrial applications, but it will improve in the near future. Rapid advances in the performances of ultrafast laser systems will further expand research in and practical applications of ultrafast laser processing.

References

1. Srinivasan R., Sutcliffe E., and Braren B. (1987). Ablation and etching of polymethylmethacrylate by very short (160 fs) ultraviolet (308 nm) laser pulses, *Appl. Phys. Lett.*, **51**, 1285–1287.
2. Küper S. and Stuke M. (1987). Femtosecond UV excimer laser ablation, *Appl. Phys.*, **B44**, 199–204.
3. Küper S. and Stuke M. (1989). Ablation of polytetrafluoroethylene (Teflon) with femtosecond UV excimer laser pulses, *Appl. Phys. Lett.*, **54**, 4–6.

4. Küper S. and Stuke M. (1989). Ablation of UV-transparent materials with femtosecond UV excimer laser pulses, *Microelectron. Eng.*, **9**, 475–480.

5. Momma C., Chichkov B. N., Nolte S., Alvensleben F., Tünnermann A., Welling H., and Wellegehausen B. (1996). Short-pulse laser ablation of solid targets, *Opt. Commun.*, **129**, 134–142.

6. Yanik M. F., Cinar H., Cinar H. N., Chisholm A. D., Jin Y. I., and Ben-Yakar A. (2004). Neurosurgery: Functional regeneration after laser axotomy, *Nature*, **432**, 822.

7. Barsch N., Korber K., Ostendorf A., and Tönshoff K. H. (2003). Ablation and cutting of planar silicon devices using femtosecond laser pulses, *Appl. Phys.*, **A77**, 237–242.

8. Nakata Y., Okada T., and Maeda M. (2002). Fabrication of dot matrix, comb, and nanowire structures using laser ablation by interfered femtosecond laser beams, *Appl. Phys. Lett.*, **81**, 4239–4241.

9. Reif J., Costache F., Henyk M., and Pandelov S. V. (2002). Ripples revisited: non-classical morphology at the bottom of femtosecond laser ablation craters in transparent dielectrics, *Appl. Surf. Sci.*, **197–198**, 891–895.

10. Wu Q., Ma Y., Fang R., Liao Y., Yu Q., Chen X., and Wang K. (2003). Femtosecond laser-induced periodic surface structure on diamond film, *Appl. Phys. Lett.*, **82**, 1703–1705.

11. Rudolph P. and Kautek W. (2004). Composition influence of non-oxidic ceramics on self-assembled nanostructures due to fs-laser irradiation, *Thin Solid Films*, **453–454**, 537–541.

12. Miyaji G. and Miyazaki K. (2006). Ultrafast dynamics of periodic nanostructure formation on diamondlike carbon films irradiated with femtosecond laser pulses, *Appl. Phys. Lett.*, **89**, 191902 .

13. Davis K. M., Miura K., Sugimoto N., and Hirao K. (1996). Writing waveguides in glass with a femtosecond laser, *Opt. Lett.*, **21**, 1729–1731.

14. Glezer E. N., Milosavljevic M., Huang L., Finlay R. J., Her T. H., Callan J. P., and Mazur E. (1996). Three-dimensional optical storage inside transparent materials, *Opt. Lett.*, **21**, 2023–2025.

15. Watanabe W., Sowa S., Tamaki T., Itoh K., and Nishii J. (2006). Three-dimensional waveguides fabricated in poly(methyl methacrylate) by a femtosecond laser, *Jpn. J. Appl. Phys.*, **45**, L765–L767.

16. Hanada Y., Sugioka K., and Midorikawa K. (2010). UV waveguides light fabricated in fluoropolymer CYTOP by femtosecond laser direct writing, *Opt. Express*, **18**, 446–450.

17. Kawata S., Sun H. B., Tanaka T., and Takada K. (2001). Finer features for functional microdevices, *Nature*, **412**, 697– 698.

18. Rudd J. V., Korn G., Kane S., Squire J., Mourou G., and Bado P. (1993). Chirped-pulse amplification of 55-fs pulses at a 1-kHz repetition rate in a Ti-Al_2O_3 regenerative amplifier, *Opt. Lett.*, **18**, 2044–2046.

19. Arai A., Bovatsek J., Yoshino F., Liu Z., Cho G. C., Shah L., Fermann M. E., and Uehara Y. (2006). Fiber chirped pulse amplification system for micromachining, *Proc. SPIE*, **6343**, 63430S .

20. Kleinbauer J., Eckert D., Weiler S., and Sutter D. H. (2008). 80 W ultrafast CPA-free disk laser, *Proc. SPIE*, **6871**, 68711B.

21. Marchese S. V, Baer C. R. E., Engqvist A. G., Hashimoto S., Maas D. J. H. C, Golling M., Suedmeyer T., and Keller U. (2008). Femtosecond thin disk laser oscillator with pulse energy beyond the 10-microjoule level, *Opt. Express*, **16**, 6397–6407.

22. Fan W. S., Storz R., Tom H. W. K., Bokor J. (1992). Electron thermalization in gold, *Phys. Rev.*, **B46**, 13592–13595.

23. Sun C. K., Vallée F., Acioli L. H., Ippen E. P., and Fujimoto J. G. (1994). Femtosecond-tunable measurement of electron thermalization in gold, *Phys. Rev.*, **B50**, 15337–15348.

24. Wellershoff S. S., Hohlfeld J., Güdde J., and Matthias E. (1999). The role of electron-phonon coupling in femtosecond laser damage of metals, *Appl. Phys.*, **A69**, S99–S107.

25. Hohlfeld J., Wellershoff S. S., Güdde J., Conrad U., Jähnke V., and Matthias E. (2000). Electron and lattice dynamics following optical excitation of metals, *Chem. Phys.*, **251**, 237–258.

26. Anisimov S. I. and Rethfeld B. (1997). Theory of ultrashort laser pulse interaction with a metal, *Proc. SPIE*, **3093**, 192–203.

27. Corkum P. B., Brunel F., Sherman N. K., and Srinivasan-Rao T. (1988). Thermal response of metals to ultrashort pulse laser excitation, *Phys. Rev. Lett.*, **61**, 2886–2889.

28. Fujita M. and Hashida M. (2004). Applications of femtosecond lasers, *Oyo Buturi*, **73**, 178–185 (in Japanese).

29. Hanada Y., Sugioka S. V., Miyamoto I., and Midorikawa K. (2005). Double-pulse irradiation by laser-induced plasma-assisted ablation (LIPAA) and mechanisms study, *Appl. Surf. Sci.*, **248**, 276–280.

30. Mao S. S., Quere F., Guizard S., Mao X., Russo R. E., Petite G., and Martin P. (2004). Dynamics of femtosecond laser interactions with dielectrics, *Appl. Phys.*, **A79**, 1695–1709.

31. Eaton S. M., Zhang H., Herman P. R., Yoshino F., Shah L., Bovatse K. J., and Arai A. Y. (2005). Heat accumulation effects in femtosecond laser-written waveguides with variable repetition rate, *Opt. Express,* **13**, 4708–4716.

32. Eaton S. M., Zhang H., Ng M. L., Li J., Chen W. J., Ho S., and Herman P. R. (2008). Transition from thermal diffusion to heat accumulation in high repetition rate femtosecond laser writing of buried optical waveguides, *Opt. Express,* **16**, 9443–9458.

33. Tamaki T., Watanabe W., Nishii J., and Itoh K. (2005). Welding of transparent materials using femtosecond laser pulses, *Jpn. J. Appl. Phys.*, **44**, L687–L689.

34. Miyamoto I., Horn A., and Gottmann J. (2007). Local melting of glass material and its application to direct fusion welding by ps-laser pulses, *J. Laser Micro/Nanoeng.*, **2**, 7–14.

35. Horn A., Mingareev I., and Werth A. (2008). Investigations on melting and welding of glass by ultra-short laser radiation, *J. Laser Micro/Nanoeng.*, **3**, 114–118.

36. Chichkov B. N., Momma C., Nolte S., von Alvensleben F., and Tunnermann A. (1996). Femtosecond, picosecond and nanosecond laser ablation of solids, *Appl. Phys.*, **A63**, 109–115.

37. Sugioka K., Gu B., and Holmes A. (2007). The state of the art and future prospects for laser direct-write for industrial and commercial applications, *MRS Bull.*, **32**, 47–54.

38. Tönshoff H. K., Ostendorf A., Nolte S. Korte F., and Bauer T. (2000). Micromachining using femtosecond laser, *Proc. SPIE*, **4088**, 136–139.

39. Emmony D. C., Howson R. P., and Willis L. J. (1973). Laser mirror damage in germanium at 10.6 μm, *Appl. Phys. Lett.*, **23**, 598–600.

40. Jain A. K., Kulkarni V. N., Sood D. K., and Uppal J. S. (1981). Periodic surface ripples in laser-treated aluminum and their use to determine absorbed power, *J. Appl. Phys.*, **52**, 4882–4884.

41. Keilmann F. and Bai Y. H (1982). Periodic surface structures frozen into CO_2 laser-melted quartz, *Appl. Phys.*, **A29**, 9–19.

42. Sakabe S., Hashida M., Tokita S., Namba S., and Okamuro K. (2009). Mechanism for self-formation of periodic grating structures on a metal surface by a femtosecond laser pulse, *Phys. Rev.*, **B79**, 033409.

43. Hashida M., Fujita M., Tsukamoto M., Semerok A. F., Gobert O., Petite G., Izawa Y., and Wagner J. F. (2002). Femtosecond laser ablation of metals: Precise measurement and analytical model for crater profiles, *Proc. SPIE*, **4830**, 452–457.

44. Yasumaru N., Miyazaki K., and Kiuchi J. (2003). Femtosecond-laser-induced nanostructure formed on hard thin films of TiN and DLC, *Appl. Phys.*, **A76**, 983–985.

45. Borowiec A. and Hauge H. K. (2003). Subwavelength ripple formation on the surfaces of compound semiconductors irradiated with femtosecond laser pulses, *Appl. Phys. Lett.*, **82**, 4462–4464.

46. Costache F., Henyk M., and Reif J. (2003). Surface patterning on insulators upon femtosecond laser ablation, *Appl. Surf. Sci.*, **208**, 486–491.

47. Shimotsuma Y., Kazansky P. G., Qiu J. R., and Hirao K. (2003). Self-organized nanogratings in glass irradiated by ultrashort light pulses, *Phys. Rev. Lett.*, **91**, 247405.

48. Hashida M., Nagashima K., Fujita M., Tsukamoto M., Katto M., and Izawa Y. (2003). Femtosecond laser ablation of metals: Characterization of new processing phenomenon and formation of nano-structures, *Proc. 9th Sym. Microjoining Assembly Technol. Electron.*, **9**, 517–522.

49. Her T. H., Finlay R. J., Wu C., Deliwala S., and Mazur E. (1998). Microstructuring of silicon with femtosecond laser pulses, *Appl. Phys. Lett.*, **73**, 1673–1675.

50. Baldacchini T., Carey J. E., Zhou M., and Mazur E. (2006). Superhydrophobic surfaces prepared by microstructuring of silicon using a femtosecond laser, *Langmuir*, **22**, 4917–4919.

51. Carey J. E., Crouch C. H., Shen M., and Mazur E. (2005). Visible and near-infrared responsivity of femtosecond-laser microstructured silicon photodiodes, *Opt. Lett.*, **30**, 1773–1775.

52. Younkin R., Carey J. E., Mazur E., Levinson J. A., and Friend C. M. (2003). Infrared absorption by conical silicon microstructures made in a variety of background gases using femtosecond-laser pulses *J. Appl. Phys.*, **93**, 2626–2629.

53. Wang F., Chen C., He H., and Liu S. (2011). Analysis of sunlight loss for femtosecond laser microstructured silicon and its solar cell efficiency, *Appl. Phys.*, **A103**, 977–982.

54. Zorba V., Stratakis E., Barberoglou M., Spanakis E., Tzanetakis P., Anastasiadis S. H., and Fotakis C. (2008). Biomimetic artificial surfaces quantitatively reproduce the water repellency of a lotus leaf, *Adv. Mater.*, **20**, 4049–4054.

55. Naya B. K., Gupta M. C., and Kolasinski K. W. (2007). Spontaneous formation of nanospiked microstructures in germanium by femtosecond laser irradiation, *Nanotechnology*, **18**, 195302.

56. Nakashima S., Sugioka K., and Midorikawa K. (2010). Enhancement of resolution and quality of nano-hole structure on GaN substrates using the second-harmonic beam of near-infrared femtosecond laser, *Appl. Phys.*, **A101**, 475–481.

57. Nakashima S., Sugioka K., Ito T., Takai H., and Midorikawa K. (2011). Fabrication of high-aspect-ratio nanohole arrays on GaN surface by using wet-chemical-assisted femtosecond laser ablation, *J. Laser Micro/Nanoeng.*, **6**, 15–19.

58. Nakata Y., Okada T., and Maeda M. (2003). Nano-sized hollow bump array generated by single femtosecond laser pulse, *Jpn. J. Appl. Phys.*, **42**, L1452–L1454.

59. Nakata Y., Okada T., and Maeda M. (2004). Lithographical laser ablation using femto-second laser, *Appl. Phys.*, **A79**, 1481–1483.

60. Nakata Y., Miyanaga N., and Okada T. (2007). Effect of pulse width and fluence of femtosecond laser on the size of nanobump array, *Appl. Surf. Sci.*, **253**, 6555–6557.

61. Nakata Y., Tsuchida K., Miyanaga N., and Furusho H. (2009). Liquidly process in femtosecond laser processing, *Appl. Surf. Sci.*, **255**, 9761–9763.

62. Nakata Y., Momoo K., Hiromoto T., and Miyanaga N. (2011). Generation of superfine structure smaller than 10 nm by interfering femtosecond laser processing, *Proc. SPIE*, **7920**, 79200B.

63. Chimmalgi A., Choi T. Y., Grigoropoulos C. P., and Komvopoulos K. (2003). Femtosecond laser apertureless near–field nanomachining of metals assisted by scanning probe microscopy, *Appl. Phys. Lett.*, **82**, 1146–1148.

64. Lin Y., Hong M. H., Wang W. J., Law Y. Z., and Chong T. C. (2005). Sub-30 nm lithography with near-field scanning optical microscope combined with femtosecond laser, *Appl. Phys.*, **A80**, 461–465.

65. Atanasov P. A., Takada H., Nedyalkov N. N., and Obara M. (2007). Nanohole processing on silicon substrate by femtosecond laser pulse with localized surface plasmon polariton, *Appl. Surf. Sci.*, **253**, 8304–8308.

66. Eversole D., Luk'yanchuk B., and Ben-Yakar A. (2007). Plasmonic laser nanoablation of silicon by the scattering of femtosecond pulses near gold nanospheres, *Appl. Phys.*, **A89**, 283–291.

67. See, for example, http://www.cmet.co.jp/en/index.html.

68. Cumpston B., Ananthavel S., Barlow S., Dyer D., Ehrlich J., Erskine L., Heikal A., Kuebler S., Lee I.-Y., McCord-Maughon D., Qin J., Röckel

H., Rumi M., Wu X.-L., Marder S., and Perry J. (1999). Two-photon polymerization initiators for three-dimensional optical data storage and microfabrication, *Nature*, **398**, 51–54.

69. Kuebler S. M., Braun K., Zhou W., Cammack J., Yu T., Ober C., Marder S., and Perry J. (2003). Design and application of high-sensitivity two-photon initiators for three-dimensional microfabrication, *J. Photochem. Photobiol.*, **158**, 163–170.

70. http://gtresearchnews.gatech.edu/newsrelease/3dmicrostructures.htm.

71. Sun H. B., Matsuo S., and Misawa H. (1999). Three-dimensional photonic crystal structures achieved with two-photon-absorption photopolymerization of resin, *Appl. Phys. Lett.*, **74**, 786–788.

72. Serbin J., Ovsianikov A., and Chichkov B. (2004). Fabrication of woodpile structures by two-photon polymerization and investigation of their optical properties, *Opt. Express*, **12**, 5221–5228.

73. Maruo S. and Inoue H. (2006). Optically driven micropump produced by three-dimensional two-photon microfabrication, *Appl. Phys. Lett.*, **89**, 144101.

74. Maruo S. and Inoue H. (2007). Optically driven viscous micropump using a rotating microdisk, *Appl. Phys. Lett.*, **91**, 084101.

75. Tian Y., Zhang Y. L., Ku J. F., He Y., Xu B. B., Chen Q. D., Xia H., and Sun H. B. (2010). High performance magnetically controllable microturbines, *Lab Chip.*, **10**, 2902–2905.

76. Wang J., He Y., Xia H., Niu L. G., Zhang R., Chen Q. D., Zhang Y. L., Li Y. F., Zeng S. J., Qin J. H., Lin B. C., and Sun H. B. (2010). Embellishment of microfluidic devices via femtosecond laser micronanofabrication for chip functionalization, *Lab Chip.*, **10**, 1993–1996.

77. Wu D., Chen Q. D., Niu L. G., Wang J. N., Wang J., Wang R., Xia H., and Sun H. B. (2009). Femtosecond laser rapid prototyping of nanoshells and suspending components towards microfluidic devices, *Lab Chip.*, **9**, 2391–2394.

78. Ovsianikov A., Malinauskas M., Schlie S., Chichkov B., Gittard S., Narayan R., Löblerc M., Sternbergc K., Schmitzc K. P., and Haverich A. (2011). Three-dimensional laser micro- and nano-structuring of acrylated poly(ethylene glycol) materials and evaluation of their cytoxicity for tissue engineering applications, *Acta Biomater.*, **7**, 967–974.

79. Farsari M. and Chichkov B. (2009). Two-photon fabrication, *Nature Photon.*, **3**, 450–452.

80. Tan D. F., Li Y., Qi F. G., Yang H., and Gong Q. H. (2007). Reduction in feature size of two-photon polymerization using SCR500, *Appl. Phys. Lett.*, **90**, 071106.

81. Tanaka T., Ishikawa A., and Kawata S. (2006). Two-photon-induced reduction of metal ions for fabricating three-dimensional electrically conductive metallic microstructure, *Appl. Phys. Lett.*, **88**, 081107.

82. Chan J. W., Huser T. R., Risbud .S. H., and Krol D. M. (2003). Modification of the fused silica glass network associated with waveguide fabrication using femtosecond laser pulses, *Appl. Phys.*, **A76**, 367–372.

83. Dharmadhikari J. A., Dharmadhikari A. K., Bhatnagar A., Mallik A., Singh P. C., Dhaman R. K. and Mathur D. (2011). Writing low-loss waveguides in borosilicate (BK7) glass with a low-repetition-rate femtosecond laser, *Opt. Comm.*, **284**, 630–634.

84. Efimov O. M., Glebov L. B., Richardson K. A., Van Stryland E., Cardinal T., Park S. H., Couzi M., and Bruneel J. L. (2001). Waveguide writing in chalcogenide glasses by a train of femtosecond laser pulses, *Opt. Mater.*, **17**, 379–386.

85. Le Coq D, Masselin P., Przygodski C., and Bychkov E. (2009). Morphology of waveguide written by femtosecond laser in As_2S_3 glass, *J. Non-Crystal. Solids*, **355**, 37–42.

86. Zhang H., Eaton S. M., and Herman P. R. (2006). Low-loss type II waveguide writing in fused silica with single picosecond laser pulses, *Opt. Express*, **14**, 4826–4834.

87. Watanabe W., Asano T., Yamada K., Itoh K., and Nishii J. (2003). Wavelength division with three-dimensional couplers fabricated by filamentation of femtosecond laser pulses, *Opt. Lett.*, **28**, 2491–2493.

88. Florea C. and Winick K. A. (2003). Fabrication and characterization of photonic devices directly written in glass using femtosecond laser pulses, *J. Lightwave Technol.*, **21**, 246–253.

89. Bricchi E., Mills J. D., Kazamsky P. G., Klappauf B. G., and Baumberg J. J. (2002). Birefringent Fresnel zone plates in silica fabricated by femtosecond laser machining, *Opt. Lett.*, **27**, 2200–2202.

90. Vella G. D., Taccheo S., Osellame R., Festa A., Cerullo G., and Laporta P. (2007). 1.5 μm single longitudinal mode waveguide laser fabricated by femtosecond laser writing, *Opt. Express*, **15**, 3190–3194.

91. Kawamura K., Hirano M., Kurobori T., Takamizu D., Kamiya T., and Hosono H. (2004). Femtosecond-laser-encoded distributed-feedback color center laser in lithium fluoride single crystals, *Appl. Phys. Lett.*, **84**, 311–313.

92. Miura K., Qiu J., Fujiwara S., Sakaguchi S., and Hirao K. (2002). Three-dimensional optical memory with rewriteable and ultrahigh density using the valence-state change of samarium ions, *Appl. Phys. Lett.*, **80**, 2263–2265.

93. Luo F. F., Qian B., Lin G., Xu J., Liao Y., Song J., Sun H. Y., Zhu B., Qiu J. R., Zhao Q. Z., and Xu Z. Z. (2010). Redistribution of elements in glass induced by a high-repetition-rate femtosecond laser, *Opt. Express*, **18**, 6262–6269.

94. Hongo T., Sugioka K., Niino H., Cheng Y., Masuda M., Miyamoto I., Takai H., and Midorikawa K. (2005). Investigation of photoreaction mechanism of photosensitive glass by femtosecond laser, *J. Appl. Phys.*, **97**, 063517.

95. Miura K., Shimotsuma Y., Sakakura M., Kanehira S., Hamabe M., and Hirao K. (2006). Three-dimensional deposition of silicon from silicate glass with dispersed metallic aluminum by a femtosecond laser, *Proc. SPIE*, **6413**, 64130K.

96. Marcinkevicius A., Juodkazis S., Watanabe M., Miwa M., Matsuo S., Misawa H., and Nishii J. (2001). Femtosecond laser-assisted three-dimensional microfabrication in silica, *Opt. Lett.*, **26**, 277–279.

97. Sugioka K., Cheng Y., and Midorikawa K. (2005). Three-dimensional micromachining of glass using femtosecond laser for lab-on-a-chip device manufacture, *Appl. Phys.*, **A81**, 1–10.

98. Sugioka K., Hanada Y., and Midorikawa K. (2010). Three-dimensional femtosecond laser micromachining of photosensitive glass for biomicrochips, *Laser Photon. Rev.*, **3**, 386–400.

99. Cheng Y., Sugioka K., Midorikawa K., Masuda M., Toyoda K., Kawachi M., and Shihoyama K. (2003). Three-dimensional Micro-optical components embedded in photosensitive glass by a femtosecond laser, *Opt. Lett.*, **28**, 1144–1146.

100. Wang Z., Sugioka K., and Midorikawa K. (2007). Three dimensional integration of microoptical components buried inside photosensitive glass by femtosecond laser direct writing, *Appl. Phys.*, **A89**, 951–955.

101. Wang Z., Sugioka K., and Midorikawa K. (2008). Fabrication of integrated microchip for optical sensing by femtosecond laser direct writing of Foturan glass, *Appl. Phys.*, **A93**, 225–229.

102. Crespi A., Gu Y., Ngamsom B., Hoekstra H. J. W. M. Dongre C., Pollnau M., Ramponi R., van den Vlekkert H. H., Watts P., Cerullo G., and Osellame R. (2010). Three-dimensional Mach-Zehnder interferometer in a microfluidic chip for spatially-resolved label-free detection, *Lab Chip.*, **10**, 1167–1173.

103. Hanada Y., Sugioka K., Ishikawa I. S., Kawano H., Miyawaki A., and Midorikawa K. (2011). 3D Microfluidic chips with integrated functional microelements fabricated by femtosecond laser for studying the gliding mechanism of cyanobacteria, *Lab Chip.*, **11**, 2109–2115.

104. Watanabe S. (2010). *Laser Precision Microfabrication*, eds. Sugioka K., Meunier M, and Pique A., Chapter 6 Laser Nanosurgery, Manupulation, and Transportation of Cells and Tissues (Springer, Heidelberg) 145–161.

105. König K., Riemann I., and Fritzsche W. (2001). Nanodissection of human chromosomes with near-infrared femtosecond laser pulses, *Opt. Lett.*, **26**, 819–822.

106. Guo S. X., Bourgeois F., Chokshi T., Durr N. J., Hilliard M. A., Chronis N., and Ben-Yakar A. (2008). Femtosecond laser nanoaxotomy lab-on-a-chip for *in vivo* nerve regeneration studies, *Nat. Meth.*, **5**, 531–533.

107. Tirlapur U. K. and König K. (2002). Cell biology—Targeted transfection by femtosecond laser, *Nature*, **418**, 290–291.

108. Csaki A., Garwe F., Steinbruck A., Maubach G., Festag G., Weise A., Riemann I., König K., and Fritzsche W. (2007). A parallel approach for subwavelength molecular surgery using gene-specific positioned metal nanoparticles as laser light antennas, *Nano Lett.*, **7**, 247–253.

109. Juhasz T., Kastis G. A., Suarez C., Bor Z., and Bron W. E. (1996). Time-resolved observations of shock waves and cavitation bubbles generated by femtosecond laser pulses in corneal tissue and water, *Laser Surg. Med.*, **19**, 23–31.

110. Nakamura K., Sora Y., Yoshikawa H. Y., Hosokawa Y., Murai R., Adachi H., Mori Y., Sasaki T., and Masuhara H. (2007). Femtosecond laser-induced crystallization of protein in gel medium, *Appl. Surf. Sci.*, **253**, 6425–6429.

111. Kaji T., Ito S., Miyasaka H., Hosokawa Y., Masuhara H., Shukunami C., and Hiraki Y. (2007). Nondestructive micropatterning of living animal cells using focused femtosecond laser-induced impulsive force, *Appl. Phys. Lett.*, **91**, 023904.

112. Yamaguchi A., Hosokawa Y., Louit G., Asahi T., Shukunami C., Hiraki Y., and Masuhara H. (2008). Nanoparticle injection to single animal cells using femtosecond laser-induced impulsive force, *Appl. Phys.*, **A93**, 39–43.

113. Kuo Y. E., Wu C. C., Hosokawa Y., Maezawa Y., Okano K., Masuhara H., and Kao F. J. (2010). Local stimulation of cultured myocyte cells by femtosecond laser-induced stress wave, *Appl. Phys.*, **A101**, 597–600.

114. Hosokawa Y., Takabayashi H., Miura S., Shukunami C., Hiraki Y., and Masuhara H. (2004). Nondestructive isolation of single cultured animal cells by femtosecond laser-induced shockwave, *Appl. Phys.*, **A79**, 4–6.

115. Dausinger F. (2002). Femtosecond pulses for medicine and production technology – overview of a German national project, *Proc. SPIE*, **4426**, 9–16.

116. http://www.amo-inc.com/products/refractive/ilasik/intralase-fs-laser.

117. Haight R., Wagner A., Longo P., and Lim D. (2005). Femtosecond laser ablation and deposition of metal films on transparent substrates with applications in photomask repair, *Proc. SPIE*, **5714**, 24–36.

118. Bauer Th. and König J. (2010). Applications and perspectives of ultrashort pulsed lasers, *Technical Digest of LPM 2010*, JSPS, 127.

119. Booth H. (2010). Laser processing in industrial solar module manufacturing, *Proc. LPM 2010*, JSPS, #10–02.

120. Hasegawa S. and Hayasaki Y. (2009). Adaptive optimization of hologram in holographic femtosecond laser processing system, *Opt. Lett.*, **34**, 22–24.

121. Obata K., Koch J., Hinze U., and Chichkov B. N. (2010). Multi-focus two-photon polymerization technique based on individually controlled phase modulation, *Opt. Express*, **18**, 17193–17200.

122. Sakakura M., Sawano T., Shimotsuma Y., Miura K., and Hirao K. (2010). Fabrication of three-dimensional 1 × 4 splitter waveguides inside a glass substrate with spatially phase modulated laser beam, *Opt. Express*, **18**, 12136–12143.

123. Englert L., Rethfeld B., Haag L., Wollenhaupt M., Sarpe-Tudoran C., and Baumert T. (2007). Control of ionization processes in high band gap materials via tailored femtosecond pulses, *Opt. Express*, **15**, 17855–17862.

124. Stoian R., Boyle M., Thoss A., Rosenfeld A., Korn G., Hertel I. V., and Campbell E. E. B. (2002). Laser ablation of dielectrics with temporally shaped femtosecond pulses, *Appl. Phys. Lett.*, **80**, 353–355.

125. Wollenhaupt M., Englert L., Horn A., and Baumert T. (2009). Control of ionization processes in high band gap material, *J. Laser Micro/Nanoeng.*, **4**, 144–151.

126. Sano T., Mori H., Sakata O., Ohmura E, Miyamoto I., Hirose A., and Kobayashi K. F. (2005). Femtosecond laser driven shock synthesis of the high-pressure phase of iron, *Appl. Surf. Sci.*, **247**, 571–576.

127. Juodkazis S., Vaalions A., Gamaly E. G., Mizeikis V., Yang W., and Rode A. (2011). Micro-explosions triggered by femtosecond laser pulses: new tool of nano-structuring, *Technical Digest of LPM 2011*, JSPS, 127.

128. Vailionis A., Gamaly E. G., Mizeikis V., Yang W., Rode A. V., and Juodkazis S. (2011). Evidence of superdense aluminium synthesized by ultrafast microexplosion, *Nat. Comm.*, **2**, 445.

Chapter 2

Lasers for Ultrafast Laser–Materials Processing

Mark Ramme,[a] Andreas Vaupel,[a] Michaël Hemmer,[a,b] Jiyeon Choi,[a,c] Ilya Mingareev,[a,d] and Martin C. Richardson[a]

[a] *Townes Laser Institute, College of Optics & Photonics: CREOL & FPCE, University of Central Florida, 4000 Central Florida Blvd, Orlando FL 32816-2700, USA*

[b] *The Institute of Photonic Sciences, Mediterranean Technology Park, Av. Carl Friedrich Gauss, Num. 3, 08860 Castelldefels (Barcelona), Spain*

[c] *Department of Laser & Electron Beam Applications, Korea Institute of Machinery & Materials (KIMM), 156, Gajeongbuk, YuseongGu, Daejeon, 305-343, Korea*

[d] *Fraunhofer-Townes Collaboration between Fraunhofer Institute for Laser Technology (Steinbachstr.15, Aachen Germany) and Townes Laser Institute, University of Central Florida*

mramme@creol.ucf.edu

Nearly all advances made in processing materials in the fs (< 100 fs) domain have so far been achieved with the use of Ti:Sapphire lasers at a wavelength of 800 nm. For longer pulses (< 1 ps), lasers operating at 1 μm have played an important role. Whereas the latter might be considered for industrial or manufacturing environments, Ti:Sapphire lasers have several disadvantages. With knowledge and expertise in the processing of many different categories of

Ultrafast Laser Processing: From Micro- to Nanoscale
Edited by Koji Sugioka and Ya Cheng

ISBN 978-981-4267-33-5 (Hardcover), 978-981-4303-69-9 (eBook)
www.panstanford.com

materials over a wide range of processing parameters, we review the different effects that can be created with ultrafast laser pulses from the viewpoint of the optimum laser performance. Beginning from an understanding of the fundamental processes involved in ultrafast laser–materials interaction, we survey the different effects that can be created, from nanostructuring, to bond-modification, refractive index modification, localized changes in density, micro-crystallization, and the various mechanisms involved in ablation, and then discuss the impact of the laser parameters used and the processing optics involved. With a knowledge of the new technologies that will influence the development of tomorrow's ultrafast lasers, those that are compact, efficient, cost-effective, and deployable in workspace environments, we identify those processing technologies that will be viable first to enter future markets.

2.1 Introduction

Ultrashort lasers became practical for material processing following the realization of efficient ultrashort pulse amplification using chirped pulse amplification (CPA) in 1985 [1]. Since then the distinctive features of the ultrashort laser materials processing have been extensively explored. In the early years extraordinary properties of ultrashort laser pulse interaction with materials were investigated by comparative studies with various pulse durations [2–8]. The first and most visible difference between laser-processed structures between ns and fs laser structuring was the lack of a heat-affected zone (HAZ). One of the initial studies [5] showed clearly the differences between using 16 ns and 300 fs excimer UV laser pulses on polymethyl-methacrylate (PMMA). The plasma screening effect is a serious factor in ns laser ablation. A plasma gradient is established early in the plasma formation preventing much of the laser energy from being deposited at the ablation surface. In fs laser interaction this inertial plasma formation is avoided and all the laser energy is deposited at the ablation surface. This feature lead to the notion that since fs ablation was dominated by electron generation on a timescale short compared to the motion of ions, it would be largely material independent. The classical thermodynamic model of UV laser ablation of dielectrics [9] was not appropriate. The breakdown threshold fluence F_{th} does not follow the empirical scaling law $F_{th} \sim \sqrt{\tau}$ dependence on pulse duration τ of pulses longer than 10 ps [3, 7]. Moreover, two different ablation regimes exist for pulses shorter than 1 ps [10]. At low fluence the etch rate is on a nanometer scale, governed by the optical skin depth while at higher fluence an "explosive" ablation regime occurs following the $\sqrt{\tau}$ law of diffusion. Compared to ns laser pulses ultrashort-pulsed lasers were initially shown to be advantageous for micromachining a variety of materials including metals, semiconductors, and dielectrics [2, 5, 7, 11]. They exhibited (i) lower ablation thresholds (J/cm^2), (ii) minimal heat affected zone, (iii) micromachining of high bandgap dielectric materials through nonlinear absorption, and (iv) deterministic machining dependent on the generation of seed electrons by avalanche ionization. These findings, so different from the interaction of conventional ns pulsed laser light with matter, ushered in the notion of a new engineering discipline. Whereas conventional laser ablation relied on the establishment of thermalized ablative regimes at material temperatures in excess of that required for melting and vaporization, this ultrafast interaction ablated material through plasma or electron-driven processes. Instead of a pool of molten material existing in the focal zone, a thin sheet of highly ionized material was created. Matter is ablated by a "cold-plasma" process in a thin sheet of thermal dense plasma, where highly energetic electrons both ablate and accelerate "cold" ions from the irradiation region. Very soon after these basic tenets of fs ablation were understood, it was clear that this form of laser–materials engineering would have special properties and lead to exciting new applications.

In this chapter we chronicle the development of the ultrashort lasers that have driven this new pathway to structuring materials with submicron accuracy. We describe the primary physical processes involved in the interaction—the interaction science, and the parametric dependencies they expose. This discussion takes into account all of those processes produced by pulses shorter in duration than ~10 ps, for beyond this timescale, thermalizing plasma processes become dominant. We give some impression of the broad range of possible applications areas. In the last decade or so we have witnessed the evolution of many of these potentially groundbreaking manufacturing techniques. In many cases the implementation of these exciting new options have stalled in implementation for reasons of cost, complexity, reproducibility or processing throughput. However we live in a fast changing world in laser development. As we formulate this overview of fs laser materials processing, we see many innovative new laser technologies emerging from many corners of laser technology, fiber lasers, new diode-pumped solid-state laser technologies, new components such as semiconductor saturable absorber mirrors (SESAMs) and grisms, and new architectures such as optical parametric chirped-pulse amplification (OPCPA), that collectively can help bypass these previous restrictions. The consequential investment unleashed by the prospects of new microscopic and submicroscopic manufacturing technologies will then propel developments in this field with ever-increasing speed. We believe we are at the cusp of this transformation in ultrafast laser technology.

2.2 Fundamental Interaction Processes: Laser Dependence

Ultrafast laser pulse-matter interaction exhibits many distinctive features that are advantageous to precision micromachining compared to cw or longer pulse lasers. Ultrafast lasers produce a deterministic ablation process with high repeatability, excellent micromachining quality such as the near absence of surface debris or recast layers and the creation of a limited HAZ. The energy loss due to thermal diffusion is negligible compared to the longer pulse laser–matter interaction regimes. Therefore the material ablation thresholds are reduced for ultrafast lasers [3, 7]. Moreover, the high intensities associated with these short pulses induce absorption mechanisms not achievable with conventional laser outputs, such as nonlinear absorption in wide bandgap transparent media. This opens the door to 3D volumetric micro-structuring by a single step process not easily available with any other existing fabrication methods. In this section, the principal mechanisms of ultrafast laser-matter interaction are discussed primarily to provide the context in terms of materials and laser parameters to provide fundamental understanding (also see Chapter 3). These mechanisms will become building blocks for the creation of novel applications utilizing upcoming ultrafast lasers.

2.2.1 Principal Physical Mechanisms in Ultrafast Laser Processing

Ultrafast lasers deliver a highly intense photon flux in extremely short time durations comparable to the elementary interaction time scales for electron collision (a few fs), electron cooling (1–10 ps), and lattice heating times (0.1~1 ns) [10, 12]. In terms of energy deposition, the light absorption in the ultrashort-pulsed regime is decoupled from energy dissipative processes as the energy deposition times is much shorter than the thermal diffusion time of most materials. Laser energy deposition occurs before any processes can transport energy to the surroundings in the vicinity of focus. The short interaction time limits the processes within an absorption volume, sometimes called a "voxel" (volume pixel) when using focusing conditions.

2.2.1.1 Reduction of the damage threshold

One principal advantage of ultrafast lasers in precision micromachining is the high machining quality owing to the nonthermal nature of the process by decoupling the energy deposition from thermal dissipation. In order to understand the process several models have been proposed. Among them, a two-temperature model, described by a non-equilibrium state of "hot" electrons combined with relatively "cold" surroundings [13, 14]. The energy loss due to thermal diffusion is negligible compared to the ns laser pulse–matter interaction regime. Therefore, the material damage thresholds are reduced [3,

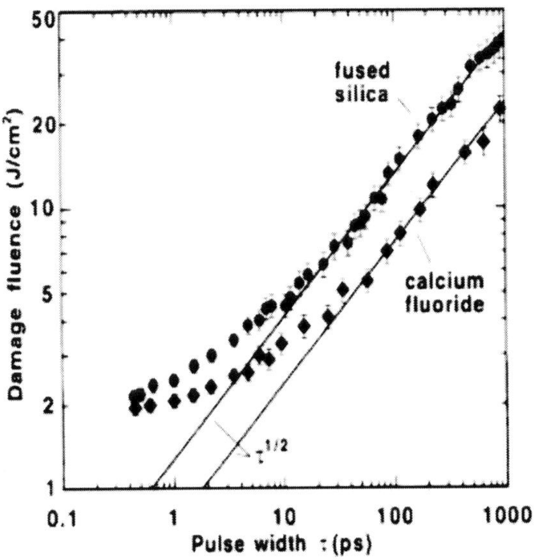

Figure 2.1 Damage fluence for fused silica and CaF_2 [7].

7]. Figure 2.1 shows the reduction of damage fluence of both fused silica and CaF_2 as the pulse width is decreased. For pulse durations of several tens of picoseconds (ps) the plot fits well the $\sqrt{\tau}$ scaling, implied by thermal diffusion [6, 7, 15]. For short pulse durations other mechanisms are dominant.

2.2.1.2 Nonlinear absorption processes

Both linear and nonlinear absorption mechanisms occur in ultrafast laser interaction with matter. For linear absorption to occur, the bandgap energy of the material must be smaller or equal to the photon energy, and Beer's Law applies:

$$I(z) = I(0)e^{-az} \tag{2.1}$$

where a is the absorption coefficient and z is the propagation distance along the axis of the incident light beam. In this case, energy deposition is not localized. Several nonlinear mechanisms occur. Both tunneling ionization and multiphoton ionization (MPI) are dominant absorption mechanisms (Fig. 2.2) in micromachining of high bandgap dielectrics including transparent optical materials with bandgap energies $E_g > \hbar\omega$. With MPI, n photons of energy $\hbar\omega$ are required to ionize atoms A with charge state q, according to

$$n\hbar\omega + A^q = A^{q+1} + e^-, \tag{2.2}$$

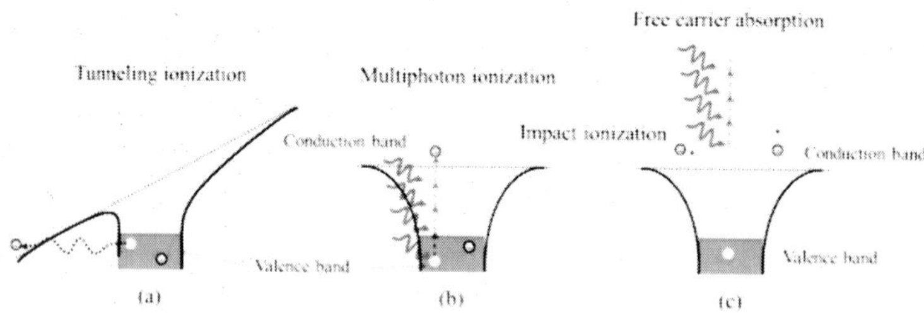

Figure 2.2 Schematic diagram of ionization mechanisms (a) tunneling ionization (b) multiphoton ionization (c) avalanche ionization consists of free carrier absorption and impact ionization.

MPI of transparent material is the dominant ionizing mechanism for intensities in the range from 10^{12} to 10^{16} W/cm^2 at 800 nm wavelength of common ultrafast laser systems. The MPI rate W is described as the differential of the free electron density,

$$W(I, \hbar\omega, E_g) = \sigma_n I(t)^n, \tag{2.3}$$

where $\sigma(n)$ is the absorption cross section of n photons [16, 17].

Tunneling ionization, characterized by the Keldysh parameter γ [18] and generally occurring when $\gamma < 1$, however is dominant if the peak laser field strength exceeds the atomic field strength,

$$\gamma = \frac{\omega}{e}\left[\frac{mcn_0\varepsilon_0 E_g}{I}\right]^{1/2} \tag{2.4}$$

where m and e are the mass and charge of the electron respectively, c is the speed of light, E_g and n_0 are the bandgap energy and the refractive index of the material of interest, respectively, ω and I are the optical frequency and intensity of the laser, respectively, and ε_0 is the permittivity in vacuum. If $\gamma > 1.5$, the ionization follows MPI, otherwise tunneling takes place. Because of its nonlinear nature, the absorption is confined only to the tightly focused volume in the medium. Avalanche ionization (Fig. 2.2) resulting from free-carrier absorption and subsequent impact ionization, also occurs when electrons are present in the conduction band. This process will continue throughout the duration of a light pulse increasing the free electron density in the focal volume [7].

2.2.2 Ablation Mechanisms and Surface Structuring

Material ablation and surface structuring were the first industrial applications of lasers (see also Chapters 5 and 13). Ultrafast laser radiation has some distinct advantages for processing many materials. The high peak intensities enable processing material such as ceramics or composite compounds that are difficult to process using conventional techniques due to their physical properties.

2.2.2.1 Surface modification thresholds

Surface modification of materials by laser light is characterized by the distinct ablation or damage threshold (J/cm^2) mentioned above. Below this threshold fluence the material is essentially unchanged. When the threshold fluence is reached, however, permanent modification can be introduced. Figure 2.3 shows an example of a modified and ablated area on the surface of a borosilicate glass after ultrashort laser irradiation. Although the material is ablated in the central region it is surrounded by a region that is photo-expanded.

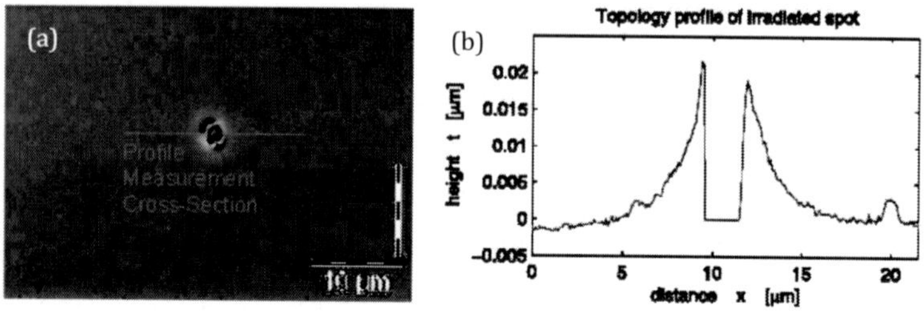

Figure 2.3 (a) 100× light microscope image of an ablation spot on borosilicate glass resulting from 250,000 fs-pulses with 85 nJ pulse energy focused by a 40× NA 0.65 microscope objective; (b) measured surface profile.

A common technique of determining the modification/ablation threshold of a material is described by Liu [19]. The surface modification/ablation threshold depends on whether the material is a metal, dielectric, or semiconductor. Consideration of this threshold for dielectric materials with ultrashort laser radiation helped redefine the approach to determining its value. With longer laser pulses (t_p > 10 ps) breakdown threshold values were determined statistically since the energy absorption process was dominated by heating of conduction-band electrons at defect sites, followed by avalanche ionization and energy transfer to the material lattice [20]. A more deterministic value can be obtained with ultrafast laser irradiation since MPI produces directly ionized conduction-band electrons without the need of a defect site [3, 6]. The same holds true for semiconductors. Whereas with longer pulses, the ablation differs from metals and dielectrics, with ultrashort pulse radiation the interaction is identical to that occurring in metals [21]. Another important factor in ablation or surface structuring is the penetration depth of the radiation from the surface into the material. For ultrashort pulses incident on transparent dielectrics, the absorption depth depends on the focusing geometry with nonlinear effects occurring where the field intensity is highest. For metals and semiconductors, the penetration depth is very shallow, given by the skin depth of the material and used radiation.

2.2.2.2 Material ablation

Material ablation occurs when matter is either directly removed by the deposited energy or indirectly by means of an established plasma on the surface. In the first case, the ionization process on the surface generates free electrons, which when ejected rapidly from the surface creating a positive space charge from the ions that are left behind. If this charge is strong enough the remaining matter will nonthermally explode, giving rise to a so-called Coulomb explosion [21]. This phenomenon can be observed only on dielectric materials since metals and semiconductors posses a natural bath of free electrons neutralizing the repulsive force of the positive charges of the remaining matter. Beyond the electronic ejection of matter from the irradiated spot, there is also a thermal process that allows melt expulsion and vaporization of matter on a much slower time scale. These slow thermal processes have shown to decrease ablation rates compared to ultrashort ablation [22, 23]. The ejection mechanisms are also different. For fs pulses the process is driven by direct solid-plasma transition [24, 25]. In the case of ns and longer pulses, the material has sufficient time to change into a molten state. The material ejection here is mainly driven by boiling and evaporation from a liquid phase [6, 24]. The short interaction time of ultrashort radiation with the matter decouples the energy deposition process and the thermal dissipation process of the deposited energy within the material lattice. Moreover, since the material state change directly from solid to plasma, there is no melt ejected at the ablation side that would redeposit on the surface. This results in an appreciable reduction of the heat-affected-HAZ surrounding the ablated region leading to smaller feature sizes [22, 24]. There is a dramatic difference in the surface ablation characteristics of dielectrics with fs and ns pulses at the same fluence (Fig. 2.4). Furthermore, if precision Gaussian profiled beams are used, by operating close to threshold, process dimensions much less than the diffraction limit can be obtained

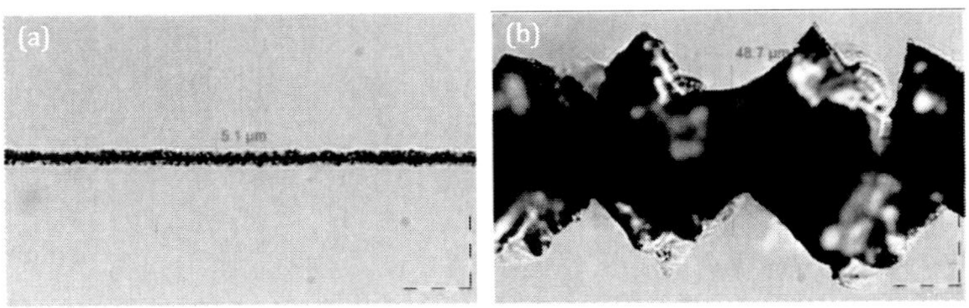

Figure 2.4 Image of an ablation track on borosilicate glass surface using a microscope objective with a NA 0.5 and approx. 100 J/cm^2, (a) fs laser radiation, (b) ns laser radiation.

[26] (Fig. 2.5). Plasma etching is rather uncommon using ultrashort lasers since an effective heating of the plasma cannot be achieved using such laser systems. The generation of a plasma on the surface can reduce the efficiency of the direct ablation processes since the plasma will shield the matter from the incident energy [27].

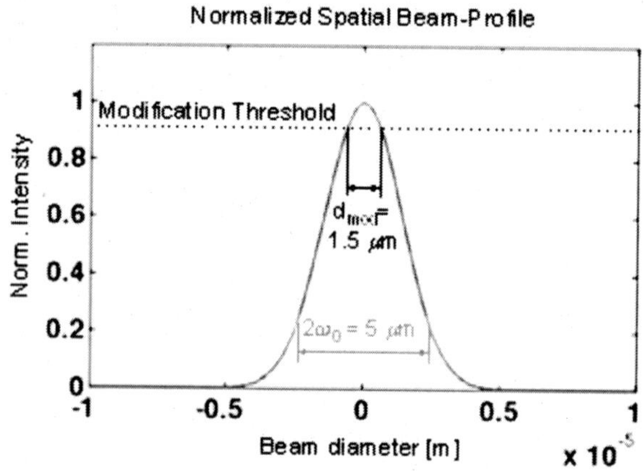

Figure 2.5 Reduced modification diameter due to modification threshold when using ultrashort-pulsed radiation.

2.2.2.3 Surface texturing

Beside ablation, surface texturing or patterning is an important field of interest since it can influence the behavior of how a material interacts with its surroundings. Nanometer-scale surface texturing, such as that found on lotus flower leaves, prohibits the adherence of water and dirt (see also Chapter 6). When irradiating surfaces with ultrashort radiation, a similar effect can be created as nanostructured ripples or laser-induced periodical surface structures (LIPSS) having surface periodicities well below the laser wavelength, as opposed to their counterparts produced by longer pulses [28–30]. The origin of the structures generated with ultrashort radiation requires further study, with several theories being developed [30–32].

2.2.3 Bulk Material Modification below the Ablation Threshold

With ultrafast laser pulse these absorption processes can lead to an interesting new set of effects in and on the surface of materials. Above a certain threshold intensity, surface ablation of the material occurs. If the material is transparent to the laser light, and it is focused inside the material, then several phenomena can occur. Depending on the nonlinear refractive index, n_2, self-focusing of the beam can occur

$$n = n_0 + n_2 I \tag{2.5}$$

At intensities inside the material sufficient for avalanche ionization to lead to catastrophic mechanical breakdown optical breakdown or a micro-explosion occurs. However, unique to ultrashort pulses, because the absorbed energy by nonlinear mechanisms is localized and occurs in times shorter than any diffusion times, material effects can be created deep inside transparent media by these same absorption processes at intensities well below the ablation threshold. In particular localized bond structure changes can be induced, or changes in the density of the material that result in changes in the refractive index. This creates the opportunity, richly exploited in many glasses, some polymers, and a few crystals, of making optical waveguides, diffractive structures, and micro-fluid structured enabled by the differing chemical etch rates of modified material.

The first reports of bulk refractive index modification in transparent media irradiated by ultrafast lasers were published 15 years ago [33, 34] (see also Chapters 8 and 9). Since then intense studies have explored the physical mechanisms that give rise to this modification in different media by many groups. Fused silica has been the medium investigated most. Positive refractive index changes Δn of $\sim 10^{-4}$ can be produced. In phosphate glasses a negative change of the same order can be created. The highest refractive index changes, approaching 10^{-2} are created in chalcogenides, such as $Ge_{23}Sb_7S_{70}$ and As_2S_3 [35].

Many photonic devices, from as simple as waveguides and waveguide arrays in different transparent dielectrics [36–38], integrated diffractive elements such as Fresnel zone plates and FBGs [39–43], three-dimensional beam splitters [44–48], to active photonic devices have been demonstrated using the direct writing technique [40, 41, 49, 50]. Since each sample material responds uniquely with positive or negative refractive index change to the fs pulse, tuning of the embedded geometries allowed to utilize a wider variety of host materials, not just dielectrics, for the introduced photonic circuits [51, 52]. Phenomena such as heat accumulation

(HA), discussed below were discovered and utilized to improve the loss characteristics of the written waveguides [53, 54].

2.2.4 Heat-Accumulation Effects in Laser Processing

An important consequence of ultrashort laser materials processing is the reduction of the HAZ due to the decoupling energy deposition and energy dissipation, as depicted in Fig. 2.6.

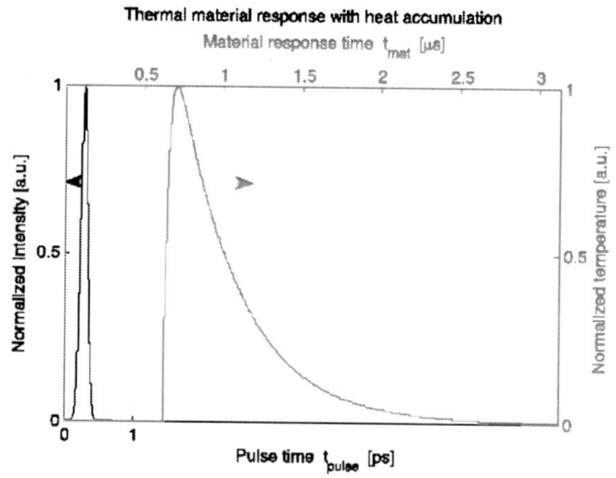

Figure 2.6 Model of energy deposition and thermal material reaction.

If the repetition rate of the laser pulses increases to the point where the duty-cycle time is shorter than the energy dissipation time then heat will build up in the focal volume. This process is referred to as HA. Depending on the pulse-to-pulse duty-cycle and the material energy dissipation time, this thermal build-up can be very significant, allowing the local temperature in the voxel to rise to several thousands of Kelvin, as reported by numerous researchers [53, 55]. Thus variation of the pulse repetition rate adds an important control parameter to the processing conditions by providing the means to control the local temperature of the material. Since many of the physical and mechanical properties of the material are temperature-dependent, this capability has a profound impact on the capabilities of ultrafast laser materials processing.

It is easy to recognize this localized heating effect during the irradiation process since it is accompanied by strong white light emission (Fig. 2.7).

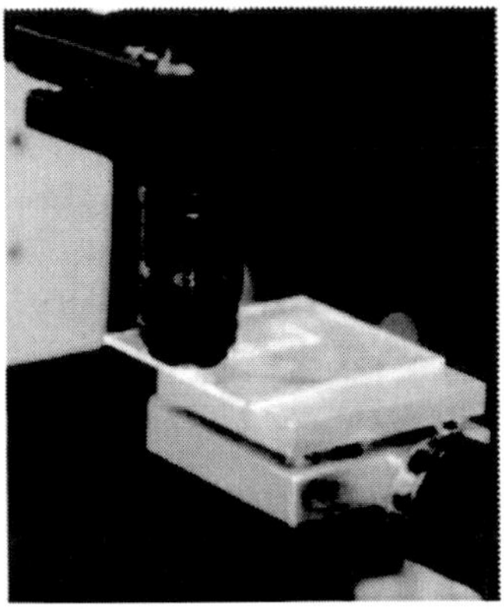

Figure 2.7 White light blackbody emission in the focal volume writing in heat-accumulation regime.

HA can be used in several processing techniques. It improves the quality of index-modified material for waveguides and laser-written optical elements [53], and can in some case increase Δn, the change in refractive index [54]. Operating the material at higher localized temperatures can dramatically increase the process rate [56]. A potential disadvantage to operating in the HA regime is the possibility of higher degrees of HAZ [53]. For the fabrication of microfluidic channels, the etch ratio may be adversely affected by the reduction or elimination of the nanogratings within the laser-written structure by this localized melting process [57].

2.3 Principal Application Areas and Their Dependence on Laser Parameters

2.3.1 Micromachining Based on Ablation

The pioneer researchers on ultrashort-pulsed micromachining focused on the rigorous measurement of the etch rate of the material removal as functions of laser parameters [3, 58]. They demonstrated rigorous study on the ablation by fs lasers. Based on their achievements, drilling has been intensively investigated: high edge quality and high aspect ratio have been demonstrated in a variety of materials including metals, ceramics, and dielectrics [22, 24]. Much work has been done in ultrashort laser development since the early experiments in ultrashort material processing, which led to much simpler to maintain laser system. With that development in mind, cutting of ultrahard materials became of interest as a new field of applications [59–61]. Nevertheless, scaling of repetition rates and average power is still necessary to deploy ultrashort pulses laser system in a bigger scheme in industrial environments. Ultrashort pulse lasers were utilized with or without mask.

2.3.2 Surface Structuring

Several novel phenomena occur when using ultrashort laser radiation for surface structuring. One of these is a ripple structure. The application fields for such nanostructures are manifold, from biotechnology and devices for directed cell-growth [62], over structuring for biocompatibility of artificial materials [63] to surface protection using structures similar to the Lotus effect [64, 65]. Moreover, the hydrophobic properties of materials can be varied widely when nanostructuring their surface. The requirements on the employed laser systems to achieve nanostructures are thereby rather relaxed, only demanding sub-ps pulses and more a few shot per spot, as shown exemplarily in Table 2.1.

The most influential parameters for the generation of such periodic surface structures are the polarization of the incident radiation, the process atmosphere the irradiation takes part in, and the overlap of the irradiation spot. These parameters determine the direction of the periodicity of the structure, the final form that the structures will take, and the homogeneity of the overall irradiated area. These factors influence the choice and operating parameters of the Therefore it becomes obvious that the processing station is an essential factor of the development process.

Beyond the purely topographical use of surface structuring there is also a growing need to modify the chemical and crystalline properties of materials to tailor their behavior toward new applications. Especially semiconductors have been shown to be quite susceptive to ultrashort laser radiation to modify their absorption behavior. It has been demonstrated that the topmost layer of silicon can be highly doped when ultrashort irradiated in a chalcogen-atmosphere, known as "black silicon" and first reported by Mazur *et al.* [66]. This process generated a new bandgap scheme in the diffusion zone of the material, a so-called subbandgap, which then exhibits increased sensitivity in the IR spectrum [67]. During the fabrication process most important is again the process atmosphere. The direct use of this technology was demonstrated when the efficiency of solar cells was increased by 50% [68].

Table 2.1 Laser systems used to generate periodic surface structures

Laser wavelength	Pulse duration	Number of shots	Reference
λ [nm]	τ [fs]		
810	120	10^3 to 10^6	[69]
800	<15	10^6 to 2.5×10^8	[70]
800	130/200	—	[71]
800	120	5000	[72]
400/800/2000[1]	100	several 10	[73]
1300/2100[1]	<50	—	[74]

2.3.3 Back-Side Structuring

A very interesting application for ultrashort pulses laser systems is the machining of back-side surfaces of transparent materials. Since the ablation is based on a nonlinear absorption process located in the focal volume of the beam, the ablation spot can be freely translated on the front as well as back-side of the sample. Shifting the laser center wavelength toward the IR-region of the spectrum, such processing can become very useful in the semiconductor industry.

2.3.4 Laser Surface Cleaning

An indirect application of ultrashort laser radiation was demonstrated when using the plasma filament generated by a 130 fs laser pulse for wafer clean [75]. The filament with a length of approximately 120 mm was translated in front of a silicon wafer to remove 100 nm sized particles from its surface. The major parameter to control in that case was the gap between the wafer and the filament, which also illustrates the need for specialized stages.

[1]Mid-IR wavelength were generated using an optically parametric amplifier (OPA).

2.3.5 3D Structuring and Microfluidics

In-volume laser processing of dielectrics and semiconductors using tightly focused ultrafast laser radiation (pulse duration $t_p <$ 10 ps) is a promising, novel technology, enabling refractive index modifications for waveguide applications, laser fusion welding of similar and dissimilar materials, and selective etching of high aspect ratio microstructures (Fig. 2.8). This versatility in combination with high precision opens new avenues to a broad variety of applications in integrated optics, microfluidics, and micromechanics.

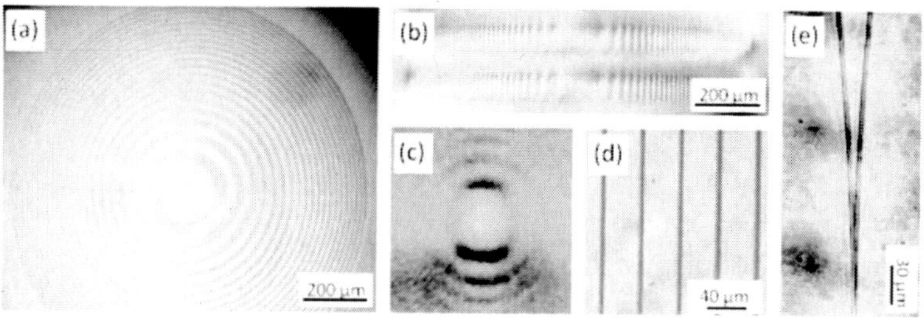

Figure 2.8 (a) Optical image of a Fresnel zone plate written in a borosilicate substrate, top view, (b) side view of FZP, (c) far field image from Telluride waveguide, (d) waveguides written in SiC, top view, (e) optical waveguide splitter written in a borosilicate glass.

2.3.5.1 Fusion welding of glass

The unique property of localized energy deposition in the volume of transparent dielectrics enables the use of the laser energy as a heat source for fusion welding of glass. Similar to the welding of plastics, glass-welding exploiting nonlinear absorption of ultrafast laser radiation is performed in the interior of a transparent material, avoiding the emission of any contaminants. Welding of various glasses has been performed exploiting nonlinear absorption of tightly focused fs- and ps-laser radiation. Ultrafast laser joining of similar and dissimilar materials can find numerous applications in fabrication of sensors, micro-electromechanical systems (MEMS) and microfluidic devices.

Stuart *et al.* reported internal modification of bulk glass by ultrashort laser pulses [6], where laser energy is absorbed by nonlinear process including MPI followed by avalanche ionization.

This unique property has also provided a new opportunity for high-repetition-rate ultrafast laser radiation to be used as a heat source for fusion welding of glass, where the interface of glass plates can be selectively melted without any absorbent giving no influence to the top and the bottom surfaces of the glass plates, if the laser energy is high enough.

Meanwhile local melting of glass without development of cracks has been reported using ultrashort laser pulses for joining fused silica and dissimilar glass [76, 77]. It was also clearly shown that laser-melted region of glass plate contains no cracks by SEM pictures [78] (Fig. 2.9). In addition, the advent of ultrashort pulse laser with high pulse-repetition rates with enough pulse energy has enabled a wide variety of internal modification by utilizing HA for internal modification of bulk glass, and also enabled the fusion welding of glass at high throughput [79]. While the process of fusion welding of glass is similar to the laser transmission welding of plastic in a sense that the welding is done interior of transparent material, and can be used even in clean room due to no emission of any contaminant, big difference is fusion welding glass is based on nonlinear process caused by ultrashort pulse laser without necessity of any absorbent [80] (Figs. 2.10 and 2.11).

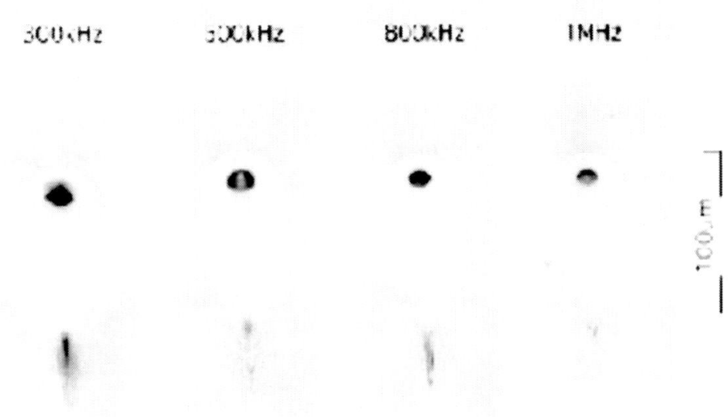

Figure 2.9 Ultrafast laser-induced modification and cracks observed in fused silica at different repetition rates (cross-sections, t_p=16 ps, NA = 0.55, P = 6 W, λ = 1064 nm) [78].

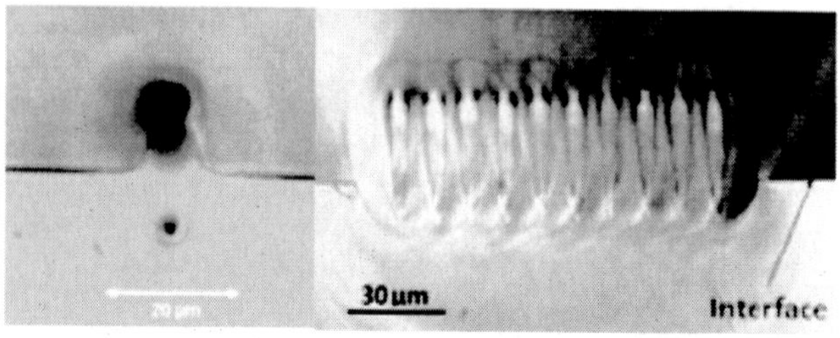

Figure 2.10 Cross sections of the weld seams using single-path and multi-path overlapping joints [80].

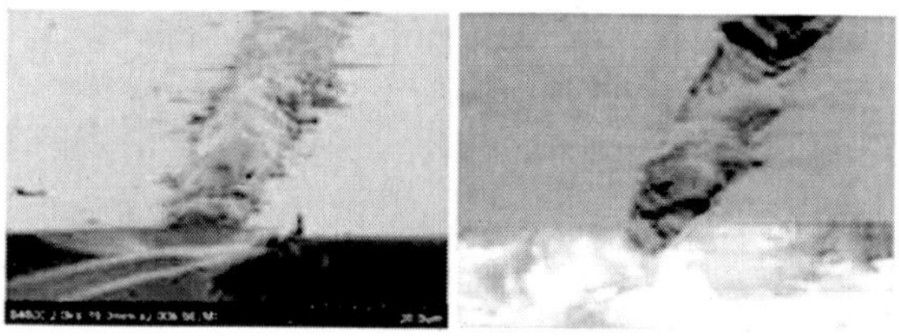

Figure 2.11 SEM images of the peeled-off weld joints [80].

Welding of glass with silicon has been reported using high-repetition rate ultrafast laser radiation [81]. The laser-induced modifications of the glass-silicon interface have been analyzed by optical microscopy at cross-sections, and by SEM after mechanical clipping of the samples (Fig. 2.12). Thereby the morphological changes in the glass sample are very similar to those observed in welding of glass with glass. However the modifications in the silicon wafer do not exceed 2–3 μm in depth and width. The mechanical clipping of the glass and silicon samples reveals large glass residues left on the surface of the silicon wafer. Narrow contours of the weld seam can be observed by optical microscopy within the glass residues. The width of these joint modifications is nearly equal to the focus spot size of the laser radiation $d_f = 3.9$ μm and indicates a large mechanical joint strength. Additionally, welding of glass with polymers [82] and metals [83] including alloys has been reported.

2.3.5.2 Selective laser etching and microfluidics

A combined approach of material modification followed by chemical etching provides the possibility of manufacturing micro channels in the volume of a material or surface structures (Fig. 2.13) for example, for microfluidic applications [84–87](see also Chapter 10). Aqueous solutions of hydro-fluoric acid (HF) as well as potassium hydroxide (KOH) have been shown to work as an etching agent [88, 89]. Etching ratios between modified and unmodified material of about 100:1 in glass, 200:1 in fused silica, and 10^4:1 in sapphire have been reported [15, 90, 91]. A combination of selectively etched micro channels and waveguides enables lab-on-a-chip devices, such as for laser induced fluorescence (LIF) detection in biomolecules or the inspection of living microorganisms. Selective etching of glass materials is usually described in terms of a density decrease and a higher etch rate for the less dense material. Similar to birefringence in fs laser written waveguides, the formation of nanoplanes leads to a dependence of the etch-rate on the incident laser polarization [92].

(a) (b)

Figure 2.12 Welding seams observed between glass and silicon wafer in (a) cross section and (b) after separation of the welded materials in top view, imaged by optical microscopy [81].

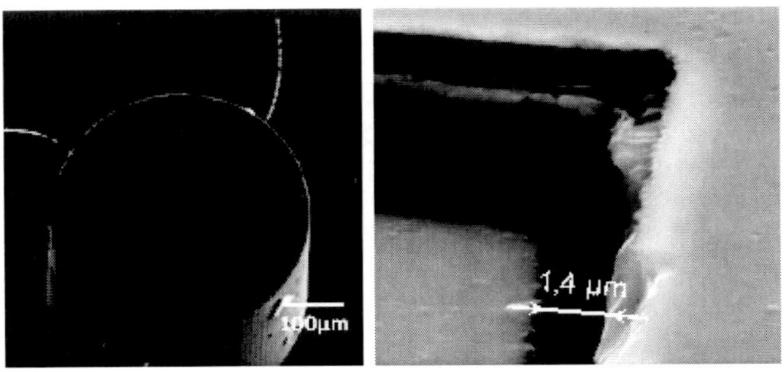

(a) (b)

Figure 2.13 Sapphire micro-cylinder produced by selective etching featuring surface roughness of the etched parts <100 nm (left). Corner of a selectively etched micro-cube in sapphire (right). (Courtesy of LLT.)

Microchannels and self-assembled nanostructures were produced in the bulk of sapphire and quartz glass by fs-laser irradiation followed by chemical etching with HF or KOH [93, 94]. After etching self-organized nanostructures, elliptical or circle microchannels are produced, depending on process parameters. By irradiating larger volumes, extruded 2D or complex 3D structures, even with internal hollow structures can be manufactured by this technique. Using a uniquely in-house developed scanning system, the manufacturing time for such parts has been decreased by a factor of more than 100 and the potential of commercially available high-repetition-rate fiber CPA (FCPA) lasers can be fully exploited. The potential of the selective etching as a very promising technique for mass production of microfluidic and micromechanical parts is determined by the very low material consumption and the high structural flexibility in three dimensions (Fig. 10.12).

Comprehensive reviews on research and development of microfluidic devices based on selected laser etching can be found for FOTURAN glass and fused silica elsewhere [95, 96].

2.3.6 3D Laser Lithography Based on Two-Photon Polymerization

Two-photon polymerization (TPA) has opened a new avenue of 3D structuring. This application was exploiting in several new fields of research including photonic crystal [97] and micromechanical elements for MEMS applications [98]. Entering to the 21st photonic era, promising novel photonic devices exploiting photonic crystal structures attracted much attention for the feasibility to control photons. To bring the available spectrum to NIR communication regime, forming submicron-width is essential. High-resolution writing optics with low power oscillator of Ti:Sapphire at 80 MHz are utilized. This method offered 3D photonic structures, which is not available with any other lithographic technique. They demonstrated that fs-laser direct writing can be a strong tool in fabricating photonic structures and microdevices. Furthermore, this method evolves as a nanostructuring tool toward fabricating metamaterials by virtue of the innovative stimulated emission detection (STED) microscopic technique for high resolution beyond diffraction limit [99, 100]. STED microscopic technique was adapted to TPA-based laser lithography to improve the resolution of optical diffraction limit. Fs laser pulses followed by cw doughnut beam has increased the resolution; about $\lambda/20$ of the writing wavelength at 800 nm could be achieved with STED inspired laser lithographic technique [101]. This method has been also applied to create metamaterials in optical regime [102].

2.3.7 Structures in Refractive Index Modification

Since 1996, when it was shown for the first time [33, 34], refractive index modification in the volume of glasses by fs laser irradiation is rapidly becoming an alternative to standard fabrication techniques for optical waveguides [36, 45, 48, 103–105] (see also Chapter 9). The possibility of processing transparent optical dielectrics enables to fabricate photonic devices such optical waveguides [36, 38], optical couplers [106], three-dimensional beam splitters [44–48], gratings [107], integrated diffractive elements such as Fresnel zone plates and FBGs [39–43] and diffractive lenses [107–110]. Refractive index increases of up to 5×10^{-3} in fused silica and 3×10^{-2} in phosphate glasses are realized [53, 111, 112]. While most of studies utilized optical glasses, several groups [37] used optical polymers demonstrating that extraordinary tubular waveguide was formed in PMMA. The size of such structures is limited to the focal volume of the incident laser radiation, and scanning the sample relative to the laser focus can provide a 3D capability for the refractive index modification technique. Thus, a broad variety of applications in telecommunication and integrated optics are enabled. Passive elements such as beam splitters and couplers, three-dimensional optics and phase plates are realized as well as signal amplifiers and waveguide lasers [43, 45, 104].

By externally modulating the laser energy and/or the writing speed, frequency selective Bragg gratings are realized [113, 114]. Depending on the pulse energy and the focal intensity distribution, nonlinear propagation effects such as self-focusing and filamentation are influencing the resulting structure shape and can lead to crack formation [115, 116]. Formation of self-assembled nanostructures (nanoplanes) perpendicular to the polarization of the incident laser radiation leads to birefringence in fs laser written waveguides [85, 117, 118]. One reason for the fs laser induced refractive index changes is the density change induced by a localized heating and fast cooling of the material. Even below the glass transition temperature, thermal expansion and induced pressure waves lead to a permanent morphological change of the material [119, 120]. At high repetition rates of the incident laser pulses, HA effects become more prominent [53]. Besides thermally induced density changes, the formation of color centers leads to a refractive index change, induced by the electron dynamics during and shortly after the pulse [33, 111]. Investigation of refractive index modifications with temporally shaped pulses (t_p < 3 ps) and fs double pulses (t_p < 200 ps) confirm a strong influence of the temporal intensity profile on the final refractive index profile and the damping of the resulting waveguides [121, 122].

Different glasses such as fused silica, phosphate, telluride, and chalcogenide glasses have been processed in order to locally change the refractive index, usually a reduction in value of the order of 10^{-4} to 10^{-1}, depending on the material used [38]. Waveguides generated in chalcogenide bulk glass and thin films have been demonstrated [123]. The unique transmittance of this type of glass in the far infrared (FIR) region of the spectrum will enable applications in biological and medical sensing. Using the available low and high-repetition-rate laser systems the influence of HA in these functional glasses on the resulting refractive index change has been investigated.

Active waveguide are also shown to be a building block of photonic light sources. Er:Yb doped phosphate glass are first used as an active medium of direct laser writing [40, 41, 49, 50, 124, 125]. More complicated photonic structures were later demonstrated such as X and Y coupler [126] and gratings [127].

Diffractive optical elements (DOEs) provide a pathway to fabrication of compact photonic devices by substituting thin diffractive optics for bulky refractive optics. Furthermore, μ-DOEs offer new functionality with regard to beam shaping techniques for laser materials processing as well as to integrated platforms toward optoelectronic sensors and such. Using direct writing techniques, volumetric DOEs (Fresnel zone plates) have been created by alternating either void formation or the modification of the local refractive index in optical materials (Fig. 2.14) [128, 129]. Various integration schemes of laser-written DOEs in photonic and optoelectronic devices have been investigated [130].

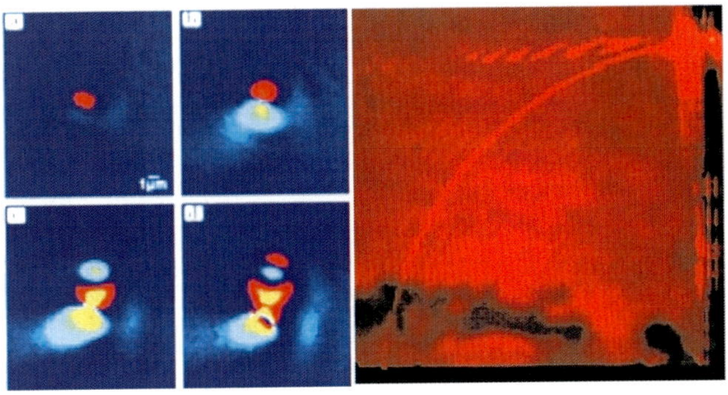

Figure 2.14 Left: Cross-sectional refractive index distribution in Yb-doped IOG glass, irradiated with increasing pulse energy. Right: Curved waveguide in D263 glass, written with a self-developed scanning system with sub-μm-resolution and scanning velocities > 100 mm/s (courtesy of LLT).

Waveguiding structures have been manufactured by fs-laser irradiation in the volume of various glasses such as fused silica, D263 and rare earth doped phosphate glasses (Schott IOG, LG760). A refractive index increase of up to 2×10^{-3}, resulting in waveguides with a numerical aperture of NA = 0.1 has been achieved in fused silica using fs-laser double pulses for irradiation [112]. Additionally, the effect of pulse energy, pulse repetition rate, and scanning velocity on the refractive index change, structure size, and shape has been investigated. *Ex situ* phase contrast microscopy and white light interference microscopy have been utilized with longitudinal and cross sections of the written structures [131].

Curved waveguides with numerical apertures of up to NA = 0.1 have been realized. Using a commercial scanning system, combined with a self-developed imaging system and a microscope objective, scanning velocities up to 100 mm/s in combination with submicron focal spot size of the laser beam have been realized [57].

2.3.8 Other Applications

The attempt to develop optical storage is using fs laser has been demonstrated [34]. Recently, generation of metallic nanoparticle in silver containing glass is demonstrated. Third harmonic generation in the interface of nanoparticle was detected by high-resolution THG microscope [132].

2.4 Ultrashort Pulse Lasers: Scientific and Commercial

Ultrafast material processing comprises applications as diverse as the probing of minuscule refractive index changes, engineering of material properties on a submicrometer scale, writing of features in materials via local refractive index modification, welding of polymers or glass, and more recently the processing of biological material [133]. For these applications, laser requirements can dramatically differ and each specialty of ultrafast material processing imposes high—and different—constraints on the laser system employed. Here we summarize the range of ultrafast laser technologies, and review how their performance characteristics impacts the material processing community.

2.4.1 Recent Developments on Ultrashort Pulse Generation

Since the first laser pulse was shot, the goal has consistently been for researchers and engineers to develop robust laser systems delivering shorter and shorter pulses with increasing pulse energy. This trend is shown in Fig. 2.15. The main driving motivation was—and still holds

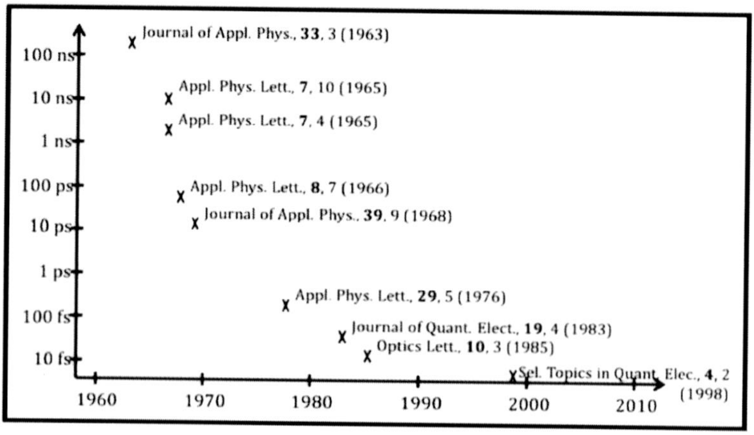

Figure 2.15 Milestones in the evolution of pulse duration from laser systems over the past five decades.

nowadays—to achieve the highest peak powers and intensities to probe physical phenomena and investigate light-matter interactions. Remarkably, the evolution of pulse duration as a function of time over the past decades shows a quasi-perfect exponential decay for wavelength in the near-IR spectral region [134, 135].

Advances in the past decade have unveiled the possibility to further scale down the pulse duration to the attosecond duration [136] by operating at wavelength in the UV and XUV spectral region. This spectro-temporal regime, currently available in a handful a laboratory worldwide will certainly enable the emergence of a new type of material processing.

2.4.1.1 Laser of the early days of material processing (70s and early 80s)

Dye lasers were widely used for the generation of fs pulses in the 70s and the 80s but were rather complex to operate. Alternative solutions relying on spectral broadening and pulse compression of Nd:YAG or Nd:glass lasers offered limited energy scaling prospective due to the limitations in power handling of the spectral broadening stages.

2.4.1.2 Broadening the spectrum... and the horizons (late 80s)

Up until the mid-80s, the main limitation toward the development of laser systems delivering ultrashort laser pulses laid in the lack of gain materials with large emission bandwidths. The mathematical constraints imposed by the time-bandwidth-product relates the duration of an optical pulse to its spectral bandwidth. Sufficient spectral bandwidth is required to support a fs optical pulse. Typically, at 800 nm central wavelength, a 10 nm spectral width results in a 100 fs pulse while 50 nm of bandwidth enables pulses as short as 20 fs. Most solid-state materials available until the mid-80s offered spectral bandwidth hardly exceeding a few nanometers prohibiting routine generation of sub-200 fs pulses. In the mid-80s, progress in material science enabled the growth of crystals such as Cr:LiSAF, Tm:YAG, Ho:YAG and Ti:Sapphire with good optical quality. This was a major breakthrough for the laser development community as for the first time spectral bandwidth extending over 100 nm could be generated and amplified. Through the development of intra-cavity dispersion compensation schemes, such as prism compressors, reliable and robust oscillators bloomed delivering hundreds of mW output power at tens of MHz repetition rate and pulse duration largely below the 100 fs landmark.

2.4.1.3 Energy scaling, pulse shortening, wavelength scaling (mid-80s to today)

The availibility of pulses with sub-100 fs pulse duration allowed a number of exciting applications in the field of material processing but the energy level, typically in the nJ regime, of these novel short pulse oscillators was unsufficient for a number of application. Again a major breakthrough occurred with the invention of the CPA (discussed in Section 2.4.2.4) technique. In this technique, a nJ pulse is temporally stretched, amplified, and temporally compressed, therefore alleviating peak power related issues upon amplification. This technique is currently widely spread and has enabled major advances largely exceeding those of material processing.

Further refinements in spectral phase management inside Ti:Sapphire oscillators have led to sub-two-cycle pulse duration of 5.4 fs from an oscillator [137]. Further refinements in pulse amplification in particular with the use of optical parametric chirped pulse amplification (OPCPA—discussed in Section 2.4.4) have led to the amplification of these ultrashort pulses to energy levels approaching the mJ energy level at kHz repetition rate [138]. Finally, the availibity of high energy pulses in the near-IR have made possible nonlinear shifting of central wavelengths to the mid-IR spectral region while preserving well beyond mJ energy at kHz repetition rate and sub-100 fs pulse duration using optical parametric amplification (OPA—discussed in Section 2.4.4) [139].

2.4.2 Techniques for Generation of Ultra Short Pulses at High Average Power

A number of modern material processing applications call for laser systems delivering high average power since both high energy (for achieving the foreseen light-matter interaction) and high repetition rate (for rapid processing) are usually needed. Designing and manufacturing such lasers is a particular challenge for the laser engineers and requires merging and selecting a number of techniques. This section offers a description of the most common techniques employed for manufacturing laser systems suitable for the demanding material processing applications.

2.4.2.1 Laser oscillators

Mode-locked oscillators can be well-operated at low average power emitting pulse trains of tens of MHz and average powers of below 5 W. The energy per pulse cannot typically be scaled by increasing the mode size in the cavity in particular in the common case of Kerr lens mode-locking (see section 4.2.2). Increasing the pulse energy from oscillators has been achieved by reducing the repetition rate since oscillators typically maintain output average power. This can be done by inserting intra-cavity Herriott-type multipass cell (MPC), which effectively increases the cavity round trip time of the oscillator by increasing the cavity length. As a result, the repetition rate is typically lowered from almost 100 MHz to several MHz and the pulse energy can be increased by a similar ratio (5–100) [140].

2.4.2.2 Mode-locking techniques

Mode-locking techniques rely on the modulation of intra-cavity losses. Contrary to Q-switching where the losses are set by an external trigger reference, in the case of mode-locking the losses are modulated at exactly the repetition rate set by the length of the oscillator cavity. This allows only a short temporal window of lasing per round trip. Short low-loss time gate initiates the pulse formation and maintains a short pulse duration. Temporal shortening of the pulse is assisted by the fact that the center of the temporal window exhibits lower losses therefore has a higher photon density and experiences more amplification. This shaping mechanism leads to shorter and shorter pulse per round trip in the transient regime setting up the mode-locking only limited in the steady state regime by the spectral extend of the emission bandwidth of the gain medium.

Active mode-locking

Active mode-locking is no longer extensively used to obtain short pulses from oscillators because of its complex implementation compared to passive mode-locking techniques. Nonetheless, it provides an easy picture of the mode-locking mechanism that is common to passive mode-locking techniques. Acousto-optical modulators (AOM) with high modulation depth are typically employed to achieve active mode-locking.

SESAM mode-locking

A SESAM is a special type of dielectric mirror in which a layer of semiconductor material is inserted in the quarter-wave stack constituting a dielectric mirror. The semiconductor layer is carefully designed such that the semiconductor bandgap matches the energy of the intra-cavity photon and such that the absorption can be bleached—or saturated—and the full mirror reflectivity retrieved, therefore constituting an ultrafast temporal gate. For a given pulse at each round trip the saturable absorber temporally gates the pulse with low loss and establishes a high cavity loss shortly after. The fast decay time of the excited electrons from the conduction band to the valence band in a semiconductor is typically much shorter (ps time scale) than the round trip time (ns time scale) of an optical pulse in a standard mode-locked cavity (several ns). The choice of the semiconductor material, its thickness, and its position in the stack depends on the cavity design, the wavelength of operation, and the desired saturation level. SESAM mode-locking is particularly suitable for oscillators with gain materials featuring low nonlinear refractive index corresponding to a low amount of Kerr lensing. SESAMs are also frequently used for architectures where Kerr lens mode-locking cannot be implemented such as fiber and thin-disk lasers.

Kerr lens mode-locking

Kerr lens mode-locking (KLM) is the method of choice to achieve the shortest pulse duration due to its almost instantaneous response time. Kerr lensing is either obtained in the gain medium (for Ti:Sapphire systems) or in an intra-cavity glass plate. The cavity is designed such that the round trip loss for cw operation is larger compared with mode-locking operation, which favors mode-locking operation once initiated. This loss modulation can be achieved by a soft-aperture: the mode-mismatch between pump beam and the resonating mode size for cw operation in the gain material decreases the available gain compared to the perfect mode-matching for mode-locked operation.

Similarly, a hard aperture can be implemented but involves the introduction of an extra intra-cavity component and more alignment sensitivity. Kerr lens mode-locking can be initiated by introducing a slight intensity disturbance in an otherwise cw lasing cavity, which establishes mode-locking operation. The amplification of the slight intensity disturbance is enhanced until an ultrashort pulse is formed and further stably maintained.

2.4.2.3 Master oscillator power amplifier

Most mode-locked oscillators cannot be scaled in output power without sacrificing parameters of the output pulse trains. To overcome this scaling problem, one can decouple the generation of the pulses and the amplification/power scaling. In the master oscillator power amplifier (MOPA) scheme, the oscillator can run at low average power while the extracted pulse train is further amplified in consecutive amplifiers. A common amplifier scheme is a multipass amplifier stage, where the pulse train is amplified through another pass through the same gain medium or a different gain medium with nonzero emission cross section at the used wavelength. Typical small signal gain coefficients can be between 1.5 and 10 per pass for solid-state amplifiers and can be higher than 30 dB for fiber amplifiers. A fluence similar to the saturation fluence of the used amplifier medium needs to be reached with the seed in order to extract energy efficiently. Another common technique is regenerative amplification, where a pulse is trapped in an amplifier resonator, kept inside until the stored energy in the gain medium is depleted and extracted afterwards. Typically, the overall gain of regenerative amplifiers can be 60 or 70 dB, amplifying sub-nJ pulses to the mJ pulse energy level. MOPAs are most times operated at moderate to high pulse energies in order to extract as much pulse energy while avoiding optical damage in the amplifiers. To achieve this, the repetition rate of the system is reduced with a pulse picker, which allows for recovery of the stored energy in the gain medium. The limit of energy scaling in MOPAs is given by the damage thresholds and the onset of nonlinear effects such as self-focusing. If fibers are employed, the typical maximum pulse energy is 50–300 μJ for 10–30 μm fundamental mode diameter and a typical damage threshold for fused silica of 30–40 J/cm^2. An increase in average power is most times possible by increasing the repetition rate of the system.

2.4.2.4 Chirped pulse amplification

Energy scaling of MOPAs is limited by the optical damage threshold for the used fs or ps pulses. In 1985 Strickland and Mourou [1] and in 1988 Maine *et al.* [141] presented the scheme of chirped-pulse amplification (CPA). The pulses from the mode-locked oscillator are commonly stretched in time from the fs domain to several hundreds of ps (a stretching ratio of 10,000). This can be done via Martinez [142], Öffner [2] or Barty [143] stretchers, which provide positive or normal dispersion. After amplification, the pulses are re-compressed in, for example, a Treacy (1969) grating compressor [144] providing the exact amount of negative—or anomalous—dispersion. During the amplification, the peak power is reduced far below the optical damage threshold while keeping the operational fluence close to the saturation fluence of the amplifier stage. The CPA technique is commonly used in solid-state, ultrafast, and fiber amplifiers. Ti:Sapphire based amplifiers can provide > 100 TW pulse [145] in laboratories around the world. Other solid-state amplifiers have been presented operating at several hundreds of watt average power and very high pulse energies. Fiber amplifiers have been presented with up to mJ pulse energy at moderate to high repetition rates [146]. Free-space stretcher and compressor offer scalability of peak power and high flexibility. However, for industrial applications simpler devices are preferred to allow full-day operation of the system. New technologies such as chirped fiber Bragg grating (CFBG), engineered photonic bandgap fibers (PBGF) and chirped volume Bragg gratings (CVBG) can be utilized for monolithic, all-fiber systems. Due to the limited fundamental mode diameter these systems are limited to the μJ level due to optical damage and onset of nonlinearities. A common approach is all-fiber stretching and amplification and free-space compression with aperture scalable grating compressors. As in most optical amplifiers, background amplified spontaneous emission (ASE) and pre-pulses can degrade the pulse contrast. Several devices, such as saturable absorbers, electro-optical gating, cross-phase modulation (XPM) and plasma mirrors have been developed to increase the contrast to tens of dB.

2.4.2.5 Merging technologies: the gateway to high-repetition-rate, few-cycle high-energy pulses

Despite the numerous advantages provided by CPA system, certain applications such as deterministic damage threshold measurement

[27], call for few-cycle pulses with high energy at high repetition rate. Few cycle pulses can hardly be delivered directly from a CPA system due to spectral narrowing upon amplification in either regenerative or multipass amplifiers [147]. A solution to this issue consist in performing spectral broadening at the output of a standard CPA or MOPA system in a hollow core fiber filled with a rare gas and compress the obtained supercontinuum with chirped mirrors. Such architecture is now rather common in research laboratories and delivers few-cycle pulses with hundreds of microjoules at kHz repetition rate, enabling material processing applications in the few-cycle regime [148].

2.4.3 Current High-Average Power, Ultrashort Laser Technology

The focus of this chapter is set on the current laser technology with an output performance at the μJ to several mJ level at high repetition rates beyond 1 kHz, pulse durations of 50 ps and below to sub-ps as well as excellent beam quality. These are constrains that material processing dictates to the utilized laser system for fast and efficient processing and best output result. In addition, the chosen laser technology should offer long-term stability, be alignment free, robust, and reliable as well as low in initial and running costs. Power scaling is in most of the following technologies achieved by increasing the cooling capability through different geometries. For a cylindrical rod or fiber structure, the surface to volume ratio is given by

$$\frac{S}{V}=\frac{2\pi RL}{\pi R^2 L}=\frac{2}{R} \tag{2.6}$$

for a given radius R and length L. Generally, the cooling efficiency is highest if the surface to volume ratio is largest. Thus, a smaller radius of the structure supports higher cooling capacity and generally higher average powers. Fiber lasers with a radius of a few micrometers and lengths of meters to kilometers are taking this ratio to the extreme.

On the other hand, power scaling is limited by the damage threshold and onset of nonlinear effects such as self-phase modulation (SPM), stimulated Brillouin scattering (SBS) and stimulated Raman scattering (SRS). The nonlinear effects can either lead to material damage or to the distortion of the ultrashort pulse. Since the limitations are peak intensity dependent, an increased mode diameter is desired. Thus, the onset of damage thresholds and nonlinear effects due to small beam sizes competes against the concept of increased heat removal capacity by decreased beam size. The following subsections describe several technologies, which operate preferable at high repetition rates, high pulse energies and/or high average powers.

2.4.3.1 Fiber amplifier system

In the simplest picture, optical fiber consists of a higher refractive index material in the center, termed the core, which is surrounded by a lower refractive index material, termed the cladding. Due to total internal reflection waveguiding is established and depending on core size, refractive index difference and used wavelength only a limited number of modes can be guided in this geometry. For a recent review about ultrafast fiber amplifier the authors would like to refer to reference [149].

The obtainable output pulse energy of conventional single-clad fiber is low due to the small mode size diameter. Limitations are given by bulk damage and onset of nonlinear effects. In addition, the coupling of pump light is limited to several Watts due to the availability of high power, single-mode pump diodes. Double-clad fiber offer the advantage of being multimode (i.e. large mode diameter) for the pump wavelength and thus ease coupling of pump light and maintain single-mode for the operational wavelength. This can be simply accomplished by introducing another cladding, referred to as outer cladding, which guides the higher order pump modes at the pump wavelength. If the core is doped with rare-earth ions such as Yb, a fraction of the pump mode overlapping with the core area is being absorbed along the fiber length. Recently, the average powers of fiber-based, ultrashort systems approached the kW-level: Eidam *et al.* presented in 2010 [150] an Yb-fiber based system with 830 W average output power and 640 fs pulse duration (Fig. 2.16). The amplifier chain supports the full repetition rate of the oscillator of 78 MHz. Pulse energies as high as 10.6 μJ with good beam quality of $M^2 < 1.4$. This new generation of kW, fs fiber systems are suitable for materials processing at high processing speed by parallel processing with multiple lines. Yb-based fiber MOPA and FCPA systems are commonly used for industrial material processing with sub-ps pulses and 1 to 100 μJ pulse energies. These systems are highly reliable, robust, and cost efficient and offer long-term hands-off operation.

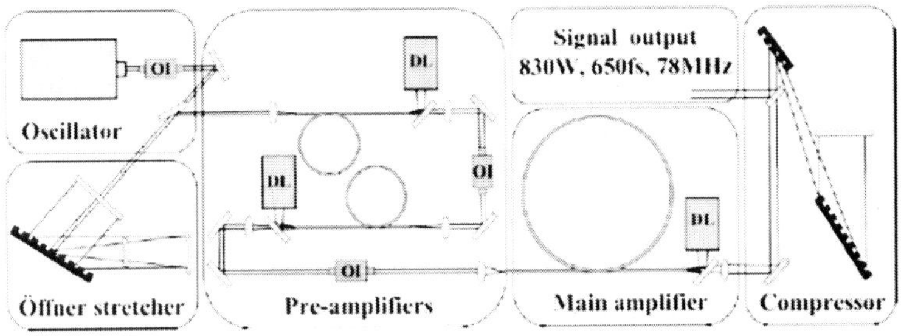

Figure 2.16 kW-level, fs system at high repetition rate and good beam quality [150].

2.4.3.2 Ti:Sapphire-based CPA systems

After its first use as lasing material in 1992, Ti:Sapphire became the work horse for most ultrafast and high energy applications. The gain bandwidth of Ti:Sapphire spans 125 nm centered at 800 nm and typical output pulse durations are 5 to 100 fs for commercially available oscillators. The CPA scheme has been extensively utilized to amplify fs pulses with Ti:Sapphire to even the PW-level with hundreds of J pulse energy but at low repetition rates. For higher repetition rates (1 to 10 kHz) the typical commercially available average power is on the order of 5 W for water-cooled systems and 10 to 20 W for cryo-cooled systems corresponding to pulse energies of <3 mJ and <10 mJ, respectively. In general, Ti:Sapphire-based amplifiers are maintenance extensive and need skilled technicians to operate on a long term basis. Ti:Sapphire requires high-quality, frequency-doubled, solid-state lasers as pump, which increases costs, complexity, and maintenance further. Due to the steady decrease in pulse duration of these amplifier systems, the phase between the electric field and the envelope of pulses with duration of less than 10 fs became an interesting feature for highly nonlinear experiments. Over the last decade technology evolved to stabilize this relationship, termed carrier-envelope phase (CEP) stabilization, for the pulse train emitted by the oscillator and for the full amplifier system (see, for example, [151]). A need set of materials processing experiments is now available, where the processing occurs due to the fully-controllable electric field rather than the pulse intensity.

2.4.3.3 Solid-state, rod-type systems

Workhorses of materials processing with ps pulse duration are solid-state MOPA system based on Nd:YVO_4, Nd:YLF, and Nd:YAG. Typical commercial average powers are on the order of 5 to 20 W at kHz to MHz repetition rates and 10 μJ to mJ pulse energies and 10 ps pulse duration. These MOPA systems are based on high average power oscillators with MHz repetition rates, pulse pickers to reduce the repetition rate and multipass amplifiers to ensure reliable and robust operation. Recently, Lührmann *et al.* [152] presented a regenerative amplifier based on diode-pumped Nd:YVO_4 as gain media operating at a pulse duration of 217 ps to 1 ns at a repetition rate of 20 kHz (see Fig. 2.17). The average power at the fundamental wavelength of 1064 nm was as high as 33.7 W corresponding to a pulse energy of 1.7 mJ while the beam quality was kept $M^2 < 1.5$. Similar systems operating at a slightly lower level average power level can be found commercially.

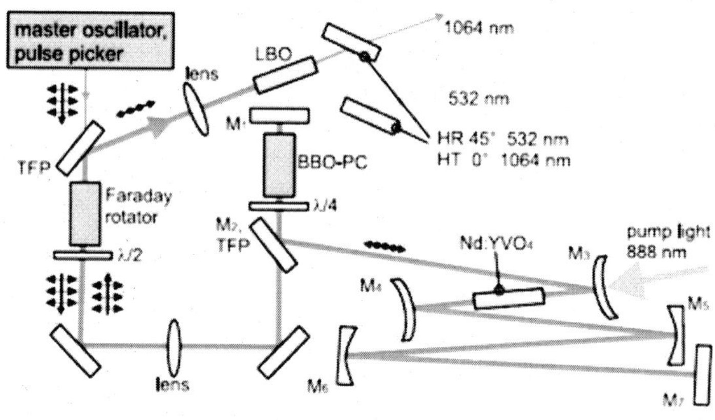

Figure 2.17 Regenerative amplifier providing 33.7 W average power [152].

2.4.3.4 Slab systems

Solid-state amplifier with rectangular aperture and large aspect ratio between height and width are commonly referred to as slab amplifiers. The slab geometry used offers several advantages for power scaling since the larger surfaces can be in contact directly with a heat sink with a large surface area. In addition, constrains on this surface (i.e. transparent, polished to optical quality, etc.) are low. The heat load induced by optical pumping is generating a heat gradient only in direction normal to these large area surfaces. Thus, a higher aspect ratio helps reduce this gradient and thus reduce thermal distortions in this direction. Meanwhile, the other direction parallel to the large surfaces shows only small thermal distortions. The beam is inserted and guided through the slab either by large external mirrors or by AR coating on the facets. Figure 2.18 shows the zigzag path geometry through a slab amplifier similar as suggested first by Kane *et al.* in 1983 [153]. The zigzag path effectively inverts the beam profile after each reflection, which leads to an inverting effect of the thermally induced optical distortions at each alternated pass. In average, thermal lensing and depolarization is highly reduced in this geometry. A challenging part in the slab geometry is the cylindrical beam shaping optics before and after the slab to establish a first Gaussian (or flattop) profile with elliptical aspect ratio and a symmetrical profile after. Slab lasers over potential to achieve sub-ps pulse duration with MHz repetition rate and kilowatt-level average power. In 2009 Russbueldt *et al.* [154] presented an Innoslab amplifier based on an Yb-doped slab with 1 mm height, 10 mm width, and 10 mm length to amplify 700 fs pulses to 400 W corresponding to 5.3 μJ pulse energy at 75.8 MHz repetition rate with a beam quality of M_x = 1.44 and M_y = 1.28. The concept of this slab design is shown in Fig. 2.19. In a recent upgrade, Russbueldt *et al.* [155] presented a similar amplifier including a single-pass, Innoslab, post-amplifier to push the average power to 1.1 kW at 20 MHz repetition rate corresponding to 55 μJ pulse energy.

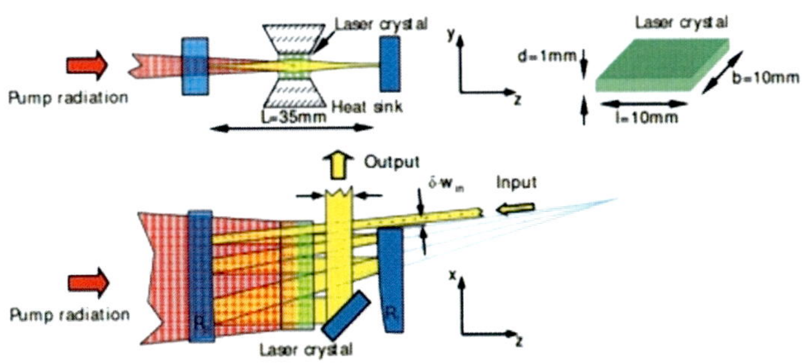

Figure 2.18 400 W slab amplifier [154].

A similar Innoslab amplifier design is commercially available from EdgeWave GmbH and is based on Nd:YAG providing 10 ps pulse duration at 1064 nm with 400 W average power at 0.4 to 15 MHz and up to 1 mJ pulse energy.

All these systems are currently limited by the accumulated B-integral and thus the onset of detrimental, nonlinear effects such as self-focusing. Recently Schulz *et al.* [156] reported on an Innoslab amplifier, which utilizes chirped-pulse amplification (CPA) to increase the obtainable pulse energy. The presented amplifier provided pulses with 830 fs duration and energies from 20 mJ at 12.5 kHz to 2 mJ at 100 kHz corresponding to an average power of 250 W or 200 W, respectively.

2.4.3.5 Thin-disk systems

After the first presentation of a thin disk laser by Giessen *et al.* in 1994 [157], several system were presented ranging from several kW cw laser to high-energy, Q-switched oscillators and high-average, power post amplifiers as well as high-average power, mode-lock oscillators. Commonly, the thin-disk is mounted on a cooling finger to provide a heat flow along to the optical axis, which causes an almost homogenous heat profile and thus highly reduces thermal lensing, and generally, higher power levels can be achieved. Innershofer *et al.* [158] presented a mode-locked oscillator providing up to 60 W average power at 34.3 MHz repetition rate with 810 fs pulse duration and 1.75 μJ pulse energy. A schematic of the setup is shown in Fig. 2.20. The output beam has an excellent quality of $M^2 < 1.1$. Later, the same research group presented a similar system [159] with extended cavity by the use of a Herriott-type multipass cell to achieve a higher pulse energy of 11.3 μJ at 4 MHz and comparable pulse duration. These systems are sutiable for materials processing at a high repetition rate and offer a simple setup and good robustness.

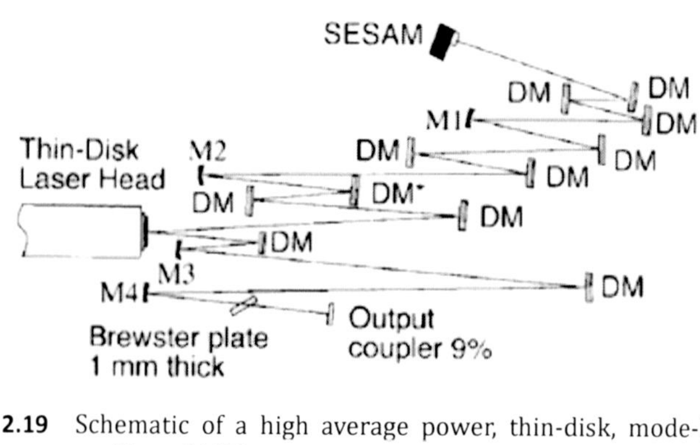

Figure 2.19 Schematic of a high average power, thin-disk, mode-locked oscillator [158].

2.4.3.6 Cryogenically cooled, ultrashort laser system

Recently, a new class of laser amplifiers evolved due to the increased interest in cryogenically cooled laser materials. The main driving force is the significant progress in high-brightness, frequency-stabilized, narrow-linewidth, diode laser, which allows efficient optical pumping of the reduced absorption linewidth under cryogenic cooling. In the case of Yb:YAG, the quasi three level system at room temperature becomes an entirely four-level system at 77 K, which permits ground-state absorption. The stimulated-emission cross section increases by a factor of 5, which effectively increases small signal gain as well as decreases the saturation fluence. The thermal conductivity is increased by factor of 2.5 to 4 depending on the Yb-dopant level [160]. On the other hand, the absorption width at 939 nm becomes only several nanometers of FWHM, which requires frequency-stabilized, narrow-linewidth pump sources. Recently Hong *et al.* [161] reported on a cryogenic system based on Yb:YAG rods at 2 kHz repetition rate. The regenerative amplifier allowed amplification of pulses with 15 ps duration and 6.5 mJ pulse energy after compression. Brown *et al.* [160] presented a system, shown in Fig. 2.20, based on several thick disks (5 mm long) in double-pass and single-pass amplifier geometry. The first amplifier consists of seven disks each pumped by two 30 W pump laser diodes, while the second amplifier consists of 8 disks each pumped by two 100 W pump diodes. A mode-locked oscillator is used to provide the 50 MHz pulse train with 12.4 ps pulse duration. After amplification, the obtained output power was as high as 758 W corresponding to 15.2 μJ pulse energy while keeping a beam quality of $M^2 < 1.3$.

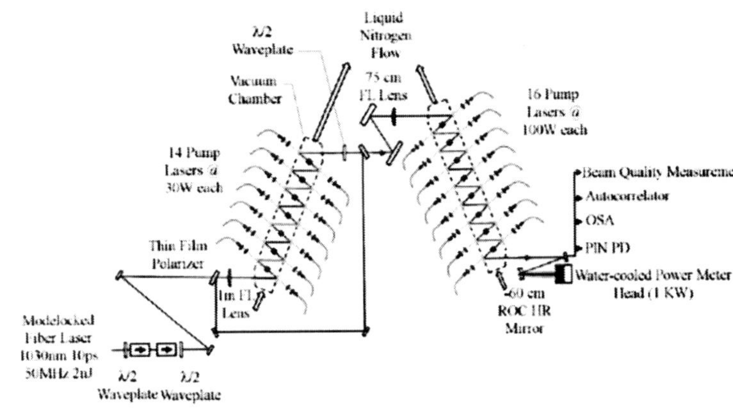

Figure 2.20 Cryogenic system generating 758 W average power [160].

2.4.4 OPA, OPCPA, and the Next Generation of Mid-IR Lasers

2.4.4.1 Optical parametric amplification (OPA)

Optical parametric amplifiers are a pathway toward wavelength conversion and tunability in the mid-IR from common laser systems such as Ti:Sapphire amplifier systems. OPAs are commercially available making it possible to significantly extend the capabilities of a Ti:Sapphire amplifier system without advanced knowledge in laser engineering and nonlinear optics. In an OPA, the output of a high energy CPA or MOPA system—typically an fs system with energy of at least a few hundreds of μJ—is focused in a solid-state material to achieve white-light generation. Portion of the continuum generated can be selected for amplification via phase-matching in staged nonlinear crystals such as BBO, KDP, DKPD, or BiBO. OPAs, nonetheless, face two major drawbacks that limit their operation for industrial purposes: poor optical long-term stability and poor efficiency. The first asks for frequent re-alignment, a regularly checked calibration and are highly dependent on the output parameters of the driving CPA or MOPA system. Such sensitivity is related to the nonlinear nature of the parametric interaction and of the white light seed generation. The optical-to-optical efficiency of common OPAs is 10% to 20% but can drop dramatically toward the spectral edges of the tuning range. This inherently implies that the OPA output performance in terms of repetition rate, pulse energy, average power, spatial profile, and stability is highly dependent on the driving laser system.

2.4.4.2 Optical parametric chirped-pulse amplification (OPCPA)

Optical parametric chirped pulse amplifier architectures address a number of the issues encountered with Ti:Sapphire CPA systems and OPA systems. It is an architecture that enables higher average power systems. Their advantages include (i) higher repetition rates than CPA systems, (ii) the generation of ultrashort laser pulses at longer wavelengths, and (iii) the elimination of the complexity of multipass Ti-sapphire amplifiers. This comes not without considerable expense at the present time mostly in the diode-pumped OPCPA pump laser. The footprint of these lasers can be significantly less than commercial CPA systems

OPCPA systems at 800 nm

OPCPA systems operating at 800 nm complement traditional Ti:Sapphire CPA systems and can provide solutions to the current limitations in repetition rate, pulse duration and spatial profile of CPA systems. Current state-of-the-art, high-repetition-rate CPA Ti:Sapphire systems seem to be limited to below 50 W of compressed average power [162]. A few systems have been reported and an asymptotical trend seems to appear at this power level, hinting at intrinsic limits of the material. In an OPCPA, thermal related problems can be suppressed. The amplification process consists of a nonlinear three-wave interaction in which the energy of a high energetic pump pulse is transferred instantaneously to a seeded signal pulse under proper phase-matching conditions without heating of gain material. This feature has the potential to increase the average power of fs laser systems at 800 nm beyond the tens of Watt level [163]. The OPCPA architecture enables the direct amplification of pulses with duration as short as a few-optical cycle. This feature is inaccessible from CPA systems due to spectral gain narrowing occurring in regenerative amplifier and multipass amplifiers upon amplification. The potential for controlling the carrier-envelope phase (CEP) of laser systems takes its full potential when dealing with few-cycle pulses. This is in particular interesting for light-matter interaction governed primarily by field-dependent processes rather than peak power dependent processes. Bringing material processing into the few-cycle regime promises the potential for unveiling new aspects of material science.

OPCPA systems for mid-IR generation

Similarly to OPAs, the OPCPA architecture offers the capability to widely tune the wavelength of operation, enabling to amplify optical pulses in spectral regions where laser gain media are either not mature or unavailable. In particular, it offers the possibility to amplify ultrashort, potentially few-cycle, pulses in the near-IR [164] to mid-IR [165] spectral region with intrinsic and robust CEP stabilization. The OPCPA systems operating in the mid-IR spectral region benefits from the same advantages as these stated for 8000 nm wavelength: repetition rate and energy scaling can be performed while maintaining beam profile due to the absence of energy storage in the amplification process.

The driving force for developing such OPCPA laser systems is currently the development of reliable sources for high-field physics experiment. Only a small set of experiments have been performed using these state-of-the-art lasers for material processing applications. It is nonetheless foreseeable that these sources will play a key role in the understanding and development of new applications in material processing. Mid-IR wavelengths enable exploring new horizons in material processing as reducing the photon energy can dramatically impact multiphoton absorption driven processes.

2.5 Irradiation Optics, Online Monitoring, and Diagnostics

As ultrafast laser processing progressively permeates the industrial sector, other important technical issues that affect processing stability speed and costs become increasingly important. Here we discuss those parts of the processing regime that have an impact on these factors. Ultimately their development will play an important role on determining the ease with which ultrafast laser materials processing enters the workplace.

2.5.1 Laser Power and Processing Speed

Apart from any considerations on the physical and mechanical effects produced on and in materials by ultrafast lasers, two of the principal drawbacks to their introduction at an industrial level have been (i) the cost, complexity and reliability of the lasers themselves, and (ii) the low component or process throughput. In overcoming these limitations, it soon becomes apparent that their origin is not necessarily the laser alone, but may also be associated with the beam scanning system, or indeed the material itself.

Firstly, we examine the limitations on the laser system. Up to the present time, lasers used for ultrafast laser materials processing have had relatively low average power, for instance a CPA system delivering 10 GW, 100 fs pulses at 1 kHz repetition rate has an average power of 1 W. GigaWatt power levels with pulse energies of 1 mJ are typical of the energies needed for ablative applications of ultrafast lasers. These power levels may be sufficient for single shot applications. At these power levels, if the beam diameter on the final element of the focusing lens is ~1 mm (typical microscope objective), the power density on these optics reaches $\sim 10^{11}$ W/cm^2, about an order of magnitude below the damage threshold of most transparent polished materials (Fig. 2.1).

The new developments in laser architecture we describe here will enable this regime to be pushed too much higher repetition rates. Commercial systems operating at the 100 W, 100 kHz level are needed. These will allow for meaningful production rates.

The HA regime has some benefit to certain processes, particular in the fabrication of optical waveguides and microfluidics, usually is effective at frequencies of 100 kHz to 1 MHz. Industrial demand in this area will focus more on laser systems that can provide the required pulse energy, usually in 1 μJ or less. With the maximum energy, or irradiance clamped to a specific value below the damage threshold, production or throughput rates can only be increased with the employment of more sophisticated irradiation regimes. We discuss these briefly in the following sections.

2.5.2 Irradiation Optics and Regimes

The discussion in the previous sections has an impact on the optical beam line used from the laser to the irradiation station and on the choice of focusing optics. Optical processing techniques such as refractive index modification depend on the irradiance (W/cm^2) [44, 130]. Here it is important to consider the effect these optics have on the pulse itself. Apart from the intrinsic damage threshold of these components and their coatings, nonlinear optical effects can change the pulse characteristics. To avoid some of these effects, simply increasing the beam cross section to lower the power density is sufficient. However the peak power can become the limiting parameter in refractive optics because self-phase-modulation and self-focusing depend solely on the power [166], not power density. These two nonlinear phenomena will influence the temporal and spatial properties of the beam, leading in the worst case to catastrophic damage of the delivery system [3, 6, 167]. Eventually this limitation forces the use of new materials with smaller nonlinear response, or the adoption of reflective optics for the beam delivery system. The pulse characteristics will also be changed by any non-uniformities in the spectral response of reflective coating and refractive components. In the extreme case of single-cycle pulses, the spectral content of the pulse can span over 300 nm of bandwidth [168]. Deforming the spectral content of the pulse will lead to a temporal broadening of the pulse and therefore reduce the peak power on the sample.

The introduction of chirp on the optics train is another important factor while designing processing systems for ultrashort laser radiation. Where it is customary to use fiber-beam-delivery systems with high-power cw-laser sources, for ultrashort pulses this method is not suitable. Beside the inability of optical fibers to withstand the high peak powers, dispersion in the medium would deform the pulse and destroy the temporal profile. Hollow fibers could be used as an alternative here as long as the wall-coating does not get damaged by the pulse peak power. More often reflective optics are used when steering an ultrashort-pulsed beam from the laser source to the sample. These are not the ideal solution for a rugged industrial processing system.

A commonly used approach to increase the throughput for industrial laser applications is the use of parallel processing. When combined with ultrashort radiation, scaling for parallel processing means increased repetition rates from a single laser source and shared beam time at the processing ends of the fabrication system. Increase repetition rate corresponds to increased average power if the rest of the laser parameters are kept constant, here mainly pulse duration and pulse energy. At that point, the power handling of the beam delivery system has to be of concern. Backside mirror cooling systems can alleviate this, but only with added expense. Switching the laser beam from one processing station to the next becomes more challenging. Two schemes are commonly used, rotating mirror or acousto-/electro-optical (AO/EO) switches. Both systems have advantages and disadvantages. Mirrors are efficient in switching the full energy content of the beam and usually do not introduce dispersion. To reduce switching times, the mirror is commonly kept small, which in return limits the maximum power. AO and EO switching are faster than mirrors but suffer significant loss. There is also the possibility of using a beam splitter to partially switch pulses from station to station. For parallel processing this scheme is rather uncommon since increased pulse repetition rates on the sample can reduce the process efficiency [169].

2.5.3 Processing Stations

Processing speed is a major enabler to gaining acceptance of many ultrafast processing techniques to the industrial market. Most experimental studies use linear translation stages stacked to provide 3D capability. These systems can have high translational accuracy (<100 nm), but at the expense of processing speed. Maximum translation speeds are a few m/s. Moreover with linear stages inaccuracies can occur because of effects of associated with the acceleration of stages and registration of the stages themselves. Acceleration effects limit the mass of the stages. Jitter, slippage, or backtalk in the stages can also decrease accuracy. In principle, scanning systems can overcome some of this issues, particularly relating to speed. The limiting factors here with respect to translation speed are the mirrors themselves inside the scanner assembly, where the focal plane would be located on a spherical surface [170]. To overcome this problem, f-theta lenses are in general used to project the focal plane into a flat plane on the sample surface. Tight focusing is however difficult using f-theta lenses. Few systems have been demonstrated using scanner assembles featuring microscope objectives to focus the beam [171]. In such cases, beam distortions are common, as well as a reduced processing area. Combining scanner and linear stage systems can compensate for this disadvantage.

2.5.4 Online Imaging, Diagnostics, and Feedback

Process control is an important factor in laser materials processing with ultrafast lasers. Online image monitoring is in many cases necessary, for instance in precision hole drilling [172]. In those applications that are dependant on the phase front and chirp of the pulse then on line characterization of the pulse length and spectral characteristics will be necessary. Techniques such as frequency-resolved optical gating (FROG) [173] or spectral phase interferometry for direct electric-field reconstruction (SPIDER) [174, 175] measure both the pulse duration and the phase of the pulse.

Finally for industrial applications feedback and control circuits must be developed to control the pulse characteristics online during the fabrication process.

2.6 Future Prospects

2.6.1 Laser Development: The Split between Shorter Pulses and Industrial Deployment

The trend shown in Fig. 2.15 clearly points toward development of shorter pulses, but how far will that development be beneficial for industrial deployment? Without doubt, for scientific research and ultrashort light-matter-interaction studies the reduction of pulse duration will continue and be an important factor to gain knowledge of involved physical processes. However, for the actual deployment of ultrashort-pulsed laser systems in industrial environments the ease of maintenance and system up-time will be the critical factors. Here the trend is pointing toward higher average powers by means of high repetition rates rather than shorter pulse durations.

Obviously, fiber lasers have been in the center of the discussion for quite some time and will continue to be in the future. Simple fiber oscillators generating short pulses with the use of SESAMs clearly mark the frontend of future industrial laser systems. The major focus must be on high repetition rates while maintaining a good beam profile and a robust, stable seed laser system. The output from such seed systems will then be amplified using various stages of amplification modules, such as either fiber amplifiers, which can be coupled directly to the seed systems, or thin disk- and slab-type amplifiers. The main interest here has to be that the repetition rate of the initial ultrashort-pulsed system is kept during the amplification process. This approach will allow for several GW of peak power while increasing the beam-on-time. Finally that enables to parallel-process or multiplex the beam between processing stations to fulfill the throughput requirements of industrial applications.

The challenge to overcome in the future to enable high-power, high-repetition-rate processing is to control the direct output of the laser source. Sequencing individual pulses into different processing stations at ns-cycle times seems to be practical already; however, doing so at pulse peak powers in the GW range is difficult. Moving mirrors that are actively cooled to withstand the overall power need to be developed. The mass of the mirror-cooling assemble has to be kept to a minimum. Solutions that incorporate the mirror mount as a head-sink are imaginable. EO and AO switching needs to employ new materials as host based on optical ceramics. Such materials are readily available for research purposes but not for commercial processing systems.

At the same time it will be necessary to introduce fast feedback systems that are capable of terminating the laser output in case of malfunctioning beam steering. Uncontrolled beam paths or even just stationary incidence of the full laser source on a single processing station will have catastrophic consequences on the beam transport and focusing optics. Electronics-based feedback systems can only partially meet this need since the temporal regime of the laser pulses is much shorter than electronic response times. Self-limiting absorbers have to be developed that (i) can sustain the peak-power during normal operation and (ii) terminate the beam, even by catastrophic destruction of the absorber itself, in the case of malfunctions to protect expensive processing optics.

It is necessary to gain further knowledge about when and how to work in the most efficient processing regime on the sample. Initial research has been conducted already [169], but a larger database of material responses is needed.

Another trend for industrial ultrashort-pulsed laser development has to be the extension of the laser wavelength range toward the mid-IR. Applications in the semiconductor industry become promising when the bandgap of Si can be approached. Today's systems often rely on OPAs and OPOs, which are inherently hard to maintain. The ease of the laser source and the increase of output powers are essential to open this promising market.

2.6.2 Future Advances in Processing Stations

With the event of high-repetition-rate laser systems the need for hybrid processing stations consisting of a scanner and a 3D axis systems is apparent. Synchronized motion of the sample and the processing beam is needed to extent the processing area.

The transition from the lab into industrial environments also requires the simplification of the human-machine interface. CAD based programming of the desired geometries has to be accompanied by converting tools to plot the final beam path on the sample. The designs of current integrated systems are nowadays drawn from CNC milling technologies. To take advantage of the full capabilities of laser material processing with new laser sources, tools have to be developed that are directly based on a massless tool machining of the sample.

2.6.3 Laser-Fabricated Photonic Devices

The applications for femtosecond laser direct writing (FLDW) seemed to be endless when this technique was first explored one-and-a-half decades ago. Up to now it has proved difficult for this technique to directly compete with photonic circuits fabricated by lithographic means when solely relying on the 3D capability of refractive index change. Very few applications can take advantage of this freedom but instead require high index changes to minimize device size. The exception here is rapid prototyping or special customized part with a small margin.

A reorientation toward applications that cannot be realized using lithographic techniques has to be made to revive the FLDW technology and make it compatible in an industrial environment. Post-processing applications to decrease the error-margin during conventional fabrication can help to drive the technology toward industrial deployment.

The extension of the laser wavelength spectrum will be a promising application field yet. To open the field of classical semiconductor materials for post-processing applications should be in the focus of the community. Basic research to understand the interaction scheme of semiconductors with ultrashort laser radiation is necessary as much as simple integration of processing systems in the semiconductor fabrication environment. From simple restructuring the surface for enhanced responses to developing entirely new, custom materials can be thought of.

2.7 Summary

From a brief summary of the principles of fundamental light-matter interaction, we have discussed the unique phenomena when irradiating matter with ultrashort pulses. We have described many material modification features generated by these pulses. The subtle improvements that can in some cases be accrued from increasing the localized material temperature by use of the HA regime is discussed. Most specifically we discuss the impacts on processing science when this technology transitions to regimes needed for industrially required process speeds and throughput requirements. These are affected most by developments in ultrafast laser technology. Here we summarize the state-of- the-art, and project future trends that will bring this technology to the factory floor. The demands high processing speeds, larger processing areas and greater precision will have on the processing stations themselves are also discussed. New technologies and components will be needed to ensure this element of ultrafast laser materials technology does not become the bottleneck to industrialization.

Finally we look to future prospects from laser development and ULMP in research and industry. The requirements on laser development are split between research seeking to employ shorter pulsed systems to gain further knowledge about physical interaction phenomena, and industrial requirements for more conservative laser sources scaled in power and repetition rate for increased processing speed.

Acknowledgments

This review was supported in part by research support in laser materials processing in the Laser Plasma Laboratory of the Townes Laser Institute provided by the National Science Foundation, DoE, ARO, the State of Florida, and the Fraunhofer–Townes Collaboration. The Laser Plasma Laboratory benefits from support and close collaboration with a number of companies, including, IMRA America, Inc., Newport Corporation., Aerotech, Inc., and Trumpf Inc.

References

1. Strickland, D, and Mourou, G (1985). Compression of amplified chirped optical pulses, *Optics Communications*, **56**, 219–221.
2. Cheriaux, G, Rousseau, P, Salin, F, Chambaret, J, Walker, B, and Dimauro, L (1996). Aberration-free stretcher design for ultrashort-pulse amplification, *Optics Letters*, **21**, 414–416.

3. Du, D, Liu, X, Korn, G, Squier, J, and Mourou, G (1994). Laser-induced breakdown by impact ionization in SiO_2 with pulse widths from 7 ns to 150 fs, *Applied Physics Letters*, **64**, 3071–3073.
4. Ihlemann, J, Wolff, B, and Simon, P (1992). Nanosecond and femtosecond excimer laser ablation of fused-silica, *Applied Physics A*, **54**, 363–368
5. Küper, S, and Stuke, M (1987). Femtosecond UV excimer laser ablation, *Applied Physics B*, **44**, 199–204.
6. Stuart, B, Feit, M, Herman, S, Rubenchik, A, Shore, B, and Perry, M (1996). Nanosecond-to-femtosecond laser-induced breakdown in dielectrics, *Physical Review B*, **53**, 1749–1761.
7. Stuart, B, Feit, M, Rubenchik, A, Shore, B, and Perry, M (1995). Laser-induced damage in dielectrics with nanosecond to subpicosecond pulses, *Physical Review Letters*, **74**, 2248–2251.
8. Tam, A, Brand, J, Cheng, D, and Zapka, W (1989). Picosecond laser sputtering of sapphire at 266 nm, *Applied Physics Letters*, **55**, 2045–2047.
9. Sutcliffe, E, and Srinivasan, R (1986). Dynamics of UV laser ablation of organic polymer surfaces, *Journal of Applied Physics*, **60**, 3315–3322.
10. Nolte, S, Momma, C, Jacobs, H, Tünnermann, A, Chichkov, BN, Wellegehausen, B, and Welling, H (1997). Ablation of metals by ultrashort laser pulses, *Journal of the Optical Society of America B*, **14**, 2716–2722.
11. Liu, X, Du, D, and Mourou, G (1997). Laser ablation and micromachining with ultrashort laser pulses, *Quantum Electronics, IEEE Journal of*, **33**, 1706–1716.
12. Sun, Q, Jiang, H, Liu, Y, Wu, Z, Yang, H, and Gong, Q (2005). Measurement of the collision time of dense electronic plasma induced by a femtosecond laser in fused silica, *Optics Letters*, **30**, 320–322.
13. Gamaly, E, Rode, A, Luther-Davies, B, and Tikhonchuk, V (2002). Ablation of solids by femtosecond lasers: ablation mechanism and ablation thresholds for metals and dielectrics, *Physics of Plasmas*, **9**, 949 http://dx.doi.org/10.1063/1.1447555.
14. Jiang, L, and Tsai, H (2005). Improved two-temperature model and its application in ultrashort laser heating of metal films, *Journal of Heat Transfer-Transactions of the Asme*, **127**, 1167–1173.
15. Juodkazis, S, Nishimura, K, Misawa, H, Ebisui, T, Waki, R, Matsuo, S, and Okada, T (2006). Control over the crystalline state of sapphire, *Advanced Materials*, **18**, 1361–1364.

16. Braunlich, P, Brost, G, Schmid, A, and Kelly, P (1981). The role of laser-induced primary defect formation in optical breakdown of NaCl, *Quantum Electronics, IEEE Journal of*, **17**, 2034–2041.

17. Kelly, P, Jones, S, Shen, X, Simpson, L, Braunlich, P, and Casper, R (1990). Measurement of the 3-photon-absorption cross-section and intrinsic optical-breakdown of Ki at 532 nm, *Physical Review B*, **42**, 11370–11372.

18. Keldysh, L (1965). Ionization in the field of a strong electromagnetic wave, *Soviet Physics JETP-USSR*, **20**, 1307–1314.

19. Liu, JM (1982). Simple technique for measurements of pulsed Gaussian-beam spot sizes, *Optics Letters*, **7**, 196–198.

20. House, R, Bettis, J, and Guenther, A (1977). Subsurface structure and laser damage threshold, *Quantum Electronics, IEEE Journal of*, **13**, 363–364.

21. Stoian, R, Ashkenasi, D, Rosenfeld, A, and Campbell, E (2000). Coulomb explosion in ultrashort pulsed laser ablation of Al_2O_3, *Physical Review B*, **62**, 13167–13173.

22. Shah, L, Tawney, J, and Richardson, MC (2001). Femtosecond laser deep hole drilling of silicate glasses in air, *Applied Surface Science*, **183**, 151–164.

23. Tuennermann, A, Nolte, S, and Limpert, J (2010). Femtosecond vs. picosecond laser material processing, *Laser Technik Journal*, **1**, 34–38.

24. Chichkov, B, Momma, C, Nolte, S, von Alvensleben, F, and Tünnermann, A (1996). Femtosecond, picosecond and nanosecond laser ablation of solids, *Applied Physics A*, **63**, 109–115.

25. Zoubir, A, Shah, L, Richardson, K, and Richardson, M (2003). Practical uses of femtosecond laser micro-materials processing, *Applied Physics A*, **77**, 311–315.

26. Smith, GN, Kalli, K, and Sugden, K (eds.) (2010). 15 *Advances in Femtosecond Micromachining and Inscription of Micro and Nano Photonic Devices* (INTECH, Croatia). 674.

27. Sanner, N, Utéza, O, Chimier, B, Sentis, M, Lassonde, P, Légaré, F, and Kieffer, JC (2010). Toward determinism in surface damaging of dielectrics using few-cycle laser pulses, *Applied Physics Letters*, **96**, 071111.

28. Borowiec, A, and Haugen, HK (2003). Subwavelength ripple formation on the surfaces of compound semiconductors irradiated with femtosecond laser pulses, *Applied Physics Letters*, **82**, 4462.

29. Costache, F, Henyk, M, and Reif, J (2002). Modification of dielectric surfaces with ultra-short laser pulses, *Applied Surface Science*, **186**, 352–357.

30. Reif, J, Varlamova, O, Varlamov, S, and Bestehorn, M (2011). The role of asymmetric excitation in self-organized nanostructure formation upon femtosecond laser ablation, *Applied Physics A*, **104**, 969–973.

31. Bonse, Jr, Rosenfeld, A, and Kruger, J (2011). Implications of transient changes of optical and surface properties of solids during femtosecond laser pulse irradiation to the formation of laser-induced periodic surface structures, *Applied Surface Science*, **257**, 5420–5423.

32. Buividas, R, Rosa, L, Šliupas, R, Kudrius, T, Šlekys, G, Datsyuk, V, and Juodkazis, S (2010). Mechanism of fine ripple formation on surfaces of (semi)transparent materials via a half-wavelength cavity feedback, *Nanotechnology*, **22**, 055304.

33. Davis, K, Miura, K, Sugimoto, N, and Hirao, K (1996). Writing waveguides in glass with a femtosecond laser, *Optics Letters*, **21**, 1729–1731.

34. Glezer, EN, Milosavljevic, M, Huang, L, Finlay, RJ, Her, TH, Callan, JP, and Mazur, E (1996). Three-dimensional optical storage inside transparent materials, *Optics Letters*, **21**, 2023.

35. Anderson, T, Petit, L, Carlie, N, Choi, J, Hu, J, Agarwal, A, Kimerling, L, Richardson, KC, and Richardson, MC (2008). Femtosecond laser photo-response of $Ge_{23}Sb_7S_{70}$ films, *Optics Express*, **16**, 20081.

36. Miura, K, Qiu, J, Inouye, H, Mitsuyu, T, and Hirao, K (1997). Photowritten optical waveguides in various glasses with ultrashort pulse laser, *Applied Physics Letters*, **71**, 3329.

37. Zoubir, A, Lopez, C, Richardson, MC, and Richardson, KC (2004). Femtosecond laser fabrication of tubular waveguides in poly(methyl methacrylate), *Optics Letters*, **29**, 1840–1842.

38. Zoubir, A, Richardson, MC, Rivero, C, and Schulte, A (2004). Direct femtosecond laser writing of waveguides in As_2S_3 thin films, *Optics Letters*, **29**, 748–750.

39. Lai, Y, Zhou, K, Sugden, K, and Bennion, I (2007). Point-by-point inscription of first-order fiber Bragg grating for C-band applications, *Optics Express*, **15**, 18318–18325.

40. Marshall, GD, Ams, M, and Withford, MJ (2006). Direct laser written waveguide-Bragg gratings in bulk fused silica, *Optics Letters*, **31**, 2690–2691.

41. Marshall, G, Dekker, P, Ams, M, and Piper, J (2008). Directly written monolithic waveguide laser incorporating a distributed feedback waveguide-Bragg grating, *Optics Letters*, **33**, 956–958.

42. Takeshima, N, Kuroiwa, Y, Narita, Y, Tanaka, S, and Hirao, K (2004). Fabrication of a periodic structure with a high refractive-index difference by femtosecond laser pulses, *Optics Express*, **12**, 4019–4024.

43. Watanabe, W, Kuroda, D, Itoh, K, and Nishii, J (2002). Fabrication of Fresnel zone plate embedded in silica glass by femtosecond laser pulses, *Optics Express*, **10**, 978–983.

44. Nolte, S, Will, M, Burghoff, J, and Tuennermann, A (2003). Femtosecond waveguide writing: a new avenue to three-dimensional integrated optics, *Applied Physics A*, **77**, 109–111.

45. Nolte, S, Will, M, Burghoff, J, and Tuennermann, A (2003). Three-dimensional structuring of glass by ultrashort laser pulses, *Proceedings of the SPIE*, **4984**, 194.

46. Will, M, Burghoff, J, Nolte, S, and Tuennermann, A (2003). Fabrication of three-dimensional photonics devices using femtosecond laser pulses, *Proceedings of the SPIE*, **4978**, 147.

47. Will, M, Burghoff, J, Nolte, S, and Tuennermann, A (2003). Femtosecond-laser-induced refractive index modifications for fabrication of three-dimensional integrated optical devices, *Proceedings of the SPIE*, **4941**, 58.

48. Will, M, Nolte, S, Chichkov, B, and Tuennermann, A (2002). Optical properties of waveguides fabricated in fused silica by femtosecond laser pulses, *Applied Optics*, **41**, 4360–4364.

49. Ams, M, Marshall, G, Dekker, P, and Piper, J (2008). Ultrafast laser written active devices, *Laser & Photonics Reviews*, **10**, 535–544.

50. Fernandez, TT, della Valle, G, Osellame, R, Jose, G, Chiodo, N, JHA, A, and Laporta, P (2008). Active waveguides written by femtosecond laser irradiation in an erbium-doped phospho-tellurite glass, *Optics Express*, **16**, 15198–15205.

51. Burghoff, J, Nolte, S, and Tünnermann, A (2007). Origins of waveguiding in femtosecond laser-structured $LiNbO_3$, *Applied Physics A*, **89**, 127–132.

52. Chan, J, Huser, T, Risbud, S, Hayden, J, and Krol, D (2003). Waveguide fabrication in phosphate glasses using femtosecond laser pulses, *Applied Physics Letters*, **82**, 2371.

53. Eaton, SM, Zhang, H, Herman, PR, Yoshino, F, Shah, L, Bovatsek, J, and Arai, A (2005). Heat accumulation effects in femtosecond laser-written waveguides with variable repetition rate, *Optics Express*, **13**, 4708–4716.

54. Eaton, SM, Zhang, H, Ng, M, Li, J, Chen, W, and Ho, S (2008). Transition from thermal diffusion to heat accumulation in high repetition rate femtosecond laser writing of buried optical waveguides, *Optics Express*, **16**, 9443.

55. Schaffer, C, Garcia, J, and Mazur, E (2003). Bulk heating of transparent materials using a high-repetition-rate femtosecond laser, *Applied Physics A*, **76**, 351–354.

56. Choi, J, *Femtosecond Laser Written Volumetric Diffractive Optical Elements and Their Applications*, 2009, doctoral dissertation (University of Central Florida, 2009).

57. Gottmann, J, and Wortmann, D (2008). High-speed manufacturing of periodical surface and in-volume nanostructures by fs-laser direct writing, *Proceedings of the SPIE*, **6879**, doi:10.1117/12.758353.

58. Pronko, P, Dutta, S, Squier, J, Rudd, J, and Du, D (1995). Machining of sub-micron holes using a femtosecond laser at 800 nm, *Optics Communications*, **114**, 106–110.

59. Klimt, B, *Industrial Picosecond Lasers for Micromachining*, rpmclasers.com.

60. Okuchi, T, Ohfuji, H, Odake, S, Kagi, H, Nagatomo, S, Sugata, M, and Sumiya, H (2009). Micromachining and surface processing of the super-hard nano-polycrystalline diamond by three types of pulsed lasers, *Applied Physics A*, **96**, 833–842.

61. Shirk, M, and Molian, PA (1998). A review of ultrashort pulsed laser ablation of materials, *Journal of Laser Applications*, **10**, 18–28.

62. Martinez, E, Engel, E, Planell, J, and Samitier, J (2009). Effects of artificial micro- and nano-structured surfaces on cell behaviour, *Annals of Anatomy—Anatomischer Anzeiger*, **191**, 126–135.

63. Koroleva, A, Schlie, S, Fadeeva, E, Gittard, SD, Miller, P, Ovsianikov, A, Koch, J, Narayan, RJ, and Chichkov, BN (2010). Microreplication of laser-fabricated surface and three-dimensional structures, *Journal of Optics*, **12**, doi:10.1088/2040-8978/12/12/124009.

64. Zorba, V, Persano, L, Pisignano, D, Athanassiou, A, Stratakis, E, Cingolani, R, Tzanetakis, P, and Fotakis, C (2006). Making silicon hydrophobic: wettability control by two-lengthscale simultaneous patterning with femtosecond laser irradiation, *Nanotechnology*, **17**, 3234–3238.

65. Zorba, V, Stratakis, E, Barberoglou, M, Spanakis, E, Tzanetakis, P, and Fotakis, C (2008). Tailoring the wetting response of silicon surfaces via fs laser structuring, *Applied Physics A*, **93**, 819–825.

66. Younkin, R, Carey, JE, Mazur, E, and Levinson, J (2003). Infrared absorption by conical silicon microstructures made in a variety of background gases using femtosecond-laser pulses, *Journal of Applied Physics*, **93**, 2626.

67. Tull, BR, Winkler, M, and Mazur, E (2009). The role of diffusion in broadband infrared absorption in chalcogen-doped silicon, *Applied Physics A*, **96**, 327–334.

68. Halbwax, M, Sarnet, T, Delaporte, P, Sentis, M, Etienne, H, Torregrosa, F, Vervisch, V, Perichaud, I, and Martinuzzi, S (2008). Micro and nano-structuration of silicon by femtosecond laser: Application to silicon photovoltaic cells fabrication, *Thin Solid Films*, **516**, 6791–6795.

69. Tomita, T, Kumai, R, Nomura, H, and Matsuo, S (2011). Surface roughness assisted 100 kHz femtosecond laser induced nanostructure formation on silicon surface, *Applied Physics A*, **105**, 89–94.

70. Straub, M, Afshar, M, Feili, D, Seidel, H, and König, K (2011). Efficient nanostructure formation on silicon surfaces and in indium tin oxide thin films by sub-15 fs pulsed near-infrared laser light, *Physics Procedia*, **12**, 16–23.

71. Bonse, Jr and Rosenfeld, A (2010). Femtosecond laser-induced periodic surface structures: recent approaches to explain their sub-wavelength periodicities, *Proceedings of the SPIE*, **7994**, 79940M-79940M–10.

72. Reif, J, Varlamova, O, and Costache, F (2008). Femtosecond laser induced nanostructure formation: self-organization control parameters, *Applied Physics A*, **92**, 1019–1024.

73. Ramme, M, Choi, J, Anderson, T, and Mingareev, I (2010). Sub-micron machining of semiconductors: femtosecond surface ripples on GaAs by 2 μm laser light, *Proceedings of the SPIE*, **7590**, doi:10.1117/12.843701.

74. Crawford, T, and Haugen, H (2007). Sub-wavelength surface structures on silicon irradiated by femtosecond laser pulses at 1300 and 2100 nm wavelengths, *Applied Surface Science*, **253**, 4970–4977.

75. Cho, S-H, Park, J-K, Chang, W-S, Kim, J-G, and Whang, K-H, Removal of nano particles on silicon wafer by self-channeled plasma filament, in *CLEO/PACIFIC RIM*2009. 1.

76. Tamaki, T, Watanabe, W, Nishii, J, and Itoh, K (2005). Welding of transparent materials using femtosecond laser pulses, *Japanese Journal of Applied Physics Part 2*, **44**, L687–L689.

77. Watanabe, W, Onda, S, Tamaki, T, and Itoh, K (2006). Space-selective laser joining of dissimilar transparent materials using femtosecond laser, *Applied Physics Letters*, **89**, 021106.

78. Miyamoto, I., Horn, A., Gottmann, J., Wortmann, D., and Yoshino, F. (2007). Fusion welding of glass using femtosecond laser pulses with high-repetition rates, *Journal of Laser Micro/Nanoengineering*, **2**, 57–63.

79. Miyamoto, I, and Herrmann, T (2007). Characteristics of internal melting of glass for fusion welding using ps laser pulses with average power up to 8W, *Proceedings of the 8th International Symposium on Laser Precision Microfabrication—LPM2007 (2007).*

80. Miyamoto, I., Horn, A., Gottmann, J., Wortmann, D., Mingareev, I., Yoshino, F., Mingareev I., Schmidt M., Bechtold P., Okamoto Y., Uno Y., Herrmann T. (2008). Novel fusion welding technology of glass using ultrashort pulse lasers. *ICALEO 2008 Proceedings, #M304.*

81. Horn, A, Mingareev, I, Werth, A, Kachel, M, and Brenk, U (2008). Investigations on ultrafast welding of glass–glass and glass–silicon, *Applied Physics A*, **93**, 171–175.

82. Watanabe, W, Tamaki, T, and Itoh, K (2008). Femtosecond laser micromachining and biological therapy, *Laser Physics*, **18**, 263–269.

83. Ozeki, Y, Inoue, T, Tamaki, T, Yamaguchi, H, Onda, S, Watanabe, W, Sano, T, Nishiuchi, S, Hirose, A, and Itoh, K (2008). Direct welding between copper and glass substrates with femtosecond laser pulses, *Applied Physics Express*, **1**, 082601.

84. Bellouard, Y, Said, A, and Bado, P (2005). Integrating optics and micro-mechanics in a single substrate: a step toward monolithic integration in fused silica, *Optics Express*, **13**, 6635–6644.

85. Hnatovsky, C, Taylor, R, Simova, E, Rajeev, P, Rayner, D, Bhardwaj, V, and Corkum, P (2006). Fabrication of microchannels in glass using focused femtosecond laser radiation and selective chemical etching, *Applied Physics A*, **84**, 47–61.

86. Matsuo, S, Shichijo, Y, Tomita, T, and Hashimoto, S (2007). Laser fabrication of ship-in-a-bottle microstructures in sapphire, *Journal of Laser Micro/Nanoengineering*, **2**, 114–116.

87. Sugioka, K, Cheng, Y, and Midorikawa, K (2005). Three-dimensional micromachining of glass using femtosecond laser for lab-on-a-chip device manufacture, *Applied Physics A*, **81**, 1–10.

88. Kiyama, S, Matsuo, S, Hashimoto, S, and Morihira, Y (2009). Examination of tching agent and etching mechanism on femotosecond laser microfabrication of channels inside vitreous silica substrates, *The Journal of Physical Chemistry C*, **113**, 11560–11566.

89. Matsuo, S, Tokumi, K, Tomita, T, and Hashimoto, S (2008). Three-dimensional residue-free volume removal inside sapphire by high-

temperature etching after irradiation of femtosecond laser pulses, *Laser Chemistry*, **2008**, 1–4.

90. Bellouard, Y, Said, A, Dugan, M, and Bado, P (2004). Fabrication of high-aspect ratio, micro-fluidic channels and tunnels using femtosecond laser pulses and chemical etching, *Optics Express*, **12**, 2120–2129.

91. Mazilu, M, Juodkazis, S, Ebisui, T, and Matsuo, S (2007). Structural characterization of shock-affected sapphire, *Applied Physics A*, **86**, 197–200.

92. Hnatovsky, C, Taylor, R, Simova, E, Bhardwaj, V, Rayner, D, and Corkum, P (2005). Polarization-selective etching in femtosecond laser-assisted microfluidic channel fabrication in fused silica, *Optics Letters*, **30**, 1867–1869.

93. Gottmann, J, Wortmann, D, and Horstmann-Jungemann, M (2009). Fabrication of sub-wavelength surface ripples and in-volume nanostructures by fs-laser induced selective etching, *Applied Surface Science*, **255**, 5641–5646.

94. Wortmann, D, Gottmann, J, Brandt, N, and Horn-Solle, H (2008). Micro- and nanostructures inside sapphire by fs-laser irradiation and selective etching, *Optics Express*, **16**, 1517–1522.

95. Osellame, R, Hoekstra, HJWM, Cerullo, G, and Pollnau, M (2011). Femtosecond laser microstructuring: an enabling tool for optofluidic lab-on-chips, *Laser & Photonics Review*, **5**, 442–463.

96. Vazquez, R, Osellame, R, Nolli, D, Dongre, C, Vlekkert, H, Ramponi, R, Pollnau, M, and Cerullo, G (2009). Integration of femtosecond laser written optical waveguides in a lab-on-chip, *Lab on a Chip*, **9**, 91–96.

97. Sun, H-B, Matsuo, S, and Misawa, H (1999). Three-dimensional photonic crystal structures achieved with two-photon-absorption photopolymerization of resin, *Applied Physics Letters*, **74**, 786–788.

98. Kawata, S, Sun, H, Tanaka, T, and Takada, K (2001). Finer features for functional microdevices, *Nature*, **412**, 697–698.

99. Hell, S, and Wichmann, J (1994). Breaking the diffraction resolution limit by stimulated-emission: stimulated-emission-depletion fluorescence microscopy, *Optics Letters*, **19**, 780–782.

100. Rittweger, E, Han, KY, Irvine, SE, Eggeling, C, and Hell, SW (2009). STED microscopy reveals crystal colour centres with nanometric resolution, *Nature Photonics*, **3**, 144–147.

101. Li, L, Gattass, RR, Gershgoren, E, Hwang, H, and Fourkas, JT (2009). Achieving $\lambda/20$ resolution by one-color initiation and deactivation of polymerization, *Science*, **324**, 910–913.

102. Fischer, J, von Freymann, G, and Wegener, M (2010). The materials challenge in diffraction-unlimited direct-laser-writing optical lithography, *Advanced Materials*, **22**, 3578–3582.

103. Osellame, R, Chiodo, N, Maselli, V, Yin, A, Zavelani-Rossi, M, Cerullo, G, Laporta, P, Aiello, L, de Nicola, S, Ferraro, P, Finizio, A, and Pierattini, G (2005). Optical properties of waveguides written by a 26 MHz stretched cavity Ti: sapphire femtosecond oscillator, *Optics Express*, **13**, 612–620.

104. Saliminia, A, Nguyen, N, Nadeau, M, and Petit, S (2003). Writing optical waveguides in fused silica using 1 kHz femtosecond infrared pulses, *Journal of Applied Physics*, **93**, http://dx.doi.org/10.1063/1.1557777.

105. Streltsov, A, and Borrelli, N (2002). Study of femtosecond-laser-written waveguides in glasses, *Journal of the Optical Society of America B*, **19**, 2496–2504.

106. Minoshima, K, Kowalevicz, A, Hartl, I, IPPEN, E, and Fujimoto, J (2001). Photonic device fabrication in glass by use of nonlinear materials processing with a femtosecond laser oscillator, *Optics Letters*, **26**, 1516–1518.

107. Katayama, S, Horiike, M, Hirao, K, and Tsutsumi, N (2002). Diffraction measurement of grating structure induced by irradiation of femtosecond laser pulse in acrylate block copolymers, *Japanese Journal of Applied Physics Part 1*, **41**, 2155–2162.

108. Bricchi, E, Mills, J, Kazansky, P, Klappauf, B, and Baumberg, J (2002). Birefringent Fresnel zone plates in silica fabricated by femtosecond laser machining, *Optics Letters*, **27**, 2200–2202.

109. Srisungsitthisunti, P, Ersoy, OK, and Xu, X (2007). Laser direct writing of volume modified Fresnel zone plates, *Journal of the Optical Society of America B*, **24**, 2090–2096.

110. Yamada, K, Watanabe, W, Li, Y, Itoh, K, and Nishii, J (2004). Multilevel phase-type diffractive lenses in silica glass induced by filamentation of femtosecond laser pulses, *Optics Letters*, **29**, 1846–1848.

111. Guizard, S, Martin, P, Petite, G, D'Oliveira, P, and Meynadier, P (1996). Time-resolved study of laser-induced colour centres in, *Journal of Physics-Condensed Matter*, **8**, 1281.

112. Wortmann, D, Ramme, M, and Gottmann, J (2007). Refractive index modification using fs-laser double pulses, *Optics Express*, **15**, 10149–10153.

113. Osellame, R, Lobino, M, Chiodo, N, Marangoni, M, Cerullo, G, Ramponi, R, Bookey, HT, Thomson, RR, Psaila, ND, and Kar, AK (2007).

Femtosecond laser writing of waveguides in periodically poled lithium niobate preserving the nonlinear coefficient, *Applied Physics Letters*, **90**, http://dx.doi.org/10.1063/1.2748328.

114. Zhang, H, Eaton, SM, Li, J, Nejadmalayeri, AH, and Herman, PR (2007). Type II high-strength Bragg grating waveguides photowritten with ultrashort laser pulses, *Optics Express*, **15**, 4182.

115. Cerullo, G, Osellame, R, Taccheo, S, Marangoni, M, Polli, D, Ramponi, R, Laporta, P, and de Silvestri, S (2002). Femtosecond micromachining of symmetric waveguides at 1.5 μm by astigmatic beam focusing, *Optics Letters*, **27**, 1938–1940.

116. Shah, L, Yoshino, F, Arai, A, Eaton, S, Zhang, H, Ho, S, and Herman, PR (2005). MHz-rate ultrafast fiber laser for writing of optical waveguides in silica glasses, *Proceedings of SPIE*, **5714**, 253.

117. Hnatovsky, C, Taylor, R, Rajeev, P, Simova, E, Bhardwaj, V, Rayner, D, and Corkum, P (2005). Pulse duration dependence of femtosecond-laser-fabricated nanogratings in fused silica, *Applied Physics Letters*, **87**, http://dx.doi.org/10.1063/1.1991991.

118. Yang, W, Bricchi, E, Kazansky, P, Bovatsek, J, and Arai, A (2006). Self-assembled periodic sub-wavelength structures by femtosecond laser direct writing, *Optics Express*, **14**, 10117–10124.

119. Kitamura, N, Fukumi, K, Kadono, K, Yamashita, H, and Suito, K (1994). Reflection spectra of dense amorphous SiO_2 in the vacuum-UV region, *Physical Review B*, **50**, 132–135.

120. Zhang, X, Xu, X, and Rubenchik, A (2004). Simulation of microscale densification during femtosecond laser processing of dielectric materials, *Applied Physics A*, **79**, 945–948.

121. Nagata, T, Kamata, M, and Obara, M (2005). Optical waveguide fabrication with double pulse femtosecond lasers, *Applied Physics Letters*, **86**, 251103.

122. Stoian, R, Boyle, M, Thoss, A, Rosenfeld, A, Korn, G, Hertel, I, and Campbell, E (2002). Laser ablation of dielectrics with temporally shaped femtosecond pulses, *Applied Physics Letters*, **80**, 353–355.

123. Ramme, M, Anderson, T, Choi, J, Ebendorff-Heidepriem, H, Monro, T, and Richardson, MC, Waveguide writing and characterization in Tellurite glass, in *CLEO/QELS*2009, 1–2.

124. Osellame, R, Taccheo, S, and Marangoni, M (2003). Femtosecond writing of active optical waveguides with astigmatically shaped beams, *Journal of the Optical Society of America B*, **20**, 1559–1567.

125. Taccheo, S, della Valle, G, Osellame, R, Cerullo, G, Chiodo, N, Laporta, P, Svelto, O, Killi, A, Morgner, U, and Lederer, M (2004). Er: Yb-doped waveguide laser fabricated by femtosecond laser pulses, *Optics Letters*, **29**, 2626–2628.

126. Homoelle, D, Wielandy, S, and Gaeta, A (1999). Infrared photosensitivity in silica glasses exposed to femtosecond laser pulses, *Optics Letters*, **24**, 1311–1313.

127. Scully, P, and Jones, D (2003). Femtosecond laser irradiation of polymethylmethacrylate for refractive index gratings, *Journal of Optics A*, **5**, S92.

128. Choi, J, and Richardson, MC, Fabrication and characterization of femtosecond laser direct written volume diffractive optical elements in fused silica, in *CLEO/QELS*2008, 1–2.

129. Siiman, LA, Lumeau, J, and Glebov, LB (2009). Phase Fresnel lens recorded in photo-thermo-refractive glass by selective exposure to infrared ultrashort laser pulses, *Optics Letters*, **34**, 40–42.

130. Choi, J, Ramme, M, and Anderson, T (2010). Femtosecond laser written embedded diffractive optical elements and their applications, *Proceedings of the SPIE*, 7589, doi:10.1117/12.843697.

131. Esser, D, Mahlmann, D, Wortmann, D, and Gottmann, J (2009). Interference microscopy of femtosecond laser written waveguides in phosphate glass, *Applied Physics B*, **96**, 453–457.

132. Lamprecht, B, Krenn, JR, Leitner, A, and Aussenegg, FR (1999). Resonant and off-resonant light-driven plasmons in metal nanoparticles studied by femtosecond-resolution third-harmonic generation, *Physical Review Letters*, **83**, 4421–4424.

133. Gattass, RR, and Mazur, E (2008). Femtosecond laser micromachining in transparent materials, *Nature Photonics*, **2**, 219–225.

134. Keller, U (2003). Recent developments in compact ultrafast lasers, *Nature*, **424**, 831–838.

135. Steinmeyer, G, Sutter, D, Gallmann, L, Matuschek, N, and Keller, U (1999). Frontiers in ultrashort pulse generation: pushing the limits in linear and nonlinear optics, *Science*, **286**, 1507.

136. Goulielmakis, E, Schultze, M, Hofstetter, M, Yakovlev, VS, Gagnon, J, Uiberacker, M, Aquila, AL, Gullikson, EM, Attwood, DT, Kienberger, R, Krausz, F, and Kleineberg, U (2008). Single-cycle nonlinear optics, *Science*, **320**, 1614–1617.

137. Morgner, U, Kärtner, F, Cho, S, Chen, Y, Haus, H, Fujimoto, J, Ippen, E, Scheuer, V, Angelow, G, and Tschudi, T (1999). Sub-two-cycle pulses

from a Kerr-lens mode-locked Ti: sapphire laser, *Optics Letters*, **24**, 411–413.

138. Adachi, S, Ishii, H, Kanai, T, Ishii, N, Kosuge, A, and Watanabe, S (2007). 1.5 mJ, 6.4 fs parametric chirped-pulse amplification system at 1 kHz, *Optics Letters*, **32**, 2487–2489.

139. Petrov, V, Noack, F, Tzankov, P, Ghotbi, M, Ebrahim-Zadeh, M, Nikolov, I, and Buchvarov, I (2007). High-power femtosecond optical parametric amplification at 1 kHz in BiB_3O_6pumped at 800 nm, *Optics Express*, **15**, 556–563.

140. Suedmeyer, T, Marchese, SV, Hashimoto, S, Baer, CRE, Gingras, G, Witzel, B, and Keller, U (2008). Femtosecond laser oscillators for high-field science, *Nature Photonics*, **2**, 599–604.

141. Maine, P, Strickland, D, Bado, P, Pessot, M, and Mourou, G (1988). Generation of ultrahigh peak power pulses by chirped pulse amplification, *Quantum Electronics, IEEE Journal of*, **24**, 398–403.

142. Martinez, O, Gordon, J, and Fork, R (1984). Negative Group-Velocity Dispersion Using Refraction, *Journal of the Optical Society of America A*, **1**, 1003–1006.

143. Lemoff, B, and Barty, C (1993). Quintic-phase-limited, spatially uniform expansion and recompression of ultrashort optical pulses, *Optics Letters*, **18**, 1651–1653.

144. Treacy, E (1969). Optical pulse compression with diffraction gratings, *Quantum Electronics, IEEE Journal of*, **5**, 454–458.

145. Kalachnikov, M, Karpov, V, Schonnagel, H, and Sandner, W (2002). 100-terawatt titanium-sapphire laser system, *Laser Physics*, **12**, 368–374.

146. Eidam, T, Rothhardt, J, Stutzki, F, Jansen, F, Hädrich, S, Carstens, H, Limpert, J, and Tünnermann, A (2011). Fiber CPA System delivering 2.2 mJ, sub 500 fs pulses with 3.8 GW Peak Power, *Advanced Solid-State Photonics*, 2, 13–16.

147. Tanabe, T, Yamanaka, M, Okamoto, T, and Kannari, F (2004). Compensation for a transfer function of a regenerative amplifier to generate accurately shaped ultrashort pulses in both the amplitude and phase, *Selected Topics in Quantum Electronics, IEEE Journal of*, **10**, 221–228.

148. Schenkel, B, Biegert, J, Keller, U, Vozzi, C, Nisoli, M, Sansone, G, Stagira, S, de Silvestri, S, and Svelto, O (2003). Generation of 3.8-fs pulses from adaptive compression of a cascaded hollow fiber supercontinuum, *Optics Letters*, **28**, 1987–1989.

149. Tuennermann, A, Schreiber, T, and Limpert, J (2010). Fiber lasers and amplifiers: an ultrafast performance evolution, *Applied Optics*, **49**, F71–F78.

150. Eidam, T, Hanf, S, Seise, E, Andersen, T, and Gabler, T (2010). Femtosecond fiber CPA system emitting 830 W average output power, *Optics Letters*, **32**, 94–96.

151. Helbing, FW, Steinmeyer, G, and Keller, U (2003). Carrier-envelope offset phase-locking with attosecond timing jitter, *Selected Topics in Quantum Electronics, IEEE Journal of*, **9**, 1030–1040.

152. Luehrmann, M, Theobald, C, Wallenstein, R, and Lhuillier, JA (2009). High energy cw-diode pumped Nd:YVO_4 regenerative amplifier with efficient second harmonic generation, *Optics Express*, **17**, 22761–22766.

153. Kane, T, Eckardt, R, and Byer, R (1983). Reduced thermal focusing and birefringence in zig-zag slab geometry grystalline lasers, *Quantum Electronics, IEEE Journal of*, **19**, 1351–1354.

154. Russbueldt, P, Mans, T, Rotarius, G, Weitenberg, J, Hoffmann, HD, and Poprawe, R (2009). 400W Yb:YAG Innoslab fs-amplifier, *Optics Express*, **17**, 12230–12245.

155. Russbueldt, P, Mans, T, Weitenberg, J, Hoffmann, H, and Poprawe, R (2010). Compact diode-pumped 1.1 kW Yb: YAG Innoslab femtosecond amplifier, *Optics Letters*, **35**, 4169–4171.

156. Schulz, M, Riedel, R, Willner, A, Mans, T, Schnitzler, C, Russbueldt, P, Dolkemeyer, J, Seise, E, Gottschall, T, and Hädrich, S (2011). Yb: YAG Innoslab amplifier: efficient high repetition rate subpicosecond pumping system for optical parametric chirped pulse amplification, *Optics Letters*, **36**, 2456–2458.

157. Giesen, A, HiigeP, H, Voss, A, Wittig, K, Brauch, U, and Opower, H (1994). Scalable Concept for diode-pumped high-power solid-state lasers, *Applied Physics B*, **58**, 365–372.

158. Innerhofer, E, Sudmeyer, T, Brunner, F, Haring, R, Aschwanden, A, Paschotta, R, Honninger, C, Kumkar, M, and Keller, U (2003). 60-W average power in 810-fs pulses from a thin-disk Yb : YAG laser, *Optics Letters*, **28**, 367–369.

159. Marchese, SV, Baer, CRE, Engqvist, AG, Hashimoto, S, Maas, DJHC, Golling, M, Suedmeyer, T, and Keller, U (2008). Femtosecond thin disk laser oscillator with pulse energy beyond the 10-microjoule level, *Optics Express*, **16**, 6397–6407.

160. Brown, DC, Singley, JM, Kowalewski, K, Guelzow, J, and Vitali, V (2010). High sustained average power cw and ultrafast Yb: YAG near-

diffraction-limited cryogenic solid-state laser, *Optics Express*, **18**, 24770–24792.

161. Hong, KH, Gopinath, JT, Rand, D, Siddiqui, AM, Huang, SW, Li, E, Eggleton, BJ, Hybl, JD, Fan, TY, and Kärtner, FX (2010). High-energy, kHz-repetition-rate, ps cryogenic Yb: YAG chirped-pulse amplifier, *Optics Letters*, **35**, 1752–1754.
162. Matsushima, I, Yashiro, H, and Tomie, T (2006). 10 kHz 40 W Ti: sapphire regenerative ring amplifier, *Optics Letters*, **31**, 2066–2068.
163. Rothhardt, J, Haedrich, S, Seise, E, Krebs, M, Tavella, F, Willner, A, Duesterer, S, Schlarb, H, Feldhaus, J, Limpert, J, Rossbach, J, and Tuennermann, A (2010). High average and peak power few-cycle laser pulses delivered by fiber pumped OPCPA system, *Optics Express*, **18**, 12719–12726.
164. Fuji, T, Ishii, N, Teisset, CY, Gu, X, Metzger, T, Baltuska, A, Forget, N, Kaplan, D, Galvanauskas, A, and Krausz, F (2006). Parametric amplification of few-cycle carrier-envelope phase-stable pulses at 2.1 μm, *Optics Letters*, **31**, 1103–1105.
165. Chalus, O, Thai, A, Bates, P, and Biegert, J (2010). Six-cycle mid-infrared source with 3.8 μJ at 100 kHz, *Optics Letters*, **35**, 3204–3206.
166. Diels, J, ed. (2006) *Ultrashort Laser Pulse Phenomena: Fundamentals, Techniques, and Applications on a Femtosecond Time Scale* (Vch Verlagsgesellschaft Mbh).
167. Soileau, M, Mansour, N, van Stryland, E, and Williams, W (1989). Laser-induced damage and the role of self-focusing, *Optical Engineering*, **28**, 1133–1144.
168. Hemmer, M, Vaupel, A, Webb, B, and Richardson, MC (2010). Multi-kHz, multi-mJ, phase stabilized, OPCPA amplifier system, *Proceedings of the SPIE*, 7578, doi:10.1117/12.842893.
169. Nolte, S, Döring, S, Ancona, A, and Limpert, J (2011). High repetition rate ultrashort pulse micromachining with fiber lasers, *Fiber Lasers and Applications*, Istanbul, Turkey, February 16, 2011.
170. Goodman, DS, ed. (2004) eds. *Geometrical Optics* (Polaroid, Cambridge, Massachusetts) 1628.
171. Horstmann-Jungemann, M, and Gottmann, J (2010). 3D micro structuring of sapphire using fs-laser irradiation and selective etching, *Journal of Laser Micro/Nanoengineering*, **5**, 145–149.
172. Doering, S, Richter, S, Nolte, S, and Tuennermann, A (2010). *In situ* imaging of hole shape evolution in ultrashort pulse laser drilling, *Optics Express*, **18**, 20395–20400.

173. Trebino, R, DeLong, KW, Fittinghoff, DN, Sweetser, JN, Krumbugel, MA, Richman, BA, and Kane, DJ (1997). Measuring ultrashort laser pulses in the time-frequency domain using frequency-resolved optical gating, *Review of Scientific Instruments*, **68**, 3277–3295.

174. Iaconis, C, and Walmsley, IA (1999). Self-referencing spectral interferometry for measuring ultrashort optical pulses, *Quantum Electronics, IEEE Journal of*, **35**, 501–509.

175. Morita, R, Hirasawa, M, Karasawa, N, Kusaka, S, Nakagawa, N, Yamane, K, Li, L, Suguro, A, and Yamashita, M (2002). Sub-5 fs optical pulse characterization, *Measurement Science and Technology*, **13**, 1710.

Chapter 3

Fundamentals of Ultrafast Laser Processing

Nadezhda M. Bulgakova[a,b]
[a] *Institute of Thermophysics SB RAS, 1 Lavrentyev Ave., Novosibirsk, 630090, Russia*
[b] *Optoelectronics Research Center, University of Southampton, University Road, Southampton, SO17 1BJ, UK*
nbul@itp.nsc.ru

When laser light of visible, near-IR or UV spectral range hits condensed matter, it interacts with the valance and/or conduction electrons of the system under action. Depending on laser intensity and irradiation geometry, the interaction can have different far-reaching consequences such as melting, ablation, changing of optical properties, mechanical and chemical transformations. Among existing laser systems, ultrafast lasers have become an extraordinary tool for processing of any kind of materials. With proper choosing the irradiation conditions, laser action allows either inducing highly localized gentle modifications or obtaining strongly damaged material sites with desired or deleterious structures such as voids, periodic nanocracks, periodic surface structures, and craters of various shapes and dimensions. This chapter presents a review on tremendous efforts of researchers in order to achieve clearer insights

Ultrafast Laser Processing: From Micro- to Nanoscale
Edited by Koji Sugioka and Ya Cheng

ISBN 978-981-4267-33-5 (Hardcover), 978-981-4303-69-9 (eBook)
www.panstanford.com

into laser–matter interactions in ultrashort irradiation regimes. The review does not pretend to completeness and aims to outline main ideas, achievements, and most intriguing findings still waiting for explanations and theoretical treatments.

3.1 Introduction

Ultrafast lasers have become an extraordinary tool for processing of any kind of materials [1–4]. By proper choosing the irradiation conditions, one can induce highly localized gentle modifications or obtain strongly damaged material sites of desired structures such as voids, periodic nanocracks, periodic surface structures, and deep craters. Because of minimized heat diffusion effects, subwavelength-diameter craters can be produced in metal films [5]. Tight focusing of infrared or visible laser pulses of sub-ps duration inside wide-bandgap dielectrics allows modification within the focal volume [6–11]. However, within the same material family, considerably different material modifications are achieved when applying similar laser pulses. Deep understanding of the processes triggered in laser-irradiated materials by femtosecond (fs) laser pulses is of vital need for further advance of laser material processing techniques.

Over the past three decades, considerable progress has been achieved in establishing the main mechanisms of ultrafast laser interaction with metals, semiconductors, and inorganic dielectrics, which include radiation absorption, leading to creation of highly non-equilibrium thermodynamic states, material heating through equilibration of electronic and ionic/atomic subsystems, and recombination of laser-excited charge carriers (in the case of bandgap materials) with generation of various defect states. Several mechanisms of fs-laser ablation have been ascertained, such as phase explosion, mechanical spallation, and coulomb explosion (mainly for dielectric materials). A lesser understanding has been achieved in respect of thermoelastoplastic processes and plastic deformations resulted from creation of high temperature (and pressure) gradients inside matter as well as of effects of heat, stress, and defect accumulation in multi-pulse irradiation regimes. Below a brief description and a short historical overview of the main approaches developed to date are presented.

The complexity of the interrelated processes involved in laser–matter interaction gives rise to elaboration of theoretical and computational descriptions of the processes upon laser–matter interaction relying on different approaches. A striking fact is that the idea of one of the oldest but important models was proposed by Kaganov *et al.* [12] before invention of lasers. The authors assumed that under extremely fast energy coupling to metal surfaces, free electrons could gain a temperature T_e while the lattice ions remained relatively cold (at a temperature T_l) and dynamics of thermalization between electrons and lattice could be described by a simple relaxation equation with a characteristic time t_r

$$\frac{\partial T_l}{\partial t} = \frac{T_e - T_l}{t_r}. \tag{3.1}$$

The two-temperature model (TTM) developed on the basis of this idea by Anisimov *et al.* [13] for description of laser-induced metal heating constitutes hitherto the heart of not only continuum modeling but also atomistic simulations. Based on the TTM, a great body of information has been accumulated that sheds light on material behavior under different conditions of ultrafast laser excitation (see Section 3.2.1).

The main premise of the TTM is that free electrons absorbing laser radiation are thermalized on a timescale shorter than the laser pulse duration that allows treating the free electron gas in terms of temperature already on irradiation. The energy exchange between the electron and lattice subsystems occurs on picosecond (ps) timescale (from several to hundred ps depending on metal properties). Hence, for sub-ps laser pulses irradiated materials are excited in a strongly non-equilibrium state with hot gas of free electrons and cold lattice (hereafter for simplicity the atomic/ionic skeleton of bulk materials is referred as lattice for both crystalline and amorphous materials). The TTM allows quick low-cost analyses based on the thermodynamic considerations how the state of the system absorbed laser energy evolves, including material heating, melting, and probable ablation routes. Moreover, an analog of the TTM is used for nonmetallic materials where thermalization within the subsystem of electrons excited to the conduction band can be delayed up to a hundred of fs after excitation [14]. During the electron non-equilibrium stage, the energy balance equation for free carrier excitation, recombination, and phonon emission can be considered in detail [15] or in a simplified form being coupled with the lattice heat diffusion equation [16, 17]. Alternatively, the electron temperature can be considered in the usual TTM as a measure of the electron average energy [18, 19].

To gain deeper insights to kinetic and thermodynamic pathways and mechanical response of bulk materials exposed to the action of ultrashort laser pulses, more sophisticated approaches have been developed. The simplest one combines the TTM with the equations of the thermoelasticity [20–24]. This approach allows revealing the influence of the laser-generated stress waves on heating dynamics, the levels of stresses that can lead to material rupture, and 3D plastic deformations in the laser affected zones. For the analysis of strongly non-equilibrium processes and dynamics of thermalization within the laser-excited electron subsystem in metals, semiconductors, or dielectrics, the kinetic approaches based on the Boltzmann or Fokker–Planck equations are utilized [14, 25–30]. However, labor-intensiveness and large computational burden limit applications of these techniques for studies of relatively low laser fluences at the regions where optical response of matter is not significantly disturbed by laser radiation.

To describe the evolution of a free electron density $n_e(t)$ in wide-bandgap dielectrics under intense laser irradiation, a very simple but intuitive rate equation can be written in the following form [31–34]:

$$\frac{\partial n_e(t)}{\partial t} = \sigma_k I^k(t) + \delta I(t) n_e(t) . \tag{3.2}$$

Here σ_k and δ are the multiphoton ionization cross section and the avalanche coefficient, respectively, k is the number of photons required for a photo-ionization event. Such equations represent a simplest way for modeling the actual experimental conditions taking into account generating nonhomogeneous profiles of the electron density in the surface layer or inside the bulk depth and changing the optical response of laser-excited matter [16–18, 31–38]. The rate equation is easily coupled with the energy balance equations, thus allowing elucidating a whole spectrum of processes occurring in bulk materials under ultrafast excitations. Similar rate equations are also used for simulations of the free carrier dynamics in semiconductors [16–18]. In recent years, new interesting approaches of treating electron excitation and free electron absorption in dielectric materials have been proposed, such as a multiple rate equation (MRE) [39, 40] or a more rigorous approach based on the time-dependent Schrödinger Eq. [41]. However, the need of self-consistent descriptions of ultrashort-laser-induced dynamics, transport, and energy of charge carriers in dielectrics and semiconductors imposes restrictions on employing these approaches in view of awkwardness and complexity of the modeling representations. The limitations are determined by the necessity of considering a surface layer of the sample of thickness on the order of the irradiation spot radius where the electron density can experience sharp gradients, by the dynamic behavior of the optical properties, etc.

The listed modeling descriptions belong to the continuum type models and their applications are restricted to irradiation regimes below material destruction or to stages of laser-excited matter evolution before rupturing/disintegration. Breaking the homogeneous nature of the irradiated matter (e.g., homogeneous nucleation of vapor phase with subsequent material disintegrations to atoms, small clusters, and droplets [42, 43] or crystal fracture [44]) is studied by the models based on hydrodynamic considerations and on atomistic (molecular dynamics) techniques. The hydrodynamic models are based on solving the one-dimensional two-temperature hydrodynamic equations with application of the realistic equations of state [45–47]. These models allow elucidating evolution of laser-excited bulk materials from the excitation stage to decomposition, though the decomposition process is considered conventionally as the emergence of alternating high- and low-density layers, which are treated as vapor and condensed phases. However, they give valuable information on material response to laser excitation at high fluences and, thus, high stresses with generation of shock wave structures.

By now atomistic modeling based on the molecular dynamics (MD) method is the most advanced technique, which gives possibility to follow not only laser-induced material spallation into layers but also disintegration of bulk matter into atoms and clusters. MD techniques possess a great potential, being, however, still strongly restricted to the description of a relatively small amount of matter compared with that usually involved in laser–solid interaction and subsequently developed laser ablation. Additionally, for strongly non-equilibrium matter containing electrons as degrees of freedom (especially for highly excited dielectrics and semiconductors), the MD technique has to be based on the many-body potential energy surface simulations taking into account occupation of the electronic levels in the valance band as well as free electron contribution [48]. So, at high levels of electronic excitations, the models based on an interatomic potential such as Lennard-Jones, Tersoff, and Stillinger-Weber do not allow reasonable description of ultrashort structural changes [48].

The following sections present the more detailed discussions of the progresses achieved with application of the above-mentioned theoretical approaches and modeling techniques in establishing fundamental mechanisms governing the behavior of matter on ultrafast laser excitation. In Section 3.2, processes initiated by ultrashort laser pulses in different materials (metals, semiconductors, inorganic dielectrics) are analyzed. Special attention is given to the two-temperature model, whose employment is still extending for studies of laser–matter interactions. The mechanisms of ultrashort laser ablation are considered in Section 3.3 with an overview of different simulation techniques for ablation studies, molecular dynamic, hydrodynamic and continuum models. Section 3.4 focuses on studies of volume modifications in transparent materials with addressing the laser pulse propagation, heat accumulation, plastic deformations in heat-affected zones (HAZ), plasma effects. The role of ambient gas in laser processing of solid surfaces is discussed in Section 3.5. Finally, Section 3.6 lists the most intriguing findings that still require explanations and adequate theoretical descriptions.

3.2 Material Excitations by Ultrashort Laser Pulses

The uniqueness of fs laser pulses for material microprocessing is provided by two important aspects of ultrashort laser interaction with matter:

- During laser energy coupling into irradiated materials, heat transfer from the coupling zone is negligible. Hence, the absorbed laser energy is localized in a region whose size can be controlled.
- Material ablation occurs after the laser pulse termination that ensures the absence of shielding of the laser beam by plasma of the ablation products.

As a result, ablation of a given amount of matter by fs laser pulses requires considerably smaller energy compared with longer laser pulses [49]. An essential feature of interaction of ultrashort laser pulses with matter is violation of thermal equilibrium between the atomic/ionic lattice and free electrons as was first noticed for metals in [12]. In this section, we consider main mechanisms of laser–matter interaction on the timescales of tens to hundreds fs for different kinds of materials at laser intensities typical for ultrashort laser processing.

3.2.1 Two-Temperature Model as a Building Block for Studies of Laser–Matter Interaction at Ultrashort Irradiation Regimes

The two-temperature model is based on the assumption of thermal equilibrium within the electronic and lattice subsystems and is expressed in the form of the heat flow equations for electrons absorbing laser radiation and lattice heated due to heat exchange with electrons [13]:

$$C_e \frac{\partial T_e}{\partial t} = \frac{\partial}{\partial z} K_e \frac{\partial T_e}{\partial z} - g(T_e - T_l) + \Sigma(z,t)\,, \tag{3.3}$$

$$C_l \frac{\partial T_l}{\partial t} = g(T_e - T_l)\,. \tag{3.4}$$

Here z is the coordinate directed from the target surface to the bulk; C_e, C_l, and K_e are the heat capacities and the thermal conductivities of electrons and lattice, respectively; g is the electron–lattice coupling constant; $\Sigma(z,t)$ is the laser energy source term. The TTM is usually written in the one-dimensional form, which is justified by the fact that in highly absorbing solids, effective absorption depths are much smaller than typical irradiation spot sizes. In metals, the laser light is absorbed predominantly by free electrons in a quasi-linear manner via the inverse bremsstrahlung process that allows the application of the Beer–Lambert law. The law states that light attenuation is proportional to the absorbance of the medium through which the light is traveling. Accordingly, the term $\Sigma(z,t)$ is written in the form $I(t,z) = (1 - R)I_0(t)\exp(-\alpha z)$ or, more generally, $I(t,z) = (1 - R(t))I_0(t)\exp(-\int_0^z \alpha(z,t)dz)$, where I and I_0 denote the local and incident laser intensities, $\alpha(z,t)$ and $R(t)$ are the absorption and reflection coefficients, which in general case can be variable, and z is the beam propagation path through the absorbing media. The temporal shape of fs laser beams can be described by a Gaussian function as $I(t)=(1-R(t))(2F_0/\tau_L)\sqrt{\ln 2/\pi}\exp(-4\ln 2(t/\tau_L)^2)$, where F_0 is laser fluence and τ_L is the laser pulse duration (full width at half maximum). Note that in Eq. (3.4), the lattice conduction term is missing as the lattice conductivity is negligible on the timescale when the thermal equilibrium between electrons and lattice is establishing after ultrafast laser excitation. It should also be admitted that on femto- and ps timescales, the lattice heat transport does not obey Fourier's empirical law, which does not account for the finite speed of the thermal energy carriers (this relates to the electrons in much lesser extent).

The validity of the TTM for treating electron–phonon coupling in laser-excited metals on sub-ps timescales has been discussed abundantly in the literature. In fact, the model does not take into account the nonthermal character of free electron behavior in the initial stages of laser excitation as was found experimentally [50–52]. Schematically, the processes triggered in metal by ultrashort laser pulses are shown in Fig. 3.1 [52]. The laser energy is absorbed by a fraction of free electrons that leads to pronounced deviation of the electron-energy distribution from a Fermi–Dirac distribution in the absorption region. The absorbed energy is transferred to the lattice through a whole sequence of the processes. The ballistic motion of the nonthermalized electrons together with their diffusive heat transfer leads to efficient transport of the energy to deeper target layers. At the same time, these electrons exchange the energy with the rest thermal bath of the non-excited electrons and, partially, with the lattice. As not all electrons are initially laser-excited, the electron–phonon scattering rate is reduced. The thermal (Fermi) electrons transfer the energy obtained from the nonthermalized electrons to lattice atoms with simultaneous diffusive energy transport toward the bulk depth. This sequence of the interconnected processes can be described in the frames of the TTM by separating the electron distribution into two parts, for highly laser-excited electrons and the Fermi component acting as a heat reservoir [51]. The evolution of this system "nonthermal electrons–Fermi electrons–lattice" can be described by supplementing the TTM with the third coupled equation for the highly excited fraction of electrons [51, 52]. The forms of the refined TTM can be found in [51, 52].

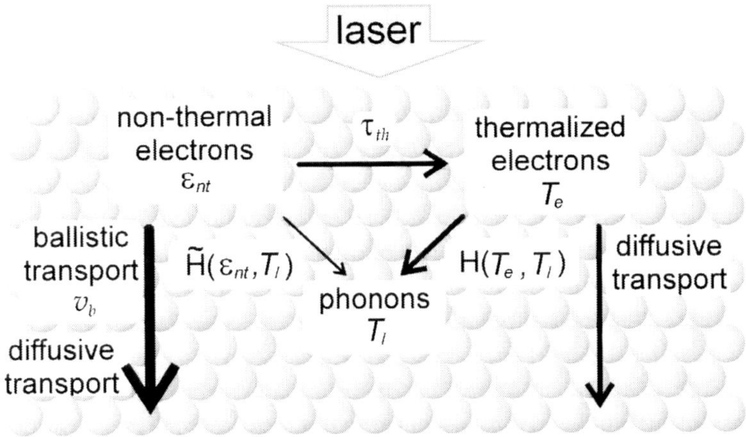

Figure 3.1 Schematic representation of energy flow in a metal after femtosecond optical excitation (extended heat bath model). The electron system is split in nonthermal and thermal baths, coupled with different strength to the phonon bath characterized by the lattice temperature T_l. Energy in the nonthermal bath ε_{nt} relaxes with a time-constant τ_{th} to the thermal bath. Heat transport of electrons is modeled as diffusive (see comments on the electron diffusivity in Section 3.2.2) and for the nonthermal bath as ballistic transport with velocity v_b which is of order of the Fermi velocity. The terms $\tilde{H}$ and H represent energy coupling to phonons from nonthermalized and thermal electrons, respectively. Courtesy of M. Wolf.

Although the experimental studies of non-equilibrium electron dynamics upon ultrafast laser excitation (either on photoelectron spectroscopy [52] or transient optical response [51]) are performed with low laser fluences, typically below 1 mJ/cm^2, they unambiguously point out that electron thermalization time decreases with increasing laser fluence. Numerical studies performed for aluminum as an example on the basis of the Boltzmann Eq. [29] show that at fluences near the damage threshold, the electron–electron thermalization time decreases to a few fs and the TTM adequately describes energy exchange in laser-excited metals. However, using the traditional TTM often encounters difficulties in view of poor agreements with the data of particular experiments on ultrafast laser excitation of metals (transient changes of optical response, melting and ablation thresholds, ablation crater depths). This stimulated extensive studies of different aspects that could influence the TTM accuracy such as ballistic range of nonthermal electrons, dependences of optical response, and the parameters of the electron gas (heat capacity, electron–phonon coupling) on the electron temperature. Main outcomes of these studies are summarized in the next Section.

3.2.2 Metals: Insights from the Two-Temperature Modeling

The TTM (Eqs. (3.3) and (3.4)), being simple at the first glance, implicates a whole range of processes that should be taken into consideration on its application. Ignoring transient changes of optical response (both varying reflectivity and absorption coefficient) of laser-excited matter and/or strong variations of the physical properties of the electron gas upon swift heating leads to a considerable mismatch between TTM results and experimental data. Changing of optical response, being especially pronounced for transition metals exhibiting $d \to p$ transitions on excitations, is usually reasonably described with the Drude model [53–57], which, however, implies a number of fitting parameters [53, 56]. According to the theory proposed by Pail Drude in 1900 [58], the complex dielectric function of metals can be expressed as

$$\varepsilon = 1 - \frac{\omega_{\mathrm{p}}^2}{\omega^2 + \dfrac{i\omega}{\tau}}\,. \tag{3.5}$$

Here $\omega_{\mathrm{p}} = \sqrt{4\pi e^2 n_{\mathrm{e}} / m_{\mathrm{e}}}$ and ω are the electron plasma frequency and the angular frequency of external (here, laser) electric field; m_{e} is the electron mass; τ is a damping time that is needed to accelerate electrons in response to the varying field and determined by the effective electron collision frequency [55, 57]. The ω_{p} and τ values are usually taken from best fitting to experimental data [53–57]. However, these values are considerably varying parameters upon excitation as they depend on the electron density, which can be changing due to $d \to p$ transitions, lattice expansion upon heating, and melting, and on the electron temperature. The most dramatic variation is experienced by the effective collision frequency of electrons, $\nu_{\rm eff} = \tau^{-1}$. Its value can be expressed as $\nu_{\rm eff} = \min[\nu_{\rm e\text{-}ph} + \nu_{\rm e\text{-}e}, \nu_{\rm c}]$ where $\nu_{\rm e\text{-}ph} = k_{\rm s}e^2T_{\rm l}/\hbar^2 v_{\rm F}$ and $\nu_{\rm e\text{-}e} = A_{\rm T}T_{\rm e}^2/\hbar T_{\rm F}$ are the electron–phonon and electron–electron collision frequencies, respectively, $T_{\rm F}$ and $v_{\rm F}$ are the Fermi temperature and velocity [54, 56]. The value $\nu_{\rm c} = \sqrt{v_F^2 + T_e/m_e}/r_0$ is the maximum possible collision frequency determined by the minimum mean free path, which cannot be smaller than the ion sphere radius r_0 [57]. The coefficients $k_{\rm s}$ and $A_{\rm T}$ are introduced to fit metal absorptivity at room temperature and to adjust for the quantitative uncertainty in the contribution of the electron–electron collisions to the effective collision frequency [57]. Hence, the application of the Drude model, being highly useful for understanding the main trends of optical response of metals to ultrafast laser excitation, implies a whole range of fitting parameters. Moreover, it is applicable only at electron temperatures below the Fermi temperature and, at high excitations when $T_{\rm e} > T_{\rm F}$, the dielectric permittivity can be described based on Coulomb collisions applicable for a dense plasma [57, 59]. Often simulations in the frames of the TTM are performed for absorbed laser energy. However, direct comparisons of such modeling data with the experiments impose difficulties in view of the above discussion.

Also, a remark should be made that transient changing of the dielectric function implies a spatiotemporal variation of the absorption coefficient. Therefore, the Beer–Lambert law should be used in the integral form as mentioned in Section 3.2.1. Additionally, the presence of nonthermal electrons leads to an enhanced (ballistic) energy transport, which can be accounted for via introducing an effective absorption coefficient $\alpha_{\rm eff} = 1/(\alpha^{-1} + l_{\rm ball})$ where $l_{\rm ball}$ is the so-called ballistic range [60–62]. In fact, the ballistic range concept is oversimplified as the $l_{\rm ball}$ value is considered a metal-specific constant, whereas it should be dependent on the excitation degree and associated change in electron collision dynamics (see discussion in [62]).

Another considerable complication for direct applications of the TTM to describe specific experimental conditions is in violent changes of the physical properties of the electron gas as noted above. Once the electron subsystem of an irradiated metal is rapidly heated, the electron–phonon coupling rate as well as the heat capacity and heat conductivity of electrons experience dramatic changes already in the course of irradiation. It is widely accepted that at relatively low laser fluences the electron–phonon coupling rate can be considered constant, while the heat capacity and conductivity of electrons can be described by linear dependences on the temperature as $C_e = A_e T_e$ and $K_e = K_{e,0} T_e / T_l$ [60, 61, 63]. However, such dependences are reasonably valid only for relatively low electron temperatures, and, hence, at low laser fluences typically well below damage thresholds of metals. A systematic analysis of the temperature dependences of the electron–lattice coupling and the electron heat capacity based on the electron density of states (DoS) has been performed for a number of metals in [64]. Its results show strong deviations of the electron gas parameters from the commonly used dependences. Moreover, the electron coupling factor, which considerably increases with the electron temperature for Al, Au, Ag, Cu, and W, drops by an order of magnitude for Ni and by several times for Pt in the T_e range of 300–10,000 K [64]. For the case of platinum, the $g(T_e)$ dependence is even more complicated. It increases approximately three times when the T_e value grows from the room temperature to ~5000 K and becomes a decreasing function with further increasing T_e [64]. A plenty of work on the DoS analysis of the electron gas parameters in laser-excited metals has been done by Zhigilei *et al.* [64, 65].

One further comment should be made on the electron heat conductivity. Based on the Drude model, it can be expressed as $K_e(T_e, T_l) = v_F^2 \tau_e(T_e, T_l) C_e(T_e)/3$. where τ_e is the electron scattering time [64, 66, 67]. The electron scattering time is determined by the sum of the electron–electron and electron–phonon scattering rates $\tau_{e\text{-}e}$ and $\tau_{e\text{-}ph}$, respectively, as $1/\tau_e = 1/\tau_{e\text{-}e} + 1/\tau_{e\text{-}ph}$. The $\tau_{e\text{-}e}$ and $\tau_{e\text{-}ph}$ values can be approximated at low temperatures as $1/\tau_{e\text{-}e} = AT_e^2$ and $1/\tau_{e\text{-}ph} = BT_l$ with the constants A and B specific for each metal [64, 66, 67], while at high electron temperatures they can behave in a more complicated manner [64]. Thus, the described set of the electron-temperature-dependent parameters makes applications of the TTM for description of the actual experimental conditions in laser-excited metals to be a difficult task, requiring in-depth knowledge of evolution of metals upon excitation.

Another important feature of ultrafast laser heating of materials is superheating of the solid phase to a temperature T_m^s (above the melting point under normal conditions, T_m) at which a catastrophic homogeneous melting occurs [68–70]. The superheating level $(T_m^s - T_m)/T_m$ can reach values of 0.4 or even higher, depending on metal properties and heating rate [68, 70]. Strong superheating observed at high laser fluences is explained by increased electronic delocalization or, alternatively, by the effect of the electronically induced pressure, which reduces the concentration of seeds for homogeneous nucleation [70]. Molecular dynamic studies [69] have revealed that due to the nucleation catastrophe, the lattice temperature drops almost to the normal melting temperature level at the melting interface (boundary of complete melting) as a result of energy expense for the latent heat of fusion (lattice destabilization). The TTM does not enable for study the mentioned superheating features. Even if the Stephan problem is coupled with the TTM to follow the solid–liquid transition as is often done, the rates of the melting process are overestimated.

The TTM gives a reasonable fit to the experimental data at weak excitations when variations of the various parameters discussed above are minimized, e.g., at low temperatures with gentle laser excitations as in studies of non-equilibrium magnetization dynamics in gadolinium [71, 72]. In the latter cases, the TTM with the simplified "low-temperature" parameters and undisturbed optical properties exhibits good agreement with pump-probe experiments at time delays >100 fs, while at smaller time delays the transient energy density of electrons is underestimated by TTM simulations [71].

The above analysis demonstrates that the TTM alone supplies us with useful but evaluative information of the electron and lattice temperature dynamics. A sharp increase in lattice temperature in a confined surface region of an irradiated sample results in formation of the pressure waves that can lead to material spallation [73, 74]. This additional aspect of the interaction of ultrashort laser pulses with matters can noticeably influence the heating and melting dynamics and melting thresholds. It is suggested that on ultrafast laser irradiation, the electron thermomechanical (blast) waves can be important in overall dynamics of heating and ablation of metals [75–78]. To follow the formation and evolution of the stress waves, the TTM is supplemented with an equation describing the dynamics of the lattice displacement u [24, 76–80]:

$$\rho \frac{\partial^2 u}{\partial t^2} = E_1 \frac{\partial^2 u}{\partial z^2} - E_2 k_\varepsilon \frac{\partial T_\mathrm{l}}{\partial z} + 2\Lambda \left(T_\mathrm{e} \frac{\partial T_\mathrm{e}}{\partial z} \right). \tag{3.6}$$

In this case, Eq. (3.4) of the TTM should be written in the following form to account for the effect of stress wave formation on the heating dynamics:

$$C_1 \frac{\partial T_1}{\partial t} = g(T_e - T_1) - \frac{C_1 \eta}{k_\varepsilon} \frac{\partial^2 u}{\partial t \partial z} . \tag{3.7}$$

In Eqs. (3.6) and (3.7), ρ is the sample density; k_ε is the linear expansion coefficient; $\eta = 3Gk_\varepsilon^2 T_{10} / C_1$ is the thermomechanical factor; the constants E_1, E_2, and G are defined by Young's modulus E and the Poisson ratio ν as $E_1 = E(1 - \nu)/(1 + \nu)(1 - 2\nu)$, $E_2 = E/(1 - 2\nu)$, and $G = E/(3(1 - 2\nu))$. In the frame of the free electron gas model, the Λ value can be considered a parameter determining relation between the electron pressure P_e and the temperature as $P_e = \Lambda T_e^{\ 2}$ and is of the order of the electron DoS at the Fermi surface [75,81]. For $T_e << T_F$, this parameter can be estimated as $\Lambda = \pi^2 N_e k_B/(3T_F)$ [78] with N_e and k_B to be the density of free electrons under normal conditions and the Boltzmann constant, respectively. The value of thermomechanical stress is defined by the lattice temperature and displacement

$$\sigma = E_1 \frac{\partial u}{\partial z} - E_2 k_\varepsilon T_1 . \tag{3.8}$$

The formation of stress waves requires of some energy expenses and hence leads to some increase in the melting threshold of the irradiated samples. However, the consideration of the stress waves becomes especially important at fluences near and above the ablation threshold of different kinds of materials as they are a driven force of the spallation mechanism and play essential role in overall dynamics of laser-induced material disintegration. Additionally, formation of the stress waves at laser beam focusing inside transparent material samples is one of the main mechanisms of material compaction exploited in the waveguide writing techniques. The ablation and material compaction mechanisms will be discussed in further sections.

The above consideration demonstrates that despite extensive studies of interaction of ultrashort pulse radiation with metals over several decades, a complete understanding of the processes in laser-excited metals has not been reached and this topic still requires much research. Especially, this concerns different aspects of non-equilibrium thermodynamics as will be additionally discussed in further sections. Turning to the bandgap materials, it is necessary to underline that the aspects considered in this section are also present on ultrafast laser excitation of semiconductors and dielectrics. Once free electron gas (dense plasma) has been excited in a bandgap material, the latter behaves like a metal with all further consequences. However, in view of more complicated nature of interaction via excitation of valance electrons to the conduction band, many of the listed processes are not considered or treated in a highly simplified manner. The following two sections focus only on peculiarities of ultrafast laser excitation of semiconductors and dielectrics without repeating the questions analyzed here.

3.2.3 Laser-Induced Processes in Semiconductors

This section will outline excitation of semiconductors by ultrashort laser pulses. We shall restrict our consideration to silicon as a typical representative of this material class excited by 800 nm fs laser pulses. Below, an example of the modeling framework is given with discussing the principal processes taking place in semiconductors under ultrashort laser irradiation. In the most semiconductors, due to the narrow bandgap, generation of free electrons occurs by one- and two-photon absorption in the wavelength range from the near-UV to near-IR region of the spectrum [16, 38]. As soon as a definite level of the electron density is reached (typically starting from approximately 10^{18}–10^{19} cm^{-3}), the collisional ionization is initiated and becomes dominating at even higher densities [16, 38]. The recombination in semiconductors is preferably of Auger (tree-particle) type [16]. However, radiative recombination is also present, especially at low free electron densities [82]. Taking into account the listed processes, the equation describing the spatiotemporal dynamics of the free carrier density for the case of silicon, which, when properly modified, can be used for other semiconductors and for other wavelengths, reads as follows:

$$\frac{\partial n_e}{\partial t} = \left[\left(\sigma_1 + \frac{1}{2}\sigma_2 I\right)\frac{I}{\hbar\omega} + \delta n_e\right]\frac{n_0 - n_i}{n_0} - R_e. \tag{3.9}$$

Here n_e, n_i, and n_0 are the electron and ion densities and the density of undisturbed lattice, respectively; I is the laser intensity; σ_1 and σ_2 are the one- and two-photon ionization coefficients, respectively (for silicon at 800 nm σ_1 = 1021 cm^{-1}, σ_2 = 10 cm/GW [83, 84]). Here we assume $n_e = n_i$, though in a surface target layer where effective photoemission occurs, this condition can be violated. The consequences of breaking quasi-neutrality on the surfaces of the laser-excited samples will be discussed in Section 3.3.3. The collisional (avalanche) electronic multiplication coefficient δ is dependent on the free electron temperature [16]. We assume that the loss term R_e is determined only by the Auger recombination process [16]. It is known that Auger recombination saturates at electronic densities approaching 10^{21} cm^{-3} due to plasma screening effects that can be described as [85]:

$$R_e = \frac{n_e}{\tau_0 + 1/Cn_e n_i}. \tag{3.10}$$

For silicon, $C = 3.8 \times 10^{-31}$ cm^6/s [16] and $\tau_0 = 6 \times 10^{-12}$ s [85]. Equation (3.9) implies indirectly that an inhomogeneous profile of the free electron density is generated [18, 86] with decreasing the density toward the bulk depth due to attenuation of the beam propagating through absorbing matter. Moreover, because of a dynamic behavior of free electron density governed by a highly nonlinear avalanche process at relatively high laser fluences, optical response of laser-excited semiconductors is strongly varying already during the laser pulse (with further changing upon melting). Spatiotemporal dynamics of the optical parameters can be described within the Drude formalism and the Fresnel formulas via the complex dielectric function $\varepsilon^*(n_e)$ whose value can be seen as a contribution of both unexcited matter and generated dense plasma [87]:

$$\varepsilon^*(n_e) \cong 1 + (\varepsilon_g - 1)\left(1 - \frac{n_e}{n_{val}}\right) - \frac{n_e}{n_{cr}}\left(1 + i\frac{1}{\omega\tau}\right)^{-1}. \tag{3.11}$$

Here ε_g is the dielectric function of unexcited material; n_{cr} is the critical density of electrons (for λ = 800 nm $n_{cr} = 1.74 \times 10^{21}$ cm^{-3}); n_{val} is the total valance-band density in the unexcited state; $\omega\tau$ is the damping factor [88], whose meaning was also discussed in Section 3.2.2. We underline that the reflection coefficient from an inhomogeneous dense plasma should be calculated within a multilayer reflection model [89, 90].

Once the absorption coefficient α_{ab} of the laser-excited semiconductor can be found via $\varepsilon^*(n_e)$ as a function of sample depth and time, the spatial and temporal distribution of a nonreflected part of the laser intensity inside the sample can be calculated as

$$\frac{\partial}{\partial x} I(z,t) = -\left(\sigma_1 \frac{n_0 - n_i}{n_0} + \sigma_2 \frac{n_0 - n_i}{n_0} I(z,t) + \alpha_{ab}(z,t)\right) I(z,t) \tag{3.12}$$

accounting for one-, two-photon ionization, and light absorption by the laser-induced free electrons in the conduction band.

To follow the dynamics of the sample heating, one can assume that strongly ionized semiconductors can be considered as dense plasmas and an analog of the TTM (Eqs. (3.3) and (3.4)) may be applied for the energy balance description either for the average electron energy [16, 17] or the electron temperature considered as a measure of the average energy under the conditions of incomplete thermalization within the electron subsystem [18] (see discussion in Sections 3.2.1 and 3.2.2). A special attention should be paid to designing the laser energy source term. It has been shown [18] that the most accurate form of the term reads as

$$\Sigma(z,t) = n_e \frac{\partial E_e}{\partial t} = \frac{\partial E_f}{\partial t} - \frac{3}{2} k T_e \frac{\partial n_e}{\partial t}. \tag{3.13}$$

Here E_f is the local energy density of the electronic subsystem and E_e is the average energy per electron ($E_e = 3kT_e/2$). The variations of the E_f value are followed via detailed consideration of the energy balance of the electron subsystem, accounting the channels for laser energy absorption and redistribution:

$$\frac{\partial E_f}{\partial t} = \left((\hbar\omega - E_g)\frac{\sigma_1 I}{\hbar\omega} + (2\hbar\omega - E_g)\frac{\sigma_2}{2}\frac{I^2}{\hbar\omega} - E_g \delta n_e \right) \frac{n_0 - n_i}{n_0} + \alpha_{ab}(z,t) I(z,t) + E_g R_e. \tag{3.14}$$

The electron–lattice coupling rate g can be defined via the characteristic electron–lattice relaxation time τ_r as $g = C_e/\tau_r$, which, considering screened electron–lattice interaction at high densities, can be taken for the case of silicon as $\tau_r = \tau_0(1 + (n_e/n_{cr})^2)$ with $\tau_0 =$ 240 fs [83].

The above formalism allows following the formation and evolution of dense plasma in laser-irradiated semiconductors and can be applied to getting insights into heating dynamics. Importantly, such features of laser-excited metals as incomplete thermalization within the electronic subsystem, superheating of the solid phase, and insufficient knowledge about behaviors of a number of physical properties stay valid for semiconductors even to a greater extent. An additional intriguing feature of ultrafast laser exposure, which is observed only for semiconductor materials, is ultrafast ("cold") melting. If a considerable fraction (~10–15%) of the valance electrons is excited to the conduction band, the lattice is destabilized and loses the crystalline order, being yet at a temperature well below the melting point [91–93]. The ultrafast order-disorder phase transition requires several hundreds of fs [93].

A simplified modeling performed for silicon within the framework described above [38] has demonstrated a reasonable agreement with the experimental observations. Simulations were performed for silicon irradiated by 800 nm laser pulses of 100 fs duration. It was assumed in the modeling that ultrafast melting is initiated when approximately 10% of the valance electrons (2×10^{22} cm^{-3}) are excited to the conduction band. In the next 400 fs, required for lattice destabilization [93], the excited region of the sample was treated within the solid dynamics and then instantaneously the silicon properties were changed to those of the liquid phase, assuming that all initial valance electrons belong now to the conduction band [94]. The results of these simplified calculations are presented in Fig. 3.2. According to the modeling, the threshold of ultrafast melting is ~0.65 J/cm^2. This value is app. three times higher than the threshold for normal (thermal) melting calculated within the same framework and agrees with the experimentally observed ultrafast phase transition [92]. The depth of nonthermally melted layer is 20–40 nm in the fluence range 1–5 J/cm^2 (Fig. 3.2a), also in reasonable correlation with observations on the basis of time-resolved X-ray diffraction [93].

It should be emphasized that thermal melting, which inevitably follows ultrafast disordering upon electron–lattice thermalization in ps timescale, occurs in a 10 times thicker layer of the surface region of the sample. The average energy of electrons immediately after the bandgap collapse (given in Fig. 3.2b in the temperature units) increases from ~0.8 to 3 eV in the fluence range of 0.65 to 5 J/cm^2. One can expect that immediately after the bandgap collapse, the distribution function of the electrons consists of a cold electron core and a high-energy tail (> 10% of electrons with $E_e \geq 10$ eV according to simulations). This will cause the second round of thermalization within the electron subsystem whether or not the electrons initially excited into conduction band have thermalized. However, such aspects of ultrafast laser excitation have not been studied yet.

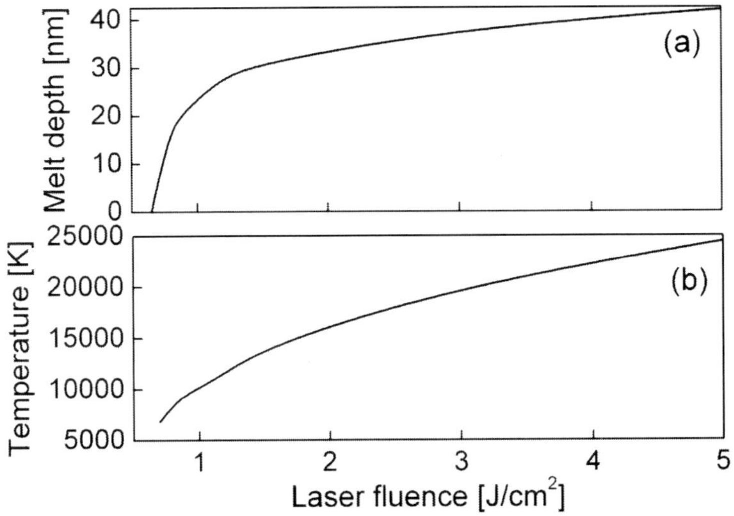

Figure 3.2 (a) The depth of the non-thermally molten layer of a silicon sample irradiated by 800 nm laser pulse of 100 fs duration as obtained by estimative simulations [38]. (b) The averaged energy of the electrons (presented in the temperature units) calculated at the time moment immediately after the bandgap collapse.

3.2.4 Photoexcitation and Relaxation of Transparent Dielectrics Irradiated by Ultrashort Laser Pulses

In spite of growing technological applications of transparent materials and their paramount promises for using in optoelectronics and photonics, the basics of laser-induced excitation of transparent crystalline materials and glasses have not sufficiently studied. This is conditioned by both the complexity of the processes in the laser-excited dielectrics and a great variety of compound materials when even a small variation in atomic composition and structure changes dramatically material response to laser excitation. In this Section the most fundamental aspects of ultrafast laser excitation specific for inorganic dielectric materials will be reviewed with an assessment of the state-of-art and further directions.

Ultrashort pulsed laser radiation applied to inorganic dielectric materials by focusing either on the surface or toward the bulk depth induces, depending on the irradiation conditions, different kinds of modifications such as formation of surface [95–99] and volume [99–102] periodic structures (nanogratings), densification and refractive index changes [6–9, 11, 103–105], formation of micro- and nanovoids [101, 106–108], phase transitions such as amorphization of crystalline materials [108], etc. Such modifications are already used in many applications in optoelectronics and photonics (optical signal transmission and optical beam manipulation, 3D writing of information) [109], and microfluidics [110] and the list of potential applications is rapidly expanding (see Chapters 8–10). Successful development of applications based on laser-induced micro- and nanomodifications of transparent materials requires deep understanding of the whole chain of the intricate processes initiated in dielectrics by fs laser pulses and extending up to millisecond timescales with formation of permanent mechanically deformed and/or chemically modified states.

The mentioned chain of the processes starts from material photoionization with creation of seed free electrons. The latter absorb laser energy and, at proper conditions, can produce secondary electrons in the collisional ionization process, thus generating electron avalanche multiplication. This results in considerable change of the optical response of the laser-irradiated region toward metallization. Laser beams focused inside the bulk of a transparent material can experience self-focusing starting from a threshold material-dependent power. However, the beam self-focusing collapse is counteracted by scattering from the laser produced plasma. In some dielectrics, the active recombination of the free electrons ("solid plasma") starts already at sub-ps timescales, while in other dielectrics, the free electron gas survives up to hundreds of ps [111]. The rapid recombination process of the solid plasma leads to swift heating of the photoexcited region that occurs at ps timescale, when heat conductions effects are negligible. As a result of fast heat release into the atomic subsystem and corresponding pressure rise, thermoelastic waves are generated, which, depending on the heating level and heat localization, can either completely dissipate or lead to significant plastic deformations of the material or even to mechanical damage in the form of micro- and nanovoids in the energy-release zone.

It should be underlined that the stress waves induce deformations not only in a hot laser-excited region but also in an extended cold zone around it. The thermomechanical effects terminate at

~10 ns after the laser pulse action when the three-dimensional pressure waves propagate a distance of a few tens of micrometers and substantially dissipate. However, under some experimental conditions the "mechanical scenario" can repeat at timescales of a few tens of microseconds when the locally released energy spreads due to heat conduction effects and reaches the regions of "cold" deformation. Softening (or even melting) of the deformed regions can cause a significant secondary redistribution of the material (as shown in [105]) and even emission of secondary thermoelastic waves. Figure 3.3 shows the characteristic timescales of the response of transparent materials on ultrashort laser excitation [90].

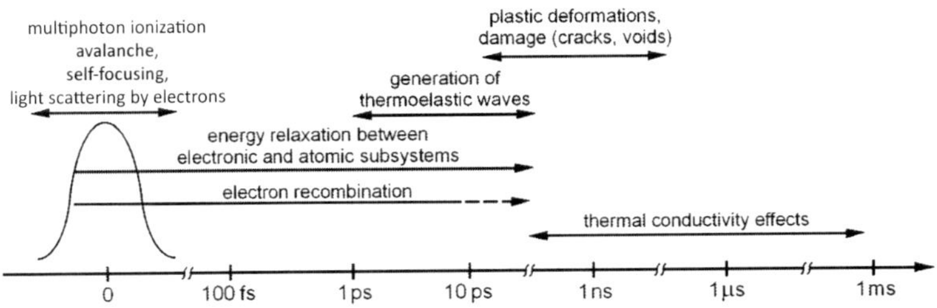

Figure 3.3 Characteristic timescales of various processes caused by irradiation of a transparent dielectric material by a femtosecond laser pulse [90].

At present, most theoretical investigations of laser excitation of wide-bandgap transparent materials are focused on the mechanisms of plasma generation and its relaxation. It has been established that the main mechanisms involved in generation of free carriers in wide-bandgap dielectrics irradiated by visible and near-IR fs laser pulses are the multiphoton and avalanche ionizations (see Eq. (3.2)) [25, 31–34]. The excitation process can conventionally be divided into two stages. At the initial stage, free electrons are generated via the multiphoton ionization while avalanche (collisional) ionization develops when a definite "seed" level of the free-electron density is reached, which is sufficient to efficiently absorb laser light. At relatively high radiation intensities, the tunneling ionization mechanism can dominate over multiphoton ionization as determined by the Keldysh parameter $\gamma = \omega(2m_{\text{eff}}E_{\text{g}})^{0.5}/(eE_{\text{L}})$ (m_{eff} and e are the effective electron mass and the unit charge, respectively; E_{L} is the electric field of the laser wave; E_{g} is the bandgap width in dielectric materials or ionization potential of individual atoms or molecules) [14, 112–114]. In fact, the Keldysh parameter is a ratio of the characteristic time that an electron takes to overcome the energy barrier (the ionization potential whose value can vary in the strong-wave field) to the electromagnetic field period. At $\gamma >> 1$, the multiphoton ionization mechanism prevails while at $\gamma << 1$ the tunneling mechanism becomes dominating. It is widely accepted that for the tunneling ionization to produce a noticeable effect, the condition $\gamma < 0.5$ should be met. However, a number of experimental facts indicate that multiphoton ionization can dominate even at $\gamma << 1$. These facts led to numerous generalizations of the Keldysh theory for photo-ionization of dielectric materials [113], as well as for atoms and molecules in the gas phase [114]. However, it must be underlined that the tunneling mechanism of ionization can play a significant role only upon focusing on the surface layer of material in the vacuum or low-pressure gas environment. On focusing into the material bulk and/or in the presence of a dense ambient gas, the attenuation of the laser beam in its way to the laser focus (see Sections 3.4.1 and 3.5) leads to the intensity constraints at levels at which the mechanism of multiphoton ionization prevails unambiguously (clamping effect) [115, 116]. It should be noted that for relatively longs wavelengths of femtosecond laser pulses, toward the mid-IR range, the rate of multiphoton ionization is strongly decreasing and tunneling ionization will inevitably play the dominant role. Although the femtosecond lasers at mid-IR wavelengths are seldom used today, this trend should be mentioned for providing more complete understanding.

The role of avalanche ionization in the breakdown of dielectrics at ultrafast irradiation regimes is still debated [117–119] from complete negation of its existence for sub-ps pulses [119] to assertion that its effect can be even more pronounced with decreasing pulse duration due to a decrease of the potential barrier in a strong laser field (cold avalanche) [120]. Electron recombination in inorganic dielectrics proceeds usually in the form of trapping in localized states ($R_e = n_e/\tau_{tr}$ where τ_{tr} is a characteristic recombination time), accompanied by creation of color centers, excitons, and other defects [88, 121, 122]. Defect generation leads to pronounced incubation effects manifested as a decrease of the damage threshold in multipulse irradiation regimes [122]. One can, however, expect that the incubation effect, i.e., accumulation of the defect states and their preferential ionization, can appear already during the pulse action, increasing by the end of the laser pulse when the pulse duration exceeds the characteristic recombination time (e.g., for fused silica where $\tau_{tr} \approx 150$ fs [123]).

Thus, the scenario of the excitation of an inorganic dielectric material can be described as following. Multiphoton ionization, whose order is determined by the ratio between the bandgap width and photon energy $\hbar\omega$, results in excitation of electrons from the valance band to the conduction band. The free electrons efficiently absorb laser radiation due to the bremsstrahlung and, if an electron has absorbed a sufficient amount of the laser energy ($>E_g$), it can collisionally ionize a neutral atom of the material. The development of avalanche (collisional multiplication of free electrons) changes the optical properties of the laser-excited region of the material, similarly to the case of semiconductors (see the previous section). For materials with sub-ps characteristic times of the electron trapping process, electrons trapped during the pulse can be re-excited into the conduction band by continuing irradiation or by electron collision. A detailed description of the trapping and re-excitation processes can be found in [123], where a set of the rate equations is presented (though neglecting the avalanche process). It has also been demonstrated [123] that the accumulating trapped electrons can contribute to the dielectric function variation that implies introduction of an additional term to Eq. (3.11).

A model of laser-induced excitation of a dielectric material and its heating in dynamics can be constructed similarly to semiconductors as discussed in Section 3.2.3. Such a model should include Eqs. (3.9) and (3.11–3.14) revised for the dielectric case and an analog of the TTM (Eqs. (3.3) and (3.4)) either for the average electron energy [16,17] or for the electron temperature as a measure of the average energy under the conditions of incomplete thermalization within the electron subsystem [18]. The rate equation should take into account the several processes. In the simplest form, it contains the multiphoton ionization and avalanche terms (see Eq. (3.2)). If solely the electron excitation dynamics is of interest, Eq. (3.2) gives the first rough estimate for the evolution of the free electron density. However, for materials where electron trapping proceeds efficiently already at sub-ps timescale, the electron recombination term is an obligatory part of the rate equation (see [111, 123, 124]). Moreover, the possibility of the re-excitation of the trapped electrons implies introducing the additional terms into the free electron rate equation (see [111, 123]). For a wide-bandgap dielectric excited with ultrashort laser pulses with focusing on the surface, the rate equation without taking into account re-excitation of the exciton states can be written as

$$\frac{\partial n_{\mathrm{e}}(z,t)}{\partial t}=(\sigma_{\mathrm{k}}I^{\mathrm{k}}+\delta In_{\mathrm{e}})\frac{(n_0-n_{\mathrm{i}})}{n_0}-\frac{n_{\mathrm{e}}}{\tau_{\mathrm{tr}}}. \tag{3.15}$$

A revised equation for laser beam attenuation upon propagation from the surface toward the bulk depth reads as

$$\frac{\partial I(z,t)}{\partial x}=-\sigma_{\mathrm{k}}(I(z,t))^{\mathrm{k}}\frac{n_0-n_i}{n_0}k\hbar\omega-a_{\mathrm{ab}}I(z,t)\ , \tag{3.16}$$

while the equation describing the variation of the local energy density of the free electron subsystem has the following form

$$\frac{\partial E_{\mathrm{f}}(z,t)}{\partial t}=((k\hbar\omega-E_{\mathrm{g}})\sigma_{\mathrm{k}}I^{\mathrm{k}}-E_{\mathrm{g}}\delta n_{\mathrm{e}}I)\frac{n_0-n_{\mathrm{i}}}{n_0}+\alpha_{\mathrm{ab}}I+E_{\mathrm{g}}R_{\mathrm{e}}. \tag{3.17}$$

Equations (3.11) and (3.13) keep their forms. However, if the generation of excitons and their re-excitation by laser radiation are of importance, the excitonic states can influence the dielectric function (see Ref. [123]).

In this section, the main processes in wide-bandgap dielectrics have been outlined for the case of laser beam focusing on the sample surface. Actually the dynamics of laser-induced excitations of dielectric materials is much richer, which motivates their utterly wide applications in various optical technologies. In Section 3.4, the peculiarities of optical material modifications by ultrashort laser pulses upon focusing in the bulk depth will be discussed. Going to the general ablation mechanisms described in the next section, we mention that by now the mechanism of Coulomb explosion (CE) has been proven for dielectric materials, while for other material kinds, CE is still being debated (see Section 3.3.3).

3.3 Studies of Laser Ablation Mechanisms in Ultrashort Irradiation Regimes

A distinctive feature of fs laser ablation, rapid energy delivery, which enables avoiding the laser-plume interaction and minimizing HAZ, allows the generation of well-defined microstructures with high quality and reproducibility [3, 49]. Very promising applications of pulsed laser ablation can also be found in the generation of high-energetic ion and cluster beams by controlling the thermodynamic paths of the materials under laser exposure [45, 125]. However, for further development of the present pulsed laser technologies, a detailed understanding of the dynamics and mechanisms of particle desorption and ablation from the irradiated surfaces is of crucial importance.

Mechanisms of laser ablation of different materials are still debated, especially for ultrashort, fs laser pulses. Among the relevant mechanisms taking place in different materials on the different timescales, one can list Coulomb explosion, phase explosion, spallation, and fragmentation into the plasma state [18, 38, 42, 111, 126–129]. Based on the present knowledge, the listed ablation mechanisms can be represented schematically as shown in Fig. 3.4 [130]. At fairly high laser fluences, an irradiated target can be divided by convention into slabs according to the laser energy coupling and ablation timescales. A superficial layer obtains a highest energy density. If for a given material, the laser pulse can generate electric field sufficient for electrostatic disintegration of the surface layer I (typically of ~3–4 nm wide) [126], Coulomb explosion occurs at the earliest stage of ablation, during first hundreds of fs [18, 131].

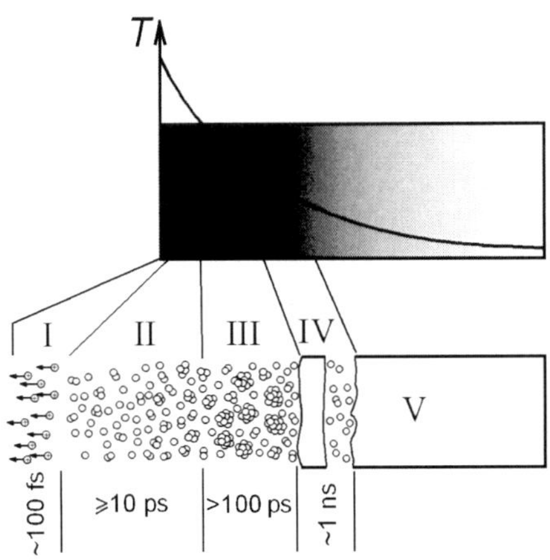

Figure 3.4 Schematic of the main ablation mechanisms in the case of their co-existence under femtosecond laser irradiation of solids [130]. I: Coulomb explosion; II: fragmentation into plasma, atomic and small cluster states; III: phase explosion; IV: spallation. V refers to a non-ablated target area. Typical timescales of manifestation of the mechanisms are indicated at the bottom.

The next target layer (II), heated up to extremely high, above-critical temperatures upon electron–lattice relaxation, experiences fragmentation into the atomic phase, which can be substantially ionized and freely expands away from the surface. This fragmentation process can occur already in first dozens of ps (though the fragmentation timescale depends on the electron–phonon coupling time inherent for a particular material). In the underlying liquid slab (III), homogeneous nucleation develops culminating in decomposition of matter into a mixture of vapor and droplets (called usually phase explosion). This explosive decomposition delayed in respect to the atomic phase ejection and can take from dozens to hundreds of ps. Finally, the thermoellastic waves generated in laser-irradiated matter can cause material spallation (IV) as a final point of the ablation process. This schematic picture shows that all listed mechanisms can plausibly exist together under particular irradiation conditions.

Generally the mechanisms depend on material kind, laser fluence, wavelength, and pulse duration. Thus, for materials with high free electron mobility, Coulomb explosion can be reached only at extreme laser irradiation conditions. For example, critical values of the electric field induced by ultrashort laser pulses were observed upon irradiation of micrometer-sized metal targets with fluence of $\sim 10^{18}$ J/cm^2 [132]. With decreasing laser fluences at such materials, the layer fragmented into the plasma state reduces until it totally vanishes and gives place to phase explosion. At even lower fluences, only spallation can occur [133–135] while in dielectric materials, in view of low electron mobility, the Coulomb explosion mechanism remains working even when ablation through the other mechanisms does not occur anymore, thus resulting in gentle, high-accuracy ablation [126, 127]. Also it should be remarked that, at relatively high laser fluences, material spallation can be suppressed by the recoil pressure of the vapor/cluster plume as was shown recently for double-pulsed laser action [46, 47]. Below the ablation mechanisms are considered in more detail as they emerge in different theoretical approaches.

3.3.1 Molecular Dynamic Studies: Spallation and Phase Explosion

In the last two decades, the classical MD method has proved to be a powerful tool for studies of laser-induced ablation of materials with providing insights into the mechanisms of ablation on the atomistic level. This method is applied to different materials under different irradiation conditions [42, 43, 73, 74, 133–142]. A detailed review on application of the MD method to metals can be found in [43]. Here we will outline only the main features of this method and discuss its advantages and limitations.

Grosso mode, the classical MD method of simulations, is based on solving of a set of the Newtonian equations of motion for each particle in a considered system of particles

$$m_i \frac{d^2\vec{r}_i}{dt^2} = \vec{F}_i, \quad i = 1,2,3,\ldots,N \tag{3.18}$$

where m_i and $\vec{r}_i$ are the mass and the position of a particle i, N is the number of particles in the system, and $\vec{F}_i$ is the force acting on this particles due to its interaction with the other particles of the system. In the general way, this force can be found if one knows the interparticle potential which is a function of position of all particles:

$$\vec{F}_i = -\vec{\nabla}_i U(\vec{r}_1, \vec{r}_2, \vec{r}_3, \ldots \vec{r}_N). \tag{3.19}$$

If the positions and the velocities of all particles are known at a time moment t_0 and the interaction potential is defined, the further evolution of the system can be followed via numerical solving the set of equations (3.18). A tremendous advantage of this method is that if an interaction potential is accurately defined, other assumptions on the thermodynamic and mechanical properties are not required. Moreover, these properties can be obtained through MD modeling, including their evolution in a strongly non-equilibrium matter on its way to equilibrium. In this sense, the MD simulation can be considered as a "computer experiment," which additionally allows the atomic-scale resolution. This makes the MD method an important predictive tool for investigations of matter evolution under extreme conditions of ultrafast energy input upon ultrashort pulse laser irradiation.

The necessity of tracking trajectories of all particles in the system imposes severe constraints on the system size and simulation timescales. The timescale factor (the ratio of the simulation time to the time of a real process) for the problems of laser–matter interaction is very large even if parallel computing is performed on a state-of-art supercomputer, making MD simulations expensive and restricting tracking time of the process under consideration.

However, the fundamental knowledge obtained via the application of the MD method to a number of realistic systems exposed to the ultrafast laser action justifies the expenses. We consider here only several examples of the MD simulations that, however, are representative of attempts to get insight to the laser-excited matter behavior on the atomic scale.

An appropriate interaction potential constitutes a heart of the MD method [48, 141, 143]. In fact, each material should be approximated by a potential that is capable of reproducing the material structure and properties, including heat capacity, thermal conductivity, elastic constants, and phase transition points. The pair-wise potentials of a Lennard-Jones, Tersoff, or Stillinger-Weber type do not describe adequately the properties of metals as they do not include the electrons, important "degrees of freedom" [48]. To solve this problem, the embedded atom methods (EAMs) have been developed [144, 145]. The EAMs include pair-wise (ion–ion) interactions as a part of the full potential, while the second part characterizes the energy of an ion surrounded by a cloud of electrons. In this case, the force (Eq. (3.19)) is written as $\vec{F} = -\vec{\nabla} E_{\text{tot}}$:

$$E_{\text{tot}} = \sum_{i} E_i; E_i = \frac{1}{2}\sum_{j \neq i} \varphi(r_{ij}) + \Phi(\rho_i);\ \rho_i = \sum_{i \neq j} f(r_{ij}). \tag{3.20}$$

Here ρ_i is the density of electrons in the ith ion location site; Φ is a function characterizing the ion energy in electronic surroundings; $\varphi(r_{ij})$ is a pair-wise interatomic potential; $f(r_{ij})$ is a contribution of jth ion into the density of the electron cloud in the ith ion location site. This method was demonstrated to be an effective tool for simulations of metals with different crystalline structure exposed to ultrafast laser excitation [42, 43, 62, 69, 73, 133, 141]. Another successful approach is based on a time-dependent, many-body potential energy surface derived from a tight-binding Hamiltonian [48, 136]. Notably, the first-principles methods based on the density functional theory do not allow performing a multi-variant computational experiment with varying laser fluence and pulse duration because of too high computational costs [48].

In view of large computational costs, it is impossible to study the evolution of laser-excited matter over the whole irradiation spot and laser affected depths. In practice, a limited region of the irradiated spot is considered with the periodic boundary conditions as illustrated in Fig. 3.5 [141]. Thus, simulating the different sights of

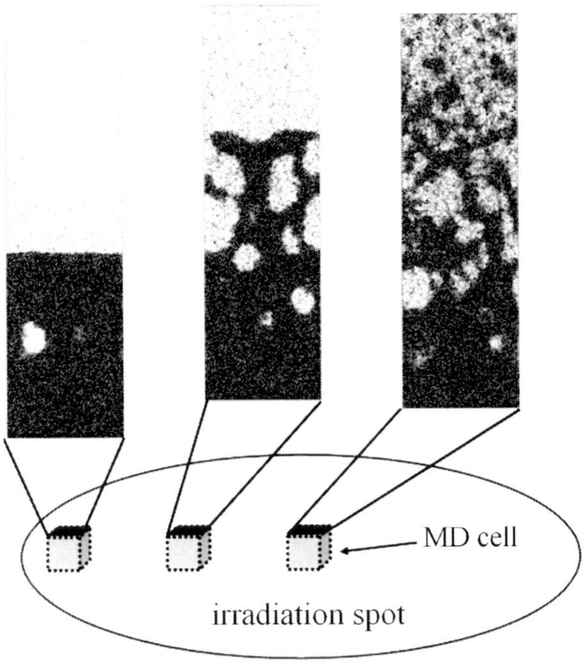

Figure 3.5 Schematic illustration of the local areas represented in MD simulations of laser ablation of different locations within a laser-irradiated spot [141]. Copyright © 2009 Authors.

the laser-irradiated spot with setting up the different levels of laser fluence according to a laser beam spatial profile, one can restore a picture of material heating, melting, and disintegration over the spot. The laser energy absorption is considered usually via the TTM (its electronic part, see Sections 3.2.1–3.2.2) with introducing the energy coupled from the electron subsystem to the lattice into the MD unit of the so-called combined TTM-MD model [42, 43, 62, 69, 73, 133, 138, 141]. In the combined model, the TTM is applied in one dimension to a whole laser-heat-affected depth of an irradiated target, while the MD method is used for getting insight about material behavior in a target surface layer of interest. In the boundary of the "superficial" MD domain in the target depth where the combined TTM-MD method turns into the pure TTM, a dynamic nonreflecting boundary condition is applied to provide free propagation of the laser-induced pressure waves [133, 141]. Another approach is setting up the absorbing boundary conditions at the bottom of the

MD domain while the laser energy absorption is considered as a sequence of discrete photons absorbed by the target according the Beer–Lambert law [134, 139, 140].

Figure 3.5 illustrates how, at relatively high laser energies, material ablation dynamics changes over the laser-irradiated spot from formation of a void in the spot periphery, which can result in spallation of a superficial layer to a considerable atomization and rupture of matter into large clusters and, finally, to a developed phase explosion with material disintegration into an atomic/nanocluster phase. The MD simulations provide rich information on the dynamic behavior of the material parameters, including temperature, pressure, density, and the evolution of matter over the phase diagram [43, 139, 141]. Figure 3.6 shows the temporal and spatial evolution of the temperature in a Ni target irradiated by 1 ps laser pulses of different intensities [43]. According to Fig. 3.5, the presented contour plots can be also considered illustrating the material behavior in the different sites of the irradiation spot. Different types of material response to laser irradiations are demonstrated, from the fast laser melting followed by resolidification (Fig. 3.6a), to spallation of a single layer of liquid material (Fig. 3.6b) and multiple layers or droplets ejection (Fig. 3.6c) and finally to explosive decomposition of an overheated surface layer (Fig. 3.6d). It should be mentioned that the laser-induced pressure behavior is not distinguished by such diversity as for the temperature evolution and the ablation dynamics [43]. The pressure evolves in the range from ~11 GPa on sudden material heating to approximately –7 GPa upon formation and propagation of an unloading wave. With increasing laser fluence, one can note a decrease of the tensile stress that is presumably conditioned by the recoil pressure of the ablation products.

As one can notice from Fig. 3.6, MD naturally reveal the occurrence of superheating of the solid state [69, 133] as mentioned in Section 3.2.2. The crystalline order is followed in simulations via the local order parameter [133], which determines the position of an atom relative to neighboring atoms. The values 0 and 1 of the local order parameter correspond to complete disorder (melt) and order (crystal), respectively. Setting a definite small value of this parameter (e.g., 0.04 as in [69]) allows fixing the superheating states. As was mentioned, superheating levels $(T_m^s - T_m)/T_m$ above 0.4 were identified in [69] that are consistent with observations [70].

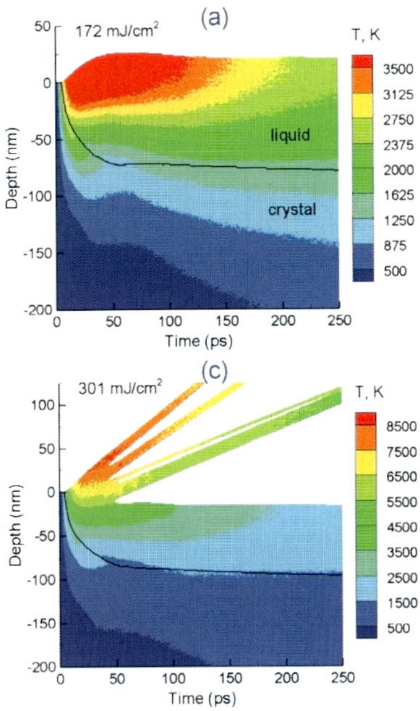

Figure 3.6 Temperature contour plots obtained by MD simulations of a bulk Ni target irradiated with 1 ps laser pulses of Gaussian temporal profiles with peak intensities at 2.5 ps from the beginning of the simulations [43]. Values of the absorbed fluence are 172 J/cm^2 (a), 193.5 J/cm^2 (b), 301 J/cm^2 (c), and 408.5 J/cm^2 (d). Laser pulses are incident from the top. Black lines separate the molten and crystalline parts of the target. Reprinted with permission from [43]. Copyright © 2009 American Chemical Society.

It must be emphasized that the described MD simulations, being two- [139, 140] or three-dimensional [42, 43, 62, 69, 73, 133, 135–138, 141, 142] in respect of material disintegration, are in fact one-dimensional as they do not account for the lateral stress evolution. Upon heating, matter confined in the irradiation spot experiences strong lateral stresses that result in the formation of micro-bumps and nano-jets [22, 146, 147] as well as in dotted transfer of matter in the laser-induced transfer schemes [148, 149]. Recently, however, first attempts of MD modeling were reported where the authors performed simulations over the whole irradiation spots [134, 150]. Although the simulations involve considerable simplifications with respect to the considered number of particles ("scaled" irradiation spots and/or film thickness), they demonstrated porosity formation upon extrusion of metal by the laser-induced lateral stress [150] and emergence of simultaneous manifestations of the different ablation mechanisms (Fig. 3.7) with emission of dislocations [134]. Note that some conclusions of 2D MD simulations may be incorrect as discussed in [43].

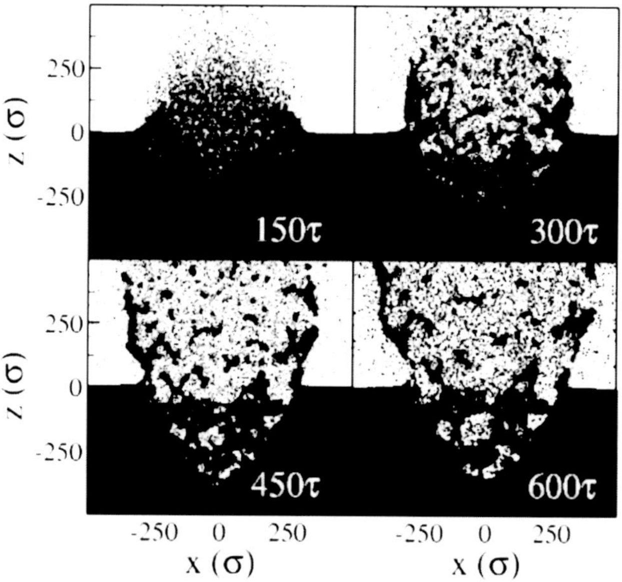

Figure 3.7 Snapshots of laser ablation of a target obtained in generic two-dimensional simulations on the basis of the Lennard-Jones potential [134]. $\tau = (\epsilon/m\sigma^2)^{0.5}$ is a characteristic timescale with ϵ and σ to be the parameters of the Lennard-Jones potential. Laser fluence is 1500ϵ with the ablation threshold between 625 and 800ϵ. The picture demonstrates differences in material decomposition in the central and edge parts of the irradiation spot. Reprinted with permission from [134]. Copyright © 2009 American Physical Society.

Upon further development of computer techniques, the MD simulation method can become one of the main tools of computer experiments in a number of research fields including laser–matter interactions. At present, large-scale applications of the MD method for studying laser excitation of bulk materials are mainly restricted to metals and molecular solids [151]. For the latter, special course-grained models have been developed, which are based not on atomic-level but on molecular representation of molecular solids, thus allowing a considerable extension of time and length scales of simulations.

It must be underlined that the model materials used for MD simulations, though reproducing to some extent the real material properties, cannot be considered an exact copy of a material. For example, the EAM applied for gold [152] gives the reduced melting point and specific heat of melting compared with the experimental values (~963 K and 8.4 kJ/mol instead of 1336 K and 12.8 kJ/mol, respectively). A qualitative nature of MD simulations should be always remembered when analyzing the simulation results in respect of the experimental situations. For strongly excited bandgap materials, existing MD technique are not applicable in view of temporal and spatial variations of the interaction potentials in the cause of the laser excitation with chemical bond breaking.

Usually MD simulations are performed for the absorbed laser energy as the optical response of materials is a function of the electron temperature in metals (see Section 3.2.2) and the free electron density in bandgap materials (Sections 3.2.3 and 3.2.4). This conditions the popularity and utilization of other methods of numerical simulations based on continuum approaches with the elements of kinetics of the free electron population (see next sections).

3.3.2 Hydrodynamic Modeling

One more approach, the hydrodynamic method of modeling based on the Euler equations (so-called ideal hydrodynamics), enables describing some important features of material ablation [45–47, 108, 153–155]. Whereas material decomposition is considered on the atomic level in the MD simulations, in one-dimensional hydrodynamic treatments, there are no clear differentiations between the vapor and droplet phases as the models do not include assumptions on kinetics of vapor phase nucleation. However, when involving criteria based on thermodynamic estimations [45] or material strength theory [155], the use of such models benefit from much smaller computational time expenses compared with MD techniques. Models based on the hydrodynamic method, being used in a two-temperature form to account for plasma state of matter and containing a multi-phase wide-range equation of state [45–47], consistently describe the dynamics of phase transformations and the evolution of the HAZ in a laser-irradiated material, though without atomic resolution of decayed matter.

The role of the main predetermined "parameter," which is played by an interaction potential in MD simulation techniques, is assigned to a reliable equation of state (EoS) in the hydrodynamic models. Such an EoS must describe the evolution of the material properties over the phase diagram from a cold state to melting, further heating, superheating, and, finally, transformation to a vapor/plasma state, including the vicinities of singularities like the thermodynamic critical point and the spinodal. Analytical forms of such EoSs do not exist and the tabulated semi-empirical EoSs represent the best choice [156].

Like in the MD models where the electron gas absorbing laser radiation is considered via the TTM, the hydrodynamic models require assumptions on the electron gas properties (thermal conductivity, heat capacity, electron lattice coupling) for solving the equation for the electron energy. Additionally, also as in both the TTM and MD simulations, the variations in the optical properties of matter under irradiation are usually treated in the frames of the Drude formalism as was discussed in Section 3.2.2.

The hydrodynamic codes are mostly applied to the cases of metal ablation [45–47, 153–155], though under some additional assumptions they can be used for describing the phase transformations in bandgap materials [108]. The calculations based on the EoS allow the determination of the state of matter in each its point at each time moment, thus following the evolution of matter over the phase diagram, including regions of coexistence of different phases. Figure 3.8 shows the thermodynamic paths of copper under double-pulse irradiation with different time separations between pulses [47]. In the phase diagram, the regions of the stable and metastable states of matter are marked by the corresponding first letters (metastable states are marked with bracketed letters), including the phase coexistence regions. Note that in the simulations [47], only overheated liquid and overcooled gas have been considered, while the superheating of the solid state is not taken into account, which is of importance for ultrashort laser irradiation [70]. The superheating of the solid state is still a challenging task for the hydrodynamic models, while in MD techniques this aspect of ultrafast laser excitation of metals is a natural outcome of modeling [69].

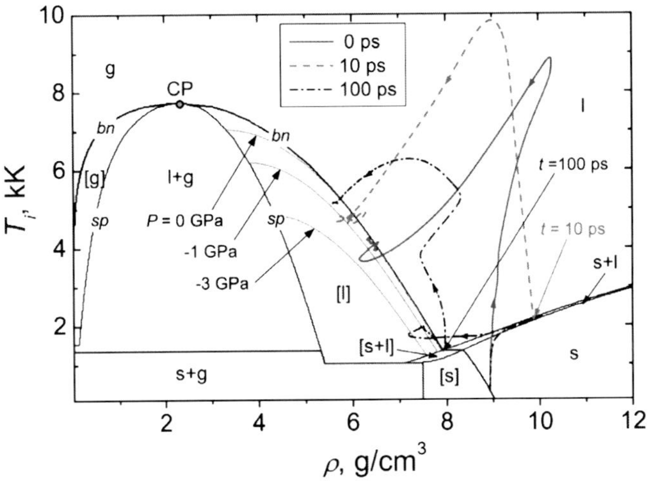

Figure 3.8 Phase diagram of copper [47]: CP: critical point; *bn*: the binodal; *sp*: the spinodal; g and [g]: stable and metastable gas respectively; l and [l]: stable and metastable liquid; s and [s]: stable and metastable solid; s+l: stable melting; [s+l]: metastable melting; s+g: solid–gas mixture. Isobars with P = 0, -1 and -3 GPa are given by solid green lines. The trajectories of the target layer located initially at 60 nm under the surface are presented by solid red line, dashed magenta line, and dash-dotted blue line respectively for 0, 10 and 100 ps separation time between pulses. Reprinted with permission from Elsevier, copyright © 2011.

By examples of Figs. 3.8 and 3.9, one can analyze thermodynamic pathways of matter upon ultrafast laser heating. The trajectories of the matter evolution over the phase diagram shown in Fig. 3.8 have been obtained for double laser pulses of 800 nm wavelength and 100 fs duration with fluence of 2 J/cm^2 each. Three trajectories are given for a copper layer located initially 60 nm deep with respect of the irradiated sample surface. First trajectory shown by the continuous red line corresponds to a zero time delay, that is, irradiation with a single 4 J/cm^2 pulse. Two other lines are for double pulses separated by 10 (dashed line) and 100 (dot-dashed line) ps. All three trajectories originate from the same point (at normal conditions). The single-pulse trajectory crosses the melting region near-isochorically and evolves in the liquid phase, reaching supercritical temperatures as a result of electron–lattice coupling. Rapid heating of the lattice (note that here the term "lattice" does not mean a crystalline order but refers to the ionic/atomic component of irradiated matter) leads to formation of the shock wave (SW, Fig. 3.9) propagating toward the target depth that is seen as some compression of matter along the trajectory. An unloading (tensile) wave (TW, Fig. 3.9), which moves from the free surface following the shock wave, leads to material expansion that results in adiabatic cooling of the expanding layer and its penetration into the metastable liquid state. In the metastable region homogeneous nucleation of the vapor phase together with the mechanical effects of the negative pressure culminate in fragmentation of matter into a droplet-vapor phase and presumably spallation of the whole layers.

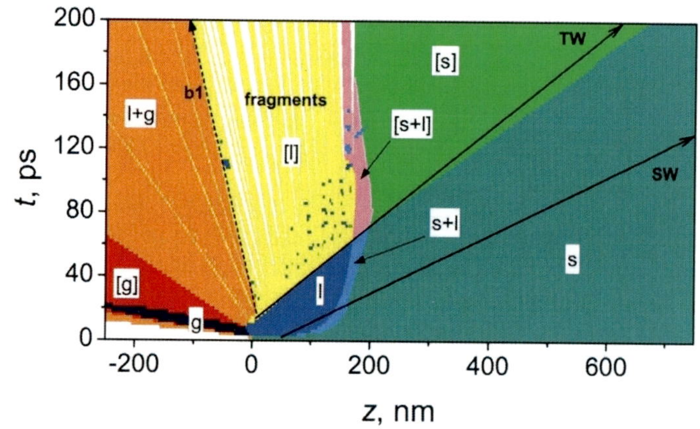

Figure 3.9 *z-t* diagram of phase states in the copper target irradiated by a single laser pulse [47]. The irradiation conditions correspond to zero time delay in Fig. 3.8. $z = 0$ corresponds to the initial position of the irradiated surface. Propagation of the shock wave (SW) and the tensile wave (TW) are shown by arrows. The boundary between the high-density (metastable liquid) and low-density (liquid – gas mixture) is denoted as b1. Reprinted with permission from Elsevier, copyright © 2011.

Under the action of double pulses, the matter evolution considerably differs. The first pulse heats matter toward the melting point. The second pulse, which acts in 10 ps after the first one, heats matter to even higher temperature than the single pulse of double fluence. However, this heating is accompanied by expansion. After heating peak, the matter again enters the metastable region though the levels of the negative pressures reached upon expansion are smaller. One can expect that mechanical fragmentation of matter should occur gentler compared with the case of zero delay.

For 100 ps pulse separation, matter has time to relax to zero pressure before the second pulse arrives, though after entering into the metastable state and experiencing a negative pressure greater than −3 GPa. This presumably causes the spallation of the target surface layer. The further evolution of matter upon irradiation by the second pulse looks intricate, which is most probably conditioned by hydrodynamic interaction of the spalled layers and fragments resulting from the first laser pulse.

This analysis has helped to reveal the mechanisms of reducing the ablation depths with increasing pulse separation between the double pulses as observed in a number of experimental studies [157, 158]. At time delays of order of a characteristic time for electron–lattice thermalization, the second laser pulse generates the second shock wave, thus reducing the amplitude of the first-pulse-generated unloading wave and, hence, somewhat suppressing material stretching and fragmentation. At large time delays, the second pulse interacts with the ablation products, reheats them, and causes partial redeposition back to the target.

It should be mentioned that the similar thermodynamic analysis can be performed in MD simulations, where, however, the state of ablated material is analyzed by averaging over all the particles or the groups of particles [139, 140]. While the molecular dynamics method enables atomistic resolution of material ablation with revealing such details as loss of the crystalline order, the nucleation of the vapor phase, the spallation of whole layers, and the decay of matter into cluster–vapor mixture, the one-dimensional hydrodynamic techniques allow making conclusions on the possible routes of matter decay via following the averaged material parameters and their evolution over the phase diagram. Yet another peculiarity of ultrafast laser–matter interaction is a possibility of electrostatic disintegration of a superficial layer of the target caused by effective photoemission of electrons and associated surface charging, so-called Coulomb explosion. To reveal the features of this ablation mechanism, the processes connected with the generation of a self-consistent electric field and field-induced charge relocation should be considered.

3.3.3 Coulomb Explosion of Dielectric Surfaces

Coulomb explosion of bulk materials is one of the electronic mechanisms of ablation manifesting itself as a swift ion emission upon laser irradiation of solids. This mechanism was widely debated for different materials during the last decade [18, 126, 127, 131, 159–165]. The main features of CE are the following. The energetic ions of different species (e.g., aluminum and oxygen in the case of ultrafast laser ablation of sapphire [126]) observed in time-of-flight signals show the same momenta and not the same energy. Also, doubly charged ions have velocities twice as high as singly charged ones (so-called momentum scaling). This indicates that the ions could be extracted and accelerated in the electric field produced in the surface layer of the target. The electric field can be generated due to intensive electron photoemission, leading to the accumulation of positive charge in a superficial target layer. Thus, the fundamental concept of Coulomb explosion is based on the fact that due to photoemission, the irradiated surface gains a high positive charge, resulting in the generation of a self-consistent electric field. Under the action of the electric field, the charge carriers (electrons and holes if the latter are mobile as in semiconductors) relocate within the target, tending to neutralize the excess charge. Additionally, a part of photoelectrons can return to the target being drawn by the field. If neutralization is not fast enough, the repulsion force between ions exceeds the lattice binding strength, thus resulting in surface layer disintegration.

It should be underlined that here the macroscopic CE is meant, that is, the ablation of at least several monolayers, but not electrostatic desorption, which can be of stochastic nature [166]. An interesting aspect of the macroscopic CE is that the exploded material leaves behind a smooth, perfect surface that makes CE attractive for nanoscale applications, e.g., ultrafine material polishing [126]. With increasing laser fluence and/or increasing number of shots, the "strong" ablation phase takes over and the irradiated surface begins to show the characteristics of "phase explosion" [126].

Coulomb explosion has been proven, both experimentally [126, 127, 161] and theoretically [18, 127, 164], to occur in wide-bandgap dielectrics at fs laser irradiation with 800 nm laser wavelength in the regimes slightly above the experimental ion emission thresholds. There are some indications of the possibility of CE in semiconductors at higher laser fluences, above the plasma formation level [160].

However, at the developed ablation regimes the ions originated from CE constitute a small (if not negligible) fraction of the ablation products (note that the CE depth saturates on increasing laser fluence [164]). These ions are masked by the thermal ones, including those accelerated by the plume double layer [130, 160]. We mention that CE has been observed in thin metal wires under ultra-intense laser irradiations when a substantial fraction of electrons carried away from the wire by the laser field [132].

In view of an intricate picture of interrelated processes involved in the generation of the electric field in laser-irradiated materials that results in controversies of CE realization in materials of different kind, attempts of numerical modeling—taking into account the most important features of the CE phenomenon—add clarity to the understanding of the physical conditions under which CE can be not only realized but also detected. Conceptually, the modeling representations can be constructed by drawing an analogy with CE of atomic clusters in strong laser fields upon ionization [167, 168]. However, for bulk materials the model of CE must include the possibility of charge relocation from the bulk depth to the ionized surface layer. In view of excessive complexity of the CE phenomenon in bulk materials, a drift-diffusion approach has been developed, which, though being simplified, unambiguously uncovers the fundamental processes and material properties responsible to explosive electrostatic decomposition of laser excited matter [18]. The approach was applied to gold, silicon, and sapphire as typical representatives of different classes of materials (metals, semiconductors, and dielectrics) for revealing the differences in charging behavior under the action of ultrashort laser pulses.

According to the observations [126], the explosive ablation caused by the Coulomb repulsion of ions in a positively charged surface layer occurs already at timescale of 100 fs to 1 ps. Thus, composing a set of equations for describing the processes involved into the generation of the electric field upon laser irradiation of solids of different kind, we limit our consideration to the time range of order of 1 ps, that is, during and shortly after the laser pulse action. Therefore, the following equations are solved self-consistently.

(1) The kinetics of charge carrier generation can be described in the frames of a rate equation similar to Eqs. (3.9) and (3.15), taking into account that due to electron photoemission, quasi-neutrality can be violated in a target surface layer and charge

carriers can relocate under the action of the electric field and/or because of the density and temperature gradients. Thus, the continuity equation in a general form can be written as

$$\frac{\partial n_x}{\partial t}+\frac{1}{q_x}\frac{\partial J_x}{\partial z}=S_x+L_x. \tag{3.21}$$

Here S_x and L_x are the source and loss terms describing the free carrier populations; q_x and n_x denote the carrier charges and densities with subscript x = e and i representing electrons and ions (holes), respectively; J_x is the carrier flux density. In metals $S_x = 0$ while the loss term L_e can take into account the electron photoemission from the surface of the irradiated sample. Alternatively, one can set $L_x = 0$ and use the electron photoemission flux density as a boundary condition [18]. For each dielectric or semiconductor material, the source and loss terms are individually constructed taking into account the photo-ionization mechanism, avalanche multiplication, electron photoemission, and recombination, which can proceed either as electron trapping with generation of the defect states or via the Auger and photo-recombination mechanisms, as already mentioned (see Sections 3.2.3 and 3.2.4; the details and comments on photoemission can be found in [18, 164]).

(2) We assume that the charge-carrier flows are caused by quasi-neutrality violation on and below the target surface due to electron photoemission and strong density and temperature gradients. The expressions for the electric current densities $J_x = q_x n_x v_x$ with v_x being the directional velocity include drift and diffusion terms and can be considered the equations of motion:

$$J_e=-en_e\mu_e E-eD_e\nabla n_e;\; J_i=|e|n_i\mu_i E-|e|D_i\nabla n_i \tag{3.21}$$

where e and $|e|$ are the electron and hole charges and μ_x is the charge carrier mobility. The time and space dependent diffusion coefficient D_x can be calculated according to the Einstein relation as $D_x = kT_x\mu_x/|e|$ with T_x representing the carrier temperature. In semiconductors the holes are mobile and their current strongly influences the charging dynamics, otherwise one can assume $J_i = 0$.

(3) The electric field generated as a result of breaking quasi-neutrality in the irradiated target (quasi-stationary with respect to the laser oscillating field) is calculated with the Poisson equation:

$$\frac{\partial E}{\partial z}=\frac{|e|}{\varepsilon\varepsilon_0}(n_\mathrm{i}-n_\mathrm{e}). \tag{3.22}$$

(4) As the diffusion coefficient is dependent on the temperature of charge carriers, the energy conservation equations have to be applied to account for the heating of electronic and lattice subsystems. We use an analog of the TTM (Eqs. (3.3) and (3.4)) where Eq. (3.3) is modified to the form:

$$C_\mathrm{e}\left(\frac{\partial T_\mathrm{e}}{\partial t}+\frac{J_\mathrm{e}}{en_\mathrm{e}}\frac{\partial T_\mathrm{e}}{\partial z}\right)=\frac{\partial}{\partial z}K_\mathrm{e}\frac{\partial T_\mathrm{e}}{\partial z}-g(T_\mathrm{e}-T_\mathrm{l})+\Sigma(z,t)\ . \tag{3.23}$$

As discussed in Sections 3.2.3 and 3.2.4, the source term in the energy equation (3.23) should be constructed to account for the energy balance of the electrons. For taking into account the electron thermionic and photoemission processes in the energy balance, we assume that each emitted electron carries away from the target an instantaneous value of the average electron energy. Thus, for the case of a silicon sample irradiated with an ultrashort laser pulse of 800 nm wavelength, Eq. (3.14) for the variations of the E_f value is rewritten in the form:

$$\begin{aligned}\frac{\partial E_\mathrm{f}}{\partial t}=&\left((\hbar\omega-E_\mathrm{g})\frac{\sigma_1 I}{\hbar\omega}+(2\hbar\omega-E_\mathrm{g})\frac{\sigma_2}{2}\frac{I^2}{\hbar\omega}-E_\mathrm{g}\delta n_\mathrm{e}\right)\frac{n_0-n_\mathrm{i}}{n_0}\\&+\alpha_\mathrm{ab}(z,t)I(z,t)+E_\mathrm{g}R_\mathrm{e}-E_\mathrm{e}PE(z,t),\end{aligned} \tag{3.24}$$

where $PE(z,t)$ is the rate of electron emission (for the both photo- and thermionic processes) [18, 131]. Note that the above assumption underestimates the energy losses due to electron emission and actual values can be higher. However, at large electron emission rates, the surface is highly charged so that the effective work function can increase. This leads to self-regulation of electron emission from laser-irradiated surfaces as discussed in [164, 169]. By simulating electron emission from copper under the action of ultrafast laser pulses, Wendelen *et al.* [170] have shown that the space charge above the irradiated surface reduces electron emission rates by several orders of magnitude.

To conclude about the possibility of CE, a criterion is required that would relate the electric field generated due to electron emission from laser-excited surfaces with the bond strength of atoms/ions in the lattice. The CE criterion can be written in the following form [18, 90, 131]:

$$E_{\text{th}}\Big|_{z=0} = \sqrt{\frac{2(\Lambda_{\text{at}} - 3kT_{\text{l}})n_0}{\varepsilon\varepsilon_0}}. \tag{3.25}$$

Here E_{th} is the threshold electric field that is necessary to initiate the CE mechanism; Λ_{at} is the binding energy in the lattice per atom/ion. Note that the condition (3.25) may be too strong for dielectrics and semiconductors as the ionized materials are characterized by the so-called softening of the bonds due to their partial rupture. Nevertheless, this condition predicts the order of magnitude of the electric field that causes the microscopic CE (ablation of at least several monolayers rather than probabilistic emission of individual ions from the surface). We mention that recent studies unambiguously show that in metals excited by ultrafast laser pulses electronic bonds are hardened [70]. It should be underlined that though the drift-diffusion approach does not take into account the electric field induced in the target surface by the cloud of emitted electrons moving away from the target (see [159]), which create a capacitor-type structure [131] and, thus, increase the calculated field amplitude, Eq. (3.25) has a general form. Also note that the electric field calculated by the Poisson Eq. (3.22) is quasi-static on the background of the electromagnetic-wave field. The latter causes oscillative displacements of free electrons with the period of the wave and, thus, cannot be responsible for macroscopic CE of solids.

The further equations that supplement the drift-diffusion approach for describing the effects of laser-induced surface charging are the expressions for laser light attenuation in a semiconductor or dielectric target (Eqs. (3.12) and (3.16)) and the Drude model, which implies calculations of spatiotemporal changes of the complex dielectric function (see comments in Section 3.2.2 and Eq. (3.11)). The simulation results for the materials of different classes can be found in [18, 38, 90, 131]. Note that the simulations were performed for the regimes of fs, near-IR laser irradiation of sapphire, silicon, and gold at threshold fluences with respect to the efficient ions emission.

Here we summarize main conclusions on the electrostatic explosive mechanism of ablation referenced as CE:

(1) This mechanism can be unambiguously detected at laser intensities *near the ablation threshold*. At high laser fluences when thermal (phase explosion) and/or mechanical (fragmentation) mechanisms of ablation dominate, the ions originating from CE constitute a small fraction of the ablation products and, hence, are masked by the thermal ions, including those accelerated in the gas phase by the double layer effects in the expending plasmas.

(2) CE is usually detected as a *momentum scaling* of emitted ions of different species in different charged states. However, a careful analysis is necessary when making conclusions on CE as the ions accelerated in the double layer are also distinguished by the same momenta but not the same energy.

(3) Although, according to modeling [38, 131], the photoemission yield is smaller for dielectrics compared with semiconductors and metals, the generated electric field is, on the contrary, largest and can exceed the CE threshold value.

(4) The most important parameter responsible for CE realization is the *mobility of free electrons* in bulk materials. Low electron mobility in dielectrics [171] conditions the possibility of generation on the irradiated surface of a self-consistent electric field comparable with the interatomic field in the lattice that results in electrostatic disintegration of a surface layer. In metals and semiconductors, mobile electrons prevent accumulation of a high positive charge in a localized surface zone and, thus, inhibit the CE mechanism of ablation.

(5) Another important parameter is the energy of excited electrons. TTM simulations show that in metals near the ablation threshold, the maximum electron temperature is usually around 1 to 1.5 eV. The energy balance considerations on the basis of the TTM approach yield the free electron temperatures of ~6 and 10 eV for semiconductors and dielectrics, respectively, in the regimes when efficient emission of ions is observed. As in dielectric materials the conduction band minimum lies close to the vacuum level, the energetic electrons leave the surface more readily compared with other material classes. However, rapid depleting of the surface layer

by electrons leads to self-regulation of the electron emission process [164].

(6) It has been shown [18] that the possibility of CE detection at the near-ablation-threshold regimes depends on laser wavelength. For example, in semiconductors irradiated with UV fs laser pulses, the absorbed laser energy is strongly localized in a thin targets layer, resulting in the lowering of the melting threshold compared with larger wavelengths and in the occurrence of phase explosion as the most probable consequence.

3.4 Volume Modifications of Transparent Materials

The previous sections presented the fundamental processes initiated in the materials by the action of ultrashort laser pulses in near-IR, visible, or UV spectral range and the effects of material ablation upon focusing laser light on the surfaces of materials of different classes. This section is devoted to transparent crystalline and amorphous materials. The irradiation of the surface of transparent materials is widely used for a variety of optical applications, including cleaning, structuring, and modification of surfaces of optical elements. Of even greater interest are the regimes of focusing ultrashort laser pulses inside the transparent crystals and glasses, leading to local modifications of the properties of the irradiated sample. The main application field here is direct writing of waveguide structures, based on a controlled change in the refractive index in laser-modified zones [6–9, 11, 103–105, 172–181]. The formation of a phase object embedded in the dielectric matrix is caused by the rearrangement of bonds in the sample with the displacement of atoms and corresponding change in density, accumulation of stresses, and appearance of color centers. Controlling modifications in laser-irradiated materials requires a detailed study of both individual laser-induced processes and their interrelations on the timescale from photoexcitation to imprinting a final 3D structure into material matrix of the original material. In the last years, considerable progress has been achieved in theoretical modeling of laser-excited processes, both during the laser pulse propagation through a transparent material sample and post-irradiation effects.

3.4.1 Propagation of Focused Laser Beams through Nonlinear Absorbing Media

Several types of modeling approaches can be listed that have been developed for the detailed study of propagation of an electromagnetic wave in transparent materials with accounting laser energy absorption. A simple analytical model [182] can be very helpful for an estimative analysis of the geometry of laser energy absorption regions and light transmission through the sample. The model relates $\partial I(z,r,t)/\partial t$ and $\partial n_e(z,r,t)/\partial t$ to I and n_e. For the regimes considered in [182], it was unambiguously shown that in transparent solids, the laser intensity is strongly clamped to the maximal reached levels of order of 5×10^{13} W/cm^2 by nonlinear absorption. Another important observation is that the absorbed laser energy is directly proportional to the pulse duration. It must be underlined that a more rigorous approach based on solving the nonlinear Schrödinger equation (NLSE) supports the latter conclusion, indicating, however, that at longer pulses the laser energy is deposited into a more localized region [183].

The models of laser light propagation in transparent media based on the NLSE are widely utilized for studying the processes of laser excitation of dielectrics in the regimes of modification. The NLSE is an asymptotic parabolic approximation of Maxwell's equations [184] applicable for describing unidirectional propagation of slowly varied envelopes of laser pulses. This equation describes the self-focusing effect, which manifests itself as a laser beam collapse at beam energies beyond a critical value particular for a Kerr medium with the positive nonlinear refractive index n_2. We note that for transparent crystals and glasses, the n_2 value is typically in the range of 10^{-16} to 10^{-14} cm^2/W (the situations when the n_2 value becomes negative, e.g., in the presence of nanoparticles embedded in glass at wavelengths near plasmon resonances [185], are not considered here). To account for additional physical effects such as a small nonparaxiality, plasma defocusing, and multiphoton ionization, the additional terms are introduced to the scalar models based on the NLSE.

A generalized NLSE that takes into account radiation losses for generation of electron plasma on the beam way and plasma-induced changing of the permittivity of the medium can be written in a cylindrically symmetric form as [183, 186–189]

$$\frac{\partial \in}{\partial z} = \frac{i}{2k_0} T^{-1} \left(\frac{\partial^2}{\partial r^2} + \frac{1}{r}\frac{\partial}{\partial r} \right) \in - \frac{ik''}{2} \frac{\partial^2 \in}{\partial t^2}$$
$$+ \frac{ik_0 n_2 T}{n} \left[(1 - f_R)|\in|^2 + f_R \int_{-\infty}^{t} R(t-\tau)|\in|^2 d\tau \right] \in \tag{3.26}$$
$$- \frac{\sigma}{2}(1 + i\omega_0 \tau_c) T^{-1} (n_e \in) - \frac{1}{2} \frac{W_{PI}(|\in|) E_g}{|\in|^2} \in$$

where $\in$ is the complex amplitude of the electric field strength of a light wave. For a Gaussian beam with cylindrical symmetry we have

$$\in(r,t,0) = \in_0 \exp\left(-\frac{r^2}{w^2} - \frac{t^2}{\tau_{las}^2} - \frac{ik_0 r^2}{2f} \right). \tag{3.27}$$

Here $\in_0^2 = 2E_{IN}/[\pi w^2 \tau_{las}(\pi/2)^{0.5}]$ is the input pulse intensity; the curvature radius f and the focusing distance d are related as $f = (d + z_f^2/d)$; z_f is the Rayleigh length; E_{IN} is the pulse energy; w is the beam waist; τ_{las} is the pulse duration (half-width determined by a decrease in the field envelope by 1/e times); $k_0 = n\omega_0/c$ and ω_0 are the wave number and the frequency of the carrier wave; n is the refractive index of the medium; c is speed of light; the parameter k'' describes the second-order group velocity dispersion; $E_g = E_{g0} + e^2\in^2/(2cn\varepsilon_0 m_r \omega_0^2)$ is the effective ionization potential in the field of a strong electromagnetic wave; m_r is the reduced mass of the electron and hole. Equation (3.26) takes into account the beam diffraction in the transverse direction, group velocity dispersion, optical Kerr effect with a term corresponding to the delayed (Raman) response of the nonlinear material (characterized by the parameter f_R), plasma defocusing, energy absorption due to the photoionization and the inverse bremsstrahlung. The operator $T = 1 + (i/\omega_0)\times(\partial/\partial t)$ describes the self-steepening effects. The inverse bremsstrahlung process is described by the Drude theory with the absorption cross section $\sigma = k_0 e^2 \omega_0 \tau_c/[n_0^2 \omega_0^2 \varepsilon_0 m_e(1 + \omega_0^2 \tau_c)]$ where the characteristic time for collisions between electrons τ_c is of order of a few fs [190].

It should be noted that the linear term in Eq. (3.26) can slightly underestimate the absorption efficiency when the free electron concentration increases as the influence of the electron concentration on the absorption cross section is not taken into account. Additionally, the possibility of multiphoton absorption by free electrons is neglected, which can be important at relatively high radiation intensities [41]. However, at laser beam focusing to the sample volume, the clamping effect limits the attainable intensities [115, 116]. The rate equation describing generation and recombination kinetics of free electrons is written similarly as Eqs. (3.2) and (3.15):

$$\frac{\partial n_e}{\partial t} = \left[W_{PI}(|\in|) + \frac{\sigma n_e}{(1+m_r/m_e)E_g}|\in|^2 \right] \frac{n_0 - n_e}{n_0} - \frac{n_e}{\tau_{tr}}. \quad (3.28)$$

This equation takes into account the production of free electrons in the processes of photoionization and avalanche as well as electron annihilation in a trapping-like process associated with local deformations of the atomic lattice (see Section 3.2.4). The rate of photoionization W_{PI} can be defined by the Keldysh formalism [112]. However, in spite of numerous generalizations of the Keldysh theory, its formalism perfectly predicts a qualitative photoionization picture, while quantitatively it often over- or underestimates the photoionization rates, especially for dielectric materials. Thus, for the case of laser excitation of the fused silica surface (τ_{las} = 150 fs, λ = 800 nm), the experimentally detected damage threshold was more than twice higher compared with the calculated critical fluence at which the maximum free-electron density reaches the critical value (3.9 J/cm^2 against 1.6 J/cm^2) [191]. Even larger disagreement between the experimental and calculated damage thresholds exists for borosilicate glass (BK7, E_{g0} = 4.2 eV). In order to overcome this problem, an experimentally determined three-photon absorption coefficient was used in simulations [105].

Numerical investigations based on the NLSE allow elucidating important features of laser pulse propagation through transparent solids such as filamentation [187, 188], clamping [90, 187, 188], strong dependence of laser energy deposition geometry on pulse duration [90, 183] for different irradiation conditions. Remarkable is the temporal dynamics of laser energy deposition into bulk dielectrics in the modification regimes as illustrated in Fig. 3.10 [183]. The figure presents the simulation results of excitation of a-SiO_2 by a Ti:sapphire laser pulse. The contour plot of the energy absorbed in a sample to the end of a single laser pulse as a result of multiphoton ionization and inverse bremsstrahlung is given in Fig. 3.10a where the inset shows the experimental image of the modified region (irradiation conditions are the same as in the simulations).

To trace the absorption process in time, the laser pulse is divided into four parts. The sequence of the absorbed energy integrated over each part of the pulse is presented in Figs. 3.10b–e. Interesting is that only a small fraction of the pulse leading edge, containing about 15% of the pulse energy, is absorbed with a high efficiency near and in front of the geometric focus. Because of strong defocusing scattering of the electron plasma generated by the pulse leading edge, the next parts of the laser pulse do not fall into the region near the geometric focus. However, as a result of the self-focusing effect, these parts are absorbed well before the geometric focus and, integrally, they generate the second region of efficient absorption (Fig. 3.10a). An important consequence of the complex correlation between self-focusing and plasma defocusing effects is that the local intensity over the whole pulse does not exceed 5×10^{13} W/cm^2 in the sample, pointing once more to unavoidable intensity clamping mentioned above. In the context of the clamping effect, the problem of the efficient delivery of laser energy into a local region within the sample still remains open.

The validity of the NLSE for ultrashort laser beams focused inside transparent crystals and glasses can be broken down in many situations, that is conditioned by neglecting some small terms upon its derivation from Maxwell's equations. The condition of a slowly varying envelope limits applications of the NLSE to relatively long laser pulses. For pulse durations of order of 10 fs and shorter, either the NLSE has to be generalized with additional terms to accounting features of such extremely short pulses or, more appropriate, the complete set of Maxwell's equations is to be used for describing light propagation through a nonlinear medium. Another strong limitation imposed on using the NLSE is the requirement of unidirectionality of the light beam. This requirement makes impossible to apply the NLSE to tightly focused beams as well as to the cases when dense plasma is generated causing light scattering to large angles. Maxwell's equations are free of the above limitations. To describe laser beam propagation through an absorbing ionizable medium, Maxwell's equations are appropriately supplemented to account for multiphoton ionization, multiphoton absorption (that is, the depletion of the laser beam due to multiphoton ionization), the Kerr effect, and plasma dispersion while the optical response of the plasma is described in the frames of a plasma fluid model [192]. Maxwell's equations and the plasma fluid equations are coupled via the free electron current. The complete set of the equations is solved using a finite-difference time-domain (FDTD) algorithm. At present, such model represents the best choice for modeling tightly focused laser beams, which can potentially generate critical and overcritical plasma inside transparent dielectric materials. In the 3D geometry, the developed model allows the elucidation of the effects of laser light polarization [192].

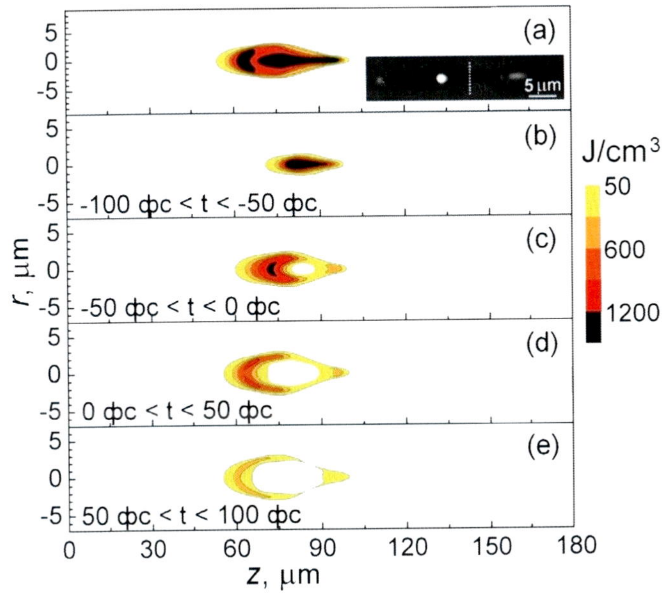

Figure 3.10 The contour plots of the absorbed laser energy obtained in simulations of excitation of a-SiO_2 by a Ti:sapphire laser pulse (λ = 800 nm, τ_{las} = 120 fs, E_{IN} = 1 μJ, w = 0.9 μm) [183]: the energy absorbed integrally by the end of the laser pulse (a); sequential energy absorption over the four time domains of the pulse (b–e). The pulse maximum corresponds to t = 0, the pulse domains are indicated in the corresponding contour plots. The geometric focal plane is at z = 90 μm. The inset shows the experimental image of the modified region, obtained by the phase contrast method (the geometric focus is shown by a white line).

Figure 3.11 shows the FDTD modeling results by Popov *et al.* [192] on free-electron plasma generation by a laser pulse focused inside a fused silica sample (800 nm, 50 fs FWHM, peak intensity at the focus

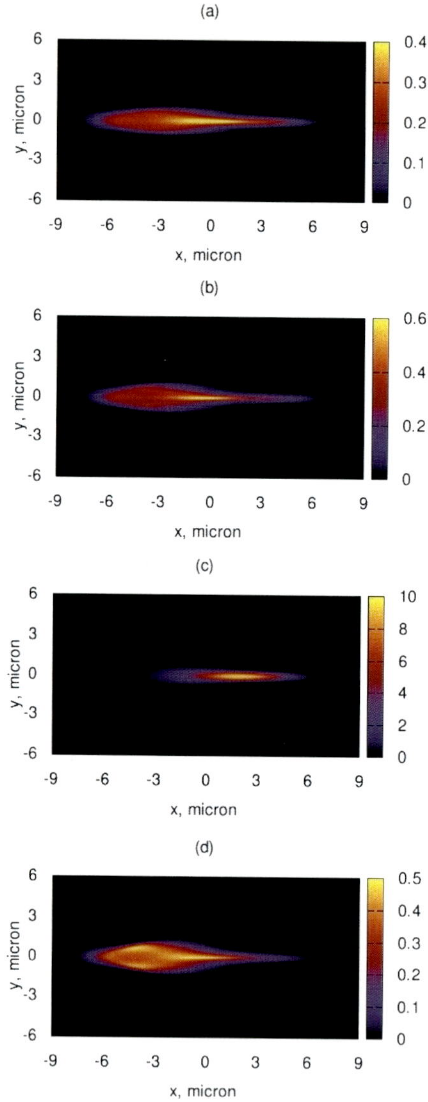

Figure 3.11 The contour plots of the generated plasma [192]. The electron density is given in units of the critical plasma density. The geometric focus is at $x = 2.75$ μm. (a) All non-linear propagation effects (plasma dispersion, multiphoton absorption (MPA), Kerr effect) are taken into account; (b) no MPA; (c) MPA only; (d) all nonlinear effects under the assumption of a collisionless plasma (damping factor is put zero). Reprinted with permission from [192]. Copyright © 2011 Optical Society of America.

of 10^{14} W/cm^2 in the absence of nonlinear effects and dispersion, NA = 0.65). The figure clearly demonstrates the importance of the different processes influencing both plasma generation and beam propagation dynamics. When all nonlinear effects are accounted for in the simulations, the resulting geometry of the laser-excited region is characterized by a narrow shape extending along the laser beam axis with the maximum of the free electron density located before the geometric focus of the laser pulse (Fig. 3.11a). When the effects of multiphoton absorption are disregarded, the maximum density of the generated electron plasma increases by ~50% staying, however, below the critical electron density value n_c ($n_c = 1.74 \times 10^{21}$ cm^{-3} for $\lambda = 800$ nm) while the shape of the laser-excited region remains similar (Fig. 3.11b). If all nonlinear effects except laser energy absorption in the multiphoton ionization process are excluded from consideration, the maximum free electron density dramatically grows approaching $10n_c$ (Fig. 3.11c). This indicates that the clamping effect is provided by plasma defocusing in a larger extent than by the multiphoton ionization process. Finally, under the assumption of a collisionless plasma (with damping factor being equal zero, Fig. 3.11d), the maximum plasma density is only slightly higher as compared with that obtained by the complete model whereas the laser-excited region becomes wider and the total number of the generated free electrons is noticeably larger [192]. It should be noted that the processes of inverse bremsstrahlung absorption and avalanche electron multiplication, which have not taken into account in the model [192], can also play a considerable role in the laser-excited plasma morphology, which calls for further studies.

3.4.2 Heat Accumulation Effects

Laser writing of local structures of modified matter in transparent materials for different application purposes is performed mainly in a gentle manner by multipulse irradiation at relatively low laser intensities in order to avoid material failure. Depending on a desired final structure, writing is performed either into a fixed volume within a sample or by moving the sample relative to the laser beam focus with different scanning speeds. Under such conditions, accumulation of laser-induced modifications from pulse to pulse is a topic of prime importance for designing structures with desired properties.

To provide crack-free laser writing of permanent structures with positive refractive index changes for waveguiding applications, laser irradiation should be applied gently in an accumulative manner avoiding conditions of material failure. Several accumulation mechanisms responsible for gentle modification of transparent materials toward waveguiding properties are established among which heat accumulations is best demonstrated [103, 174, 193–201]. At relatively large repetition rates, the energy absorbed at the focal volume from each pulse has no time to diffuse out before the subsequent pulse, thus forming a point source of heat. The process of heat accumulation upon waveguide writing can be controlled by several means that includes *variations of pulse energy, pulse repetition rate, scanning speed, and focusing conditions.* A simple analysis of the heat accumulation effect can be performed on the basis of the heat flow equation. The characteristic time of heat propagation by the distance r can be estimated as $t_{\text{heat}} \sim r^2(\rho c/\lambda)$ where ρ, c, and λ are the material density, heat capacity, and thermal conductivity, respectively. For fused silica, a heat wave propagates by ~1 μm during 1 μs, while the complete dissipation of heat from the focal volume of few micrometer size takes times of at least several dozens of microseconds (note that, according to the heat flow equation, dissipation is determined by both the thermal conductivity and the temperature gradient and is slowing down with decreasing of the latter). Hence, for unmoved silicate glass samples at a pulse energy sufficient for noticeable heating of a localized focal zone, pulse repetition rate of order of 100 kHz and higher is required for pronounced accumulation of heat [195]. For fixed laser pulse energy, the higher the repetition rate, the fewer pulses are required for considerable accumulative heating of the sample within the focal volume. The morphologies of laser-induced modifications with and without heat accumulation are completely different [199]. The structures produced at low repetitions rates are strongly localized. In fact, only a zone absorbing laser energy is modified while surrounding material looks essentially unchanged. At high repetition rates when the conditions for heat accumulation are realized, an extended volume around the light-absorbing region shows signs of modification where one can recognize two distinct zones, the interior with indications of material melting and a halo whose outer boundary is presumably determined by reaching the softening temperature upon laser exposure (see Fig. 1 in Ref. [199]).

Translating the sample relative to the laser pulse imposes further restrictions on the minimal repetition rate. The requirement here is the number of pulses coupling with the same sample volume upon translation, which should be sufficient for heat accumulation inducing local modification. Increased laser pulse energy or a higher repetition rate results in larger radius of heat modified zone and allow faster translating for producing a similar final structure that is beneficial for practical applications [9]. This process must be thoroughly controlled as excessive heat accumulation causes material failure with the formation of random or organized voids [201]. On the other hand, such regimes are exploited for writing of the microfluidic channels for biosensing and biomicrochip applications [110, 176, 202]. However, in such regimes heat accumulation cannot be considered separately from the formation of highly stressed states of matter resulting in stress wave emission with possible consequences in the form of plastic deformations (see the next section). Another parameter considerably influencing the modification structure is focusing. For example, a laser pulse with the energy of only 20 nJ applied at tight focusing (0.65 NA) induces a similar refractive index change (though in a smaller volume) as the 1 μJ pulse with loose focusing (~0.1 NA) [193]. Hence, a tighter focusing, being energetically advantageous compared with loose focusing conditions, results also in the miniaturization of written structures.

In spite of evidence that accumulative processes play a key role in glass modification, an agreement on the accumulative modification mechanism has not been reached yet. Thus, glass modification is attributed to melting and further rapid solidification at a higher density due to rapid cooling [193, 196]. In other studies, heat modification is stated to occur upon exceeding the glass softening point [174, 198]. In multipulse irradiation regimes, a considerable role is played by the accumulation of various defect states and color centers. However, there is evidence that the color centers cannot produce permanent changes of refractive index as they are easily annealed at moderate temperatures [203], while the modification structures survive upon annealing up to 1100 K [204]. Another mechanism proposed for explaining glass aging or densification by ultraviolet-induced rearranging of bond in the material [205, 206] is also suggested to be responsible for glass modification in the waveguide writing regimes with near-IR lasers (though through multi-photon ionization) [204]. Recently, Benayas *et al.* [200] performed a detailed comparative study of waveguide writing in thermal (heat accumulation) and nonthermal (low-repetition rates when heat accumulation is avoided) regimes. Nonthermal accumulation regimes can be attributed to gentle glass rebonding from pulse to pulse like in [204–206]. It has been shown that the thermal regime provides better waveguiding structures (more symmetrical mode and lower propagation losses).

It must be remarked that in multipulse irradiation regime each subsequent pulse arrives to a slightly modified matter compared with the previous one. Thus, a gradual accumulation of the defect states with larger excitation cross sections (see Section 3.2.4) makes excitation easier for the subsequent pulses. On the other hand, the refractive index is also varying due to defect generation [123], densification, and stress accumulation, thus affecting laser pulse propagation through the modified volume. However, such aspects of laser micromachining still require further studies.

3.4.3 Plastic Deformations

Heat accumulation within a confined volume inevitably leads to the formation of the stress state of matter that can culminate in the generation of the stress waves, which, being strong enough, can modify the surrounding matter through inducing irreversible deformations. Although the detailed studies of stress effects in glass modification do not abound, a definite level of understanding has already been achieved, which can be summarized as follows.

Conventionally, the stress formation regimes in transparent materials upon ultrafast laser modification can be divided into three types, in compliance with the modification mechanisms discussed in the previous Section. The first group relates to relatively low intensities when plasma breakdown conditions are not reached. In such "nonthermal" regimes [199, 200, 205], free electrons are created only via multiphoton ionization without avalanche multiplication and with minor inverse bremsstrahlung absorption, and, hence, the material is not heated noticeably and the generation of stress waves is not expected. The mechanism of material modification is analogous to that proposed for explaining densification by UV-induced rearranging of bond in the material [205, 206]. However, due to the densification of a localized volume within the laser focus region, the surrounding matter experiences stretching. Thus, around the compacted zone, a region of enhanced tensile stress is formed. As a result, in this region stress-induced birefringence is observed whose refractive index change can be described as a mutual contribution of density variations due to material stretching and elastic contribution [207, 208]. Such nonthermal modification regimes without heat accumulations are characterized by a steady-state stress accumulation.

On the contrary, in the thermal regime of modification, the stress levels induced by each laser pulse are high and the stress evolution cannot be considered quasi-stationary. When the energy of an ultrashort laser pulse focused inside a bulk transparent material is efficiently absorbed in a localized volume through the creation of microplasma, the matter in the absorption zone is driven into a strongly non-equilibrium stressed state with large temperature (and, hence, pressure) gradients. Thus, heating by only 500 K in a spherical volume of 5 μm radius induces instantaneous temperature gradients of $\sim 10^5$ K/mm. For materials with low heat conductivity, the formation of such gradients culminates in emission of a stress wave due to sudden expansion of matter within the hot zone. The stress wave can induce material modification within a radius where the wave amplitude is sufficiently high to exceed the material yield strength. Even cold solid glasses can thereby be deformed while, upon softening or close to the melting point where solids become ductile, deformation occurs more easily. Depending on material response on the induced stress, the thermal regime can be divided into two types, without and with material failure. When the generated stress does not exceed the ultimate strength of material, the emitted stress wave does not burst the matter but leads to material redistribution with formation of regions with enhanced and reduced density. By appropriately chosen conditions, one can control the redistribution process. Thus, even glasses with a high linear expansion coefficient, whose natural response to sudden localized heating is the formation of a rarefied core with a narrow shell of densified material, can be forced to form distinct compacted zones suitable for waveguiding [105].

Presumably, stress-induced material redistribution in the regimes without material failure can be realized when the temperature in the laser-excited zone exceeds the glass softening point with possible melting in the interior. At sufficiently large tensile stresses upon sudden heating and expansion, a softened/molten matter can experience rupture with the formation of voids or void chains [106–108, 201, 209–212]. This happens when the laser-induced stress levels exceed the ultimate material strength, which is considerably reduced upon softening and melting. It must be noted that in the liquid phase, the cavitation of bubbles can be followed by collapsing (see comprehensive analyses in [36, 213, 214]). Note that upon both bubble cavitation and collapsing, shock wave emission occurs, which implies the possibility of observing a series of stress waves after a single laser pulse. Formalisms reported in [213] for liquids can also be applied to bubble/void formation in glass materials though with accounting for solidification.

The generation of stress waves upon ultrafast laser excitation of transparent materials is widely supported by direct pump-probe measurements [215–217]. It is demonstrated that the stress wave is generated in several hundred ps after the laser pulse action [217]. Measurements of laser-induced birefringence show strong dependence of shear stress frozen into matter on laser beam energy [218] and the local stress values can exceed 1 GPa level [219].

3.4.4 Volume Nanogratings

Under intermediate irradiation conditions between the nonthermal and thermal regimes of modification of transparent materials, an intriguing phenomenon of self-assembled volume nanograting formation was recently discovered [100] (Fig. 3.12), which provoked strong interest for studies of underlying physical mechanisms as well as for potential applications [99, 101, 102, 220–228]. In less than a decade of studying this intriguing phenomenon, accumulative efforts of different research groups over the world have demonstrated that self-organized volume nanogratings (NGs) can be controllably produced, erased, and rewritten in fused silica glass. Basically, the NGs are the layered structures created in the laser-affected volume inside a number of transparent materials that either consist of the alternating oxygen-deficient and oxygen-excessive layers [100] or represent the nanoslits periodically spaced with subwavelength period [101, 224]. They are formed as a result of accumulative action of thousands linearly polarized laser pulses focused inside the material bulk, and the NG layers are always perpendicular to the light polarization vector. It is intriguing that only three materials, fused silica, sapphire, and TeO_2, allow the inscription of the NG structures in their bulk. Although the mechanisms of NG formation are still debated and intrinsically unclear, many important applications of these amazing structures have been proposed for the development of various integrated optical and microfluidic devices and rewritable 3D memory storage [220–229].

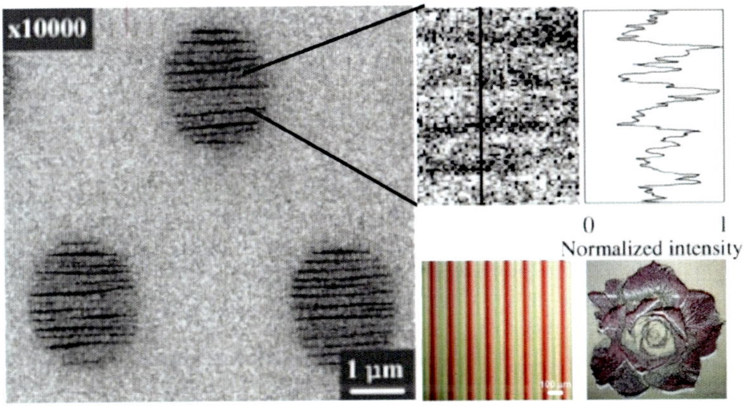

Figure 3.12 Self-organized nanogratings written by femtosecond laser irradiation inside fused silica samples [100, 220]. *On the left:* backscattering electron images of fused silica surface polished close to the focal spot. *On the right:* Auger spectrum map and line scan of oxygen (top) showing its redistribution within the irradiated region. Note that the silicon content does not experience spatial modulations. Polarization diffraction grating and birefringent rose printed in silica glass by laser induced self-organized nanograting (bottom). Courtesy of P.G. Kazansky.

The following main features of the NG formation process have been revealed to date. The grating period, usually in the range of 100–300 nm, decreases with laser exposure time at a fixed pulse energy, while it increases with pulse energy for a fixed number of applied laser pulses [100]. At increased laser energy and/or increased pulse duration, the second periodicity within the NG structure along the laser wave vector is observed with spacing of approximately λ/n_0 [221, 224]. It is intriguing that when translating the sample so that the laser beam focus moves from the bulk to the surface, the NGs are transformed from the volume to the surface periodic structures without changing periodicity [226].

Several mechanisms of volume NG formation have been proposed, which are attributed mainly to the plasma effects such as the generation of plasma standing waves excited by the polarized light [100, 101] and nanoplasma self-organization [102, 224], while defect formation can play a key role [99]. Further investigations are required to reveal the true mechanisms of the striking phenomenon of NG inscription in the bulk transparent materials. If the plasma effects are responsible for NG formation, a question arises why these structures are observed only in few selected materials, while plasma can be created in any transparent crystal or glass upon ultrafast laser excitation. Answering this question can have important consequences for ultrafast laser micromachining of different materials. The fact that the volume NGs turn into the surface ripples upon moving the laser focus toward the surface can indicate to a definite similarity in the formation mechanisms of the volume and surface periodic structures. A detailed review on the volume and surface periodic structures can be found in [99]. New indications to both the mentioned similarity and importance of the plasma effects have recently been obtained in experiments on laser imprinting of the complex polarization states of the laser wave into fused silica glass [230] and zinc acetate thin films [231]. These works additionally demonstrate great potentials of the NG discovery for cutting-edge technologies.

3.5 Influence of Ambient Gas on Ultrashort Laser Processing of Surfaces

As was already mentioned, due to rapid energy delivery to materials by ultrashort laser pulses in the regimes of surface irradiation, the laser–plume interaction is completely avoided and HAZ are minimized. This makes ultrashort laser pulses exclusively attractive for ultrafine surface processing [3], allowing the generation of well-defined microstructures with high quality and reproducibility. However, since in the majority of technological applications, material processing by fs laser pulses is carried out in an ambient environment (air, inert or reactive gases, liquid media), the questions arise about ambient medium ionization and how the ambient plasma can intervene in the process. Plasma etching of different surfaces is a widely studied phenomenon due to its extensive usage in plasma-chemical technologies [232]. Even plasma of inert gases becomes an etching agent when its energetic ions bombard the contiguous surfaces. Plasma of highly reactive gases leads to strong corrosion of surfaces. Liquid media such as water and organics used for laser-assisted nanoparticle production [233, 234] gain high reactive properties when reaching their near-critical thermodynamic states [235].

A strong negative influence of the presence of air on laser micro-processing of surfaces with fs laser pulsed compared with the vacuum conditions has been proved in a number of studies [236–241]. Laser-drilled holes and laser-ablation craters in air are usually surrounded by an extended area with clear signs of modification and droplets of redeposited material, while in vacuum the surface around the laser-processed spot is not damaged and much cleaner from re-deposit. Additionally, upon fs laser ablation, gas environment has been reported to affect the dynamics of heating of metallic targets [241–243]. It was demonstrated that the thermal energy retained in the target after irradiation can be enhanced above the natural absorptivity of the target material. In the past, a similar effect was also observed for longer nano- and microsecond laser pulses and is usually explained by energy transfer to the sample from both ablation plasma and laser-induced plasma created in ambient gas. The effect of gas breakdown near irradiated surfaces is well-established for the cases of μs [244–247], ns [247, 248], and ps [249, 250] laser pulses. Gas ionization can be caused by photoelectrons, avalanche ionization, multiphoton ionization, and due to microparticles suspended in ambient gas in multi-pulse irradiation regimes. However, for fs laser pulses, ambient air plasma can be produced efficiently only due to the multiphoton ionization process. In fact, cross sections of electron scattering by molecules of atmospheric gases (O_2, N_2, Ar) are of order of 10^{-15} cm^2. The velocities of electrons with the energies of several electron-Volts are of order of 10^8 cm/s. Correspondingly, characteristic time between collisions of such electrons with gas molecules at normal conditions is several ps and, hence, the avalanche process cannot develop [251, 252].

Recently, several models were elaborated in order to get insight into the consequences of ambient gas ionization for surface microprocessing with ultrashort laser pulses. The models, though simplified so far, are based on the hydrodynamics [237, 251, 252] and laser-induced gas ionization/recombination kinetics [130] considerations. It has been shown that already at moderate laser intensities slightly above the ablation threshold, atmospheric air is considerably ionized in front of the irradiation spot on a metal target, in particular due to interference of the incident and reflected parts of the beam. Even in the absence of ablation, this instantaneous ionization leads to the emission of the strong shock wave propagating radially from the irradiation spot [251, 252]. The expansion of the ablation products contributes to further strengthening of the shock wave and, hence, to increasing the ionization degree behind the wave [237]. At relatively tight focusing, the shock wave propagates quasi-spherically [237, 251, 252], while for long focuses the shock wave structure can be more complicated due to air ionization along the beam path [236, 253] (Fig. 3.13) as supported by observations [236].

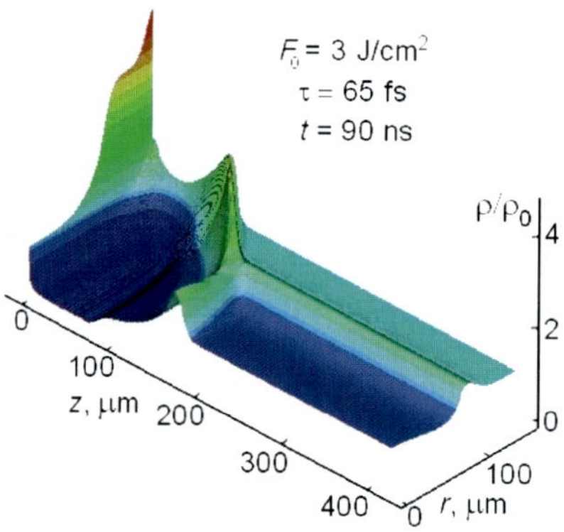

Figure 3.13 An example of a shock wave pattern (density distribution) formed in air at fs laser irradiation of a metallic sample [253].

Although much more studies are required to elucidate different aspects of air ionization impact on fs-laser micromachining of surfaces, a picture of "air-plasma assisted" surface microprocessing can be sketched as shown in Fig. 3.14 [130]. As a result of gas ionization in the focusing zone in front of the irradiated sample, a strong shock wave is generated and propagates quasi-spherically away from the irradiation spot. The target surface in contact with the compressed plasma in the shock wave is exposed to fluxes of energetic ions, electrons, and neutral molecules. This can result in substantial modification of the surface via mechanical sputtering and chemical etching. The involved chemical processes depend on both the ambient gas composition and the kind of the target material. Learning of the laser-induced plasma etching can be based on the well-developed plasma-chemical theories. However, a number of other aspects are to be understood, in particular, the role of recombinative and bremsstrahlung radiations of expanding air plasma in target heating and etching.

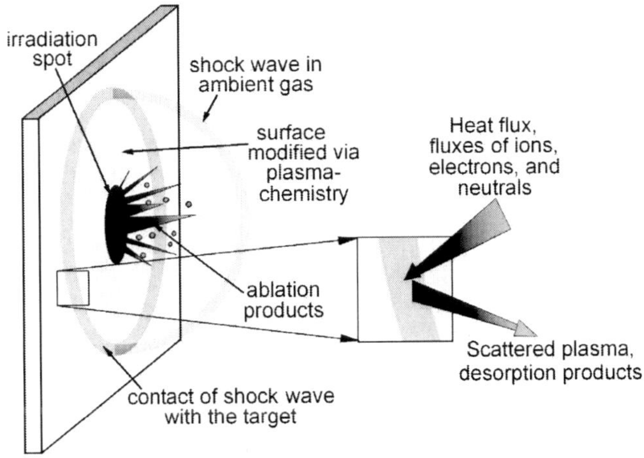

Figure 3.14 Schematic representation of the processes occurring under femtosecond laser ablation of a target in the presence of a relatively dense ambient gas [130]. Plasma formation in the near-surface region results in formation of a strong shock wave. At a contact of the shocked plasma with the target, fluxes of the charged particles interact mechanically and chemically with the surface molecules. This can result in surface modifications in the form of roughening, etching, oxidation, porosity development.

3.6 Conclusions

Fifty years after their invention, lasers have become an indispensable tool of the contemporary technologies. This review aimed to outline tremendous efforts made by researchers to get deep insights into mechanisms and dynamics of interaction of ultrashort laser pulses with inorganic materials (metals, semiconductors, and dielectrics) in order to establish principles of controllable exploiting of lasers in different technologies. The review demonstrates that in spite of outstanding achievements, far not all aspects of ultrafast laser excitation of materials of different classes have been understood. This is conditioned by extremely wide multidisciplinarity of the topic, which involves solid-state physics, optics, electrodynamics, plasma physics, thermodynamics, physical chemistry, mechanics, and hydrodynamics. Here we mention only some effects that are awaiting convincing explanations and/or adequate descriptions.

(1) Although ultrafast laser excitation of metals is most extensively studied, Section 3.2.2 underlines considerable difficulties in the treatment of the experimental observations. In the first place, this is caused by poor knowledge of dynamics of optical response of metals to laser excitation in strongly non-equilibrium regimes.

(2) The similar problems arise for bandgap materials, whose behavior under the action of ultrashort laser pulses, however, is much more complicated compared with metals. Here the ultrafast melting, long thermalization time within the electronic subsystem, as well as accumulation, decay, and re-excitation of various defect states can be mentioned. Additionally, the electronic mechanism of ablation (so-called Coulomb explosion) is still debated for semiconductor and metallic materials.

(3) In fact, when matter is heated by ultrashort laser, it enters a highly non-equilibrium thermodynamic state (so-called warm dense matter) from where it tends to evolve toward equilibrium. Usually, this evolution is treated on the basis of the concepts of equilibrium thermodynamics as non-equilibrium thermodynamics is still a poorly developed field. In this respect, the great potential can be referred to the methods of molecular dynamics (Section 3.3.1), which, however, presently are cumbersome and can be applied only to particular materials and mainly on nanoscale.

(4) As outlined in Section 3.4.1, the theory of laser pulse propagation through transparent dielectric materials when dense plasmas are created and strong plasma scattering of the light as well as reflection can develop is still in its infancy. The processes of harmonic generation that have not been touched here can, nevertheless, be also mentioned.

(5) Transparent crystals and glasses propose new intriguing puzzles, some of which are mentioned in Section 3.4. The mystery of nanograting formation with the inscription of light polarization both within the bulk and on its surface, "quill" writing, formation of micro- and nanovoids and void chains still require further studies. Nanoparticle formation by ultrashort laser ablation and laser-induced nanoparticle precipitation in glasses with dissolved metals deserve a separate review.

This list can be further continued though it is already clear that the phenomenon of ultrafast laser excitation of different materials is extremely rich from the fundamental point of view and opens new exciting possibilities for its application in material processing, optoelectronics, photonics, new material synthesis, and other field.

References

1. Kaiser, W., ed. (1993) Ultrashort Laser Pulses: Generation and Applications (Springer-Verlag, Berlin).
2. Diels, J.-C., and Rudolph, W. (1996) Ultrashort Laser Pulse Phenomena: Fundamentals, Techniques, and Applications on a Femtosecond Time Scale (Academic, New York).
3. Bäuerle, D. (2000) *Laser Processing and Chemistry,* 3rd Ed. (Springer, Berlin, Heidelberg).
4. Phipps, C. (2007) Laser Ablation and Its Application (Springer).
5. Pronko, P. P., Dutta, S.K., Squier, J., Rudd, J. V., Du, D., and Mourou, G. (1995). Machining of sub-micron holes using a femtosecond laser at 800 nm, *Opt. Commun.,* **114**, 106–110.
6. Davis, K. M., Miura, K, Sugimoto, N., and Hirao, K. (1996). Writing waveguides in glass with a femtosecond laser, *Opt. Lett.,* **21**, 1729–1731.
7. Gattass, R. R., and Mazur, E. (2008). Femtosecond laser micromachining in transparent materials, *Nat. Photon.,* **2**, 219–225.
8. Wang, Z., Sugioka, K., Hanada, Y., and Midorikawa, K. (2007). Optical waveguide fabrication and integration with a micro-mirror inside photosensitive glass by femtosecond laser direct writing, *Appl. Phys. A,* **88**, 699–704.
9. Shah, L., Arai, A., Eaton, S., and Herman, P. (2005). Waveguide writing in fused silica with a femtosecond fiber laser at 522 nm and 1 MHz repetition rate, *Opt. Express,* **13**, 1999–2006.

10. Bhardwaj, V. R., Simova, E., Corkum, P.B., Rayner, D.M., Hnatovsky, C., Taylor, R.S., Schreder, B., Kluge, M., and Zimmer., J. (2005). Femtosecond laser-induced refractive index modification in multicomponent glasses, *J. Appl. Phys.*, **97**, Paper 083102.

11. Yang, W., Corbari, C., Kazansky, P. G., Sakaguchi, K., and Carvalho, I. C. S. (2008). Low loss photonic components in high index bismuth borate glass by femtosecond direct writing, *Opt. Express*, **16**, 16215–16226.

12. Kaganov, M. I., Lifshitz., I. M., and Tanatarov, M. V. (1957). Relaxation between electrons and the crystalline lattice, *Sov. Phys. JETP*, **4**, 173–178.

13. Anisimov, S. I., Kapeliovich, B. L., and Perel'man, T. L. (1974). Electron-emission from surface of metals induced by ultrashort laser pulses, *Sov. Phys. JETP*, **39**, 375–377.

14. Kaiser, A., Rethfeld, B., Vicanek, M., and Simon, G. (2000). Microscopic processes in dielectrics under irradiation by subpicosecond laser pulses, *Phys. Rev. B*, **61**, 11437–11450.

15. Yoffa, E. J. (1980). Dynamics of dense laser-induced plasmas, *Phys. Rev. B*, **21**, 2415–2425.

16. Van Driel, H. M. (1987). Kinetics of high-density plasmas generated in Si by 1.06– and 0.53-μm picosecond laser pulses, *Phys. Rev. B*, **35**, 8166–8176.

17. Mao, S. S., Mao, X.-L., Greif, R., and Russo, R. E. (1998). Simulation of infrared picosecond laser-induced electron emission from semiconductors, *Appl. Surf. Sci.*, **127–129**, 206–211.

18. Bulgakova, N. M., Stoian, R., Rosenfeld, A., Hertel, I. V., Marine, W., and Campbell, E. E. B. (2005). A general continuum approach to describe fast electronic transport in pulsed laser irradiated materials: a problem of Coulomb explosion, *Appl. Phys. A*, **81**, 345–356.

19. Jiang, L., and Tsai, H.-L. (2008). A plasma model combined with an improved two-temperature equation for ultrafast laser ablation of dielectrics, *J. Appl. Phys.*, **104**, Article 093101.

20. Yilbas, B. S. (2002). Short-pulse laser heating of gold-chromium layers: termo-elasto-plastic analysis, *J. Phys. D: Appl. Phys.* **35**, 1210–1217.

21. Chen, J. K., Beraun, J. E., and Tham, C. L. (2004). Ultrafast thermoelasticity for short-pulse laser heating, *Int. J. Eng. Sci.*, **42**, 793–807.

22. Meshcherryakov, Y. P., and Bulgakova, N. M. (2006). Thermoelastic modeling of microbump and nanojet formation on nanosize gold films under femtosecond laser irradiation, *Appl. Phys. A*, **82**, 363–368.

23. Tzou, D. Y., and Pfautsch, E. J. (2008). Ultrafast heating and thermomechanical coupling induced by femtosecond pulses, *J. Eng. Math.*, **61**, 231–247.

24. Shugaev, M. V., and Bulgakova, N. M. (2010). Thermodynamic and stress analysis of laser-induced forward transfer of metals, *Appl. Phys. A*, **101**, 103–109.

25. Arnold, D., and Cartier, E. (1992). Theory of laser-induced free electron heating and impact ionization in wide-band-gap solids, *Phys. Rev. B*, **46**, 15102–15115.

26. Dobryakov, A. L., Farztdinov, V. M., Lozovik, Y. E., and Marowsky, G. (1999). Laser-induced nonequilibrium electron distribution in metals on a femtosecond time scale, *Phys. Scripta*, **60**, 572–578.

27. Hüttner, B. (1999). Thermodynamics and transport properties in the transient regime, *J. Phys.: Condens. Matter*, **11**, 6757–6777.

28. Apostolova, T., and Hahn, Y. (2000). Modeling of laser-induced breakdown in dielectrics with subpicosecond pulses, *J. Appl. Phys.*, **88**, 1024–1034.

29. Rethfeld, B., Kaiser, A., Vicanek, M., and Simon. G. (2002). Ultrafast dynamics of nonequilibrium electrons in metals under femtosecond laser irradiation, *Phys. Rev. B*, **65**, Paper 214303.

30. Kang, K. G., Lee, S. H., Choi, Y. K., Ryou, H.-S., Lee, J. S., and Park, S. (2006). Fokker–Planck approach to laser-induced damage in dielectrics with subpicosecond pulses, *Nanoscale Microscale Thermophys. Eng.*, **10**, 217–232.

31. Epifanov, A. S. (1975). Avalanche ionization in solid transparent dielectrics induced by strong laser radiation pulses, *Sov. Phys. JETP*, **35**, 897–903.

32. Gorshkov, B. G., Epifanov, A. S., and Manenkov, A. A. (1979). Avalanche ionization produced in solids by large radiation quanta and relative role of multi-photon ionization in laser-induced breakdown, *Sov. Phys. JETP*, **49**, 309–315.

33. Stuart, B. C., Feit, M. D., Rubenchik, A. M., Shore, B. W., and Perry, M. D. (1995). Laser-induced damage in dielectrics with nanosecond to subpicosecond pulses, *Phys. Rev. Lett.*, **74**, 2248–2251.

34. Stuart, B. C., Feit, M. D., Herman., S., Rubenchik, A. M., Shore, B. W., and Perry, M. D. (1995). Nanosecond-to-femtosecond laser-induced breakdown in dielectrics, *Phys. Rev. B*, **53**, 1749–1761.

35. Mao, S. S., Quéré, F., Guizard, S., Mao, X., Russo, R. E., Petite, G., and Martin, P. (2004). Dynamics of femtosecond laser interactions with dielectrics, *Appl. Phys. A*, **79**, 1695–1709.

36. Vogel, A., Noack, J., Hüttman, G., and Paltauf, G. (2005). Mechanisms of laser nanosurgery of cells and tissues, *Appl. Phys. B*, **81**, 1015–1047.

37. Mero, M., Rudolph, W., Ristau, D., and Starke, K. (2005). Scaling laws of femtosecond laser pulse induced breakdown in oxide films, *Phys. Rev. B*, **71**, Paper 115109.

38. Bulgakova, N. M., Burakov, I. M., Meshcheryakov, Y. P., Stoian, R., Rosenfeld, A., and Hertel, I. V. (2007). Theoretical models and qualitative interpretations of fs laser material processing, *J. Laser Micro/Nanoeng.*, **2**, 76–86.

39. Rethfeld, B. (2004). Unified model for the free-electron avalanche in laser-irradiated dielectrics, *Phys. Rev. Lett.*, **92**, Paper 187401.

40. Christensen, B.H., and Balling, P. (2009). Modeling ultrashort-pulse laser ablation of dielectric materials, *Phys. Rev. B*, **79**, Paper 155424.

41. Bachau, H., Belsky, A. N., Bogatyrev, I. B., Gaudin, J., Geoffroy, G., Guizard, S., Martin, P., Popov, Y. V., Vasil'ev, A. N., and Yatsenko, B. N. (2010). Electron heating through a set of random levels in the conduction band of insulators induced by femtosecond laser pulses, *Appl. Phys. A*, **98**, 679–689.

42. Zhigilei, L. V. (2003). Dynamics of the plume formation and parameters of the ejected clusters in short-pulse laser ablation, *Appl. Phys. A*, **76**, 339–350.

43. Zhigilei, L. V., Lin, Zh., and Ivanov, D. S. (2009). Atomistic modeling of short pulse laser ablation of metals: connections between melting, spallation, and phase explosion, *J. Phys. Chem. C*, **113**, 11892–11906.

44. Kubota, Y., Matsumoto, R., and Nakagaki, M. (2007). Molecular dynamics analysis on crack growth behavior in single and nano-crystalline Fe by the use of FS-2NNMEAM hybrid potential, *Key Eng. Mat.*, **340–341**, 985–990.

45. Colombier, J. P., Combis, P., Rosenfeld, A., Hertel, I. V., Audoard, E., and Stoian, R. (2006). Optimized energy coupling at ultrafast laser-irradiated metal surfaces by tailoring intensity envelopes: consequences for material removal from Al samples, *Phys. Rev. B*, **74**, Paper 224106.

46. Povarnitsyn, M. E., Itina, T. E., Khishchenko, K. V., and Levashov, P. R. (2009). Suppression of ablation in femtosecond double-pulse experiments, *Phys. Rev. Lett.*, **103**, Paper 195003.

47. Povarnitsyn, M. E., Itina, T. E., Levashov, P. R., and Khishchenko, K. V. (2011). Simulation of ultrashort double-pulse laser ablation, *Appl. Surf. Sci.*, **257**, 5168–5171.

48. Jeschke, H. O., and Garcia, M. E. (2003) *Nonlinear Optics, Quantum Optics and Ultrafast Phenomena with X-rays*, ed. Adams, B. W., Chapter 7 "Ultrafast structural changes induced by femtosecond laser pulses" (Kluwer Academic Publishers, Boston/Dordrecht/London) 175–214.

49. Chichkov, B. N., Momma, C., Nolte, S., von Alvensleben, F., and Tünnermann, A. (1996). Femtosecond, picosecond and nanosecond laser ablation of solids, *Appl. Phys. A*, **63**, 109–115.

50. Fann, W. S., Storz, R., Tom, H. W. K., and Bokor, J. (1992). Direct measurement of nonequilibrium electron-energy distributions in subpicosecond laser-heated gold films, *Phys. Rev. Lett.*, **68**, 2834–2837.

51. Sun, C.-K., Vallée, F., Acioli, L. H., Ippen, E. P. and Fujimoto, J. G. (1994). Femtosecond-tunable measurement of electron thermalization in gold, *Phys. Rev. B*, **50**, 15337–15348.

52. Lisowski, M., Loukakos, P. A., Bovensiepen, U., Stähler, J., Gahl, C., and Wolf, M., (2004). Ultra-fast dynamics of electron thermalization and transport effects in Ru (001), *Appl. Phys. A*, **78**, 165–176.

53. Ordal, M. A., Bell, R. J., Alexander Jr., R.W., Long, L.L., and Querry, M. R. (1985). Optical properties of fourteen metals in the infrared and far ionfrared: Al, Co, Cu, Au, Fe, Pb, Mo, Ni, Pd, Pt, Ag, Ti, V, and W, *Appl. Opt.*, **24**, 4493–4499.

54. Hohlfeld, J., Grosenick, D., Conrad, U., and Matthias, E. (1995). Femtosecond time-resolved reflection second-harmonic generation on polycrystalline copper, *Appl. Phys. A*, **60**, 137–142.

55. Agranat, M. B., Andreev, N. E., Ashitkov, S. I., Veisman, M. E., Levashov, P. R., Ovchinnikov, A. V., Sitnikov, D. S., Fortov, V. E., and Kishchenko, K. V. (2007). Determination of the transport and optical properties of a nonideal solid-density plasma produced by femtosecond laser pulses, *JETP Lett.*, **85**, 271–276.

56. Ping, Y., Hanson, D., Koslow, I., Ogitsu, T., Prendergast, D., Schwegler, E., Collins, G., and Ng, A. (2008). Dielectric function of warm dense gold, *Phys. Plasmas*, **15**, Paper 056303.

57. Kirkwood, S. E., Tsui, Y. Y., Fedosejevs, R., Brantov, A. V., and Bychenkov, V. Y. (2009). Experimental and theoretical study of absorption of femtosecond laser pulses in interaction with solid copper targets, *Phys. Rev. B*, **79**, Paper 144120.

58. Drude, P. (1900). Zur Elektronentheorie der metalle, *Annal. Phys.*, **306**, 566–613.

59. Lee, Y. T., and More, R. M. (1984). An electron conductivity model for dense plasmas, *Phys. Fluids*, **27**, 1273–1286.

60. Wellershoff, S.-S., Hohlfeld, J., Güdde, J., and Matthias, E. (1999). The role of electron–phonon coupling in femtosecond laser damage of metals, *Appl. Phys. A*, **69**, S99–S107.

61. Hohlfeld, J., Wellershoff, S.-S., Güdde, J. Conrad, U., Jähnke, V., and Matthias, E. (2000). Electron and lattice dynamics following optical excitation of metals, *Chem. Phys.*, **251**, 237–258.

62. Thomas, D. A., Lin, Zh., Zhigilei, L. V., Gurevich, E. L., Kittel, S., and Hergenröder, R. (2009). Atomistic modeling of femtosecond laser-induced melting and atomic mixing in Au film—Cu substrate system, *Appl. Surf. Sci.*, **255**, 9605–9612.

63. Corkum, P. B., Brunel, F., and Sherman, N. K. (1988). Thermal response of metals to ultrashort-pulse laser excitation, *Phys. Rev. Lett.*, **61**, 2886–2889.

64. Lin, Zh., Zhigilei, L. V., and Celli, V. (2008). Electron-phonon coupling and electron heat capacity of metals under conditions of strong electron–phonon nonequilibrium, *Phys. Rev. B*, **77**, Paper 075133.

65. http://www.faculty.virginia.edu/CompMat/electron-phonon-coupling/

66. Ashfold, N. W., and Mermin, N. D. (1976) *Solid State Physics* (Holt, Rinehart and Winston, New York).

67. Wang, X. Y., Riffe, D. M., Lee, Y.-S., and Downer, M. C. (1994). Time-resolved electron-temperature measurement in a highly excited gold target using femtosecond thermionic emission, *Phys. Rev. B*, **50**, 8016–8019.

68. Rethfeld, B., Sokolowski-Tinten, K., von der Linde, D., and Anisimov, S. I. (2002). Ultrafast thermal melting of laser-excited solids by homogeneous nucleation, *Phys. Rev. B*, **65**, Paper 092103.

69. Ivanov, D. S., and Zhigilei, L. V. (2007). Kinetic limit of homogeneous melting in metals, *Phys. Rev. Lett.*, **98**, Paper 195701.

70. Ernstorfer, R., Harb, M., Hebeisen, C. T., Sciaini, G., Dartigalongue, T., and Miller, R. J. D. (2009). The formation of warm dense matter: experimental evidence for electronic bond hardening in gold, *Science*, **323**, 1033–1037.

71. Bovensiepen, U. (2007). Coherent and incoherent excitations of the Gd(0001) surface on ultrafast timescales, *J. Phys.: Condens. Matter*, **19**, Paper 083201.

72. Melnikov, A., Prima-Garcia, H., Lisowski, M., Gießel, T., Weber, R., Schmidt, R., Gahl, C., Bulgakova, N. M., Bovensiepen, U., and Weinelt, M. (2008). Nonequilibrium magnetization dynamics of gadolinium studied by magnetic linear dihroism in time-resolved 4f core-level photoemission, *Phys. Rev. Lett.*, **100**, Paper 107202.

73. Leveugle, E., Ivanov, D. S., and Zhigilei, L. V. (2004). Photomechanical spallation of molecular and metal targets: molecular dynamics study, *Appl. Phys. A*, **79**, 1643-1655.

74. Inogamov, N. A., Zhakhovskii, V. V., Ashitkov, S. I., Petrov, Y. V., Agranat, M. B., Anisimov, S. I., Nishihara, K., and Fortov, V. E. (2008). Nanospallation induced by an ultrashort laser pulse, *J. Exp. Theor. Phys.*, **107**, 1–19.

75. Falkovsky, L. A., and Mishchenko, E. G. (1999). Electron–lattice kinetics of metals heated by ultrashort laser pulses, *J. Exp. Theor. Phys.*, **88**, 84–88.

76. Chen, J., Tomov, I. V., Elsayed-Ali, H. E., and Rentzepis, P. M. (2006). Hot electrons blast wave generated by femtosecond laser pulses on thin Au(111) crystal, monitored by subpicosecond X-ray diffraction, *Chem. Phys. Lett.*, **419**, 374–378.

77. Lee, Y.-M., and Tsai, T.-W. (2008). Effect of interfacial contact conductance on thermo-elastic response in a two-layered material heated by ultrafast pulse-laser, *J. Phys. D*, **41**, Paper 045308.

78. Tzou, D. Y., and Pfautsch, E. J. (2008). Ultrafast heating and thermomechanical coupling induced by femtosecond lasers, *J. Eng. Math.*, **61**, 231–247.

79. Yilbas, B. S. (2002). Short-pulse laser heating of gold-chromium layers: thermo-elasto-plastic analysis, *J. Phys. D*, **35**, 1210–1217.

80. Chen, J. K., Beraun, J. E., and Tham, C. L. (2003). Investigation of thermal response caused by pulse laser heating, *Numer. Heat Trans. A*, **44**, 705–722.

81. Matsui, M., and Shima, N. (2003). Electronic thermal pressure and equation of state of gold at high temperature and high pressure, *J. Appl. Phys.*, **93**, 9679–9682.

82. Dumke, W. P. (1957). Spontaneous radiative recombination in semiconductors, *Phys. Rev.*, **105**, 139–144.

83. Sjodin, T., Petek, H., and Dai, H.-L. (1998) Ultrafast carrier dynamics in silicon: a two-color transient reflection grating study on a (111) surface, *Phys. Rev. Lett.* **81**, 5664–5667.

84. Choi, T. Y., and Grigoropoulos, C.P. (2002). Plasma and ablation dynamics in ultrafast laser processing of crystalline silicon, *J. Appl. Phys.* **92**, 4918–4925.

85. Bok, J., and Combescot, M. (1981). Comment on the evidence for a self-confined plasma in laser annealing, *Phys. Rev. Lett.*, **47**, 1564.

86. Agranat, M. B., Ashitkov, S. I., Anisimov, S. I., Ovchinnikov, A. V., Shvartsburg, A. B., Sitnikov, D. S., and Fortov, V. E. (2009). Formation of absorbing heterogeneous plasma layer by femtosecond laser-induced melting and ablation of silicon, *Appl. Phys. A*, **94**, 879–887.

87. Sokolowski-Tinten, K., and von der Linde, D. (2000). Generation of dense electron-hole plasmas in silicon, *Phys. Rev. B*, **61**, 2643–2650.

88. Jones. S.C., Braunlich, P., Casper, R.T., Shen, X.-A., and Kelly, P. (1989). Recent progress on laser-induced modifications and intrinsic bulk damage of wide-gap optical materials, *Opt. Eng.*, **28**, 1039–1068.

89. Furman, Sh., and Tikhonravov, A. V. (1992). Basics of Optics of Multilayer Systems (Editions Frontières, Gif-sur-Yvette).

90. Bulgakova, N. M., Stoian, R., and Rosenfeld, A. (2010). Laser-induced modification of transparent crystals and glasses, *Quantum Electron.*, **40**, 966–895.

91. Shank, C. V., Yen, R., and Hirlmann, C. (1983). Femtosecond-time-resolved surface structural dynamics of optically-excited silicon, *Phys. Rev. Lett.*, **51**, 900–902.

92. Sokolowski-Tinten, K., Bialkowski, J., and von der Linde, D. (1995). Ultrafast laser-induced order-disorder transitions in semiconductors, *Phys. Rev. B*, **51**, 14186–14198.

93. Rousse, A., Rischel, C., Fourmaux, S., Uschmann, I., Sebban, S., Grillon, G., Balcou, P., Förster, E., Geindre, J. P., Audebert, P., Gauthier, J. C., and Hulin, D. (2001). Non-thermal melting in semiconductors measured at femtosecond resolution, *Nature*, **410**, 65–68.

94. Li, K. D., and Fauchet, P.M. (1987). Drude parameters of liquid silicon at the melting temperature, *Appl. Phys. Lett.*, **51**, 1747–1749.

95. Temple, P.A., and Soileau, M. J. (1981). Polarization charge model for laser-induced ripple patterns in dielectric materials, *IEEE J. Quantum Electron.*, **17**, 2067–2072.

96. Emel'yanov, V. I., Konov, V. I., Tokarev, V. N., and Seminogov, V. N. (1989). Formation of periodic surface ripples under the action of pulsed carbon-dioxide laser-radiation on fused-silica, *J. Opt. Soc. Am. B*, **36**, 640–647.

97. Reif, J., Costache, F., Henyk, M., and Pandelov, S. (2002). Surface morphology after femtosecond laser ablation of insulators, *Proc. SPIE*, **4760**, 980–985.

98. Das, S. K., Sagan, Z., Rosenfeld, A., and Grunwald, R. (2010), Extended-area nanostructuring of TiO_2 with femtosecond laser pulses at 400 nm using a line focus, *Nanotechnology*, **21**, Paper 155302.

99. Her, T.-H. (2011). *Comprehensive Nanoscience and Nanotechnology* **4**, eds. Andrews, D.L., Scholes, G.D. and Wiederrecht, G.P., Chapter 4.10 "Femtosecond-laser-induced periodic self-organized nanostructures" (Elsevier, Academic Press) 277–314.

100. Shimotsuma, Y., Kazansky, P.G., Qiu, J. R., and Hirao, K. (2003). Self-organized nanogratings in glass irradiated by ultrashort light pulses, *Phys. Rev. Lett.*, **91**, Paper 247405.

101. Shimotsuma, Y., Hirao, K., Qiu, J. R., and Kazansky, P.G. (2005). Nano-modification inside transparent materials by femtosecond laser single beam, *Mod. Phys. Lett. B*, **19**, 225–238.

102. Bhardwaj, V. R., Simova, E., Rajeev, P. P., Hnatovsky, C., Taylor, R. S., Rayner, D. M., and Corcum, P. B. (2006). Optically produced arrays of planar nanostructures inside fused silica, *Phys. Rev. Lett.*, **96**, Paper 057404.

103. Schaffer, C. B., Brodeur, A., Garcia, J. F., and Mazur, E. (2001). Micromachining bulk glass by use of femtosecond laser pulses with nanojoule energy, *Opt. Lett.*, **26**, 93–95.

104. Florea, C., and Winick, K. A. (2003). Fabrication and characterization of photonic devices directly written in glass using femtosecond laser pulses, *J. Lightwave Technol.*, **21**, 246–253.

105. Mermillod-Blondin, A., Burakov, I. M., Meshcheryakov, Y. P., Bulgakova, N. M., Audouard, E., Rosenfeld, A., Husakou, A., Hertel, I. V., and Stoian, R. (2008). Flipping the sign of refractive index changes in ultrafast and temporally shaped laser-irradiated borosilicate crown optical glass at high repetition rate, *Phys. Rev. B*, **77**, Paper 104205.

106. Glezer, E. N., and Mazur, E. (1997). Ultrafast-laser driven micro-explosions in transparent materials, *Appl. Phys. Lett.*, **71**, 882–884.

107. Watanabe, W., and Itoh, K. (2002). Motion of bubble in solid by femtosecond laser pulses, *Opt. Express*, **10**, 603–608.

108. Juodkazis, S., Nishimura, K., Tanaka, S., Misawa, H., Gamaly, E. G., Luther-Davies, B., Hallo, L., Nicolai, P., and Tikhonchuk, V. T. (2006). Laser-induced microexplosion confined in the bulk of a sapphire

crystal: evidence of multimegabar pressures, *Phys. Rev. Lett.*, **96**, Paper 166101.

109. Qiu, H.R., Miura, K., and Hirao, K. (2008). Femtosecond laser-induced modifeatures in glasses and their applications, *J. Non-Cryst. Solids*, **354**, 1100–1111.

110. Sugioka, K., Hanada, Y., and Midorikawa, K. (2010). Three-dimensional femtosecond laser micromachining of photosensitive glass for biomicrochips, *Laser Photon. Rev.*, **4**, 386–400.

111. Burakov, I. M., Bulgakova, N. M., Stoian, R., Rosenfeld, A., and Hertel. I. V. (2005). Theoretical investigations of material modification using temporally shaped femtosecond laser pulses, *Appl. Phys. A*, **81**, 1639–1645.

112. Keldysh, L. V. (1965). Ionization in the field of a strong electromagnetic wave, *Sov. Phys. JETP*, **20**, 1307–1314.

113. Gruzdev, V. E. (2007). Photoionization rate in wide band-gap crystals, *Phys. Rev. B*, **75**, Paper 205106.

114. Mishima, K., Hayashi, M., Yi, J, Lin, S. H., Selzle, H. L., and Schlag, E. W. (2002). Theoretical studies of the long-range Coulomb potential effect on photoionization by strong lasers, *Phys. Rev. A*, **66**, Paper 053408.

115. Becker, A., Akozbek, N., Vijayalakshmi, K., Oral, E., Bowden, C. M., and Chin, S. L. (2001). Intensity clamping and re-focusing of intense femtosecond laser pulses in nitrogen molecular gas, *Appl. Phys. B*, **73**, 287–290.

116. Sun, Q., Liang, F., Vallee, R., and Chin. S. L. (2008). Nanograting formation on the surface of silica glass by scanning focused femtosecond laser pulses, *Opt. Lett.*, **33**, 2713–2715.

117. Lenzner, M., Krüger, J., Sartania, S., Cheng, Z., Spielmann, C., Mourou, G., Kautek, W., and Krausz, F. (1998). Femtosecond optical breakdown in dielectrics, *Phys. Rev. Lett.*, **80**, 4076–4079.

118. Petrov, G.M., and Davis, J. (2008). Interaction of intense ultra-short laser pulses with dielectrics, *J. Phys. B*, **41**, Paper 025601.

119. Quéré, F., Guizard, S., and Martin, P. (2001). Time-resolved study of laser-induced breakdown in dielectrics, *Europhys. Lett.*, **56**, 138–144.

120. Rajeev, P.P., Gertsvolf, M., Corcum, P. B., and Rayner, D. M. (2009). Field dependent multiphoton ionization in wide band gap crystals, *Phys. Rev. Lett.*, **102**, Paper 083001.

121. Tanimura, K., Fujiwara, H., and Suzuki, T. (1996). Time resolved spectroscopy for radiation damage processes induced by electronic excitation in insulators, *Nucl. Instrum. Meth. B*, **116**, 26–32.

122. Krüger, J., Lenzner, M., Martin, S., Lenner, M., Spielman, C., Fiedler, A., and Kautek, W. (2003). Single- and multi-pulse femtosecond laser ablation of optical filter materials, *Appl. Surf. Sci.*, **208**, 233–237.

123. Martin, P., Guizard, S., Daguzan, Ph., Petite, G., D'Oliveira, P., Maynadier, P., and Pedrix, M. (1997). Subpicosecond study of carrier trapping dynamics in wide-band-gap crystals, *Phys. Rev. B,* **55**, 5799–5810.

124. Mero, M., Liu, J., Rudolph, W., Ristau, D., and Starke, K. (2005). Scaling laws of femtosecond laser pulse induced breakdown in oxide films, *Phys. Rev. B*, **71**, Paper 115109.

125. Stoian, R., Mermillod-Blondin, A., Bulgakova, N. M., Rosenfeld, A., Hertel, I. V., Spyridaki, M., Koudoumas, E., Tzanetakis, P., and Fotakis, C. (2005). Optimization of ultrafast laser generated low-energy ion beams from silicon targets, *Appl. Phys. Lett.*, **87**, Paper 124105.

126. Stoian, R., Ashkenasi, D., Rosenfeld, A., and Campbell, E. E. B. (2000). Coulomb explosion in ultrashort pulsed laser ablation of Al_2O_3, *Phys. Rev. B*, **62**, 13167–13173.

127. Stoian, R., Rosenfeld, A., Ashkenasi, D., Hertel, I. V., Bulgakova, N. M., and Campbell, E. E. B. (2002). Surface charging and impulsive ion ejection during ultrashort laser ablation, *Phys. Rev. Lett.*, **88**, Paper 097603.

128. Bulgakova, N. M., and Bourakov, I. M. (2002). Phase explosion under ultrashort pulsed laser ablation: modeling with analysis of metastable state of melt, *Appl. Surf. Sci.*, **197**, 41–44.

129. Hergenröder, R., Miclea, M., and Hommes, V. (2006). Controlling semiconductor nanoparticle size distribution with tailored ultrashort pulses, *Nanotechnology,* **17**, 4065–4071.

130. Bulgakova, N. M., Bulgakov, A. V., Zhukov, V. P., Marine, W., Vorobyev, A. Y., and Guo, Chunlei. (2008). Charging and plasma effects under ultrashort pulsed laser ablation, *Proc. SPIE*, **7005**, Paper 70050C.

131. Bulgakova, N. M., Stoian, R., Rosenfeld, A., Hertel, I. V., and Campbell, E. E. B. (2004). Electronic Transport and consequences for material removal in ultrafast pulsed laser ablation of materials, *Phys. Rev. B*, **69**, Paper 054102.

132. Borghesi, M., Romagnani, L., Schiavi, A., Campbell, D. H., Haines, M. G., Willi, O., Mackinnon, A. J., Galimberti, M., Gizzi, L., Clarke, R. J., and Hawkes, S. (2003). Measurement of highly transient electrical charging

following high-intensity laser-solid interaction, *Appl. Phys. Lett.*, **82**, 1529–1531.

133. Ivanov, D. S., and Zhigilei, L. V. (2003). Combined atomistic-continuum modeling of short-pulse laser melting and disintegration of metal films, *Phys. Rev. B*, **68**, Paper 064114.

134. Bouilly, D., Perez, D., and Lewis, L. J. (2007). Damage in materials following ablation by ultrashort laser pulses: a molecular-dynamic study, *Phys. Rev. B*, **76**, Paper 184119.

135. Demaske, B. J., Zhakhovsky, V. V., Inogamov, N. A., and Oleynik, I. I. (2010). Molecular dynamics simulations of femtosecond laser ablation and spallation of gold, *AIP Conf. Proc.*, **1278**, 121–130.

136. Jeschke, H. O., Garcia, M. E., Lenzner, M., Bonse, J., Krüger, J., and Kautek, W. (2002). Laser ablation thresholds of silicon for different pulse durations: theory and experiment, *Appl. Surf. Sci.*, **197**, 839–844.

137. Xu, X., Cheng, C., and Chowdhury, I. H. (2004). Molecular dynamics study of phase change mechanisms during femtosecond laser ablation, *J. Heat Transfer*, **126**, 727–734.

138. Nedialkov, N. N., Atanasov, P. A. Amoruso, S., Bruzzese, R., and Wang, X. (2007). Laser ablation of metals by femtosecond pulses: theoretical and experimental study, *Appl. Surf. Sci.*, **253**, 7761–7766.

139. Lewis, L. J., and Perez, D. (2009). Laser ablation with short and ultrashort laser pulses: basic mechanisms from molecular-dynamics simulations, *Appl. Surf, Sci.*, **255**, 5101–5106.

140. Perez, D., Beland, L. K., Deryng, D., Lewis, L. J., and Meunier, M. (2008). Numerical study of the thermal ablation of wet solids by ultrashort laser pulses, *Phys. Rev. B*, **77**, Paper 014108.

141. Zhigilei, L. V., Leveugle, E., Ivanov, D. S., Lin, Zh., and Volkov, A. N. (2010). *Nanosized Material Synthesis of High-Power Energy Fluxes on Matter*, Chapter 7 "Molecular dynamics simulations of short pulse laser ablation: mechanisms of material ejection and particle generation" (IT SB RAS, Novosibirsk, Russia), 147–220 (in Russian).

142. Urbassek, H. M., and Rosandi, Y. (2010). Insight from molecular dynamics simulations into ultrashort-pulse laser ablation, *Proc. SPIE*, **7842**, Paper 784214.

143. Frenkel, D., and Smit, B. (1992). Understanding Molecular Simulations: From Algorithms to Applications (Academic, San Diego).

144. Daw, M. S., and Baskes, M. I. (1984). Embedded-atom method: derivation and application to impurities, surfaces, and other defects in metals, *Phys. Rev. B*, **29**, 6443–6453.

145. Foiles, S. M., Baskes, M. I., and Daw, M. S. (1984). Embedded-atom-method functions for the fcc metals Cu, Ag, Au, Ni, Pd, Pt, and their alloys, *Phys. Rev. B*, **33**, 7983–7991.

146. Korte, F., Koch, J., and Chichkov, B. N. (2004). Formation of microbumps and nanojets on gold targets by femtosecond laser pulses, *Appl. Phys. A*, **79**, 879–881.

147. Kuznetsov, A. I., Koch, J., and Chichkov, B. N. (2009). Nanostructuring of thin gold films by femtosecond lasers, *Appl. Phys. A*, **94**, 221–230.

148. Banks, D. P., Grivas, Ch., Mills, J. D., Eason, R. W., and Zergioti, I. (2006). Nanodroplets deposited in microarrays by femtosecond Ti:sapphire laser-induced forward transfer, *Appl. Phys. Lett.*, **89**, Paper 193107.

149. Kuznetsov, A. I., Koch, J., and Chichkov, B. N. (2009). Laser-induced backward transfer of gold nanodroplets, *Opt. Express*, **17**, 18820–18825.

150. Ivanov, D. S., Rethfeld, B., O'Connor, G. M., Clynn, T. J., Volkov, A. N., and Zhigilei, L. V. (2008). The mechanism of nanobump formation in femtosecond pulse laser nanostructuring of thin metal films, *Appl. Phys. A*, **92**, 791–796.

151. Zhigilei, L. V., Leveugle, E., Garrison, B. J., Yingling, Y. G., and Zeifman, M. I. (2003). Computer simulations of laser ablation of molecular substrates, *Chem. Rev.*, **103**, 321–347.

152. Lin, Zh., and Zhigilei, L.V. (2006). Time-resolved diffraction profiles and atomic dynamics in short-pulse laser-induced structural transformations: molecular dynamic study, *Phys. Rev. B*, **73**, Paper 184113.

153. Colombier, J.-P., Combis, P., Bonneau, F., Le Harzic, R., and Audouard, E. (2005). Hydrodynamic simulations of metal ablation by femtosecond laser irradiation, *Phys. Rev. B*, **71**, Paper 165406.

154. Chimier, B., Tikhonchuk, V. T., and Hallo, L. (2007). Heating model for metals irradiated by a subpicosecond laser pulse, *Phys. Rev. B*, **75**, Paper 195124.

155. Povarnitsyn, M. E., Itina, T. E., Khishchenko, K. V., and Levashov, P. R. (2007). Multi-material two-temperature model for simulation of ultra-short laser ablation, *Appl. Surf. Sci.*, **253**, 6343–6346.

156. Levashov, P. R., and Khishchenko, K. V. (2007). Tabular multiphase equations of state for metals and their applications, *AIP Conf. Proc*, **955**, 59–62.

157. Semerok, A., and Dutouquet, C. (2004). Ultrashort double pulse laser ablation of metals, *Thin Solid Films*, **453**, 501–505.

158. Noël, S., and Hermann, J. (2009). Reducing nanoparticles in metal ablation plumes produced by two delayed short laser pulses, *Appl. Phys. Lett.*, **94**, Paper 053120.

159. Gamaly, E. G., Rode, A. V., Ludier-Davies, B., and Tikhonchuk, V. T. (2002). Ablation of solids by femtosecond lasers: ablation mechanism and ablation thresholds for metals and dielectrics, *Phys. Plasmas*, **9**, 949–957.

160. Roeterdink, W. G., Juurlink, L. B. F., Vaughan, O. P. H., Dura Diez, J., Bonn, M., and Kleyn, A. W. (2003). Coulomb explosion in femtosecond laser ablation of Si(111), *Appl. Phys. Lett.*, **82**, 4190–4192.

161. Stoian, R., Rosenfeld, A., Hertel, I. V., Bulgakova, N. M., and Campbell, E. E. B. (2004). Comment on Coulomb explosion in femtosecond laser ablation of Si(111) Appl. Phys. Lett. 82, 4190 (2003), *Appl. Phys. Lett.*, **85**, 694–695.

162. Costache, F., and Reif, J. (2004). Femtosecond laser induced Coulomb explosion from calcium fluoride, *Thin Solid Films*, **453–454**, 334–339.

163. Dachraoui, H., Husinsky, W., and Betz, G. (2006). Ultra-short laser ablation of metals and semiconductors: evidence of ultra-fast Coulomb explosion, *Appl. Phys. A*, **83**, 333–336.

164. Bulgakova, N. M., Rosenfeld, A., Ehrentraut, L., Stoian, R., and Hertel, I. V. (2007). Modeling of electron dynamics in laser-irradiated solids: progress achieved through a continuum approach and future prospects, *Proc. SPIE*, **6732**, Paper 633208.

165. Kaplan, A., Lenner, M., and Palmer, R. E. (2007). Emission of ions and charged clusters due to impulsive Coulomb explosion in ultrafast laser ablation of graphite, *Phys. Rev. B*, **76**, Paper 073401.

166. Kanasaki, J., and Tanimura, K. (2002). Laser-induced electronic desorption of Si atoms from Si(111)-(7×7), *Phys. Rev. B*, **66**, Paper 125320.

167. Ditmire, T., Springate, E., Tisch. J. W. G., Shao, Y. L., Mason, M. B., Hay, N., Marangos, J. P., and Hutchinson, M. H. R. (1998). Explosion of atomic clusters heated by high-intensity femtosecond laserpulses, *Phys. Rev. A*, **57**, 369–382.

168. Dermota, T. E., Zhong, Q., and Castleman, Jr., A. W. (2004). Ultrafast dynamics in cluster systems, *Chem. Phys.*, **104**, 1861–1886.

169. Riffe, D. M., Wang, X. Y., Downer, M. C., Fisher, D. L., Tajima, T., Erskine, J. L., and More, R. M. (1993). Femtosecond thermionic emission from metals in the space-charge-limited regime, *J. Opt. Soc. Am. B*, **10**, 1424–1435.

170. Wendelen, W., Autrique, D., and Bogaerts, A. (2010). Space charge limited electron emission from a Cu surface under ultrashort pulsed laser irradiation, *Appl. Phys. Lett.*, **96**, Paper 051121.

171. Miotello, A., and Dapor, M. (1997). Slow electrons impinging on dielectric solids. 2. Implantation profiles, electron mobility, and recombination processes, *Phys. Rev. B*, **56**, 2241–2247.

172. Streltsov, A. M., and Borelli, N. F. (2001). Fabrication and analysis of a directional coupler written in glass by nanojoule femtosecond laser pulses, *Opt. Lett.*, **26**, 42–43.

173. Watanabe, W., Asano, T., Yamada, K., Itoh, K., and Nishii, J. (2003). Wavelength division with three-dimensional couplers fabricated by filamentation of femtosecond laser pulses, *Opt. Lett.*, **28**, 2491–2493.

174. Eaton, S. M., Zhang, H. B., Herman, P. R., Yoshino, F., Shah, L., Bovatsek, J., and Arai, A. Y. (2005). Heat accumulation effects in femtosecond laser-written waveguides with variable repetition rate, *Opt. Express*, **13**, 4708–4716.

175. Suzuki, K., Sharma, V., Fujimoto, J. G., and Ippen, E. P. (2006). Characterization of symmetric [3×3] directional couplers fabricated by direct writing with a femtosecond laser oscillator, *Opt. Express*, **14**, 2335–2343.

176. Sun, H. Y., He, F., Zhou, Z. H., Cheng, Y., Xu, Z. Z., Sugioka, K., and Midorikawa, K. (2007). Fabrication of microfluidic optical waveguides on glass chip with femtosecond laser pulses, *Opt. Lett.*, **32**, 1536–1538.

177. Nejadmalayery, A. H., and Herman, P. R. (2007). Rapid thermal annealing in high repetition rate ultrafast laser waveguide writing in lithium niobite, *Opt. Express*, **15**, 10842–10854.

178. Jang, W. J., Kazansky, P. G., and Svirko, Y. P. (2008). Non-reciprocal ultrafast laser writing, *Nat. Photon.*, **2**, 99–104.

179. Ams, M., Marshall, G. D., Dekker, P., Dubov, M., Mezentsev, V. K., Bennion, I., and Withford, M. J. (2008). Investigation of ultrafast laser-photonic material interactions: challenges for directly written glass photonics, *IEEE J. Select. Topics Quantum Electron.*, **14**, 1370–1381.

180. Mauclair, C., Cheng, G., Huot, N., Audouard, E., Rosenfeld, A., Hertel, I. V., and Stoian, R. (2009). Dynamic ultrafast laser spatial tailoring for parallel micromachining of photonic devices in transparent materials, *Opt. Express*, **17**, 3531–3542.

181. Ferrer, A., de la Cruz, A. R., Puerto, D., Gawelda, W., Valles, J. A., Rebolledo, M. A., Berdejo, V., Siegel, J., and Solis, J. (2010). *In situ* assessment and

miniaturization of nonlinear propagation effects for femtosecond-laser waveguide writing in dielectrics, *J. Opt. Soc. Am. B*, **27**, 1688–1692.

182. Rayner, D. M., Naumov, A., and Corkum, P. B. (2005). Ultrashort pulse non-linear optical absorption in transparent media, *Opt. Express*, **13**, 3208–3217.

183. Burakov, I. M., Bulgakova, N. M., Stoian, R., Mermillod-Blodin, A., Audouard, E., Rosenfeld, A., Husakou, A., and Hertel, I. V. (2007). Spatial distribution of refractive index variations induced in bulk fused silica by single ultrashort and short laser pulses, *J. Appl. Phys.*, **101**, Paper 043506.

184. Temam, R., and Miranville, A. (2005) *Mathematical Modeling in Continuum Mechanics* (Cambridge University Press, UK).

185. Ganeev, R. A., Ryasnyansky, A. I., Stepanov, A. L., and Usmanov, T. (2004). Saturated absorption and nonlinear refraction of silicate glasses doped with silver nanoparticles at 532 nm, *Opt. Quantum Electron.*, **36**, 949–960.

186. Gaeta, A. L (2000). Catastrophic collapse of ultrashort pulses, *Phys. Rev. Lett.*, **84**, 3582–3585.

187. Couairon, A., Sudrie, L., Franco, M., Prade, B., and Mysyrowicz, A. (2005). Filamentation and damage in fused silica by tightly focused femtosecond laser pulses, *Phys. Rev. B*, **71**, Paper 125435.

188. Couairon, A., and Mysyrowicz, A. (2007). Femtosecond filamentation in transparent media, *Phys. Rep.*, **441**, 47–189.

189. Kandidov, V. P., Shlenov, S. A., and Kosareva, O. G. (2009). Filamentation of high-power femtosecond laser radiation, *Quantum Electron.*, **39**, 205–228.

190. Sun, Q., Jiang, H., Liu, Y., Wu, Z., Yang, H., and Gong, Q. (2005). Measurement of the collision time of dense electronic plasma induced by a femtosecond laser in fused silica, *Opt. Lett.*, **30**, 320–322.

191. Winkler, S. W., Burakov, I. M., Stoian, R., Bulgakova, N. M., Husakou, A., Mermillod-Blondin, A., Rosenfeld, A., Ashkenasi, D., and Hertel, I. V. (2006). Transient response of dielectric materials exposed to ultrafast laser radiation, *Appl. Phys. A*, **84**, 413–422.

192. Popov, K. I., McElcheran, C., Briggs, K., Mack, S., and Ramunno, L. (2010). Morphology of femtosecond laser modification of bulk dielectrics, *Opt. Express*, **19**, 271–282.

193. Schaffer, C. B., García, J. F., and Mazur, E. (2003). Bulk heating of transparent materials using a high-repetition rate femtosecond laser, *Appl. Phys. A*, **76**, 351–353.

194. Gattass, R. R., Cerami, L. R., and Mazur, E. (2006). Micromachining of bulk glass with bursts of femtosecond laser pulses as variable repetition rate, *Opt. Express*, **14**, 5279–5284.

195. Zhang, H., Eaton, A. M., Li, J., and Herman, P. R. (2007). Heat accumulation during high repetition rate ultrafast laser interaction: Waveguide writing in borosilicate glass, *J. Phys. Conf. Ser.*, **59**, 682–686.

196. Miyamoto, I., Horn, A., and Gottmann, J. (2007). Local melting of glass material and its application to direct fusion welding by ps-laser pulses, *J. Laser Micro/Nanoeng.*, **2**, 7–14.

197. Sakakura, M., Shimizu, M., Shimotsuma, Y., Miura, K., and Hirao, K. (2008). Temperature distribution and modification mechanism inside glass with heat accumulation during 250 kHz irradiation of femtosecond laser pulses, *Appl. Phys. Lett.*, **93**, Paper 231112.

198. Eaton, S. M., Zhang, H., Ng, M. L., Chen, W.-J., Ho, S., and Herman, P. R. (2008). Transition from thermal diffusion to heat accumulation in high repetition rate femtosecond laser writing of buried optical waveguides, *Opt. Express*, **16**, 9443–9458.

199. Shimizu, M., Sakakura, M., Ohnishi, M., Shimatsuma, Y., Nakaya, T., Miura, K., and Hirao, K. (2010). Mechanism of heat-modification inside a glass after irradiation with high-repetition rate femtosecond laser pulses, *J. Appl. Phys.*, **108**, Paper 073533.

200. Benayas, A., Silva, W. F., Ródenas, A., Jacinto, C., Vázquez de Aldana, J., Chen, F., Tan, Y., Thompson, R. R., Psaila, N. D., Reid, D. T., Torchia, G. A., Kar, A. K., and Jaque, D. (2011). Ultrafast laser writing of optical waveguides in ceramic Yb:YAG: a study of thermal and non-thermal regimes, *Appl. Phys. A*, **104**, 301–309.

201. Bellouard, Y., and Hongler, M.-O. (2011). Femtosecond-laser generation of self-organized bubble patterns in fused silica, *Opt. Express*, **19**, 6807–6821.

202. Liao, Y., Ju, Y., Zhang, L., He, F., Zhang, Q., Shen, Y., Chen, D., Cheng, Y., Xu, Z., Sugioka, K., and Midorikawa, K. (2010). Three-dimensional microfluidic channel with arbitrary length and configuration fabricated inside glass by femtosecond laser direct writing, *Opt. Lett.*, **35**, 3225–3227.

203. Will, M., Nolte, S., Chichkov, B. N., and Tünnermann, A. (2002). Optical properties of waveguides fabricated in fused silica by femtosecond laser pulses, *Appl. Opt.*, **41**, 4360–4364.

204. Homoelle, D., Wielandy, S., Gaeta, A. L., Borrelli, N. F., and Smith, C. (1999). Infrared photosensitivity in silica glasses exposed to femtosecond laser pulses, *Opt. Lett.*, **24**, 1311–1313.

205. Borrelli, N. F., Smith, C., Allan, D. C., and Seward III, T. P. (1997). Densification of fused silica under 193-nm excitation, *J. Opt. Soc. Am. B*, **14**, 1606–1615.

206. Borrelli, N. F., Smith, C. M., and Allan, D. C. (1999). Excimer-laser-induced densification in binary silicate glasses, *Opt. Lett.*, **24**, 1401–1413.

207. Gusarov, A. I., and Doyle, D. B. (2000). Contribution of photoinduced densification to refractive-index modulation in Bragg gratings written in Ge-doped silica fibers, *Opt. Lett.*, **25**, 872–874.

208. Burghoff, J., Hartung, H., Nolte, S., and Tünnermann, A. (2007). Structural properties of femtosecond laser-induced modifications in $LiNbO_3$, *Appl. Phys. A*, **86**, 165–170.

209. Gamaly, E. G., Juodkazis, S., Nishimura, K., Misawa, H., Luther-Davies, B., Hallo, L., Nicolai, P., and Tikhonchuk, V. T. (2006). Laser–matter interaction in the bulk of a transparent solid: confined microexplosion and void formation, *Phys. Rev. B*, **73**, Paper 214101.

210. Sun, H., Song, J., Li, C., Wang, X., Cheng, Y., Xu, Z., Qiu, J., and Jia, T. (2007). Standing electron plasma wave mechanism of void array formation inside glass by femtosecond laser irradiation, *Appl. Phys. A*, **88**, 285–288.

211. Terakawa, M., Toratani, E., Shirakawa, T., and Obara, M. (2010). Fabrication of a void array in dielectric materials by femtosecond laser micro-processing for compact photonic devices, *Appl. Phys. A*, **100**, 1041–1047.

212. Beresna, M., Gecevičius, M., Bulgakova, N. M., and Kazansky, P. G. (2011). *Twisting light with micro-spheres produced by ultrashort light pulses*, *Opt. Express*, **19**, 18989–18996.

213. Vogel, A., Busch, S., and Parlitz, U. (1996). Shock wave emission and cavitation bubble generation by picosecond and nanosecond optical breakdown in water, *J. Acoust. Soc. Am.*, **100**, 148–165.

214. Vogel, A., and Venugopalan, V. (2003). Mechanisms of pulsed laser ablation of biological tissues, *Chem. Rev.*, **103**, 577–644.

215. Horn, A., Kreutz, E. W., and Poprawe, R. (2004). Ultrafast time-resolved photography of femtosecond laser induced modifications in BL7 glass and fused silica, *Appl. Phys. A*, **79**, 923–925.

216. Sakakura, M., and Terazima, M. (2004). Oscillation of the refractive index at the focal region of a femtosecond laser pulse inside a glass, *Opt. Lett.*, **29**, 1548–1550.

217. Sakakura, M., Terazima, M., Shumotsuma, Y., Miura, K., and Hirao, K. (2007). Observation of pressure wave generated by focusing a femtosecond laser pulse inside a glass, *Opt. Express*, **15**, 5674–5686.

218. Poumellec, B., Sudrie, L., Franco, M., Prade, B., and Mysyrowicz, A. (2003). Femtosecond laser irradiation stress induced in pure silica, *Opt. Express*, **11**, 1070–1079.

219. Morikawa, J., Orie, A., Hashimoto, T., and Juodkazis, S. (2010). Thermal and optical properties of the femtosecond-laser-structured and stress-induced birefringent regions in sapphire, *Opt. Express*, **18**, 8300–8310.

220. Beresna, M., and Kazansky, P. G. (2010). Polarization diffraction grating produced by femtosecond laser nanostructuring in glass, *Opt. Lett.*, **35**, 1662–1664.

221. Kazansky, P. G., Yang, W., Bricchi, E., Bovatsek, J., Arai, A., Shimotsuma, Y., Miura, K., and Hirao, K. (2007). "Quill" writing with ultrashort light pulses in transparent materials, *Appl. Phys. Lett.*, **90**, Paper 151120.

222. Wortmann, D., Gottmann, J., Brandt, N., and Hoen-Solle, H. (2008). Micro- and nanostructures inside sapphire by fs-laser irradiation and selective etching, *Opt. Express*, **16**, 1517–1522.

223. Kazansky, P. G., and Shimotsuma, Y. (2008). Self-assembled sub-wavelength and form birefringence created by femtosecond laser writing in glass: properties and applications, *J. Ceram. Soc. Japan*, **116**, 1052–1062.

224. Taylor, R., Hnatovsky, C., and Simova, E. (2008). Applications of femtosecond laser induced planar nanocracks inside fused silica glass, *Laser Photon. Rev.*, **2**, 26–46.

225. Cheng, G., Mishchik, K., Mauclair, C., Audouard, E., and Stoian, R. (2009). Ultrafast laser photoinscription of polarization sensitive devices in bulk silica glass, *Opt. Express*, **17**, 9515–9525.

226. Hörstmann-Jungemann, M., Gottmann, J., and Wortmann, D. (2009). Nano- and microstructuring of SiO_2 and sapphire with fs-laser induced selective etching, *J. Laser Micro/Nanoeng.*, **4**, 135–140.

227. Richter, S., Heinrich, M., Döring, S., Tünnermann, A., and Nolte, S. (2011). Formation of femtosecond laser-induced nanogratings at high repetition rates, *Appl. Phys. A*, **104**, 503–507.

228. Yu, X., Liao, Y., He, F., Zeng, B., Cheng, Y., Xu, Z., Sugioka, K., and Midorikawa, K. (2011). Tuning etch selectivity of fused silica irradiated by femtosecond laser pulses by controlling polarization of the writing pulses, *J. Appl. Phys.*, **109**, Paper 053114.

229. Beresna, M., Gecevičius, M., and Kazansky, P. G. (2011). Polarization sensitive elements fabricated by femtosecond laser nanostructuring of glass, *Opt. Mat. Express*, **1**, 783–795.

230. Hnatovsky, C., Shvedov, V., Krolikowski, W., and Rode, A. (2011). Revealing local field structure of focused ultrashort pulses, *Phys. Rev. Lett.*, **106**, Paper 123901.

231. Li, X., Feng, D., Sun, Z., and Xu, Z. (2011). Fabrication of complex periodic micro/nanopatterns by multiple interfered femtosecond laser pulses on thin solid films, *Micro Nano Lett.*, **6**, 177–180.

232. Winters, H. F., and Coburn, J. W. (1992). Surface science aspects of etching reactions, *Surf. Sci. Rep.*, **14**, 161–269.

233. Compagnini, G., Scalisi, A. A., Puglisi, O., and Spinela, C. (2004). Synthesis of gold colloids by laser ablation in thiol-alkane solutions, *J. Mater. Res.*, **19**, 2795–2798.

234. Besner, S., Kabashin, A. V., and Meunier, M. (2007). Two-step laser ablation-based method for the synthesis of stable and ultra-pure gold nanoparticles in water, *Appl. Phys. A*, **88**, 269–272.

235. Kruusing, A. (2004). Underwater and water-assisted laser processing: Part 2—Etching, cutting and rarely used methods, *Opt. Laser Eng.*, **41**, 329–352.

236. McDonald, J. P., Ma, S., Pollock, T. M., Yalisove, S. M., and Nees, J. A. (2008). Femtosecond pulsed laser ablation dynamics and ablation morphology of nickel based superalloy CMSX-4, *J. Appl. Phys.*, **103**, Paper 093111.

237. Lehecka, T., Mostovych, A., and Thomas, J. (2008). Long duration light emission from femtosecond laser-target interactions, *Appl. Phys. A*, **92**, 727–741.

238. Weck, A., Crawford, T. H. R., Wilkinson, D. S., Haugen, H. K., and Preston, J. S. (2008). Laser drilling of high aspect ratio holes in copper with femtosecond, picosecond and nanosecond pulses, *Appl. Phys. A*, **90**, 537–543.

239. Zhang, N., Zhu, X., Yang, J., Wang, X., and Wang, M. (2007). Time-resolved Shadowgraphs of material ejection in intense femtosecond laser ablation of aluminum, *Phys. Rev. Lett.*, **99**, Paper 167602.

240. Bulgakova, N. M., and Bulgakov, A.V. (2008). Comment on Time-resolved Shadowgraphs of material ejection in intense femtosecond laser ablation of aluminum, *Phys. Rev. Lett.*, **101**, Paper 099701.

241. Hu, W., Shin, Y. C., and King, G. (2011). Early-stage plasma dynamics with air ionization during ultrashort laser ablation of metal, *Phys. Plasmas*, **18**, Paper 093302.

242. Vorobyev, A. Y., and Guo, C. (2005). Direct observation of enhanced residual thermal energy coupling in solids in femtosecond laser ablation, *Appl. Phys. Lett.*, **86**, Paper 011916.

243. Vorobyev, A. Y., and Guo, C. (2006). Enhanced energy coupling in femtosecond laser metal interactions at high intensities, *Opt. Express*, **14**, 13113–13119.

244. Barchukov, A. I., Bunkin, F. V., Konov, V. I., and Prokhorov, A. M. (1973). Low-threshold breakdown of air near a target by CO_2 radiation, and associated large recoil momentum, *JETP Lett.*, **17**, 294–296.

245. Pirri, A. N., Root, R. G., and Wu, R. K. S. (1978). Plasma energy transfer to metal surfaces irradiated by pulsed lasers, *AIAA J.*, **16**, 1296–1304.

246. McKay, J. A., Dleach, B. R., Nagel, D. J., Schriempf, J. T., Hall, R. B., Pond, C. R., and Manlief, S. K. (1979). Pulsed-CO_2-laser interaction with aluminium in air—Thermal response and plasma characteristics, *J. Appl. Phys.*, **50**, 3231–3240.

247. Kabashin, A. V., Nikitin, P. I., Marine, W., and Sentis, M. (1998). Experimental study of spontaneous electric field generated by a laser plasma, *Quantum Electron.*, **28**, 24–28.

248. Ageev, V. P., Barchukov, A. I., Bunkin, F. V., Konov, V. I., Puzhaev, S, B., Silenok, A. S., and Chapliev, N. I. (1979). Heating of metals by the pulsed CO_2-laser radiation, Sov. *J. Quantum Electron.*, **9**, 43–47.

249. Mao, S. S., Mao, X., Grief, R., and Russo, R. E. (2000). Initiation of an early-stage plasma during picosecond laser ablation of solids, *Appl. Phys. Lett.*, **76**, 31–33.

250. Klimentov, S. M., Kononenko, T. V., Pivovarov, P. O., Garnov, S. V., Konov, V. I., Prokhorov, A. M., Breitling, D., and Dausinger, F. (2001). The role of plasma in ablation of materials by ultrashort laser pulses, *Quantum Electron.*, **31**, 378–382.

251. Bulgakova, N. M., Zhukov, V. P., Vorobyev, A. Y., and Guo, C. (2008). Modeling of residual thermal effect in femtosecond laser ablation of metals: role of a gas environment, *Appl. Phys. A*, **92**, 883–889.

252. Zhukov, V. P., and Bulgakova, N. M. (2009). Role of ambient gas in heating of metal samples by femtosecond pulses of laser radiation, *Thermophys. Aeromech.*, **16**, 165–176.

253. Bulgakova, N. M., Bulgakov, A. V., Zhukov, V. P., Rosenfeld, A., Grebing, C., Steinmeyer, G., and Bonse, J. (2009). Fs-laser processing of metals in air: Pulse duration effects and the role of air ionization, *Abstracts 10th Intern. Conf. Laser Ablation*, COLA 2009, 146.

Chapter 4

Spatial and Temporal Manipulation of Ultrafast Laser Pulses for Micro- and Nano-Processing

Yoshio Hayasaki and Satoshi Hasegawa
Center for Optical Science and Education (CORE), Utsunomiya University, 7-1-2 Yoto, Utsunomiya 321-8585, Japan
hayasaki@opt.utsunomiya-u.ac.jp

Temporal manipulation of ultrashort laser pulses allows efficient injection of energy to a material to perform laser processing with high precision and high energy-use efficiency. Spatial manipulation is also effective in laser material processing with high throughput and high energy-use efficiency regardless of the irradiation energy. In the entire laser processing system, from the laser source to the target material, temporal pulse manipulation allows spectral dispersion control of the system and photon–matter interactions, and spatial pulse manipulation allows efficient photon delivery to the target. Both types of manipulation are mathematically identical and share many similarities in their implementations, except for the difficulty of direct time-domain control of ultrashort pulses. Recently spatiotemporal pulse manipulation techniques that simultaneously

Ultrafast Laser Processing: From Micro- to Nanoscale
Edited by Koji Sugioka and Ya Cheng

ISBN 978-981-4267-33-5 (Hardcover), 978-981-4303-69-9 (eBook)
www.panstanford.com

control the temporal and spatial properties of pulses have been studied. With this technique, total optimization of the pulse properties greatly enhances the processing capabilities.

4.1 Introduction

Femtosecond laser processing provides high spatial resolution and causes reduced thermal destruction of a target as a result of the ultrashort interaction time between the laser pulse and the material. Therefore, femtosecond laser processing is an attractive technology in the field of nanofabrication and is widely used to fabricate next-generation optical devices. In particular, when a femtosecond laser pulse is tightly focused in a transparent material, the intensity in the focal volume becomes high enough to cause nonlinear absorption, which leads to optical breakdown and microplasma generation. Due to the nonlinearity of this process, the energy deposition is confined to the focal volume inside the transparent material, where scattering damage and a refractive index change are induced. For these reasons, femtosecond laser processing is an important tool for developing three-dimensional (3D) optical devices such as 3D optical memories, optical waveguides, diffractive optical elements (DOEs), and photonic crystals.

However, wide practical application of this technique is still limited by the high photon cost in comparison with nanosecond and continuous-wave laser processing. To fabricate devices, a huge number of processing points are required. One strategy to enhance the appeal of femtosecond laser processing is to improve the throughput by making efficient use of the laser pulse energy. Spatial and temporal manipulation of ultrafast laser pulses is effective in achieving this.

Temporal manipulation of ultrafast laser pulses enables efficient injection of energy, resulting in laser material processing with high precision, high throughput, and high energy-use efficiency. There are two methods of temporally controlling pulses: a temporal-domain technique and a temporal-frequency–domain technique. In the temporal manipulation of pulses, spatial control techniques are used because it is difficult to control an ultrafast pulse directly in the time domain. One technique is to use an optical delay line that regulates the optical path length. Another is to use a pulse shaper composed of two diffraction gratings, 4*f*-optics, and a spatial light modulator located at the frequency plane. Some spectral dispersion control techniques that are implemented with gratings and prisms can also be used.

Spatial manipulation of ultrafast laser pulses is also effective in achieving laser material processing with high throughput and high energy-use efficiency. Laser processing based on spatial pulse manipulation is classified into two methods according to the pulse shape. One method uses a spatially deformed pulse that has a continuous spatial shape, for example, linear, rectangular, or triangular. The other uses parallel pulses with multiple pulse peaks. Like the temporal techniques, the spatial techniques for pulse manipulation include a spatial-domain technique and a spatial-frequency–domain technique. The spatial-domain technique is typically implemented with beam dividing optics and an optical mask. The spatial-frequency–domain technique is implemented with a computer-generated hologram displayed on a spatial light modulator to generate desired 2D and 3D arrangements of focused pulses. Recently, techniques that simultaneously control the temporal and spatial properties of pulses have also been demonstrated.

In this chapter, we discuss temporal, spatial, and spatiotemporal control of ultrafast laser pulses that is useful for material laser processing.

4.2 Fundamentals of Ultrafast Pulse Control

Here we give a fundamental mathematical description of temporal and spatial manipulation of ultrashort laser pulses and describe their optical implementations.

4.2.1 Mathematical Description of Ultrafast Laser Pulse

Assuming a scalar light field (a linearly polarized light field), the complex amplitude of an optical pulse traveling along the *z*-axis on a plane $(x, y, 0)$ at time t is written as

$$e(x,y,t)=\int g(x,y,\omega)\exp[-i\omega t+i\phi_t(x,y,\omega)]d\omega\,, \tag{4.1}$$

where $g(x, y, \omega)$ and $\phi(x, y, \omega)$ are the amplitude and phase of each component with a temporal angular frequency ω at position (x, y). The amplitude $g(x, y, \omega)$ is written as

$$g(x,y,\omega)=\iint a(\omega_x,\omega_y,\omega)\exp[i(\omega_x x+\omega_y y)+i\phi_s(\omega_x,\omega_y,\omega)]d\omega_x,d\omega_y, \tag{4.2}$$

where $a(\omega_x, \omega_y, \omega)$ and $\phi_s(\omega_x, \omega_y, \omega)$ are the amplitude and phase of each component with a spatial angular frequency ω. $\phi_t(x, y, \omega)$ and $\phi_s(\omega_x, \omega_y, \omega)$ are redundant duplicate description but are introduced for ease of illustration of the optical phenomena. From Eqs. (4.1) and (4.2), the complex amplitude is

$$e(x,y,t)=\iiint a(\omega_x,\omega_y,\omega)\exp[i(\omega_x x+\omega_y y-\omega t)+\phi(\omega_x,\omega_y,\omega)]d\omega_x d\omega_y d\omega \cdot \tag{4.3}$$

The descriptions of the temporal and spatial components of the optical pulse are identical. Therefore, control of the temporal and spatial shapes of the optical pulse can be performed with similar optical arrangements and techniques.

4.2.2 Analogy between Control of Time- and Space-Varying Signals

To control a time-varying signal, the simplest and most fundamental optical system is an interferometer with an optical delay line, as shown in Fig. 4.1(a). First, it is assumed that the laser pulse has spatially uniform intensity and phase. In the temporal domain, the two pulses coming from both arms interfere, and the interference signal at the detector has lower contrast with increasing optical path difference $\Delta l = c\Delta t = l_2 - l_1$ when Δl is within the coherence length, which is related to the spectral width, whereas the two pulses do not interfere when Δl is sufficiently larger than the coherence length. In Eq. (4.1), $\phi_t(\omega) = l\omega/c = kl$, where $k = 2\pi/\lambda$, and the interference signal for Δl is

$$\begin{aligned}\left\langle |e_2(t)+e_1(t)|^2 \right\rangle &= \left\langle \left| \int g(\omega)\exp(ikl_2-i\omega t)d\omega+\int g(\omega)\exp(ikl_1-i\omega t)d\omega \right|^2 \right\rangle \\ &= \int g(\omega)^2 |\exp(ikl_2)+\exp(ikl_1)|^2 d\omega \\ &= \int 2g(\omega)^2[1+\cos(k\Delta l)]d\omega \\ &= \int 2g(\omega)^2[1+\cos(\omega\Delta t)]d\omega \end{aligned} \tag{4.4}$$

where $\langle * \rangle$ is the time average, and the spatial coordinates are omitted. This is the well-known low-coherence interference signal.

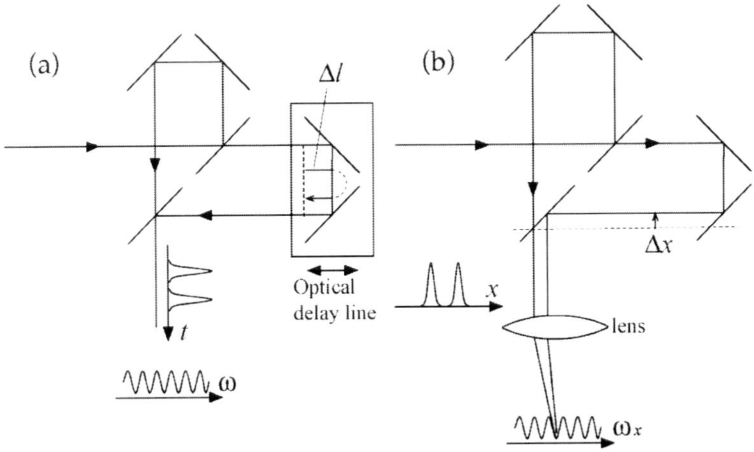

Figure 4.1 (a) Fringes in temporal frequency domain. (b) Fringes in spatial frequency domain.

In the temporal-frequency domain, fringes are also observed:

$$\begin{aligned}\left\langle \left|\Im[e_1(t)+e_2(t)]\right|^2 \right\rangle &= \left|g(\omega)\exp(ikl_1)+g(\omega)\exp(ikl_2)\right|^2 \\ &= 2g(\omega)^2[1+\cos(k\Delta l)] \\ &= 2g(\omega)^2[1+\cos(\omega\Delta t)]\end{aligned} \tag{4.5}$$

where $\Im$ is the Fourier transform. The fringes in the temporal frequency domain have a period proportional to the time difference (the path difference). This phenomenon is frequently used as an easy method for matching two optical path lengths.

Now consider the spatial domain. Two spatially divided pulses are obtained by introducing a lateral shift Δx of one arm of the interferometer. The two pulses interfere when the distance between them is within the beam diameter. This is shearing interference. The change in interference with distance is different between the time and space domains because a femtosecond laser pulse has low temporal coherence but high spatial coherence, but the analogy is an important and useful one. In the spatial frequency domain, the interference fringes are observed as

$$\left\langle \left| \Im[g(x)+g(x+\Delta x)] \right|^2 \right\rangle = \left| G(\omega_x)\exp(-i\omega_x x_1) + G(\omega_x)\exp(-i\omega_x x_2) \right|^2$$
$$= 2G(\omega_x)^2[1+\cos(\omega_x \Delta x)]. \tag{4.6}$$

The fringes in the spatial frequency domain have a period proportional to the spacing between the pulses.

We confirmed this analogy between the temporal and spatial signal manipulations. Although two divided pulses are considered above, in practice, the number of the divided pulses is large, as described in Sections 4.2.5 and 4.2.6.

4.2.3 Analogy between Control in Temporal Frequency Domain and Spatial Frequency Domain

For control in the temporal frequency domain, when a beam in one arm of the interferometer is given a frequency shift, the interference signal is temporally modulated. As shown in Fig. 4.2(a), if the frequency shift has a single frequency, the observed time-domain signal has a sinusoidal shape. This is the well-known heterodyne beat signal.

In a similar way, replacing time with space now, a beam in one arm of the interferometer is given a spatial frequency shift instead of a temporal frequency shift, which is easily implemented by tilting a mirror, as shown in Fig. 4.2(b). As a result, we observe fringes in the space-domain signal. This is just well-known interference fringes.

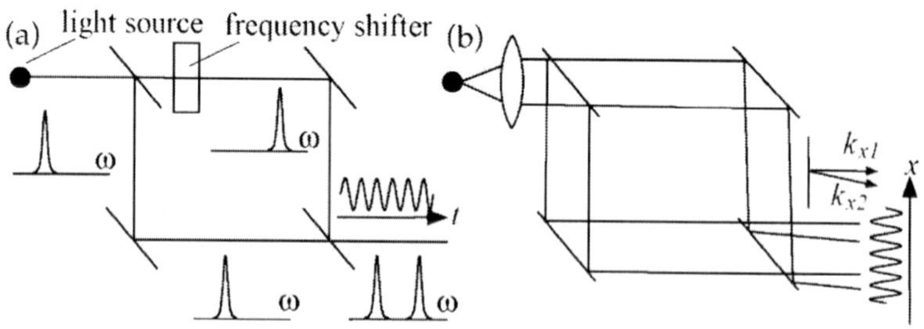

Figure 4.2 (a) Fringes in time-domain signal. (b) Fringes in space-domain signal.

4.2.4 Linear Filtering

Linear filtering is a well-known method for controlling not only electrical signals but also optical signals. Optical signals in the spatial frequency domain have been employed, in addition to those in the time frequency domain. Since the 1960s, spatial filtering has been studied in the field of optical computing, and numerous techniques have been developed. Linear filtering for optical temporal control was first used in 1974, and after the development of short pulse lasers, its importance increased in optical telecommunications and ultrafast control of material excitations and so on.

The filtering process is described in either the time domain or the frequency domain. In the time domain, the filter is characterized by its impulse response function $h(t)$. The output $e_{out}(t)$ in response to an input $e_{in}(t)$ is given by the convolution

$$e_{out}(t) = e_{in}(t) * h(t) = \int e_{in}(t')h(t-t')dt', \tag{4.7}$$

where * denotes convolution. This equation is extended to a spatiotemporal convolution as follows:

$$\begin{aligned} e_{out}(x,y,t) &= e_{in}(x,y,t) * h(x,y,t) \\ &= \iiint e_{in}(x',y',t')h(x-x',y-y',t-t')dx'dy'dt'. \end{aligned} \tag{4.8}$$

If the input is a delta function, the output is $h(x,y,t)$. Therefore, a sufficiently short pulse and/or a sufficiently small beam can be changed to an arbitrary desired pulse shape and/or beam shape (a spatiotemporal shape) with a linear filter.

Equation (4.8) in the frequency domain can be rewritten to take account of the frequency response of the system:

$$E_{out}(\omega) = E_{in}(\omega)\,H(\omega), \tag{4.9}$$

where E_{out}, E_{in}, and H are the Fourier transforms of e_{out}, e_{in}, and h, respectively. The Fourier transform of e_{in} is written as

$$E_{in}(\omega) = \Im(e_{in}) = \int e_{in}(t)\exp(-i\omega t)dt. \tag{4.10}$$

The output is calculated by the inverse Fourier transform of Eq. (4.9):

$$e_{out}(t) = \Im^{-1}[E_{in}(\omega)H(\omega)]. \tag{4.11}$$

Extension to spatial and spatiotemporal forms is trivial.

4.2.5 Temporal Pulse Manipulation with Linear Filtering (Pulse Shaping)

A direct optical implementation of Eq. (4.7) on picosecond and femtosecond time scales is not easy. A chirped mirror is a good example of a device for controlling an ultrafast signal in the time domain. In practice, however, the flexibility of a chirped mirror is low, so that Eq. (4.9) is normally implemented with a pulse shaper arrangement composed of two gratings, two positive lenses (or two concave mirrors), and a complex amplitude filter, as shown in Fig. 4.3. An incident optical pulse is spatially dispersed by the first grating and the first lens, effectively performing the calculation in Eq. (4.10). An amplitude and phase mask is placed at the focal plane of the first lens, which is a frequency plane (the Fourier plane). The dispersed light is spectrally modulated by the mask, performing the calculation in Eq. (4.9). The second lens and second grating perform the inverse Fourier transform in Eq. (4.11). Finally, an arbitrary pulse shape is obtained [1,2].

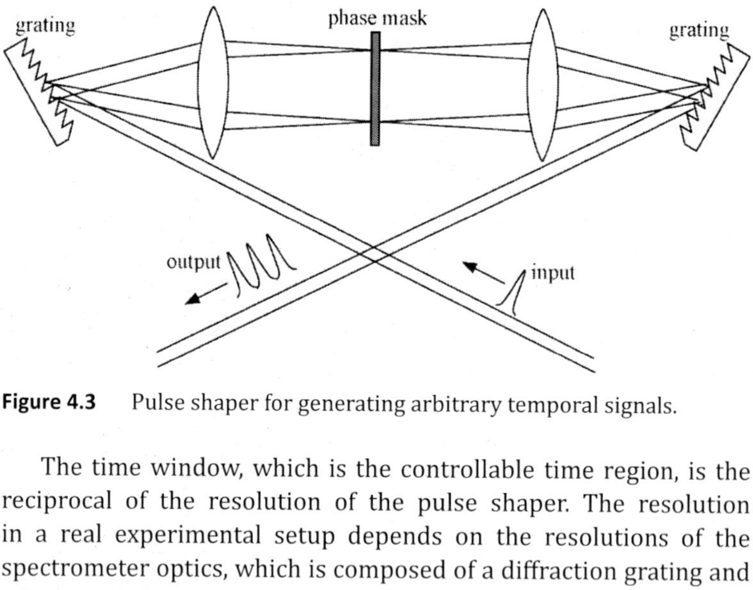

Figure 4.3 Pulse shaper for generating arbitrary temporal signals.

The time window, which is the controllable time region, is the reciprocal of the resolution of the pulse shaper. The resolution in a real experimental setup depends on the resolutions of the spectrometer optics, which is composed of a diffraction grating and a lens, and the mask. Similarly, the resolution of the time-domain signal is the reciprocal of the spectral window width.

4.2.6 Spatial Pulse Manipulation with Linear Filtering (Computer Generated Hologram)

An optical apparatus for calculating Eq. (4.9) in the spatial domain is shown in Fig. 4.4. It is composed of two lenses and an amplitude and/or phase mask. The first lens forms a collimated beam from a point source; in other words, it calculates the spatial Fourier transform of a spatial delta function. At the Fourier plane, the mask spatially modulates the beam (the spatial calculation of Eq. (4.9)). The spatially modulated beam forms an arbitrary spatial pattern via a spatial Fourier transform (Eq. (4.11)) by the second lens. This technique, employing a computer-generated hologram (CGH), is well known.

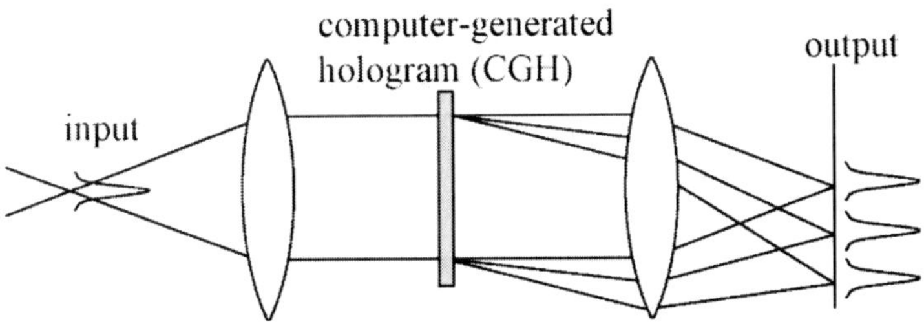

Figure 4.4 Computer-generated hologram (CGH) for generating arbitrary spatial patterns.

The optical calculation processes in both optical apparatuses shown in Figs. 4.3 and 4.4 are mathematically identical, the only difference being the dimension; therefore, the method of designing a mask (a frequency filter) is exactly the same.

4.2.7 Frequency Filter Optimization

It is easy to design a frequency filter that modulates the complex amplitude of a lightwave by using a Fourier transform (FT), but modulation of the amplitude is inefficient in terms of energy consumption; therefore, a phase-only filter is frequently employed. Designing a phase-only filter requires iterative optimization to obtain the desired output.

Some optimization methods for phase filters are described briefly here; more details are given in the literature [3–12]. One of the most well-known CGH optimization methods is the iterative Fourier transform (IFT) method (or Gerchberg–Saxon method) [3], which can be easily implemented with a fast Fourier transform (FFT) and constraint conditions in both the signal plane and the frequency plane. The calculation time with the IFT method is independent of the number of diffraction points. Consequently, the method is effective if there are a large number of diffraction points.

Other optimization methods have also been developed, such as the direct binary search method [4], the simulated annealing algorithm [5, 6], the error diffusion method [7], the optimum rotation angle (ORA) method [8], genetic algorithms [9, 10], the lens grating method [11], and the center (offset) phase search method [12]. Simulated annealing and genetic algorithms are universal methods that are applicable to a wide range of optimization problems. They are useful in cases where we know the estimation function of the target system that we desire to optimize but do not know the operation principle. The ORA method gives a CGH that generates diffraction peaks with high uniformity. The calculation time for the method depends on the number of diffraction points. The lens grating method can be used to design multiple diffraction peaks with a small computational cost. With the offset phase search method, when used in combination with the lens grating method, the calculation time increases with an increase in the number of diffraction points but produces diffraction peaks with high uniformity.

4.3 Temporal Pulse Shaping

In this section, we focus on temporal manipulation of ultrashort laser pulses, and especially its application to material processing. Two or more temporally separated ultrashort pulses independently propagate in a material if photon absorption is sufficiently small. However, in a material that exhibits photon absorption under irradiation with ultrashort pulses, phenomena induced by the first pulse have some effect on the second pulse.

The phenomena that an ultrashort pulse induces are, in sequence, excitation of electrons, diffusion of electrons, recombination, deposition of energy into the lattice, ultrafast (supersonic) thermal expansion, and ensuing compression of the surrounding material. Heating and expansion result in cavitation and void formation with a pressure wave. Subsequent thermal diffusion, re-flow and re-solidification follow. The entire phenomena up to the resulting final structure involves various effects depending on the properties of the material and the pulse parameters, including the energy, duration, repetition, frequency, focusing conditions, and complex amplitude (described by spectral amplitude and phase) of the pulses.

Pulse duration and chirping in ultrafast laser processing should be discussed as the simplest aspects of temporal pulse shaping because they have important effects. Since they have been described in detail in the literature [13–20], here we focus on temporal pulse shaping. First, we discuss ultrafast laser processing using double pulses produced with a beam splitter, relatively complex pulses formed by a pulse shaper, and pulses generated by an adaptive scheme. Adaptive pulse control can be achieved by two different schemes. One involves controlling the shape of the pulse after it passes through the optics. In this case, temporal pulse manipulation compensates for internal imperfections in the system and external disturbances to the system. The other scheme involves controlling the phenomena that occur after irradiating the target with the pulse. The total system, including the optics and the target, is optimized as the result that the manipulated pulse has a complex structure derived from the spectral–temporal complexity of the light–matter interactions.

4.3.1 Double Pulse Method

The simplest type of temporally shaped laser pulse is formed by double-pulse irradiation. It is easily implemented with the interferometer-like optical setup shown in Fig. 4.1(a). The time delay is controlled by axially translating the mirror in one of the two arms. There have been several studies where various materials were irradiated with a pair of femtosecond pulses with a wavelength of 800 nm and a pulse duration of ~100 fs, such as silicon [21–23], Ti [24], Al [25–27], Cu [25–27], Ni [28], steel [25, 29], and wide-bandgap dielectrics [30–33]. In these reports, the observed phenomena are related to the interaction of the second pulse and the plasma formed by the first pulse. The plasma has two effects on the second pulse. One is to increase the absorption of the second pulse, which leads to enhanced ablation. The other is to be blocked the second pulse by the plasma with high reflectivity, which, conversely, leads to reduced ablation. The balance between these effects is strongly dependent on the pulse parameters for each material and it is complex. Therefore, the optimum pulse parameters of the double-pulse method should be determined by a lot of iterative laser processing trials.

4.3.2 Material Processing Using Temporally Shaped Pulses Formed by a Pulse Shaper

A pulse shaper easily controls the temporal pulse shape so that the effect of the temporal shape of the pulse in material processing can be investigated. When dielectrics and semiconductors are irradiated with double and triple pulse trains with subpicosecond separation, shaped-pulse irradiation leads to improved processing quality through carrier generation and trapping, electron–phonon coupling, and local heating caused by the first pulse and successive enhanced absorption and edge-defined ablation [34, 35]. Temporal pulse shaping can also optimize the axial size of refractive index changes and microdamaged structures [36, 37]. Adaptive optimization with in-situ observation of the laser processing plays an important role, as described in the next section. Spatial aberrations are dynamically corrected with the same scheme [38].

A temporally asymmetric pulse can produce a nanostructure much smaller than the diffraction limit on a dielectric surface, as shown in Fig. 4.5 [39, 40]. The balance between photo-ionization and

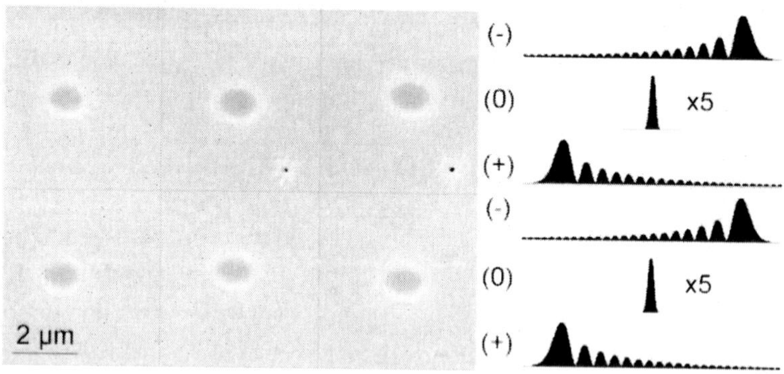

Figure 4.5 SEM images of fabricated structures in fused silica for various temporal asymmetries of the laser pulse. Top row: surface in focal plane; lower row: surface 1 μm below focal plane. The large structure in all areas is due to the unshaped pulses (0), and their size is close to the diffraction limit. The appearance of the small structure for positive cubic phase (+) in the middle column defines the threshold for $\varphi_3 = +6 \times 10^5$ fs^3. For negative cubic phase (−), $\varphi_3 = -6 \times 10^5$ fs^3, i.e., a time-reversed pulse [39]. (Courtesy of Thomas Baumert.)

electron–electron impact ionization controls the localized generation of electrons with dependence on the pulse shape. The cubic chirped pulses control the two different bases of electron generation with a positive or negative chirp. This result suggests the possibility of tailoring the temporal pulse control depending on the material, to produce the desired surface nanostructures.

4.3.3 Adaptive Control Based on Temporal Pulse Shaping

Adaptive feedback control for temporal pulse shaping is applied to two different kinds of schemes in systems where a material is irradiated with an ultrashort pulse. A pulse shaper is a valuable tool for effective excitation of materials in many systems. As described in the previous section, it is an important technique for obtaining higher quality material processing. One scheme is to perform optimization that obtains a desired pulse shape using the pulse itself as a target. The other scheme is to perform optimization of the pulse shape to produce a desired response in the material to the laser pulse.

4.3.3.1 Minimization of pulse duration

In material processing, it is important to obtain a pulse with the minimum duration from a pulse that is given a spectral dispersion, especially as it decreases the threshold processing energy. The first experimental demonstrations of this by using a pulse shaper have been reported [41–45]. As shown in Fig. 4.6, the optical setup is composed of a pulse shaper and a photomultiplier tube (PMT) for detecting the time-integrated second harmonic signal, and

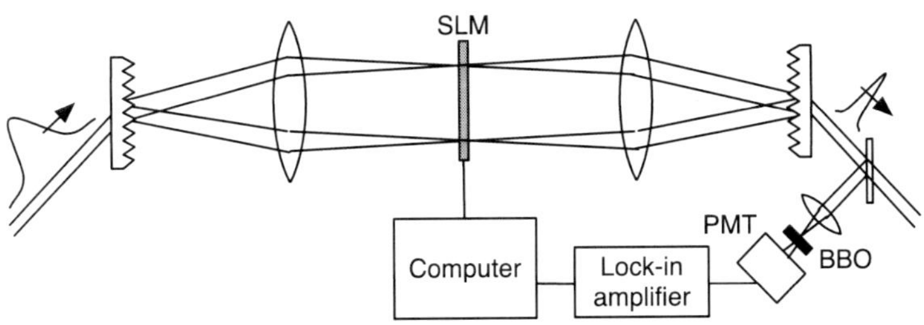

Figure 4.6 Adaptive control system for minimizing pulse duration.

a computer minimizes the pulse duration by using a simulated annealing algorithm [41, 42] or a genetic algorithm [43]. Adaptive shaping of a pulse propagating in a single-mode optical fiber has been demonstrated [44]. The adaptive pulse compression is performed with the same optical setup, involving two-photon absorption in a photodetector for estimating the pulse duration [45].

4.3.3.2 Optimization of material's response

In the optimization process, the response of the material after it is irradiated with a laser pulse is measured, and the pulse is modified by the pulse shaper based on the measurement results. Through iterative modification, the total system, including the laser source, the optical system, and the material, is optimized to maximize the desired phenomenon that is induced by the laser pulse.

A feedback control circuit composed of a femtosecond laser, a pulse shaper, a reflection time-of-flight mass spectrometer, and a computer were used to optimize the branching ratios of different organometallic photodissociation reaction channels with an evolutionary algorithm [46]. A similar experimental system was also applied to the optimization of Si ablation [47] and the control of laser-induced breakdown in Ar and Xe by measuring light emission from plasma [48], as well as filamentation in water [49] and in air [50]. For the problem of molecular alignment, pulse shaping based on evolutionary algorithms has been applied [51]. The study also demonstrated that a covariance matrix adaptation evolutionary strategy (CMA-ES) had better performance than other evolutionary algorithms. Adaptive pulse shape control has been applied to ultrafast laser processing. The axial size of the processing area in fused silica was minimized by pulse shaping with a phase-contrast image of a permanent fabricated structure, as shown in Fig. 4.7 [52]. Adaptive pulse shape control has been used to find the pulse parameters that maximize a region of positive refractive index change in BK7 glass [53]. The pulse had an envelope of several picoseconds with many femtosecond spikes.

As demonstrated in the studies described above, adaptive schemes are promising to search for suitable pulse parameters for fabricating a desired structure.

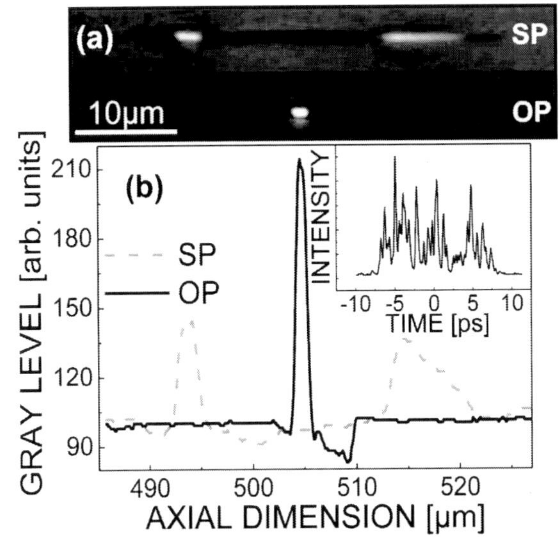

Figure 4.7 (a) Structures at a depth of 500 μm induced by a short pulse (SP) and an optimal pulse (OP) with the same energy. (b) The axial cross sections of the corresponding refractive index change. The dark color denotes a positive index variation. The inset shows the OP shape [52]. (Courtesy of Razvan Stoian.)

4.4 Spatial Pulse Shaping for Parallel Material Processing

Femtosecond laser processing can make 3D structures with high resolution. The resolution can be smaller than $\lambda/10$ in glass and smaller than $\lambda/20$ in two-photon polymerization. Therefore, a vast number of processing points are required to fabricate 2D and 3D structures with millimeter scale. For instance, if a 1 mm^3 volume of material is processed at intervals of 1 μm, the number of processing points is 10^9. If a laser with a repetition rate of 1 MHz is used, laser processing would take 17 minutes.

There are two possible methods to improve the processing throughput. One is to employ beam scanning. Although this method can achieve arbitrary large-scale patterning, for high throughput it requires a high-repetition laser, a high-speed optical switching device, and a high-speed scanning device. The other method is to use spatial pulse shaping.

In this section, we will describe material processing with spatially-shaped femtosecond laser pulses. From an engineering perspective, spatial shaping techniques are indispensable to obtain sufficient processing throughput and to achieve the required energy-use efficiency.

4.4.1 Requirements of Spatial Pulse Shaping and Its Advantages

4.4.1.1 Parallel pulses and deformed pulse

We will mainly discuss parallel pulses, but the discussion is also applicable to a deformed pulse if an effective parallelism, P_{df}, of the deformed pulse is defined as $P_{df} = F_s/F_{df}$, where $F_s = E_s/(\pi R_s^2)$ is the fluence of an original single pulse with energy E_s and beam radius R_s, and $F_{df} = E_{df}/S_{df}$ is the fluence of the deformed pulse with energy E_{df} and area S_{df}. The multiple parallel pulses are generated from a complex amplitude wavefront with a discrete distribution, but the deformed pulse requires a complex amplitude wavefront with a continuous distribution. Consequently, the design methods for both parallel pulses and a deformed pulse have a difference, but have same advantages. The question of which is better in each application, especially in combination with beam scanning, is an interesting one, but there are few studies on such combined use at present.

4.4.1.2 Pulse energy and duration

From an industrial viewpoint, it is very important to consider how to deliver all of the laser power to the material spatially and temporally. Figure 4.8 shows the number of pulse shots per second on the horizontal axis and pulse energy on the vertical axis. In the figure, the lower-case letters *a*, *b*, and *c* indicate lines, and the upper-case letters *A*, *B*, and *C* indicate points. To simplify the discussion, it is assumed that we have a 1 W, 1 kHz regenerative-amplified Ti:sapphire femtosecond pulsed laser including a 1 W, 80 MHz Ti:sapphire femtosecond pulsed laser. Line *a* indicates a laser power of 1 W, corresponding to a photon use efficiency of 100%, and line *b* indicates 10 mW and 1%, respectively. The rectangle region with point *A* at its upper right corner corresponds to the pulse parameters that are feasible with the 1 kHz pulsed laser. The rectangle region with point *B* at its upper right corner corresponds to the pulse parameters that are feasible with the 80 MHz pulsed laser.

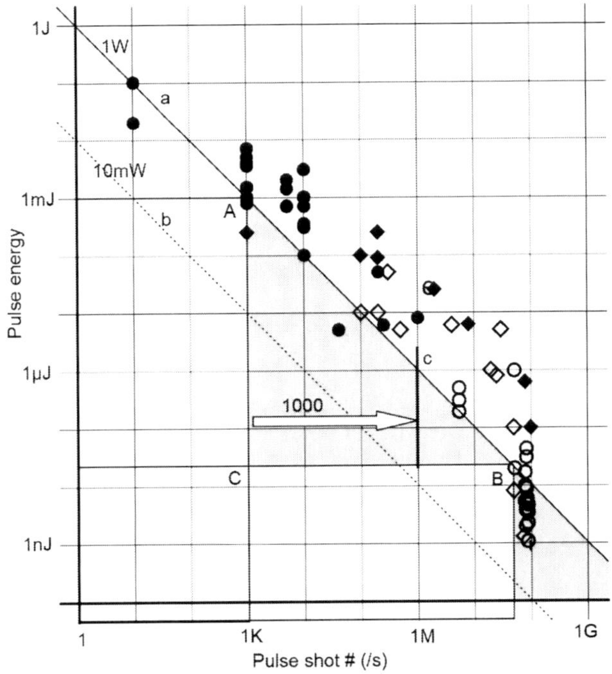

Figure 4.8 Relation between energy per pulse and number of pulse shots per second. The filled circles, open circles, filled rhombuses, and open rhombuses indicate regenerative-amplified Ti:sapphire femtosecond lasers, mode-locked Ti:sapphire lasers, picosecond lasers, and fiber lasers that are commercially available (more than 100 mW) as of September 2010, respectively.

4.4.1.3 High throughput

The gray region in the triangle *ABC* in Fig. 4.8 corresponds to the pulse parameters that are not realizable using the two lasers above. Parallel processing can realize the pulse parameters. If 1000 parallel pulses at most can be generated, the laser parameters on the left side of the vertical line *c* are available. Namely, we can obtain a processing throughput of 1 MHz by using the 1 kHz pulsed laser.

4.4.1.4 High energy-use efficiency

When a material is irradiated with 1 kHz pulses with an attenuating pulse energy of 100 nJ, the energy-use efficiency is extremely low at 0.01%. If 1000 parallel pulses are applied, the energy-use efficiency increases to 10%. Parallel processing allows use of a variety of suitable pulse parameters with little reduction in the energy-use efficiency.

4.4.1.5 High uniformity

Laser pulses have inherent variation in the pulse energy depending on the laser performance. Therefore, there is a corresponding variation of the structures (size and depth) in the laser processing. In particular, the variation is pronounced near the threshold energy E_{th}, because the theoretical curve of the fabrication diameter D as a function of the pulse energy E:

$$D = \omega\sqrt{2\ln\left(\frac{E}{E_{th}}\right)}, \tag{4.12}$$

where $\omega = 1.0\lambda/NA$ is the full-width at half maximum of the Airy disk, and E_{th} is the threshold energy. In parallel processing for fabrication of nanostructures, high energy uniformity in the parallel pulses is extremely important.

4.4.1.6 Variable pattern processing

Use of a spatial light modulator (SLM) allows reconfigurable spatial pulse shaping, which enables novel laser processing schemes. When parallel processing forms different patterns in a time region that is comparable to the response time of the SLM, the processing throughput N_{tp} [pulses/s] is the product of the frame rate N_{SLM} [frames/s] of the SLM and the number of parallel pulses (parallelism) N_P [pulses], or $N_{tp} = N_{SLM}N_P$. Parallel laser processing uses 2π-phase modulation, which gives 100% energy-use efficiency; however, since the typical frame rate of a nematic liquid crystal SLM is less than 30 Hz, in this case, the maximum energy-use efficiency is 3% when the laser repetition rate is N_{rep} = 1 kHz. If a point is processed with N_{shot} shots and irradiation energy E_{ir}, then the energy-use efficiency is improved to $E_{ir}N_PN_{SLM}N_{shot}/(E_{laser}\,N_{rep})$, where E_{laser} is the pulse energy. The total irradiation energy per pulse, $E_{ir}N_P$, does not exceed the pulse energy on the sample plane, which is the product of E_{laser} and the transmittance of the optical system T_{os} ($E_{ir}N_P \leq E_{laser}\,T_{OS}$), and the product of the number of shots N_{shot} and the frame rate N_{SLM} does not exceed the laser repetition rate N_{rep} ($N_{SLM}N_{shot} \leq N_{rep}$). In practical applications, it is critical to properly set the four parameters E_{ir}, N_p, N_{SLM}, and N_{shot} according to the material to be processed.

4.4.1.7 Instantaneous processing

There are many difficulties in processing at the proper position when a target moves and deforms. In particular, external vibrations are a serious problem in micro- and nano-fabrication. Thus, the laser processing system requires a feedback control system composed of a position detection system and a beam steering system [54]. A system that allows instantaneous parallel processing reduces the need to repeat position and shape measurements.

4.4.2 Passive Optical Components for Spatial Pulse Shaping

Spatial pulse shaping is implemented with passive optical components, such as apertures (optical masks), beam splitters, lens arrays, and DOEs. These will be described in the following.

4.4.2.1 Apertures (optical masks)

Controlling the amplitude of light using an optical mask having a circular aperture, a slit, or a specific pattern causes loss of light but is an easy and effective implementation. The optical mask is placed at the conjugate image plane, the spatial frequency plane, or the pupil plane. When the optical mask is placed at the Fourier plane, the beam diameter and the Rayleigh length of the focused pulse are controlled, as shown in Fig. 4.9 [55, 56]. In ordinary focusing, such

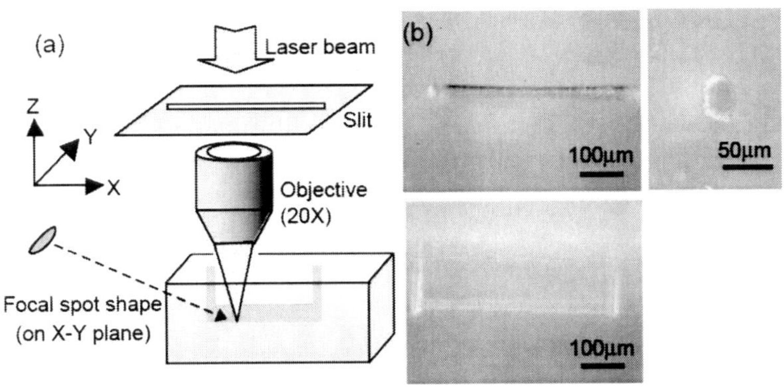

Figure 4.9 (a) Schematic diagram of the experimental setup for controlling the focus volume. (b) Photomicrographs of fabricated bridge-like microstructures fabricated with a slit of 0.2 mm width [55]. (Courtesy of Ya Cheng and Koji Sugioka.)

as focusing of a Gaussian pulse by an objective lens, the focusing volume has different lengths in the lateral and axial directions, and laser processing produces a fabricated structure that depends on the focusing volume. Therefore, in the fabrication of an optical waveguide, the beam is reduced in size in one direction by an aperture, and the difference between the lateral and axial lengths is minimized. The same effect can be achieved with a cylindrical lens [57]. A dark-ring mask and an annular mask at the pupil plane of an objective lens can be used to obtain smaller fabricated structures in glass based on the principle of superresolution [58].

4.4.2.2 Beam splitters

Beam splitters can easily form parallel pulses without introducing large changes in the spatial and temporal properties of the pulses. The method allows highly flexible pulse manipulation because of the ease of inserting other optical components. There is no fundamental limit to the number of divided pulses, but a high degree of parallelism increases the costs of the optical components and decreases the system stability. One important optical arrangement for parallel processing is interference exposure for making periodic structures [59–61]. A combination of this arrangement with a special prism has been employed to minimize the pitch [62].

4.4.2.3 Lens arrays

Lens arrays can easily form many focal spots at regular intervals [63,64] and are useful components for fabricating identical structures in parallel. The intensity of the focal spots is proportional to the incident laser power entering the pupils of the individual lenses; therefore, the uniformity of the illumination light on the lens array is extremely important for fabricating uniform parallel structures.

4.4.2.4 Diffractive optical elements

Diffractive optical elements are effective for producing arbitrary spatial patterns. In laser processing systems, DOEs are used for beam combining and beam branching. Beams from three diffraction gratings with different directions, used instead of a prism combiner, have been made to interfere on a sample for fabricating a periodic structure via photopolymerization [65]. DOEs inherently have spatial dispersion, and since ultrashort pulses have broadband, so this optical system is suitable for pulse lengths down to subpicosecond level [66]. An optical system composed of a diffractive beam splitter and telescope optics interferes among ultrashort pulses without spectral dispersion in a wide area, as described in Section 4.4.4.1. Another approach for compensating for spectral dispersion is to use a combination of two DOEs to perform parallel multifocal processing with femtosecond pulses [67–70]. The spatial spectral dispersion along the radial direction (Fig. 4.10(c)) can be minimized by the dispersion compensation method, as shown in Fig. 4.10(d). A diffractive equivalent axicon lens has been used to fabricate a microchannel with a high aspect ratio, as shown in Fig. 4.11 [71–74]. The large-volume ablation of glass using femtosecond Bessel pulses was explained by rapid expansion due to heating, with subsequent cavitation.

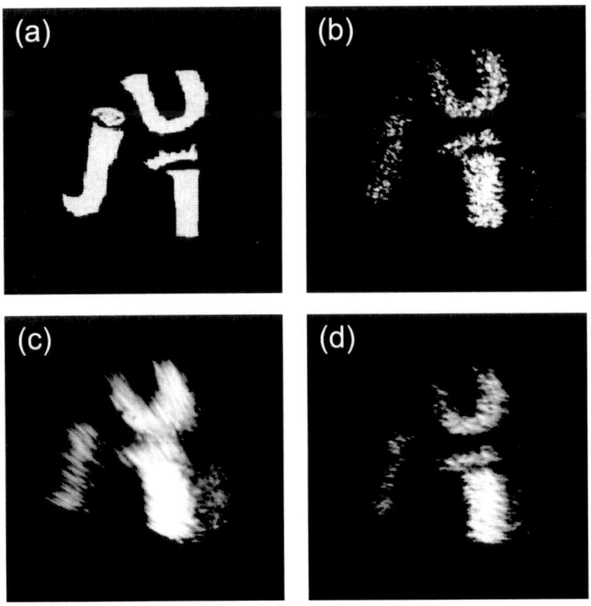

Figure 4.10 (a) Computational reconstruction of a CGH. (b) Optical reconstruction using monochromatic laser light. Femtosecond laser reconstructions (c) without and (d) with spatial-chirp correction [70]. (Courtesy of Lluís Martínez-León and Gladys Mínguez-Vega.)

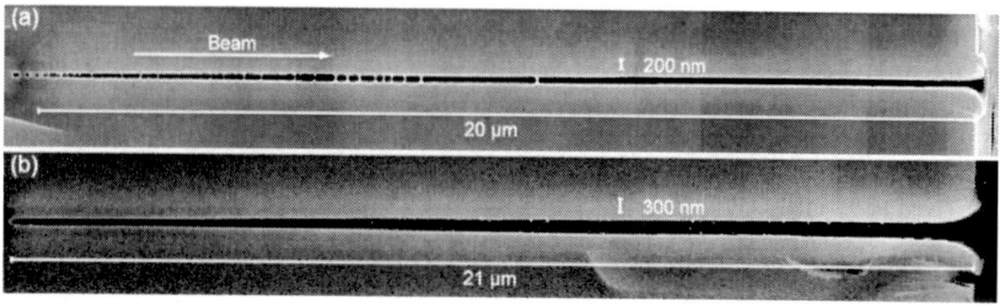

Figure 4.11 Microchannels fabricated by Bessel pulses with energies of (a) 0.65 μJ and (b) 0.85 μJ, respectively [74]. (Courtesy of John Dudley and Francois Courvoisier.)

4.4.3 Active Optical Components for Spatial Pulse Shaping: Spatial Light Modulators

4.4.3.1 Liquid crystal spatial light modulators

Liquid crystal spatial light modulators (LCSLMs), originally developed as a key device to actively and dynamically control light in displays and optical information processing, have recently been used in imaging systems, optical tweezers, and laser processing systems. A nematic LCSLM performs phase modulation with low loss of light and is frequently used as a device for modulating spectral phase in a pulse shaper and as a device for displaying DOEs, including CGHs. An LCSLM has a spatial frequency response; therefore, the phase value $\varphi(\mathbf{r})$ at position $\mathbf{r}$, calculated on a computer, should be given to the LCSLM with compensation of the spatial frequency response. Thus, the calculated phase value $\varphi(\mathbf{r})$ is transformed to an actual phase $\varphi^{(a)}(\mathbf{r})$ on the LCSLM as follows:

$$\varphi^{(a)}(\mathbf{r}) = \Im^{-1}\left\{\frac{\Im[\varphi(\mathbf{r})]}{1+(\nu/\nu_h)^2}\right\}, \tag{4.13}$$

where $\Im$ and $\Im^{-1}$ are the Fourier transform and its inverse, ν is the spatial frequency, and ν_h is the spatial frequency where the modulation transfer function is the half maximum. The derivation of Eq. (4.13) is described in Refs. 75 and 76. Figure 4.12 shows a computer simulation of the reconstruction of a CGH displayed on an LCSLM with a spatial frequency response, compensation in optical reconstruction of the CGH, and application to holographic laser processing [76].

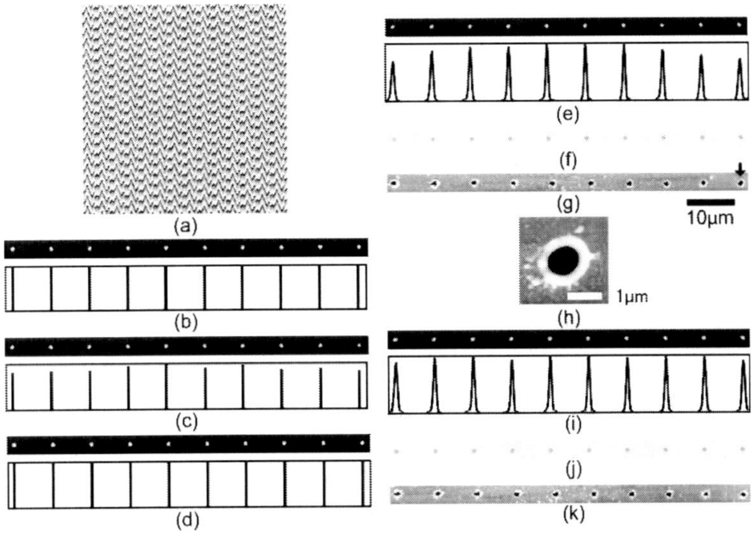

Figure 4.12 (a) A CGH designed to generate ten diffraction peaks. Computer reconstructions of the CGH (b) without and (c) with the spatial frequency response of the LCSLM taken into account, and (d) computer reconstruction of a CGH designed to include compensation of the spatial frequency response. The respective diffraction efficiencies were U = 95.8%, 70.0%, and 69.2%, respectively. The respective uniformities were η = 99.97%, 82%, and 99.86%, respectively. (e) Optical reconstruction of the CGH without compensation. U was 74% and η was 60%. The uniformity was smaller than that of the computer reconstruction because the actual diffusion length l was larger than that used in the computer simulation (ν_h = 1/16 μm^{-1}). (f) Transmission microscope image and (g) AFM image of the area in a glass microscope cover slip processed by a single femtosecond pulse. The uniformity of the diameter, $U_d = d_{min}/d_{max}$, was 76%. (h) The processed structure consisted of a cavity surrounded by a ring-shaped protrusion, debris, and small droplets on the glass surface. (i) Optical reconstruction of the CGH with compensation. U was 91% and η was 61%. (j) Transmission microscope image and (k) AFM image of the processed area. U_d improved to 84% by introducing compensation of the spatial frequency response. [76]

4.4.3.2 Deformable mirrors

Deformable mirrors were originally developed for adaptive optics in astronomy, and small-scale devices were developed for use in observation of the human retina. Recently, commercial products have become available at a relatively low price. The number of pixels is not so large compared with an LCSLM, so it is difficult to generate a complex optical pattern, but they can be used for spatial deformation of pulses [77] and aberration compensation in laser processing systems. These devices are effective for real-time, adaptive wavefront control because the response speed is faster than that of nematic LCSLMs.

4.4.4 Optical Setups

Optical setups for parallel laser processing are classified into three types according to the positional relationship between the target and the device used for spatial pulse control (a hologram in this case), as shown in Fig. 4.13. These setups are described in the following sections.

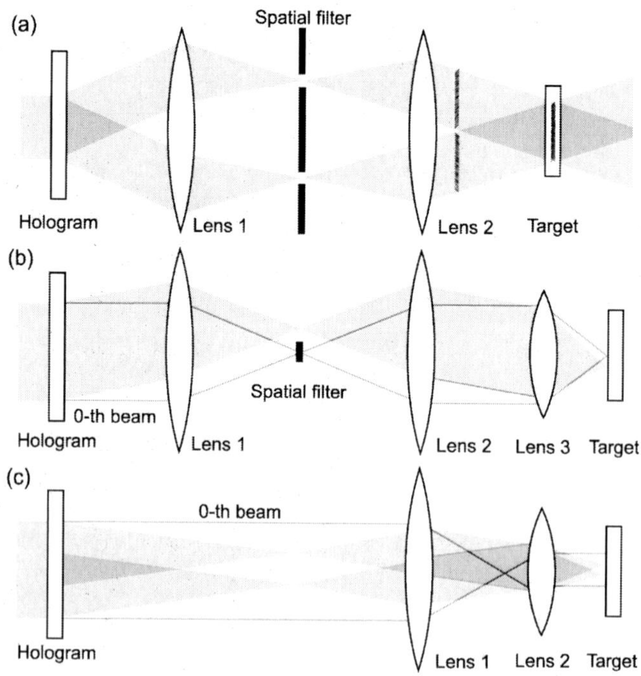

Figure 4.13 Three optical setups for parallel laser processing, with a target placed at (a) the image plane, (b) the Fourier plane, and (c) the Fresnel plane of the hologram.

4.4.4.1 Image plane

When the target is placed at the image plane of the hologram [78], as shown in Fig. 4.13(a), the irradiation beam is divided into multiple beams by a diffractive beam splitter, which is a kind of hologram. The diffracted beams form an interference pattern on the target through a 4*f* imaging system. Selecting the necessary diffracted beams with a spatial frequency filter placed at the Fourier plane allows variable interference patterning [79–83]. An advantage of this method is its suitability for processing a periodic structure over a large area. The spatial frequency filter is space-invariant; that is to say, the effect it has on the incident light does not differ from location to location. If a space-variant structure is required, the desired period and direction are applied at the desired position in the diffractive beam splitter.

4.4.4.2 Fourier plane

When the target is placed at the Fourier plane of the hologram, as shown in Fig. 4.13(b), a Fourier transform hologram is used. A high-quality hologram can be designed with low computational cost by using optimization techniques. The zero-order beam is obstructed with a spatial filter because the zero-order beam causes undesired processing. Shift-variant parallel femtosecond laser processing with a fixed hologram [84] and variable holograms [85] have been demonstrated with this optical system. Some groups have demonstrated parallel femtosecond laser processing [86–99]. Other demonstrations that have been reported include

internal 3D processing in glass using multi-shot or single-shot irradiation [84], high-speed processing in combination with a galvanometer scanner (Fig. 4.14) [89, 91], parallel two-photon polymerization [87, 88, 93, 94], waveguide fabrication (Fig. 4.15) [94–96], phase gratings [97], 3D processing with aberration correction (Fig. 4.16) [98], and antireflection surface structuring [99]. Once parallel processing has been sufficiently improved to achieve faster processing, it is important how to determine the scanning speed and the number of parallel processing points according to the irradiation energy, the repetition rate and power of the laser, and an application.

In the optical system using the Fourier CGH, the relation between the pulse diffraction position r, defined as a distance from the optical axis, and a spatial frequency component of the CGH, v, is as follows.

When light with center wavelength λ_c is diffracted by fringes with spatial frequency ν on the LCSLM, the diffraction angle θ is given by $\theta = \sin^{-1}(\nu\lambda)$. Then, the diffraction position r on the focal plane (the Fourier plane) of the lens with a focal length F is

$$r = F\tan[\sin^{-1}(\nu\lambda_c)] = F\left[\nu\lambda_c + \frac{(\nu\lambda_c)^3}{2} + \cdots\right]. \tag{4.14}$$

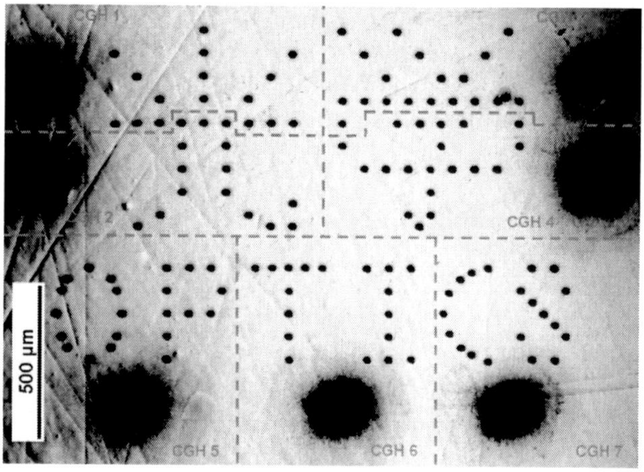

Figure 4.14 A pattern of holes forming the word 'OPTICS' (bottom) and the corresponding Japanese Kanji characters (top), completed by using seven CGHs applied at appropriate positions using a galvanometer scanner. The three large holes formed by the zero-order diffracted beams at the bottom are easily eliminated by a filter [91]. (Courtesy of Zheng Kuang and Ken Watkins.)

If $\nu\lambda_c$ is sufficiently smaller than 1 ($\sim 2 \times 10^{-2}$ in a real system), then r is approximated as

$$r = F\nu\lambda_c M \tag{4.15}$$

where M is a newly introduced magnification given by the reduction optics to match the size of the imaged CGH to the pupil of the objective lens. The Fourier CGH has spatial spectral dispersion, so that the focal spot has spatial broadening Δr given by

$$\Delta r = F\nu\Delta\lambda M = r\Delta\lambda/\lambda_c, \tag{4.16}$$

where $\Delta\lambda$ is the spectral width. Δr depends on only r if the light source is fixed (λ_c and $\Delta\lambda$ are fixed).

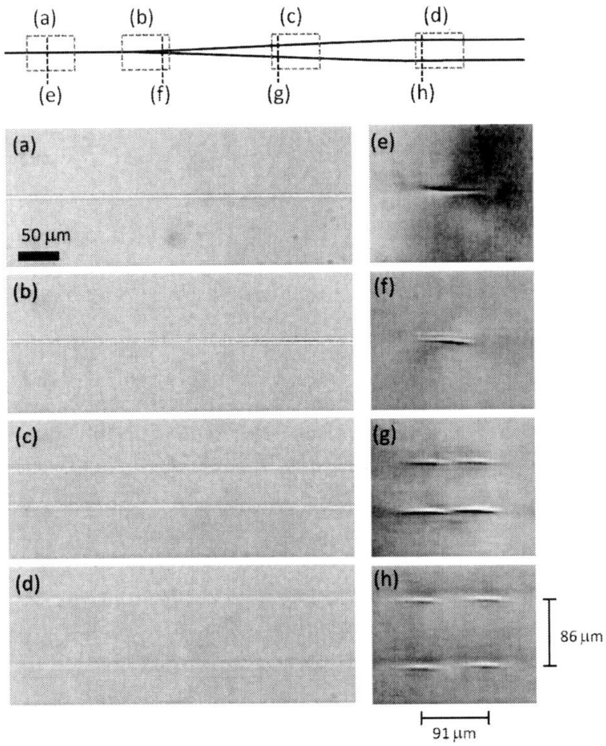

Figure 4.15 Schematic diagram (top) and optical microscope images of a 1×4 splitter. (a) Single-straight region, (b) branching region, (c) separating region, and (d) four-straight region. (e–h) Cross-sectional images at corresponding regions in (a–d) [96]. (Courtesy of Masaaki Sakakura and Kiyotaka Miura.)

A CGH with $N \times N$ pixels is displayed on an area of the LCSLM with dimensions $W \times W$. The maximum frequency of the CGH is $\nu_{max} = N/(2W)$. The CGH is imaged on the pupil plane of the objective lens (OL; Lens 3) with magnification $M = F_2/F_1$, where F_2 and F_1 are the focal lengths of Lens 1 and Lens 2, respectively. The side length of the OL's pupil plane is $W_{pupil} = WM$. Therefore, the minimum step size in the spatial frequency domain is $1/W_{pupil} = 1/(WM)$. Consequently, using Eq. (4.15), the equivalent pixel size p_s on the sample plane is geometrically calculated as

$$p_s = \frac{\lambda_c F_{OL}}{W_{pupil}} = \frac{\lambda_c F_{OL} F_1}{W F_2}. \tag{4.17}$$

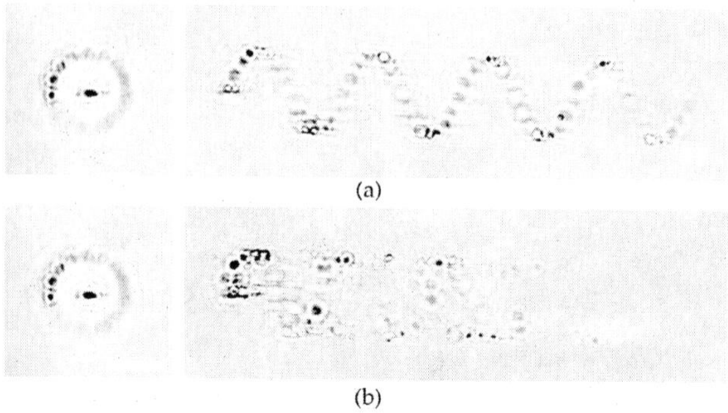

Figure 4.16 Helical structures written into fused silica using a single hologram (a) with and (b) without correction of the depth-dependent spherical aberration. The scale bar corresponds to 10 μm [98]. (Courtesy of Alexander Jesacher.)

The focal spot diameter, d_{airy}, at the sample plane is calculated from the full width at half maximum (FWHM) of the airy disk pattern:

$$d_{airy} = 1.03\frac{\lambda_c}{NA_{eff}}, \quad (4.18)$$

where $NA_{eff} = R_{pupil}/F_{OL}$ is the effective numerical aperture in the experimental system with the effective beam radius $R_{pupil} = W_{pupil}/2$ at the pupil of the OL. The spot diameter d_{airy} is almost equal to $2p_s$. It is important that we select M so as to adjust the size of the imaged CGH to the diameter of the OL's pupil, $W_{OL} = 2NA_{OL}F_{OL}$. Then NA_{eff} is nearly equal to NA_{OL}, and therefore, the focal spot diameter is minimized.

4.4.4.3 Fresnel plane

When the target is placed at the Fresnel plane of the hologram, as shown in Fig. 4.13(c), a Fresnel transform hologram can be used, as is done for focal spot shaping [100]. The zero-order beam is background light of the diffraction peaks. In the case of a transparent material, the femtosecond laser processing is based on a multi-photon process, so that the background light does not contribute to the processing. To change the axial position (depth) without any mechanical movements and to realize 3D parallel processing with a single shot, a Fresnel transform hologram is required [12, 101–103]. The focus diameter of the hologram lens is the same as that given by Eq. (4.18). The maximum numerical aperture of the hologram lens is determined by the pixel size *p*, namely, $NA^{(\max)} = \lambda_c/(2p)$. The minimum focus diameter is almost the same as the pixel size *p*. In Fig. 4.17(c), the focal spot is reduced to the diffraction limit with a set of reduction optics.

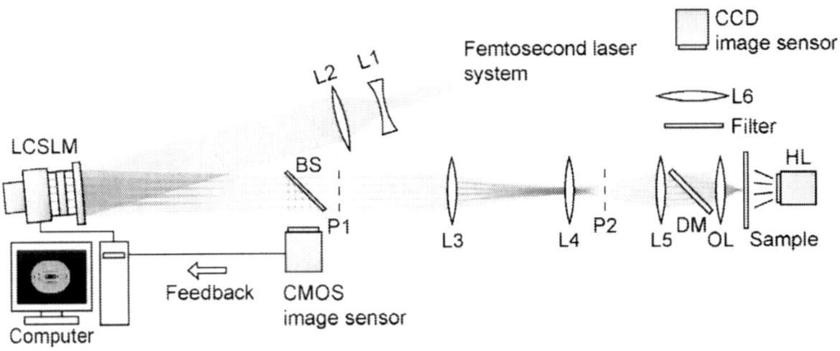

Figure 4.17 Holographic femtosecond laser processing system with Fresnel CGH. A complementary metal-oxide semiconductor (CMOS) image sensor detects the reconstructed CGH displayed on the LCSLM, and the CGH was recalculated in the computer according to the measured values [102].

4.4.5 Adaptive Optimization of CGH in Laser Processing System

The optimized CGH had high-uniformity diffraction peaks in a computer reconstruction. In actual optical reconstruction, however, the uniformity was decreased due to the spatial properties of the optical system. To obtain a CGH with high uniformity in the optical system, the CGH was optimized using its optical reconstruction. In the adaptive optimization, the optically reconstructed CGH that had previously been sufficiently optimized in a computer was measured by an image sensor, as shown in Fig. 4.14. The CGH was recalculated from the measured values to obtain the desired pattern. These steps were iteratively performed until a satisfactory output was obtained. The fabrication uniformity, U_d, reached 0.84 for nine parallel fabricated structures on a glass surface [102]. Furthermore, with second harmonic optimization, 18 parallel fabricated structures were formed on a glass surface with uniformity U_d = 0.88. In the parallel laser processing, the minimum average pit diameter was 271 nm when the mean fluence of the beams was 0.88 J/cm^2 [104].

4.5 Spatiotemporal Pulse Manipulation

4.5.1 Spatiotemporal Dispersion Control

In the holographic technique, a femtosecond laser pulse diffracted by the CGH has reduced peak intensity and extended pulse width at the focal point due to wavelength dispersion in the CGH. Some wavelength dispersion compensation techniques have been proposed, including hybrid diffractive-refractive lenses [105], a pair of DOEs [67, 106], and a combination of a diffraction grating and diffractive lens for multifocal processing [70].

4.5.2 Spatiotemporal Focusing

Refractive spatial and temporal focusing [107–111] forms an ultrashort pulse only at the focal plane, thus reducing background excitation in multi-photon microscopy. A combination of refractive spatiotemporal focusing and pulse shaping for optimization of the phase distribution with an SLM has also been proposed [110]. Simultaneous spatiotemporal focusing for axial scanning has been investigated by adjusting the group velocity dispersion [111], and multifocal temporal focusing using an Echelle grating has also been proposed [112]. This approach was used to fabricate hollow microfluidic channels with a circular cross-section in fused silica by spatiotemporally focusing a femtosecond laser beam [113].

4.5.3 Diffractive Spatiotemporal Focusing

A diffractive spatiotemporal lens that compensates for the wavelength dispersion in addition to achieving spatial and temporal focusing has also been proposed [114]. The technique using a diffractive lens realizes not only spatiotemporal focusing and wavefront optimization, but also focusing with straightforward reconfigurability and easy additive phase modulation, such as producing multiple focal points to compensate for the wavelength dispersion simultaneously.

4.5.4 Spatiotemporal Pulse Shaping

Until now, most of the demonstrations of spatiotemporal pulse shaping have involved two-photon microscope observation of biological cells [115, 116]. There are few demonstrations applied to material processing. However, a spatiotemporal lens is extremely promising for material processing because it performs temporal pulse shaping and spatial pulse shaping simultaneously with a relatively simple optical system [117]. Thus, the laser processing system can apply a pulse with the minimum duration and the minimum diameter to a material even if the system and the material have internal and external disturbances in real environments.

4.6 Conclusions and Future Perspectives

In this chapter, we have described techniques for temporal and spatial manipulation of ultrashort laser pulses and showed that they have the same underlying mathematical basis and similar optical implementations. Techniques developed for temporal manipulation can be applied to spatial manipulation, and vice versa. Furthermore, simultaneous techniques for spatially and temporally maximizing the peak intensity can be understood as extensions of these techniques.

Temporal manipulation and spatial manipulation have slightly different roles in laser material processing. Temporal manipulation is used to control the photon–material interaction to improve the processed morphology and throughput. Our understanding of the whole picture of the relation between the pulse shape and the material excitation efficiency is still at an early stage.

The photon–material interaction depends not only on the temporal shape of the laser pulse but also on the oscillation direction, namely, the polarization. Vector pulse shaping control [118] is important. At present, it is mainly used for scientific applications, but in the near future, vector pulse shaping control is expected to enhance ultrafast laser processing.

Spatial pulse manipulation is used to perform laser material processing with high throughput and high energy-use efficiency regardless of the irradiation energy by effectively delivering photons from a laser source to a target material. Spatiotemporal pulse manipulation, in which the temporal and spatial properties of the pulse are simultaneously controlled, has the possibility of further enhancing the photon–material interactions because a series of phenomena induced by laser pulses includes spatial phenomena such as diffusion of electrons and phonons, pressure wave propagation, and resulting spatial distributions of heating, softening, and structure formation.

From an engineering viewpoint, optimization of the temporal properties and spatial properties of the entire laser processing system, from the laser source to the target material, is important. Spectral dispersion control and aberration control are required to ensure the desired pulse shape at the laser irradiation point. The emerging developments in these areas are expected to further improve the processing capability.

References

1. Weiner, A. M. (1995). Femtosecond pulse shaping and processing, *Prog. Quant. Electr.,* **19**, 161–237.
2. Weiner, A. M. (2000). Femtosecond pulse shaping using spatial light modulators, *Rev. Sci. Instrum.,* **71**, 1929–1960.
3. Gerchberg, R. W., and Saxton, W. O. (1972). A practical algorithm for the determination of the phase from image and diffraction plane pictures, *Optik,* **35**, 237–246.
4. Seldowitz, M. A., Allebach, J. P., and Sweeney, D. W. (1987). Synthesis of digital holograms by direct binary search, *Appl. Opt.,* **26**, 2788–2798.
5. Kirpatrick, S., Gelatt, C. D., and Vecchi M. P. (1983). Optimization by simulated annealing, *Science,* **220**, 671–680.
6. Kim, M. S., Feldman, M. R., and Guest, C. C. (1989). Optimum encoding of binary phase-only filters with a simulated annealing algorithm, *Opt. Lett.,* **14**, 545–547.
7. Barnard, E. (1988). Optimal error diffusion for computer-generated holograms, *J. Opt. Soc. Am. A,* **5**, 1803–1817.
8. Bengtsson, J. (1994). Kinoform design with an optimal-rotation-angle method, *Appl. Opt.,* **33**, 6879–6884.
9. Goldberg, D. E. (1989). *Genetic Algorithms in Search, Optimization, and Machine Learning* (Addison-Wesley, Reading, MA)
10. Yoshikawa, N., Itoh, M., and Yatagai, T. (1995). Quantized phase optimization of two-dimensional Fourier kinoforms by a genetic algorithm, *Opt. Lett.,* **20**, 752–754.
11. Liesener, J., Reicherter, M., Haist, T., and Tiziani, H. J. (2000). Multi-functional optical tweezers using computer-generated holograms, *Opt. Commun.,* **185**, 77–82.

12. Hasegawa, S., Hayasaki, Y., and Nishida, N. (2006). Holographic femtosecond laser processing with multiplexed phase Fresnel lenses, *Opt. Lett.,* **31**, 1705–1707.

13. Stuart, B. C., Feit, M. D., Herman, S., Rubenchik, A. M., Shore, B. W., and Perry, M. D. (1996) Nanosecond-to-femtosecond laser-induced breakdown in dielectrics, *Phys. Rev. B,* **53**, 1749–1761.

14. Chichkov, B. N., Momma, C., Nolte, S., von Alvensleben, F., Tunnermann, A. (1996). Femtosecond, picosecond and nanosecond laser ablation of solids, *Appl. Phys. A*, **63**, 109–115.

15. Lenzner, M., Krüger, J., Sartania, S., Cheng, Z., Spielmann, Ch., Mourou, G., Kautek, W., and Krausz, F. (1998). Femtosecond optical breakdown in dielectrics, *Phys. Rev. Lett.*, **80**, 4076–4079.

16. Tien, A.-C., Backus, S., Kapteyn, H., Murnane, M., and Mourou, G. (1999). Short-pulse laser damage in transparent materials as a function of pulse duration, *Phys. Rev. Lett.*, **82**, 3883–3886.

17. Rosenfeld, A., Lorenz, M., Stoian, R., and Ashkenasi, D. (1999). Ultrashort-laser-pulse damage threshold of transparent materials and the role of incubation, *Appl. Phys. A*, **69**, S373–S376.

18. Jasapara, J., Nampoothiri, A. V. V., Rudolph, W., Ristau, D., and Starke, K. (2001). Femtosecond laser pulse induced breakdown in dielectric thin films, *Phys. Rev. B*, **63**, 045117.

19. Mero, M., Liu, J., Rudolph, W., Ristau, D., and Starke, K. (2005). Scaling laws of femtosecond laser pulse induced breakdown in oxide films, *Phys. Rev. B*, **71**, 115109.

20. Louzon, E., Henis, Z., Pecker, S., Ehrlich, Y., Fisher, D., Fraenkel, M., and Zigler, A. (2005). Reduction of damage threshold in dielectric materials induced by negatively chirped laser pulses, *Appl. Phys. Lett.*, **87**, 241903.

21. Choi, T. Y., Hwang, D. J., and Grigoropoulos, C. P. (2002). Femtosecond laser induced ablation of crystalline silicon upon double beam irradiation, *Appl. Surf. Sci.*, **197–198**, 720–725.

22. Zhang, Z., Van Rompay, P. A., and Pronko, P. P. (2003). Ion characteristics of laser-produced plasma using a pair of collinear femtosecond laser pulses, *Appl. Phys. Lett.,* **83**, 431–433.

23. Hu, Z., Singha, S., Liu, Y., and Gordon, R. J. (2007). Mechanism for the ablation of Si<111> with pairs of ultrashort laser pulses, *Appl. Phys. Lett.,* **90**, 131910.

24. Scuderi, D., Albert, O., Moreau, D., Pronko, P. P., and Etchepare, J. (2005). Interaction of a laser-produced plume with a second time delayed femtosecond pulse, *Appl. Phys. Lett.,* **86**, 071502.

25. Harzic, R. L., Breitling, D., Sommer, S., Föhl, C. , König, K., Dausinger, F., and Audouard, E. (2005). Processing of metals by double pulses with short laser pulses, *Appl. Phys. A,* **81,** 1121–1125.

26. Amoruso, S., Bruzzese, R., Wang, X., and Xia, J. (2008). Ultrafast laser ablation of metals with a pair of collinear laser pulses, *Appl. Phys. Lett.,* **93**, 191504.

27. Povarnitsyn, M. E., Itina, T. E., Khishchenko, K. V., and Levashov, P. R. (2009). Suppression of ablation in femtosecond double-pulse experiments, *Phys. Rev. Lett.,* **103**, 195002.

28. Donnelly, T., Lunney, J. G., Amoruso, S., Bruzzese, R., Wang, X., and Ni, X. (2009). Double pulse ultrafast laser ablation of nickel in vacuum, *J. Appl. Phys.,* **106**, 013304.

29. Liebig, C. M., Srisungsitthisunti, P., Weiner, A. M., and Xu, X. (2010). Enhanced machining of steel using femtosecond pulse pairs, *Appl. Phys. A*, **101**, 487–490.

30. Chowdhury, I. H., Wu, A. Q., Xu, X., and Weiner, A. M. (2005). Ultra-fast laser absorption and ablation dynamics in wide-band-gap dielectrics, *Appl. Phys. A,* **81**, 1627–1632.

31. Chowdhury, I. H., Xu, X., and Weiner, A. M. (2005). Ultrafast double pulse ablation dynamics of fused silica, *Appl. Phys. Lett.,* **86**, 151110.

32. Nagata, T., Kamata, M., and Obara, M. (2005). Optical waveguide fabrication with double pulse femtosecond lasers, *Appl. Phys. Lett.,* **86**, 251103.

33. Jiang, L., and Tsai, H.-L. (2005). Repeatable nanostructures in dielectrics by femtosecond laser pulse trains, *Appl. Phys. Lett.,* **87**, 151104.

34. Stoian, R., Boyle, M., Thoss, A., Rosenfeld, A., Korn, G., Hertel, I. V., and Campbell, E. E. B. (2002). Laser ablation of dielectrics with temporally shaped femtosecond pulses, *Appl. Phys. Lett.,* **80**, 353–355.

35. Spyridaki, M., Koudoumas, E., Tzanetakis, P. Fotakis, C., Stoian, R., Rosenfeld, A., and Hertel, I. V. (2003). Temporal pulse manipulation and ion generation in ultrafast laser ablation of silicon, *Appl. Phys. Lett.,* **83**, 1474–1476.

36. Stoian, R., Boyle, M., Thoss, A., Rosenfeld, A., Korn, G., and Hertel, I. V. (2003). Dynamic temporal pulse shaping in advanced ultrafast laser material processing, *Appl. Phys. A,* **77**, 265–269.

37. Burakov, I. M., Bulgakova, N. M., Stoian, R., Rosenfeld, A., and Hertel, I. V. (2005). Theoretical investigations of material modification using

temporally shaped femtosecond laser pulses, *Appl. Phys. A,* **81**, 1639–1645.

38. Mauclair, C., Mermillod-Blondin, A., Huot, N., Audouard, E., Stoian, R. (2008). Ultrafast laser writing of homogeneous longitudinal waveguides in glasses using dynamic wavefront correction, *Opt. Express*, **16**, 5481–5492.

39. Englert, L., Rethfeld, B., Haag, L., Wollenhaupt, M., Sarpe-Tudoran, C., and Baumert, T. (2007). Control of ionization processes in high band gap materials via tailored femtosecond pulses, *Opt. Express,* **26**, 17855–17862.

40. Englert, L., Wollenhaupt, M., Haag, L., Sarpe-Tudoran, C., Rethfeld, B., and Baumert, T. (2008). Material processing of dielectrics with temporally asymmetric shaped femtosecond laser pulses on the nanometer scale, *Appl. Phys. A,* **92,** 749–753.

41. Meshulach, D., Yelin, D., and Silberberg, Y. (1997). Adaptive ultrashort pulse compression and shaping, *Opt. Commun.,* **138**, 345–348.

42. Yelin, D., Meshulach, D., and Silberberg, Y., (1997). Adaptive femtosecond pulse compression, *Opt. Lett.,* **22**, 1793–1795.

43. Baumert, T., Brixner, T., Seyfried, V., Strehle, M., and Gerber, G. (1997). Femtosecond pulse shaping by an evolutionary algorithm with feedback, *Appl. Phys. B,* **65**, 779–782.

44. Omenetto, F. G., Taylor, A. J., Moores, M. D., and Reitze, D. H. (2001). Adaptive control of femtosecond pulse propagation in optical fibers, *Opt. Lett.,* **26**, 938–941.

45. Siegner, U., Haiml, M., Kunde, J., and Keller, U. (2002). Adaptive pulse compression by two-photon absorption in semiconductors, *Opt. Lett.,* **27**, 315–317.

46. Assion, A., Baumert, T., Bergt, M., Brixner, T., Kiefer, B., Seyfried, V., Strehle, M., and Gerber, G. (1998). Control of chemical reactions by feedback-optimized phase-shaped femtosecond laser pulses, *Science,* **30**, 919–922.

47. Shverdin, M. Y., Goda, S. N., Yin, G. Y., and Harris, S. E. (2006). Coherent control of laser-induced breakdown, *Opt. Lett.,* **31**, 1331–1333.

48. Dachraoui, H., and Husinsky, W. (2006). Thresholds of plasma formation in silicon identified by optimizing the ablation laser pulse form, *Phys. Rev. Lett.,* **97**, 107601.

49. Heck, G., Sloss, J., and Levis, R. J. (2006). Adaptive control of the spatial position of white light filaments in an aqueous solution, *Opt. Commun.,* **259**, 216–222.

50. Ackermann, R., Salmon, E., Lascoux, N., Kasparian, J., Rohwetter, P., Stelmaszczyk, K., Li, S., Lindinger, A., Wöste, L., Béjot, P., Bonacina, L., and Wolf, J. -P. (2006). Optimal control of filamentation in air, *Appl. Phys. Lett.*, **89,** 171117.

51. Siedschlaga, C., Shirb, O. M., Bäckb, T., and Vrakkinga, M. J. J. (2006). Evolutionary algorithms in the optimization of dynamic molecular alignment, *Optics Commun.*, **264**, 511–518.

52. Mermillod-Blondin, A., Mauclair, C., Rosenfeld, A., Bonse, J., Hertel, I. V., Audouard, E., and Stoian, R. (2008). Size correction in ultrafast laser processing of fused silica by temporal pulse shaping, *Appl. Phys. Lett.*, **93**, 021921.

53. Mermillod-Blondin, A., Burakov, I. M., Meshcheryakov, Y. P., Bulgakova, N. M., Audouard, E., Rosenfeld, A., Husakou, A., Hertel, I. V., and Stoian, R. (2008). Flipping the sign of refractive index changes in ultrafast and temporally shaped laser-irradiated borosilicate crown optical glass at high repetition rates, *Phys. Rev. B,* **77,** 104205.

54. Takita, A., Hayasaki, Y., and Nishida, N. (2006). Femtosecond laser processing system with target tracking feature, *J. Laser Micro/Nanoengineering,* **1**, 288–291.

55. Cheng, Y., Sugioka, K., Midorikawa, K., Masuda, M., Toyoda, K., Kawachi, M., and Shihoyama, K. (2003). Control of the cross-sectional shape of a hollow microchannel embedded in photostructurable glass by use of a femtosecond laser, *Opt. Lett.*, **28**, 55–57.

56. Ams, M., Marshall, G. D., Spence D. J., and Withford, M. J. (2005). Slit beam shaping method for femtosecond laser direct-write fabrication of symmetric waveguides in bulk glasses, *Opt. Express,* **13**, 5676–5681.

57. Cerullo, G., Osellame, R., Taccheo, S., Marangoni, M., Polli, D., Ramponi, R. Laporta, P., and De Silvestri, S. (2002). Femtosecond micromachining of symmetric waveguides at 1.5 mm by astigmatic beam focusing, *Opt Lett.,* **27**, 1938–1940.

58. Merano, M., Boyer, G., Trisorio, A., Chériaux, G., and Mourou, G. (2007). Superresolved femtosecond laser ablation, *Opt. Lett.,* **32**, 2239–2242.

59. Kawamura, K., Ogawa, T., Sarukura, N., Hirano, M., and Hosono, H. (2000). Fabrication of surface relief gratings on transparent dielectric materials by two-beam holographic method using infrared femtosecond laser pulses, *Appl. Phys., B.* **71**, 119–121.

60. Wu, L., Zhong, Y., Chan, C. T., Wong, K. S., and Wang, G. P. (2005). Fabrication of large area two- and three-dimensional polymer photonic crystals using single refracting prism holographic lithography, *Appl. Phys., Lett.,* **86**, 241102.

61. Lei, M., Yao, B., and Rupp, R. A. (2006). Structuring by multi-beam interference using symmetric pyramids, *Opt. Express,* **14**, 5803–5811.

62. Bloomstein, T. M., Marchant, M. F., Deneault, S., Hardy, D. E., and Rothschild, M. (2006). 22-nm immersion interference lithography, *Opt. Express,* **14**, 6434–6443.

63. Matsuo, S., Juodkazis, S., and Misawa, H. (2005). Femtosecond laser microfabrication of periodic structures using a microlens array, *Appl. Phys. A,* **80**, 683–685.

64. Kato, J., Takeyasu, N., Adachi, Y., Sun, H.–B., and Kawata, S. (2005). Multiple-spot parallel processing for laser micronanofabrication, *Appl. Phys. Lett.,* **86,** 044–102..

65. Shishido, A., Diviliansky, I. B., Khoo, I. C., Mayer, T. S., Nishimura, S., Egan, G. L., and Mallouk, T. E. (2001). Direct fabrication of two-dimensional titania arrays using interference photolithography, *Appl. Phys. Lett.,* **79**, 3332 –3334.

66. Klein-Wiele, J.-H., and Simon, P. (2003). Fabrication of periodic nanostructures by phase-controlled multiple-beam interference, *Appl. Phys. Lett.,* **83**, 4707–4709.

67. Amako, J., Nagasaka, K., and Kazuhiro, N. (2002). Chromatic-distortion compensation in splitting and focusing of femtosecond pulses by use of a pair of diffractive optical elements, *Opt. Lett.,* **27**, 969–971.

68. Mínguez-Vega, G., Lancis, J., Caraquitena, J., Torres-Company, V., and Andrés, P. (2006). High spatiotemporal resolution in multifocal processing with femtosecond laser pulses, *Opt. Lett.,* **31**, 2631–2633.

69. Mínguez-Vega, G., Tajahuerce, E., Fernández-Alonso, M., Climent, V., Lancis, J., Caraquitena, J., and Andrés, P. (2007). Dispersion-compensated beam-splitting of femtosecond light pulses: Wave optics analysis, *Opt. Express,* **15**, 278–288.

70. Martínez-León, L., Clemente, P., Tajahuerce, E., Mínguez-Vega, G., Mendoza-Yero, O., Fernández-Alonso, M., Lancis, J., Climent, V., and Andrés, P. (2009). Spatial-chirp compensation in dynamical holograms reconstructed with ultrafast lasers, *Appl. Phys. Lett.,* **94**, 011104.

71. Amako, J., Sawaki, D., and Fujii, E. (2003). Microstructuring transparent materials by use of nondiffracting ultrashort pulse beams generated by diffractive optics, *J. Opt. Soc. Am. B,* **20**, 2562–2568.

72. Courvoisier, F., Lacourt, P.-A., Jacquot, M., Bhuyan, M. K., Furfaro, L., and Dudley, J. M. (2009). Surface nanoprocessing with nondiffracting femtosecond Bessel beams, *Opt. Lett.,* **34**, 3163–3165.

73. Bhuyan, M. K., Courvoisier, F., Lacourt, P.-A., Jacquot, M., Furfaro, L., Withford, M. J., and Dudley, J., M. (2010). High aspect ratio taper-free microchannel fabrication using femtosecond Bessel beams, *Opt. Express,* **18**, 566–574.

74. Bhuyan, M. K., Courvoisier, F., Lacourt, P. A., Jacquot, M., Salut, R., Furfaro, L., and Dudley, J. M. (2010). High aspect ratio nanochannel machining using single shot femtosecond Bessel beams, *Appl. Phys. Lett.,* **97**, 81–102.

75. Chaen, K., Takahashi, H., Hasegawa, S., and Hayasaki, Y. (2007). Display method with compensation of the spatial frequency response of a liquid crystal spatial light modulator for holographic femtosecond laser processing, *Optics Commun.,* **280**, 165–172.

76. Takahashi, H., Hasegawa, S., and Hayasaki, Y. (2007). Holographic femtosecond laser processing using optimal-rotation-angle method with compensation of spatial frequency response of liquid crystal spatial light modulator, *Appl. Opt.,* **46**, 5917–5923.

77. Thomson, R. R., Bockelt, A. S., Ramsay, E., Beecher, S., Greenaway, A. H., Kar, A. K., and Reid, D. T. (2008). Shaping ultrafast laser inscribed optical waveguides using a deformable mirror, *Opt. Express,* **16**, 12786–12793.

78. Maznev, A. A., Crimmins, T. F., and Nelson, K. A. (1998). How to make femtosecond pulses overlap, *Opt. Lett.,* **23**, 1378–1380.

79. Sun, H. B., Xu, Y., Juodkazis, S., Sun, K., Watanabe, M., Matsuo, S., Misawa H., and Nishii, J. (2001). Arbitrary-lattice photonic crystals created by multiphoton microfabrication, *Opt. Lett.,* **26**, 325–327.

80. Kondo, T., Matsuo, S., Juodkazis, S., and Misawa, H. (2001). Femtosecond laser interference technique with diffractive beam splitter for fabrication of three-dimensional photonic crystals, *Appl. Phys. Lett.,* **79**, 725–727.

81. Nakata, Y., Okada, T., and Maeda, M. (2002). Fabrication of dot matrix, comb, and nanowire structures using laser ablation by interfered femtosecond laser beams, *Appl. Phys. Lett.,* **81**, 4239–4242.

82. Y. Li, Watanabe, W., Yamada, K., Shinagawa, T., Itoh, K., Nishii, J., and Jiang, Y. (2002). Holographic fabrication of multiple layers of grating inside soda-lime glass with femtosecond laser pulses, *Appl. Phys. Lett.,* **80**, 1508–1510.

83. Venkatakrishnan, K., Sivakumar, N. R., Hee, C. W., Tan, B., Liang, W. L., and Gan, G. K. (2003). Direct fabrication of surface-relief grating by interferometric technique using femtosecond laser, *Phys. A,* **77**, 959–963.

84. Kuroiwa, Y., Takeshima, N., Narita, Y., Tanaka, S., and Hirao, K. (2004). Arbitrary micropatterning method in femtosecond laser microprocessing using diffractive optical elements, *Opt. Express,* **12**, 1908–1915.

85. Hayasaki, Y., Sugimoto, T., Takita, A., and Nishida, N. (2005). Variable holographic femtosecond laser processing by use of a spatial light modulator, *Appl. Phys. Lett.,* **87**, 031101.

86. Lee, G. J., Lee, Y. P., Cheong, H., Yoon, C. S., Oh, C. H., Kim, E. K., Son, Y. D., and Jang, J. (2006). Femtosecond-laser surface structuring of amorphous and crystalline silicon, *J. Korean Phys. Soc.,* **48**, 1268–1272.

87. Kelemen, L., Valkai, S., and Ormos, P., (2007). Parallel photopolymerisation with complex light patterns generated by diffractive optical elements, *Opt. Express,* **15**, 14488–14497.

88. Takahashi, H., Hasegawa, S., Takita, A., and Hayasaki, Y. (2008). Sparse-exposure technique in holographic two-photon polymerization, *Opt. Express,* **16**, 16592–16599.

89. Kuang, Z., Perrie, W., Leach, J., Sharp, M. Edwardson, S. P., Padgett, M., Dearden, G., and Watkins, K. G. (2008). High throughput diffractive multi-beam femtosecond laser processing using a spatial light modulator, *Appl. Surf. Sci.,* **255**, 2284–2289.

90. Yamaji, M., Kawashima, H., Suzuki, J., and Tanaka, S. (2008). Three dimensional micromachining inside a transparent material by single pulse femtosecond laser through a hologram, *Appl. Phys. Lett.,* **93**, 041116.

91. Kuang, Z., Liu, D., Perrie, W., Edwardson, S., Sharp, M., Fearon, E., Dearden, G., and Watkins, K. (2009). Fast parallel diffractive multi-beam femtosecond laser surface micro-structuring, *Appl. Surf. Sci.,* **255**, 6582–6588.

92. Jenness, N. J., Hill, R. T., Hucknall, A., Chilkoti, A., Clark, R. L. (2010). A versatile diffractive maskless lithography for single-shot and serial microfabrication, *Opt. Express,* **18**, 11754–11762.

93. Obata, K., Koch, J., Hinze, U., and Chichkov, B. N. (2010). Multi-focus two-photon polymerization technique based on individually controlled phase modulation, *Opt. Express,* **18**, 17193–17200.

94. Mauclair, C., Cheng, G., Huot, N., Audouard, E., Rosenfeld, A., Hertel, I. V., and Stoian, R. (2009). Dynamic ultrafast laser spatial tailoring for parallel micromachining of photonic devices in transparent materials, *Opt. Express,* **17**, 3531–3542.

95. Sakakura, M., Sawano, T., Shimotsuma, Y., Miura, K., and Hirao, K. (2009). Parallel drawing of bent optical waveguides by using a spatial light modulator, *Jpn J. Appl. Phys.*, **48**, 126507.
96. Sakakura, M., Sawano, T., Shimotsuma, Y., Miura, K., and Hirao, K. (2010). Fabrication of three-dimensional 1 × 4 splitter waveguides inside a glass substrate with spatially phase modulated laser beam, *Opt. Express*, **18**, 12136–12143.
97. Liu, D., Kuang, Z., Perrie, W., Scully, P. J., Baum, A., Edwardson, S. P., Fearon, E., Dearden, G., and Watkins, K. G. (2010). High-speed uniform parallel 3D refractive index micro-structuring of poly(methyl methacrylate) for volume phase gratings, *Appl. Phys. B*, **101**, 817–823.
98. Jesacher, A., and Booth, M. J. (2010). Parallel direct laser writing in three dimensions with spatially dependent aberration correction, *Opt. Express*, **18**, 21090–21099.
99. Imamoto, H., Kanehira, S., Wang, X., Kametani, K., Sakakura, M., Shimotsuma, Y., Miura, K., and Hirao, K. (2011). Fabrication and characterization of silicon antireflection structures for infrared rays using a femtosecond laser, *Opt. Lett.*, **36**, 1176–1178.
100. Sanner, N., Huot, N., Audouard, E., Larat, C., Huignard. J. P., and Loiseaux, B. (2005). Programmable focal spot shaping of amplified femtosecond laser pulses, *Opt. Lett.*, **30**, 1479–1481.
101. Hasegawa, S., and Hayasaki, Y. (2007). Holographic femtosecond laser processing with multiplexed phase Fresnel lenses displayed on the liquid crystal spatial light modulator, *Opt. Rev.*, **14**, 208–213.
102. Hasegawa, S., and Hayasaki, Y. (2009). Adaptive optimization of hologram in holographic femtosecond laser processing system, *Opt. Lett.*, **34**, 22–24.
103. Hasegawa, S., and Hayasaki, Y. (2009). Performance analysis of adaptive optimization of multiplexed phase Fresnel lenses, *Jpn. J. Appl. Phys.*, **48**, 09LE03.
104. Hasegawa, S., and Hayasaki, Y. (2011). Second harmonic optimization of computer-generated hologram, *Opt. Lett.*, **36**, 2943–2945.
105. Stone, T., and George, N. (1988). Hybrid diffractive-refractive lenses and achromats, *Appl. Opt.*, **27**, 2960–2971.
106. Minguez-Vega, G. Lancis, J. Caraquitena, J., and Torres-Company, V. (2006). High spatiotemporal resolution in multifocal processing with femtosecond laser pulses, *Opt. Lett.*, **31**, 2631–2633.
107. Oron, D., Tal, E., and Silberberg, Y. (2005). Scanningless depth-resolved microscopy, *Opt. Express*, **13**, 1468–1476.

108. Zhu, G. van Howe, J. Durst, M. Zipfel, W., and Xu, C. (2005). Simultaneous spatial and temporal focusing of femtosecond pulses, *Opt. Express,* **13**, 2153–2159.

109. Tal, E., Oron, D., and Silberberg, Y. (2005). Improved depth resolution in video-rate line-scanning multiphoton microscopy using temporal focusing, *Opt. Lett.,* **30**, 1686–1688.

110. Oron, D., and Silberberg, Y. (2005). Spatiotemporal coherent control using shaped, temporally focused pulses, *Opt. Express,* **13**, 9903–9908.

111. Durst, M. E., Zhu, G., and Xu, C. (2006). Simultaneous spatial and temporal focusing for axial scanning, *Opt. Express,* **14**, 12243–12254.

112. Vaziri, A., and Shank, C. V. (2010). Ultrafast widefield optical sectioning microscopy by multifocal temporal focusing, *Opt. Express,* **18**, 19645–19655.

113. He, F., Xu, H., Cheng, Y., Ni, J., Xiong, H., Xu, Z., Sugioka, K., and Midorikawa, K. (2010). Fabrication of microfluidic channels with a circular cross section using spatiotemporally focused femtosecond laser pulses, *Opt. Lett.,* **35**, 1106–1108.

114. Kimura, K., Hasegawa, S., and Hayasaki, Y. (2010). Diffractive spatiotemporal lens with wavelength dispersion compensation, *Opt. Lett.,* **35**, 139–141.

115. Papagiakoumou, E., de Sars, V., Emiliani, V., and Oron, D. (2009). Temporal focusing with spatially modulated excitation, *Opt. Express,* **17**, 5391–5401.

116. Papagiakoumou, E., de Sars, V., Oron, D., and Emiliani, V. (2008). Patterned two-photon illumination by spatiotemporal shaping of ultrashort pulses, *Opt. Express,* **16,** 22039–22047.

117. Piestun, R., and Miller, D. A. B. (2001). Spatiotemporal control of ultrashort optical pulses by refractive–diffractive–dispersive structured optical elements, *Opt. Lett.,* **26**, 1373–1375.

118. Sato, M., Suzuki, T., and Misawa, K. (2009). Interferometric polarization pulse shaper stabilized by an external laser diode for arbitrary vector field shaping,*Rev. Sci. Instrum.,* **80**, 123107.

Chapter 5

Surface Patterning, Drilling, and Cutting

Andreas Ostendorf and Benjamin Schöps
Laseranwendungstechnik, Ruhr Universität Bochum,
Universitätsstraße 150 44780 Bochum, Germany
andreas.ostendorf@rub.de

Surface patterning is one of the most important applications of ultrashort laser pulses used in micromachining. The transition from bulk machining to surface patterning will be described in more detail. In particular, absorption, reflection, and heat conduction will be analyzed with respect to ablation regimes. Characteristic effects such as beam filamentation and plasma interaction specific for ultrashort laser pulses will be explained. Surface patterning has become an emerging technology in the field of selective ablation of thin-film layers, which are used in modern industries such as flat-panel displays and integrated solar cell manufacturing. Therefore, part of this chapter focuses on the wavelength- and intensity-dependent deposition of the laser energy in a small surface layer without influencing the underlying substrate. Also the optical setup plays an important role not only in the accuracy but also in throughput. Differences between focusing optics and image or mask projection will lead to different results. Bulk machining can be used,

Ultrafast Laser Processing: From Micro- to Nanoscale
Edited by Koji Sugioka and Ya Cheng

ISBN 978-981-4267-33-5 (Hardcover), 978-981-4303-69-9 (eBook)
www.panstanford.com

on the other hand, to generate more complex three-dimensional structures. Finally, some details about the beam guiding and forming will be provided, which are specific for pico- and femtosecond laser pulse ablation.

In the second part, more information on the laser drilling process will be provided. Here, the different drilling strategies will be compared with respect to their specific application fields. For drilling with high aspect ratios, an effect of the polarization can be observed. It will be shown how this effect can be used to drill perfect circular or shaped holes. This technology requires a special optical setup that is given in more detail here. Although laser cutting is regarded quite important it can be described as an extension of the through-hole drilling process. Basically, some additional parameters will be taken into account, such as the support by a process gas. The characterization is focused on the edge roughness and the heat-affected zone.

5.1 Introduction

Microprocessing with short laser pulses, which are generated by switching the resonator quality (q-switching), has been investigated already shortly after the invention of the laser in 1960. Numerous textbooks are available describing in detail the different processes and applications [7, 31, 35, 40]. Also comprehensive models, simulating the thermal behavior and the material´s response are available, which illustrate the influence of different parameters on the ablation process [2] or provide a deeper understanding of material removal processes in metals [39]. However, it took some decades until extremely short pulses were investigated in material processing application. This is even more surprising since the main principle of generating pico- and femtosecond laser pulses was invented in the 1960s, too. In 1985, Mourou and coworkers developed the principle of Chirp-Pulse-Amplification (CPA) [36], which allowed the upscaling of pulse energies now sufficiently high to ablate materials. This was an important milestone that triggered many further research groups to develop material ablation processes on the basis of femtosecond laser pulses. Mourou´s group, among others, early revealed that the process characteristics are different compared with longer pulses, e.g., the modification thresholds are more pronounced and are independent on the pulse duration in the very short pulse regime [37, 8]. The unique properties of the short pulses have shown to be very attractive in microfabrication since—although still a thermal process—the negative thermal load to the surrounding material could be minimized compared with all other laser processes. Also, the high intensities allowed the simultaneous absorption of more than one single photon to overcome the bandgap in insulating and semiconducting materials, which has been carried out before only by ultraviolet laser photons. Many research groups in the meantime have shown that multiphoton absorption allows the structuring of nearly all kinds of material with the same precision [30, 21]. This property has also been unique compared with all other laser processes. Even more appealing is the fact that all these processes became possible by an infrared laser source since these types of lasers are often more reliable and less expensive than ultraviolet laser systems, also with respect to the optical components. Materials processing in the form of surface and thin-film patterning as well as bulk ablation for drilling and cutting are just at the edge of being transferred to industrial applications. In order to choose the right laser source with the best available parameters, a sound understanding of the relations between the incident pulses and the material properties is necessary and will be highlighted in this chapter.

5.2 Surface Patterning

5.2.1 Motivation

Surface Patterning by laser radiation has become a topic of growing interest. Although lasers suffer from the fact of being a pointwise process by nature, and the efficient processing of surface areas by serial processing takes rather long times, people have started to use this process in an increasing number of applications. Besides the flexibility in terms of materials, surface topographies, multidimensional structuring, and the absence of chemicals beam manipulation systems have been developed, which allow the parallel and economic patterning. Application areas of surface patterning are the selective ablation of thin films, e.g., in solar cells or flat panel displays; ablation of self-assembled monolayers (SAMs); tuning of the wettability; biofunctionalization of surfaces; nanostructuring, e.g., for photonic devices, improving tribological or optical properties; increase of effective surface areas, e.g., in batteries or for catalytic surfaces; and microfluidics or improving the susceptibility in analytics, e.g., tailored surfaces for surface-enhanced Raman spectroscopy (SERS). In the field of biofunctionalization, for example, the macroscopic structures determine the interaction with tissue, and the microscopic topography is responsible for the orientation effect of the cell's cytoskeleton, whereas nanostructures interact with the subcellular structures, i.e., protein interaction and cell signaling and adhesion. Thus, for many applications, hierarchical structuring covering orders of magnitude is important, which can basically be fulfilled by laser pulses.

5.2.2 Characteristics of Thin Film Ablation

Structuring of thin films by ultrashort laser pulses can be regarded as another subclass in the area of surface patterning. Micromachining of thin films finds many applications in different sectors e.g., processing and repairing of photomasks for the semiconductor lithography, structuring of thin-film transistors and pixels in the flat panel display industry or the integrated processing of serial interconnections in large area coated thin-film solar modules. Figure 5.1 shows the different processing steps (P1, P2, and P3) that are applied in thin-film solar modules (a:Si cells, CIGS cells) today. In thin-film processing, femtosecond laser pulses are perfect tools to achieve superior results compared with alternative lasers or even other processes due to the specific interaction with matter. Especially the well-defined ablation or modification thresholds in combination with wavelength and single or multiphoton absorption can lead to the selective removal of layers without disturbing the underlying substrate or other layers. Also the ultrashort pulse duration provides some advantages in terms of heat conduction. Thermal effects in the surrounding areas can be minimized, which is an important requirement both in the vertical and in the lateral direction.

Most of the films cover a thickness range of a few tens to a few hundreds of nanometers. Thus, for many materials, the thickness is in the same range compared with the optical penetration depth

$$l_{\text{opt}} = \frac{\lambda}{4\pi\kappa} \tag{5.1}$$

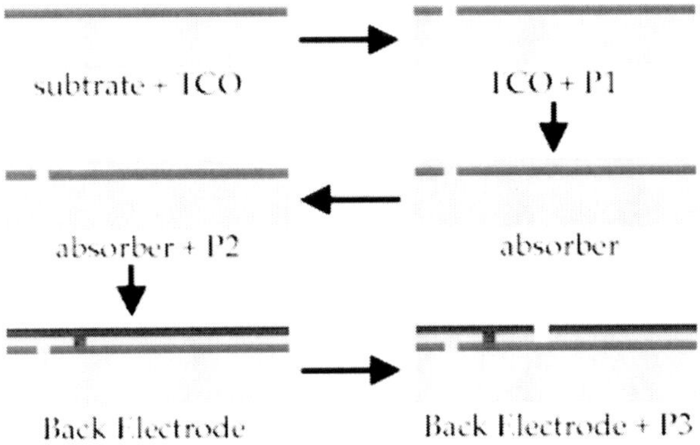

Figure 5.1 Process requirements in the integrated serial connection of thin-film solar cells.

with λ is the wavelength and κ the extinction coefficient, and also it is similar to the electron–phonon coupling length. The latter is important for pulses in the ultrafast regime. When applying laser pulses in the nanosecond regime the thermal penetration length is represented by the well know relation

$$l_{\text{th}} = \sqrt{Dt_{\text{p}}} \quad (5.2)$$

where D is the heat diffusivity of the material and t_{p} the laser pulse duration. This is a consequence of the heat diffusion equation in the equilibrium state. In the case of femtosecond laser pulses, the electrons and the lattice cannot be described in an equilibrium state, i.e., the temperatures can be significantly different for the two systems and the coupling factor, represented by the electron–lattice or electron–phonon coupling time τ_{e}, generally describes the interaction [22]. When the initially free electrons absorb the laser pulse energy usually within a few femtoseconds, they increase their kinetic energy. As a consequence, they can ballistically diffuse deeper into the material and—dependent on the electron–lattice coupling properties—can gradually transfer their energy to the lattice by phonon excitation. This mechanism can be described similarly by

$$l_{\text{th}} = \sqrt{D\tau_{\text{e}}} \quad (5.3)$$

If the film thickness is smaller than the characteristic penetration length, the electrons rapidly reach the film-substrate interface, where they transfer their energy to the lattice ions. In this case, the damage threshold is significantly lowered. Drawing the fluence damage threshold for the ablation of metallic films on insulating substrates versus the film thickness will result in two regimes. Krueger [20] carried out investigations for the ablation of gold films on BK7-substrates when using a CPA amplified Ti-Sapphire laser (Fig. 5.2). For extremely thin films, the ablation threshold scales linearly with the thickness and from a certain characteristic value for the thickness the threshold stays constant.

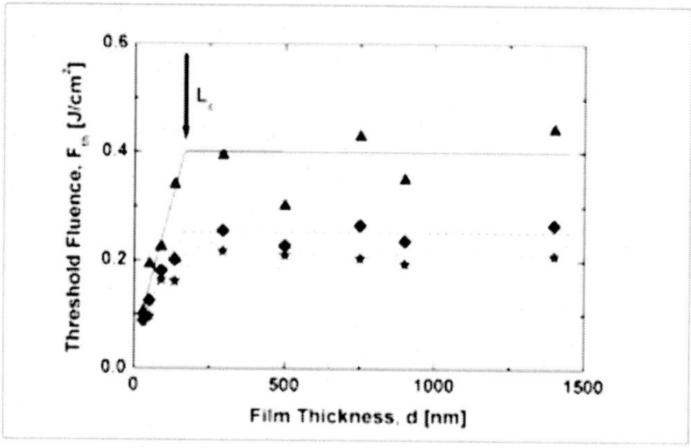

Figure 5.2 Ablation threshold vs. gold film thickness (stars: N = 10000; diamond: N = 1000; triangle: N = 100 pulses; substrate: BK7; wavelength 800 nm, t_p = 28 fs) [20].

The characteristic length L_c ≈180 nm in Fig. 5.2 represents an important parameter in the absorption process. It has to be noted that the optical penetration depth for the above-mentioned constellation can be given with 12 nm, i.e., an order of magnitude less than L_c. In this case, L_c represents the electron–lattice coupling length. Basically, a number of electrons can ballistically travel up to L_c in depth. For thicknesses above L_c the film behaves like a bulk material [9].

5.2.3 Patterning with High Repetition Rate Lasers

There have been many attempts to gain speed in laser surface machining processes. The most promising approach is the use of high-repetition-rate lasers. Besides amplified laser systems also high-repetition-rate oscillators can be used to structure thin films due to the low damage threshold fluence. For these high pulse frequencies, however, it is important to estimate the heat accumulation effect. The authors in [10] have calculated the heat accumulation in the case of a succession of laser pulses under some simplifications. They assumed that the maximum temperature $T_{\max}$ is reached at the end of a single pulse of duration t_p and the temperature at the beginning of the following laser pulse after a cooling phase during the time interval t_{pp} between two pulses is calculated by

$$T_{\min} = T_{\max}\sqrt{t_p / t_{pp}}\,. \tag{5.4}$$

The factor between the previous maximum and the following minimum therefore is described by

$$\alpha = \sqrt{t_p / t_{pp}} \tag{5.5}$$

i.e., for the nth pulse assuming a linear heating process $(T_{\min})_n = \alpha\ (T_{\max})_n$. The average surface temperature can then be calculated by

$$\tilde{T}_n = \frac{1}{n(t_p + t_{pp})}\int_0^{n(t_p+t_{pp})} T(0,t)dt = 2\alpha\frac{\left(1-\frac{2}{3}\alpha\right)}{(1+\alpha^2)}\frac{1}{n}\sum_{i=1}^{n}\tilde{T}_{\max,i} \tag{5.6}$$

For $n >> 1$ and $\alpha << 1$, the temperature can be approximated according to [10] as follows:

$$\tilde{T}_n \cong 2\alpha T_{\max,1} = 2T_{\max,1}\sqrt{t_p / t_{pp}} \tag{5.7}$$

Even for repetition rates in the range of 100 MHz the ratio α for femtosecond laser pulses is in the range of 1% or below, i.e., the target cools down between the pulses leading to only a small average temperature increase compared with the temperature rise during the pulse itself.

High-repetition-rate experiments, as described in [38], have been performed on a 300 nm gold film, which has been sputtered on top of a 250 μm-thick silicon substrate. The influence of the fluence can be seen in Fig. 5.3 for a feed rate of 4 mm/s, pulses of 13 MHz repetition rate at 214 fs, a wavelength of 515 nm, helium as a process gas and a theoretical spot diameter of 1 μm. Obviously, for low fluencies, extremely narrow cuts can be generated. The smallest widths in Fig. 5.3 are in the order of 110 nm with a heat-affected zone of about 800 nm. This demonstrates, that by adjusting the fluence, high-repetition-rate femtosecond lasers can be a very productive tool in surface patterning.

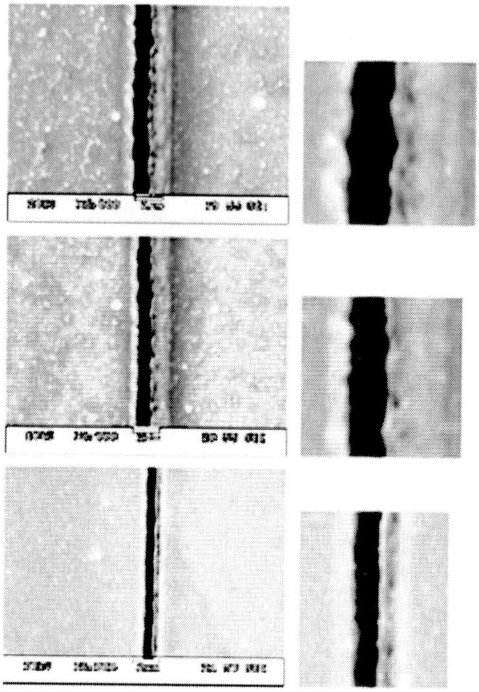

Figure 5.3 SEM photos with different fluencies at a repetition rate of 13 MHz: (a) 3.340 J/cm^2, (b) 2.672 J/cm^2, (c) 2.042 J/cm^2 [38].

5.2.4 Beam Focusing and Forming

Many applications require specific patterns to be generated on the surface or for the ablation of a thin film. Small spot sizes in the range of a few ten μm can be generated by means of a conventional lens or if smaller features are necessary sophisticated focusing objectives can be used. More complex structures are usually manufactured by a mask projection system where an image of the mask is transferred by an optical system down to the workpiece. From the scientific point of view, there are more possibilities such as near field optics and diffractive optics, but these have not yet been used in many applications and, hence, will not be described in the following. When using a simple refractive lens, one has to compensate for the pulse front distortion and the group velocity dispersion, which is characteristic for ultrashort laser pulses and which can lead to significant temporal pulse extension. The use of achromatic lenses can reduce this effect drastically. Thus, the design of the appropriate optical system is crucial for ultrashort pulse laser ablation. To avoid dispersion effects, reflective optics can be used, e.g., a Schwarzschild objective, which consists of two spherical mirrors and which projects a real image into infinity.

In Fig. 5.4, it is obvious that the reflective Schwarzschild objective nearly perfectly transmits a Gaussian beam. This can be advantageous when machining extremely small structures. However, there are some drawbacks due to the characteristics of the intensity on the edges of the profile. For ultrashort laser pulses almost all materials exhibit a threshold for melting and a higher threshold for evaporation. Between these two thresholds, a clear melting zone exists. Applying this concept to the precise ablation by Gaussian beams one can see a significant amount of melt can occur around the central evaporation zone (see Fig. 5.5).

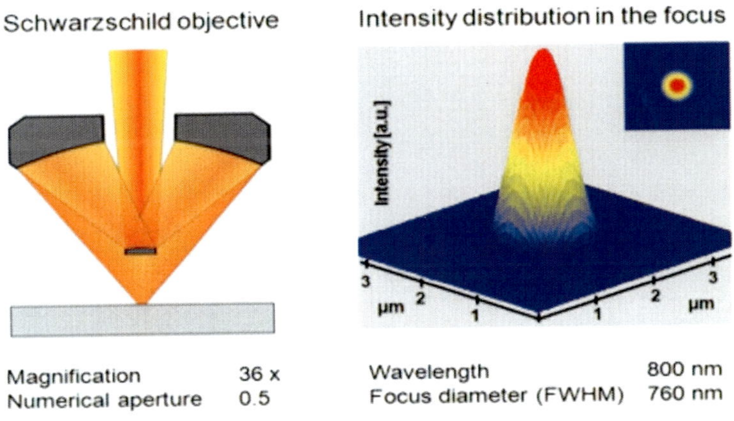

Figure 5.4 Principle of a Schwarzschild objective in focusing Gaussian beams with only very small distortion [19].

To overcome this problem, mask projection can be used. Besides the possibility to demagnify the mask down to a factor 100 or even more, homogeneous flat top profiles can be generated on the workpiece. However, one has to be careful with the optical setup since image projection often makes use of intermediate focusing. Here, intensities can reach levels where self-focusing and plasma generation can take place and disturb the projection path. Due to diffraction and interference effects a homogeneously illuminated

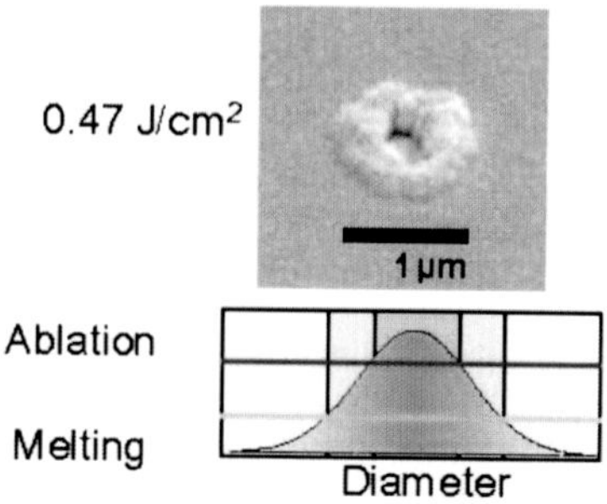

Figure 5.5 SEM picture of surface modification of a 100 nm chrome layer generated by 1000 laser pulses at a fluence of 0.38 J/cm^2 [19].

mask can result in a more or less flat-top profile in the image plane. A characteristic of a flat-top profile is also the steeper edges of the profile. Returning to the concept of melting and evaporation threshold, one can easily derive from Fig. 5.6 that the melting zone will be decreased to a minimum and the ablation area becomes larger and clearer. Therefore, for ultraprecise structuring the rims have to be taken into account and mask projection obviously reduces the negative effect of a molten zone.

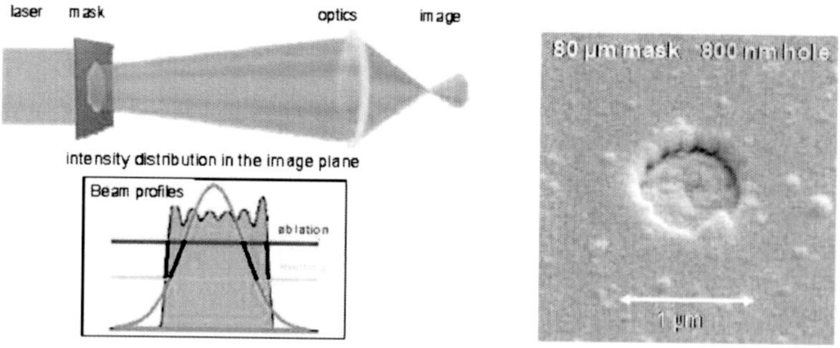

Figure 5.6 Maskprojection [19].

5.2.5 Applications

5.2.5.1 Thin-film solar modules

Thin-film solar cells are possible candidates for the next generation of photovoltaic modules due to their simple and cost-effective production. As already depicted in Fig. 5.1 an important processing step is to subdivide the large coated area into smaller cells and to seriously interconnect them. Lasers can be used to selectively ablate the material layers at different positions and in combination with subsequent coating steps to guarantee the reliable interconnection without shortcuts. Ultrashort pulse lasers are interesting tools because their energy can be locally deposited in a very small dimension without influencing environmental material. The substrates for the thin-film modules can either be glass or polymer foils. Glass is mostly used for inorganic cells such as a:Si and CIGS cells because of the high temperatures of the coating process. Recently there have also been attempts to use flexible stainless steel foils or even polymer foils to make these types of cells. Organic polymer cells that are mostly based on low-temperature solution processing are often based on flexible polymer substrates such as PET foils. Nanosecond pulsed lasers are intensively investigated to ablate thin layers from a glass substrate and many of these laser sources are already used in industry. More challenging, however, is the selective ablation of thin films from PET foils because of the soft material properties of polymers compared with glass. Light usually hits the solar cell from the substrate side, i.e., a transparent electrode is often directly coated on the substrate and has to be removed selectively (Fig. 5.7) in the P1 step.

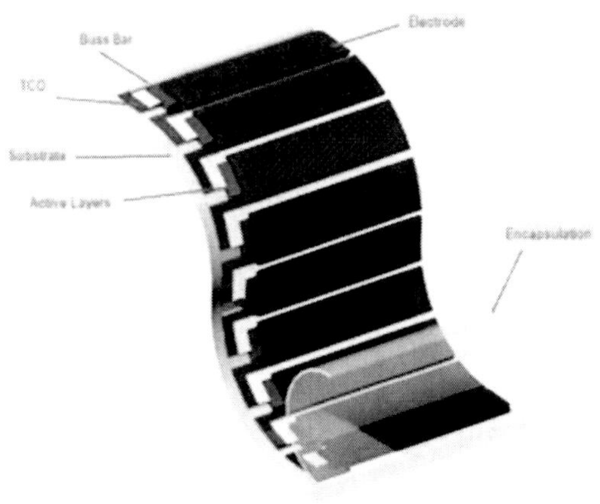

Figure 5.7 Typical structure of an organic solar cell with integrated serious connection.

As an electrode material, often indium-tin-oxide (ITO) with a thickness in the range of 100 nm is used, which is an inorganic material exhibiting a high melting and evaporation temperature. In Fig. 5.8, some ablation experiments using 10 ps laser pulses at different wavelengths are shown for frontside ablation, i.e., the laser beam hits the target from the ITO layer side. It has to be noted that the film is much thinner than the optical penetration, and the laser pulse thus propagates deeper into the material. Obviously, for the glass substrate, thermal effects can be observed and the process can generally be described by the characteristics described above. For the PET substrate, however, the process behaves differently. Due to the smaller stiffness of the substrate material (acoustic impedance mismatch for shockwaves) and the smaller adhesion forces between the substrate and the ITO layer (differences in the thermal expansion coefficients), thermomechanical processes dominate the ablation. Evaporation becomes less important; rather delamination and spallation are the driving forces for the ablation. As a consequence, it is important to exactly know the substrate properties and how the interaction with short laser pulses take place. The resulting processes can be completely different when changing the parameters.

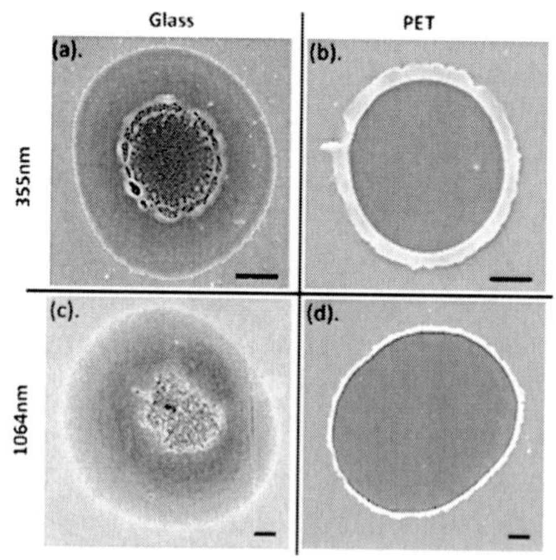

Figure 5.8 Ablation experiments (100 nm ITO layer on glass and PET foil), (a–c) were on glass substrate. (b–d) were on PET substrate. The scale bar is 2 μm. The fluencies used were: (a). 8.7Fth; (b). 3.5Fth; (c). 3.4Fth; (d). 8.9Fth. Fth is the removal threshold: 0.48 J/cm^2 for glass and 355 nm, 0.57 J/cm^2 for glass and 1064 nm, 0.14 J/cm^2 for PET and 355 nm, 0.4 J/cm^2 for PET and 1064 nm).

5.2.5.2 Self-assembled monolayers

Self-assembled monolayers, as shown in Fig. 5.9, are ultrathin organic films, which provide a variety of opportunities to tailor the wettability and to control the chemical reactivity and resistance [23]. Therefore, SAMs are often applied as ultrathin resists or functional coatings. Femtosecond laser patterning of SAMs is a promising approach to further explore the potential of SAMs. On one hand, femtosecond-laser processing can be used as a maskless noncontact direct-writing technique with the potential for nonlinear processing providing a means for sub-wavelength patterning. On the other hand, because of their ultrathin nature, SAMs can be processed with single laser pulses allowing high feed rates. The monomolecular thickness also allows for well-defined irradiation and burr-free patterning of the coating and avoids bubble and particle formation. The most common SAM systems are by today silane-based monolayers on quartz glass or surface-oxidized silicon substrates and thiol-based monolayers on gold-coated silicon substrates [24].

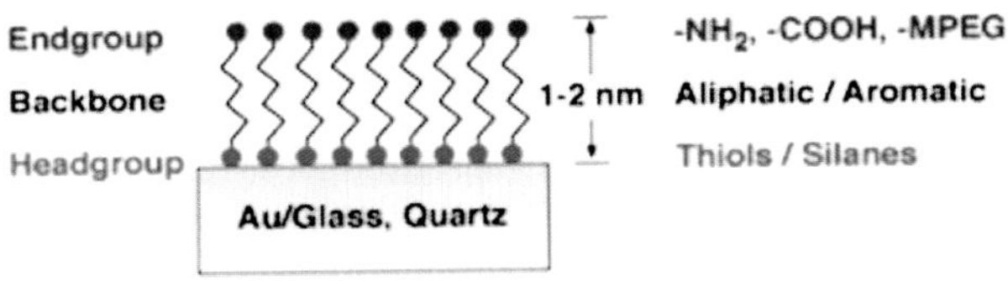

Figure 5.9 Schematic structure of SAMs. The basic SAM structure consists of three groups of molecules: the head group, the backbone, and the end group [24].

Patterning of SAMs with femtosecond laser pulses is possible in two different ways: Direct excitation of the adsorbed molecules within the monolayer or indirectly via the excitation of the substrate. Examples for direct absorption include the photoexcitation via single or multiphoton processes and field-induced processes. Substrate-mediated photochemical or photoelectrochemical processes comprise the absorption of laser light in the substrate generating excited charge carriers, such as hot electrons, which subsequently interact with the adsorbed molecules. Alternatively, the excited electrons eventually scatter inelastically with the substrate lattice, which results in a temperature rise at the surface leading to photothermal reaction pathways.

In Fig. 5.10, results of the patterning of octadecylsiloxane (ODS) monolayers on quartz glass substrates by femtosecond pulses with

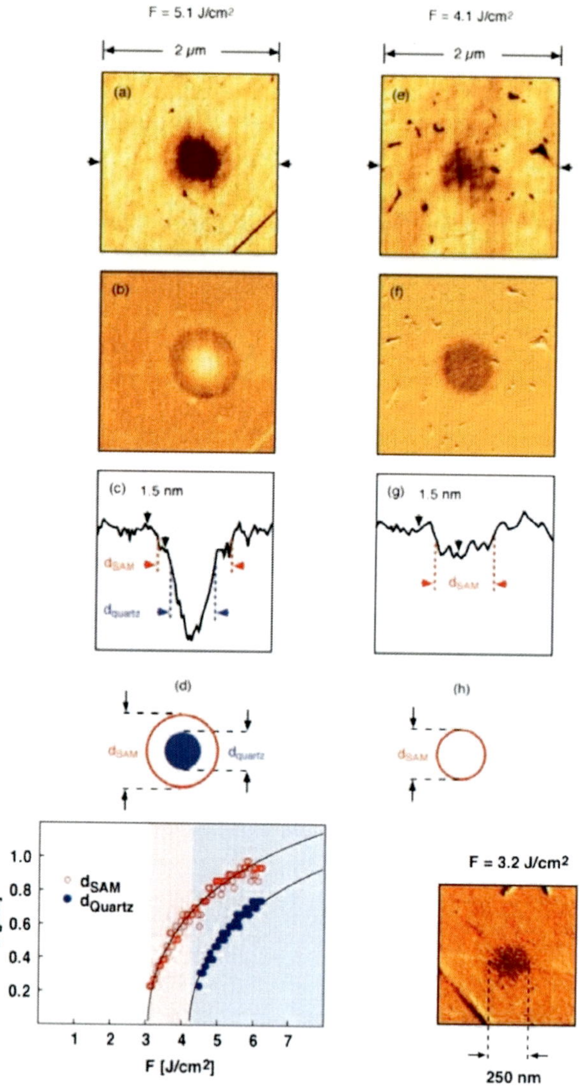

Figure 5.10 Patterning of ODS monolayers on quartz glass with single laser pulses at λ = 800 nm, t_p < 30 fs and distinct fluences f: (a–d) F = 5.1 J/cm^2 and (e–h) F = 4.1 J/cm^2. The data display: (a, e) the topography, (b, f) the friction contrast, (c, g) height profiles. (d, h) schematic presentations of the diameters d_{SAM} and d_{Quartz}, where monolayer decomposition and substrate ablation are observed, respectively. a) dependence of the diameters d_{SAM} and d_{Quartz} on the laser pulse fluence. (b) friction contrast image of a sub-wavelength structure at F = 3.2 J/cm^2 [11].

different fluencies are depicted. With a linear photodissociation energy E_D of 6 eV of the monolayer multiphoton absorption is realistic and machining beyond the diffraction limit becomes possible. For lower laser fluences a decomposition of the monolayer without affecting the substrate can be observed. A plot of the crater diameters over the fluences yields critical threshold values for the monolayer decomposition $F_{th,SAM}$ = 3.1 J/cm^2 and for substrate ablation $F_{th,quartz}$ = 4.2 J/cm^2. This window offers the possibility for the selective processing of the monolayers. Moreover, in this regime, sub-wavelength patterning beyond $\lambda/3$, i.e., around 250 nm diameter, is possible [11].

5.2.5.3 Nanotexturing, microbumps, microjets

A further approach to modify or texture surfaces or thin films is the generation of hollow bumps and jet-like structures on metal films, which have recently been investigated in more detail. In Fig. 5.11, the effect of applying femtosecond laser pulses with different pulse energies on a gold film of 60 nm thickness is nicely demonstrated [19].

In these experiments, the laser-induced dynamics is explored in a very small volume of the molten material. Gold has been chosen here as the material because of its rather weak electron–phonon coupling behavior compared with transition metals such as Cr, Mo, W, and Fe. As a consequence, the energy transfer to the lattice is much slower and the molten phase exists longer for noble metals such as gold. The key to achieving the jet-like structures is making use of the well-defined modification window that exists between the melting threshold and the ablation threshold with respect to the laser fluencies. In this regime, laser surface structuring or texturing without ablation becomes possible. If the pulse energy is gradually increased at a specific threshold bumps start to be formed on the surface. After further increasing the fluences, nanojets can be observed originating on the bumps. By excess of a fluence of 2.5 J/cm^2, the process becomes unstable and the microbump or the nanojet structures or both are destroyed. Measurements have proven that the bumps are hollow inside and the thickness is determined by the thickness of the molten layer.

In order to clarify the mechanisms responsible for the formation of these microbumps, it has been shown that the adhesion interface between the gold layer and the glass substrate is not the key point for stripping off the material although in the experiments described in Fig. 5.11 the optical penetration length (about 100 nm) in gold was in the range of the thickness of the metal layer (60 nm). However, the same behavior could be observed at much thicker gold coatings of a few μm, although the bumps and jets exhibited a smaller size in this case. Hence, there is some influence of the layer thickness, but the general formation cannot be explained by it. One can assume different mechanisms such as thermo-elastic or plastic deformation of the thin layer, Marangoni convection, and an expansion of the film due to the pressure of some small amount of evaporating material.

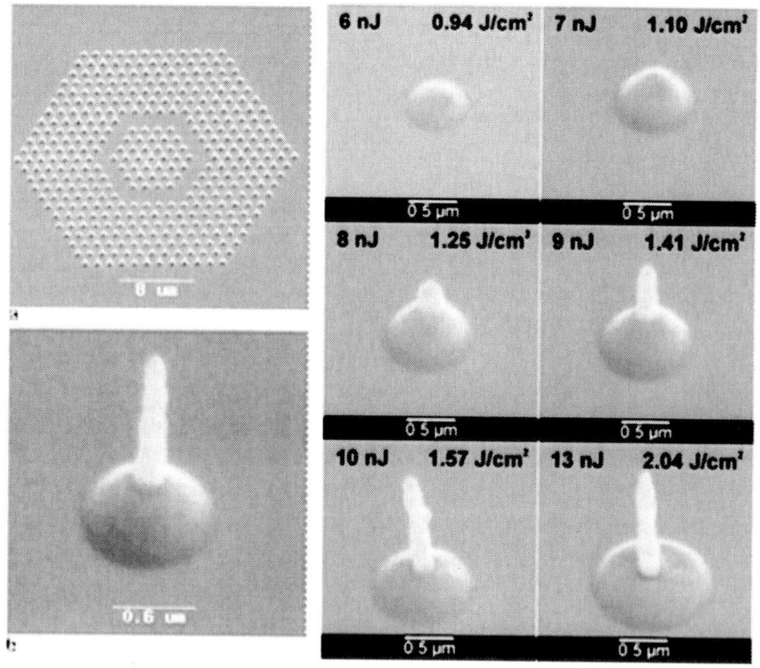

Figure 5.11 Left: Array of nanojets fabricated in a 60 nm-thick gold film with femtosecond laser pulses (a) and a single nanojet in detail (b), right: SEM images of sub-wavelength surface structures produced in a 60 nm gold layer by tightly focused femtosecond laser pulses [17].

In [14], a detailed investigation on the formation mechanisms based on a large-scale molecular dynamics simulation (MDS) combined with the two-temperature-model (TTM) has been described. It was shown that the fast heating in the center of the spot during the pulse duration leads to a partial inertial stress confinement because the heating time is shorter than the time needed for the film to freely expand. This leads to high compressive stresses during the first several picoseconds in the simulation. The relaxation of the compressive stresses then leads to the expansion in both the radial direction and the direction normal to the surface.

5.2.5.4 Black silicon

Laser patterning of silicon is one of the most attractive techniques in the field of laser-based surface structuring. Besides laser cutting and drilling of silicon, which finds applications in the electronics packaging and solar cell industry, surface patterning opens new routes in the interaction of light with silicon. Interesting phenomena can be observed when femtosecond laser pulses from an amplified laser system target a surface of an n-type silicon wafer under sulfur hexafluoride (SF_6) atmosphere (see also Chapter 6). After thermal annealing in a vacuum condition, the surface was covered with microstructures that could be adjusted in size by varying the laser fluence, the number of pulses, the SF_6 gas pressure, and the pulse duration. In Fig. 5.12, an SEM photo is shown with cones that were both 2–3 μm tall and laterally spaced. Chemical investigations revealed that the surface layer of the cones was heavily doped with sulfur. Sulfur is often used as n-dopant and thus by femtosecond laser patterning in SF_6-atmosphere an n/n+ heterojunction is produced between the original n-type crystalline substrate and the sulfur-doped disordered surface region.

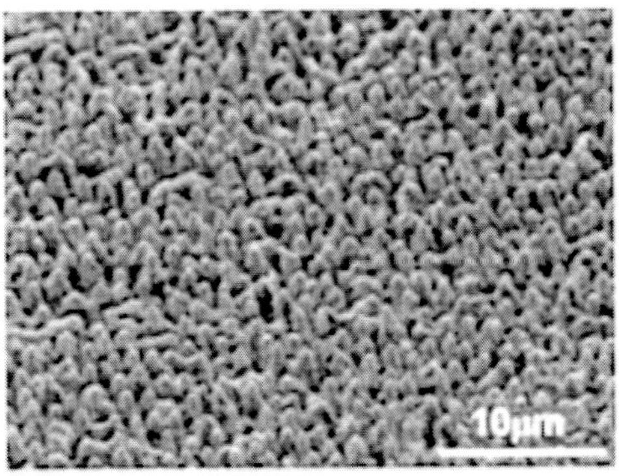

Figure 5.12 Scanning electron micrographs of microstructured Si surface [13].

In contrast to flat silicon surfaces, these surfaces exhibit low reflectance and high absorption across a much broader wavelength spectrum extending to 2 μm. These properties could extend the application potential for Si-based optoelectronic devices. Manifold applications can be envisaged in infrared imaging, microbolometers, and biomedical and chemical sensors. In [13], a microstructured photodiode has been evaluated providing a very high responsivity over a much broader spectral range compared with commercial photodiodes. Even at the well-established telecommunication wavelengths at 1.3 μm and 1.5 μm, the microstructured photodiode still exhibited photoresponse.

5.2.5.5 Wettability properties

Femtosecond laser patterning is a powerful process able to control the wettability properties of surfaces. Especially the possibility of inducing superhydrophilic and superhydrophobic behavior opens up a wide range of applications, e.g., lab-on-a-chip devices, biomedical scaffolds in tissue engineering, self-cleaning of surfaces, textiles and applications in the automotive industry. By hierarchical structuring, i.e., machining of micrometer and sub-μm structures by femtosecond laser pulses, it becomes possible to technically imitate the superhydrophobic lotus-effect. Especially in n-type silicon surfaces, femtosecond laser pulses from a CPA amplifier system are able to generate quasi-ordered arrays of cone-shaped tips of micrometer size under an SF_6-atmosphere, as shown in the previous subsection. Simultaneously, femtosecond laser irradiation may induce a disordered roughness on the spikes with feature sizes in the submicron radius range [41]. Experiments revealed that by changing the laser fluence, the surface topology can be controlled. By varying parameters other than the laser fluence (e.g., number of laser pulses) contact angles as high as 160° were measured. The impact of this topology change can then be assessed by water contact angle measurement and investigation of the partial penetration of the liquid within the two-scale surface structure. In [41], it is even shown that by properly texturing the silicon surface with a gradient of wetting surface topologies, a further gradient of surface tension can be realized. Hence, a spontaneous net motion of liquids can be induced. It even becomes possible to drive droplets to ascend a structured silicon surface, tilted at any angle.

At a certain fluence, the micrometer-sized spikes become more and more regular, and while in the lower fluence range the spikes are only small and not clearly separated at fluences around 1 J/cm^2, the cones are becoming more pronounced and spatially separated with base diameters around 8 μm and cone heights around 15 μm. Also with increasing fluence, more pronounced sub-μm features with dimensions from tens to a few hundreds of nanometers on the spikes walls appear at higher fluences. In this regime, the density of cones is in the range of 10^6/cm^2. The relating contact angle is the range of 130°.

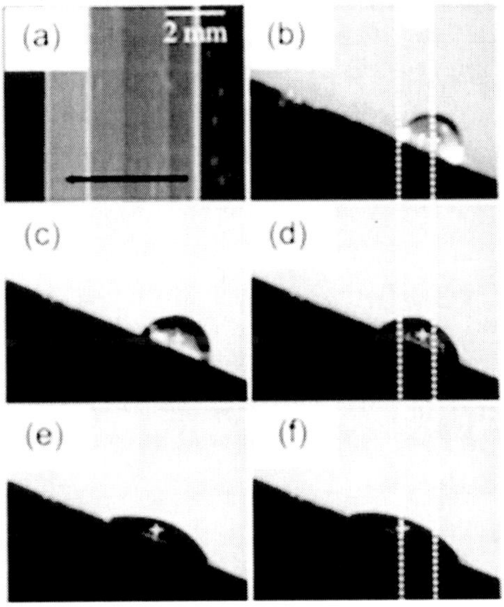

Figure 5.13 (a) Optical microscope picture of the silicon surface structured with a wettability gradient. The four laser fluences used were 1.1, 0.78, 0.56 and 0.33 J/cm^2, respectively. The arrow indicates the direction of increasing hydrophilicity (decreasing laser fluence). Photograph of a water drop ascending the same area tilted at 25°. Time from drop deposition: 0 (b), 2.4 (c), 5.0 (d), 5.8 (e), and 7.6 s (f). The shift of the center of mass from the initial to the final position is marked by the vertical white-dashed lines.

In terms of applications, there is a growing interest of surfaces with controlled wettability in the field of biomedical implants. In tissue engineering, for example, chemical and physical properties of the scaffolds strongly affect the adhesion of cells, differentiation, motility, and the specific function. Besides surface chemistry, topological features with structure sizes in μm and sub-μm range can stimulate, suppress, and control tissue development. For example, microstructured implant surfaces may help optimize the interaction between the electrode surface of a cochlea implant and neuronal cells in the inner ear [32]. Microstructures that reduce postoperative fibrotic tissue growth could help promote efficient electric stimulation of the nerve. Femtosecond laser-generated grooves with gradual geometrical parameters and contact angles, respectively, have been produced both on the platinum electrode and the silicone carrier material. By ablating grooves in the platinum material with a depth of 1 μm and a gradually increasing width from 4 to 7 μm, it is possible to selectively minimize the fibroblast cell growth while simultaneously guiding neural-like cell outgrowth. Another interesting example is described in [1], where a femtosecond laser is used to pattern bioresorbable polymers. The materials were thin films of aliphatic polyester, which can be used in scaffolds in artificial tissue engineering, drug delivery systems, surgical instruments, or microimplants. The authors demonstrated that micropatterning using femtosecond laser is possible while maintaining the structural integrity of the substrate. *In vitro* degradation test have shown nearly the same behavior of the patterned surfaces and the virgin material.

5.3 Bulk Micromachining

5.3.1 Motivation

Bulk micro-machining can be used to generate more complex three-dimensional structures. Ultrashort laser pulses are increasingly used to remove 3-D structures out of a bulk material. In thin-film ablation and in surface patterning, mostly low or medium fluences are used in order to avoid thermal destruction. In bulk processing, however, higher pulse fluences are being used. However, the quality of the walls is important, and in many applications, too, the quality of the ablation ground is an important aspect. Smooth ablation grounds can be achieved using again low laser fluencies for the complete process. This leads to low process speeds in comparison to hole drilling or cutting. The highest resolution of the machined structures can reach submicron dimension, dependent on the optical setup and the pulse energy, as described earlier.

To create extensive 2D structured surfaces via femtosecond laser pulses a direct writing process is applied. Because these structures mainly consist of linear grooves a periodic movement of the workpiece or a scanning system is used for their manufacturing. By the choice of laser parameters different dimensions and wall angles can be achieved.

5.3.2 Characteristics of Bulk Machining

In the femtosecond time regime, the pulse energy is absorbed in a thin, superficial layer. The thickness of this layer corresponds to the optical penetration depth l_{opt} (skin depth, ~10 nm). The classical heat conduction theory, based on the assumption that a material can be characterized by one temperature only, is no longer valid: thermal diffusion effects through the lattice of the solid-state material are almost irrelevant. Instead, interactions between electrons and lattice have to be taken into account, and temperatures of both the electrons and the lattice have to be treated separately. At very high fluencies ($I > 10^{16}$ W/cm^2), electrons located within the skin depth are heated up to extremely high temperatures, and additionally, overheated electrons are generated with energies up to the MeV range. Subsequent diffusion of the hot electrons transmits the major part of the pulse energy into areas underneath, which leads to higher ablation rates per pulse compared with ablation mechanisms of long pulses.

Momma *et al.* [26] provided a detailed analytical study on ablation of metals. Using femtosecond lasers, this process is characterized by rapid overheating and thermalization of the electrons within the optical penetration depth. As electrons have a low thermal capacity compared with the lattice, they are rapidly heated beyond the Fermi-level to very high, transient temperatures, forcing an extreme non-equilibrium state between the electron and lattice system. Assuming the electron thermal capacity to be described by $C_e = 3N_e k_B/2$, where N_e is the electron number density, k_B is the Boltzmann constant, and the electron thermal conductivity given by $k_e = C_e v_F^2 \tau/3$, with v_F being the Fermi velocity and the electron relaxation constant τ described in a first approximation by $\tau = a/v_F$, the electron diffusion coefficient D determined by $D = k_e/C_e$ remains temperature independent, in order to be applied with the differential equation system below, modeled for the temperature prediction of the electrons (T_e) and of the lattice (T_l) along the solid plane normal (z) and the time scale (t)

$$\frac{\partial T_e}{\partial t} = D\frac{\partial^2}{\partial z^2}T_e - \frac{T_e - T_I}{\tau_e} + \frac{IA\alpha}{C_e}e^{-\alpha z} \tag{5.8}$$

$$\frac{\partial T_I}{\partial t} = \frac{T_e - T_I}{\tau_I} \tag{5.9}$$

with the initial and boundary values

$$T_e(z,t=-\infty) = T_I(z,t=-\infty) = T_0 \approx 0 \tag{5.10}$$

$$\left.\frac{\partial T_e}{\partial z}\right|_{z=0} = \left.\frac{\partial T_e}{\partial z}\right|_{z=\infty} = 0 \tag{5.11}$$

The expression for the analytical solution of T_I for a pulse with the fluence $I(t)$, and the absorption A is then approximately given by

$$T_I \approx \frac{F_a}{C_I}\frac{1}{l^2-\delta^2}[le^{-z/l} - \delta e^{-z/\delta}] \tag{5.12}$$

where F_a is the absorbed fluence, C_I the lattice thermal capacity, the characteristic electron diffusion length $l = \sqrt{D\tau_a}$ with the ablation interval τ_a that is independent of the pulse length in this time regime, and the optical penetration length $\delta = l_{opt} = 1/\alpha$. Based on this analytical solution, two extreme situations can be derived for the lattice temperature distribution:

$$T_I \approx \frac{F_a}{C_I\delta}e^{-z/'} \qquad (l \ll \delta) \tag{5.13}$$

and

$$T_I \approx \frac{F_a}{C_I l}e^{-z/l} \quad (l \gg \delta) \tag{5.14}$$

The first equation is valid for negligible electron diffusion ($l << \delta$), while the latter one is valid in case that the electron diffusion length is considerably greater than the optical penetration depth ($l >> \delta$).

Assuming that the conditions for significant ablation are fulfilled if the absorbed energy density is larger than the solid enthalpy of evaporation $C_l T_l \geq \rho\Omega$, with the density ρ, and specific enthalpy of evaporation Ω, the following expressions for the ablation depth can be derived from the shown equations:

$$L \approx \delta \cdot \ln\left(\frac{F_a}{F_{th}^{\delta}}\right) \quad (l \ll \delta) \tag{5.15}$$

and

$$L \approx l \cdot \ln\left(\frac{F_a}{F_{th}^{l}}\right) \quad (l \gg \delta) \tag{5.16}$$

with the respective ablation thresholds determined by $F_{th}^{\delta} = \rho\Omega\delta$ and $F_{th}^{l} = \rho\Omega l$.

According to the previously mentioned simplified, theoretical model, the equations yield two expressions for the ablation depth per pulse as logarithmic functions of the laser fluence, i.e., the pulse energy. Equation (5.15) is the characteristic expression for the ablation depth per pulse for a mechanism without a significant heat transfer beyond the optical penetration depth. Equation (5.16) describes the ablation mechanism that is characterized by a significant heat transfer beyond the optical penetration depth due to hot electron diffusion. Both effects (Fig. 5.14) are experimentally verifiable for a variety of materials, including metals with excellent thermal conductivities, such as copper [26].

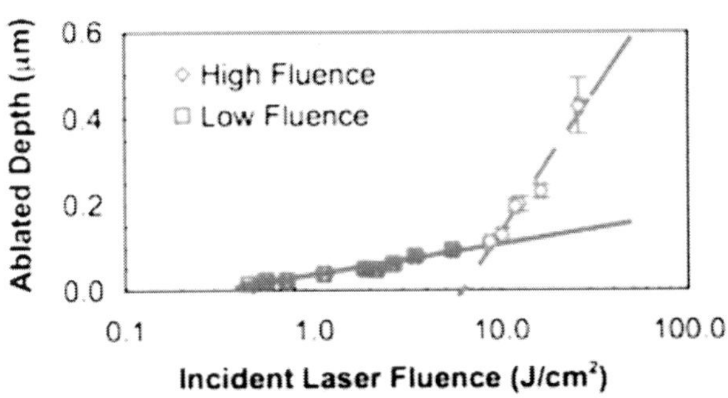

Figure 5.14 Maximum depth of femtosecond laser ablation crater as a function of the laser fluence using the example of ablation the nickel based superalloy CMSX-4 with 150 fs laser pulses of 780 nm from a Ti-sapphire laser. [25].

5.3.3 Differences in the Absorption between Metals and Dielectrics

Ablation of dielectrics with femtosecond lasers is also characterized by nonthermal breakdown of the solid material. However, the free conduction band electrons (CBE), essential for the high-intensity energy diffusion, first need to be generated. CBE generation can be described by two processes: (i) collisional (avalanche) ionization and (ii) multiphoton ionization. The dominance of either of these competing processes is determined by the form of energy gain, which is highly dependent on the experimental environment. A detailed study on this topic is presented in [10]. If sufficient CBE densities can be achieved, ablation proceeds in a similar way to the mechanism observed in short pulse ablation of metals are observed. As a part of the pulse energy is needed for CBE generation, the available pulse energy for hot electron diffusion, the effective ablative power, is reduced. This, in return, decreases the hot CBE maximal diffusion length. Thus, for predictions of the potential ablation depth of dielectrics, Eq. (8) is more favorable. The mechanisms of CBE generation are manifold and also influence the ablation thresholds. Typically, a significant dependency of the pulse length on the breakdown threshold in dielectrics can be observed (as reported in [31, 37]) and have to be considered for correct determination.

5.3.4 Plasma Influence, Self-Focusing, Filamentation

The interaction of long (nanosecond) laser pulses and ultrashort (femtosecond) laser pulses with solid materials is illustrated in Fig. 5.15 in principle. When using long pulses with sufficient fluences ($I > 10^{10}$ W/cm^2), particles, clusters, ions, and electrons are ejected from the surface, which can be regarded as a plasma plume formed above the workpiece. This laser-induced plasma reduces the amount of radiation that contributes to interaction with the solid-state material significantly, which is often referred to as "plasma shielding." On the contrary, the influence of the plasma plume is negligible when applying ultrashort laser pulses. Due to the very small spatial expansion of the plasma plume during the pulse duration, the pulses interact directly with the material surface.

Self-focusing is a process based on the optical Kerr effect. Under intense exposure the refractive index of a material can be denoted as $n = n_0 + n_2 I$, where the second-order term n_2 represents the nonlinear Kerr effect. If $n_2 > 0$, which is valid for many media, including air, the refractive index on the optical axis is increased since for Gaussian beams the highest intensity is on the optical axis with decreasing values for increasing radii. As a consequence, the Gaussian beam leads to the formation of a focusing lens. Above a certain power level, this effect becomes very distinct and further amplified because when the beam is focused by this effect the intensity even becomes higher on the optical axis resulting in a very strong focusing effect. This effect can be triggered although the second-order refractive index in many media is very low (e.g., in air).

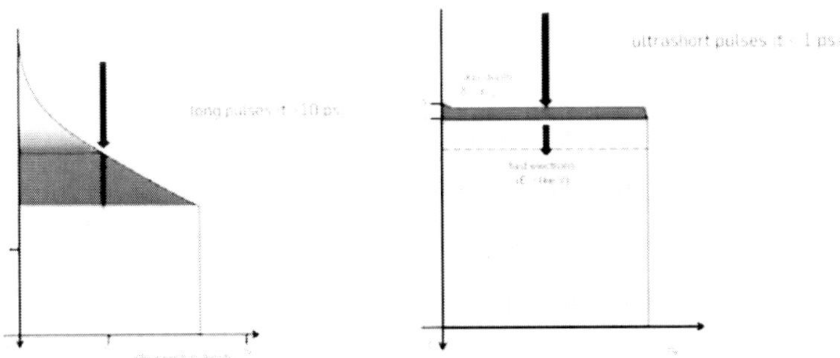

Figure 5.15 Illustration of the interaction of long and ultrashort laser pulses with solids. The laser radiation can only propagate in the plasma if the electron number density n_e is below the critical value n_c [27].

Once self-focusing is achieved, several nonlinear processes commence and interact with each other resulting in the phenomenon of beam filamentation. Filamentation can be used in many applications such as in LIDAR measurements. However, in the precise processing of materials, it often leads to disturbed features; therefore, filamentation needs to be avoided. It should be noted, however, that there is one application in the bulk modification of transparent media where long channels have to be machined, where filamentation can be used as a superior method. The extremely high intensities originating from self-focusing can trigger multiphoton ionization resulting in the formation of a quasi-transparent plasma with $n_e << n_c$. In this regime the refractive index can be calculated by the Drude model resulting in values below 1.0, i.e., the plasma behaves like a defocusing lens. In summary, the combination of self-focusing and defocusing by plasma generation leads to the effect of beam filamentation [5]. Beam filaments can be very stable and for temporally chirped pulses they can comprise a length of more than a kilometer. The energy of a beam filament propagates in the surrounding of the filament, called a "photon bath" or an "energy reservoir," which can have a radius of 1 mm or above. It has to be noted that these effects can be observed in air already for femtosecond pulses with pulse energies around 100 mJ focused down to a few 100 μm spot diameter. Thus, for precise and efficient machining of many materials, plasma formation, self-focusing, and filamentation have to be avoided. Process gases with a very high ionization potential can be used in this case to suppress these effects. Helium, for example, has an ionization energy E_I that is almost double to that of air (E_I(He) = 25 eV; E_I(Air) = 15 eV).

5.3.5 Applications

5.3.5.1 Drilling

Mechanical drilling tools are commonly used for drilling holes in metals down to approximately 200 μm in diameter at depths of approximately 1 mm. For smaller structure sizes or irregular shapes, electron beam, electro discharge machining (EDM) or conventional lasers are used. Conventional lasers such as CO_2, Nd:YAG lasers affect materials based on localized heating. All of these microfabrication techniques heat the material to the melting or boiling point, resulting in thermal stress to the remaining material and often a heat-affected zone can be observed afterward. There are manifold problems in achieving higher precision using these techniques. The removal of material without significantly transferring heat into the surrounding areas is an advantage that can only be assigned to ultrashort laser pulses (picosecond and femtosecond pulses) [4, 12, 28, 33, 37]. The important shape characteristics of a laser-drilled hole are entrance and exit diameters, roundness, wall roughness, and the hole profile.

There are three laser drilling techniques used in industrial applications, which are also relevant to femtosecond laser drilling. These techniques are single pulse drilling, percussion drilling, and trepanning (Fig. 5.16). Single pulse drilling is used when a high processing speed is necessary and relatively low quality can be accepted. If a larger depth is required (more material needs to be removed) more pulses are necessary. This also leads to an increased accuracy of the process by removing a smaller volume per pulse. During trepanning, the focused laser beam moves along a circular path relative to the workpiece. Trepanning is the standard technique for drilling larger holes with diameters up to several millimeters [6].

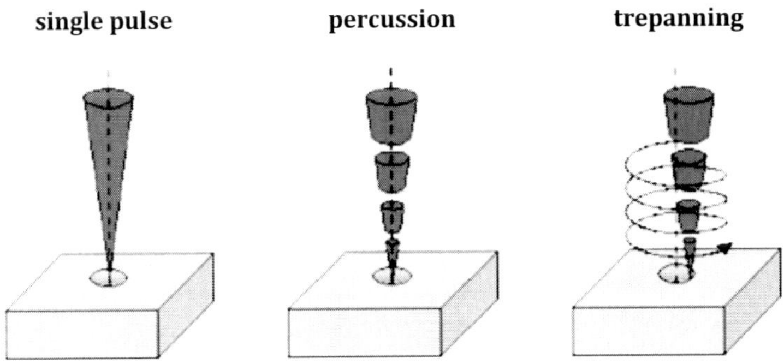

Figure 5.16 Laser drilling techniques.

Applying the trepanning technique leads to excellent results concerning the roundness of the micro holes. The roundness of the hole is less dependent on the beam profile than when simply focusing the beam without rotation, due to the fact that the focused beam moves in circles. Figure 5.17 demonstrates the difference in hole quality between percussion and trepanning results in femtosecond laser pulse drilling.

The main advantages of material processing using femtosecond laser pulses are (i) efficient, fast and localized energy deposition, (ii) well-defined damage and ablation thresholds, and (iii) minimal thermal and mechanical damage of the substrate material. These advantages make high-quality drilling of solids at relatively low laser fluences, close to the ablation threshold, possible.

However, for many practical applications, when high processing speed is required, much higher laser fluences well above the ablation threshold are necessary. At these laser fluences, the energy coupled into a workpiece and the corresponding thermal load is high (according to Eq. 5.14). The ablation depth per pulse is determined by the energy transfer into the target, due to the electron thermal conduction and/or the generated shock wave. This is similar to ablation using nanosecond and longer laser pulses. Therefore, in the high fluence regime, considerable advantages for material processing with femtosecond lasers are not usually expected. They can probably drill more efficiently and produce deeper holes, due to the following reasons: lower thermal losses and negligible hydrodynamic expansion of the ablated material during the femtosecond laser pulse. With the latter, "plasma shielding"-induced radiation losses can be avoided.

For deep drilling of metals, which is necessary in many industrial applications, the most important criteria are the hole geometry and quality. The use of more expensive femtosecond lasers could be justified if they could fabricate holes with a special geometry, superior quality, and high reproducibility. It has been demonstrated that femtosecond lasers are close to this goal. They have the potential to provide a simple processing technique that does not require any additional post-processing or special gas environment. This can be considered a real advantage of using femtosecond lasers.

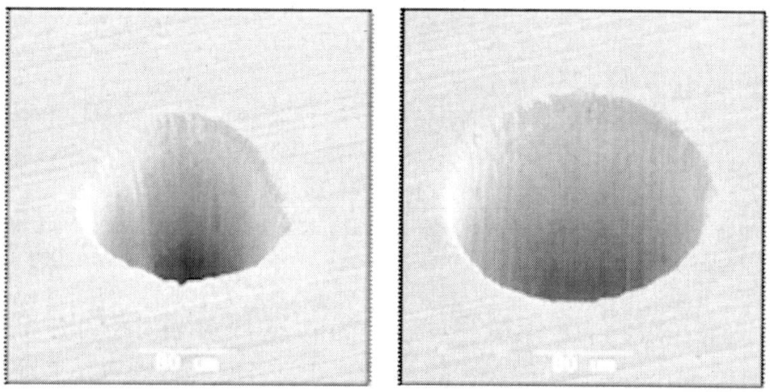

Percussion drilling Trepanning

Figure 5.17 Femtosecond laser drilling in stainless steel using percussion (left) and trepanning technique (right).

The question that needs to be answered at this point is, why, even at high laser fluences, can high-quality holes be drilled? The answer to this question is quite simple. High intensities are required to rapidly drill a through-hole. After the through-hole is drilled, the high intensity part of the laser pulse propagates directly through the hole without absorption. Only at the edge of the laser pulse, at laser intensities close to the ablation threshold, a small amount of material is removed. Starting from this point, the interaction of femtosecond laser pulses with the workpiece occurs in the low-fluence regime, where all advantages of femtosecond lasers can be realized (according to Eq. (5.13)). This can be considered subsequent low-fluence post-processing, which leads to an excellent hole quality, or a low-fluence finishing [16]. The high quality of the single holes and high reproducibility is demonstrated in Fig. 5.18.

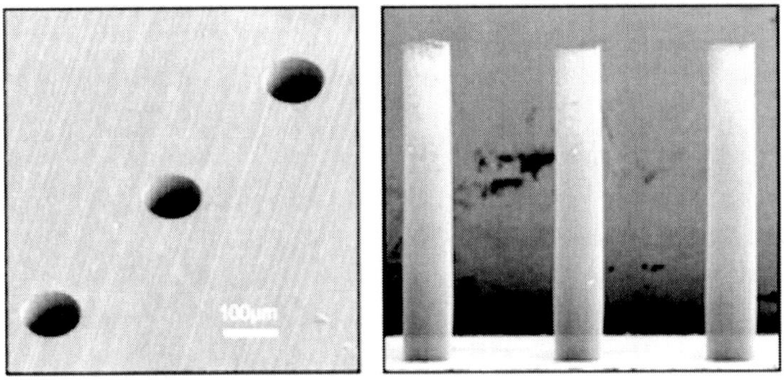

Figure 5.18 Scanning microscope image of holes drilled in 1 mm-thick stainless steel plate using femtosecond laser pulses, and their replicas.

When high aspect ratio and multipulse drilling are used for drilling holes, a strong influence of the beam polarization can be observed as shown in Fig. 5.19. This is due to polarization-dependent reflections at the walls of the hole, which might lead to non-uniform intensity distributions resulting in hole geometries with strong noncircularities. Experimentally, when a through-hole in a 200 μm stainless steel foil is drilled using 170 fs pulses, the backside of the substrate shows a bulge overlapped by a circular exit hole. This bulge is perpendicular to the direction of the beam polarization. By reflection calculations on the side walls with increasing depths using the Fresnel formulas for s- and p-polarized light, it becomes clear that the intensity distribution on the ground of the blind hole becomes gradually asymmetric with an elongation perpendicular to the polarization vector [29]. To overcome this disturbing effect, either a conversion of the linear polarization into circular polarization by a quarter-wave plate can be used or a fast mechanically rotating half-wave plate can be implemented in the beam guiding system. The latter has proven to be more robust since the quarter-wave plate has to be adjusted very carefully in order to fully compensate the effect. Even a slightly elliptical polarization can be observed in the exit hole geometry.

Summing up, femtosecond laser material processing at high laser fluencies has to be considered a practical tool for high-quality, deep drilling of metals. Thanks to recent advances in lasers concerning ease of use, reliability, and overall lifetime of femtosecond laser sources in the market, femtosecond laser-machined parts can be expected in the near future.

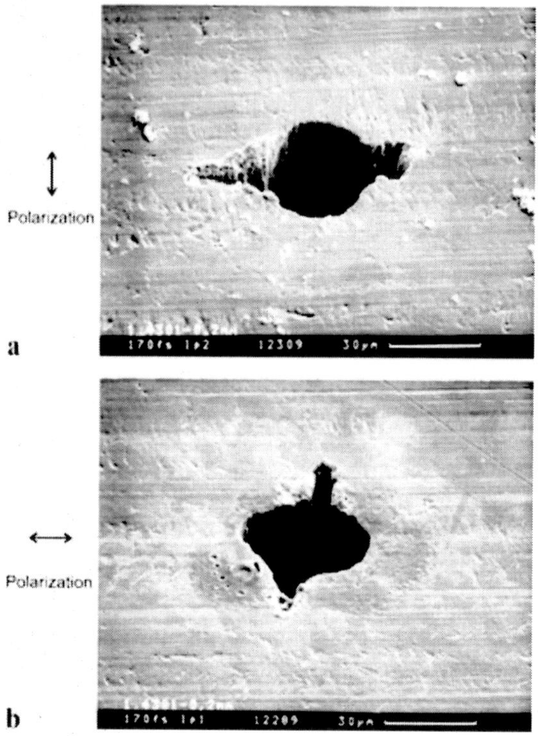

Figure 5.19 Exit hole of a 200 μm-thick stainless steel plate (170 fs pulses at a wavelength of 780 nm, $F = 2.5$ J/cm^2). The orientation of the linearly polarized laser light was perpendicular to the bulges [29].

5.3.5.2 Cutting

Today, lasers are well-established and very flexible tools for producing various microstructures. Besides marking, drilling, and welding (using longer pulses or cw-lasers, of course) applications, cutting is the most frequently applied laser machining process.

For metals, the laser-cutting process can be generally performed by melting, processes based on exothermal oxygen reaction, and by sublimation [3]. The main advantage of sublimation cutting is that melting and heat-affected zones can be minimized or even avoided. The disadvantage is, however, that the cutting speed is relatively low due to the high evaporation enthalpy of metals. This is the major reason why current industrial laser cutting is predominantly based on melting and oxygen cutting processes. Oxygen cutting has the advantage that high machining speeds can be realized, and the disadvantage is that oxidation occurs. Both processes lead to high burr formation and depositions on the material surface. In very precise applications, the burr and the depositions must be removed in further production steps.

Cutting with conventional laser sources (e.g., Nd:YAG), which produce pulses with durations in the range of nanoseconds to milliseconds, has several limitations. First of all, the thermal load in the material due to heat conduction is relatively high, which results in large heat-affected zones and melting. The melting again leads to a significant burr formation at the cutting edges and to a deposition of solidified droplets sticking on the surface. In many applications, both have to be removed via several subsequent treatments. Due to the thermal load during laser cutting, the minimum producible structure size is limited. Moreover, a variety of materials cannot be structured using this conventional technique. Thus, there is a demand for an alternative, less invasive laser machining technique.

Such a technique is femtosecond laser machining. With femtosecond lasers, almost all materials (e.g., metals, ceramics, glass, polymers, organic tissue) can be structured with minimum or no thermal damage. Materials that otherwise could not be structured with conventional (laser) techniques can be processed. The cuts are very smooth and completely burr-free, and the surface is free of depositions. Figure 5.20 shows three examples. Except for a simple ultrasonic treatment in alcohol, no postprocessing was applied here. The stent structure at the top left of Fig. 5.20 has very smooth struts, which cannot be machined using more conventional laser techniques. The potential of femtosecond laser machining technique is even more pronounced in the case of extremely sensitive and delicate materials. The SEM images in the top right the bottom of Fig. 5.20 show cuts in a PVDF foil and PMMA, respectively. Both polymers are optically transparent and very heat-sensitive materials. At present, no other suitable machining process for cutting and structuring of these polymer materials achieves the quality of femtosecond laser machining. For many applications, the nonthermal nature of the femtosecond ablation process is necessary to ensure that the material properties are maintained, even adjacent to the cut area.

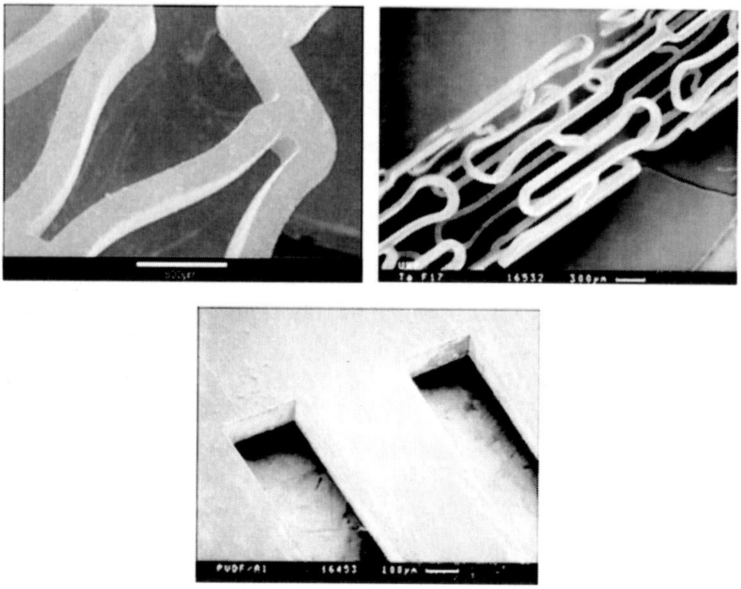

Figure 5.20 Femtosecond laser cutting of a tantalum stent (top), PVDF foil (middle) and PMMA (bottom).

Where conventional (laser) techniques are not suitable for cutting heat-sensitive materials, or for very high precision, the use of lasers producing ultrashort pulses with pulse durations in the femtosecond regime is a promising possibility. With this kind of laser, nearly all kinds of material can be treated with minimum mechanical and thermal damage.

5.4 Conclusion and Future Perspectives

Femtosecond laser ablation exhibits major advantages over long pulse laser structuring. Besides the superior flexibility in terms of machining nearly all kinds of materials especially the precise control of the thermal processes allow the structuring of surfaces as well as the deep machining of bulk materials. The short time frame of the pulse allows almost the complete absorption of the laser energy by the electron system while the transfer into the lattice takes place on scales much longer than the duration of the pulse itself. Consequently, the thermal penetration length is mainly determined by the electron–lattice interaction coefficient, i.e., how deep the strongly heated electrons can diffuse into the bulk. If this length is longer than the film to be structured, the underlying substrate hinders the further penetration into the bulk. This has some consequences for the optimal choice of ablation parameters as the ablation threshold scales with the thickness of the film. Also, for structures with higher aspect ratio, the unique physical effects can be explored. The non-equilibrium two-temperature (electrons and lattice) behavior leads to the formation of different ablation regimes. For higher fluences, the process is characterized by thermal penetration and, thus, is mostly driven by thermal effects similar to long pulses. This regime allows the effective ablation of rather large volumes but with reduced precision. For the low fluence regime, less thermal effects can be observed and the ablation rate is characterized primarily by the optical penetration length. Gaussian beams with their characteristic profile provide both, a high intensity in the center and lower intensities on the edges, which allows both the effective material removal in the center and the smoothening of the walls by the low fluence regime on the edges. This is a unique behavior characteristic for laser pulses in the femtosecond range.

In future, higher overall efficiencies and process speeds can be expected. New femtosecond laser sources with average output power in the kilowatt range will offer ultrahigh speed machining with the same precision. These sources allow together with a special optical system the cost-effective patterning of large surfaces, which is not realistic with repetition rates in the kilohertz range because a focused laser beam still is a serial process based on a spot-by-spot technology. Also the use of temporally and spatially designed pulses offers new opportunities to adjust the beam properties in an optimal way to the specific materials response. Several research groups today already use tailored pulse trains to excite, modify, and remove the material in an optimized way.

References

1. Aguilar, C.A., Lu, Y., Mao, S. Chen (2005) Direct micro-patterning of biodegradable polymers using ultraviolet and femtosecond lasers, *Biomaterials*, **26**, 7642–7649, DOI: 10.1016/j.biomaterials.2005.04.053.

2. Anisimov, S.I., Bityurin, N.M., Luk'yanschuk, B.S. (2003) Photoexcited Processes, Diagnostics and Applications, In *Models for Laser Ablation* (ed. Peled A.) (Kluwer, Dordrecht, The Netherlands) 121–159.
3. Beyer, E., Märten, O., Behler, K. *et al.* (1985). Laser cutting, *Laser Optoelektronik*, **3**, 282–290.
4. Chichkov, B. N., Momma, C., Nolte, S., von Alvensleben, F. Tünnermann, A. (1996). Femtosecond, picosecond and nanosecond laser ablation of solids, *Appl. Phys. A*, **63**, 109.
5. Couairon, A., Mysyrowicz, A. (2007). Femtosecond filamentation in transparent media, *Phys. Rep.*, **441**, 2–4, 47–189, DOI:10.1016/j.physrep.2006.12.005.
6. Dausinger, F. (2001). *Drilling of High Quality Micro-Holes*, Laser Manuf., **22**, 25–27.
7. Dowden, J. (2009) *The Theory of Laser Materials Processing* (Springer Dordrecht, the Netherlands).
8. Du, D., Liu, X., Korn, G., Squier, J., Mourou, G. (1994). Laser-induced breakdown by impact ionization in SiO_2 with pulse widths from 7 ns to 150 fs, *Appl. Phys. Lett.*, **64**, 23, 3071–3073.
9. Forster, M., Égerházi, L., Haselberger, C., Huber, C., Kautek, W. (2011). Femtosecond laser interaction with pulsed-laser deposited carbon thin films of nanoscale thickness, *J. Appl. Phys. A*, **102**, 1, 27–33, DOI: 10.1007/s00339-010-6013-5.
10. Gamaly, E. G., Rode, A. V., Luther-Davies B. (1999). Ultrafast ablation with high-pulse-rate lasers. Part I: theoretical considerations, *J. Appl. Phys.*, **85**, 4213–4221 DOI:10.1063/1.370333.
11. Hartmann, N., Franzka, S., Koch, J., Ostendorf, A., Chichkov, B. N. (2008). Subwavelength patterning of alkylsiloxane monolayers via nonlinear processing with single femtosecond laser pulses, *Appl. Phys. Lett.*, **92**, 223111-1–3, DOI:10.1063/1.2939585.
12. Horwitz, J. S., Krebs, H.-U., Murakami, K. Stuke, M. (Eds.) (1999). *Proceedings of the 5th International Conference on Laser Ablation (COLA '99).*
13. Huang, Z., Carey, J. E., Liu, M., Guo, X., Mazur, E., Campbell, J. C. (2006). Microstructured silicon photodetector, *Appl. Phys. Lett.*, **89**, 033506, DOI:10.1063/1.2227629.
14. Ivanov, D. S., Rethfeld, B., O'Connor, G., Glynn, T. J., Volkov, A. N., Zhigilei, L. V. (2008). The mechanism of nanobump formation in femtosecond pulse laser nanostructuring of thin metal films, *Appl. Phys. A*, **92**, 791–796, DOI: 10.1007/s00339-008-4712-y.

15. Jia, T.-Q., Chen, H., Zhang, Y.-M. (2000). Photon absorption of conduction-band electrons and their effects on laser-induced damage to optical materials, *Phys. Rev.* 61, **24**, 16522–16529.

16. Kamlage, G., Bauer, T. Ostendorf, A., Chichkov, B. N. (2003). Deep drilling of metals by femtosecond laser pulses, *Appl. Phys. A*, **77**, 307–310.

17. Kautek, W., Krüger, J., Lenzner, M., Sartania, S., Spielmann, C. Krausz, F. (1996). Laser ablation of dielectrics with pulse durations between 20 fs and 3 ps, *Appl. Phys. Lett.*, **69**, 3146.

18. Koch, J., Korte, F., Bauer, T., Fallnich, C., Ostendorf, A., Chichkov, B. N. (2005). Nanotexturing of gold films by femtosecond laser-induced melt dynamics, *Appl. Phys. A*, **81**, 325–328, DOI: 10.1007/s00339-005-3212-6.

19. Korte, F., Adams, S., Egbert, A., Fallnich, C., Ostendorf, A., Nolte, S., Will, M., Ruske, J.-P., Chichkov, B.N., Tuennermann, A. (2000). Sub-diffraction limited structuring of solid targets with femtosecond laser pulses, *J. Opt. Expr.*, **7**, 41–49.

20. Krüger, J., Duffta, D., Kotera, R., Hertwig, A. (2007). Femtosecond laser-induced damage of gold films, *Appl. Surf. Sci.*, **253**, No. 19, 7815–7819, DOI: 10.1016/j.apsusc.2007.02.164.

21. Lenzner, M., Krüger, J., Sartania, S., Cheng, Z., Spielmann, C., Mourou, G., Kautek, W., Krausz, F. (1980) Femtosecond optical breakdown in dielectrics, *Phys. Rev. Lett.*, **80**, 4076–4079.

22. Lin, Z., Zhigilei, L.V. (2008) Electron–phonon coupling and electron heat capacity of metals under conditions of strong electron–phonon nonequilibrium, *Phys. Rev. B*, **77**, 075133, DOI: 10.1103/PhysRevB.77.075133.

23. Love, J.C., Estroff, L.A., Kriebel, J.K., Nuzzo, R.G., Whitesides, G.M. (2005) Self-assembled monolayers of thiolates on metals as a form of nanotechnology, *Chem. Rev.*, **105**, 1103–1169.

24. Mathieu, M., Hartmann, N. (2010). Sub-wavelength patterning of organic monolayers via nonlinear processing with continuous-wave lasers, *New J. Phys.*, **12**, 125017, DOI: 10.1088/1367-2630/12/12/125017.

25. McDonald *et al.* (2008). Femtosecond pulsed laser ablation dynamics and ablation morphology of nickel based superalloy CMSX-4, *J.Appl. Phys.*, **103**, 093111.

26. Momma, C. *et al.* (1996). Short-pulse laser ablation of solid targets, *Opt. Commun.*, **129**, 134–142.

27. Momma, C., Nolte, S., Chichkov, B.N., (1997). Precise micromachining with femtosecond laser pulses, *Laser Optoelektronik*, **29**, No. 3, 82–89.

28. Nolte, S., Momma, C., Jacobs, H., Tünnermann, A., Chichkov, B. N., Wellegehausen, B., Welling, H., (1997). Ablation of metals by ultrashort laser pulses, *J. Opt. Soc. Am. B*, **14**, 2716.

29. Nolte, S., Momma, C., Kamlage, G., Ostendorf, A., Fallnich, C., von Alvensleben, F., Welling, H. (1999). Polarization effects in ultrashort-pulse laser drilling, *Appl. Phys. A*, **68**, 563–567, DOI: 10.1007/s003399900013.

30. Perry, M.D., Stuart, B.C., Banks, P.S., Feit, M.D., Yanovsky, V., Rubenchick, A.M. (1999). Ultrashort-pulse laser machining of dielectric materials, *J. Appl. Phys.*, **85**, 6803–6810.

31. Poprawe, R. (2011) *Tailored Light 2, Laser Application Technology*, 1st Ed. (Springer Heidelberg, Dordrecht, London, New York, Germany).

32. Reich, U., Mueller, P.P., Fadeeva, E., Chichkov, B.N., Stoever, T., Fabian, T., Lenarz, T., Reuter, G., (2008). Differential Fine-Tuning of Cochlear Implant Material–Cell Interactions by Femtosecond Laser Microstructuring, *J Biomed Mat Res B: Appl Biomat*, 87B, 146–153, DOI: 10.1002/jbm.b.31084.

33. Simon, P. Ihlemann, J., (1996). Machining of submicron structures on metals and semiconductors by ultrashort UV-laser pulses, *Appl. Phys. A*, **63**, 505.

34. Soileau, M.J., Williams, W. E., van Stryland, T.F., *et al.* (1982). Temporal dependence of laser induced breakdown in NaCl and SiO_2, In: *Laser-Induced Damage in Optical Materials*, Benett, H. E., Guenther, A. H., Milam, D.; *et al.* (eds), *Natl. Bur. Stand. (U.S.) Spec. Publ.*, **669**, 387–405.

35. Steen, W.M. Mazumder, J. (2010) *Laser Material Processing* (Springer London, Dordrecht, Heidelberg, New York).

36. Strickland, D., Mourou, G., (1985) Compression of amplified chirped optical pulses, *Opt. Comm.*, **56**, 219–221.

37. Stuart, B.C., Feit, M.D., Herman, S., Rubenchik, A.M., Shore, B.W., Perry, M.D., (1996). Optical ablation by high-power short-pulse lasers, *J. Opt. Soc. Am. B*, **13**, 459.

38. Tan, B., Dalili, A., Venkatakrishnan, A. (2009). High repetition rate femtosecond laser nano-machining of thin films, *Appl. Phys. A*, **95**, 537–545, DOI: 10.1007/s00339-008-4938-8.

39. Von Allmen, M. (1976). Laser drilling velocity in metals, *J. Appl. Phys.*, **47**, 12, 5460–5463.

40. Von Allmen M., Blatter, A. (1995) *Laser-Beam Interactions with Materials: Physical Principles and Applications* (Springer Berlin, Germany).

41. Zorba, V., Persano, L., Pisignano, D., Athanassiou, A., Stratakis, E., Cingolani, R., Tzanetakis, P., Fotakis, C. (2006). Making silicon hydrophobic: wettability control by two-lengthscale simultaneous patterning with femtosecond laser irradiation, *Nanotechnology*, **17**, 3234–3238, DOI:10.1088/0957-4484/17/13/026.

Chapter 6

Ultrafast Laser-Assisted Surface Micro- and Nanostructuring

Emmanuel Stratakis,[a] Ekaterina V. Barmina,[b] Panagiotis A. Loukakos,[a] Georgy A. Shafeev,[b] and Costas Fotakis[a]

[a]*Institute of Electronic Structure and Laser, Foundation for Research & Technology—Hellas, IESL-FORTH, P.O. Box 1527, Heraklion 711 10, Greece*

[b]*Wave Research Center, A.M. Prokhorov General Physics Institute, Russian Academy of Sciences, 38, Vavilov Street, Moscow 119991, Russian Federation*

stratak@iesl.forth.gr, fotakis@iesl.forth.gr

This chapter summarizes the characteristics and formation mechanisms of different kinds of micro- and nanoscale structures on the surface of metallic, semiconducting and dielectric materials, fabricated using ultrashort laser pulses. Two distinct approaches are reviewed: laser surface modification in controlled gas and liquid media. Emphasis will be placed on the exploitation of the materials and structures attained to modern applications for which there has been an increasing demand over the previous years.

6.1 Introduction

Controlling the interactions of laser beams with matter is crucial

Ultrafast Laser Processing: From Micro- to Nanoscale
Edited by Koji Sugioka and Ya Cheng

ISBN 978-981-4267-33-5 (Hardcover), 978-981-4303-69-9 (eBook)
www.panstanford.com

for the success and scalability of material-processing applications. The optimal interplay between the laser irradiation and material parameters allows controlled modification of surface properties at different length scales adding a new dimension to materials properties optimization. For a given wavelength, the laser pulse duration is a critical parameter and processing by ultrashort (i.e., sub-picosecond) laser pulses opens new exciting possibilities. In comparison to picosecond (ps) and longer laser pulses, ultrafast laser processing provides two major advantages: (a) a net decrease of the ablation threshold for identical laser wavelength and focusing conditions [1] and (b) an important decrease of the heat affected zone that can greatly reduce the extent of collateral damage to the surrounding area [2], allowing the production of sub-wavelength nanostructures. Such characteristics are due to the rapid energy deposition inside the target material and the subsequent non-linear processes induced at the focal point. Experimentally, it has been shown that the internal thermalization of the electron distribution following excitation by a femtosecond (fs) laser pulse takes place within one ps [3]. Thermalization between the electron subsystem and the lattice is usually longer and is of the order of several ps, depending on the electron–phonon coupling strength [4]. Thus, ultrafast laser excitation generates a hot electron gas which is in a highly non-equilibrium state with the lattice. As a consequence, direct irradiation of materials by ultrafast laser pulses induces modifications leading to complex micro- and nanoscale surface features, which are often found to have unique properties and cannot be realized by other, non-laser based techniques.

Laser micro- and nanostructuring of materials is important in many scientific, technological and biological applications, such as the fabrication of opto- and nanoelectronic devices, information storage systems, control over the mechanical and optical properties of solids, and microfluidics and biomedical engineering [5–10]. This chapter will summarize the fabrication, and formation mechanisms of different kinds of micro- and nanoscale structures on the surface of metallic, semiconducting and dielectric materials fabricated by means of ultrafast laser processing. Two distinct approaches are reviewed, including laser surface modification in controlled gas and liquid media, respectively. Emphasis will be placed on the exploitation of the materials and structures attained to modern applications for which there has been an increasing demand over the previous years. Besides presenting recent advances achieved by these techniques, the chapter will also delineate existing limitations and discuss emerging possibilities and future prospects.

6.2 Fabrication of Surface Micro/Nanostructures by Ultrafast Laser Processing in Gas Media

In the past decade, a number of research efforts have been concerned with the effect of fs laser pulses on the surface morphology of various types of materials. Various textures can be accurately produced by controlling processing parameters such as beam intensity, spatial and temporal profile, wavelength, and processing environment (background gas or liquid). The macroscopic dimensions of the surface features are generally defined by the shape and size of the beam. However, secondary microscale and even nanoscale features can form in and around the irradiated region due to a variety of mechanisms, including post-ablation melting and resolidification or splashing of a liquid surface due to the recoil pressure. Such secondary characteristics can be important in determining surface properties functionality for a desired application.

In general, structures that develop on solid surfaces under the action of laser light can be classified into coherent and non-coherent ones. Coherent structures are directly related to the coherence, the wavelength and the polarization of the laser light. These structures have a common origin: the oscillating radiation field on the material surface, which is generated by the interference between the incident laser beam and scattered/excited surface waves. The spatial periods of such structures are therefore proportional to the laser wavelength, while their orientation is perpendicular to the electric vector of the incident light. Non-coherent structures, however, are not directly related to any spatial periodicity of the energy input caused by interference phenomena; their period is related to the laser-beam intensity and the ambient gas pressure, when such is used, rather than the laser wavelength and polarization. They appear at higher intensities than coherent structures and exhibit a wave-like topography with a spatial period much larger than the incident laser wavelength.

Figure 6.1 presents a scanning electron microscopy (SEM) image of a metallic and a semiconducting surface after exposure to several

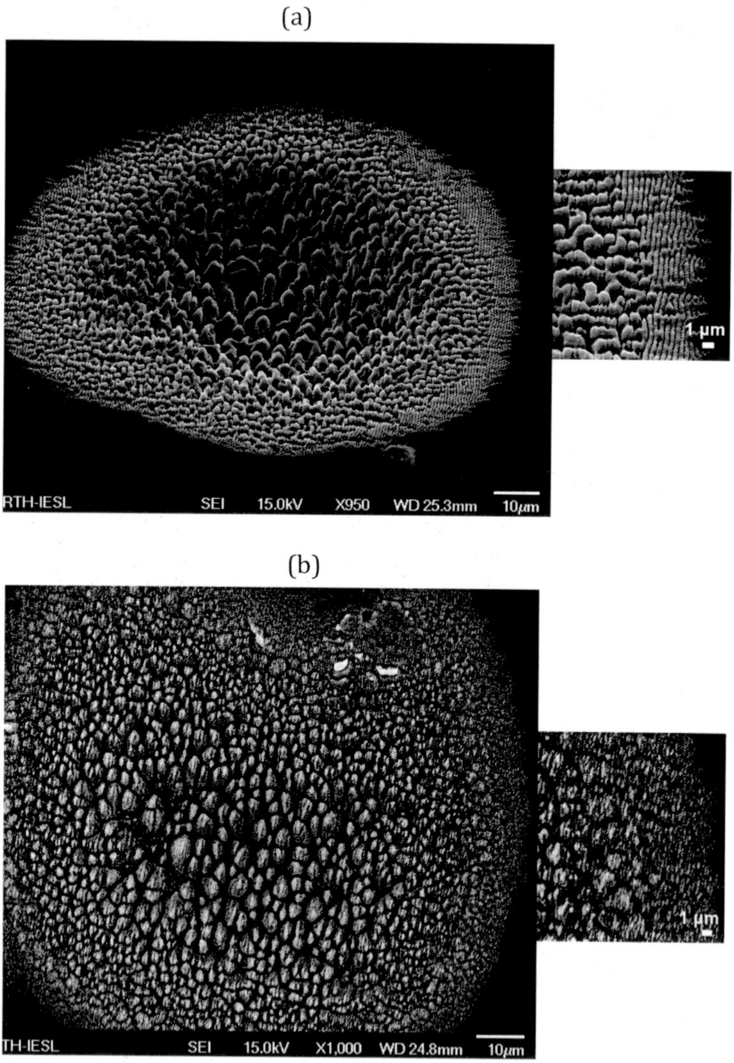

Figure 6.1 SEM images of a semiconducting (a) and a metallic (b) surface after exposure to several 800 nm/fs laser pulses exhibiting a Gaussian beam profile. Submicron ripples, bumps and microspikes are present in both single spots. The insets show high-magnification pictures of the spots' periphery exhibiting submicron ripples and bumps.

fs laser pulses exhibiting a Gaussian beam profile. On a single spot, features of much different sizes are attained. The central part of the irradiated zone contains only microstructures. At the same time, nanostructures are seen in the peripheral parts of the ablation zone, where the laser energy density is substantially lower. Submicron surface structures develop either below ablation threshold when there is a melt in the affected zone or in the ablation regime when the incident intensity slightly exceeds the ablation threshold. An attempt to study the morphology of laser- processed metallic surfaces was made by Vorobyev and Guo [11]. In their experiments, the size and shape of surface nanostructures depended not only on the incident laser intensity and number of pulses but also on the properties of the material and the initial structure of the surface. Adjusting the incident intensity and number of pulses, they were able to produce nano-, micro- and macrostructures and various combinations of these on metal surfaces. There were optimal laser processing conditions for each type of structure, in particular for pure surface nanostructuring, i.e., for producing only nanometre-sized surface structures. In general, nanostructures in the form of nanoparticles or nanometric ripples are produced for low fluences and pulse numbers, while microstructures prevail at high number of pulses at a constant fluence. In the following section, we discuss how fs laser processing can be used to produce changes to the surface morphology at the nano- and/or micro-length scales in order to optimize material performance for potential applications.

6.2.1 Laser-Induced Periodic Surface Structures

6.2.1.1 Characteristics of LIPSS formation

The formation of nanometre-sized parallel periodic fringes is an interesting effect of the interaction of ultrashort laser pulses with semiconductor and other materials surfaces. Such laser-induced periodic surface structures (LIPSS) are usually called ripples. LIPSS were first observed by Birnbaum on various semiconductor surfaces [12]. Since then, LIPSS on solid materials have been studied extensively. The spatial period of classic LIPPS is found to be close to the laser wavelength, while their orientation is always found to be perpendicular to the incident light polarization. However, recent reports indicate that ultrashort fs laser pulses can give rise to the formation of either low-spatial-frequency LIPSS (LSFL) or high-spatial-frequency LIPSS (HSFL) depending on the irradiation conditions. The period, Λ, of fs laser-induced LSFL in different materials is smaller, while that of HSFL is much smaller than the laser wavelength, λ. According to the ratio (Λ/λ) in normal incidence, LSFL fall in the window $0.4 < \Lambda/\lambda < 1$, while HSFL fall in the range $\Lambda/\lambda < 0.4$. In general, LIPSS appear in all materials surfaces, including conducting, semiconducting, dielectrics and polymers. The HSFL are observed in the case of multi-photon fs irradiation at fluences close to the ablation threshold, while the LSFL that frequently coexist with HSFL, tend to form at slightly higher fluence. This is clearly shown in Fig. 6.2 showing both LSFL (central area) and HSFL (periphery) on the same spot of the laser treated ZnO surface. Table 6.1 shows the period of LIPSS formed on different types of materials upon fs laser irradiation with a fluence at the respective ablation thresholds [13]. Such materials include certain metals and alloys, transparent dielectrics (SiO_2, LiF, MgF_2, Al_2O_3, ZBLAN), a non-transparent

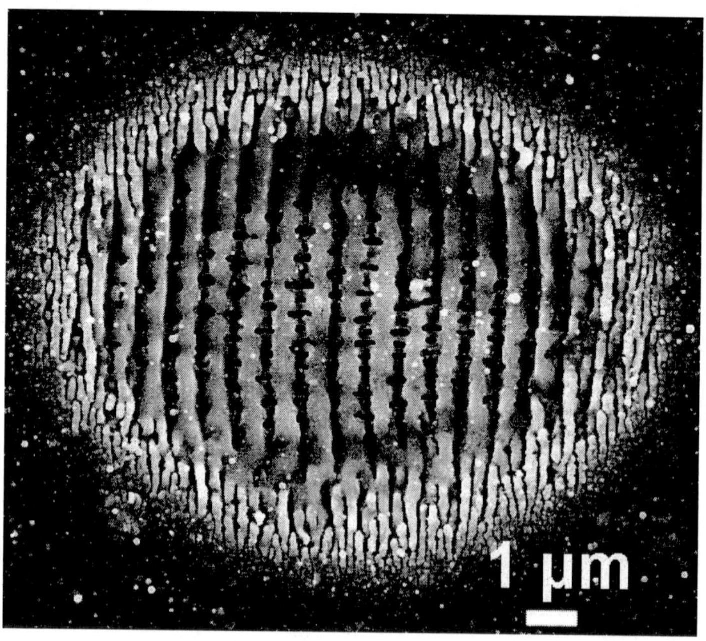

Figure 6.2 SEM image of a ZnO surface treated with several 800 nm/fs laser pulses. LSFL (central area) and HSFL (periphery) ripples are present on the same spot.

Table 6.1 Spatial period of LIPSS formed on different materials after irradiation with 100 fs Ti:Sapphire laser source at 1 kHz rep. rate

Material	Laser wavelength (nm)	Ripple period (nm)
SiO_2(1)	800	189
LiF(1)	800	215
LiF(1)	400	139
MgF_2(1)	800	235
$Al2O_3$(1)	800	276
ZBLAN(1)	800	312
PTFE(1)	800	338
Si(1)	800	532
Al(2)	800	540
Al+Pt film(2)	800	590
Cu(2)	800	600
Cu+Pt film(2)	800	670
WC-Co(3)	800	630
Ti(C,N)(3)	800	620
TiN(3)	800	600

Source:

(1) Ralph Wagner, Jens Gottmann, *J. Phys*, Conference Series 59 (2007) 333–337.

(2) T. Nakashima, T. Sano, S. Nishiuchi, and A. Hirose: Influence of thin metallic film on formation of periodic nanostructure on metals induced by femtosecond laser pulses, *The 6th Asia Pacific Laser Symposium* (Technical Digest), Nagoya, Japan (2008) 127.

(3) G. Dumitru, V. Romano, H. P. Weber, M. Sentis, and W. Marine: Femtosecond ablation of ultrahard materials, *Appl. Phys. A*, 74, (2002) 729.

polymer (PTFE) and a non-transparent semiconductor (Si). Ripples structures with spacing significantly smaller than the irradiation wavelength are observed on all the samples. The clarity of the contour and edges of the linear structures depends on the material. Ripple spacings from 139 nm to 532 nm are detected depending on the material and the wavelength. The direction of the resulting grooves is perpendicular to the polarization of the laser radiation. In most cases, ripples extend coherently regardless of the orientation of the scanning over many partially overlapping laser pulses; hence, a change of the scanning direction does not change the orientation of the ripples. A scanning perpendicular to the polarization results in long parallel grooves. Finally, the LIPSS period depends on the applied wavelength. The purpose of this section is to summarize specific features of the formation of LSFL and HSFL on the surfaces of different materials under variable fs laser irradiation parameters, including fluence, number of pulses, wavelength and light polarization. Additionally, the formation of different nanostructures such as nanoholes and nanoparticles on materials surfaces will be presented.

The initiation and evolution of surface LIPSS shows similar characteristics in all materials surfaces. Periodic surface structures appear only after a number of consecutive pulses, which is found to depend on the type of material, laser fluence, number of laser pulses, and wavelength as reported by Borowiec and Haugen [14] and Vorobyev *et al.* [15]. They reported that after the first few pulses, the surface shows nanoprotrusions with few nanocavities. At a certain pulse number, a nanoscale periodic pattern starts to appear over the initially formed random nanocavities and nanoprotrusions. The initiation of ripples on a metallic surface was investigated by scanning helium ion microscopy by Veld and Veer [16]. As shown in Fig. 6.3a it has been observed that spherical bubbles comprising a mixture of gas and liquid with a diameter in the range of 10–100 nm were initially formed. Upon increasing the laser fluence, bubble formation and initiation of pre-ripples followed by a transformation towards regular course ripples with an orientation perpendicular to laser polarization direction occurs. Then as the laser shot increases, random laser-induced surface structures entirely fill the ablated area. At high pulse numbers, LIPSS starts to disappear gradually in the central spot area. Corresponding AFM images and section profiles of the entire ablated area on stainless steel surface is shown in Fig. 6.3b.

Figure 6.4 shows AFM and SEM images of the modification induced on a Si surface using different number of pulses, *N*, of a 100 fs/800 nm laser with fluence close to the damage threshold. It is observed that the initiation and evolution of ripples on the silicon surface by fs laser pulses are different from that in metals confirmed by several research groups [15–17]. At low *N*, no ripples are formed and several rings are observed on the laser-irradiated area. The ring formation on the laser-irradiated area has been studied by von der Linde *et al.*, who found that the ablation during the ring formation

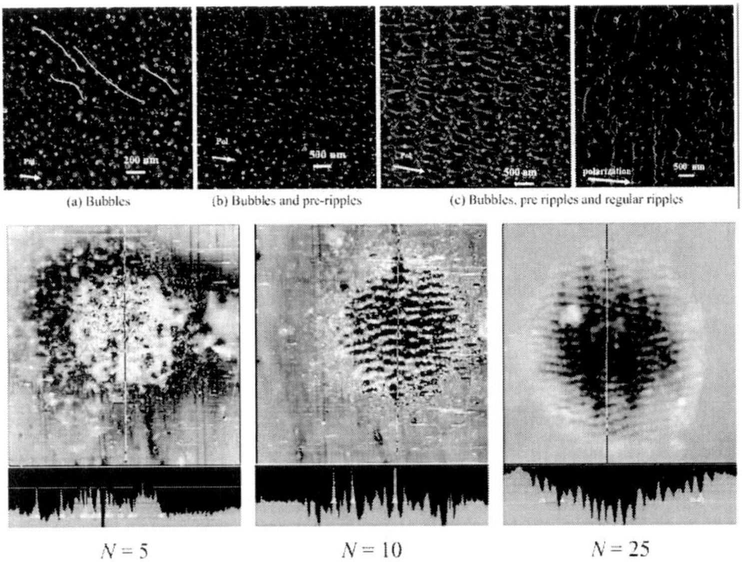

Figure 6.3 (Top) Initiation of LIPSS on steel by femtosecond laser pulses (From B. H. Veld and H. van der Veer, Initiation of femtosecond laser machined ripples in steel observed by scanning helium ion microscopy (SHIM), *J. Laser Micro/Nanoeng.* (JLMN), 5, (2010) 28). (Bottom) AFM images of section profile of the entire ablated area for stainless steel at E_p = 1.04 μJ and different number of laser pulses (From L. Qi, Microstructures fabrication using femtosecond laser, Phd. Thesis, Department of Mechanical Engineering, Chubu University).

was characterized as a rapid thermal process and the characteristic states of the conversion of solid material into hot fluid matter [18]. No particles are found around the laser-irradiated area during the ring formation. At higher N, the laser-irradiated area is characterized by rings or ripples (Fig. 6.4a). Comparing ring and ripple formation, two distinct differences occur. The first one involves a significant difference in the ablation depth; for rings an ablation depth of several nanometre or tens of nanometre occurs, while for ripples a depth of several hundreds of nanometres is observed. Second, a large number of particles are observed around the ripples, whereas no particles are found around the rings. At $N > 25$, the ripples are formed on the entire ablated area and are shown in Fig. 6.4b. Then as the laser shots increases, laser-induced surface structures entirely fill the ablated area. Upon further increasing N LIPSS start to disappear gradually in the central spot area, while at large *N* LIPSS disappear and microgrooves and microcones are formed within the ablated area. During the formation of LIPSS on the Si surface by fs laser pulses, large numbers of particles are observed around the LIPSS area, the density of which increases with the number of pulses. The diameter of such particles is initially around 100 nm and increases as *N* is increased.

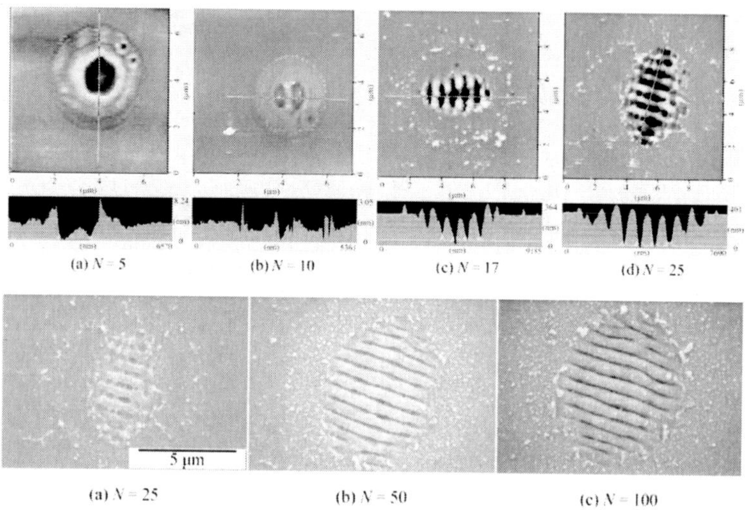

Figure 6.4 (Top) AFM images of the evolution of the morphology of the laser-irradiated area by femtosecond laser pulses on Si at E_p = 0.158 μJ and different number of laser pulses; (Bottom) Evolution of LIPSS on the ablated areas at E_p = 0.18 μJ and different number of laser pulses. (From L. Qi, Microstructures fabrication using femtosecond laser, Phd. Thesis, Department of Mechanical Engineering, Chubu University).

A systematic study of the dependence of ripple period on pulse number for various materials involving dielectrics (ZnO, ZnSe, SiC, and diamond), semiconductors (Si and GaAs), and conductors (graphite and Brass (CuZn)) is presented in Fig. 6.5 [19]. In general, when laser fluence and *N* are appropriate, regular LSFL can be observed in the centre (small *N*) or the periphery (large *N*) of the crater with a direction always perpendicular to laser polarization. In addition, as a whole, their period decreases while *N* is increasing. At large *N*, Λ deviates from λ greatly, and the decreasing rate becomes slower or

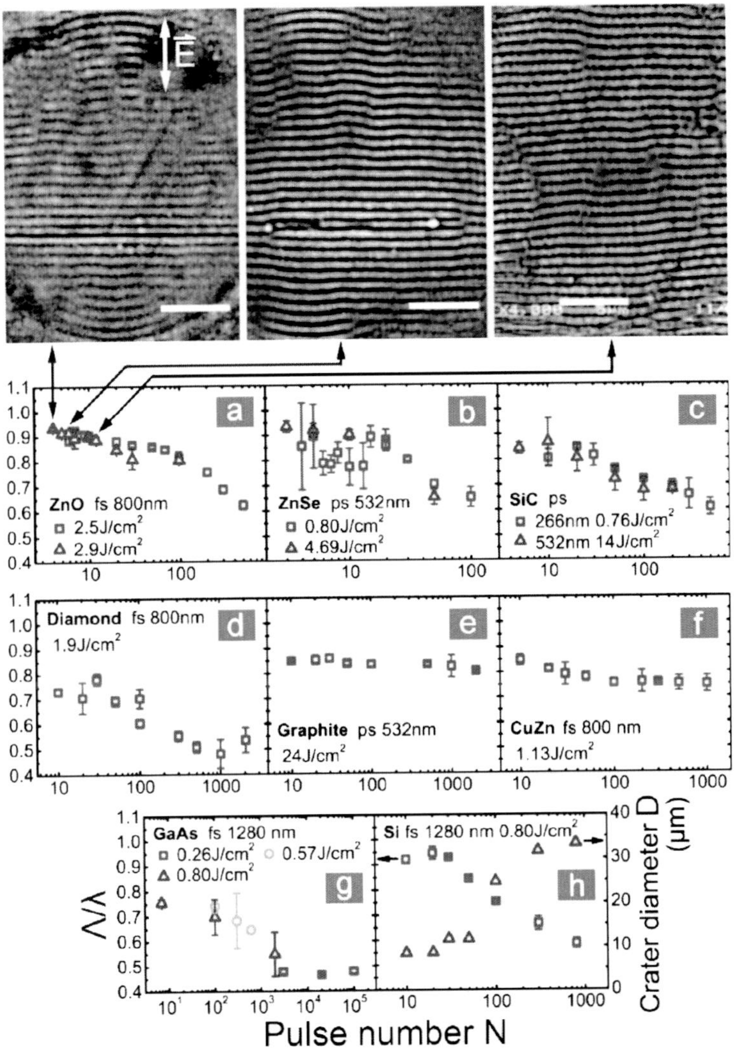

Figure 6.5 Characteristics of near-wavelength ripples on various materials irradiated by short-pulse laser ablation. Λ/λ as a function of *N* for (a) ZnO, (b) ZnSe, (c) SiC, (d) diamond, (e) graphite, (f) CuZn, (g) GaAs, and (h) Si with irradiation conditions indicated in each figure. In panel (h), Λ/λ is measured in the periphery area, and crater diameter D as a function of *N* is also exhibited. SEM images for the typical ripples of ZnO irradiated by different pulse numbers are shown (scale bar, 5 μm; double arrow, laser polarization). From M. Huang, F. Zhao, Y. Cheng, N. Xu, and Z. Xu, *ACS Nano*, 3 (2009) 4062.

saturates. Interestingly, the Λ decreasing phenomenon is far more prominent in the cases of dielectrics and semiconductors than the cases of conductors. Such properties are not strongly influenced by the laser pulse duration since similar characteristics are observed in the cases of ps laser ablation.

Huang *et al.* [20] have systematically studied the laser fluence dependence of the morphological and optical characteristics of large-area ripples produced by fs laser irradiation on GaAs, Si and Cu. It is observed that as the laser fluence increases, the feature size of the ripples and different kinds of nanostructures formed increases following a specific law. In particular, by the Fourier analysis of the treated areas, it is revealed that, in a large laser fluence range, the spatial frequencies, f, of such structures are varying non-continuously following the sequence of $2f$, f, $f/2$, $f/4$, and $f/8$. Furthermore, for a certain kind of structure, as laser fluence increases, the feature size also increases and gradually approaches a plateau value. A typical example is LSFL, whose period is getting close to laser wavelength and saturates at high laser fluences. Figure 6.6 shows the Fourier analysis results for GaAs being qualitatively the same in other materials as well.

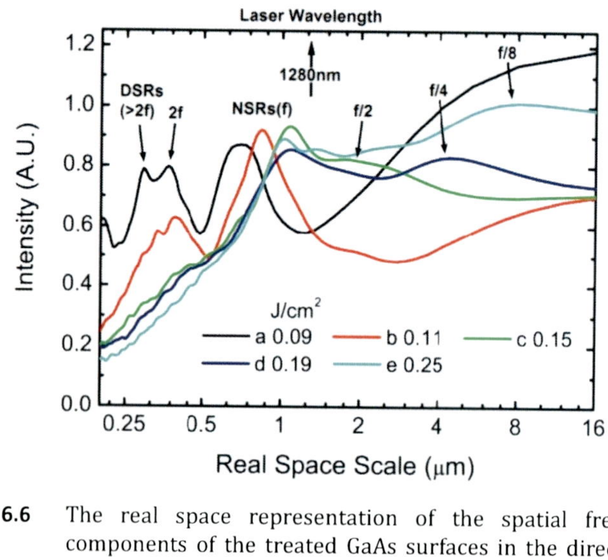

Figure 6.6 The real space representation of the spatial frequency components of the treated GaAs surfaces in the direction of ripple wave vector as a function of the incident laser fluence. (From M. Huang, F. Zhao, Y. Cheng, N. Xu, and Z. Xu, *ACS Nano*, 3 (2009) 4062.)

From the FFT images of the laser treated areas, we can transform the spatial frequency components in the direction of ripple wave vector to the real space. The principal frequency components are tagged in the corresponding positions, from which we can intuitively see that the components of HSFL, LSFL, $f/2$, $f/4$, and $f/8$ have the real space scales of about 0.3, 1, 2, 4, and 8 μm, respectively. Clearly, the real space representation of the frequency components of HSFL and LSFL is consistent with the morphological characteristics of the direct SEM observation. From Fig. 6.6, it is easy to see that with increasing laser fluence the dominant spatial scales of the laser-induced structures shift from a few hundred nanometres to a few micrometres. In detail, a phenomenon for the LSFL can be inspected—the peak of LSFL moves towards the large scale as laser fluence increases and tends to approach a maxima that is related to laser wavelength of 1280 nm.

Heitz *et al.* have reported the formation of ripples on bulk polymeric surfaces exposed to UV fs excimer-laser pulses [21]. In particular, irradiation of PET and PI surfaces with polarized fs KrF-laser light at fluences above threshold results in the formation of coherent ripple structures, which are superimposed by surface structures obtained also with nanosecond (ns) pulses. The ripples have a period of 100 nm to 250 nm and are oriented parallel or perpendicular to the polarization of the laser light. The orientation of the ripples depends only on the polarization of the laser light, while the period does not depend on the laser fluence. The above results complied with the experiments performed using ns pulses indicating that on polymers, ripples parallel to laser polarization are dominant. This is in contrast to the most results on other inorganic materials, where ripples are mostly oriented perpendicular to laser polarization at angles near normal incidence. However, Heitz *et al.* have demonstrated for the first time that upon fs irradiation of some polymers, structure ripples perpendicular to polarization can dominate. On the other hand, ablation experiments performed on PI foils with 800 nm fs laser pulses revealed the formation of ripples that are always parallel to the laser linear polarization [22]. Figure 6.7 (top-left) shows a SEM picture of a cavity generated with 50 pulses of linearly polarized radiation. At the bottom of the crater highly oriented ripples with a period of ~800 nm were observed (comparable to the laser wavelength). The ripple orientation was always parallel to the vector of the electric field. This was verified by rotating the sample in steps of 20° around the beam axis. In the outer region the ripples are very smooth and straight. In the centre of the hole each ripple splits into two parts, showing a small groove in the middle (see Fig. 6.7, bottom-left). The wall of the cavity has a rough appearance without ripples.

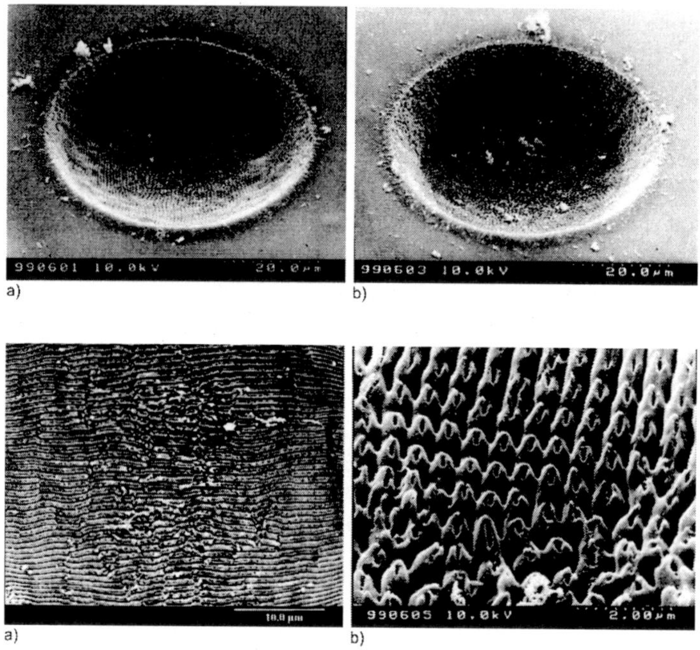

Figure 6.7 (Top) SEM pictures of the irradiated areas on Polyimide, obtained with 50 laser pulses: (a) Linear polarization (vector of electric field is parallel to the periodic structure); (b) circular polarization of the laser pulses (150 fs, 800 nm, 1.3 J/cm^2). The structures formed show the same period in both cases but differ in direction. (Bottom) SEM picture of the central part of the craters in greater detail: (a) Linear polarization and (b) circular polarization. The formation of small grooves and holes can be noticed in the parallel (a) and submicron cone structures. From S. Baudach, J. Bonse, W. Kautek, *Appl. Phys. A* 69 (Suppl.), S395–S398 (1999).

As pointed out above in this section, LIPSS are only formed by linear polarized light, while their direction always depends on the incident polarization. Using circularly polarized light, sub-wavelength particle structures analogously can be formed on material surface with similar irradiation condition [23, 24]. Irradiation of some polymers, i.e., PI, with circularly polarized light gives rise to a ripple pattern that is radially oriented (Fig. 6.7, top-right) [21]. A weak formation of separated domains can be observed, which are distinguished by slightly different ripple orientations. The ripples have the same period as those produced with linearly polarized light. The morphology changes to an array of small cones (Fig. 6.7, bottom-right) towards the centre of the cavity bottom. At the top of each cone, a small hole can be seen.

Fowlkes *et al.* [25] have irradiated Si with ns UV (248 nm) laser pulses. They observe nanoparticles formed on the Si surface that exhibit both short-range and long-range order. While they don't have an explanation for the short-range order they observed similarities for the long-range order with the LSFL, and therefore they suggest a formation mechanism analogous to the interference of the incident laser with a scattered wave. This is supported by the observation that the nanoparticle line spacing follows the law $\Lambda = \lambda/(1 - \sin\theta)$.

6.2.1.2 Proposed mechanisms for LIPSS formation

Birnbaum [12] was the first to observe the formation of a laser-induced periodic surface structure (LIPSS) after irradiation of semiconductor surfaces with a CW Ruby laser. He explained this ripple formation as the result of light diffraction effects at the focal region. The dips are produced at the region where removal of material is higher due to the higher local intensity of the electric field. Conversely the heights are the regions where less removal of the material has occurred due to the minima of the electric field. His model is based on basic optics principles, and it gave a semiquantitative description of the ripple periodicity.

Later on, Bastow [26] studied the ns laser irradiation of a metal surface and found micro craters as the result of the first stages of irradiation. Upon continuing irradiation, he observed ripple formation. He reported that the cause of the periodic structures was unipolar arcs that are created between the metal surface and the hot electron plasma above the surface. To explain the ordering of the arc sites, it is assumed that under intense radiation all points on the surface are equally likely to be arc sites, and that the positions at which arcs initiate are determined by points of high potential on the face of the plasma cloud nearest to the metal surface. He suggested that these points of high potential are the result of instability within the plasma, the wave nature of which leads to the ordered spacing of the arc sites.

Emmony *et al.* [27] used Ge as the target and suggested that a scattering centre on the surface might be responsible for the observed ripples. The interference of an incident wave with waves scattered by any kind of irregularity will give fringes with spacing equal to the laser λ under normal incidence and spacing equal to $\Lambda = \lambda/(1 \pm \sin\theta)$ depending on the direction of the scattered wave relative to the incident laser beam when the angle of incidence is other than the normal to the surface. They later explained their observations in terms of the appearance of electron avalanches around low ionization energy centres, close to or at the surface [28]. Radiation doublets might be formed at the interface, constructive interference of which would then be the cause of the periodic structures. Specifically, they observe both a ripple structure with wavelength close to the laser wavelength and a new additional ripple structure with periodicity perpendicular to the previous one and a wavelength much smaller than that of the laser. They suggest that the near field of the induced current doublet gives rise to the fine ripple pattern, while a surface wave produces new avalanches and/or increased damage at a spacing close to the laser wavelength.

Leamy *et al.* [29], based on these findings, further suggested that as a consequence of the interference fringes, the melting threshold was periodically exceeded leaving alternating regions of different crystallinity. As they stated, the intensity variations will produce periodic melting or erosion of the surface if the power density is near the melting or plasma threshold, respectively. They concluded that the ripples are produced by intensity oscillations about a threshold that is exceeded in the centre of the spots, and they suggested that ripple formation occurs when the melting threshold is periodically exceeded.

Maracas *et al.* [30] argued that their observations on laser annealing of GaAs could not support an interference model where the incident beam interferes with a surface scattered wave. Instead, they suggested that their observations were consistent with a non-linear interaction taking place in the sample between the several electric fields of the present axial modes. This interaction would produce a standing acoustic wave pattern that would drive the molten material to nodal lines of this pattern.

Later, according to Becker, Walser, and coworkers [31, 32], who have modified this idea and adapted it to the case of Si, the mechanism is resonant absorption of photons by plasmons at the surface of small metallic spheres, or droplets, of a critical size. The droplets would be produced by thermal fluctuations driving the excited solid locally above the melting temperature.

Several authors have suggested that surface plasmons or surface polaritons could produce the ripples. For example, Van Vechten [33] first proposed that ripples may occur as a consequence of the condensation of plasmons in regions with high carrier density. Similar qualitative ideas were proposed by Rendell and Ngai [34]. Keilmann and Bai [35], working in the infrared reststrahlen region of quartz, produced evidence for a dispersive behaviour of the ripple period and attributed that phenomenon to the action of surface polaritons. They were also the first to provide photographic evidence with the appearance of ripples with a spatial period far from the laser wavelength and close to half of it. They suggested that this kind of ripples with extraordinary lower wavelength than that of the laser could possibly also be ascribed to surface polaritons.

Guosheng *et al.* [36] summarized the early results of the ripple formation by laser irradiation: The ripples are almost always perpendicular to the incident laser field. At normal incidence, the period of the ripples Λ is almost equal to the laser wavelength λ. For oblique incidence and p-polarization, it is $\Lambda = \lambda/(1 \pm \sin\theta)$. For oblique incidence with s-polarization, the spacing remains equal to the laser λ. For circular polarization, no ripples are observed. Ripples can be produced by a single or many laser shots. If the surface is being displaced during multiple-pulse irradiation then there are two possibilities: (1) if the displacement is parallel to the incident polarization, the ripples can extent coherently over many laser shots and (2) for displacements perpendicular to the laser polarization, the ripples in one spot have, in general, no phase relationship with the ripples in an adjoining spot. Ripples are independent of the atmosphere and of the crystallographic direction of the surface if it is crystalline. The ripples appear at intensities close to the melting threshold.

They believe that surface plasmon-polaritons are not the cause of the surface ripples because first, in semiconductors extremely large number of carriers are required for $\omega_p > \omega$, a necessary condition for plasmons to occur and second, the wavelength of the plasmon would need to be quite different from that of the laser. Therefore, they developed a theory for the interpretation of the periodic structures formed on surfaces (metallic and semiconductor) following laser irradiation. The theory predicts the observed spacing, polarization and growth properties of the ripples. This model is based on the Fourier components of a random surface disturbance which scatters the incident light into directions along the surface with a velocity close to the speed of light. The interference of this diffracted optical wave with the incident beam then gives rise to optical interference fringes which can reinforce the initial disturbance. The ripples can grow with illumination by increasing the number of laser shots as the selection process slowly eliminates all the Fourier components except the one with the right spacing, which grows and finally becomes visible. The ripples also propagate coherently after successive and partially overlapping laser shots since ripples in the first spot scatter the incident wave in the next spot and intensity fringes are produced with correct spacing. Therefore, the ripples propagate coherently with the ripples produced in the first shot acting as a seed grating for the subsequent laser shots.

Sipe, Young and coworkers [37, 38] performed experiments [38] and developed a theory[37] to predict an inhomogeneous absorption of light on a surface that exhibits roughness. They found strong peaks in the deposition in Fourier space, which led to predictions of induced fringe patterns with spacing and orientation dependent on the angle of incidence and polarization of the damaging beam. At this point, the occurrence of the fringes had been attributed to the prevailing model of the interference of the incident beam with a "surface-scattered wave" originating at a scratch [27, 39]. Although this picture of the damage formation has been appealing because this simple picture is able to predict the experimental results, it is not physically consistent since, first, the damage patterns are presumably produced by the absorption of energy within the material, while it is not clear why the surface-scattered wave responsible should propagate with a wavelength of λ as opposed to λ/n, where n is the refractive index of the material and second, at normal-incidence excitation, the scattered wave would have to be longitudinally polarized; such waves do not satisfy the Maxwell equations. In fact, plane waves of any polarization, propagating parallel to an interface, do not satisfy the Maxwell equations.

Sipe, Young and coworkers calculated the generation of non-radiative field structures that they call "radiation remnants", which replace the "surface-scattered" waves. These are not radiation fields in the sense that they can have either transverse or longitudinal polarization but are similar to radiation fields in that they satisfy the dispersion relations in vacuum and in the underlying medium. Additionally, feedback should be introduced in a complete theory [40]. The theory predicts that the periodic surface damage can be produced in the vicinity of defects; however, a microscopically rough, albeit macroscopically smooth, surface is sufficient for the damage to occur.

Sipe, Young and coworkers then performed experiments on Ge, Si, Al and brass [38] and studied the induced periodic surface structures in the Fourier space with the help of a diffracted weak cw laser source. All materials contained similar structures. Whereas periodic ripple patterns oriented perpendicular to the polarization at near-normal incidence are commonly reported, the diffraction patterns reveal that, in fact, there exists a continuous distribution of periodic structure oriented at all angles with respect to the polarization. At near-normal incidence there are two dominant sets of "fringes" running perpendicular to the polarization with the known spacing $\lambda/(1 \pm \sin\theta)$, while for a p-polarized beam incident at >35°, there exist three dominant periodic structures: two run perpendicular to the polarization with spacing as above and one is oriented parallel to it with spacing $\lambda/\cos\theta$, where θ is the incidence angle with respect to the surface normal. For s-polarized light incident at $\theta > 35°$, there are two dominant patterns which form a cross-hatched pattern with axes oriented at 45° to the plane of incidence. Sipe, Young and coworkers found that both the initial and laser-induced surface roughness play important roles in the evolution of the damage. They observed fringes even after a single laser shot in a narrow annular region around the perimeter of the beam. Irradiation of the spot with successive laser pulses results in fringes that are formed not only in the narrow annular region but extending towards the center of the spot. At the same time, pits with diameter (1–3) μm are being developed in the central region of the spot with the irradiation by successive laser pulses. After 30 pulses, the entire surface is covered with fringes. For the case of metals, in contrast to semiconductors and dielectrics, the authors concluded that the excitation of surface plasmons rather than the generation of "radiation remnants" suffices to explain the ripple formation.

Ozkan *et al.* [41] reported the observation of ripples on diamond surfaces with periods far from the laser wavelength (HSFL) and close to $\lambda/2$; however, the first observation seems to be by Keilmann and Bai [35]. Although they attributed this to interference between the incident and the scattered fields, the ripples fail to follow the

$\Lambda = \lambda/(1 \pm \sin\theta)$ law. Borowiec *et al.* [14] studied the irradiation of InP, GaP, GaAs, InAs and Si with fs laser pulses at wavelengths 2100, 1300 and 800 nm in order to study the influence of the either below-gap or above-gap excitation of the semiconductors, or, in other words, the influence of the transparency or opacity in the ripple formation. They claimed to be among the first [23, 41–44] to observe a new family of ripples with spatial period well below the laser λ although the mention of such "fine" structures dates back to Kailmann and Bai [35]. They found that when working in the transparency regime, the high-spatial-frequency-LIPSS (HSFL) are emerging, while when working in the opacity regime the low-spatial-frequency-LIPSS (LSFL) are favoured. The fluence levels are close to the single-pulse ablation thresholds. The formation of HSFL is independent of the crystal orientation relative to the laser polarization. They suggested that because of this insensitivity, second-harmonic generation in the bulk of undamaged semiconductors does not play a key role, despite the approximate correspondence of some of the HSFL periods and laser second harmonic wavelengths. However, this is not excluded for the laser modified material.

Bonse *et al.* [45] irradiated InP (bandgap = 1.344 eV @ 300 K) with varying number of pulses and confirmed the appearance of two different kinds of ripples. For low pulse number, they found the LSFL with spacing close to the laser wavelength, and for pulse number 5–30 they observed the additional formation of the HSFL with spacing close to half of that of the laser. Both kinds of ripples have the same orientation, i.e., vertical to the laser polarization. For pulse number greater than 50, Bonse *et al.* found the formation of micrometre-spaced grooves which are oriented perpendicular to the ripple. They stated that the LSFL are a universal phenomenon that is observed in various absorbing materials such as metals, ceramics and semiconductors observed at fluences slightly above the single-pulse ablation threshold. The new HSFL with wavelengths as low as $\lambda/6$ had been observed by various groups at fluences below the single-pulse ablation threshold [38, 40, 46, 47]. Their nature is still very controversially discussed in the literature. After the first laser pulse, Bonse *et al.* did not observe any periodic structures, but after the second pulse, some periodic surface features were present in their 2D Fourier transforms analysis. Hence, they concluded that the first laser pulse generates a roughened surface leading to some scattering of the subsequent pulses. Additionally, they found the

HSFL, which have roughly half the period of the LSFL. They pointed out that the novelty in these HSFL forms—observed previously by, for instance, Borowiec and Haugen [14]—was that here they observed them for a laser wavelength where the InP was opaque, while in the other case, InP was transparent to the used laser wavelength. Because they observed that the lateral period of the HSFL is half of the LSFL and that HSFL form at the pulse number window where the LSFL are most pronounced, they suggested that the HSFL are created by a non-linear response of the surface structured with the LSFL, i.e., the involvement of a second-harmonic generation at the rough surface. The conventional theory of the surface scattered wave first suggested by Emmony *et al.* [27] and then improved by Sipe *et al.* [37] and Guosheng *et al.* [36] accurately described their observed LSFL but was not able to describe the HSFL and the groove structures.

Huang *et al.* [48] irradiated the surface of ZnO with fs laser pulses at 800 nm wavelength and alternated the pulses with crossed polarization. They managed to construct a cross-ripple configuration with 290 nm size and a square geometry on the surface of ZnO, which belongs to the family of the HSFL. They found that the chopper frequency that alternates the irradiation of the two beams is critical to the quality of the orthogonal structures. They additionally found that the ability of the formation of a uniform submicron square structure by this technique exhibits a non-interference effect, i.e., the two orthogonal subwavelength LIPSSs formed by the beams with orthogonal polarizations are isolated form each other. They attributed this to the spatial and the temporal non-interference of the alternating laser beams. Later on, they studied the deep sub-wavelength ripple formation (HSFL) in diamond and graphite [49]. They suggested that inside the grooves, the light intensity is enhanced via cavity-mode excitation in the condition of transverse-magnetic wave irradiation. The field enhancement acts as a feedback mechanism for the growth and polarization dependence of the nano-gratings (HSFL). Huang *et al.* further suggested that surface plasmons are responsible for the formation of the initial seed deep-subwavelength apertures in the view of recent models of interference between the light and plasma waves [50] and growth of nanoplasma for the formation of planar nanostructures inside fused silica by localized nanoscale explosions [51]. Similar ripples were observed in both the transparent diamond and the opaque graphite showing universality. This is supported by results from other groups; therefore they concluded that multiphoton absorption is not a necessary condition to form HSFL. These results are reported for Si [47, 52–54] InP [45] and Cu [55]. The material properties at the excited state instead of the normal state should therefore be considered.

With regard to LSFL, Huang *et al.* [19] further studied dielectrics (ZnSe, ZnO, SiC), semiconductors (Si, GaAs) and conductors (CuZn, Graphite). According to them, the ripples result from the initial direct SP-laser interference and the subsequent grating-assisted SP-laser coupling and that additionally to HSFL, the LSFL cannot be adequately described by the simplified scattering model [37]. They proposed that the irradiated surface by short laser pulses should exhibit metallic behaviour irrespective of the initial metallic, semiconducting or dielectric nature of the surface. Therefore the surface may satisfy the metallic conditions ($\varepsilon < -1$)) for launching SPs. Thus, the high-excited characteristics of the surface-laser-induced SPs are responsible for the formation of LSFL, and the initial SP-laser interference and the subsequent SP-laser coupling can provide a complete picture for understanding a variety of phenomena about LSFL. Huang *et al.* observed that Λ/λ decreases with increasing pulse number N and therefore at high N the deviation of Λ from λ is large. The decrease of Λ/λ with N is more prominent for dielectrics and semiconductors than in conductors. This is attributed to grating-assisted SP-laser coupling. If the phase-matching conditions hold, then the initial fringes formed by previous pulses act as a positive feedback to later ones. Hence, the grooves will grow with N and become deeper with the resonant wavelength displaying a red shift.

Bonse and co-workers [56, 57] irradiated Si with fs laser pulses to study the LSFL. They claimed that no satisfactory explanation has been presented why the experimentally observed fs-laser-generated LSFL in Si typically are $0.62<\Lambda<\lambda$ and why the occurrence of the LSFLs is restricted to a certain fluence range only [56]. Their results suggest that the early stage of LSFL formation in fs laser-irradiated Si is dominated by the excitation of surface plasmon polaritons (SPPs). These can be observed already for single pulse irradiation and they act as a seed for the subsequent feedback phase typically involved in the formation of LIPSSs. They are based on the model by Sipe *et al.* [37]. Their calculations support the experimental findings for the polarization dependence and that there exists a narrow fluence range around the Drude plasma resonance for the appearance of LIPSS. As they point out, the excitation of SPPs is allowed in Si at the critical density where the dielectric permittivity changes sign thus allowing SPP excitation. This is already possible by the first pulse and by interference of the incident beam with the surface near electromagnetic field generated by the SPP which was excited, for example, via coupling to a surface defect already present during the fs laser pulse.

The initial periodic surface energy deposition will act as a feedback for the ripples strengthened by the next laser pulses [57]. A decrease in the LSFL period with increasing pulse number is attributed to (i) changes in the local fluence which changes the carrier density and therefore the SPP frequency; (ii) the grating-assisted surface plasmon-laser coupling that will change as the grooves deepen thus changing the resonant wavelength of the SPP; (iii) changes in the local angle at the border region of the ablated spot [45].

Sakabe *et al.* [58] propose a different interpretation of the grating features produced by irradiation of metal surfaces by fs laser pulses. They interpret the grating formation in terms of parametric decay of laser light to surface plasma waves. They claim that with the exception of self-organized nanogratings in glass materials [50] where the cause for the gratings is considered to be laser-induced plasma waves in bulk plasma, the physical mechanism for the self-formation of periodic gratings by fs laser irradiation has remained unexplained. They perform experiments on Cu surfaces and they observe (1) the interspaces are $<0.85\lambda$, (2) the interspaces depend on the laser energy fluence, (3) the interspaces become shorter in a discontinuous manner near the ablation threshold and the fluence decreases, and (4) the gratings are perpendicular to the laser polarization. They claim that none of these features can be observed for ns-pulse laser ablations.

They propose the following: (1) Plasma waves are induced on the surface by the fs laser pulse; (2) spatially localized ion clouds Coulomb-explode to vacuum, and consequently the thin layer is ablated, and the interspacing of the grating is printed at the stage of the first several pulses (it is considered that the number of pulses depends on the laser fluence, and at certain high values for the fluence, only a single pulse is involved in this and the following processes); and (3) for the subsequent pulses, the electric field is enhanced near the initially printed structures, and the near field ablated the surface, thus deepening the structures. These do not occur for ns pulses because the surface obscured by the generated plasma which is opaque to the trailing photons of the laser pulse. For fs time windows, the plasma is still in close contact with the solid-state matter.

The same group in a subsequent work [59] extended their findings and their model in [58] into additional metals (Al, Au, Cu, Ti, Pt, Mo, and W). For the self-formation of periodic gratings on metal surfaces, the interspaces of the periodic structures depend on laser fluence. This dependence is the same for all metals, although the range of laser fluence in which the structures are formed differs between metals. The laser fluence dependence can be explained by the generation of a plasma wave through the parametric decay of laser light [58]. This indicates that the formation of periodic structures depends not only on metal properties but also on the electron density of plasma produced on a surface by fs laser pulses. They [59] mention that the spacing of grating structures for transparent materials are much shorter than the spacing for metals; therefore, the mechanisms proposed for transparent materials cannot be applied to metals.

Although the process by which the generation of the surface-plasma wave leads to the formation of periodic structures is not yet fully understood, they assume the following: when the plasma wave travels slowly with respect to the laser pulse duration, locally charged areas are periodically generated on the surface and positively charged areas are exploded toward free space by the Coulomb repulsive force (Coulomb explosion); consequently, a thin layer is ablated and the grating pattern is printed. The wave number (wavelength) of the surface-plasma wave induced on the surface then depends on only the plasma frequency (electron density) of the surface plasma [58].

Huang *et al.* [20] studied the ripples in GaAs, Si and brass and found that as the laser fluence increases, the spatial frequencies appear to follow a sequence of $2f$, f, $f/2$, $f/4$, and $f/8$, where f is the fundamental frequency corresponding to the near-subwavelength ripples. In their opinion, the new frequency components of $f/2$, $f/4$ and $f/8$ originate in the second-order, fourth-order and eighth-order grating coupling, respectively. Also, they observe that there exists a certain fluence window where each kind of structure appears: HSFLs tend to appear at fluences approaching the ablation threshold, whereas LSFLs are apt to appear at fluences higher than the ablation threshold. Han and Qu [60] irradiated Si with fs laser pulses. They observed ripples and suggested that the origin of the ripples lies in the interference of laser with surface plasmons. They found a decrease of the spatial period of the ripples with increase pulse number and attributed this to the decrease of the surface plasmon wavelength with the decreasing electron number density in the plasma in the perimeter of the laser damaged region, thus confirming the model and results of Huang *et al.* [20]. They also observed nanoparticles and debris with diameter 50–500 nm and explained this by a shock wave formation.

6.2.1.3 Applications of femtosecond laser nanostructured surfaces

The regular nanostructuring of various materials has been gaining widespread attention owing to the increasing application of nanostructures in various fields for controlling the optical, electronic, wetting, and biological properties of matter [6, 61–64].

Guo *et al.* were the first to introduce the concept of the alteration of the optical properties of metallic surfaces with ultrafast laser nanostructuring. Using a fs laser surface they turned highly reflective metals into highly absorptive, creating the so-called "black metals" [64]. Figure 6.8 (top) shows the morphology of laser-treated Al samples that appears pitch black and grey, respectively. However, it is demonstrated that when the experimental conditions are modified this technique also allows the creation of metallic surfaces that exhibit various colours at different viewing angles as well. Figure 6.8 (middle) shows such colours and Fig. 6.8c the respective reflectance spectra of Al surfaces prepared under different laser fluences, and scanning speeds. As shown in the respective SEM images, pronounced nanostructures covered LIPSS are covering the treated areas. The LIPSS period is about 540 nm, and this value is significantly less than the laser wavelength used in the experiment (800 nm). We note that producing LIPSSs is a very versatile way in modifying optical properties since the period of the LIPSS can be controlled. For linearly polarized incident laser light, the LIPSS period d is given by $d = \lambda/(\eta \pm \sin\theta)$ with $g//E$, where λ is the incident light wavelength, $\eta = \mathrm{Re}[\varepsilon/(\varepsilon + 1)]^{1/2}$ is the real part of the effective refractive index of the air–metal interface for surface plasmons, ε is the dielectric constant of the metal, θ is the incidence angle of the laser light, g is the grating vector, and E is the electrical field vector of the incident wave. This equation shows that the period of LIPSSs can be varied by changing the laser wavelength, the incidence angle, or the real part of the effective refractive index. By applying the fs laser blackening technique directly to a tungsten incandescent lamp filament, Guo *et al.* achieved a remarkable brightening of the tungsten lamp and enhanced its emission efficiency to approach 100% [65]. The uniform increase of absorptance of the laser structured black tungsten promises to produce extremely efficient ultrabroadband light sources with flat spectral response.

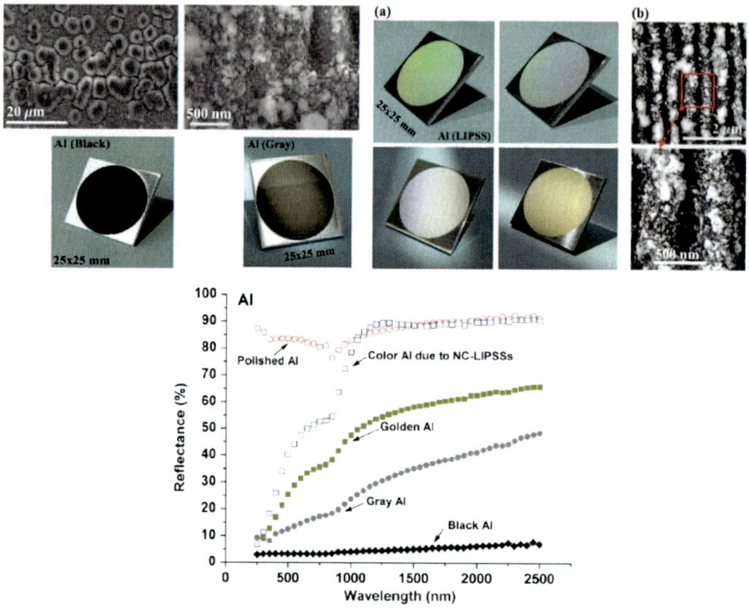

Figure 6.8 (Top) SEM images showing micro- and nanoscale features of laser treated black and grey Al samples, respectively. (Middle) Colour Al due to femtosecond laser-induced nanostructures covered LIPSSs. (a) Photographs show that the same Al sample exhibits various colours depending on the viewing angle. (b) SEM images showing LIPSSs on the colour Al sample in (a). (Bottom) Wavelength dependent reflectance of polished, golden, colour, grey, and black Al. (From A. Y. Vorobyev and C. Guo, *Appl. Phys. Lett.* 92, 041914 (2008)).

The periodic morphology of fs laser nanostructured surfaces combined with a metallic particles coating was used to provide surface-enhanced Raman scattering (SERS) active substrates [66, 67]. SERS signals from molecules adsorbed on such surfaces show a spatially uniform enhancement factor of approximately 10^7.

Spectroscopic characterization of these substrates suggests their potential for use in few or single-molecule Raman spectroscopy.

It has been shown that the efficiency of a bare silicon photovoltaic cell was improved when LSFL ripples (700–900 nm spacing) were generated on the surface after irradiation with fs laser pulses [68]. The performance could be further enhanced with the fabrication of HSFL periodic nanostructures on surfaces involved in solar cell fabrication.

Nanostructuring of materials is highly desirable for various biomedical applications. Proliferation of biological cells proceeds more efficiently if the surface has a certain degree of roughness. It was found that the optimal roughness lies in the range of several hundreds of nm. In this sense, the formation of nanostructures upon fs laser processing ideally suits the needs for fabrication of efficient culture substrates or medical implants. Figure 6.9 shows the example of fibroblast cells cultured on a fs laser structured nanorippled Si surface.

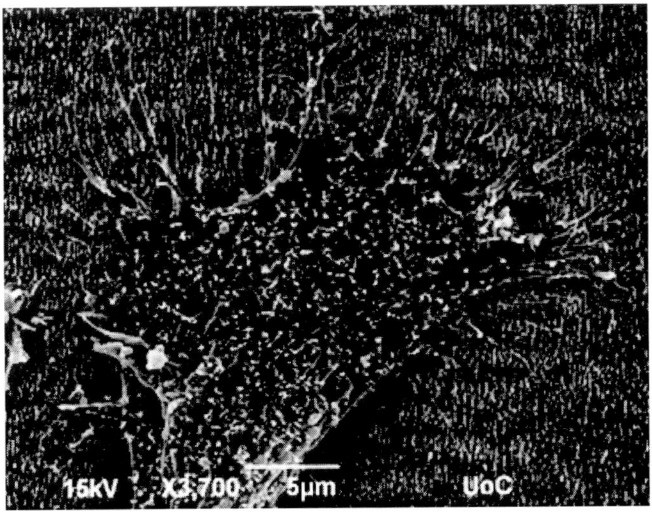

Figure 6.9 SEM image of a fibroblast cell cultured on a nanorippled Si substrate produced by 200 fs, 800 nm laser ablation of Si.

6.2.2 Formation of Micro and Micro/Nano Conical Structures

The formation of conical microstructures with laser irradiation was first reported in the mid 1980s by Rothenberg and Kelly, who produced micron-sized irregular cones protruding above the surface of Al_2O_3 using ns UV laser pulses [69]. Such laser-induced "sputter-cones" were initially an unwanted byproduct of laser sputtering of targets in pulsed laser deposition experiments. Sputter-cones produced in air or vacuum atmosphere were irregular, disordered across the surface and blunt. In the mid- to late 1990s, a few research groups began to examine more exotic structure formation on silicon following laser irradiation [70–72].

Mazur *et al.* have first reported the formation of regular arrays of conical microstructures on Si in 1998 [73–74]. It is shown that irradiation of a Si surface with a pulsed fs laser in halogen-containing ambient gases (the first report was for sulphur hexaflouride (SF_6)) results in a quasi-ordered array of micrometre-sized, conical structures. The microstructures develop upon irradiation with hundreds of pulses and form without the use of masks, while their morphology exhibits spontaneous ordering and sharpness unmatched by other laser structuring methods at the time. The size and shape of such surface microstructures was found to be a function of the laser fluence, wavelength and pulse duration as well as the background gas chemistry and pressure. Furthermore, it is observed that the grey mirror finish of the silicon wafer changed to a deep matte black colour (Fig. 6.10) wherever the laser irradiated the surface. This material is subsequently called "black Si". Following this report, similar structures have also been created in a number of semiconducting [75, 76] and metallic surfaces [77, 78]. On the contrary, microstructures formed in the presence of non-halogen-

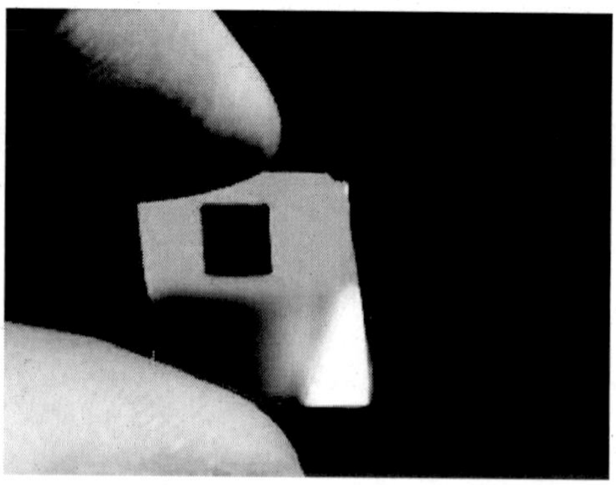

Figure 6.10 Photograph of black silicon area patterned on a planar Si wafer by fs laser structuring.

containing gases, such as N_2 or He, tend to be less conical and more blunt and irregular in shape.

Stratakis *et al.* were the first to report that under certain irradiation and ambient atmosphere parameters, the microcone morphology exhibits a secondary roughness at the nanoscale (see Fig. 6.11) [79–81]. They underlined that the morphology of dual rough micro/nanostructured black silicon and metallic surfaces mimics that of natural biological surfaces [82, 83]. Such hierarchical surfaces have been proved to be much desirable for a number of applications, including optoelectronics, microfluidics and tissue engineering [6].

(a) Si

(b) Ti

Figure 6.11 SEM images of micro/nano spikes on (a) Si and (b) Ti surfaces fabricated by 800 nm/180 fs laser structuring. The micro and nanoscale structure of a single cone is shown in the insets.

6.2.2.1 Characteristics of micro/nanostructures formation

This section summarizes the main characteristics of the formation of micro/nanostructures on materials surfaces using fs laser pulses under variable irradiation parameters. In particular, the effect of laser fluence, pulse number and duration, light polarization and ambient gas will be analysed.

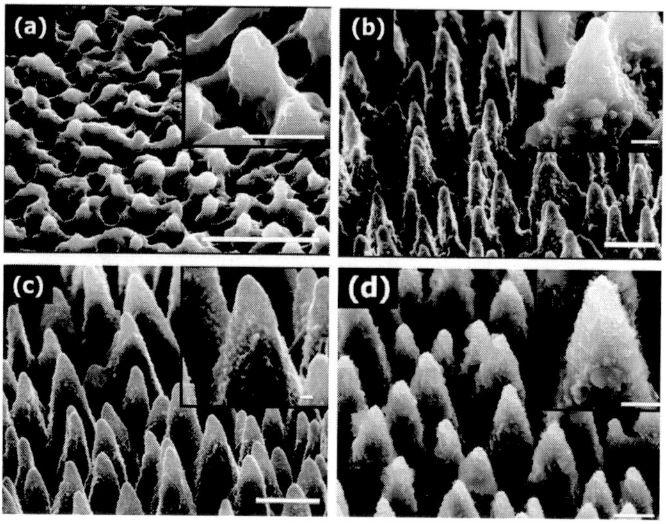

Figure 6.12 Side-view scanning electron microscope image (45°) of Si surfaces structured by 800 nm, 180 fs irradiation at different laser fluences. (a) 0.37 J/cm^2, (b) 0.78 J/cm^2, (c) 1.56 J/cm^2, and (d) 2.47 J/cm^2 (scale bar 5 μm). Higher magnifications of the obtained structures are shown in the corresponding insets (scale bar 1 μm).

The textured Si surfaces realized by employing the same number of laser pulses at different irradiation fluences are presented in Fig. 6.12. Increasing the incident energy per unit area causes remarkable changes in the structure's shape, dimension and density. At low fluences, laser heating induces melting of the surface producing a rippled landscape, with structures not completely physically separated. Upon increasing fluence, conical morphology is promoted on the surface, with spikes becoming more pronounced and spatially separated. In the low-fluence regime, the average inter-spike spacing, base diameter and height increase with laser fluence, whereas their density decreases (Fig. 6.13). For large fluence values (above

~1.0 J/cm^2), the spike density reaches a plateau, while the height stabilizes around 10 μm. Besides directly affecting the micrometre-scale surface topology, increasing fluence is also crucial to induce a more pronounced sub-micrometre decoration on the spikes walls. In particular, the protrusions with size from tens to a few hundreds of nanometres, which provide the double length-scale pattern on the Si surface, become more evident as the laser fluence increases (Fig. 6.12d). The micrometre-scale features generated by the spikes landscape together with the nanoscale features generated by the surface prongs on the cones result into a significant increase of the overall roughness. Similar textures can practically be observed upon increasing the number of pulses for a constant fluence above threshold. It should be outlined that the reactive gas plays a distinct role in the fabrication process, since it determines the sharpness of the microstructures obtained. Indeed, the formation of second-length scale roughness on the cones surface becomes more pronounced with increasing ambient gas pressure. Furthermore the number density of structures created was greatest in SF_6, slightly more than Cl_2, but approximately twice that of N_2 and air.

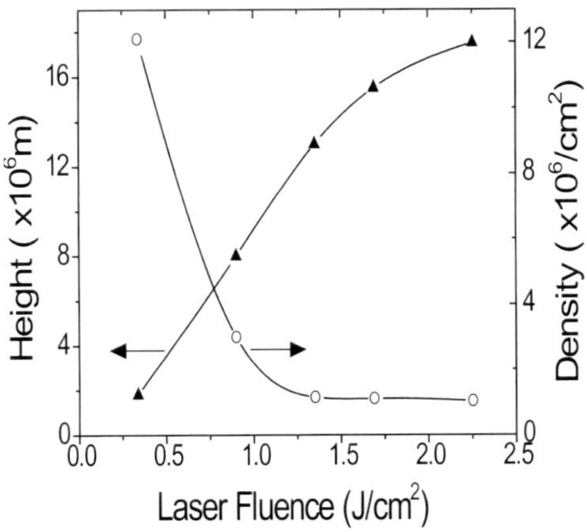

Figure 6.13 Dependence of microcones height and density on the incident laser fluence. Irradiation is performed by a 800 nm/180 fs laser source.

The pulse duration also has a noticeable effect on the resulting morphology. It is found that despite similar near-unity broadband absorption, processing in SF_6 with fs pulses produced significantly different structures on the surface of silicon than ns and ps pulses [84]. The fs-formed structures are about 10 μm tall with their tips level with the original surface, indicating that ablation and etching dominate the formation. Furthermore, the secondary nanoscale features are much more pronounced. The ns-formed structures are smoother but eight times taller at ~80 μm, and protrude from the original surface indicating that material deposition played a part in the growth process. Both cases produced structures with a crystalline silicon core covered with a highly disordered layer of nanocrystallites, nanopores, and sulphur impurities. However, this layer was much thinner and sparser on the ns-formed structures [85].

Two additional laser characteristics that affect microstructure morphology are the polarization and laser propagation direction. Figure 6.14a shows that the base of each microstructure is elliptical, with a long and a short axis; the long axis is perpendicular to the polarization of the laser light. This shape is a result of the higher absorption of p-polarized light over s-polarized light [86]. With the initial micro-bead pattern established, laser light hitting the side of a bead along the horizontal direction (for the scanning electron micrographs in this is the left and right side of the beads) is p-polarized light and light hitting the sides of a bead in the vertical direction (top and bottom) strikes as s-polarized light. Therefore higher absorption, and subsequently more ablation, occurs along the sides in the horizontal direction, where the incident light is p-polarized. Figures 6.14a–c shows top views of the microstructures produced at different laser polarizations. When the polarization is rotated by 90°, the long axis of the microstructure is also rotated

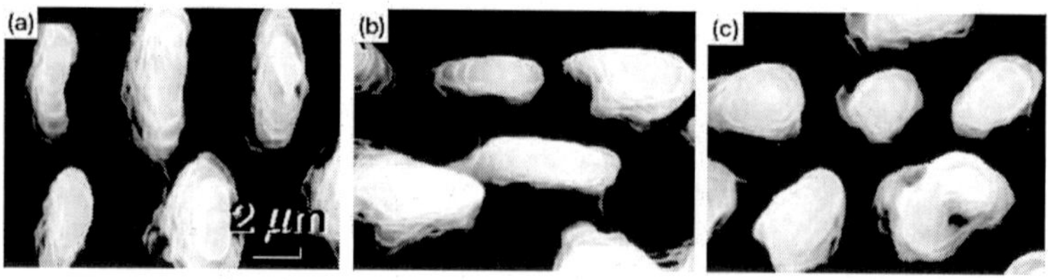

Figure 6.14 Morphology following irradiation with (a) horizontally polarized, (b) vertically polarized, and (c) circularly polarized laser pulses. (From Femtosecond-laser Microstructuring of Silicon for Novel Optoelectronic Devices, J. E. Carey, Phd Thesis, Harvard University.)

by 90°. For circularly polarized light there is no preferential direction for ablation and the base of the microstructures becomes circular. All the above results correspond to normal incidence of the irradiation beam. However, it is found that microspikes align with the direction of laser-beam propagation with little dependence on the crystallographic orientations [87].

6.2.2.2 Proposed mechanisms for micro/nano cones formation

Her *et al.* [73] were the first to report a different-than-the-usual LIPSS-like surface structure produced by irradiation with lasers. This was about the regular, self-organized sharp microconical spikes on Si surfaces following irradiation by fs laser pulses. Her et al. found that the spikes depend on the atmosphere suggesting that chemical reaction is essential to their formation; however, they stated that the mechanisms are unknown. The tips of the cones are at the same height as the surface of the surrounding undamaged Si. They continued with further studies on the formation of μ-cones on Si following fs laser irradiation [74]. They found that the base of the spikes is asymmetric with the long axis perpendicular to the laser polarization. The spikes have the same crystallinity with the bulk. The spikes point along the laser direction even when at different than the normal angle of incidence and independent on the crystallographic direction. At the centre of the crater, the spikes are wider, taller and less dense than at the edges. The spike density reduces with laser fluence and increases with pulse duration up to 1 ps. Because more Si is removed in areas where more energy is deposited, a difference in light absorption on different sides of the spikes would explain the asymmetric shape of their bases. Because absorption of p-light is larger than s-light in oblique incidence the sides of the oriented parallel to the laser polarization absorb less light than the other sides. Therefore the bases are elliptic with the long axis perpendicular to the laser polarization.

Later, Dolgaev and co-workers [78, 88] created similar self-organized micro spikes on Si surfaces with ns laser pulses. They suggested that the starting inhomogeneity for the development of the microstructures is the capillary waves on the molten solid surface. They were based on the observation of a solidified wave-like structure at the periphery of the laser-damaged area (that resembles the ripples in the annular region of Si). The origin of these waves according to the authors requires further investigations. In addition to the capillary wave, the authors suggested that mechanisms such as Benard cells, Rayleigh–Taylor and Kelvin–Helmholtz instabilities should be additionally considered.

These structures cited above are self-organized in the sense that no spatial attenuation or modification of the laser beam shape is required. All authors agree that there the explanation of the formation of such self-assembled micrometre-sized structures is still at its infancy. However, it seems that the situation is slightly simplified in the case that the micrometre structures are produced with a spatially modulated laser beam. A few works are reviewed below.

Chiba *et al.* [89] were the first to explain the formation of microcones on Si following masked laser irradiation of the Si surface. Although they did not perform numerical calculations, they suggest with a very good detailed picture of the formation mechanism that it is the anomalous density behaviour of Si when changing from the liquid to the solid phase and vice versa that is responsible for the formation of the micro cone. Initially, at the centre of the laser-irradiated area, a small molten pool appears and it progressively becomes larger to reach the size of the laser beam. As soon as the laser irradiation is terminated the molten pool begins to cool beginning at its outskirts. As a result, a round characteristic depression is formed close to the periphery of the irradiated area. As the molten pool now becomes smaller, the liquid Si in the container rises from its edge due to surface tension, because the volume of liquid Si is greater than the container capacity, the curvature depending on the volume of liquid Si and container capacity. Finally, all molten Si congeals and a unique conical protrusion forms on the Si surface. Wysocki *et al.* [90] used a regular lattice of SiO_2 microspheres on a quartz support to construct in a single laser shot arrays of Si micro-cones. They also explained the formation of the microcones by the anomalous behaviour of the density of solid and liquid Si. Resolidification starts at the edge of the molten zone and proceeds towards the centre. In contrast to the usual behaviour of materials with melting, the density of liquid Si is higher than that of solid Si. Thus, the volume of Si increases during solidification. As a consequence, during cooling the liquid Si is squeezed radially to the centre and forms a protrusion. As a result, a solid cone surrounded by a ring-shaped trench is formed. Finally, Moening and Georgiev [91] constructed micrometre-sized cones on

Si surfaces by using a masked laser beam by imaging pinholes on the focal area and confirmed the formation mechanism already proposed [89, 90] which is due to the anomalous density behaviour of Si when changing phases. Therefore, all authors tend to agree on a single explanation of the formation of isolated microconical arrays by using spatially modulated laser beam combined with optical imaging techniques.

6.2.2.3 Applications of micro/nanostructured surfaces

Materials' texturing to create micro/nanoscale surface structures has tremendous technological importance. The surface area enhancement attained due to the formation of various types of micro/nanostructures have important applications in a wide range of research and development activities, including optoelectronics, spectroscopy, microfluidics and tissue engineering.

In Section 6.2.2.1, it is stated that multiscale texturing remarkably affects the optical properties of irradiated materials. Following irradiation, the initial mirror-like surface is turned deep black, which is a clear indication that laser texturing alters its optical properties. Figure 6.15 (top-left) shows absorptance measurements of silicon microstructured by fs laser irradiation in SF_6 atmosphere at different fluences. For low fluences, an increased absorptance is observed for both the below-bandgap and above-bandgap wavelengths. However, for the highest laser fluences used, microstructured silicon has a drastically decreased absorptance over the entire measured spectrum. The increased absorption in the visible region is consistent with the gradual change of the surface shading from light grey to black as shown in Fig. 6.15 (top-right). Figure 6.15 (bottom) presents the corresponding spectra for irradiation in a variety of atmospheres. In all cases, a significant enhancement of the absorptance, for wavelengths above the bandgap (250 nm–1.1 μm), is observed. This can be attributed to the microstructure's ability to trap light and reduce reflections. On the other hand, beyond the band edge (1.1 μm–2.5 μm), the absorptance of N_2-, Cl_2-, and air-processed samples decreased continuously following the behaviour of pristine surface, while SF_6- and H_2S-processed samples remained at about 90% absorbing. It is therefore evident that sulphur appears to be a key ingredient in producing a surface that has near-unity infrared absorptance. Such absorptance spectrum is attributed to

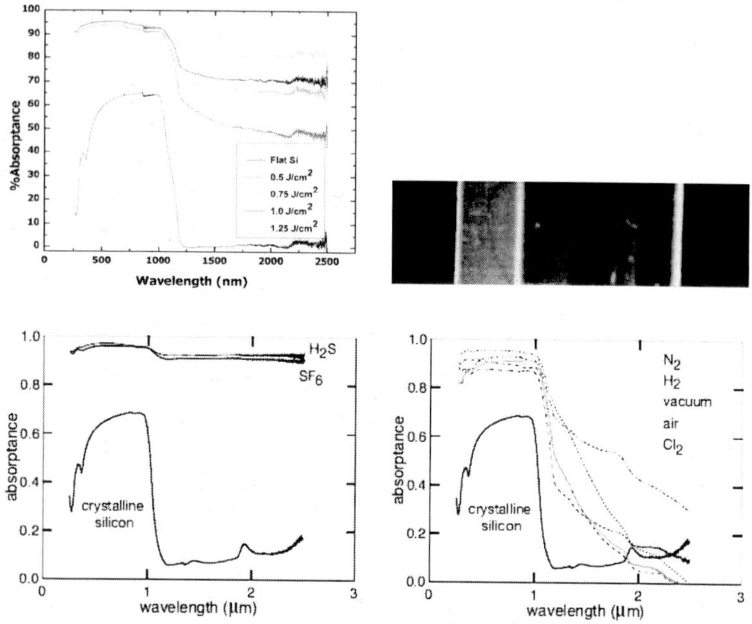

Figure 6.15 (Top left) Absorptance of laser fabricated Si structures as a function of the wavelength for different fluences; (top right) SEM image of the corresponding surface—the irradiation fluence is increased from left to right. All the samples were fabricated using 2000 pulses in a 500 Torr SF_6 atmosphere. (Bottom) Absorptance spectra of silicon irradiated in different environments: SF_6 and H_2S (left) and N_2, H_2, Cl_2, air and vacuum (right). In all cases the absorptance of the unstructured Si substrate is also shown for comparison.

structural disorder in the laser modified surface brought about by the trapping of defects during resolidification. The analysis of the surface with Rutherford backscattering spectrometry shows that the surfaces irradiated in SF_6 contain a large concentration of sulphur that exceeds the solid solubility by several orders of magnitude. Based on these results, the high concentration of sulphur atoms in the polycrystalline surface layer created a densely populated donor impurity band in the bandgap of silicon. The promotion of electrons from this impurity band into the conduction band allows for the increased absorption of infrared light [86]. Along with exploring the structure and composition of the surface layer, researchers discovered that the absorption of infrared light decreases after thermal annealing. Following irradiation, the sulphur atoms were not in the most energetically favourable equilibrium configuration, but they were trapped in this configuration as a result of the extremely fast resolidification rate. Therefore, when annealing is performed, the sulphur–silicon network relaxes to a more favourable configuration where the sulphur atom was no longer optically active.

The above optical properties provide the possibility to extend the spectral range of Si-based optoelectronic devices to the infrared spectral region as well. Therefore black Si could be exploited for numerous applications, including infrared light detection [92] and photovoltaics [86]. Figure 6.16 (top) shows schematic diagram of a microstructured sample with electrical contacts used to obtain current–voltage (*I*–*V*) measurements through the sample plane. Immediately after irradiation, the structured area behaves as a resistor; on the contrary, upon increasing annealing temperature, a diodic behaviour is observed as shown in Fig. 6.16 (bottom left). Researchers proposed that following irradiation, the incorporated sulphur atoms did not contribute additional free carrier electrons and therefore the electrical response is determined by the c-Si substrate. However, after annealing, the coordination of the sulphur atoms changes and their extra valence electrons cause an increase of the free carrier concentration in the surface layer. Furthermore, the annealing process removes lattice defects that act as traps for charge carriers. As a result the surface layer becomes heavily n-doped, confirmed by Hall measurements. The free carrier concentration gradient formed between the surface layer and the bulk builds up a static electric field that forms a diode junction. Under reverse bias, this junction was found to be highly responsive to visible and infrared light compared to commercial Si photodetectors (Fig. 6.16 (bottom right)). Under zero bias, the junction behaved as a photovoltaic device with a maximum conversion efficiency of ~1.5 %. Finally the microconical self-organized structures are found to exhibit improved ultrafast optoelectronic properties [93] that may prove to be important in applications such as efficient photovoltaic elements where the characteristic surface morphology may enhance the efficiency in the trapping/collection of the dissociated exciton that was initially created by the sun irradiation.

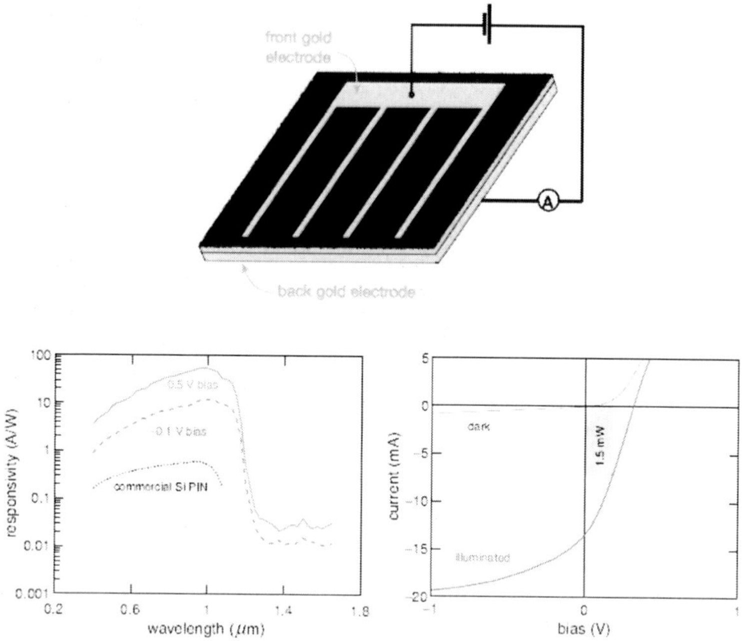

Figure 6.16 (Top) Cartoon depicting the electrode configuration used to measure *I–V* curves of sulphur-doped femtosecond laser-irradiated silicon. (Bottom) Responsivity versus wavelength of incident light for sulphur-doped femtosecond laser-irradiated silicon at different reverse bias voltages. Right: Current versus voltage graph showing the photovoltaic properties of sulphur-doped femtosecond laser-irradiated silicon when illuminated with 100 mW/cm^2 of solar radiation (AM 1.0). The lower fourth quadrant shows a maximum power generation of 1.5 mW, resulting in a conversion efficiency of ~1.5 %. Silicon substrate is Si(111), n-type (ρ = 800–1200 Ω-cm) for the photodiode and p-type (ρ=1–10 Ω-cm) for the solar cell. Data obtained from J. E. Carey, Femtosecond-Laser Microstructuring of Silicon for Novel Optoelectronic Devices, PhD thesis, Harvard University, 2004; J. E. Carey, C. H. Crouch, M. Shen, and E. Mazur, Visible and near-infraredresponsivity of femtosecond-laser microstructured silicon photodiodes, *Opt. Lett.* 30, 1773–1775, 2005; B. R. Tull, Femtosecond Laser Ablation of Silicon: Nanoparticles, Doping and Photovoltaics, PhD thesis, Harvard University, 2007].

The three-dimensional, quasi-periodical arrays of micro-spikes produced upon ultrafast laser structuring of Si show great potential as electron field emission cathodes in integrated circuits, flat-panel displays and microwave power amplifiers [94, 95]. The high aspect ratio and sharp tips, as well as their altered work function due to introduced impurities, are expected to be important for field emission applications. Furthermore, the technique used to fabricate these laser-induced silicon structures is fairly simple, relative to the traditionally used complex lithographic processes, considering that they do not require further processing steps beyond laser irradiation. Zorba *et al.* have demonstrated that such conical structures provide a significant enhancement of field emission performance at low applied voltages, due to the geometrical field enhancement from the high aspect ratio spikes [82]. It is concluded that the field emission properties of such arrays is directly dependent on the morphological characteristics of the micro/nano cones that can be preferentially tuned by varying the irradiation parameters. The optimum performance achieved showed a reduction of the field emission threshold field to values of 2.5 V/μm for Si irradiated by 15 ns laser pulses. Stratakis *et al.* have shown that the emission performance can be further improved by the growth of nano-sized secondary emitters onto the laser fabricated microconical arrays [96]. They demonstrated an efficient methodology for the fabrication large scale regular arrays of 3D CNW field emitters, produced by chemical vapour deposition (CVD) of CNWs on forests of micro-conical Si spikes (CNW/μSi). Figure 6.17 depicts corresponding SEM images of CNW/μSi field emission cathodes after the CVD process showing that nanowalls follow the surface and decorate the microspikes forming a flower-like coating. As a result, the final surface exhibits a 3D structure comprising arrays of CNW-spikes with hierarchical micro- and nanomorphology. As shown in Fig. 6.17, the FE performance of 3D CNW structures is measured to be by far superior to that of planar CNW mats and comparable to that reported for optimized CNT-based emitters. The ability of tuning the field emission properties through laser fabricated micro-spike arrays combined with the comparatively high emission efficiency achieved makes them excellent candidates to be used for electron source technologies.

In the last decade, it has become more and more evident that ultrashort laser-fabricated surfaces exhibit remarkable wetting properties that can be potentially exploited for numerous applications,

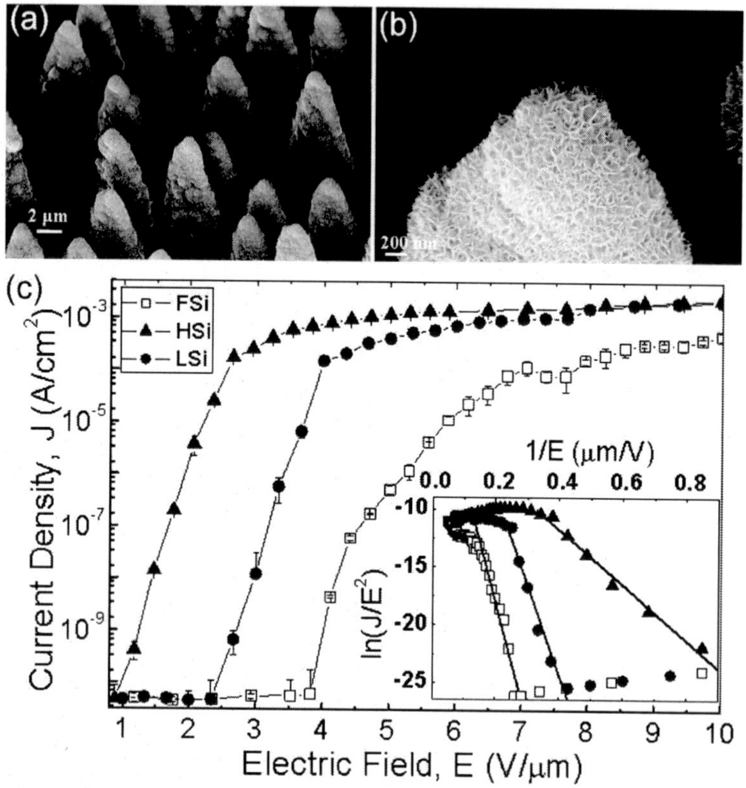

Figure 6.17 (a) Three-dimensional carbon nanowall (CNW) field emitters formed by depositing CNWs on Si microspikes produced by fs laser structuring in reactive gas atmosphere(2.25 J/cm^2, 500 pulses @1 kHz, 500 Torr SF_6); (b) higher magnification image of the structures shown in (a); (c) Plot of *J–E* FE characteristics for the different cathodes studied. Error bars indicate current fluctuations after several conditioning runs. The solid lines are guides to the eye. The corresponding Fowler–Nordheim (FN) [ln(*J*/E^2) vs E^{-1}] plots are shown in the inset. The solid lines represent linear fits to the FN equation.

including self-cleaning surfaces, biological scaffolds, microfluidics, lab-on-chip devices, waterproof coatings for automotive and aerospace vehicles and textiles [97, 98]. More important, these wetting properties can be tuned with a large degree of control by altering the surface topology through proper choice of processing parameters. Among the advantages of this technique, is that it can be applied in a wide range of materials, such as polymers, ceramics, or metals. This possibility of tailoring the wettability of materials within a wide range, from superhydrophilicity to superhydrophobicity, as will be shown in the following, is raising an increasing interest for the above applications.

In Section 2.2.1 it is shown that fs laser structuring in reactive gas atmosphere is capable to prepare artificial surfaces that possess hierarchical micro- and nanostructures. The height of the micro-scale conical features along with the number of nanoscale secondary protrusions decorating the cones increase with laser fluence, resulting in a significant increase of the overall roughness. Structuring at multiple length scales is proved to be an important mean to control the wetting properties of materials surfaces. Figure 6.18 presents the fluence dependence of the wetting angle of Si surface resulting from the engraved topology. Tuning of the wetting response of Si, from initially hydrophilic to hydrophobic, is achieved through this process without any additional surface coating. Similar observations were reported for other types of laser treated surfaces, including metallic [99] and polymeric ones [100]. The ability to tune the wetting properties can be exploited for inducing spontaneous

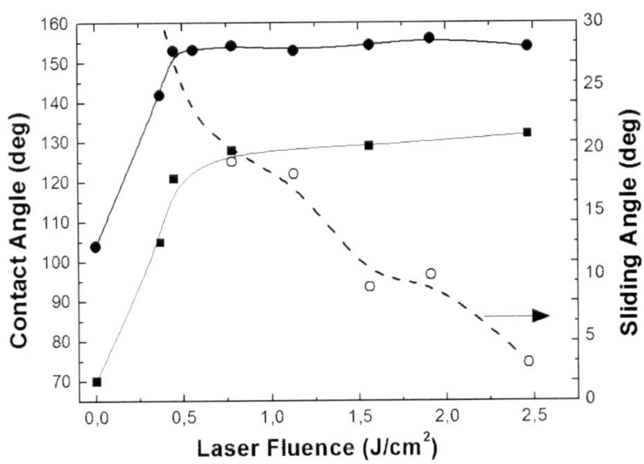

Figure 6.18 Static contact angle measurements (■) of a water drop on structured Si surfaces plotted as a function of laser fluence. Static contact angle (•) and sliding angle (○) measurements of a water drop on structured Si surfaces terminated with hydrophobic DMDCS monolayers plotted as a function of laser fluence. For fluences below 0.5 J/cm^2 the sliding angles are higher than the 30° limit of our measurement setup.

motion of liquids onto different types of surfaces. For example, by varying laser fluence, surface wettability gradients can be generated to drive microfluidic flows [80]. Further reduction of the surface free energy and thus upgrading hydrophobicity into superhydrophobicity is achieved by coating the dual-rough patterned surfaces by DMDCS organosilane monolayers. Figure 6.18 also shows the corresponding evolution of the CAs and sliding angles measured on DMDCS terminated flat and microstructured Si surfaces as a function of laser fluence.

Water droplets leave the structured area at tilt angles lower than 10° for samples treated at fluences higher than ~1.5 J/cm^2. Interestingly, these high-fluence samples exhibit the most pronounced second length-scale surface roughness. Samples in the mid fluence range (0.7 – 1.5 J/cm^2) do not meet both criteria for superhydrophobicity; they exhibit high sliding angles although they show similar contact angle to their high-fluence counterparts. At fluences lower than 0.7 J/cm^2, water drops remain pinned on these surfaces, at the highest sliding angles utilised. Evidently, contact angle hysteresis is a very important parameter when droplet motion on a surface is considered. Motion initiation, at very low inclination angles (5°) on the high-fluence superhydrophobic surfaces, is translated into a small gravitational force required to initiate motion. In other words the friction between the droplet and structured surface is accordingly low.

The artificial surfaces prepared at high fluences can quantitatively mimic both the structure and the water repellent characteristics of several natural surfaces exhibiting special water repellent properties, most notably that of the lotus (*nelumbo nucifera*) leaf (Fig. 6.19). The effectiveness of these natural textures is due to the multiscale nature of the features that ranges from the nano- to the microscale. Laser texturing can mimic these multiscale structures (Fig. 6.19) and their superhydrophobic properties. Low frictional motion is a key feature behind the unique lotus self-cleaning property and it is therefore desirable in any corresponding application, as it provides for efficient motion of water droplets even when long travel distances are necessary.

Laser micro/nanostructured surfaces can be used as templates to realize functional responsive surfaces with wetting properties that can be controllably altered on demand. Functionalization of an artificially roughened surface with a responsive coating is expected

Figure 6.19 (a) Top left: Picture of a water droplet on an artificial structured silicon surface (dark area); Top right: Static contact angle measurement of a water droplet of 0.78 mm radius on that surface; the contact angle is $154° \pm 1°$; Bottom left: SEM image of the artificial surface comprising protrusions with conical or pyramidal asperities; Bottom right: High-magnification SEM image of a single protrusion depicting nanostructures of sizes up to few hundred nanometres on the slopes of the protrusions. The surface was structured in the presence of 500 Torr SF_6 at a laser fluence of 2.47 J/cm^2 with an average of 500 pulses. (b) Top left: Picture of water droplets on a *Nelumbo nucifera* (Lotus) leaf; Top right: Static contact angle measurement of a water droplet of 0.78 mm radius on the Lotus leaf surface; the contact angle is $153° \pm 1°$; Bottom left: SEM image of the leaf surface; Bottom right: High-magnification SEM image of a single papillose depicting branch like protrusions with sizes of about 150 nm. (From Ref. 82.)

to enhance this switching effect and result in a surface that can potentially switch between hydrophobicity or super-hydrophobicity and hydrophilicity or super-hydrophilicity in response to appropriate external stimuli; the possibility of multi-responsive surfaces has been explored as well [101]. Indeed the creation of responsive surfaces via proper functionalization of fs laser micro/nanopatterned Si substrates has been demonstrated. Depending on the functional coating deposited onto those surfaces, different functionalities are attained, including photo- [102], electro-[103] and pH- responsiveness [104]. In all cases we show that the principal effect of hierarchical roughness is to cause an amplification of the response to the external stimulus. In particular: (a) The electrowetting on dielectric (EWOD) properties of black Si surfaces with different micro- and nanomorphological characteristics were investigated and compared to planar Si ones [103]. The fabricated EWOD devices exhibits remarkable CA reversibility in air environment and may constitute dual rough black Si an important material in integrated microfluidic systems; (b) pH-responsive surfaces were developed [104], which can reversibly switch between superhydrophilic at low pH and superhydrophobic and water repellent at high pH (Fig. 6.20). The surfaces were developed by "grafting from" a pH-sensitive polymer brush onto fs laser micro/nanostructured substrates. The responsive behaviour is due to the combined effect of the hierarchical micro- and nanoroughness and the hydrophobicity of the functionalizing polymer brush; (c) A photoresponsive surface can be achieved by depositing a conformal ZnO nanograin coating on the laser structured microconical surfaces through pulsed laser deposition (PLD) [102]. Following deposition, a significant enhancement of the nanoscale roughness is attained (Fig. 6.21). The micrometre scale spikes have been decorated by nano-sized ZnO protrusions, resulting in a hierarchically rough surface. As presented in Fig. 6.21, these surfaces exhibit a significant photo-induced transition to super-hydrophilicity, reaching a nearly zero contact angle in short time.

Using fs laser microstructuring novel surface patterns can be created that transforms regular metallic and Si surfaces to superwicking [105, 106]. A typical surface pattern created following fs laser treatment is shown in Fig. 6.22. The surface has multiple parallel microgrooves with a period of 100 μm that correspond to the vertical step between two horizontal scanning lines. A closer examination of the surface structures shows a combination of porous nanostruc-

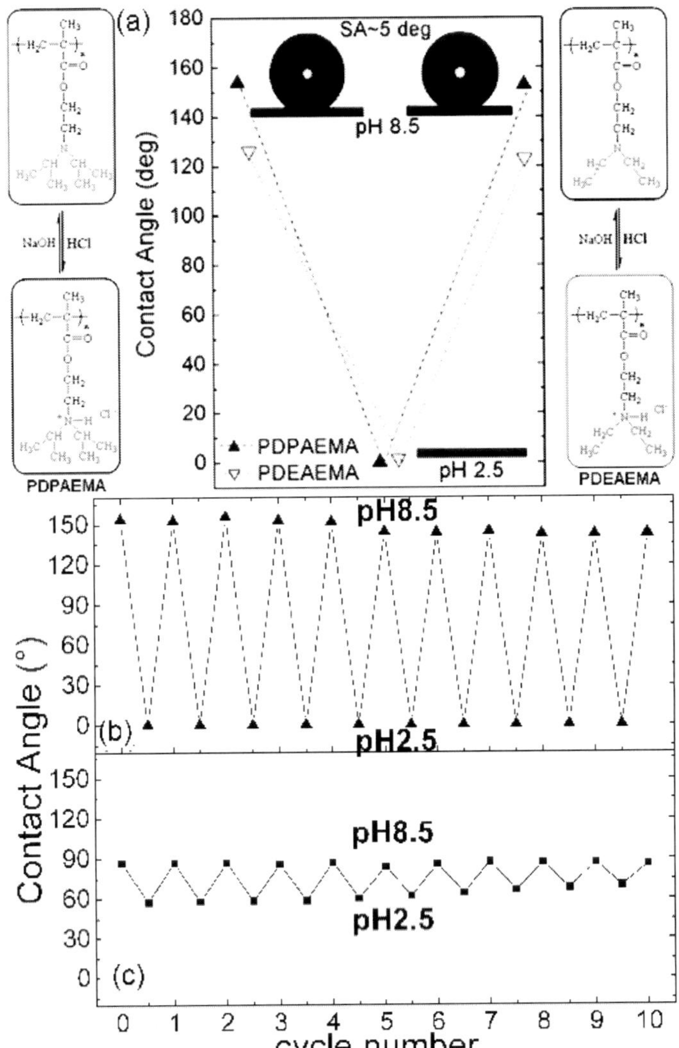

Figure 6.20 (a) Characteristic images of water droplets residing on the PDPAEMA functionalized hierarchically structured surface following immersion at pH 8.5, pH 2.5 (complete wetting) and again at pH 8.5. The scheme shows the protonation/deprotonation process of PDPAEMA (b) Average contact angle values of water drops residing on the PDPAEMA functionalized hierarchically structured surface following successive immersions at pH 8.5 and pH 2.5. (c) The same response on the planar Si surface (from Ref. 104).

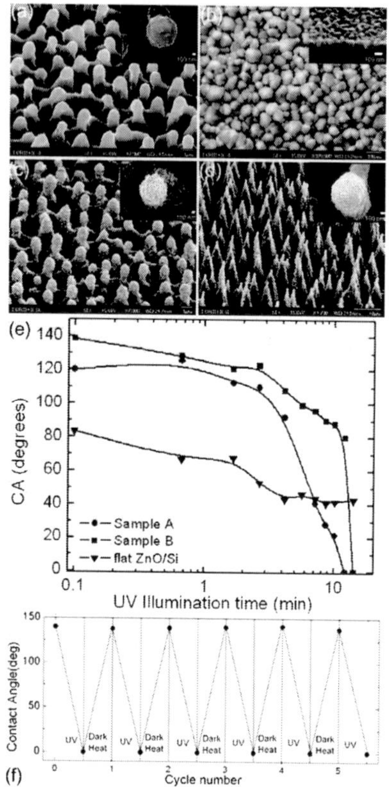

Figure 6.21 (a) Side SEM view of Si surfaces structured by fs irradiation at laser fluence of 0.17 J/cm^2. The inset shows a higher magnification of the top of a single micro-cone; (b) Top SEM view of a nanograined ZnO film prepared by PLD on a flat Si substrate. A cross-sectional image of the film is shown in the inset; (c) Side SEM view of a ZnO coated Si surface structured by fs irradiation at laser fluence of 0.17 J/cm^2. Higher magnification of the top of a single micro-cone (the scale bar is 100 nm), shown in the inset, reveals the double scale roughness of the structures; (d) The same as in (a) but at laser fluence of 2.1 J/cm^2. (From Ref. 102); (e) Dependence of the water contact angle on the UV illumination for the Sample A (ZnO nanostructured thin film on the Si microstructure prepared at 0.21 J/cm^2), Sample B (ZnO nanostructured thin film on the Si microstructure prepared at 1.1 J/cm^2) and a ZnO nanostructured thin film. (f) Reversible switch from hydrophobicity to super-hydrophilicity for the Sample B, under the alternation of UV irradiation and thermal heating at 200°C for 1 hour. Sample A exhibits similar response.

tures and fine microstructures superimposed on both ridges and valleys of the grooved pattern. It is demonstrated that liquid runs vertically uphill against gravity on such structured surfaces. More interestingly, the liquid can propel itself against gravity and accumulate to an elevated level above the liquid reservoir. Due to the strong wicking effect, a vertically standing structured metal or Si surface will constantly remain wet as long as its bottom edge is immersed in the liquid reservoir. Furthermore, the fs laser surface structuring technique provides a controllable way to modify the directionality of surface wetting.

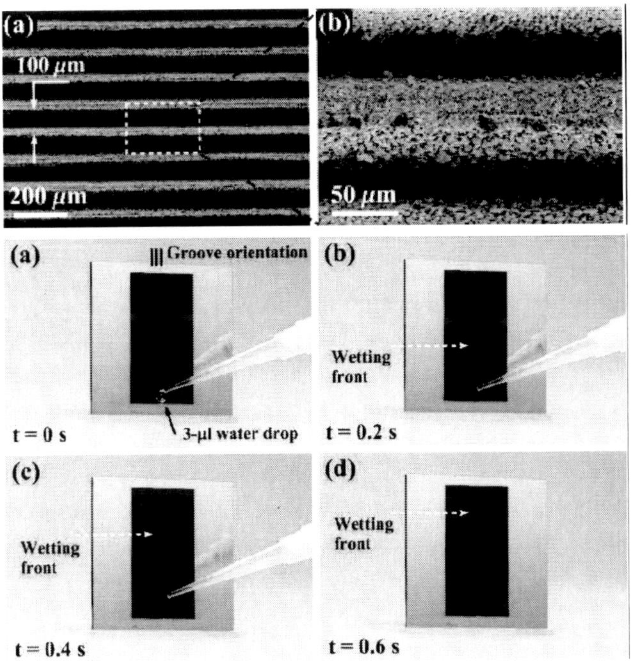

Figure 6.22 (Top) SEM images showing typical Structure of parallel grooves patterned on the laser-treated surfaces. (Bottom) Dynamics of water running uphill on a vertically standing silicon sample with vertically oriented microgrooves (from Refs. 105 and 106).

Vertes *et al.* reported that laser fabricated micronical arrays are efficient soft laser desorption/ionization substrates (SLDI) for small molecule mass spectrometry [107]. Mass spectrometric analysis of peptides and synthetic polymers has been demonstrated.

The desorption/ionization threshold fluence and efficiency were strongly dependent on the surface morphology that was determined by the medium used in microstructuring. Microconical arrays were capable of desorbing/ionizing peptides of up to 6000 Da molecular mass and exhibited low femtomole detection limits. Depending on the laser power, microcones behaved as a SLDI substrate for intact peptide analysis (low fluence) or produced structure-specific ions through ISD for top-down peptide sequencing (high fluence). The mechanism of ion formation under these conditions was attributed to the thermal, optical, and electron emission properties of the silicon microcolumns as well as to thermal and chemical confinement effects. This new matrix-free SLDI substrate offers several advantages over existing systems. Laser structured microconical arrays are more versatile than desorption/ionization on silicon method that used nanoporous silicon from galvanostatic etching in the sense that varying the laser fluence enables interchanging between molecular ion detection and structure-specific fragmentation. It is more robust than silicon nanowires as excessive laser exposure does not destroy the structure. In principle, the LISMA surface is reusable and can even be regenerated by further laser processing.

The implementation of laser engineered surfaces for the development of tissue scaffolds is additionally investigated [108, 109]. The ability to tailor their surface morphology and chemistry could be beneficial for the use of such structures as model surfaces for the systematic exploration of the role of 3D micro/nano morphology and/or surface energy on cell adhesion and growth. In Section 6.2.1.1, it was presented that direct-laser writing enables simultaneous structuring at both the micro- and the nano length scales. The variation of laser fluence caused remarkable changes in the morphological features of the spikes such as height (microscale roughness), sharpness, porosity, nanoscale roughness and wettability. Consequently, the etched substrates produced should allow more 3D free space perpendicular to the culture plane and should also provide physical cues to facilitate cell adhesion and spreading.

Cell (Fibroblasts NIH/3T3) adhesion experiments were performed on laser patterned Si surfaces exhibiting gradient roughness ratios, in order to investigate whether such surfaces were able to modulate cellular responses. The corresponding flat surfaces were tested as control samples. Fluorescence microscopy images (Fig. 6.23, top) show that the number of attached cells per unit area ini-

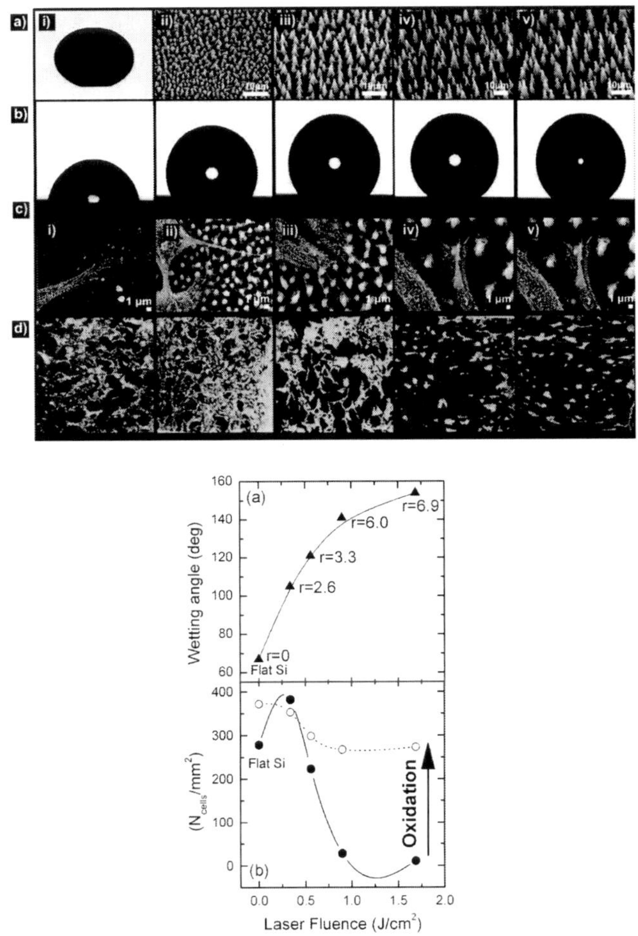

Figure 6.23 (Top) (a) Picture of a polished Si wafer (i) and side SEM views of the as-prepared Si spikes surfaces structured at four different laser fluencies (ii) 0.34 J/cm^2, (iii) 0.56 J/cm^2, (iv) 0.90 J/cm^2, (v) 1.69 J/cm^2; (b) Photographs of water droplets on the patterned Si surfaces (c) SEM micrographs of fibroblast cells adhering to the surfaces; (d) Confocal laser microscopy pictures of fibroblast cells cultured for three days on the respective surfaces. (Bottom) (a) Relation between the roughness ratio (r) and wetting angle on the irradiation fluence used to pattern the different substrates for fibroblast cells culture; (b) Dependence of fibroblast cells density on the laser fluence and thus on surface roughness and wettability for the flat and patterned Si surfaces after 72 h incubation. (From Ref. 108.)

tially increases as the roughness ratio and CA is increased and then decreases denoting that the dependence of fibroblast adhesion on surface wettability is non-monotonic, in accordance to previous theoretical models. The results presented in Fig. 6.23 (bottom) suggest the existence of a switching function for the structured Si surface characteristics, from fibroblast adhesive to fibroblast repulsive, when a critical combination of roughness and thus wettabilty, is attained. Further support to this statement is coming form the results obtained after substrates oxidation, where the substrate with the highest roughness is converted from ultrahydrophobic to ultrahydrophilic. This transition is again followed by a corresponding enhancement in cell spreading and the surface switches from cell-phobic to cell-philic. It is thus concluded that the sharp transition from superhydrophobicity to superhydrophilicity is accompanied by a similar transition in cell adherence as well. This sharp transition between the two extreme wetting states may be potentially a useful tool towards controlling cell behaviour on culture substrates.

Besides fibroblasts, Si microstructured substrates have been utilized to enable primary neurons to grow in 3D, without using synthetic extracellular matrix (ECM) coatings, or chemotropic factors [109]. Neurons depend on surface support more than other cells, but readily adhere to surfaces coated with ECM proteins. The challenge to comprehend the multitude of cues neurons receive, process and direct, in vivo, and to simulate 3D cultures in vitro, has been addressed, however, the results are limited. The importance of three-dimensional neuronal cultures can be realised if one thinks that the dimensionality of the network has a strong impact on its connectivity, and therefore plays an important role for its possible behaviour. Primary neuronal cells were cultured on laser patterned Si substrates, in the absence of chemotropic factors or synthetic extracellular matrix. Details of the substrates preparation and culture process can be found elsewhere. Similar to fibroblasts neuron cell attachment was favoured on more hydrophilic surfaces. Microstructured roughness in the form of spikes promoted the adhesion of single or clusters of neurons to the substrate (Fig. 6.24a). In contrast, only few cells survived on a flat Si substrate used as a control that verified the role of roughness in absence of synthetic ECM. The cells developed into a dynamic cellular aggregate, with long neurites sprouting over the spikes' surface (Fig. 6.24b). The integrity and uniform sprouting observed suggest the biocompatibility of the neuronal population with the surface. The protrusions of the neurolemma grow over the spikes, engulfed their top and incorporated them to form a web (Fig. 6.24c). Moreover, a network of nanoscale thin neuritic extensions was formed along the third dimension (-z axis) utilising the space towards the base of the spikes (Fig. 6.24d). In conclusion, ultrafast laser etched surfaces, with the capacity to exquisitely control the size of pores, the micro/nanotopography and the surface wettability represent a promising culture scaffold that might enable a multi-parametric assessment of the various factors that affect cell behaviour with far-reaching implications for tissue engineering.

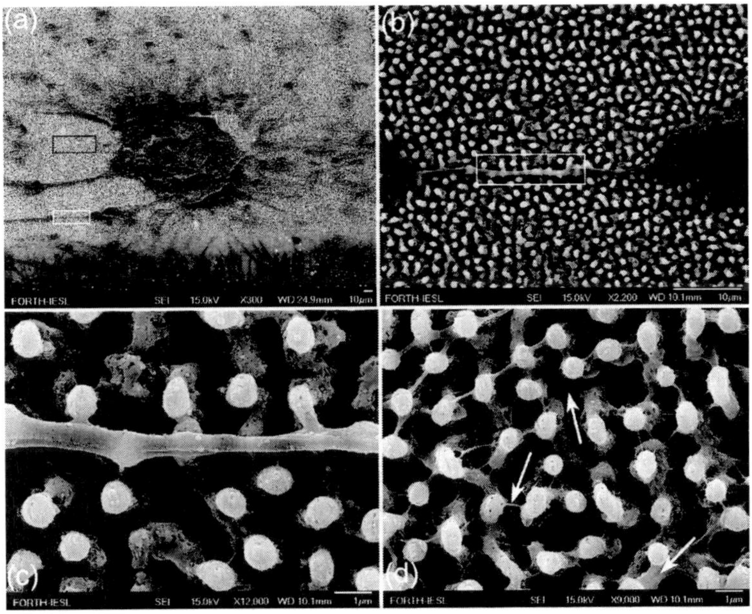

Figure 6.24 (a) Neuronal cluster on the Si spikes area; (b) Detail corresponding to white lined inset of (a), showing a long neurite that has attached and grown over the spikes; (c) Detail corresponding to white lined inset of (b), showing protrusions of neurolemma growing over and engulfing the top of the spikes; (d) Detail corresponding to black lined inset of (a), showing the 3D web of cytoplasmic processes growing along the direction vertical to the culture plane. The arrows indicate how multiple processes may initiate from one neurite. (From Ref. 109.)

6.3 Fabrication of Surface Micro/Nanostructures by Ultrafast Laser Processing in Liquid Media

Phase transitions at the solid-liquid interface, exposed to short laser pulses at an energy density sufficient to melt the solid, can be traced owing to modifications of the solid surface induced by the laser pulses. In case of a metallic target, the laser radiation is absorbed by free electrons, and the lattice temperature starts to increase due to electron–phonon relaxation process. If the absorbed energy is sufficiently high, the metal target melts, so that a layer of the liquid that surrounds it is heated up due to heat transfer from the metal. As a result of the high pressure of the adjacent medium that contacts the melt, the latter can be modified. In particular, the liquid vapours that surround the molten layer induce in it viscous flows giving rise to the formation of various structures.

The characteristic thickness of the modified layer of the solid target strongly depends on the melt thickness, and therefore, on both the laser fluence and its duration. The thickness of the molten layer h_m can be estimated by the heat diffusion length during the laser pulse as follows: $h_m \sim (at_p)^{1/2}$, where t_p stands for pulse duration, and a stands for heat diffusion coefficient of the solid. This estimation is valid only for laser fluence close to the melting threshold of the solid. If the duration of the laser pulse is less than the time of electron–phonon relaxation, then the heating of the lattice occurs within the depth of the absorption of laser radiation. For typical metals the mean free path of excited electrons during the relaxation process is too short, and the melt thickness does not exceed a fraction of micrometre even for ns laser pulses. As a result, this layer of material may be re-distributed into one or another kind of nanostructures (NS) due to the recoil pressure of the liquid medium adjacent to it.

The melting of the target surface is a necessary condition for structures formation; however, the adjacent medium is responsible for the actual type of structures. The morphology of the structures that are left in the target surface after melt solidification can provide valuable information about the phase transitions taking place at the solid-liquid interface. In a sense, the observed NS are the "fingerprints" of nanoscale inhomogeneities of the medium in its supercritical state that surrounds the target.

In this section, we shall distinguish between the two types of surface morphology that is observed under laser exposure of solids. The interaction of short-pulsed laser radiation with conducting solids is accompanied by various instabilities, which may lead to the formation of surface textures. In general, these are divided into large- and small-scale textures. The large-scale textures result from the capillary motion of the melt, produced by laser irradiation of the target. Under ordinary conditions, their characteristic length scale is tens of micrometres. On the other hand, the size of the finer surface structures is of the order of the laser wavelength, and they originate from the interference of the incident laser radiation and the surface electro-magnetic wave (SEW) generated on the of solid surface. The resulting interference pattern is stationary because the waves are in phase, and the corresponding surface morphology is determined by a variety of processes (including oxidation, sublimation, and others), whose rate is temperature dependent. Typical morphology of these structures consists of straight periodic ripples, and their period scales with the laser wavelength.

Laser ablation of solids in liquids also leads to surface texturing of the target; however, in this case different mechanisms prevail, owing to fact that the liquid adjacent to the melted surface undergoes a phase transition. Short laser pulses give rise not only to superheated liquid but also to a transient zone of elevated pressure close to the target, which may bring the surrounding medium to a supercritical state. As a result of the interaction of the pressure wave with the melt layer on the target surface, its morphology is changed. This surface profile is frozen upon cooling before being smeared by the surface tension of the melt. Typically the structures of this type have circular symmetry. Depending on the experimental conditions, such as laser peak power, target material, surrounding liquid, etc., these two types of structures – ripples and spherical features may coexist. Here we shall designate the spherical structures as NS.

6.3.1 Morphology of NS under Laser Ablation of Planar Surfaces

Nanostructure imaging using scanning probe microscopes, e.g., atomic force microscope (AFM) indicates that they are densely packed nano-cones. This is illustrated in Fig. 6.25 for the case of NS grown on an Ag target under its ablation in water with ps laser pulses

[110]. Note that both the lateral dimensions of NS and their period are much smaller than the laser spot size on the target, which in typical experimental conditions is of order of hundreds of micrometres. The estimated density of NS in Fig. 6.25 amounts to 10^{10} cm^{-2}. Therefore, the expanding vapours of the liquid that surrounds the target are unstable. Pressure difference appears within initially smooth vapour pocket above the molten layer of the target. This pocket is split into periodic cells, and the pressure difference within these cells pushes the melt from high to low pressure areas. Solidified NS on the target are just the imprints of those cells in the adjacent to target medium in which the phase transition takes place.

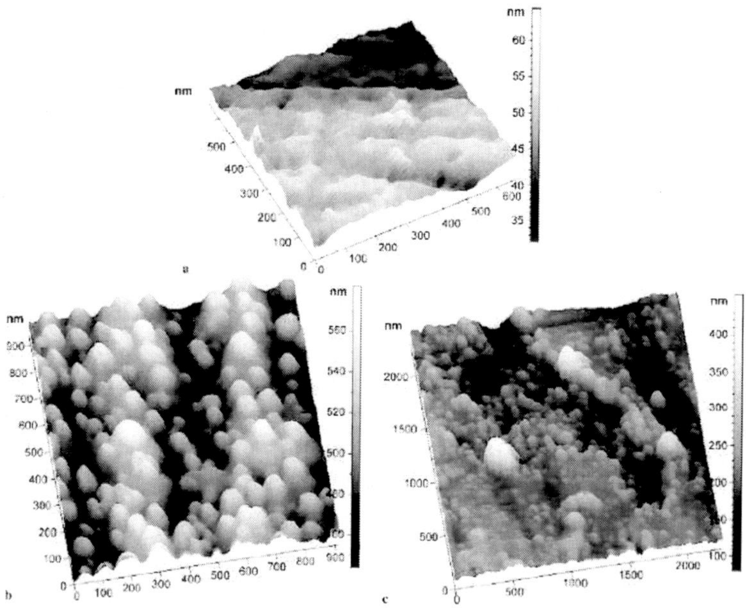

Figure 6.25 AFM view of NS on an Ag plate generated by its ablation in water under exposure to a 350 ps Nd:YAG laser radiation. Pristine surface (a), NS after laser ablation in water (b), concentration of NS in micro-depressions of the relief (c).

This type of NS is observed under ablation of solids with sufficiently short laser pulses. The upper limit of the laser pulse duration at which the NS are observed is around 300–400 ps. Longer pulses do not favour the formation of NS in the whole range of laser fluencies, from melting threshold to intense ablation. On the other

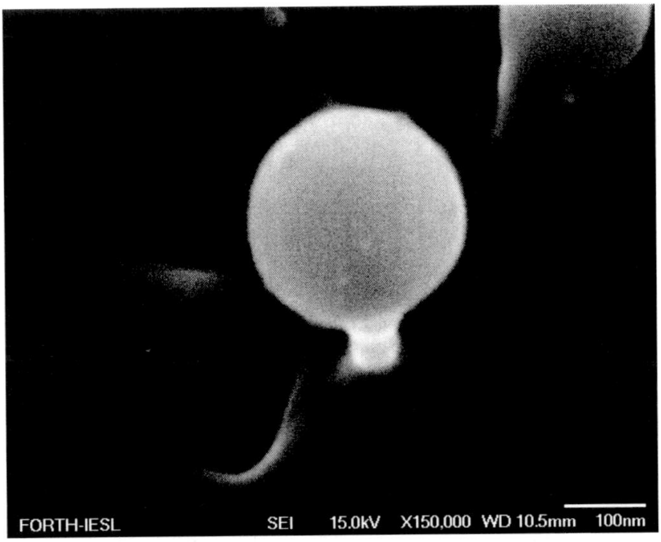

(a)

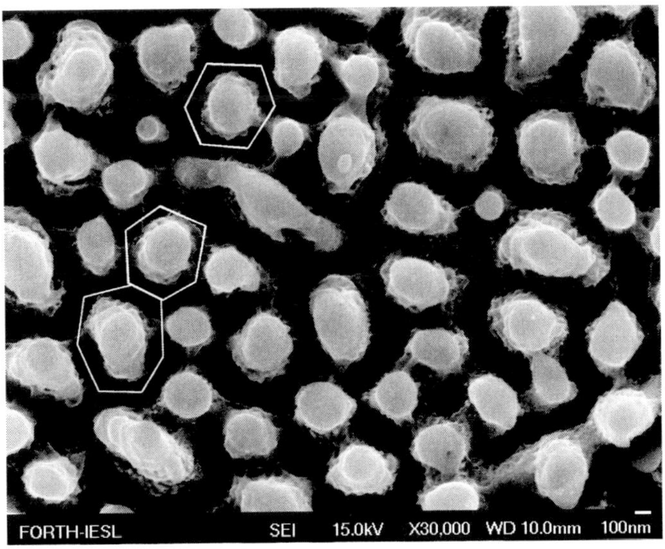

(b)

Figure 6.26 (a) Enlarged view of a single NS on Ta produced by its ablation in water with 350 ps laser pulses of a Nd:YAG lasers. FE SEM, scale bar denotes 100 nm. (b) Top FE SEM view of NS on Ta generated by its ablation in water using 5 ps, 248 nm laser radiation. The white polygonal figures are just the guides for an eye.

hand, NS are readily observed under ablation with shorter laser pulses down to fs ones. The laser wavelength needed for NS formation is not very important as soon as metal targets are considered. This is because the optical constants of most metals are nearly the same in the range of wavelengths where lasers usually emit.

As one can observe from tilted SEM micrographs (Fig. 6.26a), the shape of NS is mushroom-like, so that their lateral size is a non-monotonous function of the coordinate [111]. Figures 6.27a–d show the FE SEM view of NS on Ag, Au, Zn, and Al targets ablated in water with ps laser radiation. Most of NS have a mushroom shape similar to that observed for Ta NS. Most of NS have a mushroom shape similar to that observed for Ta NS.

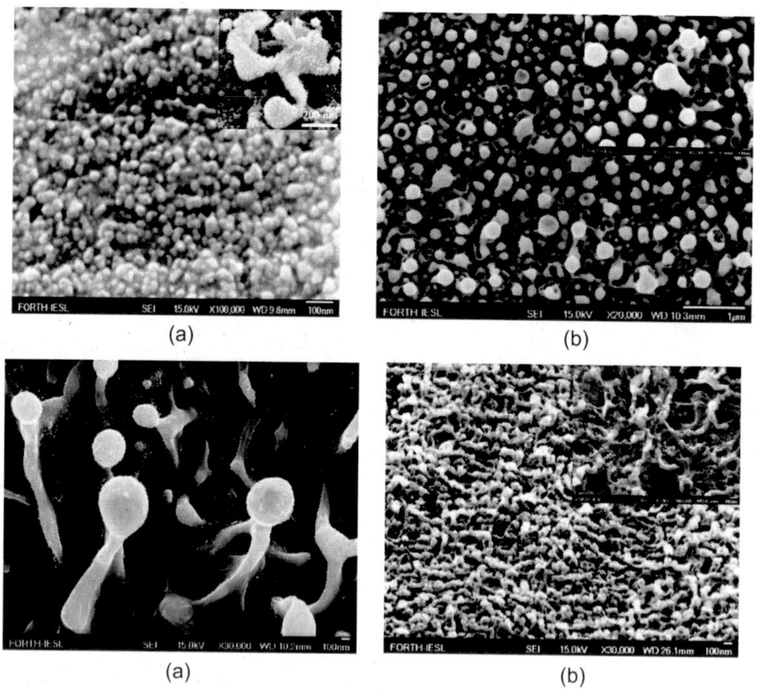

Figure 6.27 FE SEM view of NS on Ag (a) and Au (b) targets produced by their ablation in water with radiation of a 5 ps KrF laser, wavelength of 248 nm. Insets show the view at different scale. Zn NS obtained by ablation of bulk Zn in ethanol with a 150 ps laser radiation, wavelength of 1064 μm (c). NS on bulk Al ablated with 100 fs pulses of a Ti: sapphire laser in ethanol (d).

Note that for metals the temperature of the adjacent to the surrounding liquid layer is around 1000 K. As soon as the shock wave propagates toward the free surface of the liquid, the pressure in it remains high, but when it reaches the liquid surface, the pressure above the target abruptly drops. Usually the shock wave in liquids propagates in it with the speed of sound (1497 m/s). This means that, for 1 mm-thick liquid layer, the time of propagation of the shock wave is of order of microseconds, which is much longer than both the pulse duration and electron–phonon relaxation time. As a result, all kinds of structures shown in Fig. 6.28 can be formed and this is corroborated by the NS images presented above. On the other hand, NS that were detached from the melt do not always take on a spherical shape. For instance, in case of Al ablation with fs laser pulses either in water or ethanol, the nanoparticles found in liquid have a shape of a drop with "tails" as shown in Fig. 6.29. It is pertinent to note that this kind of NPs morphology was not observed so far on any other metals under their laser ablation in liquids [112]. This unique morphology can be due to the interplay of the melt viscosity and cooling rate of ejected nano-drops.

Figure 6.28 Possible types of NS morphology. The last right scheme corresponds to detachment of the molten drop that becomes a nanoparticles suspended in the surrounding liquid.

The spherical structure presented in Fig. 6.26 has been formed by the melt propulsion under the action of the pressure difference within the surrounding liquid. This is supported by the observation that each NS is surrounded by a system of pits that are almost situated around the NS, shown in Fig. 6.26b. Thus, the melt is pushed from the pits towards the centre, where the NS is formed [111]. In this context, the formation of NS can be interpreted as the result of the work of pressure. This work is spent for the increase of the interface length against the surface tension of the melt. In view of this, the elementary work dA needed to increase a circular interface length of $2\pi R$ by dR can be expressed as follows:

$$dA = \sigma \tag{6.1}$$

where σ stands for the coefficient of surface tension of the melt. The total work, A, for the generation of a spherical NS of radius R can be found as follows:

$$A = 2\int_0^R \sigma 2\pi R dR = 2\pi\sigma R^2 \tag{6.2}$$

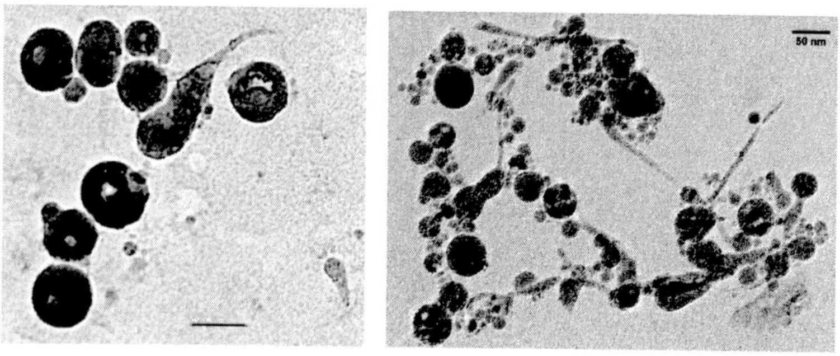

Figure 6.29 Transmission Electron Microscope view of nanoparticles of aluminium (left) and titanium (right) generated by ablation of a bulk Al target in ethanol using 100 fs Ti:sapphire laser pulses at a wavelength of 800 nm. The bar denotes 50 nm.

On the other hand, this work is performed by a force equal to $S\Delta p$, where S is the surface of the NS and Δp is the pressure difference. As this force acts on a distance $l = 2R$ its corresponding work is

$$A = Fl = \pi R^2 \Delta p 2R \tag{6.3}$$

Equalizing the above two expressions one finds the following estimation for the pressure difference:

$$\Delta p = \frac{\sigma}{R} \tag{6.4}$$

Substituting the parameters for NS on Ta shown above, with R = 100 nm, one obtains Δp = 400 atm. This is the order of magnitude of the pressure difference existing above the melt surface, applied on a very short distance, of the order of NS period. It is important that the estimated value for Δp exceeds the critical pressure for H_2O. Furthermore, the temperature of the liquid layer adjacent to the melt is close to the melting temperature of the target, equal to 3000°C in case of Ta. As a result, the medium that surrounds the target (H_2O) is in its supercritical state. Many compounds among them is water, in supercritical state are known for their enhanced chemical activity.

Although H_2O is neutral to many solids in normal conditions, so that their solubility in it is negligible, it is a good solvent to many metals and dielectrics in its supercritical state. In actual experimental conditions, dissolution of the target in supercritical H_2O should take place. The dissolved target material sediments upon pressure and temperature drop that follow the laser pulse, thus the formation of NS might be due to condensation of this material from the super-saturated solution. On the other hand, dissolution of solids in supercritical H_2O is accompanied by chemical interaction causing the formation of corresponding hydroxides. In this case, however, the re-condensed material should not be as compact as it is actually shown by the corresponding SEM images of the Ta target. The effect of supercritical dissolution of Ta on the morphology of NS has been tested studying its ablation in aqueous solutions of the base KOH.

Figures 6.30 and 6.31 show the size distribution of NS on Ta generated by its ablation in water with 1.06 μm laser radiation at different pulse durations. It is clear that the lateral size of NS on Ta increases with the increase of the pulse duration. This fact is also corroborated by comparison of lateral size of Ag NS generated with 350 (Fig. 6.25) and 5 ps (Fig. 6.27a) laser radiation. Furthermore, the lateral size of NS decreases with the decrease of laser fluence towards the melting threshold. It is pertinent to note that in all cases the size distribution function has two maxima, the first one at small sizes (50–200 nm) and the second in the vicinity of the laser wavelength, either at 250 or 1000 nm (see Fig. 6.30 and 6.31). This means that the interference of the laser radiation with surface electromagnetic wave (SEW) does influence the formation of NS, though in case of ps ablation in liquids regular periodic ripples are rarely observed. In contrast, in the case of fs laser exposure in liquids, periodic ripples dominate over self-organized NS. This is observed in almost all solids, e.g., for Ge and Ni, and for Ta ablation in water it is illustrated in Fig. 6.32. It is evident that periodic ripples are the main feature of the ablated surface, whereas mushroom-like NS, typical of longer laser pulses, are situated mostly on top of them. The formation of ripples is most probably due to the same reason as for "usual" mushroom-like NS. Namely, melting of the surface occurs in the maxima of the interference pattern, and then the melt is pushed out by the recoil pressure of vapours of the surrounding liquid. Simultaneous observation of periodic ripples and spherical NS on the same target indicates different mechanisms of their formation.

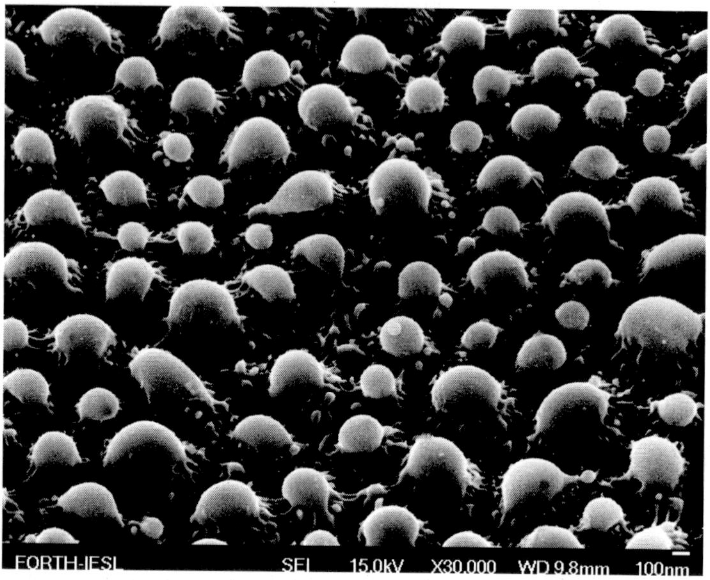

(a)

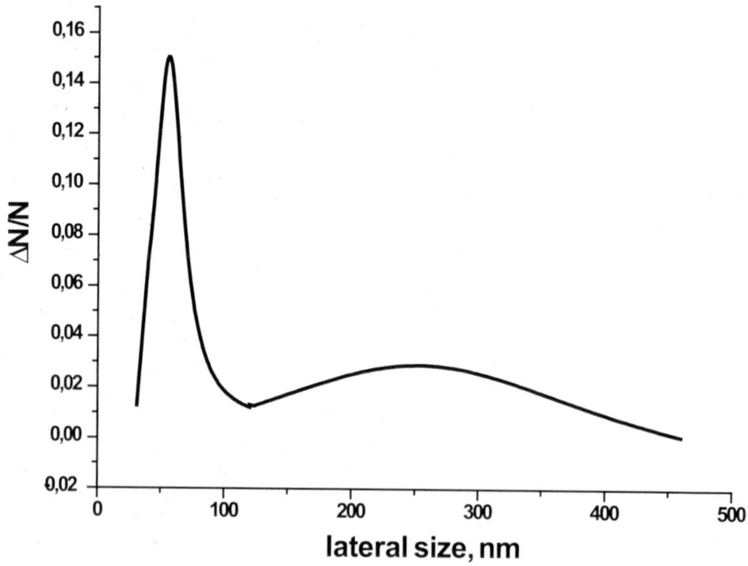

(b)

Figure 6.30 FE SEM view of NS on Ta produced by its ablation in water with a 5 ps laser pulses at wavelength of 248 nm (a). Distribution of lateral size of NS on Ta in this conditions (b).

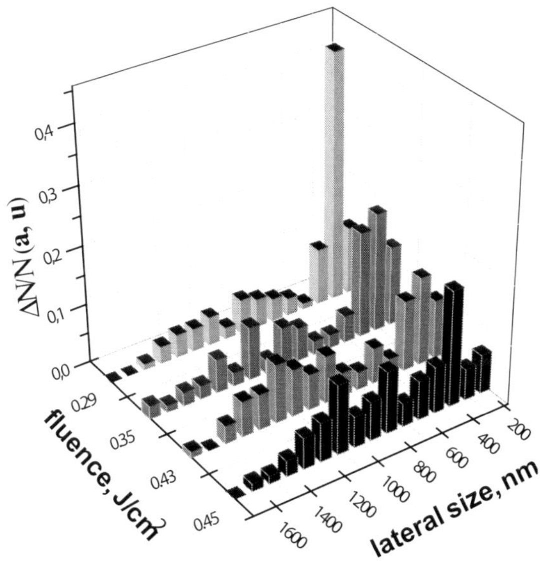

Figure 6.31 Distribution of the lateral sizes of NS on Ta with laser fluence as a parameter. Ablation in water, Nd:YAG laser, pulse duration of 350 ps at wavelength of 1.06 μm.

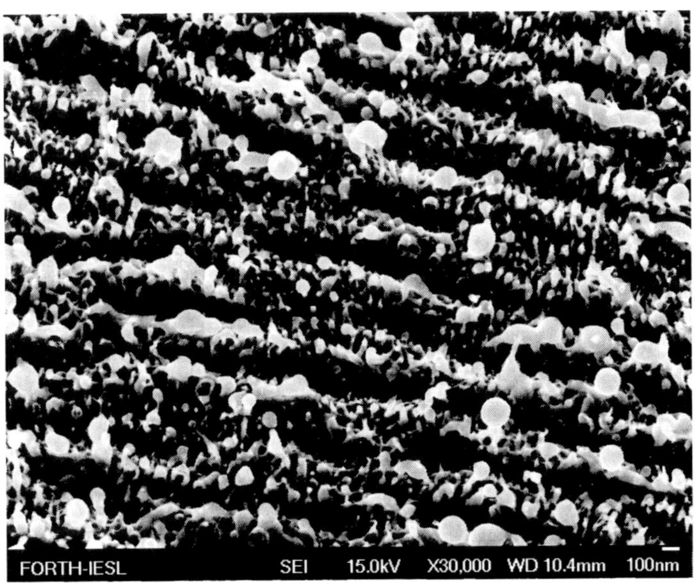

Figure 6.32 FE SEM view of a Ta surface ablated in water with 100 fs radiation of a Ti:sapphire laser at wavelength of 800 nm.

6.3.2 Growth of NS on Pre-Patterned Surfaces

Any real surface is not perfectly smooth and is characterized by a certain degree of surface roughness. The generation of NS is sensitive to its initial value. The first realization of NS on Ag under its ablation in liquids showed that some initial surface roughness is required for their formation [110]. It is observed that the initial mean roughness should be at least 50 nm, and no NS formation occurred on optically polished targets. In the latter case, a deep crater (a groove when the laser beam scans the substrate) appears, upon increasing the laser fluence. Furthermore it was found that NS are predominantly located close to the edges or protrusions of the relief. This was attributed to a weaker thermal contact of these features with the substrate, so they are the first to melt under the laser pulse. The above findings were later confirmed for other metals as well. A distinct property of small surface protrusions whose size lies in the nanometre range is the depression of their melting temperature compared to the bulk target material. As a result, nano-sized protrusions may melt, while the rest of the target remains solid at certain laser fluence. The melting temperature drop is related to the higher fraction of surface atoms of NS, which are disordered due to the large number of dandling bonds. Theoretically, it is expressed by the following relation [113]

$$T_{\mathrm{m}} = T_{\mathrm{m}}^{\infty} \exp\left(-\frac{4\delta}{\delta + 2R}\right) \tag{6.5}$$

where T_{m}^{∞} stands for the melting temperature of a bulk solid and δ is the Tolman constant. The physical meaning of this constant is the thickness of a superficial layer with distorted lattice; $\delta = 6d$, where d is the lattice parameter of a bulk solid. In other words it the thickness of the disordered external layer of the surface that can be considered "liquidified". R is the radius of a nanoparticle, which in our case is that of a local nano-protrusion. The absolute value of depression of the melting temperature depends on the nature of the target material, the maximal depression is observed for Au.

The influence of the initial target morphology on the properties of NS was studied using specially designed Ni targets pre-patterned by electron beam lithography. A layer of a photoresist (PMMA) was deposited by spin-coating on the surface of a polished Ni foil and then it was exposed to a focused electron beam. This layer was then developed in such a way that the exposed areas were dissolved. Finally, galvanic Ni deposition was carried out and the non-exposed areas of the photoresist were removed. The resulting surface comprised periodic grooves and its profile is presented in Fig. 6.33. The effect of target pre-structuring on NS formation strongly depends on the laser pulse duration. In case of fs pulses, the influence of initial roughness on the growth of NS is illustrated in Fig. 6.34. The grey square shown in the figure is the pre-patterned Ni surface, whereas black areas correspond to pristine Ni. The laser beam scans the surface from top to bottom. The beam intensity profile is slightly asymmetric, and only its central part produces NS on the pristine Ni surface being visible as bright areas. Then the beam crosses the pre-

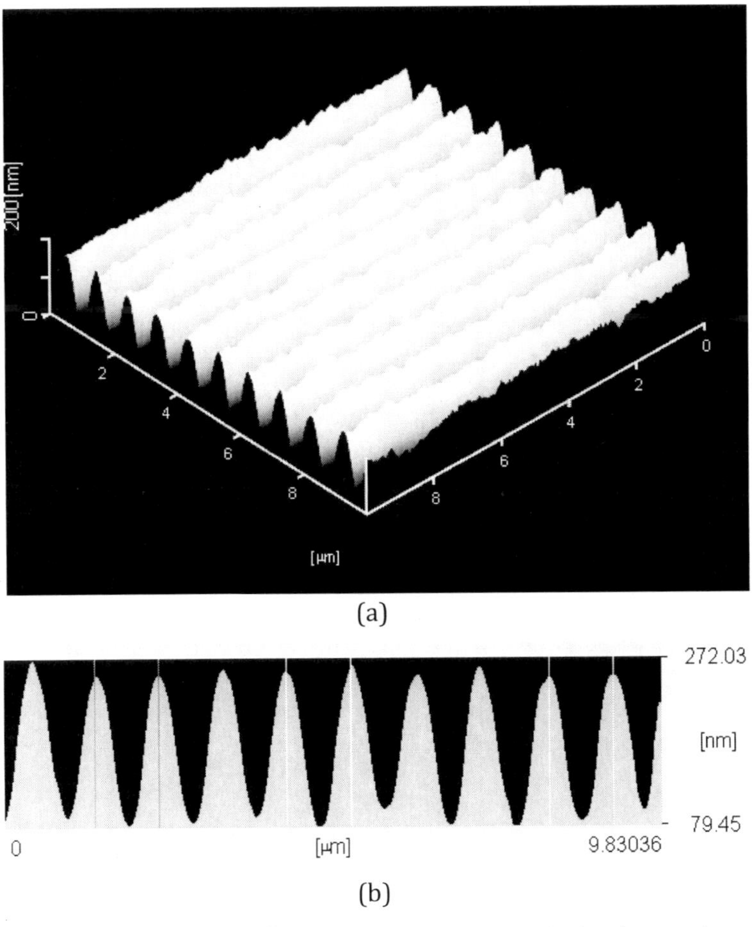

Figure 6.33 AFM view of a Ni target pre-patterned via electron beam lithography. 3D view (a), cross-sectional view (b).

Figure 6.34 Influence of the initial roughness on the growth of NS on Ni ablated into ethanol using a 50 fs, 800 nm Ti:sapphire laser. The side of the pre-patterned square is 2 mm long.

patterned area of the substrate (clear square) while the area where NS are formed rapidly expands almost over the whole square. Later when the beam leaves the pre-patterned square, the area with NS remains as wide as on the square itself. This means that the threshold laser fluence needed to produce NS is much smaller on the pre-patterned areas than on the smooth one. As soon as NS are created in the surface, they are self-sustained and can be generated at much lower fluence. This result is coherent with previous observations for generation of NS on Ag. A detailed view of NS formation on pre-patterned Ni surface is presented in the following set of FESEM images [10]. At low laser fluence only sporadic pits are observed on the pristine Ni surface. However, as soon as the laser beam reaches the initial pattern, morphology changes become more pronounced. As can be clearly seen in the enlarged view of Fig. 6.35, the surface dam-

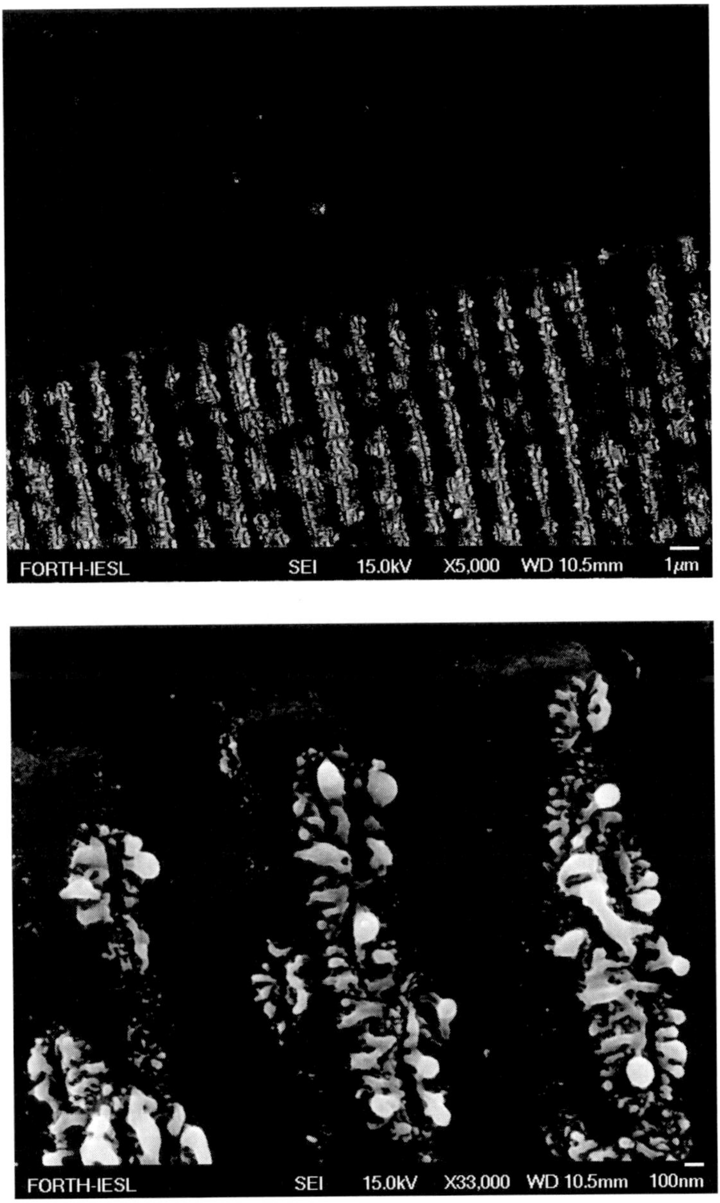

Figure 6.35 Formation of NS on the pre-patterned Ni foil under exposure to femtosecond Ti:sapphire laser radiation in ethanol. Low fluence of 4.5 J/cm^2. The frontier between pristine and pre-patterned areas (top), enlarged view (bottom).

age is located predominantly in the summits of the initial structures. It looks more like explosion rather than melting calling in mind the so called phase explosion. Upon further increase of the laser fluence both smooth and pre-patterned areas of the target are covered by NS (Fig. 6.36 left). Finally, at much higher fluences the pre-patterned areas can be seen only due to the higher average size of NS compared to that on the pristine Ni surface (Fig. 6.36, right). As shown in Fig. 6.37, the effect of pre-structuring of the target is less pronounced when longer laser pulses are used. The lateral size of initial structures is much less than the lateral dimensions of NS on a smooth surface, and the initial relief is just decorated with NS. For low fluences the edges of the initial structures are decorated with large NS, while their size in the valleys is lower. It is interesting that NS are aligned along the edges of the initial stripes. This can be understood as the depression of the melting temperature on the edges of initial relief. At higher fluences NS on top of the stripes tend to coalesce forming thus elongated nano-entities.

The surface with ripples can also be considered a pre-patterned substrate. Ripples are observed as a surface relief if some target material was ablated into the surrounding liquid. The highest depth of the ripples relief corresponds to the maximum of intensity of the interference pattern of SEW with the incident laser radiation. The recoil vapour pressure of the surrounding liquid under subsequent laser exposure pushes the molten target material from minima to maxima of the ripples relief. That is why spherical NS are located on top of ripples.

6.3.3 Optical Properties of Metallic NS

The visible manifestation of NS formation after laser irradiation of a metallic target is the coloration of exposed areas. A representative example is shown in Fig. 6.38, which shows the yellow coloured Al surface obtained after its ablation by fs pulses in water. This coloration should be distinguished from surface oxidation, since some oxides have absorption bands in the visible. However, a thin oxide layer cannot be responsible for intense coloration of the target. The optical properties of NS are closely related to the spectral features of corresponding nanoparticles (NP). This is due to the fact that electrons are confined within NS just like in NP [114, 115]. The interesting feature of the reflectivity of a surface with NS is that the

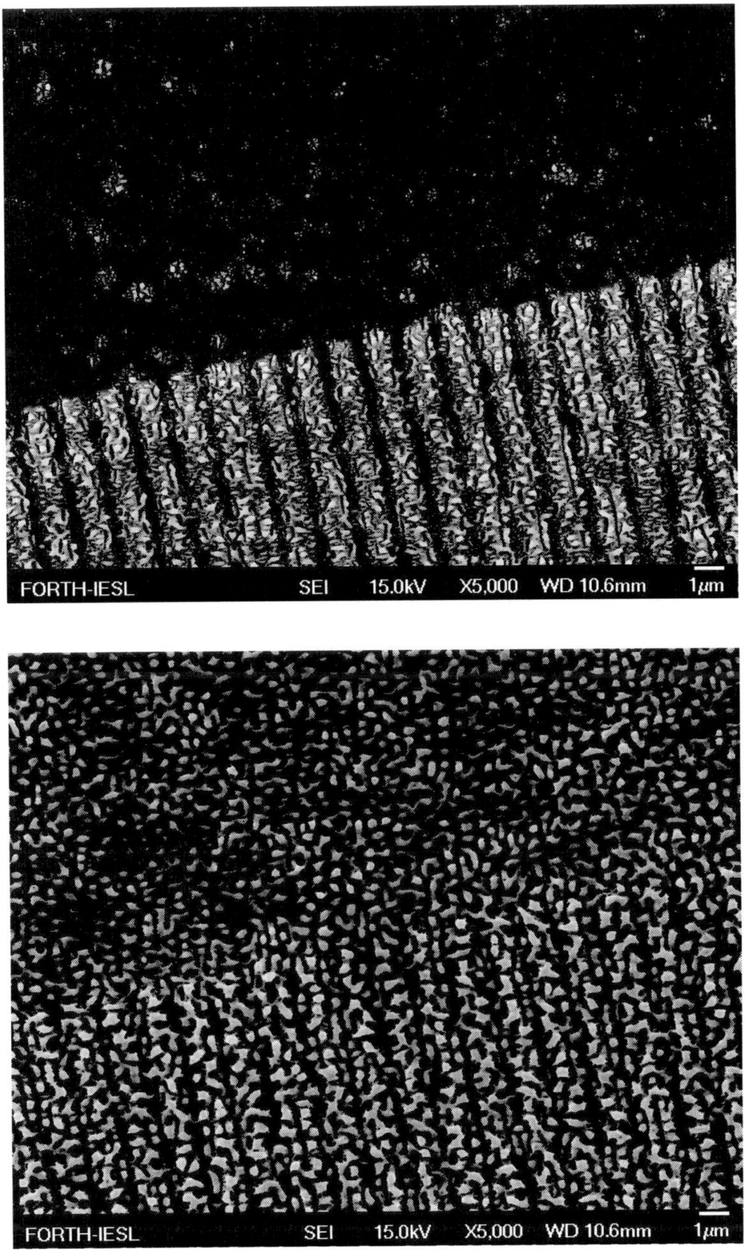

Figure 6.36 Formation of NS on the pre-patterned Ni foil under exposure to femtosecond Ti:sapphire laser radiation in ethanol. Laser fluence of 4.5 (top) and 18 J/cm^2 (bottom).

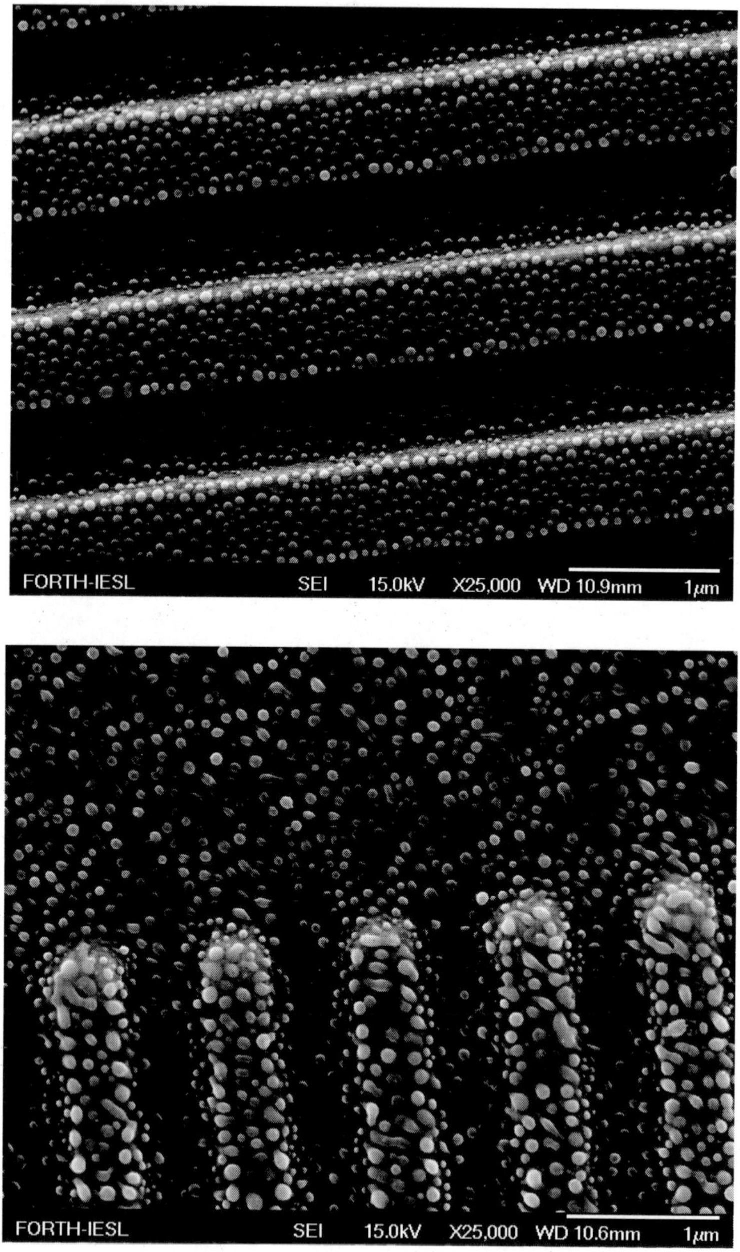

Figure 6.37 NS formation on pre-patterned Ni target under exposure with a 5 ps laser pulses in ethanol, wavelength of 248 nm. The laser fluence was 0.3 and 0.9 J/cm^2, respectively.

colour is visible only in case of mirror-like reflection. This confirms the participation of electrons in NS to the observed coloration.

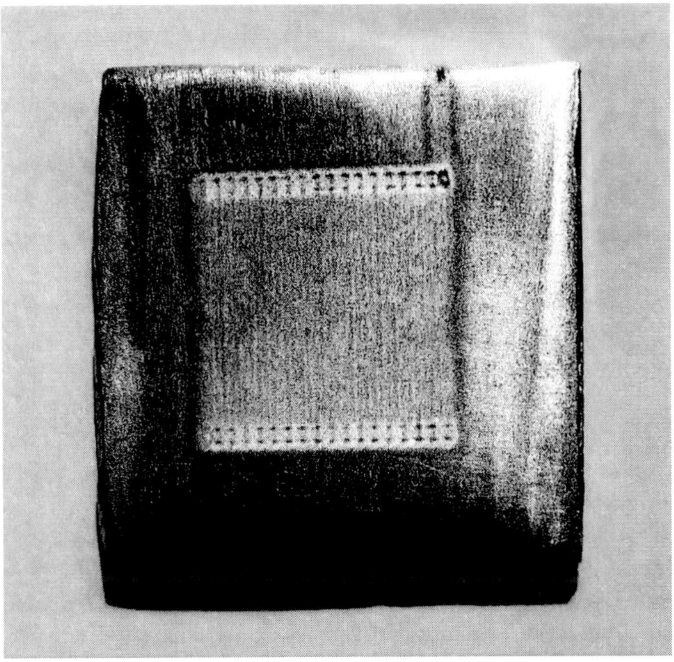

Figure 6.38 Macro view of an Al target exposed in water to radiation of 180 fs Ti:sapphire laser. The yellow square at the centre corresponds to the exposed area comprising NS. Lateral dimensions of the target are 3 × 3 cm^2.

The formation of Ag and Au NS under laser ablation of corresponding metallic targets in liquids is extensively studied, as the plasmon resonance of NS from these two metals lies in the visible. Therefore the successful formation of such NS is extremely important for optoelectronic applications. Laser exposure of Ag in liquids leads to significant modifications of the plasmon spectrum of this metal. Coloration of Al surface exposed to fs laser radiation has been reported by C. Guo *et al.* [116]. It is observed that after ablation at sufficiently high laser fluence in air, the target takes on a gold-yellow tint. Later, this coloration has been attributed to the formation of NS on Al surface under its ablation with fs laser pulses either in air or in liquid environment [114]. The authors attribute the yellow coloration of Al to the formation of periodic ripples decorated with

NS. The spectral reflectivity of the laser-treated sample is measured at near-normal incidence with poor resolution, so the UV features of Al NS presented in the present work were not observed. Coloration is ascribed either to different ripple periods or just to grooving the Al surface with various periods.

Figure 6.39 shows the absorption spectra of an Ag surface before as well as after its exposure in water with ps laser pulses. The spectrum of the initial surface shows a peak at 315 nm corresponding to the anticipated surface plasmon oscillations of electrons in the bulk Ag (spectrum 1). This spectral feature remains in the laser-exposed surface though it is widened and shifted to higher frequencies due to the NS formation (spectrum 2) that enhance the damping of plasmonic oscillations. At the same time NS bring new absorption bands and an additional wide peak appears in the near-UV region of spectrum centred at 370–380 nm. This peak shifts to the visible (spectrum 3 in Fig. 6.39, left) after sample storage for several days in air. The wing of this peak protrudes to the visible range of spectrum resulting in yellow-gold coloration of the laser-exposed areas. The liquid takes on a yellow colour as well, indicating the formation of nano-sized particles dispersed in the liquid. The theoretical position of the peak maximum of plasmon resonance for Ag NPs is situated exactly at 400 nm, provided that they are suspended in a liquid with refractive index $n = 1.33$ [117]. This peak is observed to shift towards UV, since the spectrum of the exposed surface is taken in air with $n = 1$ [110]. These spectral peaks around 400 nm indicate the formation of surface nanostructures with lateral dimensions comparable to those of Ag nanoparticles dispersed in the liquid [118]. Nanostructures of this size were indeed observed in the exposed areas of an Ag target, as shown in the previous section. Therefore the intensity of NS coloration depends on the environment. Figure 6.40 shows the macro view of nanostructured Ag substrate in air and in water. The coloration is more intense in aqueous medium, since it has higher refractive index. In water, the UV plasmon peak is red-shifted to the visible. Laser exposure of an Ag target in ethanol, under identical experimental conditions, leads to a similar pronounced spectral peak, though centred at 430 nm (Fig. 6.39, right, spectrum 2). On the contrary, an Au surface with NS formed under its ablation in water looks like bulk copper. The spectrum of NS on a bulk Au target is shown in Fig. 6.41 that compares absorption spectra of pristine and laser-exposed areas of target [119].

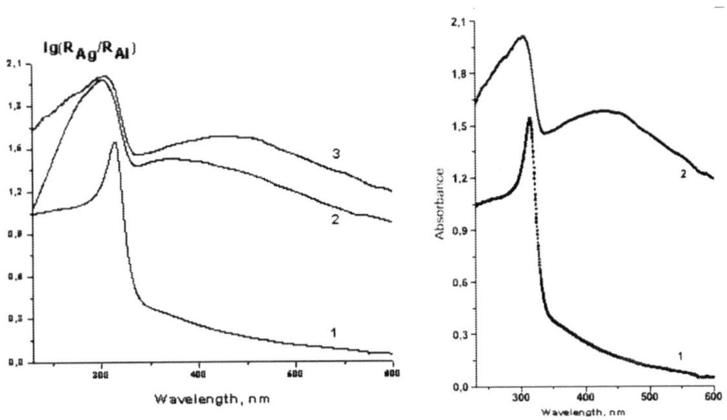

Figure 6.39 Modifications of reflectivity of an Ag target exposed (left) in water to radiation of a Nd:YAG laser, pulse duration of 350 ps. Pristine Ag surface (1), after laser exposure (2), and after storage in air for several days (3). Reference sample is bulk Al. (right) in ethanol under otherwise identical conditions.

(a) (b)

Figure 6.40 Macro view of the pattern with NS on a bulk Ag target. In air (a) and in water (b). The pattern was created by ablation with 200 fs radiation of a Ti:sapphire laser in aqueous environment.

Exposure of an Al target at short laser pulses in liquid results in the yellowish coloration of the metal, being visible at angles close to mirror reflection (Fig. 6.38). The coloration is most pronounced in the case of exposure into ethanol, where it appears just after only a few pulses at a fluence of as low as 0.05 J/cm^2. At lower fluences, virtually no changes of the Al surface are observed, even with elevated number of laser shots. This colour change is permanent, for instance, the target may be wiped by a wet tissue without any change of the colour and it is observed independently of the purity of Al target with Al content ranging from 75% through 95%. Hence, this coloration should be assigned to structuring of the Al surface at the nanometre scale similarly to NS on both Ag and Au [110, 119].

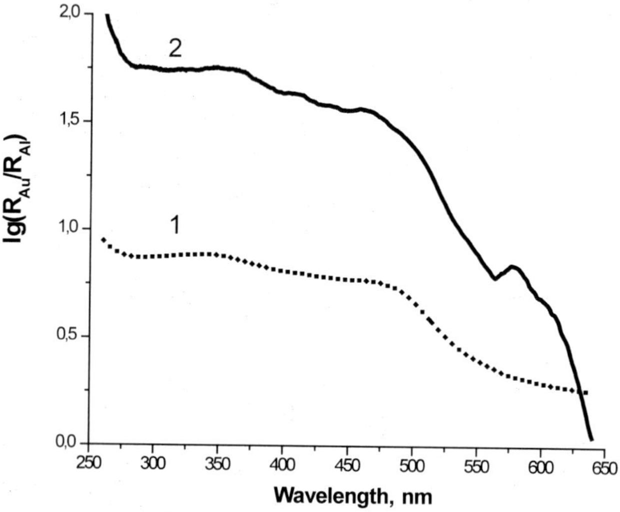

Figure 6.41 Absorptivity of the pristine Au target (1) and Au target with NS formed via its ablation in water with a 350 ps Nd:YAG laser (2). Spectra were taken with bulk Al as a reference.

Figure 6.42 presents the absorption spectra of Al surfaces structured with different laser sources. In the blue and near UV region the absorption of the exposed samples exceeds that of the initial surface, while in the NIR range, the samples ablated into liquid show lower absorption. In all cases, the absorption maximum is situated around 300 nm, which is justified by the yellow coloration of the corresponding exposed areas. This peak is shifted to lower wavelengths upon aging. When the Al target is exposed into liquids the latter takes on a yellow colour as well, indicating the formation of Al particles dispersed in the liquid. As shown in the spectrum of Fig. 6.42, these NP dispersions show a characteristic absorption peak close to 300 nm. The lateral size of NS on Al produced by its ablation in water with 350 ps laser radiation is 200 – 300 nm. The coloration of the exposed surface in this case is not visible at all. Comparison of laser ablation of bulk Al target in air and liquids has been reported recently [114]. It is found that coloration of Al is gained under laser action in both media. However, in case of laser exposure in air the target surface is covered by significant amount of sputtered material. NS generated in air are less compact presumably due to higher amount of oxide. The similarity of the spectrum under ablation of Al in air and in liquids suggests that the mechanism governing the NS formation is the same, independently on the medium. In all cases NS are formed due to spatial instability of the evaporation and appear as a result of action of recoil pressure onto a thin molten layer on Al. When the exposure is performed in liquids this pressure is the pressure of vapour of the surrounding liquid. In case of exposure in air this pressure is due to fast evaporation of the molten layer itself. That is why the laser fluence required to produce NS on Al in liquid is several times lower than in case of air exposure. Nanostructures in this case are formed owing to recoil pressure of metal vapours. Apparently, this process is also characterized by evaporation instability leading to formation of self-organized nano-sized areas within the evaporation cloud.

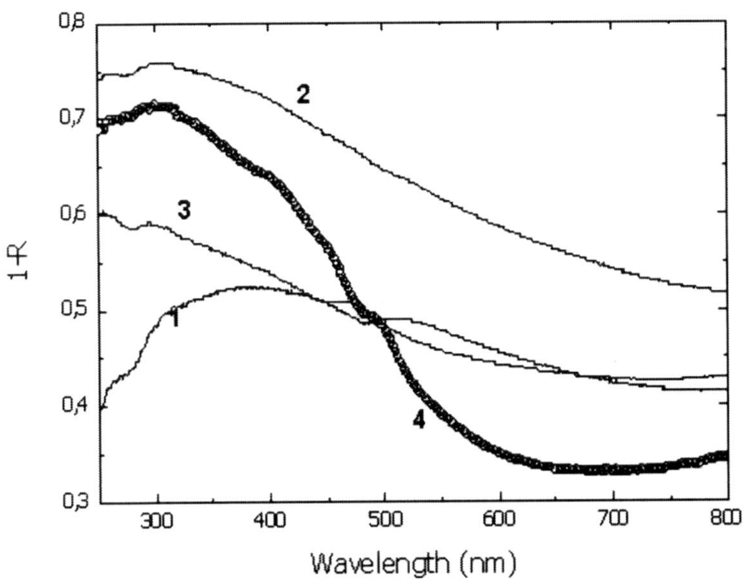

Figure 6.42 Absorption spectra of the initial Al surface (1) and of those exposed to the radiation of a 5 ps laser at 248 nm in water (2) and in ethanol (3). The curve 4 corresponds to exposure of Al in water to 180 fs-Ti:sapphire laser.

The optical absorption characteristics of an Ag surface decorated with laser-generated NS originate from the combination of plasmon resonance of electrons in bulk and nanostructured Ag. The peak at 370 nm is attributed to the generated Ag nanostructures due to the confinement of plasmon oscillations inside them [110]. This confinement occurs provided that the mean size exceeds the electron mean free path. Furthermore, the shape of the plasmon resonance peak of bulk Ag is modified by the generated nanostructures since the sharp plasmon peak at 315 nm is broadened and shifted to shorter wavelengths. The observed peak broadening evidences the enhanced scattering of electrons in the skin layer of the target, most likely due to presence of nanostructures.

The position of absorption peak of Ag nanospikes produced by ablation in ethanol can be explained by the pyrolysis of ethanol on Ag. This leads to the formation of products with high molecular mass, as already reported for the ablation of a brass target in ethanol [120]. Most probably these products are polyethylene of low molecular masses. Their deposition on Ag nanospikes also alters the refractive index of the surrounding medium and accounts for the observed red shift of the plasmon resonance.

The variation of the spectral reflectivity of the laser exposed Ag surface upon aging is due to the oxidation of nanostructures by air oxygen. Indeed, the nanospikes have a high specific surface and react easily with oxygen even at ambient temperature. The most common Ag oxide is Ag_2O, which has a higher refractive index in the visible. Partial oxidation of nanospikes alters the medium around them, and the plasmon oscillations frequency shifts to the red (lower frequencies). For similar reasons, the coloration of the laser-exposed Ag plate is more pronounced when the target is immersed in a liquid. Indeed, the surrounding liquid has a higher refractive index than air, which causes the red shift of its absorption peak. The position of the peak at 370 nm for a dry Ag surface with laser-generated nanospikes agrees with the calculated position of the plasmon resonance of Ag nanoparticles in vacuum [117]. On the other hand, it is known that any surface exposed to air humidity is covered by several monolayers of water, and these layers may shift the plasmon resonance. In the absence of any adsorbates, the absorption peak of Ag nanostructures should be shifted to higher frequencies.

The optical characteristics of an Al surface after its ablation in liquids can be again explained by a plasmon resonance absorption mechanism. The theoretical position of plasmon resonance of Al NPs of 10 nm in diameter in water was calculated in ref. 117 and the maximum of absorption lies around 200 nm (more precisely – in vacuum UV region). However, it is red-shifted for NP of higher diameters [121]. Also, oxidation of Al NP would also cause a red shift of their plasmon resonance, since its oxide, Al_2O_3, has higher refractive index in the UV range than water. It should be noted that alumina itself has no absorption in the range of study since its absorption only commences from 250 nm and even shorter wavelength, depending on its impurities. The presence of aluminium oxide is indicated by fluorescence measurements performed on the exposed surfaces. This oxide layer may be formed by fast oxidation of the irradiated surface upon solidification and efficiently passivates the surface of NS against further oxidation, so they are chemically stable to provide permanent coloration of Al surface. No fluorescence is detected in the case of a pristine Al target [115]. It is pertinent to note that the most probable effect caused by nanostructuring of a metallic surface is its yellow coloration. Plasmon frequencies for majority of metallic NPs are situated in UV with the exception of only three metals, Ag, Au, and Cu. Therefore, nanostructured metallic surface may show enhanced absorption in the blue region of spectrum, which is closer to UV. Enhanced absorption in the blue region corresponds to yellow coloration of the surface that is what is observed for Al and Ag. Of course, quadruple plasmon resonance of electrons in NS is usually shifted to the red region compared to the dipole one. However, the intensity of this absorption band is much weaker.

6.3.4 NS on Non-Metallic Surfaces

It has been recently reported that 800 nm, 100 fs laser irradiation of Si surfaces immersed in water gives rise to the formation of high-density regular arrays of nanometre-scale rods that are much smaller than the laser wavelength [122]. The authors believe that the water environment helps faster cooling of the laser-heated areas, while it is evident that this is of least importance and may be significant only for reduction of the average temperature of the substrate.

Nanotexturing of virtually all materials by laser ablation in liquids can be achieved using a recent approach suggested in ref. [123] for nanotexturing of both Ge and GaAs by laser ablation in air. The approach involves double exposure of the same area of the target with a 90° rotation of the target between exposures. The first exposure of a sample with linearly polarized laser beam produces straight ripples in the sample surface whose orientation is perpendicular to the polarization of the laser beam. The second exposure produces the SEW that propagates across previously generated relief. This technique has been used for nano-texturing of Si in ethanol with a Ti:Sapphire fs laser. At fluences in the range 2–4 J/cm^2, straight ripples form with a spacing of about 120 nm (Fig. 6.43). The long axis of the ripples is perpendicular to the laser polarization. If the ripples shown in Fig. 6.43 are irradiated for a second time in water with a polarization rotated by 90° so that it becomes parallel with the long axis of the ripples, the ripples break up into a surface that is uniformly covered with nanometre-scale rods. In general, NS generated on Si under its ablation in water with fs laser radiation show the same morphology as Ta NS obtained with fs pulses. Namely, periodic ripples coexist with spherical NS situated on top of the ripples (see Fig. 6.32). The spacing between ripples is much less than the laser wavelength. Spherical NS are located on top of ripples and look brighter.

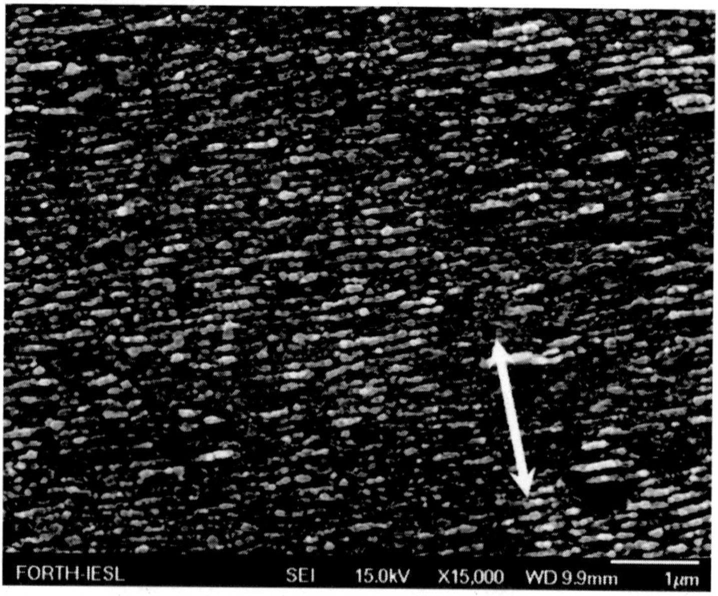

Figure 6.43 Single crystal Si ablated in water with femtosecond laser radiation at fluence of 2 J/cm^2. The arrow indicates the direction of laser beam polarization.

Nanotexturing of single crystal SiC by laser ablation in liquids has been also realized. Silicon carbide is a wide bandgap semiconductor with E_g = 3.2 eV for 4H type. The coupling of laser radiation for sub-bandgap laser photons is achieved via multi-photon excitation. The melting point of SiC is associated with its decomposition for Si and C, so the melt in proper sense does not exist. The absence of the melt makes impossible the formation of mushroom-like NS described above for other materials. The dominating feature of the relief of SiC subjected to laser exposure in liquids are straight grooves that may be associated with the interference of the incident laser wave with the induced SEW. The typical FE SEM view of the SiC surface ablated in either water or ethanol with a beam of a Ti:sapphire fs laser is presented in Fig. 6.44. The period of the grooves does not depend on the nature of the liquid and is about 155 nm.

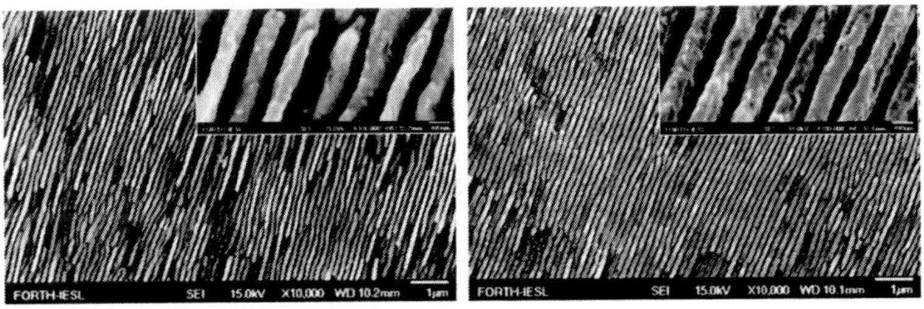

(a) (b)

Figure 6.44 FE SEM view of a single crystal 4H SiC ablated in water (a) and in ethanol (b) with the help of radiation of a Ti:sapphire laser with pulse duration of 200 fs. Scale bar in the insets denotes 100 nm.

This is much lower than the laser wavelength of 800 nm even if one takes into account that the wavelength of radiation is *n* times less, where *n* is the refractive index of either water or ethanol with respect to air. Apparently, the grooves were formed in the maxima of the interference pattern of the SEW with the laser beam. The material from these areas was decomposed and removed from the surface by the recoil pressure of the vapours of the liquid that surrounds the target. Therefore, the NS similar to those observed on metallic targets cannot be produced. SEW cannot propagate along the surface of SiC since this material is dielectric. One may suggest, however, that the surface of SiC becomes conductive under excitation of electrons from the valence to conduction band of SiC due to multi-photon absorption. The energy of the laser photon of a Ti:sapphire laser is 1.55 eV, therefore the absorption of at least two photons is needed to overcome the SiC bandgap of 3.2 eV. The result of the double exposure on SiC single crystal with 90° sample rotation between exposures is presented in Fig. 6.45. One can see that the surface of the SiC crystal is textured into regular squares of about 120 by 120 nm separated with grooves. The role of the liquid as the ablation medium is to provide the removal of the decomposition products from the ablated areas. Indeed, ablation of a SiC single crystal in air results in the formation of both Si oxide and Si nanoclusters that are left in the target surface. Other products of SiC decomposition are gaseous carbon oxides.

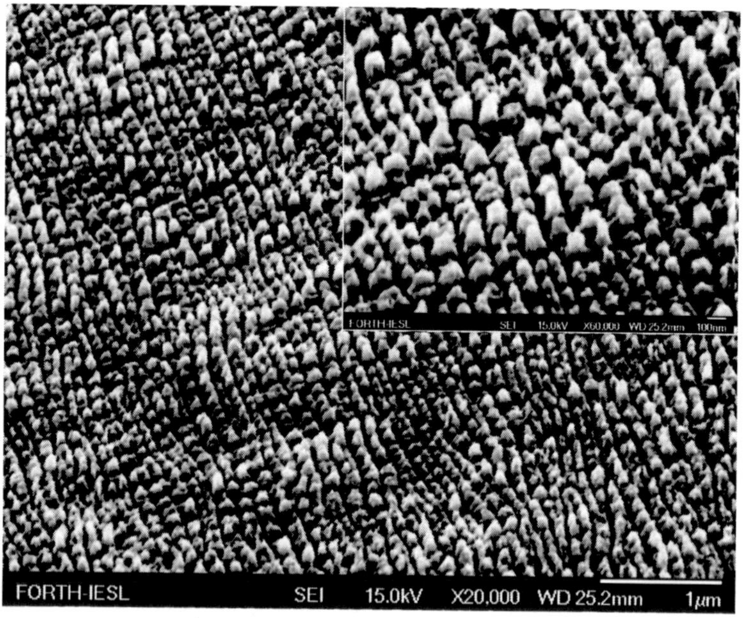

Figure 6.45 Nanotexturing of a single crystal SiC by double exposure in ethanol to radiation of a femtosecond Ti:sapphire laser. The scale bar in the inset denotes 100 nm.

6.3.5 Applications of NS

Besides evident fundamental interest, NS that are formed under laser-induced phase transitions at the solid-liquid interface possess promising potential for different applications. Nano-sized metal features are responsible for so called SERS due to local amplification of the electromagnetic waves in their vicinity. As a result, the intensity of a pumping laser beam is enhanced giving rise to a corresponding amplification of the intensity of the weak Raman-shifted scattering signal. The total enhancement factor amounts to 10^5 to 10^6.

Silver and gold are the most popular metals used in SERS experiments. The SERS experiments are usually carried out with freshly prepared colloidal metallic NP in absence of surface-active substances. However, the measurements can be conducted only once, since the same colloidal solution of NPs cannot be used twice due to absorbed organic molecules. Silver and Au NS generated via laser ablation at the solid-liquid interface (see Figs. 6.27a,b), can be potentially be a promising alternative for studying SERS activity since they offer the advantageous capability of performing multiple measurements on the same nanostructured substrate. Figure 6.46 presents experiments carried out on a freshly-prepared NS on bulk Ag plate with the help of its ablation in water with a 90 ps Nd:YAG laser. The probe molecule used was acridine dissolved in water [124].

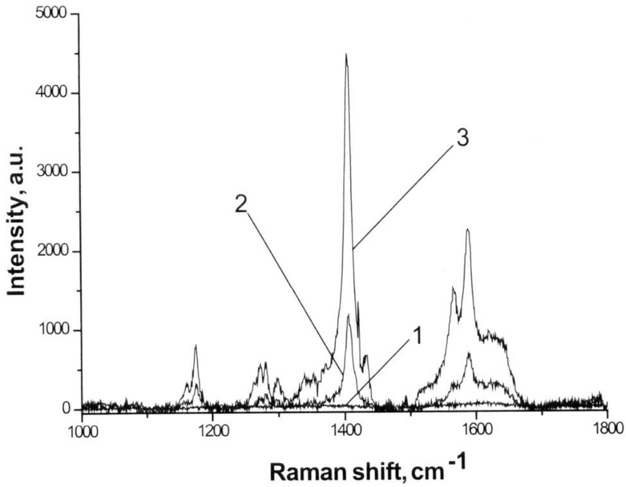

Figure 6.46 Raman spectra of acridine molecules (concentration of 10^{-5} M/l). Excitation by a cw Ar ion laser, wavelength of 514 nm, acquisition time of 30 s. NS on Ag were generated by exposure under water to the radiation of a 90 ps Nd:YAG laser. Initial Ag surface (1), NS generated with 2000 laser shots (2) and 6000 laser shots (3).

Nanostructuring of some metals is highly desirable for various medical applications. Proliferation of biological cells is more efficient if the surface has a certain value of roughness. It was found that the optimal roughness lies in the range of several hundreds of nm. In this sense formation of NS under laser ablation of solids in liquids ideally suits the needs for fabrication of efficient medical implants. Tantalum is perfectly biocompatible with living cells owing to its high chemical stability. However, its high cost limits its use as an implant. More common biocompatible materials are Ti alloys. Figure 6.47 shows the example of successful nanostructuring of a medical Ti alloy by its exposure to short laser pulses in ethanol. One can see that NS on Ti alloy consist of periodic ripples decorated on top with spherical NS typical of other solids. Therefore, the initial experiments on nanostructuring of biocompatible materials were promising in this respect.

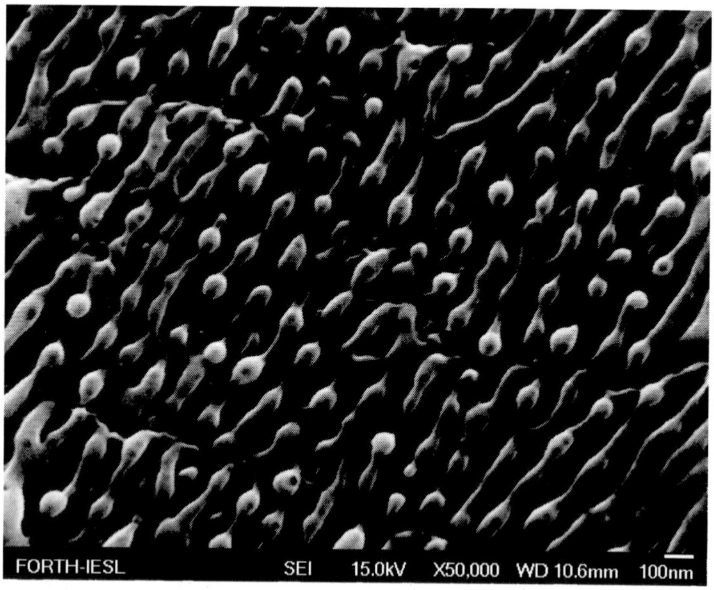

Figure 6.47 Nanostructuring of medical Ti alloy with the help of its ablation in ethanol with a 5 ps radiation of a KrF laser at 248 nm.

Another promising field of applications is the control over wettability of solid surfaces via laser-assisted generation of NS. This approach is very advantageous since the rate of NS generation is very high and can be carried out in the "laser writing" regime.

Laser nanotexturing of metals can find applications for the increase of emission properties of cathodes made of this metal. Nanoprotrusions with small radius of curvature serve as an efficient emitter of electrons since the electric field near them is higher than on a flat surface. Therefore, the potential barrier for electrons near these protrusions is lower, and the efficient work function of nanostructured surface is lower than that of a flat one. Figure 6.48 shows the image of W cathode before and after laser ablation in ethanol with pulse duration of 180 fs [125]. The surface of the Tungsten (W) target is covered by ripples with period about 350 nm. The surface protrusions are situated on top of periodic ripples.

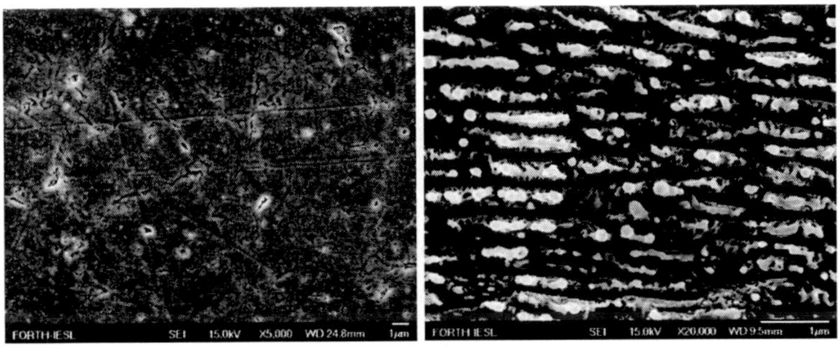

(a) (b)

Figure 6.48 FESEM image of the initial surface of W cathode (a) and NS with ripples (b) on it after ablation in ethanol. NS are located on top of ripples. Pulse duration of 180 fs, wavelength of 800 nm.

Their lateral size is between 50 and 150 nm. The cathode made of W was subjected to double exposure with linear polarization of the laser radiation. The sample was turned for 90° between the exposures [123]. The morphology of W cathode after double laser exposure in ethanol is shown in Fig. 6.49. Average lateral size of squared texture is about 150–250 nm. Spherical NS are located on top of square areas. Their size varies from 50 to 200 nm, and density of NS is 8×10^8 per cm^2. The work function of cathode with thermal emission was measured. Figure 6.50 shows the dependence of the work function of both initial and nanotextured cathodes on temperature. The work function of nanotextured cathode decreases by 2.5% with the increase of temperature. Initial cathode does not show such changes. Also presented data indicate the decrease of

(a)

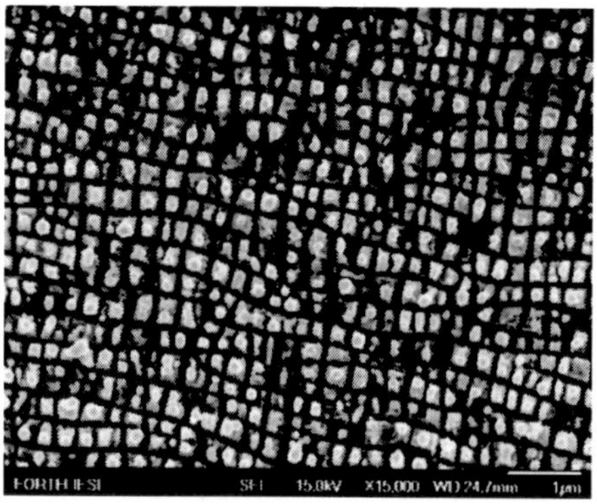

(b)

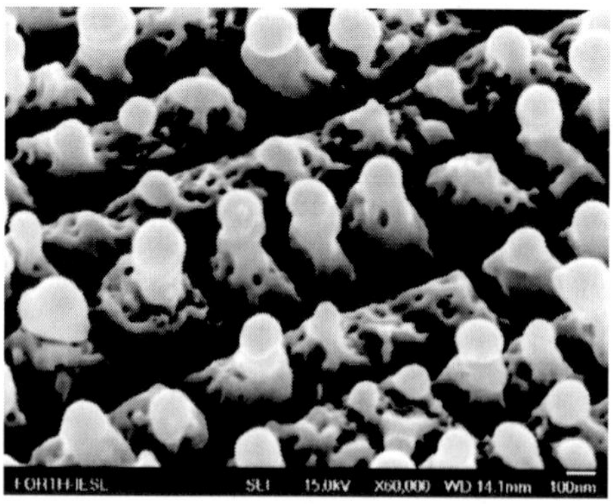

(c)

Figure 6.49 FE SEM view of W cathode: (a) general view, (b) top view of two-dimensional array obtained by double laser exposure in ethanol with sample rotation by 90°, (c) view tilted for 25°. Scale bar denotes 100 μm (a), 1 μm (b), and 100 nm (c).

work function for the structured surfaces of the cathode by 8% with respect to the initial sample. It is about 0.3 eV. Since the density of the electron current depends on the work function exponentially, this result is very promising. Nanostructuring of W allows working at lower temperature at the same emission current that allows increasing the lifetime of the device.

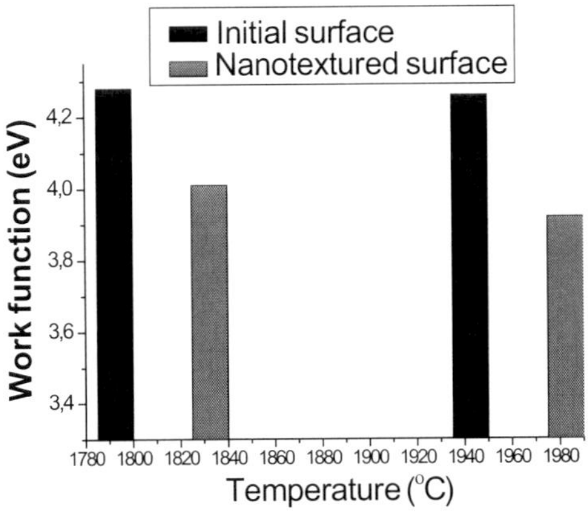

Figure 6.50 Dependence of the work function on temperature for initial and nanotextured W cathode. Ti:sapphire laser, wavelength of 800 nm, pulse duration of 180 fs, double laser exposure in ethanol.

6.4 Summary and Future Perspectives

The present article is focused on the application of ultrafast pulsed laser processing of surfaces for materials' surface modification at multiple length scales. Two distinct approaches are reviewed: laser surface structuring in controlled gas and liquid media. Ultrafast laser processing presents distinct advantages for practical applications, since: (a) it is versatile and material independent and (b) it is rapid, easily adaptable and scalable through parallel processing and (c) it could be readily adapted and integrated to mass production lines in industry. In particular, the use of focused fs laser pulses to materials texturing offers new opportunities in photonics industry. Although fs laser processing and inscription has been studied for several decades, recent significant improvements in the types of fs sources available have accelerated the acceptance of these lasers as a strong option for commercial fabrication in manufacturing and photonics industry. The lasers available today offer vastly improved peak powers and reliability offering the possibility to open the door of the fs processing technology into a range of engineering applications. Furthermore, high repetition rate fs laser systems are extensively developed and attract much attention as new light sources due to the potential increase of the fabrication rate or production throughput. In view of the above, ultrafast lasers present unique capabilities for the realization of novel surfaces, opening new opportunities for innovation and exploitation in materials industry. Therefore, the wealth of arising possibilities and the number of new approaches prescribe a future where control of materials characteristics at different length scales and subsequent applications can be accomplished with a level of sophistication that we cannot presently envisage.

References

1. B. C. Stuart, *et al., J. Opt. Soc. Am. B* 13, 459 (1996).
2. B. N. Chichkov, *et al., Appl. Phys. A-Mater.* 63, 109 (1996).
3. W. S. Fann, R. Storz, H. W. K. Tom, and J. Bokor, *Phys. Rev. B* 46, 13592 (1992).
4. J. Hohlfeld, *et al., Chem. Phys.* 251, 237 (2000).
5. K. Sugioka, *et al.* (eds.), *Laser Precision Microfabrication*, Springer Series in Materials Science 135, DOI 10.1007/978-3-642-10523-4-7.
6. E. Stratakis, E. Ranella, and C. Fotakis, *Biomicrofluidics* 5, 0113411 (2011).
7. A. V. Kabashin and M. Meunier, in *Recent Advances in Laser Processing of Materials* (Amsterdam: Elsevier) 1 (2006).
8. B. G. Arnold and A. Pique. *MRS Bull.*, 32, 9 (2007).
9. P. N. Prasad *Introduction to Biophotonics* (Boston: Wiley-Interscience) (2003).
10. G. A. Shafeev, in: *Lasers in Chemistry, Volume 2: Influencing Matter* (M. Lackner, ed.), Wiley VCH Verlag GmbH&Co, KGaA, Wienheim, 713–741 (2008).
11. A. Y. Vorobyev and C. Guo. *Phys. Rev. B* 72, 195422 (2005); A. Y. Vorobyev, C. Guo *Appl. Surf. Sci.*, 253, 7272 (2007); A. Y. Vorobyev, C. Guo *Opt. Express* 14 (6), 2164 (2006).

12. M. Birnbaum, *J. Appl. Phys.* 36, 3688 (1965).
13. R. Wagner, and J. Gottmann, *J. Phys.: Conf. Ser.* 59, 333–337 (2007).
14. A. Borowiec and H. K. Haugen, *Appl. Phys. Lett.*, 82, 4462 (2003).
15. A. V. Vorobyev, V. S. Makin, and C. L. Guo, *J. Appl. Phys.*, 101, 034903 (2007).
16. B. H. Veld and H. van der Veer, *J. Laser Micro/Nanoeng. (JLMN)*, 5, 28 (2010).
17. L. T. Qi, K. Nishii, and Y. Namba, *Opt. Lett.*, 34, 1846 (2009).
18. K. Sokolowski-Tinten, J. Bialkowski, A. Cavalleri, D. von der Linde, A. Oparin, J. Meyer-ter-Vehn, and S. I. Anisimov, *Phys. Rev. Lett.*, 81, 224 (1998).
19. M. Huang, F. Zhao, Y. Cheng, N. Xu, and Z. Xu, *ACS Nano* 3, 4062 (2009).
20. F. Huang, Y. Zhao, N. Cheng, Xu, and Z. Xu, *Opt. Express* 18, A600 (2010).
21. J. Heitz, E. Arenholz, D. Bauerle, R. Sauerbrey, and H. M. Phillips, *Appl. Phys. A* 59, 289–293 (1994).
22. S. Baudach, J. Bonse, and W. Kautek, *Appl. Phys. A* 69 [Suppl.], S395–S398 (1999).
23. N. Yasumaru, K. Miyazaki, and J. Kiuchi , *Appl. Phys. A* 76, 983 (2003).
24. O. Varlamova, F. Costache, J. Reif, and M. Bestehorn, *Appl. Surf. Sci.* 252, 4702–4706 (2006).
25. J. D. Fowlkes, A. J. Pedraza, D. A. Blom, and H. M. Meyer III, *Appl. Phys. Lett.* 80, 3799 (2002).
26. T. J. Bastow, *Nature* 222, 1058 (1969).
27. D. C. Emmony, R. P. Howson, and L. J. Willis, *Appl. Phys. Lett.* 23, 598 (1973).
28. L. J. Willis and D. C. Emmony, *Opt. Laser Technol.* 7, 222 (1975).
29. H. J. Leamy, G. A. Rozgonyi, T. T. Sheng, and G. K. Celler, *Appl. Phys. Lett.* 32, 535 (1978).
30. G. N. Maracas, G. L. Harris, C. A. Lee, and R. A. McFarlane, *Appl. Phys. Lett.* 33, 453 (1978).
31. R. M. Walser, M. F. Becker, J. G. Ambrose, and D. Y. Sheng, in: *Laser and Electron-Beam Solid Interactions and Materials Processing* (J. F. Gibbons, W. Hess, and T. Sigmon, eds.), Elsevier, New York, 177 (1981).
32. M. F. Becker, R. M. Walser, Y. K. Jhee, and D. Y. Sheng, *Proc. Soc. Photo-Opt. Instrum. Eng.* 322, 93 (1982).

33. J. A. Van Vechten, *Solid State Commun.* 39, 1285 (1981).
34. R. W. Rendell and K. L. Ngai, *Bull. Am. Phys. Soc.* 27, 235 (1982).
35. F. Keilmann and Y. H. Bai, *Appl. Phys. A* 29, 9 (1982).
36. Z. Guosheng, P. M. Fauchet, and A. E. Siegman, *Phys. Rev. B* 26, 5366 (1982).
37. J. E. Sipe, J. F. Young, J. S. Preston, and H. M. van Driel, *Phys. Rev. B* 27, 1141 (1983).
38. J. F. Young, J. S. Preston, H. M. van Driel, and J. E. Sipe, *Phys. Rev. B* 27, 1155 (1983).
39. M. Oron and G. Sorensen, *Appl. Phys. Lett.* 35, 782 (1979).
40. P. A. Temple and M. J. Soileau, *IEEE J. Quant. Elec. QE*-17, 2067 (1981).
41. A. M. Ozkan, A. P. Malshe, T. A. Railkar, W. D. Brown, M. D. Shirk, and P. A. Molian, *Appl. Phys. Lett.* 75, 3716 (1999).
42. H. Varel, M. Waehmer, A. Rosenfeld, D. Ashkenazi, and E. E. B. Campbell, *Appl. Surf. Sci.* 127–129, 128 (1998).
43. Q. Wu, Y. Ma, R. Fang, Y. Liao, Q. Yu, X. Chen, and K. Wang, *Appl. Phys. Lett.* 82, 1703 (2003).
44. F. Costache, M. Henyk, and J. Reif, *Appl. Surf. Sci.* 208–209, 486 (2003).
45. J. Bonse, M. Munz, and H. Sturm, *J. Appl. Phys.* 97, 013538 (2005).
46. J. Bonse, H. Sturm, D. Schmidt, and W. Kautek, *Appl. Phys. A* 71, 657 (2000).
47. F. Costache, S. Kouteva-Arguirova, and J. Reif, *Appl. Phys. A* 79, 1429 (2004).
48. M. Huang, F. L. Zhao, T. Q. Jia, Y. Cheng, N. S. Xu, and Z. Z. Xu, *Nanotechnology* 18, 505301 (2007).
49. M. Huang, F. Zhao, Y. Cheng, N. Xu, and Z. Xu, *Phys. Rev. B* 79, 125436 (2009).
50. Y. Shimotsuma, P. G. Kazansky, J. Qiu, and K. Hirao, *Phys. Rev. Lett.* 91, 247405 (2003).
51. E. N. Glezer and E. Mazur, *Appl. Phys. Lett.* 71, 882 (1997).
52. O. Varlamova, F. Costache, J. Reif, and M. Bestehorn, *Appl. Surf. Sci.* 252, 4702 (2006).
53. G. Daminelli, J. Krueger, and W. Kautek, *Thin Solid Films* 467, 334 (2004).
54. R. Le Harzic, H. Schmuck, D. Sauer, T. Anhut, I. Riemann, and K. Koenig, *Opt. Express* 13, 6651 (2005).

55. A. Weck, T. H. R. Crawford, D. S. Wilkinson, H. K. Haugen, and J. S. Preston, *Appl. Phys. A* 89, 1001 (2007).
56. J. Bonse, A. Rosenfeld, and J. Krueger, *J. Appl. Phys.* 106, 104910 (2009).
57. J. Bonse and J. Krueger, *J. Appl. Phys.* 108, 034903 (2010).
58. S. Sakabe, M. Hashida, S. Tokita, S. Namba, and K. Okamuro, *Phys. Rev.* B 79, 033409 (2009).
59. K. Okamuro, M. Hashida, Y. Miyasaka, Y. Ikuta, S. Tokita, and S. Sakabe, *Phys. Rev. B* 82, 165417 (2010).
60. Y. Han and S. Qu, *Chem. Phys. Lett.* 495, 241 (2010).
61. A. Y. Vorobyev and C. L. Guo, *Appl. Phys. A* (Rapid Comm.), 86, 321 (2007).
62. A. S. Mahmood, M. Sivakumar, K. Venkatakrishnan, and B. Tan, *Appl. Phys. Lett.*, 95, 034107 (2009).
63. M. Silvennoinen, J. Kaakkunen, K. Paivasaari, P. Vahimaa, and T. Jaaskelaninen, *J. Laser Micro/Nanoeng. (JLMN)*, 5, 97 (2010).
64. A. Y. Vorobyev and C. Guo, *Phys. Rev. B* 72, 195422 (2005); *Opt. Photon. News* 18, 43 (2007).
65. A. Y. Vorobyev, V. S. Makin, and Chunlei Guo, *Phys. Rev. Lett.* 102, 234301 (2009).
66. D. E. Diebold, N. H. Mack, Stephen K. Doorn, and Eric Mazur, *Langmuir* 25, 1790–1794 (2009).
67. C.-H. Lin, L. Jiang, Y.-H. Chai, H. Xiao, S.-J. Chen, and H.-L. Tsai, *Opt. Express* 17(24), 21588.
68. K. A. Al-Naimee, *Indian J. Sci. Technol.* 3, 308–310 (2010).
69. J. E. Rothenberg and R. Kelly, *Nucl. Instrum. Meth. Phys. Res. B* 1, 291 (1984).
70. F. Sanchez, J. L. Morenza, R. Aguiar, J. C. Delgado, and M. Varela, *Appl. Phys. Lett.* 69, 620 (1996).
71. W. Kautek and J. Kr¨uger, in: *Material Science Forum* (D. B. Chrisey and G. K. Hubler, eds.), Trans Tech Publications Inc., Switzerland, vol. 173–174, 17 (1995).
72. A. J. Pedraza, J. D. Fowlkes, and D. H. Lowndes, *Appl. Phys. Lett.* 74, 2322 (1999).
73. T.-H. Her, R. J. Finlay, C. Wu, S. Deliwala, and E. Mazur, *Appl. Phys. Lett.* 73, 1673 (1998).

74. T.-H. Her, R. J. Finlay, C. Wu, and E. Mazur, *Appl. Phys. A* 70, 383 (2000).
75. E. Spanakis, J. Dialektos, E. Stratakis, V. Zorba, P. Tzanetakis, and C. Fotakis, *Phys. Stat. Sol. (c)*, 5 (10), 3309 (2008).
76. B. Nayak, M. Gupta, and K. Kolasinski, *Nanotechnology* 18, 195302 (2007).
77. B. K. Nayak and M.Gupta, *Optics Lasers Eng.* 48, 940–949 (2010).
78. S. I. Dolgaev, S. V. Lavrishev, A. A. Lyalin, A. V. Simakin, V. V. Voronov, and G. A. Shafeev, *Appl. Phys. A* 73, 177 (2001).
79. V. Zorba, E. Stratakis, M. Barberoglou, E. Spanakis, P. Tzanetakis, and C. Fotakis, *Appl. Phys. A* 93, 819 (2008).
80. V. Zorba, L. Persano, D. Pisignano, A. Athanassiou, E. Stratakis, R. Cingolani, P. Tzanetakis, C. Fotakis, *Nanotechnology* 17, 3234 (2006).
81. M. Barberoglou, V. Zorba, E. Stratakis, E. Spanakis, P. Tzanetakis, SH. Anastasiadis, C. Fotakis, *Appl. Surf. Sci.* 255, 5425 (2009).
82. V. Zorba, E. Stratakis, M. Barberoglou, E. Spanakis, P. Tzanetakis, S. H. Anastasiadis, C. Fotakis, *Adv. Mater.* 20, 4049 (2008).
83. E. Stratakis and V. Zorba, in *Nanomaterials for the Life Sciences Vol. 7: Biomimetic and Bioinspired Nanomaterials* (C. S. S. R. Kumar, ed.) Wiley-VCH Verlag GmbH & Co. KGaA, Weinheim, 2010.
84. V. Zorba, *et al., Appl. Phys. Lett.* 88, 081103 (2006); C.H. Crouch, J.E. Carey, J.M. Warrender, M.J. Aziz, E. Mazur, F.Y. Génin, *Appl. Phys. Lett.* 84, 1850 (2004).
85. V. Zorba, N. Boukos, I. Zergioti, and C. Fotakis, *Appl. Optics* 47, 1846 (2008).
86. J. E. Carey, *Femtosecond-laser Microstructuring of Silicon for Novel Optoelectronic Devices*, Phd Thesis, Harvard University (2004).
87. T. H. Her, *Femtochemistry at Gas/Solid Interfaces*. PhD thesis, Harvard University, 1998.
88. S. I. Dolgaev, N. A. Kirichenko, A. V. Simakin, and G. A. Shafeev, *Quantum Electron.* 34, 771 (2004).
89. T. Chiba, R. Komura, and A. Mori, *Jpn. J. Appl. Phys.* 39, 4803 (2000).
90. G. Wysocki, R. Denk, K. Piglmayer, N. Arnold, and D. Baeuerle, *Appl. Phys. Lett.* 82, 692 (2003).
91. J. P. Moening and D. G. Georgiev, *J. Appl. Phys.* 107, 014307 (2010).
92. J. E. Carey, C. H. Crouch, M. Shen, and E. Mazur, *Opt. Lett.* 30, 1773–1775 (2005).
93. E. Magoulakis, *et al., Appl. Phys. A* 98, 701 (2010).

94. V. Zorba, P. Tzanetakis, C. Fotakis, E. Spanakis, E. Stratakis, D. G. Papazoglou, I. Zergioti, *Appl. Phys. Lett.* 88, 081103 (2006).

95. A. V. Karabutov, V. D. Frolov, E. N. Loubnin, A. V. Simakin, and G. A. Shafeev, *Appl. Phys. A* 76, 413–416 (2003).

96. E. Stratakis, R. Giorgi, M. Barberoglou, T. Dikonimos, E. Salernitano, N. Lisi, and E. Kymakis, *Appl. Phys. Lett.* 96, 043110–043112 (2010).

97. R. Blossey, *Nat. Mat.* 2, 301 (2003).

98. M. Callies and D. Quere, *Soft Mater.* 1, 55 (2005).

99. A.-M. Kietzig, S. G. Hatzikiriakos, and P. Englezos, *Langmuir* 25 (8), 4821 (2009).

100. T. Lippert and J. T. Dickinson, *Chem. Rev.* 103, 453–485 (2003).

101. F. Xia, H. Ge, Y. Hou, T. Sun, L. Chen, G. Z. Zhang, and L. Jiang, *Adv. Mater.* 19, 2520 (2007).

102. E. L. Papadopoulou, M. Barberoglou, V. Zorba, A. Manousaki, A. Pagkozidis, E. Stratakis, and C. Fotakis, *J. Phys. Chem. C* 113, 2891 (2009).

103. M. Barberoglou, V. Zorba, A. Pagozidis, C. Fotakis, and E. Stratakis, *Langmuir* 26, 13007–13014 (2010).

104. E. Stratakis, A. Mateescu, M. Barberoglou, M. Vamvakaki, C. Fotakis, and S. H. Anastasiadis, *Chem. Commun.* 46, 4136–4138 (2010).

105. A. Y. Vorobyev and C. Guo, *Appl. Phys. Lett.* 94, 224102 (2009).

106. A. Y. Vorobyev and C. Guo, *Optics Express* 18, 6455 (2010).

107. Y. Chen and A. Vertes, *Anal. Chem.* 78, 5835–5844 (2006).

108. A. Ranella, M. Barberoglou, S. Bakogianni, C. Fotakis, and E. Stratakis, *Acta Biomater.* 6(7), 2711–2720 (2010).

109. E. L. Papadopoulou, A. Samara, M. Barberoglou, A. Manousaki, S. N. Pagakis, E. Anastasiadou, *et al.*, *Tissue Eng. Part C Methods* 16(3), 497–502 (2010).

110. E. V. Zavedeev, A. V. Petrovskaya, A. V. Simakin, and G. A. Shafeev, *Quantum Electron.* 36(10) 978 (2006).

111. E. V. Barmina, M. Barberoglou, V. Zorba, A. V. Simakin, E. Stratakis, C. Fotakis, and G. A. Shafeev, *Quantum Electron.* 39, 89–93 (2009).

112. G. A. Shafeev, in: *Lasers in Chemistry, Volume 2: Influencing Matter* (M. Lackner, ed.), Wiley VCH Verlag Wienheim, 713–741, ISBN: 978-3-527-31997-8 (2008).

113. S. S. Rekhviashvili and E. V. Kishtikova, *Techn. Phys. Lett.* 32(10), 50 (2006).

114. E. Stratakis, V. Zorba, M. Barberoglou, C. Fotakis, and G. A. Shafeev, *Appl. Surf. Sci.* 255, 5346–5350 (2009).

115. E. Stratakis, V. Zorba, M. Barberoglou, C. Fotakis, and G. A. Shafeev, *Nanotechnology* 20, 105303–105307 (2009).

116. A. Y. Vorobyev and C. Guo, *Appl. Phys. Lett.* 92, 041914 (2008).

117. J. A. Creighton and D. G. Eadon, *J. Chem. Soc. Faraday Trans.*, 87, 3881 (1991).

118. G. A. Shafeev, in: *Nanoparticles: New Research* (S. L. Lombardi, ed.), 1–37, ISBN: 978-1-60456-704-5, Nova Science Publishers, Inc (2008).

119. S. Lau Truong, G. Levi, F. Bozon-Verduraz, A. V. Petrovskaya, A. V. Simakin, and G. A. Shafeev, *Appl. Surf. Sci.* 254, 1236 (2007).

120. P. V. Kazakevich, V. V. Voronov, A. V. Simakin, G. A. Shafeev, *Quantum Electron.* 34(10), 951–956 (2004).

121. B. S. Luk'yanchuk and M. I. Tribelsky, Collection of papers devoted to memory of Prof. M. N. Libenson. The St.-Petersburg Union of the Scientists, Russia, 101–117 (2005).

122. M. Shen, J. E. Carey, C. H. Crouch, M. Kandyla, H. A. Stone, E. Mazur, *Nano Lett.* 8(7), 2087 (2008).

123. E. Freysz, J. Degert, N. Lascoux, and G. Shafeev, in: *Conference on Lasers and Electro-Optics/Quantum Electronics and Laser Science and Photonic Applications Systems Technologies*, Technical Digest (Optical Society of America), paper JTuC45 (2008).

124. S. Lau Truong, G. Levi, F. Bozon-Verduraz, A. V. Petrovskaya, A. V. Simakin, and G. A. Shafeev, *Appl. Phys. A* 89(2), 373 (2007).

125. E.V. Barmina, A.A. Serkov, E. Stratakis, C. Fotakis, V. N. Stolyarov, I. N. Stolyarov, and G.A. Shafeev *Appl. Phys. A* 106, 1–4 (2012).

Chapter 7

Nanoablation Using Nanospheres and Nanotips

Mitsuhiro Terakawa and Minoru Obara
Department of Electronics and Electrical Engineering, Keio University, 3-14-1 Hiyoshi, Kohoku-ku, Yokohama 223-8522, Japan
minoruobara@aol.com

The use of an enhanced near field scattered around nanospheres and nanotips is an attractive method to strengthen the ability of laser processing, which overcomes its diffraction limit. In this chapter, we describe nanoablation processing using nanospheres and nanotips excited by pulsed laser irradiation. The combination of a near field and femtosecond laser illumination can demonstrate non-thermal and ultrafast near-field ablation processing on various kinds of materials, including transparent materials via multiphoton absorption processes. The technology of enhanced near field induced in the vicinity of nanospheres excited by femtosecond laser provides periodic nanohole arrays in air with high throughput. The near-field coupling effect between adjacent nanospheres governs the magnitude of the field enhancement and the near-field distribution

Ultrafast Laser Processing: From Micro- to Nanoscale
Edited by Koji Sugioka and Ya Cheng

ISBN 978-981-4267-33-5 (Hardcover), 978-981-4303-69-9 (eBook)
www.panstanford.com

on the material substrate to be processed. Nanoablation using small tips whose sizes are smaller than the incident wavelength is also described in this chapter. Finally, we discuss the nanoheater effect of the plasmonic nanoprocessing. The processing technology using near field around nanospheres and nanotips is smart and promising to fabricate surface nanostructures.

7.1 Introduction

The demand for the simple and robust nanoprocessing technology is increasing [1–3], and at the emphasis is on developing smart optical, electro-optical, electrical, mechanical, and biological devices. From the viewpoint of practical applications, nanopatterning of subwavelength structures on a substrate surface is a key requirement for potential applications such as antireflection surface structures, nanotribology of surface friction, structural colors, and SERS (surface enhanced Raman scattering) template. The utilization of an enhanced near field has a potential to provide high-precision processing to meet the demands.

The basic physics of the near field dates back to the Mie theory in 1908, which describes the interaction between the electromagnetic field and metallic spheres with dimensions smaller than the incident wavelength [4]. Recent development of commercial and also new robust methods for fabrication of nanospheres and other nanostructures has increased the possibility of nanoprocessing by use of the near field [5, 6]. Pulsed laser irradiation of nanostructures such as nanospheres and nanotips excites enhanced electromagnetic field in the nanospace below the diffraction limit of the far field. Besides the optical properties observed in the far field, the nature of the near field defines an exponential decrease of the optical intensity with the distance from the surface of the nanostructures, enabling precise processing in sub-micro to nanoscale. With dielectric spheres, underlying physics for enhanced near field under the spheres shifts from micro-lens effect to Mie resonance scattering domain governed mainly by the size parameter [7–9]. The strong enhanced near field is generated around the Mie resonance dielectric spheres, allowing the fabrication of nanoholes with a diameter less than 100 nm on the substrate surface under the spheres. The near field generated around the metallic spheres is strongly enhanced due to the resonant electromagnetic coupling effect between the incident laser and collective electron oscillation, i.e., plasmon resonance [10–12]. It is notable that femtosecond laser is an innovative tool to be used in this method, which can significantly contribute to a strong localization of the laser-matter interaction. The merged method of a near field and femtosecond laser can realize non-thermal, and intense near-field ablation nanoholes processing on various kinds of materials. Enhanced near field at the nanometer scale is also achievable around the tip for scanning tunneling microscope (STM) [13] and atomic force microscope (AFM) [14] by laser irradiation. Near-field scanning optical microscope (NSOM) [15] also allows nanoablation by using near field in the tip aperture.

In this chapter, we describe nanoablation processing using nanospheres and nanotips excited by pulsed laser irradiation. Section 7.2 starts with the basic physics of the enhanced near field induced in the vicinity of nanospheres, followed by the characteristics of the near field and its ablation properties. Nanoablation using small tips whose sizes are smaller than the incident wavelength is stated in Section 7.3. Finally, we discuss the nanoheaters, nanosphere-laser pulse interaction focusing on optical absorption, and nanosphere heating in Section 7.4.

7.2 Nanoablation Using Nanosphere

The processing technique with the enhanced near field in the vicinity of nanospheres is an attractive method for high-throughput nanopatterning with a simple experimental setup. This technique has been investigated extensively after the publication of the paper presenting the formation of hillocks whose size is smaller than the incident wavelength [16]. The mechanism of the formation of the subwavelength structure was found to be the enhanced near field by the spheres. As will be described in detail in Section 7.2.3, in the case of 400 nm excitation wavelength irradiation to 200 nm dielectric spheres with refractive index of 2.7 placed on a substrate, the corresponding values of effective numerical aperture (NA) can be calculated for 2.33 on SiO_2 substrate and 3.42 on Si substrate, respectively [17]. These values using a near-field nanolens are much higher than the highest NA value of 1.44 at 193 nm for liquid immersion photolithography of ULSI using conventional far-field optics. Another promising type of nanosphere used for nanoprocessing is a metallic nanosphere. In this section, the essential theory in the enhanced near field by both types of spheres will be described. Subsequently, studies on nanoablation by using metallic nanospheres and that by dielectric nanospheres will be reviewed, respectively.

7.2.1 Plasmon Polaritons and Mie Scattering Near Field

Nanospheres generating a near field are classified broadly into two categories: dielectric nanospheres (e.g., polystyrene, silica, etc.) and metallic nanospheres (e.g., gold, silver, etc.). Both types of nanospheres generate an intense near field mainly by the Mie scattering, whereas the underlying processes of near-field generation are different. Free electron oscillation (plasmons) inside the metallic nanospheres generates a highly enhanced near field, while the dominant factor in nanopatterning by using dielectric nanosphere is scattered light passing through the nanospheres.

The internal and external electromagnetic field of spheres can be expressed by the Mie coefficient. Based on the Mie theory, Messinger *et al.* obtained the scattering efficiency formula in the near-field zone [18]. The derivation of scattering efficiency and near-field efficiency from the Mie coefficient is described in Refs. [19, 20]. Near-field efficiency is defined as the value that the scattering cross section of the near-field generation around a sphere (C_{NF}) divided by the geometrical cross section of the sphere (πR^2), that is, the enhancement factor of the near-field intensity compared to the incident laser intensity. The near-field efficiency formula can be written as follows:

$$\begin{aligned} Q_{\mathrm{NF}} &= \frac{C_{\mathrm{NF}}}{\pi R^2} \\ &= 2\{|a_n|^2\left[(n+1)\left|h_{n-1}^{(2)}(k_0R)\right|^2 + n\left|h_{n+1}^{(2)}(k_0R)\right|^2\right] \\ &\quad + (2n+1)|b_n|^2\left|h_n^{(2)}(k_0R)\right|^2\} \end{aligned} \tag{7.1}$$

where R is the radius of the sphere, k_0 is the wave number in the surrounding medium, $h^{(2)}(x)$ is spherical Hankel function of the second kind, a_n and b_n are the Mie coefficients. Near-field efficiency Q_{NF} is squared amplitude of electric field in the vicinity of the sphere surface. Q_{NF} is larger than scattering efficiency in the near-field zone, and decreases rapidly with increasing the distance from the sphere surface. Figure 7.1 shows dependence of near-field efficiency Q_{NF} on the size parameter $\alpha = 2\pi R/\lambda$ for metallic and dielectric spheres. The Q_{NF} was calculated for gold, polystyrene (PS), and SiO_2 spheres. The incident wavelength is set to 820 nm. The refractive indices for PS and SiO_2 are 1.578 and 1.453, respectively. Dielectric spheres show saturation of Q_{NF} at around $\alpha = 5$ and large vibrations can be observed for Q_{NF} for size parameter higher than 7. The vibration is attributable to the induction of enhanced near field affected by the reflection at the sphere surface such as whispering-gallery mode. In the case of gold spheres, the sphere has the specific resonant size for the wavelength.

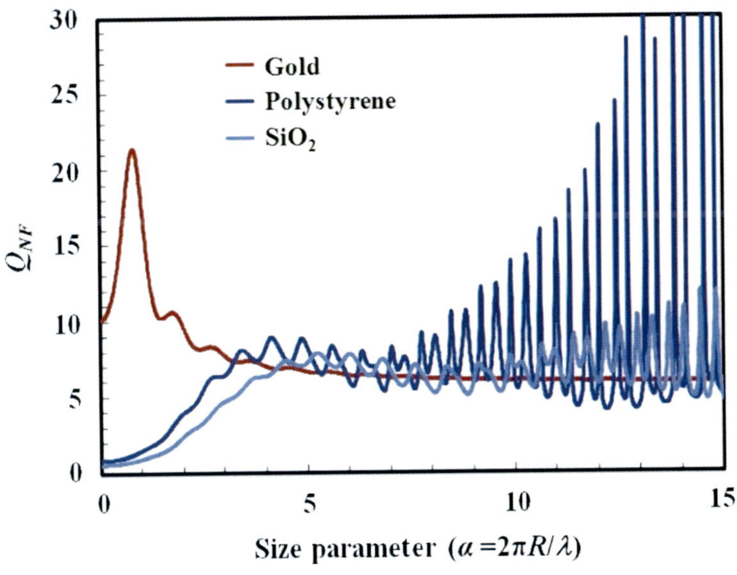

Figure 7.1 Dependence of near-field efficiency on size parameter for gold, polystyrene, and SiO_2 spheres.

7.2.2 Nanoablation Using Metallic Nanosphere

Nanopatterning mediated with surface plasmon polaritons in a metallic sphere is based on the electric field enhancement and localization on the substrate surface due to the capacitive interaction of the near field around the sphere and the image charge induced in the substrate. The processing characteristics can be discussed by the numerical calculation of optical field using the finite-difference time-domain (FDTD) method [21, 22] and the simulation result is validated by the experiment.

Figure 7.2 represents the optical field intensity distribution in the vicinity of the gold sphere with a diameter of 200 nm (a) in vacuum and (b) on a silicon substrate. The illumination laser wavelength of an 800 nm Ti:sapphire femtosecond laser with a linearly polarized light with the *E* vector is applied parallel to the x axis. In vacuum, the induced near-field intensity has the highest value on the equator plane (Fig. 7.2(a)), which is the evidence for the electric dipole excitation at 800 nm. When the sphere is placed on the substrate, the near-field distribution is drastically changed to give a strong localized and enhanced intensity distribution in the vicinity of the contact point of the sphere and the substrate (Fig. 7.2(b)). The enhanced field intensity inside the Si substrate will result in the substrate modification with a spatial dimension at the nanometer scale.

Figure 7.3 shows SEM images of nanoholes experimentally produced by a single Ti:sapphire femtosecond laser pulse with an 800 nm wavelength and pulse width of 150 fs (FWHM). The laser irradiation is directed normally to the silicon substrate. The holes are ablated by using (a) circularly polarized and (b) linearly polarized laser irradiation. The nanohole shows radial symmetry when the circular polarization is used, while an elongated-shape hole along the *E* vector direction is observed for the linear polarization. These shapes are well matched with the optical field intensity distribution inside the substrate from the top view of the substrate surface, as shown in the insets in Fig. 7.3. The wavelength dependence of the near-field intensity inside the Si substrate is shown in Fig. 7.4 [23]. The resonant wavelength does not show a significant change with changing the sphere size if the sphere is placed on the substrate.

The distributions of the optical field intensity around the gold sphere on (a) gold (metal), (b) silicon (semiconductor), and (c) soda lime glass (dielectric) substrate are shown in Fig. 7.5 [12, 24]. Due to the induction of image charge on the substrate surface and the evanescent properties of the near field, the high optical field enhancement is localized around the contact point between

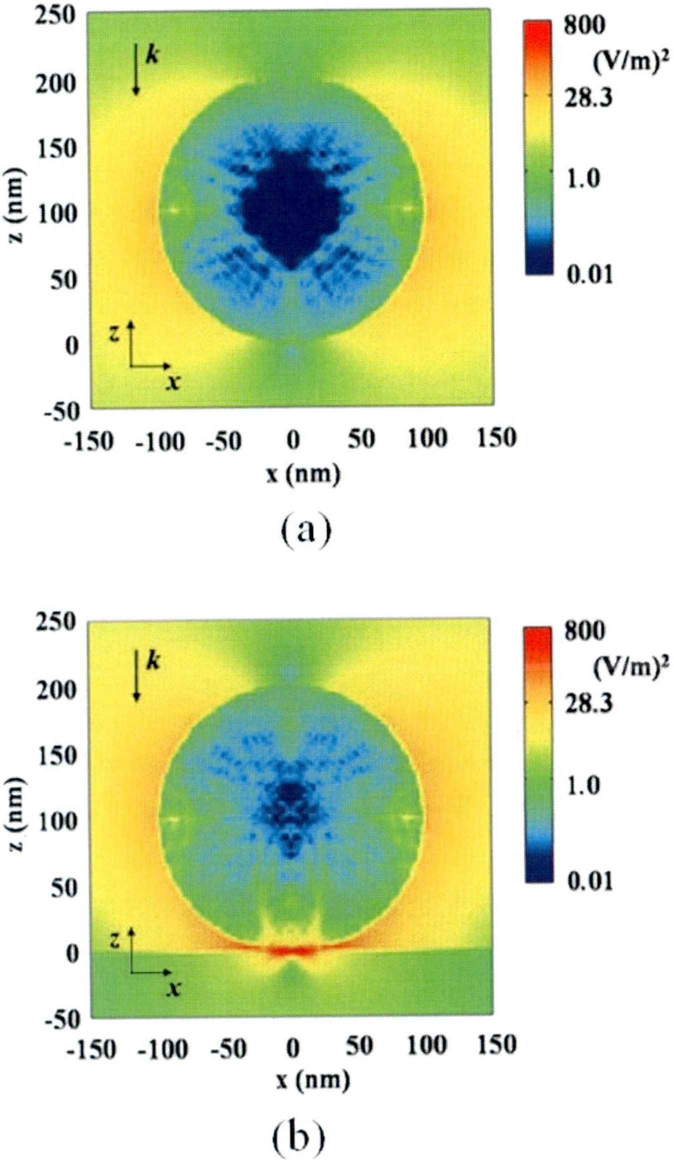

Figure 7.2 Optical field intensity ($|E/E_0|^2$) distribution for 200 nm gold sphere: (a) in vacuum and (b) on silicon substrate [23]. The 800 nm incident polarization is linear and the E vector is parallel to the x axis.

(b)

Figure 7.3 SEM images of the fabricated nanohole with 200 nm gold sphere on silicon substrate [23]. The pulse duration is 150 fs, laser fluence is 0.2 J/cm^2 and the hole is ablated by a single shot irradiation. (a) and (b) show the hole obtained with circular polarization and linear polarization, respectively. The inset shows the optical field intensity distribution inside the Si substrate ($z = -1.25$ nm). The definition of the axis is described in Fig. 7.2.

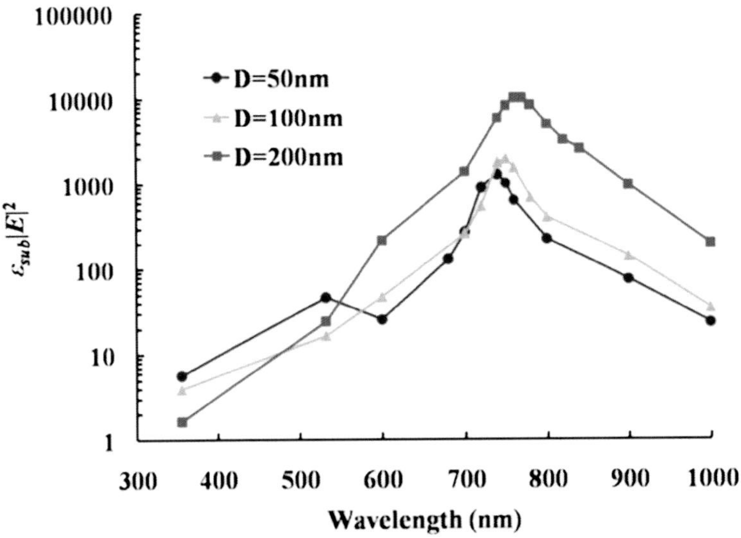

Figure 7.4 Wavelength dependence of the enhancement factor of the optical intensity inside a silicon substrate for several diameters of the gold spheres [23]. The field intensity is at $z = -0.625$ nm for $D = 50$ nm, $z = -1.25$ nm for $D = 50$ nm and $z = -2.5$ nm for $D = 200$ nm. The enhancement factor is calculated as $\varepsilon_{\text{sub}}|E|^2$ divided by $\varepsilon_{\text{air}}|E_{\text{inc}}|^2$, where ε_{sub} is the permittivity of the Si substrate, ε_{air} is the permittivity of air, and E_{inc} is the incident optical field.

the sphere and the substrate for the gold and silicon substrates. Figure 7.6 shows the calculated depth distributions of the optical field in the different substrates [24]. The dependence of the field distribution corresponds to the point on the surface where the field has a maximum value. As can be seen in Fig. 7.6, the optical field drops rapidly to the initial incident value as the decay depth is in the order of tens of nanometers. The theoretically obtained field enhancement factor suggests that the processing characteristics are strongly dependent on the optical properties of the substrate materials.

Figure 7.7 shows SEM images of holes produced on (a) Au, (b) Si and (c) SiO_2 substrate when the gold spheres on the substrate were

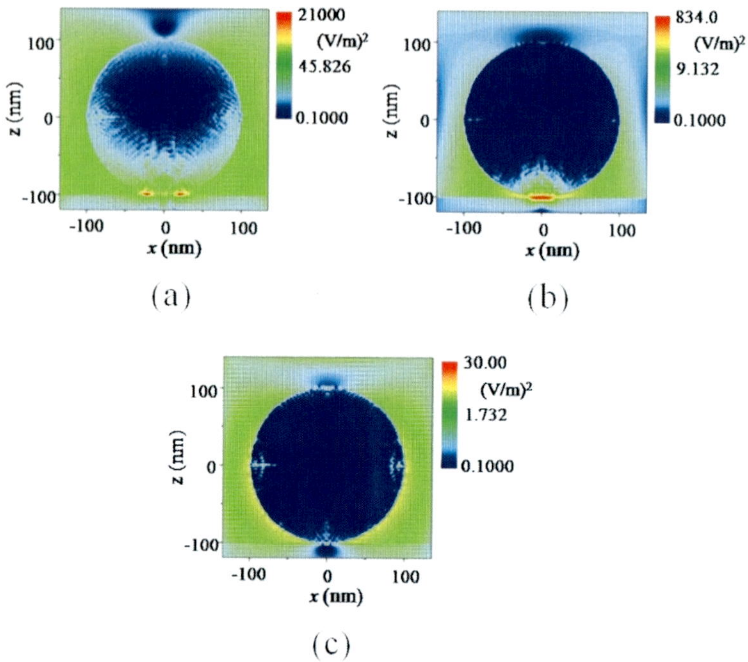

Figure 7.5 Optical field intensity ($|E/E_0|^2$) distribution on an x–z plane for simulation system with gold nanosphere placed on: (a) gold, (b) silicon, and (c) glass substrate, respectively [12]. The intensity distribution is shown on the plane through the center of the sphere and the contact point between the sphere and the substrate. The substrate surface is at z = –100 nm. The incident optical wave is circularly polarized in x–y plane, and it propagates along the –z direction. The incident optical field intensity is 1 (V/m)^2.

irradiated by a single laser shot. The laser fluence was 0.1, 0.13, and 6.3 J/cm^2, respectively [25]. In all cases, the incident fluence is lower than the threshold for bare surface modification by the single shot experiment. In the case of gold and silicon surfaces, the laser fluence is about two times lower than the bulk modification threshold. In the case of glass substrate, the nanohole formation is clearly observed at the laser fluence of about only 1.1 times lower than the threshold. The optical field distribution on the substrate surface is shown as the insets, which confirms the strong enhancement factor on a gold and silicon substrate and a weak enhancement factor on the glass substrate, being consistent with the experimental results.

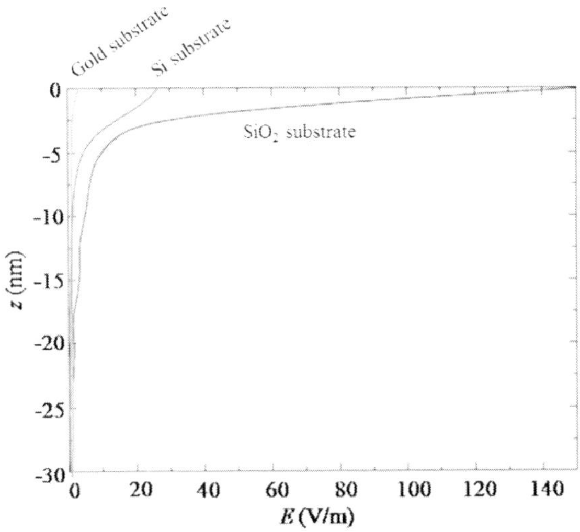

Figure 7.6 Depth distributions of the electric field for three different substrate materials [24]. The sphere has a diameter of 200 nm and the wavelength of the incident radiation is 800 nm. The position of the contact point between the sphere and the substrate is $z = 0$.

Aiming at a potential application of parallel and reproducible technique for arbitrary surface nanostructuring, the discussion on the enhanced near field of an arrayed sphere system is necessary. Due to the capacitance effect, a strong near field can be developed in the intersphere gap of closely spaced spheres. The intersphere interaction (plasmon coupling) would affect both the spatial distribution and the maximal magnitude of the optical near field. The scattering cross section of the sphere would also affect the optical near field due to the presence of the adjacent spheres. Thus, the properties of the near field on the substrate are governed not only by the interaction between the sphere's charge and the image charge induced in the substrate, as is the case of the isolated sphere, but also by the capacitive coupling between the charges of adjacent spheres.

Figure 7.8 shows the near-field intensity distribution on the gold substrate surface on which gold nanospheres with a diameter of 200

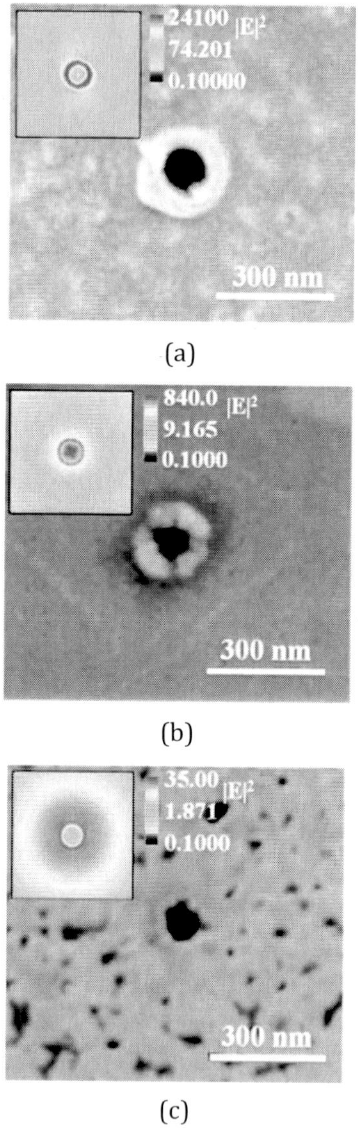

Figure 7.7 SEM images of holes produced in (a) Au, (b) Si and (c) SiO_2 substrates when the substrates with deposited nanospheres are irradiated by a single laser shot [25]. The laser fluence is 0.1, 0.13, and 6.3 J/cm^2, respectively. Laser radiation has circular polarization. The distribution of the near optical field intensity on the substrate surface calculated by FDTD simulation is also shown. The spatial scale of the intensity distribution is equivalent to the SEM image.

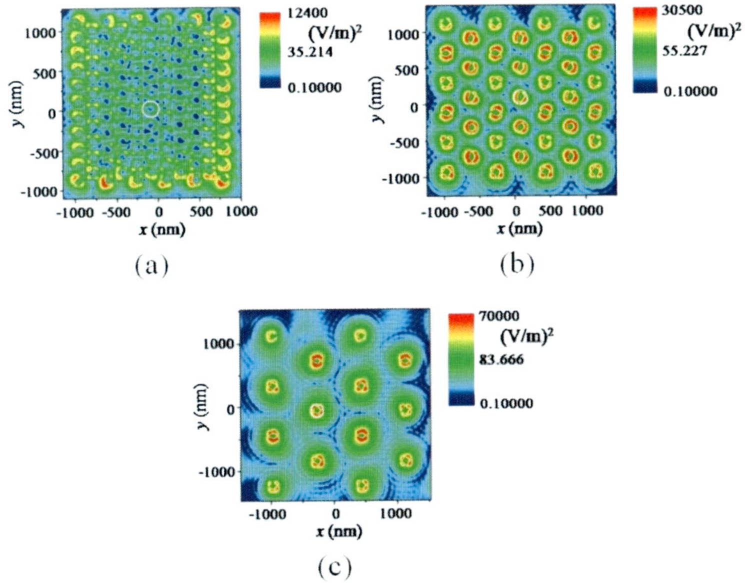

Figure 7.8 Calculated optical near-field intensity ($|E/E_0|^2$) distribution on gold substrate surface when gold nanospheres with diameter of 200 nm are arrayed hexagonally and irradiated by optical wave at a wavelength of 800 nm [25]. The field distribution is shown for three intersphere distances: (a) 200 nm (touching spheres), (b) 400 nm and (c) 785 nm. The white line represents the sphere delineation in (a).

nm are arrayed hexagonally, and irradiated by a single laser pulse at wavelength of 800 nm. The intensity distribution is shown for the three different intersphere distances: (a) 200 nm (contact mode), (b) 400 nm and (c) 785 nm. The results indicate that the field intensity is enhanced under the spheres, but both the magnitude and the spatial distribution on the gold substrate surface are strongly dependent on the intersphere distance. The lowest value for the near-field intensity is obtained for the system of no intersphere distance (Fig. 7.8(a)). In this case, the spatial distribution of the near-field intensity is not homogeneous, but the intensity is higher under the spheres located at the edge of the aggregation, and it is about an order of magnitude higher than that under the spheres located at the center of the arrayed sphere system. This intensity under the spheres increases with increasing the intersphere distance. In the case of 400 nm intersphere distance, the optical intensity distribution under all spheres is similar to all spheres in the figure.

The dependence of the maximal enhancement factor on intersphere distance is shown in Fig. 7.9. The calculation system is composed of gold nanospheres of 200 nm diameter on a Si substrate, which is irradiated with incident plane wave of an 800 nm wavelength. The incident optical field is assumed to be 1 V/m, and the incident wave is linearly polarized. An isolated sphere shows a maximal enhancement factor of 500 and the maximal enhancement factor for the sphere arrays is much smaller than that for the isolated sphere. The intensity increases monotonically with the increase of the intersphere distance [26].

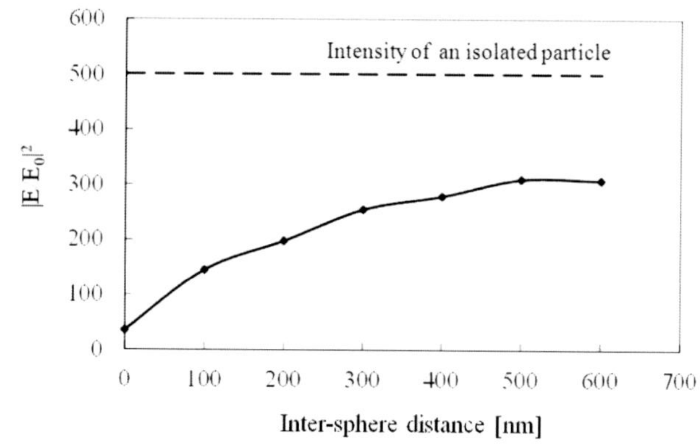

Figure 7.9 Dependence of intersphere distance on the maximal field intensity on the Si substrate in an infinite 2D gold nanosphere array [26]. The dashed line shows the intensity on the Si substrate in the isolated gold sphere case.

To solve the above technical issue, it was found that the change of the irradiation conditions by means of the angle of incidence gives another option for laser nanopatterning with tunable performance characteristics. Enhanced optical intensity comparable to that of the isolated sphere can be induced on the substrate surface by oblique incidence irradiation of the laser to the arrayed spheres due to the image charge interaction with the substrate [26]. The p-polarized wave incidence with oblique irradiation to the substrate surface can realize an uniform nanohole array patterning when gold nanospheres are closely and periodically ordered (Fig. 7.10).

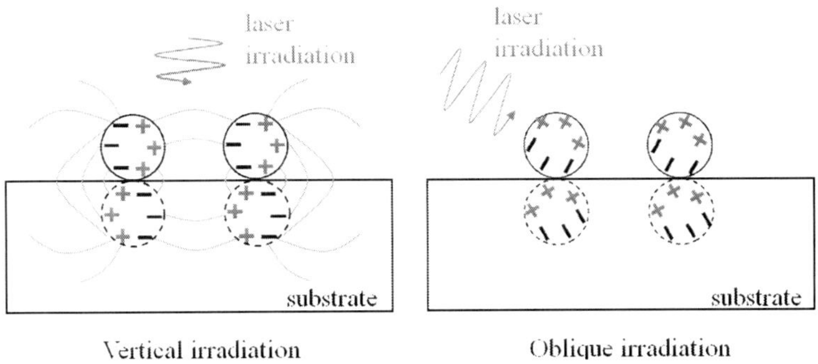

Figure 7.10 Conceptual diagrams of the intersphere interaction (plasmon coupling) with vertical irradiation and that of decreasing intersphere interaction with oblique irradiation.

7.2.3 Nanoablation Using Dielectric Nanosphere

The dominant factor in nanopatterning using dielectric sphere is scattered light passing through the spheres governed by the Mie scattering. Figure 7.11 shows optical intensity distribution under the polystyrene (PS) spheres of diameters of (a) 200, (b) 450, and (c) 820 nm calculated by FDTD method [27]. In the calculation, a plane wave with a wavelength of 820 nm was irradiated normally to the PS sphere placed on a Si substrate. Solid and dashed lines show the intensity distribution in the x and y directions, respectively. The dotted line shows the sphere size. The incident light is linearly polarized to the y direction. As shown in Fig. 7.11, the intensity distribution is a near Gaussian profile for any sphere size. The focusing radius becomes smaller as the sphere size is decreased.

For the nanoablation using a dielectric sphere, the size parameter ($\alpha = 2\pi R/\lambda$, R: radius of the sphere, λ: incident wavelength) in the Mie scattering theory is a critical parameter for the generation of near-field light. Here we describe the relationship between the size parameter and the nanohole size fabricated by femtosecond laser processing. Figure 7.12 shows the dependence of the near-field efficiency Q_{NF} on the size parameter for four wavelengths: 800 nm (fundamental wave of Ti:sapphire femtosecond laser), 400 nm (SHG), 263.5 nm (THG), and 200 nm. The near-field efficiency Q_{NF} was calculated at the position of the sphere surface. The near-field efficiencies with 800 nm and 400 nm seem to be similar, indicating that

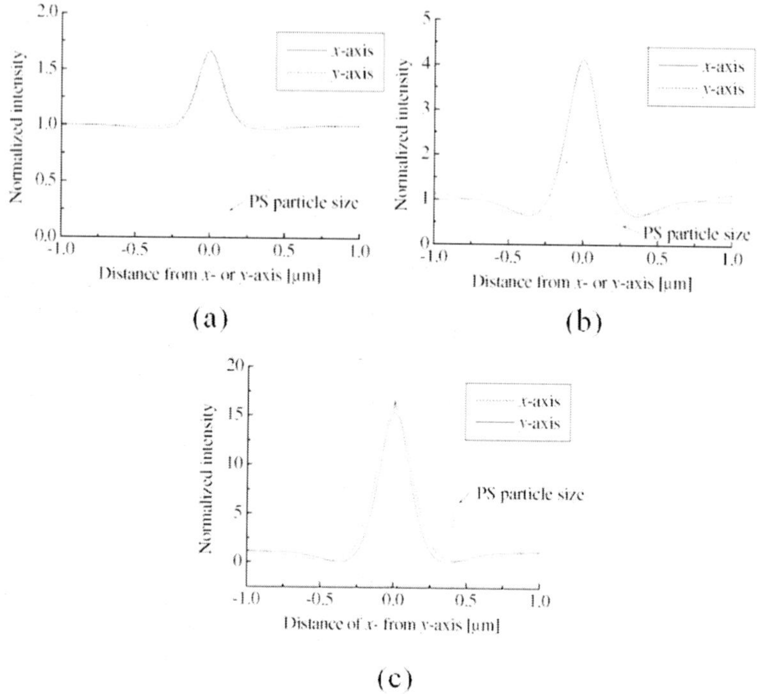

Figure 7.11 Calculated optical intensity distribution in the vicinity of contact point between PS sphere and Si surface along *x*- and *y*-axis [27]. The sphere diameters are (a) 200, (b) 450, and (c) 820 nm. The 820 nm incident light (plane wave) is linearly polarized to the *y*-axis. The laser intensity is normalized by the incident laser intensity.

a similar near-field efficiency can be achieved at both wavelengths if the ratio of the incident wavelength to the sphere size kept constant. At the incident wavelength of 263.5 nm, a different tendency from those of 800 nm and 400 nm is shown. The near-field efficiency decreases with increasing the size parameter due to the optical absorption by the PS sphere itself, which can be understood by the increase of the extinction coefficient k at shorter wavelengths. In the case of the incident wavelength of 200 nm, the near-field efficiency shows a quite different behavior. The near-field efficiency reaches a maximum value at around a size parameter of 1, and it rapidly decreases with increasing the size parameter. This result indicates that PS behaves similarly to small metallic spheres, that is, the near-

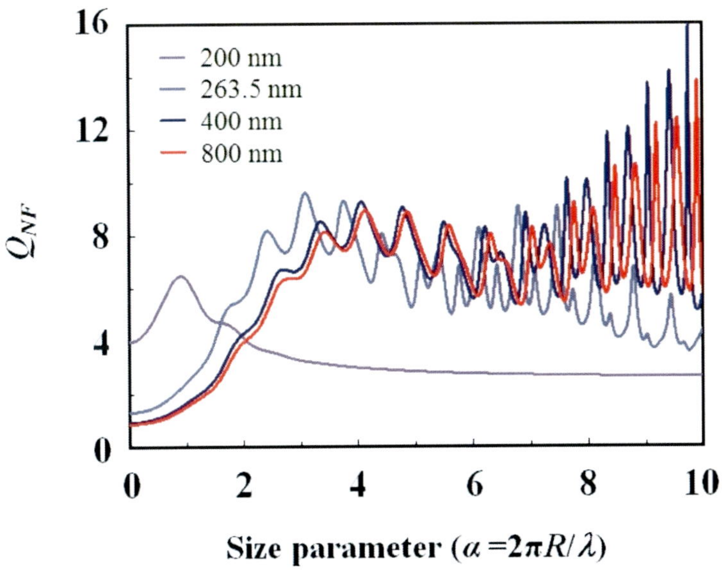

Figure 7.12 Dependence of the near-field efficiency on the size parameter for four wavelengths [8]. The near-field efficiency Q_{NF} is starting from $\alpha = 0.01$.

field efficiency shows a similar tendency depending on the extinction coefficient. Figure 7.12 indicates the size parameter suitable for efficient nanoprocessing is around $\alpha \sim 3.14$ ($=\pi$) for the incident wavelength of 800 nm, 400 nm, and 263.5 nm. Figure 7.13 shows the cross-sectional image of the optical field intensity distribution inside the silicon substrate for the size parameter around 3.14. The white dashed lines delineate the surface of the PS sphere, and the contour lines show the optical field intensity distribution inside the silicon substrate. As shown in Fig. 7.13 (a), the optical near field is focused deeply in the silicon substrate. Therefore, the incident optical field intensity is sufficiently enhanced by the micro-lens effect. Figure 7.13 (b) shows the optical field intensity distribution in the silicon substrate with 400 nm irradiation. Since the extinction coefficient k increases, the optical field intensity decreases drastically. Figure 7.13 (c) shows the optical field intensity distribution in the silicon substrate with 263.5 nm irradiation. The highest near-field efficiency in the three figures is induced by 263.5 nm radiation. However, as can be seen in Fig. 7.13 (c), little penetration of the optical field into the

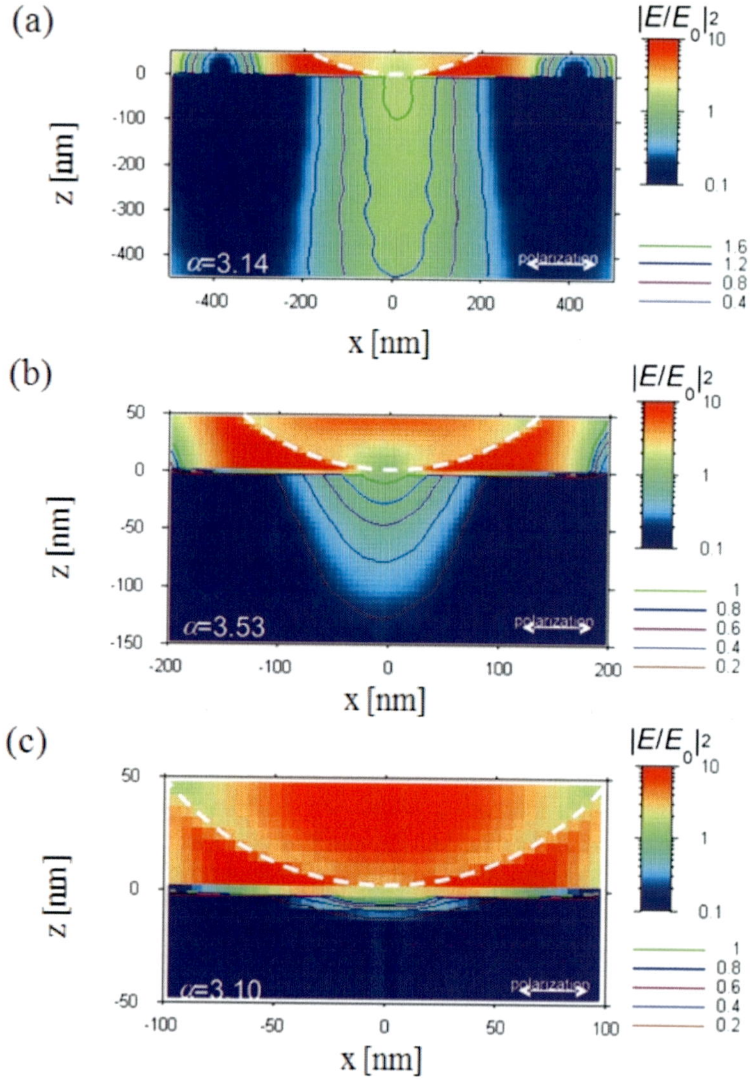

Figure 7.13 Optical field intensity distribution in x–z plane under PS sphere and inside silicon substrate [8]. White dashed line shows the PS sphere surface. The Si surface is at $z = 0$ nm. (a) Incident wavelength λ is 800 nm, the PS diameter d is 800 nm. $n = 3.673$ and $k = 0.005$. (b) $\lambda = 400$ nm, $d = 450$ nm. $n = 5.570$ and $k = 0.387$. (c) $\lambda = 263.5$ nm, $d = 260$ nm. $n = 1.764$ and $k = 4.278$.

Si substrate is observed and the field is mostly reflected off from the substrate surface. It is because the extinction coefficient $k = 4.278$ of the silicon substrate at 263.5 nm has a similar value to metals. The use of 263.5 nm wavelength can provide a smaller hole processing and the patterned hole depth becomes shallow.

The arrangement of dielectric nanospheres on a substrate is technically easy by using a wet pulling up process. Using metallic nanospheres, the near-field intensity, which is several hundred times higher than that of the incident wave, can be acquired. However, the hexagonally periodic arrangements on a substrate are difficult because metallic nanospheres are easily aggregated. Even if metallic nanospheres can be arranged periodically, the optical intensity on the substrate strongly decreases due to the plasmonic interaction between the neighboring nanospheres, as described in Section 2.2. Near field generated around dielectric nanospheres is less affected by the arrangement, thus, having the advantage in nanopatterning with high throughput for large surface.

Figures 7.14 (a) and (b) show the SEM images of the Si substrate surface after irradiating a femtosecond laser pulse of 800 nm and 400 nm wavelength to hexagonally arrayed PS spheres shown in Fig. 7.14 (c) and (d), respectively. The laser fluence was 288 mJ/cm^2 at 800 nm and 57 mJ/cm^2 at 400 nm. The ablation threshold fluence for a bare silicon substrate without PS spheres was 252 mJ/cm^2 at 800 nm and 57.8 mJ/cm^2 at 400 nm. Both Fig. 7.14 (a) and (b) show elliptical shapes. The experimentally fabricated nanohole pattern is well explained by the calculated elliptical shape with the long axis perpendicular to the polarization of the incident light [9]. The circular polarization pulse gave a circular hole patterning [28].

The diameter (short axis) and the depth of the patterned nanoholes can be controlled by changing the sphere size. Figure 7.16 shows the dependence of the diameter and the depth of patterned nanoholes on the laser fluence for the two wavelengths. Red and blue dashed lines in Fig. 7.16 indicate the ablation threshold of the bulk silicon without PS sphere mediation. The diameter and the depth are increased linearly with the increase of the fluence. This experimental result seems to be similar to that with small gold spheres [11]. The maximum nanohole depth fabricated by the 800 nm laser is about 100 nm, which is two times deeper than that at 400 nm wavelength. According to the calculation results shown in Figs. 7.13 (a) and (b), the nanohole depth is dependent upon the energy transfer efficiency into the silicon substrate, which is evidenced by the experiment. With the 800 nm laser, the patterned hole diameter experimentally ranged from 100 to 250 nm and the depth ranged from 20 to 100 nm. With the 400 nm SHG wave, the patterned hole diameter ranged from 50 to 200 nm and the depth ranged from 10 to 60 nm.

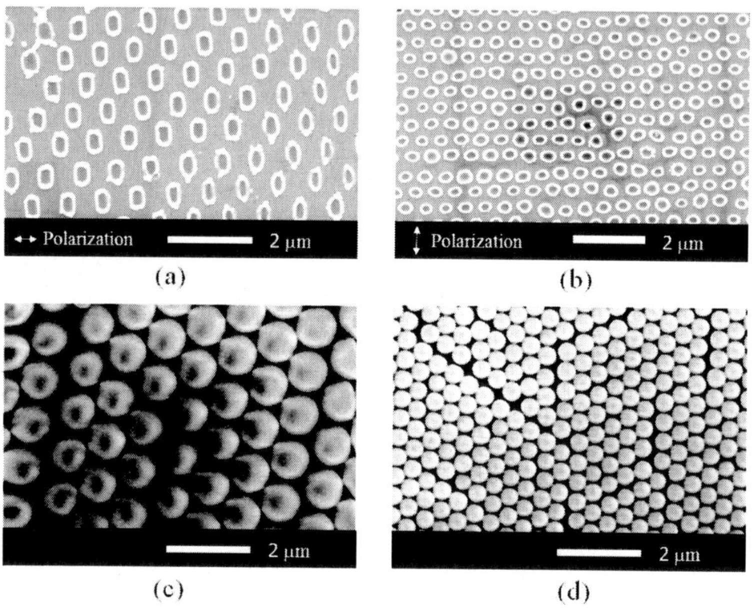

Figure 7.14 (a) and (b) SEM images of the Si surface after femtosecond laser irradiation. (a) 800 nm with fluence of 288 mJ/cm^2, (b) 400 nm with fluence of 57 mJ/cm^2. (c) and (d) SEM images of ordered monolayer PS spheres in hexagonal array on silicon substrate. (c) The diameter is 800 nm. (d) The diameter is 450 nm. Figures 7.15 (a) and (b) show the AFM images of the patterned Si surface after irradiating a single laser pulse with linear polarization at 800 nm to the arrayed spheres shown in Fig. 7.14 (c) and (d), respectively. The fluence is 570 mJ/cm^2 at 800 nm and 290 mJ/cm^2 at 400 nm. The depth of the nanohole is observed to be quite shallow in both systems, as predicted by the 3D FDTD simulation as shown in Fig. 7.13. With the shorter incident wavelength, the extinction coefficient of silicon reduces the incident energy transfer into the Si substrate. Consequently, the fabricated hole depth becomes shallower.

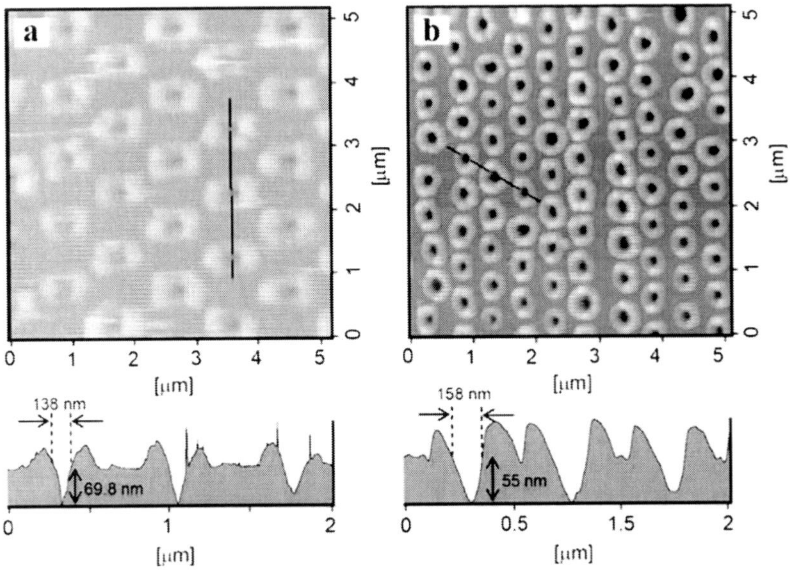

Figure 7.15 AFM images of the Si surface after laser irradiation. Cross-sectional images in the bottom figures correspond to the positions of the black lines in the top figures [8]. (a) 800 nm laser with fluence of 570 mJ/cm^2 was irradiated to the structure shown in Fig. 7.14 (c), and (b) 400 nm laser with fluence of 290 mJ/cm^2 was irradiated to the structure shown in Fig. 7.14 (d).

Nano-texturing has also been demonstrated by use of a dielectric sphere and oblique incidence laser beam. Li *et al.* reported the use of SiO_2 spheres of 500 nm in diameter with enhanced near-field excited by a illumination of a 248 nm nanosecond laser pulse (Fig. 7.17) [29]. The angle of incidence controls the position of the intensity peak point in the radial direction. Many textures can be fabricated with a single shot of laser pulse by use of a self-assembled sphere array.

High-permittivity dielectric sphere (dielectric sphere with high refractive index) with small size parameter provides a strong near-field enhancement mediated with the Mie resonance scattering. Although the near-field properties with dielectric spheres less affected by adjacent spheres than metallic spheres, near-field properties are affected by interaction not only with surface charge induced in the spheres but also with electromagnetic field inside

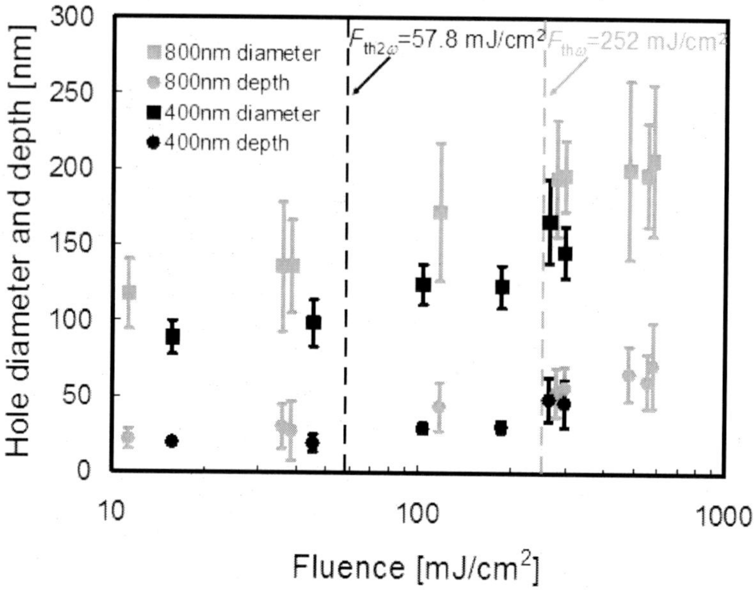

Figure 7.16 Plots of diameter and depth of nanoholes as a function of laser fluence [8]. The red dashed line represents the ablation threshold fluence for bare silicon with 800 nm femtosecond laser pulse. The blue dashed line represents the ablation threshold fluence for bare silicon with 400 nm femtosecond laser pulse.

the spheres. Theoretical calculation clarified that the near-field distribution on the substrate are affected notably for small spheres by the interaction with neighboring spheres [7, 30, 31]. Fabrication of high-density 2D periodic nanohole arrays with a clear boundary can be realized by selecting optimal high-permittivity spheres for the hexagonally close-packed sphere arrays even in the small size parameter domain. Figure 7.18 represents evaluated factors for intensity enhancement when the hexagonally close-packed sphere array with the diameter of 200 nm is placed on the Si substrate, as a function of sphere's refractive index (λ = 400 nm). The enhancement factor of the intensity inside substrate beneath the sphere–substrate contact point is shown by the blue line. Enhancement ratio of the intensity beneath the sphere–substrate contact points divided by that at the gap of the contact points, which is evaluated as a guidepost for avoiding the joining of neighboring nanoholes, is shown by the

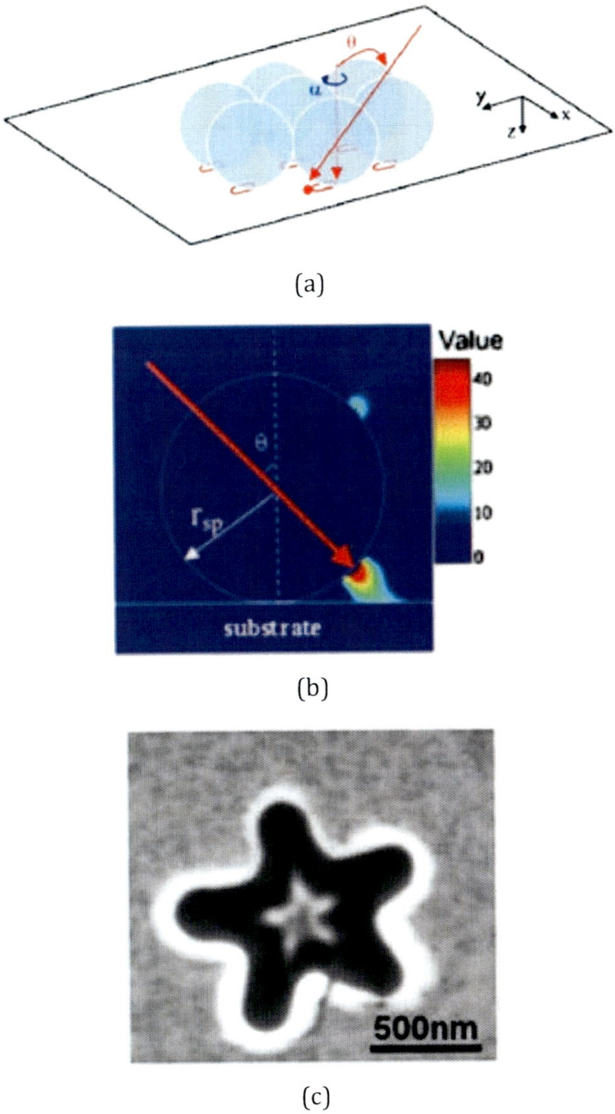

Figure 7.17 Nano-texturing by use of a dielectric sphere and an oblique incidence laser beam. (a). Schematic diagram of the experimental configuration. (b) Calculated Poynting vector intensity distribution for oblique incidence laser beam to SiO_2 sphere of 500 nm in diameter on SbTe substrate. (c) SEM image of star-shaped texture produced by scanning beams with designed angles. Reproduced with permission from [29].

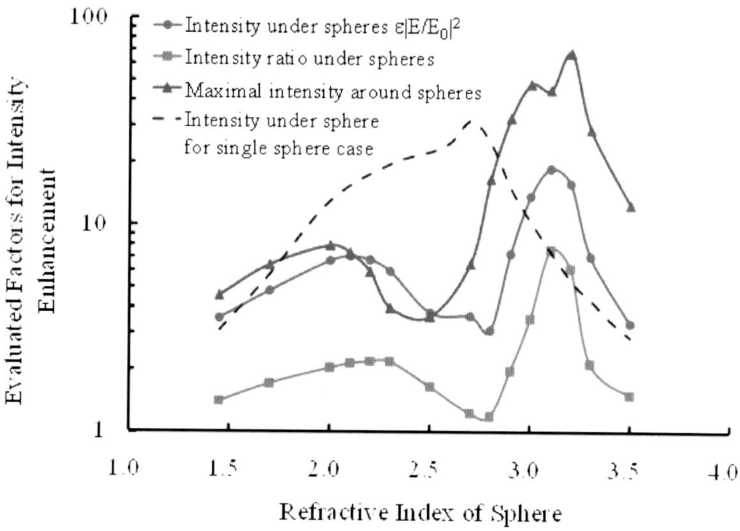

Figure 7.18 Evaluated factors for intensity enhancement when the hexagonally close-packed sphere array with the diameter of 200 nm is placed on Si substrate, as a function of sphere's refractive index (λ = 400 nm) [31].

green line. Enhancement factor of maximal intensity around sphere array above the substrate, which is commonly evaluated for the research of surface-enhanced Raman scattering (SERS), is shown by the red line. Enhancement factor of the intensity inside the substrate beneath the sphere–substrate contact point for the single sphere system is also shown by the gray dashed line. Each factor of 2D arrays takes a peak value with refractive indices around 2.1 and 3.1, shifted from the optimal refractive index of 2.7 for the single sphere system. Thus, the optimal refractive index for nanohole array fabrication is determined by the near-field resonance excitation condition for 2D sphere arrays, different from that for the isolated single sphere system. When the refractive index of the sphere is 3.1, the intensity enhancement factor under spheres is two times larger than the single sphere system. Figure 7.19 shows intensity distribution inside the Si substrate on *xz* plane with 400 nm wavelength irradiation to four systems of spheres: (a) a single sphere of 200 nm amorphous TiO_2 (n = 2.66 + 0.024i), (b) the 2D arrayed sphere of 200 nm amorphous TiO_2 spheres, (c) the 2D arrayed 200 nm ZnO spheres (n = 2.18), and (d) the 2D arrayed 200 nm polycrystalline rutile TiO_2 spheres

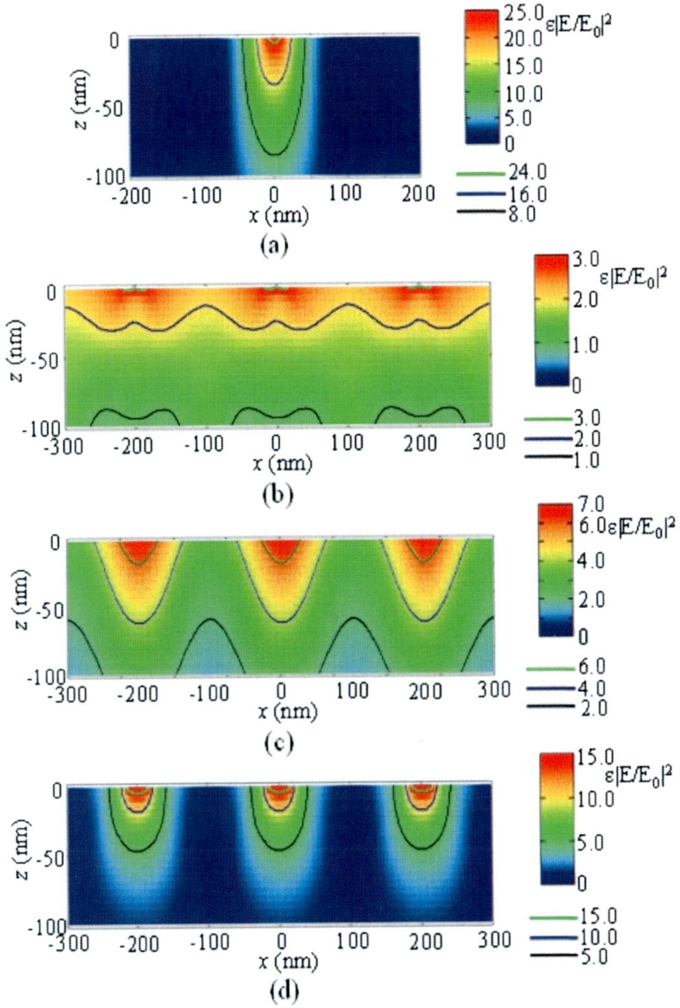

Figure 7.19 Intensity distribution inside Si substrate on *xz* plane when applying a single sphere of 200 nm amorphous TiO_2 sphere and 2D sphere array of 200 nm amorphous TiO_2 spheres, 200 nm ZnO spheres and 200 nm poly-rutile TiO_2 spheres with irradiation wavelength of 400 nm [31]. (a) single sphere of 200 nm amorphous TiO_2 sphere ($n = 2.66 + 0.024i$) on Si substrate, (b) 2D sphere array of 200 nm amorphous TiO_2 spheres ($n = 2.66 + 0.024i$) on Si substrate, (c) 2D sphere array of 200 nm ZnO spheres ($n = 2.18$) on Si substrate, (d) 2D sphere array of 200 nm poly-rutile TiO_2 spheres ($n = 3.182 + 0.00145i$) on Si substrate.

(n = 3.182 + 0.00145i). In the case of amorphous TiO_2 sphere, the intensity enhancement in the 2D array system is much lower than that in the single sphere system. ZnO and poly-rutile TiO_2 have refractive indices near the peak refractive indices for sphere array shown in Fig. 7.18. Intensity hotspots with a clear boundary are obtained under the spheres-substrate contact points with the ZnO sphere array and the poly-rutile TiO_2 sphere array. The fabrication ability of 2D nanohole array with a clear boundary can be achieved by smart selection of appropriate sphere materials.

Unlike metallic spheres, the dielectric spheres are capable of being used for processing of various materials, including transparent materials. There is a crucial issue with the processing method using metallic spheres, which is attributable to the surface plasmons that near-field enhancement factors are strongly dependent on the substrate materials. The near-field enhancement factor becomes quite low when metallic spheres are placed on the low refractive index substrates due to their weak plasmon coupling effect to the substrate surfaces. For example, when a gold sphere is irradiated by 800 nm Ti: sapphire femtosecond laser, the near-field enhancement is strong on the Si substrate (n = 3.68); while enhancement factors become by orders of magnitude lower than that on Si substrate and clear nanoholes are not obtainable when the sphere is placed on the low-refractive index substrates such as SiO_2 (n = 1.45), Al_2O_3 (n = 1.76), and ZnO (n = 1.94).

Simulated optical intensity enhancement factor and focused spot diameter (FWHM)) on the SiO_2 substrate surfaces as a function of refractive index of sphere with various extinction coefficients (k) are shown in Fig. 7.20. Large peak values of enhancement factor are obtained at a refractive index of 2.7 on the substrate. When a refractive index of sphere is 2.7, the magnetic quadrupole (TE2) mode resonance occurs. This mode gives a forward-directed near-field distribution pattern that can induce a hot spot on a substrate, resulting in the largest enhancement factor and nearly smallest spot diameter among the same size dielectric spheres on the substrate.

The corresponding values of numerical aperture (NA) with a sphere's refractive index of 2.7 calculated from spatial resolution in air are 2.33 on SiO_2. These values using near-field nanolens are much higher than the highest NA value of 1.44 at 193 nm for liquid immersion photolithography using conventional far-field optics.

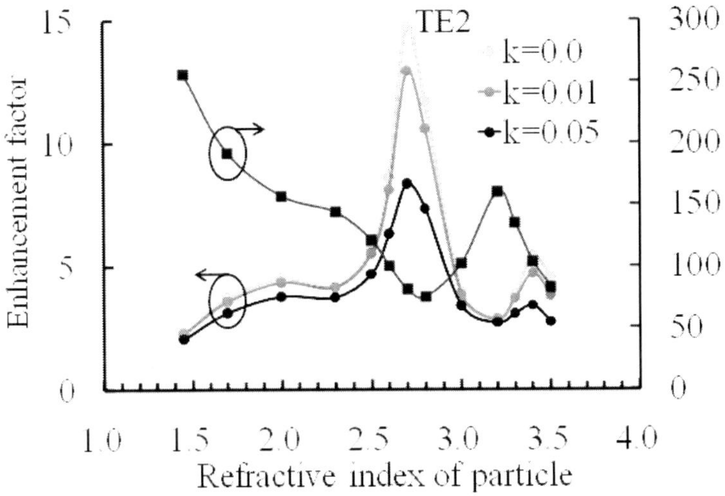

Figure 7.20 Simulated optical intensity enhancement factor and spot diameter (FWHM) with a 200 nm dielectric sphere deposited on SiO_2 substrate surfaces as a function of refractive index of sphere (n) with various extinction coefficients (k) [32].

Experimental validation was performed by using a sphere with optimal refractive index obtained from the theoretical study. Figure 7.21 shows SEM images of nanoholes fabricated by an irradiation of a 400 nm femtosecond laser pulse from the second harmonics of Ti:sapphire femtosecond laser to a 200 nm amorphous TiO_2 sphere (n = 2.66 + 0.024i) deposited on the SiO_2 substrate (E_g ~ 5 eV). A clear circular nanohole at the diameter of 90 nm was fabricated on the low-refractive-index SiO_2 substrate even with a laser fluence of 1.4 J/cm^2, which is lower than a half the ablation threshold of the bare substrate (3.8 J/cm^2). Larger nanohole diameter of 120 nm was obtained by applying a higher laser fluence of 3.5 J/cm^2. A further increase in the laser fluence up to 4.7 J/cm^2 resulted in a fabrication of 130 nm nanohole and the debris around hole was removed.

7.3 Nanoablation Using Nanotip

Nanoablation using small tips, whose sizes are much smaller than the incident wavelength, can be classified into two categories: a method using an apertureless-pointed tip (Fig. 7.22 (a)) and the other one using an apertured tip (Fig. 7.22 (b)). The former method

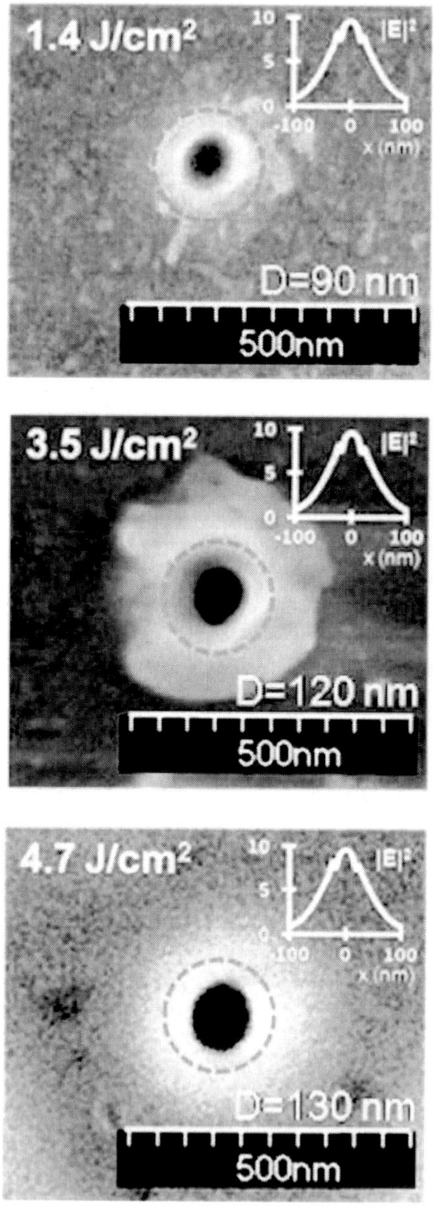

Figure 7.21 SEM images of fabricated nanoholes using a 200 nm amorphous TiO_2 sphere on SiO_2 substrate surfaces with different laser fluencies. The dashed line shows delineation of the applied sphere [32].

uses metallic tips of scanning tunneling microscope (STM) or atomic force microscope (AFM). The latter method uses an apertured tip used for near-field scanning optical microscope (NSOM). In addition to the applicability to the 2D nanohole arrays fabrication, this method has advantage in writing an arbitrary pattern using 2D scanning method [33, 34]. Figure 7.23 shows nanocharacters and nanocurve fabricated by using NSOM tip and a pulsed laser irradiation.

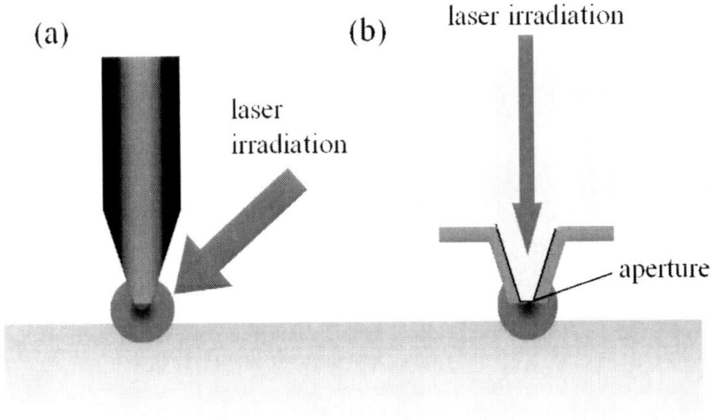

Figure 7.22 Schematic of the nanoablation using nanotip: (a) method using an apertureless-pointed metal tip, (b) method using an apertured tip with optical fiber.

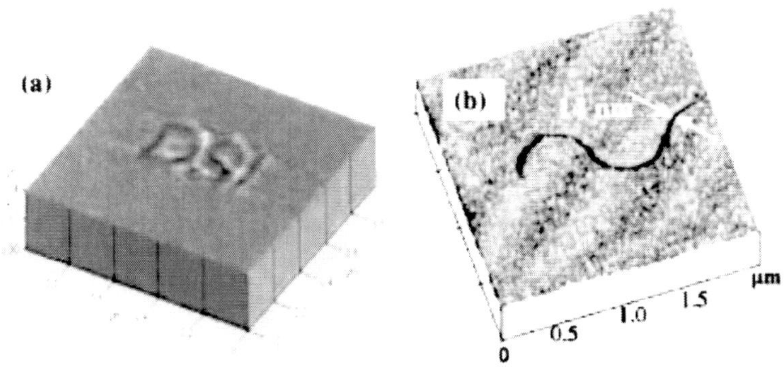

Figure 7.23 AFM images of (a) nanocharacters and (b) nanocurve written on the thin-film surfaces by the laser SPM combination technique. Reproduced with permission from [2].

The nanoablation using the apertureless tip is governed by the optical field within the tip-substrate gap. In 1996, Jersh and Dickmann reported the nanoablation of gold film by using a near field excited by the irradiation of 5 ns laser pulses at 532 nm wavelength to the STM tip [13]. As is the case for nanoablation using nanospheres, the nanoablation is achieved by an enhanced near field generated between the gap of a tip and a substrate surface. The enhanced near field generated in the gap depends on the geometry of the tip-substrate system, incident wavelength, incident angle, and materials of the tip and the substrate. Wang *et al.* showed the optical-field intensity distribution on a substrate surface for four tip-substrate systems: dielectric tip (Si) and dielectric substrate (Si) system, dielectric tip (Si) and metallic substrate (Au) system, metallic tip (Au) and dielectric substrate (Si) system, and metallic tip (Au) and metallic substrate (Au) system [35]. The system of plasmonic tip and plasmonic substrate showed the highest optical-field intensity among the four systems, which is similar to the result obtained in the nanoablation using an isolated gold nanosphere on the gold substrate, as shown in Fig. 7.5. The use of femtosecond laser was also investigated for the excitation of near field in this method [36]. Ultrafast laser irradiation results in the optical field enhancement factor of 150, causing the thin-materials to be removed under the tip. The apertured tip used for NSOM, whose diameter is in the range of nanometers, is also investigated for nanoablation (Fig. 7.24). In this method, the laser beam is guided through the optical fiber and irradiated a substrate surface as an evanescent wave around the aperture. Ablation characteristics are strongly dependent on the distance between the tip and the sample surface in these methods. An increase in the distance between the tip and sample surface resulted in the incapability of nanoablation. Lin *et al.* investigated the dependence of light intensity distribution under the tip on the distance [15]. At distances shorter than 10 nm, a double-peak intensity distribution generates. In addition, with the increase of distance, a double-peak profile is replaced by a single-peak, and a peak intensity decreases. These properties indicate that extremely precise apparatus is necessary for the control of the distance. The nanoablation using a nanotip provides the processing in a space far below the diffraction limit of the far field in air. However, a nanomachining throughput is quite low due to its scanning speed in a several μm/s range at a 10–30 nm line scale. Another drawback is an optical damage of the tip caused by the high near-field intensity at the tip.

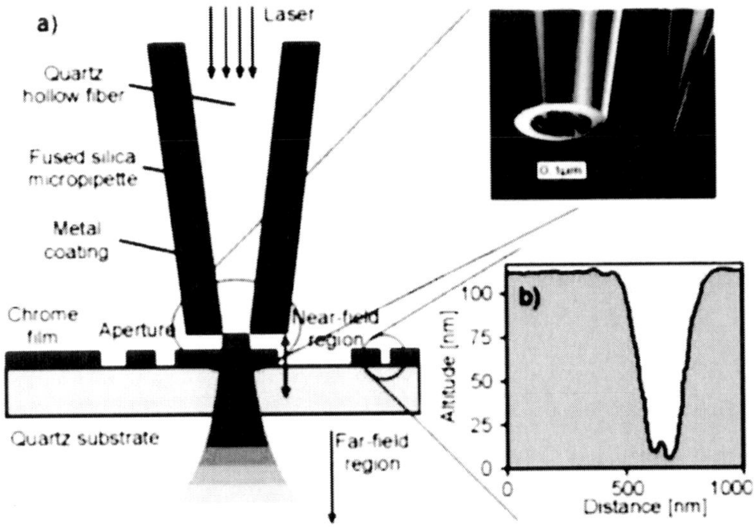

Figure 7.24 (a) Schematic diagram of a laser NSOM combination for surface nanostructuring and (b) AFM image of a cross section of an ablated groove on a Cr thin film. Reproduced with permission from [1].

7.4 Nanolens or Nanoheater?

In this section, we will describe a more detailed view for the metallic (plasmonic) nanosphere and laser pulse interaction focusing on the energy absorption and nanosphere heating (nanoheater effect). Many studies have dealt with the sphere heating and the near-field properties separately, which have limited the overall understanding of the interaction, where generally both effects play an important role. Figure 7.25 represents the cross-sectional spectra of a single gold nanosphere with radii of 20, 40, and 100 nm [37]. The clearly expressed spectral peaks in the absorption cross sections are related to the plasmon excitation in the nanospheres. The maximal values are observed in the range of about 525–550 nm for spheres with radius R = 20 and 100 nm. The absorption cross sections at wavelength of 800 nm are about two orders of magnitude lower than the optimal ones for spheres with R = 20 and 40 nm. Although the increase of the sphere dimensions results in the broadening of the plasmon absorption band and a shift of the maxima to the longer wavelengths, the cross section for sphere with R = 100 nm is about 20 times smaller than the peak value.

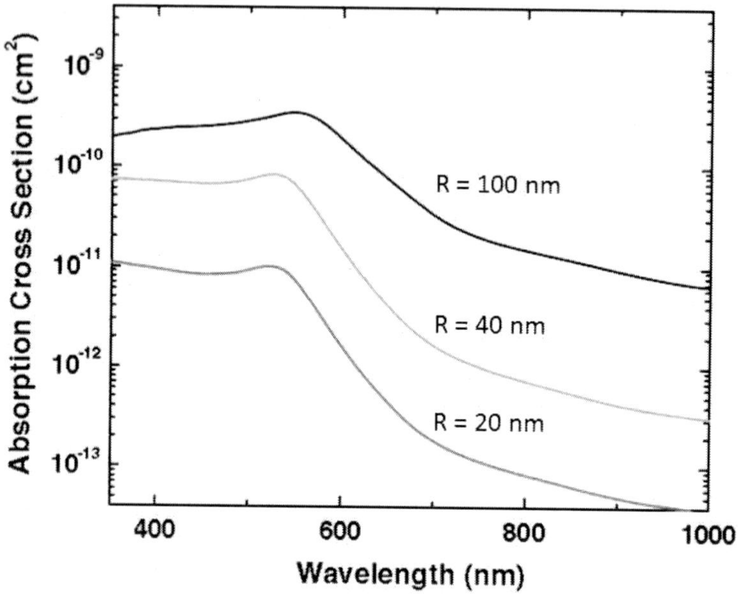

Figure 7.25 Absorption cross section for single gold nanospheres in vacuum with radii of 20, 40, and 100 nm [37].

Using the absorption cross sections, the temperature of the spheres is obtained by using the two-temperature diffusion model. The temperature evolution of the sphere during and after the interaction with the laser pulse is traced by one-dimensional two-temperature model [38]:

$$C_e \frac{\partial T_e}{\partial t} = -\gamma(T_e - T_i) + S \tag{7.2}$$

$$C_i \frac{\partial T_i}{\partial t} = \gamma(T_e - T_i) \tag{7.3}$$

$$S = IC_{abs}/V_p \tag{7.4}$$

where, T_e and T_i are the electron and lattice temperatures, S and I are the laser energy source and laser intensity (time dependent), respectively; V_p = $4/3\pi R^3$ is the sphere volume; C_i, C_e, and γ are the lattice and electron system heat capacities, and electron–lattice coupling constant, respectively. In this model, the spatial

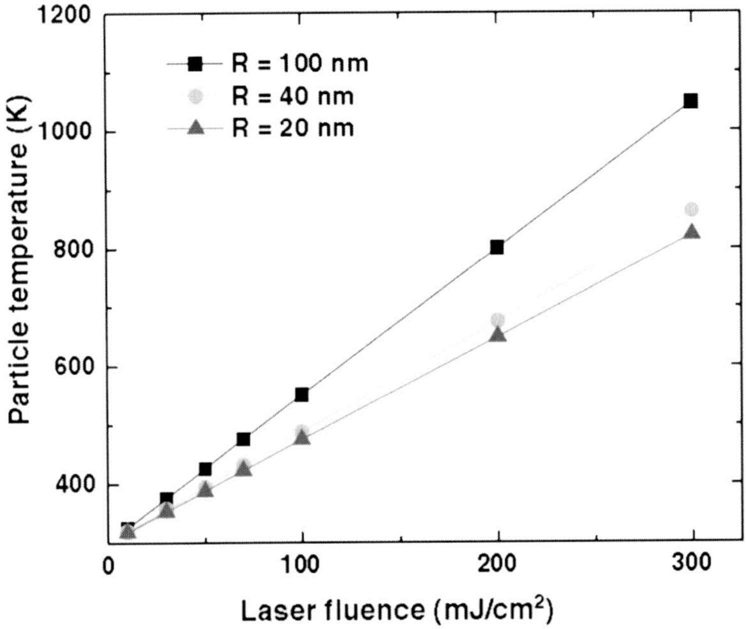

Figure 7.26 Calculated dependence of the maximal temperature reached in the nanospheres with different diameters as a function of the applied laser fluence [37]. The incident irradiation is a single laser pulse with duration of 100 fs at wavelength of 800 nm. The data are obtained from Eqs. 7.2–7.4 using results shown in Fig. 7.25.

distribution in the sphere heating is neglected, and it is assumed to be homogeneous. The values of the maximal temperature reached in the nanospheres when they are irradiated by a single 100 fs laser pulse as a function of the incident laser fluence are shown in Fig. 7.26. For comparison, the value of the maximal temperature obtained at the highest absorption (at λ = 550 nm) for R = 100 nm sphere exceeds 5 × 10^3 K at fluence of 100 mJ/cm^2. One of the promising applications of nanoheater is biophotonics. Metallic nanospheres are used for cell processing, as cancer cell photothermal therapy and cell membrane permeabilization (Fig. 7.27) [39]. It should be noted that for biophotonic applications, such as cell membrane modification, the necessary increase of the temperature should be up to about 315 K (42°C) [40] and kept for a certain period of time, especially when CW (continuous wave) or high repetition rate (tens of MHz) laser

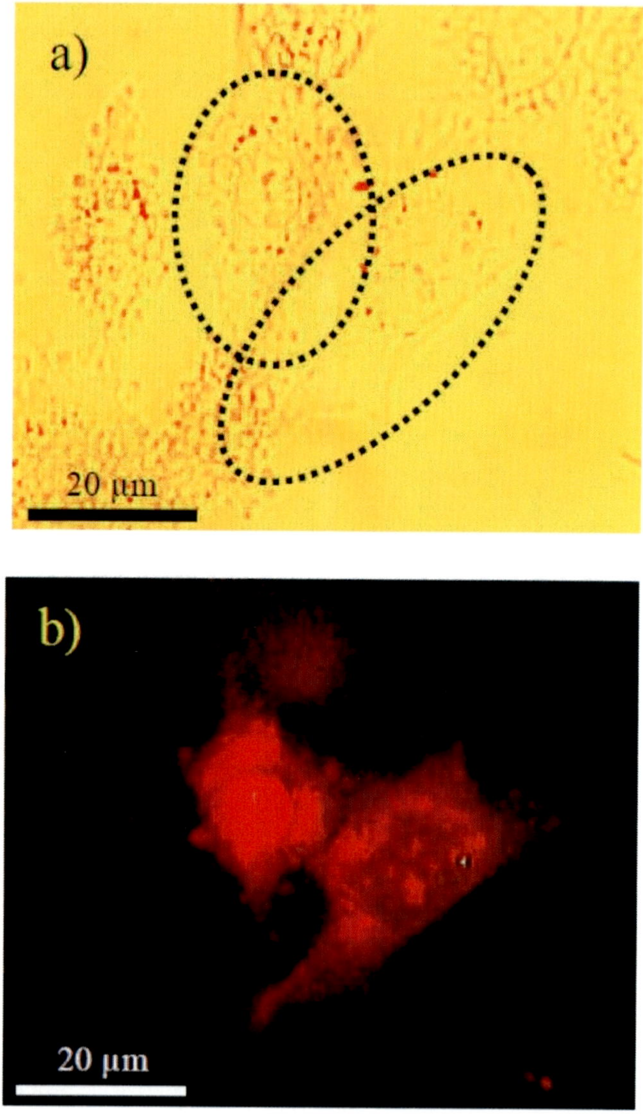

Figure 7.27 Perforated granulosa cells by using 150 nm gold spheres irradiated by a femtosecond laser pulse: (a) bright field image of cells which perforated cells are highlighted with dashed circles and (b) fluorescence image of the cells. Cells were illuminated by a 120 fs laser pulse in presence of the gold spheres located close to the cell membrane. Reproduced with permission from [39].

systems are used. For more strong interaction, when irreversible cell modification is based on the cavitation bubble formation (as in tumor cell killing), the necessary temperature should be at least 470 K [41–43]. When ultrashort laser pulses are used, such heating is realized in nearly isochoric conditions, which will result in the development of significant tensile stress [44] in the vicinity of the heating point. This will lead to the formation of a cavitation bubble without the need of heat accumulation with only a single pulse. From Fig. 7.26, it is seen that a temperature of 470 K is achievable by the gold nanospheres with the sizes assumed in this study after a single laser pulse, at fluences of about 70 mJ/cm^2 for R = 100 nm and at 100 mJ/cm^2 for spheres with R = 20 and 40 nm. It should also be mentioned that experimentally obtained threshold for biological cell modification [45] using direct femtosecond laser (far field) pulse irradiation (without nanosphere mediation) is estimated to be 500 mJ/cm^2 approximately.

7.5 Conclusions and Future Perspectives

In this chapter, experimental and theoretical investigation on nanoablation processing mediated with nanospheres and nanotips excited by pulsed laser irradiation was reviewed. The use of a near field generated around nanospheres is an attractive method to extend the industrial ability of femtosecond laser processing from laboratory to industry, providing precise non-thermal processing that overcomes the far-field diffraction limit. The method is capable of fabrication of the clear nanoholes on various advanced substrates, including semiconductors, metals, and dielectrics for photonics and electronics industries. The processing performance characteristic is well controlled by the sphere size, laser fluence, illumination wavelength, and angle of incidence. The near field generated around the gold spheres is strongly enhanced, while the near field generated around the dielectric sphere is less affected by the sphere arrangement and processing substrate materials. A further investigation on highly enhanced near field generated around an arrayed nanospheres, i.e., the properties of the near field combining both advantages of metallic and dielectric spheres, will open up the advanced industrial nanotechnology of nanoablation using nanospheres. For nanoablation using nanotips, nanomachining throughput and avoiding optical damage of the tip will be the next challenges and necessary to be technically solved. It is confirmed that the multidisciplinary and horizontally merged knowledge reviewed in this chapter gives us not only the understanding of the underlying processing physics involved in the interaction but also the thrust for developing the surface nanostructures for smart optical, electro-optical, electrical, tribology, and biological devices.

Acknowledgments

We are very much grateful to Dr. T. Sakai, Dr. T. Miyanishi, Dr. Y. Tanaka, Mr. A. Zenidaka, and Mr. G. Obara for their excellent work done at Keio University. We are also very much grateful to Prof. P. Atanasov and Prof. N. Nedyalkov of Institute of Electronics, Bulgarian Academy of Sciences.

References

1. Shalaev, V. M. (2007). Optical negative-index metamaterials, *Nat. Photon.*, **1**, 41–48.
2. Chong, T. C., Hong, M. H., Shi, L. P. (2010). Laser precision engineering: from microfabrication to nanoprocessing, *Laser Photon. Rev.*, **4**, 123–143.
3. He, Y., Fan., C., Lee, S. (2010). Silicon nanostructures for bioapplications, *Nano Today*, **5**, 282–295.
4. Mie, G. (1908). Beitrige zur Optik trUber Medien, speziell kolloidaler Metallosungen, *Ann. Phys. (Leipzig)*, **25**, 376–445.
5. Abe, K., Hanada, T., Tanigaki, N., Takiguchi, H., Nagasawa, H., Nakamoto, M., Yamaguchi, T., Yase, K. (1998). Two-dimensional array of silver nanoparticles, *Thin Solid Films.*, **31**, 327–329.
6. Zheng, X., Xu, W., Corredor, C., Xu, S., An, J., Zhao, B., Lombardi, J. R. (2007). Laser-induced growth of monodisperse silver nanoparticles with tunable surface plasmon resonance properties and a wavelength self-limiting effect, *J. Phys. Chem.*, **111**, 14962–14967.
7. Huang, S. M., Wang, Z. A., Sun, Z., Wang, Z. B., Luk'yanchuk, B. (2009). Theoretical and experimental investigation of the near field under ordered silica spheres on substrate, *Appl. Phys. A*, **96**, 459–466.
8. Sakai, T., Tanaka, Y., Nishizawa, Y., Terakawa, M., Obara, M. (2010) Size parameter effect of dielectric small particle mediated nano-hole

patterning on silicon wafer by femtosecond laser, *Appl. Phys. A*, **99**, 39–46.

9. Sakai, T., Miyanishi, T., Nedyalkov, N., Nishizawa, Y., Obara, M. (2009). Nano-dimple processing of silicon surface by femtosecond laser irradiation with dielectric particle templates in the Mie scattering domain, *J. Phys. D*, **42**, 025502/1–7.
10. Atanasov, P. A., Takada, H., Nedyalkov, N. N., Obara, M. (2007). Nanohole processing on silicon substrate by femtosecond laser pulse with localized surface plasmon polariton, *Appl. Surf. Sci.*, **253**, 8304–8308.
11. Eversole, D., Luk'yanchuk, B., Ben-Yakar, A. (2007). Plasmonic laser nanoablation of silicon by the scattering of femtosecond pulses near gold nanospheres, *Appl. Phys. A*, **89**, 283–291.
12. Atanasov, P. A., Nedyalkov, N. N., Sakai, T., Obara, M. (2007). Localization of the electromagnetic field in the vicinity of gold nanoparticles: Surface modification of different substrates, *Appl. Surf. Sci.*, **254**, 794–798.
13. Jersch, J., Dickmann, K. (1996). Nanostructure fabrication using laser field enhancement in the near field of a scanning tunneling microscope tip, *Appl. Phys. Lett.*, **68**, 868–870.
14. Huang, S. M., Hong, M. H., Lu, Y. F., Luk'yanchuk, B. S., Song, W. D., Chong, T. C. (2002). Pulsed-laser assisted nanopatterning of metallic layers combined with atomic force microscopy, *J. Appl. Phys.*, **91**, 3268–3274.
15. Lin, Y., Hong, M. H., Wang, W. J., Wang, Z. B., Chen, G. X., Xie, Q., Tan, L. S., Chong, T. C. (2007). Surface nanostructuring by femtosecond laser irradiation through near-field scanning optical microscopy, *Sens. Actuators A*, **133**, 311–316.
16. Lu, Y. F., Zhang, L., Song, W. D., Zheng, Y. W., Luk'yanchuk, B. S. (2000). Laser writing of a subwavelength structure on Silicon (100) surfaces with particle-enhanced optical irradiation, *JETP Lett.*, **72**, 658–661.
17. Tanaka, Y., Obara, G., Zenidaka, A., Terakawa, M., Obara, M. (2010). Femtosecond laser near-field nanoablation patterning using Mie resonance high dielectric constant particle with small size parameter, *Appl. Phys. Lett.*, **96**, 261103/1–3.
18. Messinger, B. J., von Raben, K. U., Chang, R. K., Barber, P. W. (1981). Local fields at the surface of noble-metal microspheres, *Phys. Rev. B*, **24**, 649–657.
19. Bohren, C. F., Huffman, D. R. (1983) *Absorption and Scattering of Light by Small Particle* (Wiley, USA).
20. van de Hulst, H. C. (1981) *Light Scattering by Small Particles* (Dover Publications, USA).

21. Yee, K. S. (1966). Numerical solution of initial boundary value problems involving Maxwell's equations in isotropic media, *IEEE T. Antennas Propag.*, **14**, 302–307.
22. Taflove, A. (2000). *Computational Electrodynamics: The Finite-Difference Time-Domain Method* (Artech House Publishers, USA).
23. Tanaka, Y., Nedyalkov, N. N., Obara, M. (2009). Enhanced near-field distribution inside substrates mediated with gold particle: optical vortex and bifurcation, *Appl. Phys. A*, **97**, 91–98.
24. Nedyalkov, N., Sakai, T., Miyanishi, T., Obara, M. (2006). Near field properties in the vicinity of gold nanoparticles placed on various substrates for precise nanostructuring, *J. Phys. D*, **39**, 5037–5042.
25. Nedyalkov, N. Atanasov, P. A., Obara, M. (2007). Near-field properties of a gold nanoparticle array on different substrates excited by a femtosecond laser, *Nanotechnology*, **18**, 305703/1–7.
26. Miyanishi, T., Tanaka, Y., Terakawa M., Obara, M. (2010). Uniform surface patterning with two-dimensional gold nanoparticle array excited by oblique femtosecond laser irradiation, *Jpn. J. Appl. Phys.*, **49**, 055001/1–7.
27. Takada, H., Obara, M. (2005). Fabrication of hexagonally arrayed nanoholes using femtosecond laser pulse ablation with template of subwavelength polystyrene particle array, *Jpn. J. Appl. Phys.*, **44**, 7993–7997.
28. Sakai, T., Nedyalkov, N., Obara, M. (2007). Positive and negative nanohole-fabrication on glass surface by femtosecond laser with template of polystyrene particle array, *J. Phys. D*, **40**, 2102–2107.
29. Li, L., Guo W., Wang, Z. B., Liu, Z., Whitehead, D., Luk'yanchuk, B. (2009). Large-area laser nano-texturing with user-defined patterns, *J. Micromech. Microeng.*, **19**, 054002/1–6.
30. Wang, Z. B., Guo, W., Luk'yanchuk, B., Whitehead, D. J. Li, L., Liu, Z. (2008). Optical near-field interaction between neighboring micro/nano-particles, *J. Laser Micro/Nanoeng.* **3**, 14–18.
31. Tanaka, Y., Obara, G., Zenidaka, A., Nedyalkov, N. N., Terakawa, M., Obara, M. (2010). Near-field interaction of two-dimensional high-permittivity spherical particle arrays on substrate in the Mie resonance scattering domain, *Opt. Exp.*, **18**, 27226–27237.
32. Tanaka, Y., Obara, G., Zenidaka, A., Terakawa, M., Obara, M. (2010). Femtosecond laser near-field nanoablation patterning using Mie resonance high dielectric constant particle with small size parameter, *Appl. Phys. Lett.*, **96**, 231103/1–3.

33. Lin, Y., Hong, M. H., Wang, W. J., Law, Y. Z., Chong, T. C. (2005). Sub-30 nm lithography with near-field scanning optical microscope combined with femtosecond laser, *Appl. Phys. A*, **80**, 461–465.

34. Hong, M. H., Huang, S. M., Luk'yanchuk, B. S., Chong, T. C. (2003). Laser assisted surface nanopatterning, *Sens. Actuators A*, **108**, 69–74.

35. Wang, Z. B., Luk'yanchuk, B. S., Li, L., Crouse, P. L., Liu, Z., Dearden, G., Watkins, K. G. (2007). Optical near-field distribution in an asymmetrically illuminated tip-sample system for laser/STM nanopatterning, *Appl. Phys. A*, **89**, 363–368.

36. Chimmalgi, A., Choi, T. Y., Grigoropoulos, C. P., Komvopoulos, K. (2003). Femtosecond laser apertureless near-field nanomachining of metals assisted by scanning probe microscopy, *Appl. Phys. Lett.*, **82**, 1146–1148.

37. Nedyalkov, N. N., Imamova, S., Atanasov, P. A., Tanaka, Y., Obara, M. (2011). Interaction between ultrashort laser pulses and gold nanoparticles: nanoheater and nanolens effect, *J. Nanopart. Res.*, **13**, 2181–2193.

38. Pustovalov, V. K., Babenko, V. A. (2005). Computer modeling of optical properties of gold ellipsoidal nanoparticles at laser radiation wavelengths, *Laser Phys. Lett.*, **2**, 84–88.

39. Schomaker, M., Baumgart, J., Ngezahayo, A., Bullerdiek, J., Nolte, I., Murua Escobar, H., Lubatschowski, H., Heisterkamp, A. (2009). Plasmonic perforation of living cells using ultrashort laser pulses and gold nanoparticles, *Proc. SPIE*, **7192**, 71920U/1–8.

40. Liu, H., Chen, D., Tang, F., Du, G., Li, L., Meng, X., Liang, W., Zhang, Y., Teng, X., Li, Y. (2008). Photothermal therapy of Lewis ling carcinoma in mice using gold nanoshells on carboxylated polystyrene spheres, *Nanotechnology*, **19**, 455101/1–7.

41. Volkov, A. N., Sevilla, C., Zhigilei, L. V. (2007) Numerical modeling of short pulse laser interaction with Au nanoparticle surrounded by water, *Appl. Surf. Sci.*, **253**, 6394–6399.

42. Zharov, V. P., Kim, J., Curiel, D. T., Everts, M. (2005). Self-assembling nanoclusters in living systems: Application for integrated photothermal nanodiagnostics and nanotherapy, *Nanomedicine*, **1**, 326–345.

43. Zharov, V. P., Letfullin, R. R., Galitovskaya, E. N. (2005). Microbubbles-overlapping mode for laser killing of cancer cells with absorbing nanoparticle clusters, *J. Phys. D*, **38**, 2571–2581.

44. Vogel, A., Noack, J., Hüttmann, G., Paltauf, G. (2007). Mechanisms of femtosecond laser nanoprocessing of biological cells and tissues, *J. Phys. Conf. Ser.*, **59**, 249–254.

45. Cho, S., Chang, W., Kim, J., Whang, K., Choi, K. Sohn, S. (2008). *In situ* observation of photo-bleaching in human single living cell excited by a NIR femtosecond laser, *Appl. Surf. Sci.*, **254**, 3370–3375.

Chapter 8

Ultrafast-Laser-Induced Phenomena in Transparent Materials

Kiyotaka Miura,[a] Kazuyuki Hirao,[a] Yasuhiko Shimotsuma,[a] and Masaaki Sakakura[b]

[a]*Department of Material Chemistry, Graduate School of Engineering, Kyoto University, Nishikyo-ku, Kyoto 615-8510, Japan*

[b]*Center for the Promotion of Interdisciplinary Education and Research, Kyoto University, Sakyo-ku, Kyoto 606-8501, Japan*

kmiura@func.mc.kyoto-u.ac.jp

Femtosecond laser (fs) has been widely used in materials microprocessing when high accuracy and small structure size are required. We demonstrate what effect the irradiation from focusing the laser pulse has on the structural changes in the laser-irradiated region and a few characteristics of the fs laser processing in transparent materials. Various interesting phenomena, e.g. space-selective refractive index change, valence state change, formation of elemental distribution, growth of nonlinear crystal, precipitation of silicon, in glass that can be accomplished with an fs laser are introduced. The photo-induced phenomena in transparent materials

Ultrafast Laser Processing: From Micro- to Nanoscale
Edited by Koji Sugioka and Ya Cheng

ISBN 978-981-4267-33-5 (Hardcover), 978-981-4303-69-9 (eBook)
www.panstanford.com

using an fs laser from the viewpoint of photonic application are also described.

8.1 Introduction

Materials processing technology by using femtosecond (fs) laser irradiation has attracted tremendous interest from both scientific and technological communities. Femtosecond laser has been widely used in materials microscopic processing when high accuracy and small structure size are required. Local structural change induced inside transparent materials using tightly focused fs laser pulses enables us to fabricate various kinds of three-dimensional (3D) structures inside glasses, polymers and inorganic crystals; this microfabrication technique with an fs laser has been used to produce small optical devices such as optical waveguides, photonic crystals, 3D optical memory and so on [1–8]. A key advantage of using fs laser, as opposed to longer pulses, for direct writing is that such pulses can rapidly and precisely deposit energy in transparent materials. When a transparent material such as glass is irradiated by a tightly focused fs laser, a photo-induced reaction should occur only near the focused part of the laser beam due to various nonlinear physicochemical reactions. In the past several years, a lot of research efforts have been devoted to the field of 3D microscopic modifications to transparent materials by using ultrafast laser.

In this chapter, we focused on the characteristics of the fs laser in transparent materials processing and various types of photo-induced phenomena with the fs laser irradiation.

8.2 Characteristics of Ultrafast Laser Processing

One of possible origins of the fs-laser-induced refractive index change is a pressure wave, which is generated as a result of the rapid relaxation of thermoelastic stress. Because a pressure wave compresses materials, the refractive index increase in an fs laser focal region might be the result of the densification by the pressure wave. Another possible origin is a rapid temperature elevation and rapid cooling. A number of studies showed that rapid cooling of a silica glass induces a glass structure of a higher fictive temperature, which has a higher density [9, 10].

8.2.1 Shock Wave Propagation

For investigating the relation between the fs-laser-induced modification and pressure wave, the observation of the pressure wave is essential. The fs-induced pressure wave inside a transparent material has been detected by ultrasonic detection [11], time-resolved phase contrast or interference microscopy [12, 13], transient lens (TrL) method [14, 15].

Juodkazis and coworkers detected the acoustic signal from the fs laser focal region inside a glass plate by ultrasonic detection technique [11]. In the detection, an acoustic transducer was attached to a glass plate (Fig. 8.1(a)), and the electric signal from the transducer is read by an oscilloscope. To detect fast pressure change, the transducer needs to have fast response time. The electric signal of the fs laser acoustic signals from a borosilicate glass is shown in Fig. 8.1(b). The first spike came from the direct heating of the transducer by the excitation laser pulse, and the second one was originated from the pressure wave. The delay time of the second pulse was determined mainly by the sound velocity and the distance between the photoexcited region and the transducer. The signals from the transducer also showed that the amplitude of the pressure wave increased as increasing excitation energy, the pressure wave

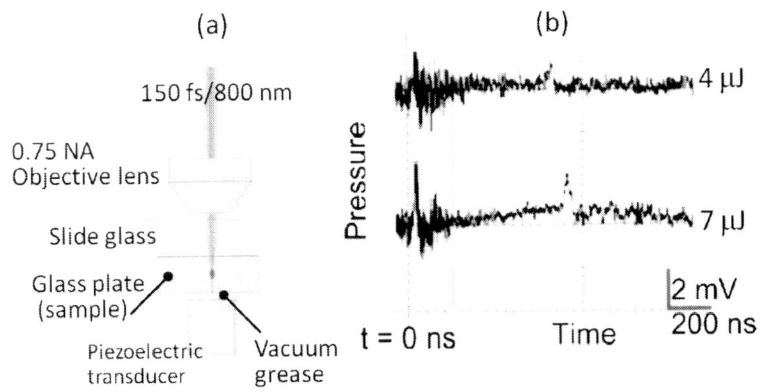

Figure 8.1 (a) Schematic illustration of experimental setup for acoustic detection of femtosecond-laser-induced breakdown inside a glass plate. (b) Waveform of acoustic signal when a single fs laser pulse was focused inside a glass.

was detected later at higher excitation energy, and there was broad compressive pressure change at higher laser pulse energy (7 μJ). The delayed time of the spike from pressure wave was attributed to the position change of the acoustic source, which means that the laser-induced plasma core in the sample was shifted in the direction to the objective lens. The broad compressive pressure change at higher excitation energy suggests that the filamentation of the excitation laser pulse became longer at higher excitation energy and it resulted in longer detection of the pressure change from the filamentation.

The ultrasonic detection does not give us direct information of the shapes of the pressure wave and laser-induced plasma. To observe the shape of the pressure wave and plasma, time-resolved optical microscopy is a powerful method. However a pressure wave does not alter optical absorption inside glasses, therefore, phase-sensitive method is necessary for detecting pressure waves. Several methods have been developed for detecting fs-laser-induced pressure wave inside a glass. Mermillod-Blondin *et al.* used time-resolved phase contrast microscopy (PCM) to observe the refractive index distribution change inside glasses after focusing of an fs laser pulse [12]. In their PCM system, the excitation laser pulse (100 fs and 800 nm) was focused inside a fused silica, and the frequency doubled laser pulse was transmitted through the photoexcited region in the perpendicular direction as the probe pulse. A phase plate was used in the PCM to detect the phase change due to the laser-induced refractive index change sensitively. Figures 8.2 (a) and (b) show the images at the optical delay of 10 ps in an optical transmission microscopy (OTM) and PCM [12]. The decrease of light due to light absorption by the laser-induced plasma was observed in the photoexcited region in the OTM image (Fig. 8.2 (a)). On the other hand, more complex change in shape was detected in the PCM image (Fig. 8.2 (b)). In the PCM image, the light increase corresponds to the refractive index decrease. The refractive index decrease due to plasma formation in the photoexcited region is surrounded by the positive refractive index change. The surrounding positive refractive index change was attributed to the temperature increase due to energy transfer from photoexcited electrons to the glass matrix. At a longer time delay, another refractive index change was observed in the PCM image. Figure 8.2(c) shows the PCM image at 800 ps. The region of positive refractive index change can be observed about 4 μm apart from the photoexcited region. This region corresponds to a pressure wave, because it propagated at a velocity of 5.75 km s^{-1}, which was the sound velocity of the glass. From the intensity change around the photoexcited region (ROI in Fig. 8.2(c)), they observed the dynamics of void formation in the photoexcited region (Fig. 8.2(d)). The density dynamics shows that the void formation started in several hundred picoseconds, which corresponds to the elastic relaxation time of the glass [16].

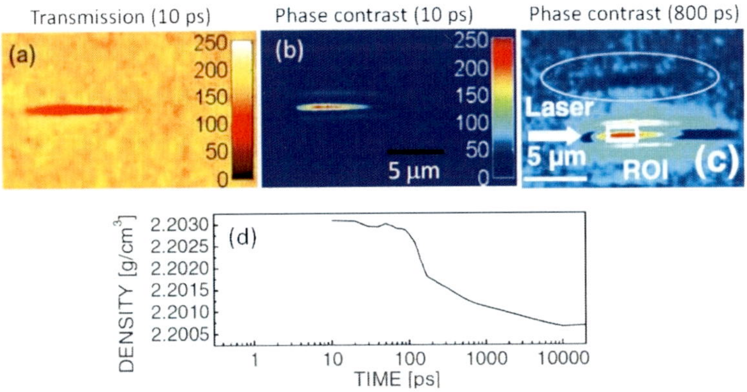

Figure 8.2 Images of the fs laser focal region. (a, b) OTM and PCM images at 10 ps, respectively. (c) PCM image at 800 ps. (d) Density dynamics in the ROI in (c).

Hayasaki *et al.* also observed a pressure wave using a pump-probe interference microscope [13]. In their microscope, a phase change due to the refractive index change around the photoexcited region could be obtained using an interferometer before the image sensor. In the interferometer, the probe beam is divided into two, and then they are mixed again on the image sensor with a small tilt. By analyzing the interference pattern between two beams, the amplitude and phase distributions in the photoexcited region can be obtained. Figure 8.3 shows time-resolved phase and amplitude images in the photoexcited region inside a borosilicate glass. In the phase images, a brighter shade indicates positive phase shift, i.e., refractive index decrease. After the photoexcitation, the refractive index at the center decreased in 1 ps, and the refractive index distribution changed gradually. The refractive index change in 1 ps was attributed to the free-carrier generation. They observed the phase jumps at the center and the adjacent region simultaneously around 100 ps and attributed them to the cavity formation. At a longer time, a pressure wave was observed (Fig. 8.3(e)) in both the phase and amplitude images.

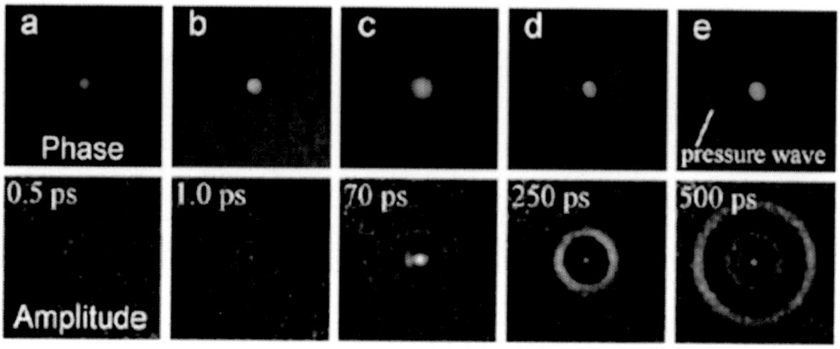

Figure 8.3 Phase and amplitude images obtained at various time-delays using a pump-probe interference microscope.

Sakakura *et al.* used a TrL method to observe the fs laser-induced dynamics inside glasses. The TrL method is one of the techniques for observing the dynamics of processes that are accompanied by a refractive index change (including change of chemical species, temperature change, density change and so on) [17–19].

Figure 8.4 shows a schematic picture of the TrL detection for observation of the fs-laser-induced dynamics inside a glass. When an fs laser pulse (pump pulse) is focused tightly inside a glass, the material around the focal region absorbs the light energy through the nonlinear photoionization and avalanche ionization. Following light absorption, the temperature in the photoexcited region is elevated and the density change is followed. The temperature change and the density change should induce the refractive index change, and the refractive index distribution works as a lens. Therefore, this laser-induced refractive index modulation is called a "transient lens". To detect the TrL, a probe beam is introduced into the photoexcited region. During the propagation of the probe beam through the TrL region, the wavefront of the light is modulated by the TrL (the lower part in Fig. 8.4). The modulation of the wavefront corresponds to the modulation of the spatial phase distribution of the probe beam. The phase modulated beam alters its spatial intensity distribution due to diffraction after the propagation. The spatial intensity distribution of the probe beam on the detection plane is called a "TrL image." In many cases, the intensity of the TrL image only at the beam center is detected. In this case, the temporal evolution of the light intensity is called "TrL signal." By analyzing TrL images or a TrL signal, the refractive index dynamics can be obtained.

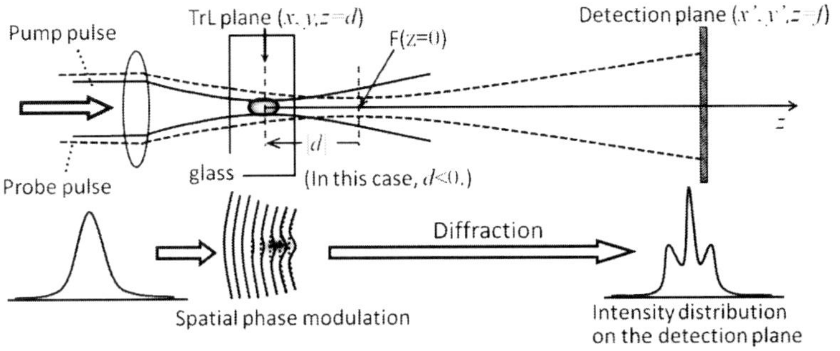

Figure 8.4 TrL method for the observation of fs-laser-induced structural change inside a glass. The lower part illustrates the deformation process of the beam profile.

The TrL images at different time delays after the photoexcitation of a soda lime glass are shown in Fig. 8.5(a). The energy of a pump pulse was 0.3 μJ. At –20 ps, the intensity distribution was Gaussian. At 0 ps, the spatial profile of the probe beam changed suddenly. This means that the lens effect appeared as a result of the photoexcitation. There was a sudden change in the beam pattern again at 1 ps. The sudden image change at 0 ps should be attributed to interactions between the glass and the light field, such as the optical Kerr effect and electron–hole plasma formation. After 1 ps, the drastic change in the beam pattern was observed until 2 ns. From these TrL images, the phase distribution changes at various time-delays can be obtained by fitting the radial intensity distribution by the relation between a phase distribution of light and the diffraction pattern on the detection plane. The results of fitting and the obtained phase distributions at various time-delays were shown in Fig. 8.5(b) and (c), respectively. The temporal evolution of the phase distribution shows that the pressure wave was generated around the photoexcited region in 200 ps and propagated away from the center at the velocity about 6 km s^{-1}. This result is consistent with the observations by the PCM and interference microscope.

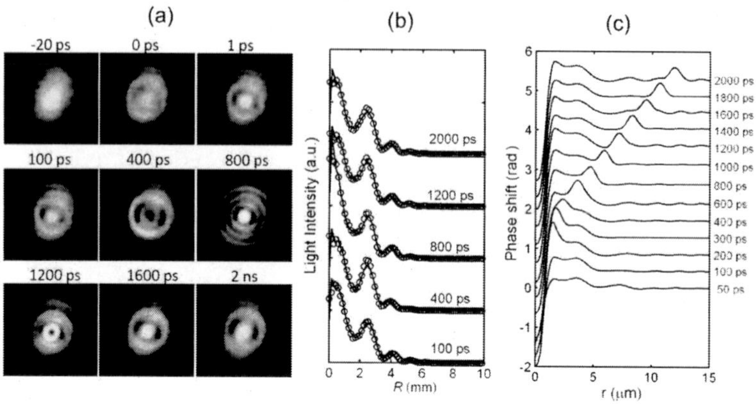

Figure 8.5 (a) TrL images at different time-delays. The pump energy was 0.3 μJ/pulse. (b) Radial distribution of light intensity in the TrL images and their fits by the diffraction equation. (c) The temporal evolution of the phase distributions obtained from the fitting. (a) TrL images at different time-delays. The pump energy was 0.25 μJ/pulse. (b) The TrL signal measured under the same conditions as for (a). (c) The TrL signal simulated using pressure wave propagation.

The pressure wave observed by the TrL measurement suggests that the temperature in the photoexcited region should increase much faster than the relaxation of the thermal stress. The generation of the thermal stress faster than the acoustic relaxation rate is called "stress confinement" [20, 21]. The acoustic relaxation time is given by D_{heat}/v_{ac}, where D_{heat} is the smallest dimension of the heated volume and v_{ac} is the sound velocity of the material. In fs laser processing inside glasses, typical values are D_{heat} = 1 μm and v_{ac} = 6 μm ns^{-1}. The calculated acoustic relaxation time is 200 ps, which means that the photoexcitation must be completed long before the elastic deformation starts to generate a large pressure wave. In other words, the pulse width of the pump pulse must be shorter than several tens of picoseconds. Therefore, the generation of a large pressure wave should be one of the characteristics of fs laser processing.

8.2.2 Thermal Diffusion

Although the elastic relaxation was completed within several nanoseconds, the temperature in the photoexcited region should still be high at the longest time-delay used in the TrL measurement system. In this longer time range, there should be slower dynamics such as thermal diffusion and viscoelastic deformation. Indeed, thermal diffusion is very important for understanding the mechanism of laser processing, because there have been a number of reports that the material deformation depends on the repetition rate of laser irradiation, an effect that must be caused by the accumulation of thermal energy. Since the deformed volume at several kilohertz is much larger than the photoexcited volume, the accumulated thermal energies should play an important role in the material deformation.

Figure 8.6(a) shows the TrL signals in a longer time range after a fs laser pulse of 0.6 μJ was focused with 20× objective lens inside a borosilicate glass (Corning 0211). The TrL signal indicates that the refractive index distribution changes after microseconds. By reproducing these signals based on thermal diffusion equation, we obtained the temporal evolutions of temperature at different positions (Fig. 8.6 (b)). When this temperature history was obtained, we assumed that the length of the heated region in the beam axis was 20 μm. The temperature at the center remains above 1500°C for 100 ns after the photoexcitation. Due to this high temperature, the glass at the center should melt, with a viscosity low enough for viscous flow to occur. We now estimate how much viscous flow occurred at the center. Using the viscosity data for Corning 0211 and Fulcher's equation [22], the viscosity at the center in 100 ns after photoexcitation should be about 10 Pa·s. According to the Stokes–Einstein (SE) relation [23], the diffusion coefficient of a spherical molecule in a continuum solvent is given by $D_{SE} = kT/6\pi\eta r$, where k is the Boltzmann constant, T temperature, η viscosity, and r the radius of the molecule. We assumed that the minimum radius of the moving component in the molten state is the bond length of B–O (~0.1 nm) [24]. Under this assumption, $D_{SE} \sim 1.4 \times 10^{-11}$ m^2s^{-1} was obtained. The diffusion length of the molecule with $D_{SE} \sim 1.4\times10^{-11}$ m^2s^{-1} in 100 ns is only 1.2 nm. This estimate means that the region undergoing viscous flow in fs-laser processing inside a glass is very small compared to the photoexcited region. According to the temporal evolution of the temperature distribution apart from the center, the temperature elevation of the material around the photoexcited region is much lower than the glass transition temperature (higher than 500°C). For example, the highest temperature that material at r = 3.0 μm reached after laser irradiation is lower than 100°C, which is too low for the glass in the region to be deformed by the thermal energy. Therefore, the observation by the TrL signal in the longer time range concludes that the thermal effect of a single irradiation on the surrounding region is very small.

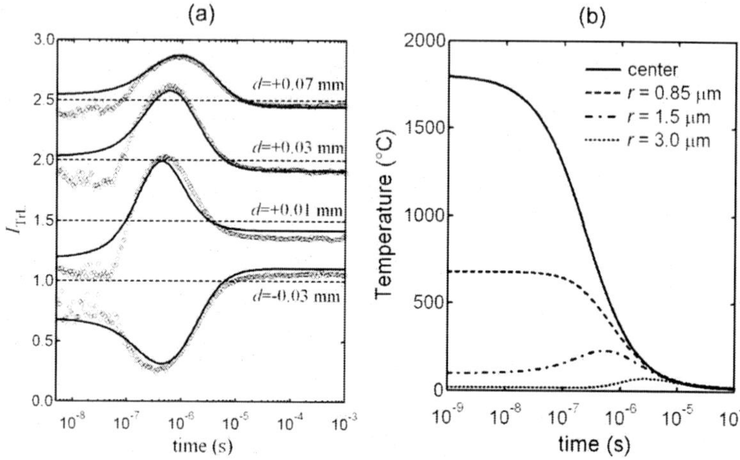

Figure 8.6 (a) TrL signals measured at different focal position mismatches with a pump energy of 0.6 μJ/pulse. The solid lines were calculated TrL signals by the temperature change shown in (b). (b) Temporal evolutions of temperature at different positions, which reproduce the TrL signals shown in (a).

8.3 Various Types of Laser-Induced Phenomena

In general, it is difficult to produce an interaction between glass and a laser when the pulse energy is not absorbed by single photon absorption process. However, an ultrashort-pulse laser can produce various types of interactions with glasses through the nonlinear optical process. In practice, white-light generation due to the self-phase modulation and second-harmonic generation were observed from the focal point of the laser beam inside the glass. After the irradiation of the fs laser, a variety of microscopic modifications were formed in the glass samples. AFM observation of a damage caused by focusing laser pulses on the surface of the glass samples are shown in Fig. 8.7(a), and the damage caused by focusing in the interior of the samples is shown in Figs. 8.7(b–d). Each sample received 200,000 pulses/spot through a microscope objective. Ablation or cracking (a) due to thermal and shockwave occurred at high power when the ultrashort laser pulse was focused on the near surface of the glass samples, but when the laser beam was focused in the interior of the glasses, no cracking was detected and only visible damage resulting from various phenomena such as the formation of color centers or lattice defects (b), local densification (c) or melting of very small area (d) were observed in the glasses. By use of the photosensitivity attributed to the nonlinear optical process, visible laser damage could only be formed inside the focused region, because nonlinearity only occurs in regions with high optical intensity above the damage threshold. Moreover, the threshold of each microscopic modification is different and dependent on the type of glass. Varying the type of glass and laser-irradiation conditions can therefore produce various types of microscopic modifications inside glass. The following describes typical examples of microscopic modifications by an fs laser.

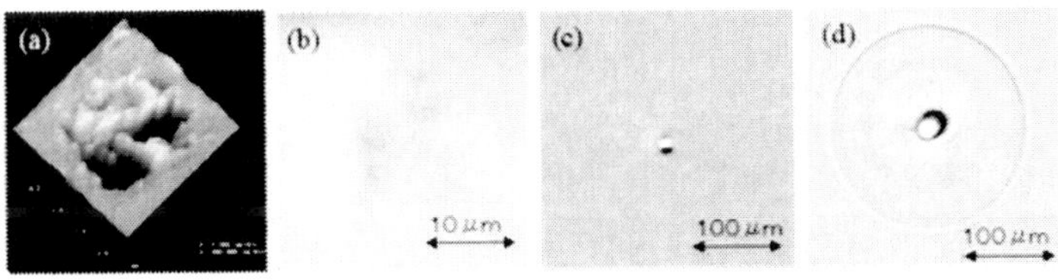

Figure 8.7 Various types of damage formed by the irradiation of the focused femtosecond laser (130 fs, 200 kHz). (a) Damage on the surface caused by 1 J pulses focused through a 10× objective lens; (b) Damage in the interior caused by 1 J pulses focused through a lens of 100 mm focal length; (c) Damage in the interior caused by 1 J pulses focused through a 10× objective lens; (d) Damage in the interior caused by 4 J pulses focused through a 50× objective lens.

8.3.1 Photo-Induced Refractive Index Change

The modifications accompanied with large refractive index change are important, because they can be applied to the fabrication of various optical devices such as waveguides, gratings or photonic crystals [1–8, 25–32]. Details of photonic device fabrication based in fs-laser-induced refractive index modification are described in Chapter 9. Figure 8.8 shows an array of the refractive index changes inside silica glass induced by a single pulse for each, where pulse energy was 0.1 μJ. These spots have diameters of roughly 400 nm, as measured with a confocal laser scanning microscope. This was significantly smaller than the focal-beam size and the wavelength of the laser (800 nm). This may be due to the self-focusing of the laser and nonlinear effect of the glass, which resulted from the interaction between the laser and the glass as described in Chapter 1. The spot size of refractive index change increases with increasing number of irradiation pulse and peak power of each pulse.

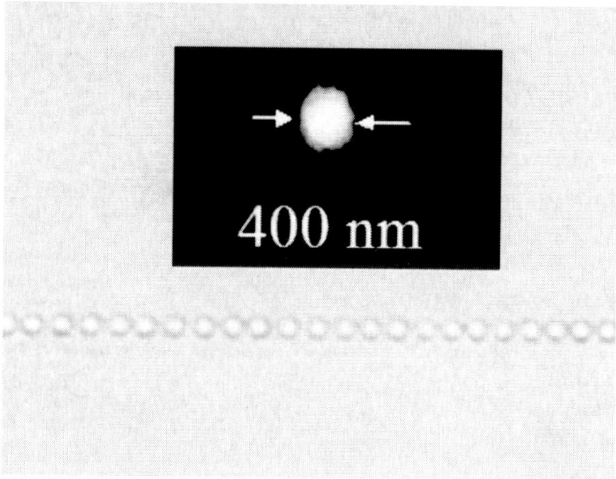

Figure 8.8 Refractive index changes inside silica glass induced by a single pulse for each.

Using an fs laser with a high repetition rate, we induced continuous refractive index changes along a path traversed by the focal point. The photos in Fig. 8.9 show typical photo-written waveguides formed inside various types of glass, such as silica, fluoride, and chalcogenide glass. The characteristics of a writing waveguide can be controlled by adjusting the writing conditions (the numerical aperture of the objective lens, the pulse energy, the pulse width, the repetition rate, the scanning speed, etc.). Figure 8.10 shows the refractive index profiles of the core of waveguides written in silica glass at various numerical apertures and average powers. The mode-field diameters and propagation modes calculated based on these refractive index profiles are shown in Table 8.1. The core diameter and the relative refractive index difference can be altered by adjusting the numerical aperture of the objective lens and the pulse energy. It is apparent, based on these results, that optical waveguides exhibiting a wide range of propagation characteristics can be written.

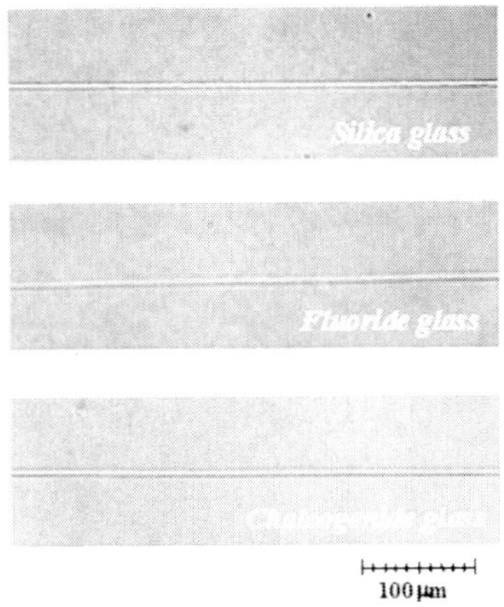

Figure 8.9 Photo-written optical waveguides formed inside various types of glass.

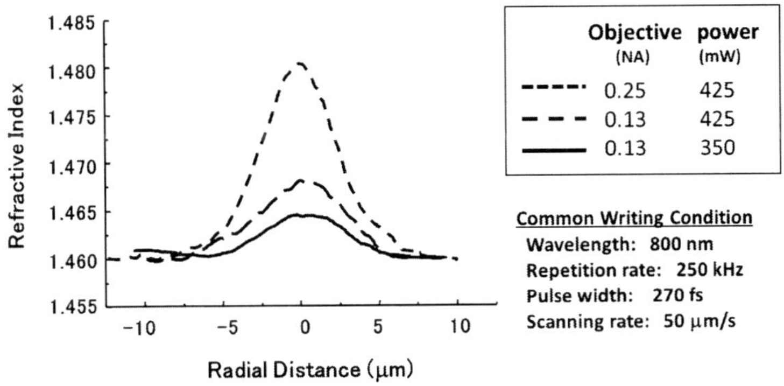

Figure 8.10 Typical photo-written waveguides formed inside silica glass, the refractive index profiles of the core of waveguides written in silica glass at various numerical apertures.

Femtosecond laser processing has often been conducted in a silica glass, because this glass is an important material for the low loss optical waveguide. Therefore, there have been number of investigations on the fs-laser-induced modification inside a silica glass. For example, Chan *et al.* observed the Raman spectrum from the region of the refractive index change inside a silica glass and found that the smaller rings of SiO_2 structure had increased [4]. Their results suggested that densification occurred after the photoexcitation by fs laser pulses. They speculated that the densification had been induced by fast cooling after the photoexcitation or shock wave generation [30]. Mysyrowicz and coworkers found two types of damage in the fs laser focal region inside a silica glass [33]. They compared the spatial dynamics of excited electrons with the damage and suggested that the damage with larger pulse energy should be associated with avalanche ionization. They also suggested that a molten phase of SiO_2 should be formed due to the high temperature after the photoexcitation based on a result from the time-resolved light transmission measurement. The authors observed the TrL signals from the fs laser focused area in a silica glass at various excitation laser pulse energies and investigated the relation between the laser-induced modification and the pressure wave generation [34].

Table 8.1 Mode-field diameters and propagation modes calculated based on refractive index profiles in Fig. 8.8

	Change distance (μm)	Refractive index difference (%)	Calculated guide mode (1.3 μm)	Calculated guide mode (1.55 μm)
1	20	1.32	LP_{01}, LP_{02}, LP_{11}	LP_{01}, LP_{11}
2	18	0.53	LP_{01}, LP_{11}	LP_{01}, LP_{11}
3	11	0.27	LP_{01}	LP_{01}

The Raman spectra from the center of the refractive index lines with various pulse energies are shown in Fig. 8.11(a). All the Raman bands observed in Fig. 8.11(a) have been assigned [35, 36]. For example, the dominant broad band around 450 cm^{-1} has been attributed to symmetric oxygen vibration of the bent Si-O-Si linkages. The sharp bands at ~490 cm^{-1} (D_1 band) and ~600 cm^{-1} (D_2 band) have been attributed to the oxygen breathing modes in four- and three-membered rings of SiO_4 tetrahedra, respectively. These bands are important for obtaining the structural information in the laser-induced modification inside a silica glass, because these bands come from the compaction of silica networks [36]. Here, the D_2 band can be separated more easily from the other bands. The intensity of the D_2 band increased with increasing laser pulse energy. The peak intensities of this band were plotted against the laser pulse energies in Fig. 8.11(b). The peak intensity of this band started to increase at about 90 nJ. The energy threshold of the pressure wave generation was about 90 nJ. The threshold of a pressure wave is nearly equal to the threshold of an increase of three-membered rings of SiO_4 tetrahedra. This suggests that the three-membered ring structure in silica networks should be induced by a pressure wave generation. After the photoexcitation, the temperature in the photoexcited region is elevated due to the energy transfer from the photoexcited electrons to the lattice in a glass, and the temperature elevation induces thermoelastic stress in the photoexcited region. The subsequent stress relaxation induces the generation of a pressure wave.

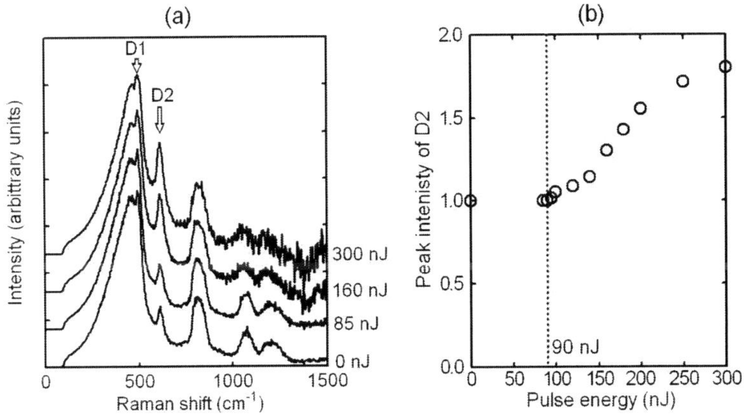

Figure 8.11 (a) Raman spectra at the center of the refractive index lines written with various pulse energies. (b) The peak intensities of the D2 bands plotted against the pulse energies. The intensities were normalized by that from the non-irradiated area.

The mechanism of the refractive index change with the densification induced by the pressure wave is shown in Fig. 8.12. Since the lifetime of photoexcited electrons in glass is short, a large amount of energy should be transferred to the lattice immediately after the photoexcitation, and a temperature increase and thermo-elastic stress should be induced in a very limited volume. The relaxation of the thermo-elastic stress produces the force driving the shock wave. This force could be a compressive acoustic wave. This shock wave propagates outward. With it, the structure of the glass is extended outside. As a result, the central density decreases. When the laser is turned off, compressive stress moves back toward the center. Therefore, a graduated high-density region is formed in the laser-focusing area. After the thermal diffusion from the irradiated region, the structural change becomes a permanent refractive index change. If the intensity of the laser pulse is strong enough, a void or a less dense region is created due to the explosive thermal expansion. If laser pulses are repeatedly irradiated, the high-temperature area rapidly expands, and, as described earlier, melting regions can be formed at arbitrary sites within the glass. With this mechanism, the laser-induced, refractive index change created by an fs laser does not depend on the type of glass.

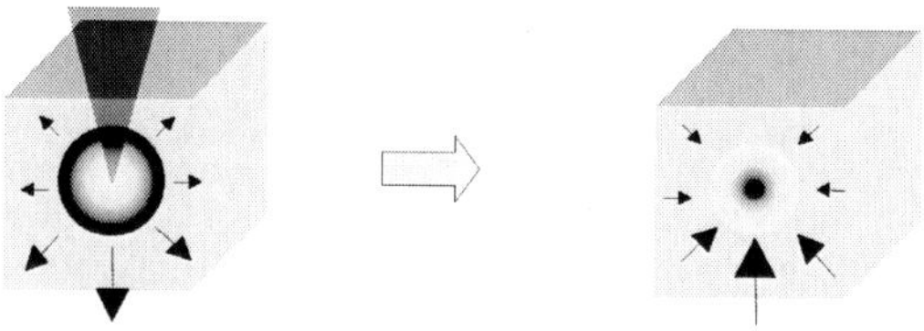

The irradiation of a single pulse	After pulse irradiation
Temperature increase and thermo-elastic stress should be induced in a very limited volume. ↓ The relaxation of the thermo-elastic stress produces the force driving the shock wave. ↓ The shock wave propagates outward. ↓ The structure of the glass is extended outside.	Compressive stress moves back toward the center. ↓ A graduated high-density region is formed in the laser- focusing area. ↓ After the thermal diffusion from the irradiated region, the structural change becomes a permanent refractive-index change.

Figure 8.12 Mechanism of the refractive index change with the densification induced by the pressure wave.

8.3.2 Valence State Change

Rare-earth ions, such as Sm^{3+} and Eu^{3+}, in glass, can be space-selectively photoreduced with an infrared fs laser [37]. Figure 8.13 shows photoluminescence spectra obtained by excitation at 488 nm for a laser-irradiated area (a) and non-irradiated area (b) in the interior of the glass. The composition of the glass samples doped with samarium ions used in this figure was $35AlF_3$-$10BaF_2$-$10SrF_2$-$20CaF_2$-$10MgF_2$-$14.5YF_3$-$0.5SmF_3$ (mol%). Samarium was present in the glass in the Sm^{3+} ionic state. Comparing (a) to (b) shows that the emission in the 650–775 nm region differed appreciably. The broad bands observed around 560, 600, and 645 nm can be attributed to the ${}^4G_{5/2} \rightarrow {}^6H_{5/2,7/2,9/2}$ transitions, respectively, of the Sm^{3+} ions. On the other hand, the emissions at 680, 700, and 725 nm are attributed to the ${}^5D_0$, ${}^7F_{0,1,2}$ transitions, respectively, of the Sm^{2+} ions. This means that laser-irradiated areas (photoreduced areas) recorded inside glass can be detected only by emissions at 680, 700, or 725 nm. Although the mechanism is not fully clear at present, we suggest the mechanism of photoreduction as follows. Active electrons and holes were created in the glass through multiphoton process. Holes were trapped in the active sites in the glass matrix while some of electrons were trapped by the Sm^{3+}, leading to the formation of hole trapped defect centers and Sm^{2+}.

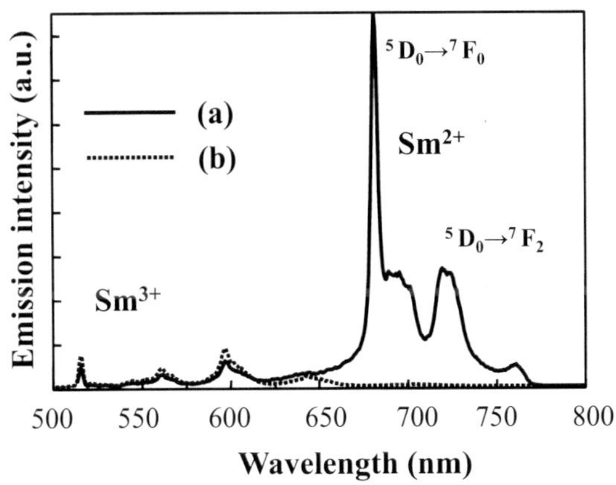

Figure 8.13 Photoluminescence spectra obtained by excitation at 488 nm from the irradiated (a) and non-irradiated (b) areas in Sm ion doped glass.

Figure 8.14 shows photoluminescence images of alphabetical characters recorded on different layers, which were observed by using a 40× objective lens and the 680 nm emission from Sm^{2+} with confocal detection implemented. By using the photoreduction of Sm^{3+} to Sm^{2+}, alphabetical characters were recorded in the form of sub-micron size bits in a 3D (layered) manner in a glass samples. Here, one recorded character consisted of 300 to 500 photoreduction bits recorded with 5000 laser shots per bit with a repetition rate of 250 kHz. The spacing between alphabetical characters was 2 μm. The bits had a diameter of 400 nm, which was significantly smaller

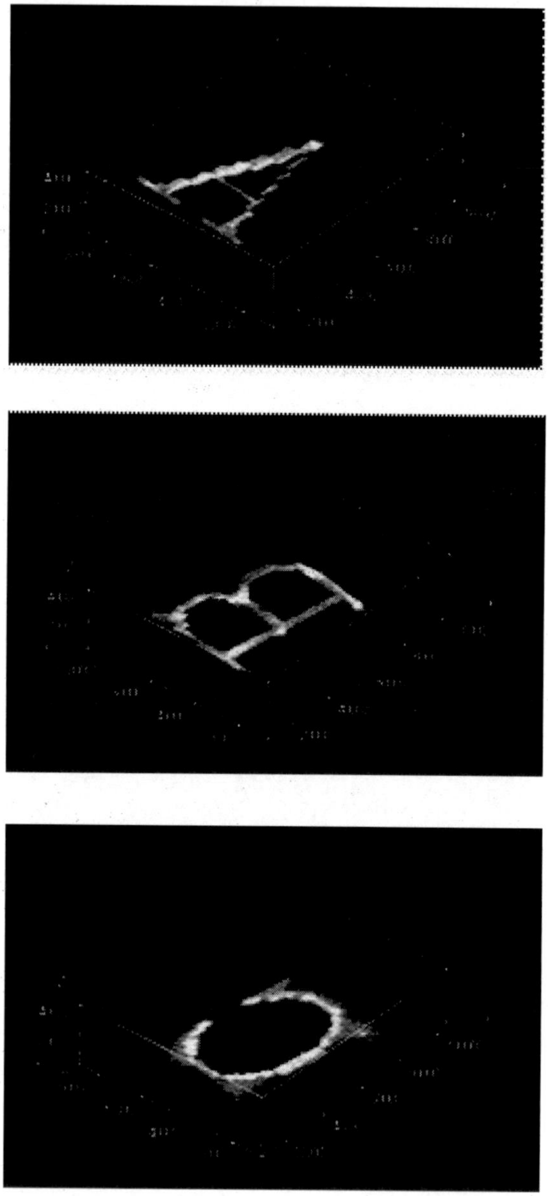

Figure 8.14 Photoluminescence images of alphabetical characters recorded on different layers, which were observed by using a 40× objective lens and the 680 nm emission from Sm^{2+} with confocal detection implemented (excitation at 488 nm, 1 mW Ar^{+} laser).

than the focal-beam size and the wavelength of the recording laser. The spacing of 2 μm between alphabetical character planes was sufficient to prevent cross-talk in the photoluminescence images. Although the 3D memory bits (photoreduced areas) were recorded with an fs laser, they could be read with a CW laser at 0.5 mW. Recording and readout in a 3D memory are possible without any influence from bits in upper or lower layers recorded previously. Although obtaining a multilayer of several hundred layers has been difficult in a conventional 3D memory, it can be achieved by applying photoreduction bits to current optical memory technology. In addition, as shown in Fig. 8.15, photoreduction bits with a 200 nm diameter could also be read out clearly by detecting the fluorescence as a signal. Since photoreduction bits can be spaced 150 nm apart on a layer in glass, a memory capacity of as high as 1 Tbits could be achieved for a glass piece with dimensions of 10 mm × 10 mm × 1 mm. Another outstanding characteristic of the photoreduction of Sm^{3+} to Sm^{2+} is that stable Sm^{2+} at room temperature can be changed to Sm^{3+} by photo-oxidation with a CW laser, such as an argon ion laser or a semiconductor laser.

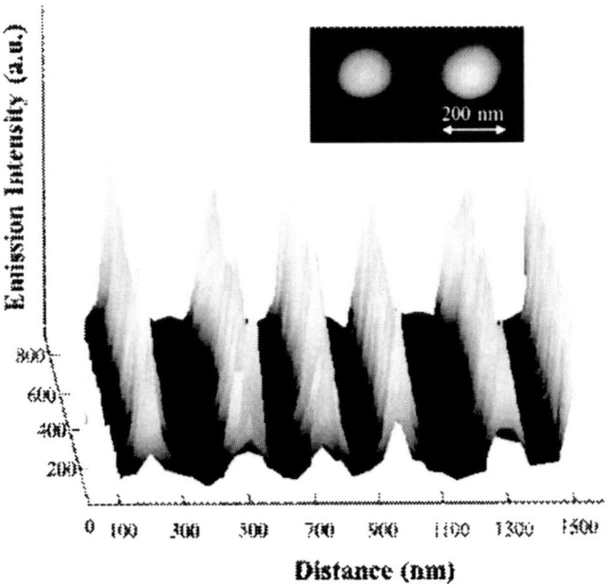

Figure 8.15 Signal read out by detecting the fluorescence for photoreduction bits with a 200 nm diameter.

8.3.3 Formation of Elemental Distribution

The composition of a glass affects its optical properties such as refractive index, light absorption, and luminescence property. Spatially selective modification of the composition in a glass would enable 3D control of such properties. In the field of fs laser processing, element redistribution induced by irradiation with fs laser pulses at high repetition rate has attracted much attention, because it has the potential to control glass composition three-dimensionally [38–41].

By irradiating a single pulse of 0.1 μJ, it was also confirmed that the temperature at the center of the laser beam reaches 3000°C or more within 1 ps, and then the temperature decreases from ~3000°C to ~300°C within 10 μs [18]. This result suggests that the heat accumulation becomes critical when the repetition rate of the irradiation is higher than 100 kHz [42–45]. When a lot of laser pulses are focused at high repetition rate (more than several hundred kilohertz), the released heat is accumulated around the photoexcited region and the surrounding glass is modified due to high temperature increase. Under such a condition, it was confirmed that the elemental distribution of a ring shaped pattern can be formed in a glass. Figure 8.16 shows an elemental distribution around the focal point after laser irradiation, as analyzed by EPMA. The relative concentration of Si and O ions increases and those of K, Na, Ca, and Zn decreases at the focal point. The increase of Si ions means a decrease in the refractive index in most glass. In the case of the phosphate glass, it was also confirmed that Ge and Ga were moved to the center in the laser-irradiated area as shown in Fig. 8.17. An increase of Ge and Ga contribute to an increase in the refractive index. By using this phenomenon, we were able to successfully write optical waveguides, where elemental distributions were continuously induced along a path traversed by the focal point.

Based on experimental results obtained for various types of glass, we found that the tendency in the elemental distribution depended on the strength of the bond between cations and oxygen ions. Strongly bonded ions such as Si or Al migrated to the center of the irradiated spot, whereas weakly bonded ions such as Na or Ca migrated to the outside.

Shimizu *et al.* studied on the formation mechanism of element distribution in glass under high-repetition-rate fs laser irradiation [43, 44]. Figure 8.18 shows the optical microscope images and

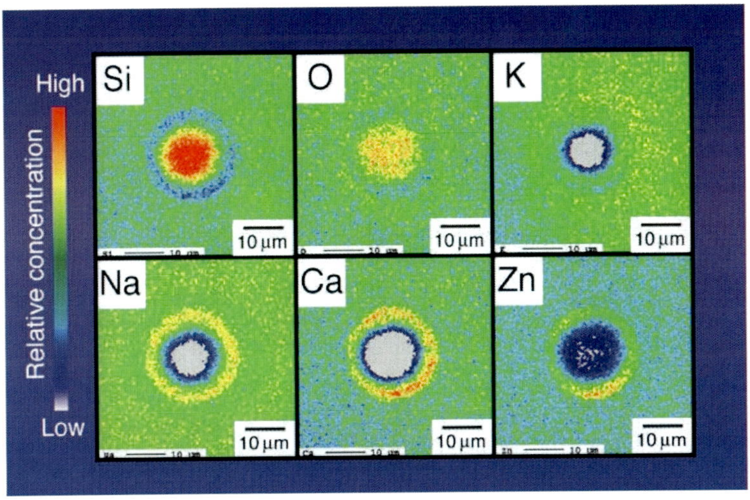

Figure 8.16 Profile of elemental concentration around the focal point in a silicate glass after laser irradiation.

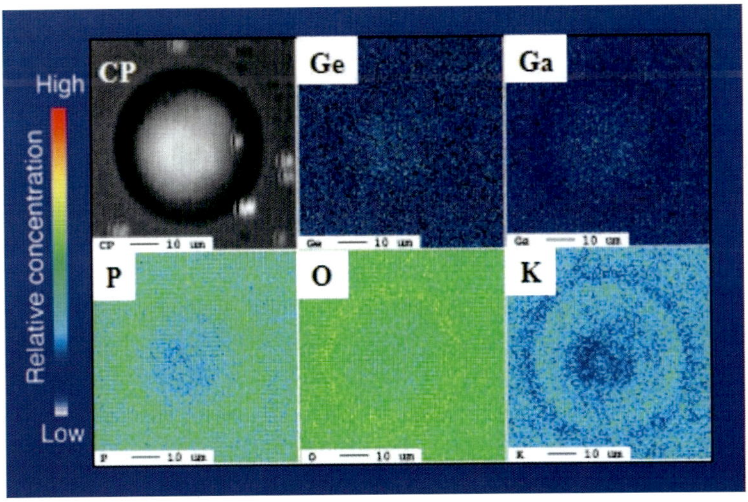

Figure 8.17 Profile of elemental concentration and refractive index distribution around the focal point in a phosphate glass after laser irradiation.

element distributions after laser irradiation. In an image (Fig. 8.18(a)), two circular boundaries are observed. During laser irradiation, the temperature increases with decreasing distance from the focal spot.

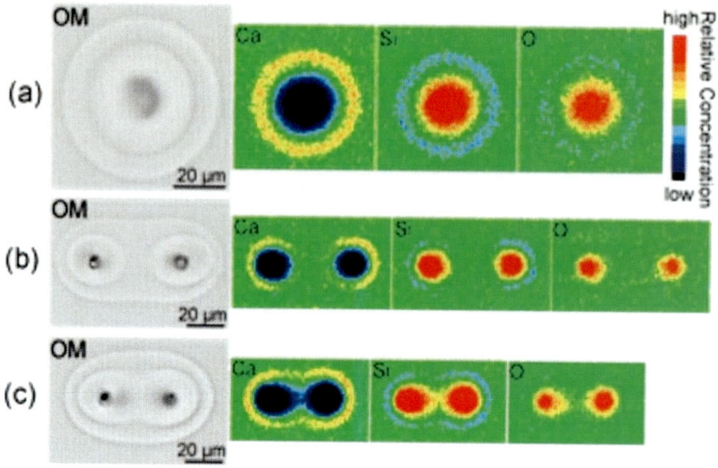

Figure 8.18 Optical microscope images and element distributions after laser irradiation. (a) One-spot irradiation. (b) Two-spot irradiation with a distance of 42.7 μm between the two focal spots. (c) Two-spot irradiation with a distance of 31.9 μm between the two focal spots.

When fs pulses are focused at a high repetition rate, the temperature increases due to accumulated heat because the following pulse come before the heat generated by the previous pulses diffuses out from the photoexcited region [42–45]. The temperature exceeded the glass transition temperature inside the outer boundary during the laser irradiation and visco-elastic deformation was induced by the stress due to thermal expansion around the photoexcited region. Inside the inner boundary, the element distribution is modified. The CaO concentration determined by a micro-Raman was more than 30 mol% at every point even after the element redistribution. This element migration should result in decrease of refractive index in the central region. Figures 8.19(a–c) show the temperature distributions for the simulations. Figures 8.19(d–i) show the simulated concentration distributions of oxides CaO and SiO_2 after 10 ms. From the compositional formula, the molar concentrations of oxides CaO and SiO_2 are equivalent to those of elements Ca and Si, respectively. All the element distributions reproduced those in the experiment of both single and two-spot irradiations. As indicated above, the order of magnitude of the heat of transport was based on assumption. The excellent qualitative correspondence between experiment and simulation indicates that this model succeeded in qualitatively modeling the formation of element distribution. This simulation includes two factors that alter the concentration distribution; one is the concentration-gradient-driven diffusion and the other is the temperature-gradient-driven diffusion. Because the former makes the concentration homogeneous, it does not contribute the inhomogeneous element distribution formed during laser irradiation. Therefore, the temperature gradient should be responsible for element migration, i.e., this phenomenon should be interpreted as thermo-migration.

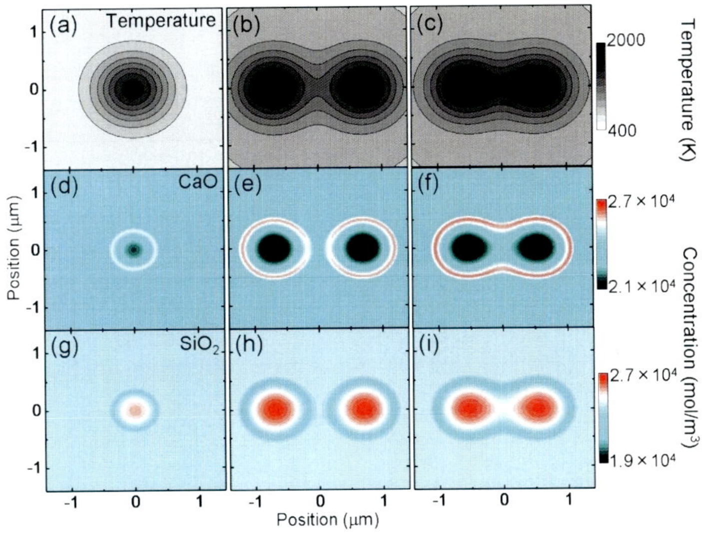

Figure 8.19 (a–c) Temperature distributions for simulation and calculated concentration distributions of (d–f) CaO and (g–i) SiO_2 after 10 ms. (a, d, g) One-spot irradiation. (b, e, h) Two-spot irradiation with a distance of 1.40 μm between the two focal spots. (c, f, i) Two-spot irradiation with a distance of 1.12 μm between the two focal spots.

8.3.4 Precipitation and Growth of Silicon

The aim of photonic integrated circuits is to attain a variety of different optical components on a single chip. A high refractive index difference is required to produce compact, integrated devices. Achieving a high refractive index difference is possible if Si structure can be formed in the glass because the respective refractive indexes of Si and conventional oxide glass are approximately 3.4 and 1.5, respectively. Silicon (Si), as a photonic medium, offers unique advantages in the fabrication of photonic integrated circuits. It is transparent in the range of optical telecommunications wavelengths (1.3 and 1.55 μm) and has a high index of refraction, which enables the fabrication of high-index-contrast sub-micrometer structures in photonic crystal devices.

It is known that oxygen-deficiency (O-deficiency) centers are formed in silica glass by irradiation with an fs laser, resulting in local changes from SiO_2 to Si-rich structures [46]. This fact suggests the possibility that Si clusters or particles can be extracted from silica or silicate glasses with an fs laser. However, very little information is currently available in the published literature on depositing Si inside a bulk glass. This is thought to be because general oxide glass cannot trap O ions in its structure, which are superfluously generated by eliminating Si–O bonds and growth of Si in an enclosed space inside the glass. To circumvent this problem, Si deposition in silicate glass by adding metallic aluminum (Al) to the starting material has been attempted by using fs laser irradiation [47].

The microphotographs in Fig. 8.20 were taken near the focal point in glass sample after irradiation with a pulse energy of 50 μJ and focused with a 50× objective lens with a numerical aperture of 0.40 on a spot for 1 s at a repetition rate of 1 kHz followed by heat treatment at 850°C. The focal spot was scanned at speed of 50 μm s^{-1} along the laser propagation direction from the bottom to the top surface. Refractive index change area with a diameter of about 10 μm was formed after laser irradiation. Micro-cracks were generated in the laser-irradiated area after heat treatment. The inset in Fig. 8.21 shows the SEM images and the graph plots the results of elemental analysis by EDS on the polished sample surface at the depth of the focal point after focused laser irradiation and subsequent heat treatment at 850°C. Silicon deposition occurred at the focused irradiation area in the glass with heat treatment (Fig. 8.21(c)). Although Silicon aggregation occurred at the focal point within a glass, complete silicon could not be observed when only laser irradiation at a repetition rate of 1 kHz was done without heat treatment (Fig.

8.21(b)). It was also confirmed that no Si precipitation occurred when heat treatment alone was done. These results indicate that the O-deficiency defects change to a structure that becomes the nucleus (i.e., seeds) of Si growth through irradiation by a laser. The Al ions can be expected to act as O-trapping centers in the silicate glass because metallic Al acts as a reducing agent in the thermite reaction. As a result, cutting the silicon–oxygen combination with laser energy and connecting the dissociated oxygen ions to the aluminum by heat treatment cause silicon deposition.

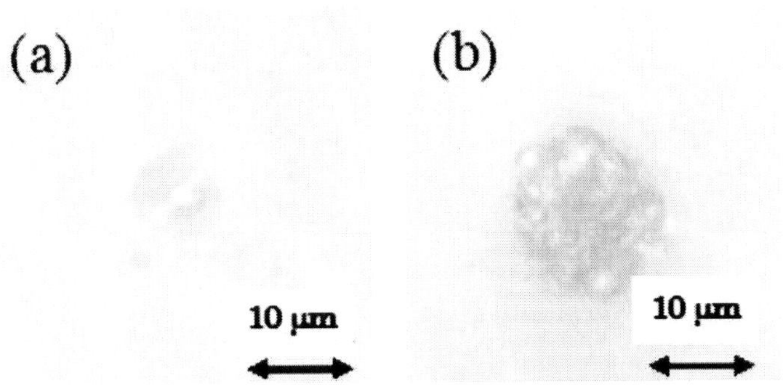

Figure 8.20 Microphotographs taken near focal point in glass sample after irradiation with a pulse energy of 30μJ and subsequent heat treatment at 850°C and focused with a 50× objective lens with a numerical aperture of 0.85 on a spot for 1 s at a repetition rate of 1 kHz. (a) After focused laser irradiation and (b) after heat treatment.

Figure 8.22 shows backscattering electron images and EPMA spectra mapping on the polished sample surface of a glass to the depth of focal spot location after irradiation with a pulse energy of 3.6 μJ and focused with a 20× objective lens with a numerical aperture of 0.40 at a repetition rate of 250 kHz. As a reference, glass prepared using Al_2O_3 (aluminum oxide) as the source material is shown together with the metallic aluminum glass. In the case of 250 kHz repetition rate, the heat caused by laser energy absorption was diffused over only ~2 μm during the interval between successive pulses (4 μs). In the case of 1 kHz (interval of pulses: 1 ms), in contrast, the heat could diffuse away out of focal volume to ~30 μm during the interval of successive pulses [41]. Therefore, heat accumulation is expected to have an effect at a repetition rate of 250

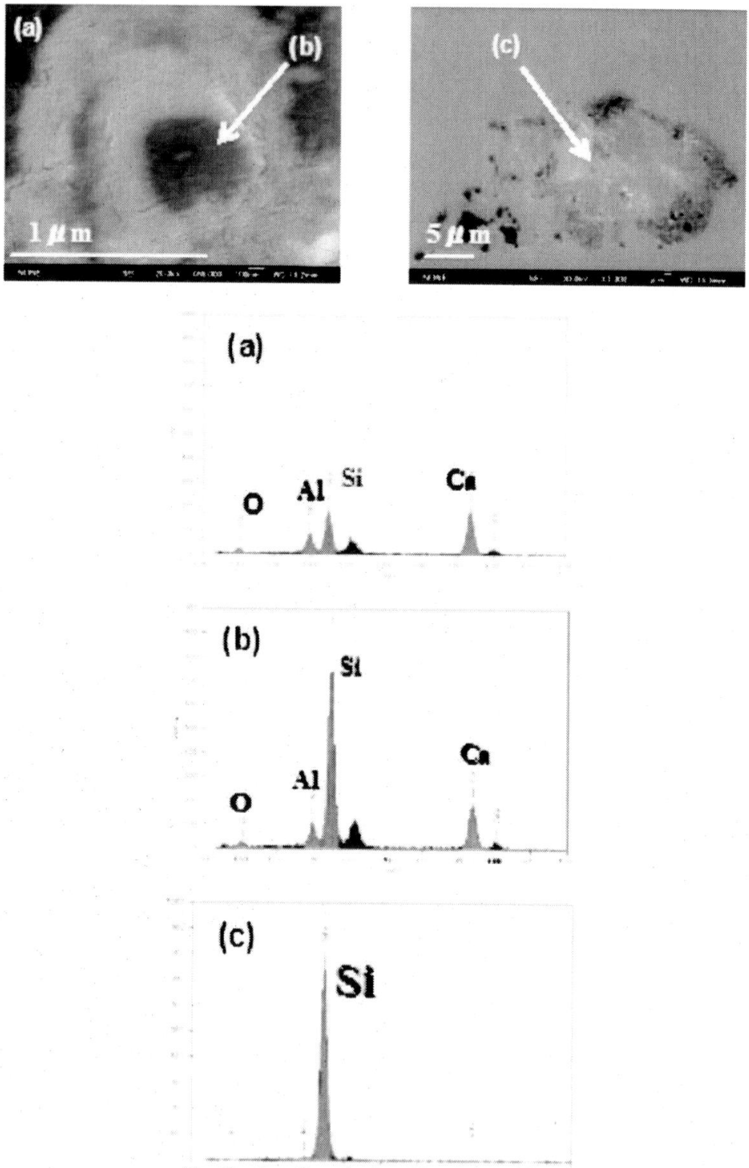

Figure 8.21 SEM image and the graph plots the results of elemental analysis by EDS of polished glass surface to depth of focal point with 1 kHz mode-locked pulses. (a) Non-irradiated, (b) after focused laser irradiation and (c) after subsequent heat treatment at 850°C.

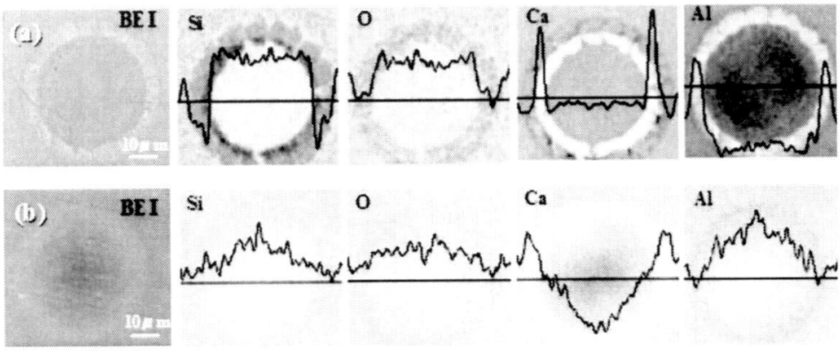

Figure 8.22 Backscattering electron images and EPMA spectra mapping on polished surface of glass to depth of focal spot after irradiation with 250 kHz mode-locked pulses. Glass prepared using (a) metallic aluminum (Al) and (b) aluminum oxide (Al_2O_3) as source material.

kHz. On both samples, the concentrations of silicon and aluminum elements are denser in the center of the irradiated area and more diffuse toward the edges. In contrast, calcium migrates to the outside from the center. However, it can be confirmed that the ion diffusion of the glass is more remarkable when metallic aluminum (Al) is added to the starting material than when Al_2O_3 (aluminum oxide) is added. Figure 8.23 shows backscattering electron images in the Si-rich region formed by the irradiation of a 250 kHz laser. TEM image observed from Si nanoparticles similar to those shown in Fig. 8.23 is shown in Fig. 8.24. The precipitation of many Si particles can be confirmed at sizes under 10 nm with a diameter of about 30 μm. The fs laser-irradiated spot is transformed to liquid or to an excited solid state at high temperature and under high pressure. Under such conditions, Si–O bonds are continuously broken and re-formed with numerous unbound atoms wandering around. Since the diffusion coefficient of O is more than twice as large as that of Si, O atoms are diffused away from the laser-irradiated areas by the generation of shockwaves, and Si-rich structures are formed at the center of the irradiated area. Furthermore, the clustering of Si is promoted by the presence of Al-rich structures in which O ions are in substoichiometric concentration. Space-selective formation of silicon structures in silicate glass could be useful both in optics, such as optical waveguides and photonic crystals, and in silicon semiconductor applications.

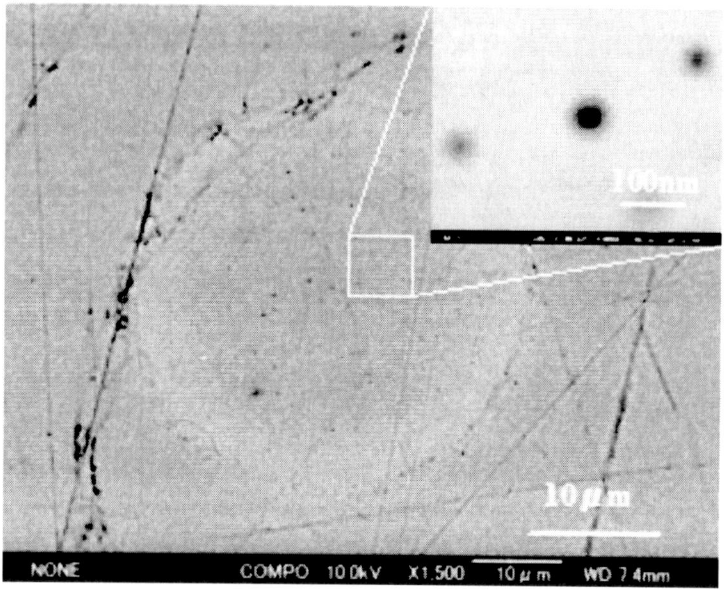

Figure 8.23 Backscattering electron images on Si rich region in Fig. 8.20 (a) by field emission scanning electron microscopy.

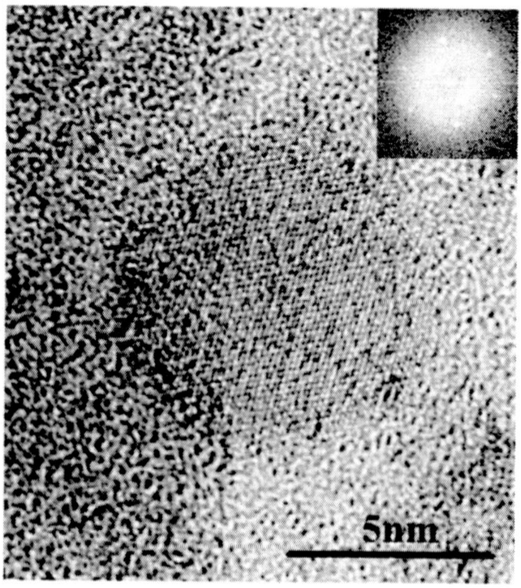

Figure 8.24 TEM images on Si nanoparticles similar to those shown in Fig. 8.23.

8.4 Final Remarks

We first focused on the characteristics of the fs laser in glass processing. According to the simulation of elastic deformation, the pressure wave is generated as the result of the stress relaxation in the photoexcited region. The investigation of the refractive index dynamics by the TrL method showed the characteristic of an fs laser processing inside a glass. The rapid temperature increase by the irradiation of fs laser pulse, which is much faster than the elastic response time of the photoexcited material, generates a strong pressure wave in the elastic relaxation process. In the case of the photoexcitation by a longer laser pulse, much more energy is needed to generate the pressure wave of the same amplitude. This means that the material deformation can be induced with a weaker photoexcitation by the fs laser irradiation. The localized photoexcitation inside a glass by the fs laser pulse should be important for preventing a random deformation due to viscous flow of heated material at high temperature.

Next, we observed various types of photo-induced phenomena, e.g., space-selective refractive index change, valence reduction of rare-earth ions, formation of elemental distribution, precipitation and growth of silicon, in glasses after the fs laser irradiation. Varying the type of glass and laser-irradiation conditions can therefore produce various types of microscopic modifications inside glass. The photo-induced phenomena in glass materials also are introduced from the viewpoint of photonic application. In particular, the use of fs laser processing to create 3D structures in the glass is technologically attractive for 3D optical devices. We are convinced that fs laser will open new possibilities in micro-optics, materials science, physics, chemistry, and bioscience.

References

1. K. M. Davis, K. Miura, N. Sugimoto, and K. Hirao (1996) Writing waveguides in glass with a femtosecond laser, *Opt. Lett.*, 21, 1729–1731.
2. K. Miura, J. Qiu, H. Inouye, T. Mitsuyu, and K. Hirao (1997) Photowritten optical waveguides in various glasses with ultrashort pulse laser, *Appl. Phys. Lett.*, 71, 3329–3331.

3. A. M. Streltsov and N. F. Borrelli (2002) Study of femtosecond-laser-written waveguides in glasses *J. Opt. Soc. Am. B*, 19, 2496.
4. J. W. Chan, T. R. Huser, S. H. Risbud, and D. M. Krol (2003) Modification of the fused silica glass network associated with waveguide fabrication using femtosecond laser pulses, *Appl. Phys. A*, 76, 367–372.
5. A. Mermillod-Blondin, I. M. Burakov, Y. P. Meshcheryakov, N. M. Bulgakova, E. Audouard, A. Rosenfeld, A. Husakou, I. V. Hertel, and R. Stoian (2008) Flipping the sign of refractive index changes in ultrafast and temporally shaped laser-irradiated borosilicate crown optical glass at high repetition rates, *Phys. Rev. B*, 77, 104205.
6. T. Toma, Y. Furuya Y, W. Watanabe, K. Itoh, J. Nishii, and K. Hayashi (2000) Estimation of the refractive index change in glass induced by femtosecond laser pulses, *Opt. Rev.*, 7, 14–17.
7. K. Miura, J. Qiu, S. Fujiwara, S. Sakaguchi, and K. Hirao (2002) Three-dimensional optical memory with rewriteable and ultrahigh density using the valence-state change of samarium ions, *Appl. Phys. Lett.*, 80, 2263–2265.
8. M. Beresna and P. G. Kazansky (2010) Polarization diffraction grating produced by femtosecond laser nanostructuring in glass, *Opt. Lett.*, 35, 1662–1664.
9. J. E. Shelby (2004) Density of vitreous silica, *J. Non-Cryst. Sol.*, 349, 331–336.
10. M. Sakakura, M. Terazima, K. Miura, and K. Hirao, (2007) Methods for observation of material deformations and energy dissipation processes during a femtosecond laser machining inside transparent materials, *Rev. Laser Eng.*, 35(2), 109–115.
11. S. I. Kudryashov, V. D. Zvorykin, A. A. Ionin, V. Mizeikis, S. Juodkazis, and H. Misawa (2008) Acoustic monitoring of microplasma formation and filamentation of tightly focused femtosecond laser pulses in silica glass, *Appl. Phys. Lett.*, 92, 101916.
12. A. Mermillod-Blondin, J. Bonse, A. Rosenfeld, I. V. Hertel, Yu. P. Meshcheryakov, N. M. Bulgakova, E. Audouard, and R. Stoian (2009) Dynamics of femtosecond laser induced voidlike structures in fused silica, *Appl. Phys. Lett.*, 94, 041911.
13. Y. Hayasaki, M. Isaka, A. Takita, and Saulius Juodkazis (2011) Time-resolved interferometry of femtosecond-laser-induced processes under tight focusing and close-to-optical breakdown inside borosilicate glass, *Opt. Express*, 19, 5725–5734.

14. M. Sakakura and M. Terazima (2005) Initial temporal and spatial changes of the refractive index induced by focused femtosecond pulsed laser irradiation inside a glass, *Phys. Rev. B,* 71, 024113-1-024113-12.

15. M. Sakakura, M. Terazima, Y. Shimotsuma, K. Miura, and K. Hirao (2007) Observation of pressure wave generated by focusing a femtosecond laser pulse inside a glass, *Opt. Express,* 15(9), 5674–5686.

16. G. Paltauf, and P. E. Dyer (2003) Photomechanical processes and effects in ablation, *Chem. Rev.*, 103, 487–518.

17. C. Hu, and J. R. Whinnery (1973) New thermooptical measurement method and a comparison with other methods, *Appl. Opt.*, 12, 72–79.

18. M. Sakakura, M. Terazima, Y. Shimotsuma, K. Miura, and K. Hirao (2007) Heating and rapid cooling of bulk glass after photoexcitation by a focused femtosecond laser pulse, *Opt. Express,* 15(25), 16800–16807.

19. M. Terazima (1995) Ultrafast transient Kerr lens in solution detected by the dual-beam thermal-lens method, *Opt. Lett.,* 20, 25.

20. M. Pons, S. Nonell, I. Garcia-Moreno, A. Costela, and R. Sastre (2002) Time-resolved thermal lens study on the heat dissipation effects in solid polymeric matrices used as laser dyes, *Appl. Phys. B.,* 75, 687–694.

21. A. Vogel, J. Noack, G. Huttman, and G. Paltauf (2005) Mechanisms of femtosecond laser nanosurgery of cells and tissues, *Appl. Phys. B,* 81, 1015–1047.

22. G. S. Fulcher (1925) Analysis of recent measurements of the viscosity of glasses, *J. Am. Ceram. Soc.*, 8, 339–355.

23. E. L. Cussler (1984) *Diffusion*, Cambridge University Press, Cambridge.

24. C. A. Coulson and T. W. Dingle (1968) The B-O bond lengths in boron-oxygen compounds, *Acta Cryst. B,* 24, 153–155.

25. Y. Shimotsuma, K. Hirao, J. Qiu, and K. Miura (2006) Nanofabrication in transparent materials with a femtosecond pulse laser, *J. Non-Crystall. Solids,* 352, 646–656.

26. J. Qiu, K. Miura, and K. Hirao (2005) Fabrication of micro-optical components by using a femtosecond laser, *SPIE,* 5350, 1–12.

27. K. Hirao and K. Miura (1998) Writing waveguides in silica-related glasses with femtosecond laser, *Jpn. J. Appl. Phys.*, 37, 49–52.

28. P. G. Kazansky, H. Inouye, T. Mitsuyu, K. Miura, J. Qiu, and K. Hirao (1999) Anomalous anisotropic light scattering in Ge-doped silica glass, *Phys. Rev. Lett.*, 82(10), 2199–2202.

29. J. Qiu, C. Zhu, T. Nakaya, J. Si, and K. Hirao (2001) Space-selective valence state manipulation of transition metal ions inside glasses by a femtosecond laser, *Appl. Phys. Lett.*, 79, 3567–3569.

30. J. Qiu, M. Shirai, T. Nakaya, J. Si, X. Jiang, C. Zhu, and K. Hirao (2002) Space-selective precipitation of metal nanoparticles inside glasses, *Appl. Phys. Lett.*, 81, 3040–3042.

31. S. Hiramatsu, T. Mikawa, Q. Ibaragi, K. Miura, and K. Hirao (2004) Laser-written optical-path redirected waveguide device for optical black-plane interconnects, *IEEE Photon. Technol. Lett.*, 16(9), 2075–2077.

32. S. Hiramatsu, K. Miura, and K. Hirao (2007) Optics backplane connectors using three-dimensional waveguide arrays, *J. Lightwave Technol.*, 25(9), 2776–2782.

33. A. Couairon, L. Sudrie, M. Franco, B. Prade, and A. Mysyrowicz (2005) Filamentation and damage in fused silica induced by tightly focused femtosecond laser pulses, *Phys. Rev. B,* 71, 125435.

34. M. Sakakura, M. Terazima, Y. Shimotsuma, K. Miura, and K. Hirao (2011) Thermal and shock induced modification inside a silica glass by focused femtosecond laser pulse, *J. Appl. Phys.*, 109, 023503.

35. P. F. McMillan, B. Piriou, and R. Couty (1984) A Raman study of pressure-densified vitreous silica, *J. Chem. Phys.*, 81, 4234–4236.

36. M. Okuno, B. Reynard, M. Shimada, Y. Syono, and C. Willaime (1999) A Raman spectroscopic study of shock-wave densification of vitreous silica, *Phys. Chem. Miner.,* 26, 304–311.

37. J. Qiu, K. Miura, and K. Hirao (2003) Space-selective microscopic modifications of glass structure by a femtosecond laser, *SPIE.*, 5061, 82–88.

38. S. Kanehira, K. Miura, and K. Hirao (2008) Ion exchange in glass using femtosecond laser irradiation, *Appl. Phys. Lett.*, 93, 023112-023112–3.

39. Y. Liu, B. Zhu, L. Wang, J. R. Qiu, Y. Dai, and H. L. Ma (2008) Femtosecond laser induced coordination transformation and migration of ions in sodium borate glasses, *Appl. Phys. Lett.*, 92, 121113-121113–3.

40. F. F. Luo, B. Qian, G. Lin, J. Xu, Y. Liao, J. Song, H. Y. Sun, B. Zhu, J. R. Qiu, Q. Z. Zhao, and Z. Z. Xu (2010) Redistribution of elements in glass induced by a high-repetition-rate femtosecond laser, *Opt. Express,* 18, 6262–6269.

41. M. Shimizu, K. Miura, N. Yasuda, M. Sakakura, S. Kanehira, M. Nishi, Y. Shimotsuma, and K. Hirao (2009) Formation of elemental distribution

in glass using thermal accumulation with femtosecond laser irradiation, *MRS Proc.*, 1230-MM07–04.

42. S. M. Eaton, H. B. Zhang, and P. R. Herman (2005) Heat accumulation effects in femtosecond laser-written waveguides with variable repetition rate, *Opt. Express,* 13, 4708–4716.

43. M. Sakakura, M. Shimizu, Y. Shimotsuma, K. Miura, and K. Hirao (2008) Temperature distribution and modification mechanism inside glass with heat accumulation during 250kHz irradiation of femtosecond laser pulse, *Appl. Phys. Lett.*, 93, 231112.

44. M. Shimizu, M. Sakakura, M. Ohnishi, Y. Shimotsuma, T. Nakaya, K. Miura, and K. Hirao (2010) Mechanism of heat-modification inside a glass after irradiation with high-repetition rate femtosecond laser pulses, *J. Appl. Phys.,* 108, 073533.

45. M. Shimizu, M. Sakakura, S. Kanehira, M Nishi, Y. Shimotsuma, K Hirao, and K Miura (2011) Formation mechanism of element distribution in glass under femtosecond laser irradiation, *Opt. Lett.*, 36, 2161–2163.

46. N. Fukata, Y. Yamamoto, K. Murakami, M. Hase, and M. Kitajima (2003) *In situ* spectroscopic measurement of defect formation in SiO_2 induced by femtosecond laser irradiation, *Phys. B*, 340–342, 986–989.

47. K. Miura, Y. Shimotsuma, M. Sakakura, S. Kanehira, M. Hamabe, and K. Hirao (2006) Three-dimensional deposition of silicon from silicate glass with dispersed metallic aluminum by a femtosecond laser, *SPIE Smart Mater., Nano, Micro-Smart Syst.*, 6413, 64130K1–13.

Chapter 9

3D Photonic Device Fabrication

Nicola Bellini, Andrea Crespi, Shane M. Eaton, and Roberto Osellame
Dipartimento di Fisica, Politecnico di Milano,
P.za Leonardo da Vinci, 32, 20133 Milano, Italy
roberto.osellame@polimi.it

Focused femtosecond laser pulses can induce permanent and localized refractive index modification of transparent materials under suitable conditions. Relative translation of the sample with respect to the laser beam allows one to literally draw optical waveguides, circuits or other complex structures in the substrate, fully exploiting the three dimensions. In this chapter, the main parameters involved in the fabrication process will be discussed. Several important devices will also be shown, which indeed exemplify the vast portfolio of applications reached nowadays by this technology by exploiting its unique capabilities.

9.1 Introduction

Focused femtosecond laser pulses yield peak intensities greater than 10 TW/cm^2, which cause strong nonlinear absorption and localized

Ultrafast Laser Processing: From Micro- to Nanoscale
Edited by Koji Sugioka and Ya Cheng

ISBN 978-981-4267-33-5 (Hardcover), 978-981-4303-69-9 (eBook)
www.panstanford.com

energy deposition in the bulk of transparent materials. After several picoseconds, the laser-excited electrons transfer their energy to the lattice, leading to a permanent material modification. Depending on the laser and material parameters, this laser modification may result in damaged and irregular scattering centres, or smooth structures with a positive refractive index alteration. Such smooth structures of positive refractive index change, suitable for waveguiding light at visible and infrared wavelengths by total internal reflection, may be formed along arbitrary three-dimensional paths by scanning the sample using computer-controlled motion stages. Additional advantages of femtosecond laser direct writing include maskless, single-step fabrication, and its ability to process many transparent materials including glasses, polymers and crystals. In this chapter, the relevant exposure parameters to produce low loss, high refractive index increase optical waveguides by femtosecond laser writing are reviewed and discussed. In particular, the chapter will focus on the possibility of using this microfabrication technology to produce important photonic devices such as directional couplers, Mach–Zehnder interferometers, Bragg grating waveguides, waveguide lasers and waveguide arrays. Exploiting the unique flexibility of the technology and its three dimensional capabilities these components can be further combined to produce more complex microsystems addressing such diverse application fields as optofluidics, optomechanics and integrated quantum optics. The chapter will review some of the most significant examples highlighting the advantages of this fabrication technology with respect to standard ones.

9.2 Waveguide Writing in Transparent Materials

9.2.1 Exposure Parameters and Considerations

As illustrated in the previous chapters, the features induced in the substrate by the action of femtosecond laser pulses depend in a nontrivial way on many irradiation parameters, such as pulse energy, pulse duration and repetition rate, translation speed of the sample, focusing conditions, as well as on material characteristics such as bandgap, thermal characteristics and crystalline or amorphous properties. Thus, a careful experimental optimization of the parameters is generally required for fabricating the optimum waveguide for a certain application.

9.2.1.1 Pulse energy and translation speed

Depending on the energy of the focused femtosecond laser pulses, three qualitatively different features have been observed in the bulk of transparent materials[1–3]: a smooth isotropic refractive index change for low pulse energies, a birefringent refractive index change for intermediate pulse energies and a void due to microexplosions for high pulse energies.

The smooth isotropic refractive index change is the most relevant for waveguide writing in glasses.[4] However, the precise microscopic mechanisms that lead to the refractive index increase are not yet fully understood and are still under investigation. Diverse interaction phenomena enter into play, including colour centre formation, densification, thermal diffusion and accumulation. The relative role of each of them is debated[5–8] and depends on the material and the exposure conditions.

In fused silica glass, for slightly higher pulse energies, the formation of birefringent nanostructures is observed, the axis of birefringence depending on the irradiation polarization.[9–11] This peculiar feature has been explored to produce waveguides and photonic devices with polarization dependent characteristics, such as structures that are able to guide only one polarization[11] and directional couplers with polarization dependent behaviour.[12]

For even higher pulse energies microexplosions are induced by the femtosecond pulses, which result in microvoid formation. This kind of modification does not present, in general, waveguiding properties. Interesting applications for this regime have been envisaged anyway in high-density three-dimensional data storage. Sub-micron size voxels (volume pixels) can be written in this way by femtosecond pulses and readout can be performed exploiting scattering or luminescence.[13,14]

For a given pulse energy, the translation speed controls the amount of total energy deposited per unit volume and typically influences the achieved refractive index change. However, since the interaction process is nonlinear, lowering (increasing) the speed with the same pulse energy is typically not equivalent to increasing (lowering) the pulse energy for a fixed speed.

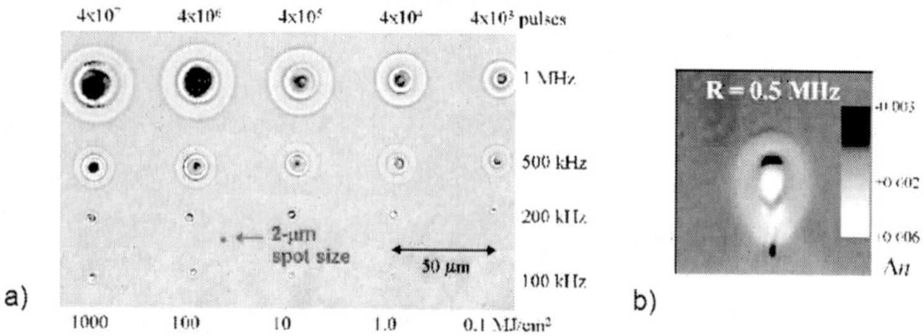

Figure 9.1 (a) Heat accumulation effects enlarge the modified region with increasing repetition rate and pulse energy.[15] (b) Refractive index profile of a waveguide cross section, written at 0.5 MHz repetition rate.[16] An overall elliptical shape can be observed: there is a balance between the isotropic heating effects and the elongated trace produced by the individual pulses.

9.2.1.2 Low vs. high repetition rate

The repetition rate of the femtosecond laser pulses is a crucial parameter which has a strong influence on the waveguide writing process, determining completely different regimes.

In the low-repetition-rate regime (1–250 kHz), typical of traditional Ti:sapphire regeneratively amplified laser systems, the material cools down to the original temperature after the irradiation of each pulse, thus the substrate modification is mainly determined by individual pulses.[15] The shape of the modified region coincides with the irradiated volume and beam-shaping techniques are required in order to obtain circular waveguide cross sections.

If the repetition rate is increased to several MHz, such that the time interval between two consecutive pulses is much shorter than the heat diffusion time (which is about 1 μs), strong cumulative thermal effects take place.[17,18] In this regime, femtosecond laser oscillators without amplification stage are typically employed, delivering low energy pulses that are focused by high-numerical-aperture oil-immersion objectives to trigger the nonlinear absorption. Due to the tight focusing conditions, the region directly irradiated by the individual pulses is very small. However, the thermal accumulation and heat diffusion generate a much larger modified region, which extends isotropically, thus proving a circular waveguide profile without the need of beam-shaping techniques.

A very interesting regime is observed at intermediate repetition rates (hundreds of kilohertz to few megahertz), in which a gradual appearance of the thermal accumulation effects occurs[15,16] (see Fig. 9.1a). To work in this repetition rate regime Yb:based laser systems are a common choice, which are capable of pulse energies up to some μJ. Tight focusing is no longer required and long-working distance objectives can be employed, enabling a full exploitation of the three-dimensional capabilities of the laser writing technology. The region directly modified by the laser pulses in this case has comparable size to the thermal affected region (see Fig. 9.1b) and the advantages of the two regimes can be combined: stronger refractive index increase, given by the action of more energetic pulses along with thermal effects which helps to make the waveguide more uniform and circular.

9.2.1.3 Longitudinal vs. transverse writing

Two different writing geometries are traditionally identified, depending on the direction of translation of the sample under the laser beam for the femtosecond irradiation: longitudinal and transverse.[4,19]

In the longitudinal geometry, the sample is translated along the beam propagation direction. As a consequence, the waveguide formed has intrinsic transverse circular symmetry, with dimensions on the order of the focal spot size. However, the overall waveguide length is limited by the working distance of the microscope objective adopted.

In the transverse geometry, the sample is translated orthogonally to the beam propagation direction. The waveguide cross section in this case becomes strongly asymmetric, since the width is approximately twice the beam waist w_0, while the depth has a dimension comparable to the (much larger) confocal parameter of the focused writing beam $2nw_0/\mathrm{NA}$ (where n is the refractive index of the glass and NA the numerical aperture of the objective). Despite this disadvantage, the gain in terms of design flexibility is enormous, since this gives complete freedom on a glass sample which is commonly few millimetres thick but some centimetres wide.

Beam-shaping techniques[20–22] have been developed to fabricate circular waveguides, which yield good mode matching with optical fibres, even in the transverse writing geometry (see Chapter 4). Briefly, these techniques enlarge the focus dimension in the plane orthogonal to the translation direction by means of a cylindrical telescope or a slit before the focusing objective. In this way, a disc-shaped focal volume is obtained, which produces circular-cross-section waveguides. However, if a high repetition rate is adopted, a circular waveguide profile may be obtained in some materials (as mentioned before), due to isotropic thermal effects, without any beam-shaping technique even in the transverse writing configuration.

Another technique for fabricating waveguides with symmetric cross section is multi-scan irradiation.[23,24] Many lines are written, each of them with elongated cross section, slightly shifted relative to each other. An overall square and symmetric cross section is thus obtained.

9.2.2 Different Materials

One of the most striking advantages of femtosecond laser waveguide writing over conventional lithographic techniques is the possibility to work on a broad range of materials with the same laser system. It would not be feasible to review here all the substrates on which waveguide writing has been successfully demonstrated; however, it may be worth summarizing briefly the main material categories.

9.2.2.1 Glasses

Glasses are the first transparent substrates on which femtosecond laser waveguide writing was demonstrated.[4] In addition, glass is the material of choice in the vast majority of the results and applications.

Fused silica is worth of particular mention. Its chemical simplicity has been the object of many studies that investigate the microscopic processes of refractive index increase due to the interaction with femtosecond pulses.[5,7,25] In addition, irradiation with ultrafast laser can produce birefringent structures[9] or regions preferentially etched by HF (hydrofluoric acid),[26] thus enabling an incredibly wide portfolio of applications. Regarding the writing parameters, thermal accumulation from high repetition rates has shown to produce irregular structures with no waveguiding properties.[27] Thus, low or intermediate repetition rates are commonly employed with this material.

Multicomponent glasses have also been employed for writing high-quality waveguides,[28] in particular borosilicate glasses[16] and phosphate glasses.[29,30]

Photosensitive glasses have also to be mentioned, since they have recently been applied for fabrication of integrated micro-optical elements.[31] In particular, void microchannels[21] and optical waveguides[32] have been demonstrated in commercially available Foturan glass from Schott Glass Corporation, which is composed of lithium aluminosilicate doped with trace amounts of silver, cerium and antimony. Further studies are required to fully understand the mechanism of refractive index change in this photosensitive Foturan glass.[33]

9.2.2.2 Crystals

Writing waveguides in ordered media, like crystals, with femtosecond laser pulses is much more critical than in amorphous media, like glasses. The modification of crystals requires a larger energy density and typically implies amorphization of the material and thus a negative refractive index change. One of the most studied crystalline substrates is lithium niobate $LiNbO_3$ (LN). This is because LN is a very important material for integrated-optical applications due to its wide transparency range and its outstanding electrooptic and nonlinear properties.

The femtosecond laser-induced mechanisms, which lead to refractive index changes useful for the fabrication of guiding devices, consist of two types; both types are well explained by the model of lattice defects at different degrees of damage.[34] For low laser fluence, point defects are created which lower the spontaneous polarization of the crystal and thus increase the extraordinary index. Anyway, lowering the pulse energy provides faint waveguides that hardly show any guiding capability. Therefore, in order to deposit a sufficient energy without introducing damage, a multiscan approach was proposed (Fig. 9.2a) and employed for the fabrication of good optical quality waveguides.[35,36] These structures are known as "type I" optical waveguides, where the guiding region corresponds to the irradiated volume. Due to the absence of induced damage, type I waveguides preserve the nonlinear properties of LN.[37] On the other hand, strong crystalline damage at high fluence leads to a volume increase that correlates with diminished refractive index. This mechanism is also responsible for the positive refractive index change in the vicinity of the structure due to stresses induced by the femtosecond pulses (Fig. 9.2b). This result can be exploited by irradiating two parallel lines, which present a negative refractive

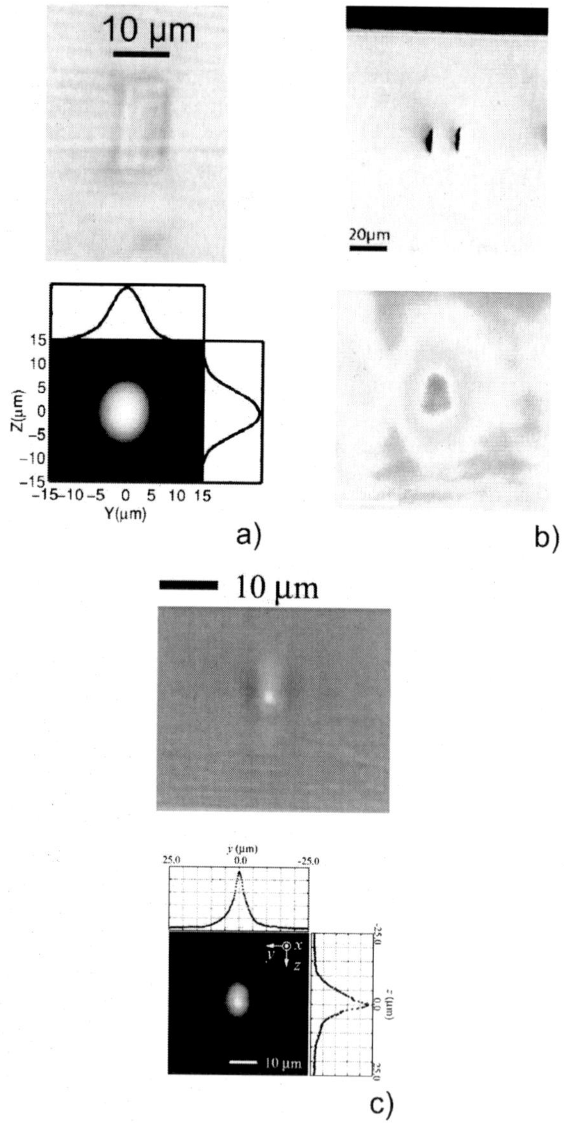

Figure 9.2 (a) LN "type I" waveguide: (top) microscope image of the waveguide end facet; (bottom) near field of the waveguide mode at 1567 nm.[37] (b) LN "type II" waveguide: (top) microscope image of the waveguide end facet; (bottom) near field of the waveguide mode at 1064 nm.[40] (c) PMMA waveguide: (top) microscope image of the waveguide end facet; (bottom) near field of the waveguide mode at 633 nm.[43] If any, lateral panels show the intensity profiles across the peak intensity.

index change, but induce a positive one in the unmodified region in between them, where light is guided. These waveguides are the so-called "type II" ones.[38–40] In Section 9.3.1.1 applications of LN type II waveguides in Mach–Zehnder interferometers will be described. Due to the general nature of the process, this discussion is also valid for other crystalline materials. Type II waveguides have been demonstrated in Yb:KGW and Yb:KYW.[41,42]

Polymeric substrates are widely used for the manufacturing of microfluidic components owing to well-developed fabrication technologies. Recently, polymer materials have also been used for fabricating photonic components due to their numerous advantages, such as their low cost, ease of manufacture and high transmission in the visible region, in addition to their good compatibility to existing microfluidic elements. In particular, polymethyl methacrylate (PMMA) is widely spread and highly employed.

First optical devices fabricated on PMMA by femtosecond laser writing were surface diffraction gratings; later, more complex diffractive devices were fabricated directly buried in the polymeric substrate.[44,45] Exploiting the 3D capability of femtosecond laser micromachining, holographic structures were also demonstrated using a Ti:sapphire 1 kHz repetition rate femtosecond laser.[46] The analysis of the diffractive efficiency of these elements permitted a deep study of the interaction between the femtosecond pulses and the modified polymer.[47–49]

The modification of the refractive index led to the fabrication of optical waveguides in PMMA and other polymers. Noteworthy waveguides on PMMA were demonstrated by using a 1 kHz repetition rate Ti:sapphire laser and transverse writing geometry.[43] A slit beam-shaping stage was introduced to control the waveguide cross section, allowing for single-mode waveguiding at 632.8 nm (Fig. 9.2c). The refractive index change and the propagation losses were estimated to be 4.6×10^{-4} and 4.2 dB/cm, respectively. A directional coupler to split the coupled beam with a 1:1 splitting ratio was also realized using the same technique. Recently, waveguides on PMMA doped with azochromophores were demonstrated.[50]

Waveguides were also fabricated in hydrogel polymers (PV2526-164) containing ~36% water by weight.[51] Although high refractive index increase was observed (0.06), larger surface roughness along the waveguides induced very large propagation loss at 632.8 nm. Such high refractive index change and losses were due to the presence of water; therefore, other experiments were performed in the same polymers with no water.

The reasons for the observed changes in refractive index are not yet clear. However, main responsible phenomena are considered to be structural changes, change in density, compression and tensile stresses.[52,53] For example, it was reported that the refractive index change in PMMA might be induced by photodecomposition; the increase in refractive index is due to the complete and partial separation of the side chain from the PMMA molecule.[48]

9.3 Photonic Devices

9.3.1 Passive Devices

Passive substrates are materials that present no optical gain, in which light can only experience attenuation along its propagation path, due to absorption, scattering, diffusion. Anyway, optical devices fabricated on these passive transparent substrates are extremely important to treat and manipulate light. To minimize the insertion losses of the device, it is fundamental to enhance as much as possible the optical quality of the guiding structures. In most cases, losses of passive devices fabricated by femtosecond laser micromachining are still slightly higher than those fabricated with standard technologies; anyway, the femtosecond laser technique has the advantage of intrinsic 3D and easy prototyping nature, thus enabling the development of a new class of devices that are not possible with 2D planar geometry.

9.3.1.1 Power routing devices

Optical power splitters are key waveguide devices in optical communication and photonic information processing. Several photonic devices can perform power splitting, including Y branches, directional couplers, and multi-mode interference waveguide devices. Moreover, such devices can be useful for several applications, such as sensing, signal multiplexing and demultiplexing, wavelength routing and light modulation. For this reason, we will discuss in detail the fabrication of these basic devices by femtosecond laser micromachining, giving more emphasis to those elements that exploit the unique 3D potential of the fabrication technique.

Power splitters A 1 × 2 Y splitter was demonstrated in 1999 using the longitudinal writing configuration in pure fused silica.[54] Infrared pulses from an amplified Ti:sapphire oscillator, at 1 kHz repetition rate, were focused using a 10× objective which allowed a 2 cm writing range. The fabricated structure was able to guide and split 514.5 nm light from an argon ion laser, with a splitting ratio close to 1:1. An analogous device was demonstrated some years later also in LN,[55] using the transverse writing configuration. For this device a 1.1:1 splitting ratio was measured, with propagation losses of 1 dB/cm and additional splitting loss of 0.8 dB, for operation at 633 nm wavelength.

In a Y splitter, power division is very sensitive to the symmetry of the two branches, which can be controlled easily when working with femtosecond laser pulses, because the same fabrication condition will produce waveguides with almost the same properties. On the other hand, using this fabrication technology the common branch needs in most cases to be irradiated twice. This may affect the waveguide properties with respect to a single scan.

Low *et al.* showed a 1 × 4 beam splitter, realized in aluminosilicate glass.[19] In this case, since four branches need to be irradiated, the common branch was subjected to four overlapped scans. However, the authors demonstrated in multi-scan straight waveguides that there was no discernible change in the waveguide diameter with increasing number of passes, up to 10. Only for higher number of passes they observed damage in the waveguide structure.

A more complex 1 × 8 splitter was demonstrated in fused silica substrate by Liu *et al.*,[56] cascading three 1 × 2 splitting sections. The splitter was single mode at 1550 nm wavelength, with well-balanced power repartition in the eight arms. The entire device was written twice, by two consecutive scans transversely offset by 1.5 μm. This mitigated additional splitting losses (reduced down to 0.5 dB) and polarization dependent losses. This double scan technique partially corrects for the asymmetry in the waveguide shape which may arise in the transverse writing configuration if no beam-shaping technique is adopted.

Even more interesting devices can be realized exploiting some peculiar capabilities of femtosecond laser microfabrication (refer to Fig. 9.3). The same femtosecond laser system, for example, can be exploited both for a gentle modification of the refractive index of the substrate, leading to waveguide formation, and for surface ablation.

Figure 9.3 (Top) Femtosecond laser written Y-splitter integrated with U-grooves for fibre alignment. Schematic and microscope pictures.[57] (Bottom) Schematic of a 1 × 3 three-dimensional splitter. In the inset near field image of the output: an equal power distribution over the three branches is appreciated.[58]

This dual capability enabled the fabrication of a 1 × 2 optical splitter,[57] integrated with U-grooves for passive fibre alignment, all realized in a single laser fabrication step on a fused silica substrate. In particular the Y splitter was irradiated with 400 nJ pulses of a Ti:sapphire laser and 50 μm/s scan speed, while the ablation was performed with much more energetic 30 μJ pulses and 500 μm/s scan speed. It is worth noting that common techniques for fabricating grooves in optical chips requires difficult post-fabrication etching processes, whereas in this case high-quality grooves for fibre alignment are realized in a straightforward fashion.

Another unique possibility of femtosecond laser waveguide writing is that it can provide arbitrary three-dimensional waveguiding paths in the bulk, which is not possible with traditional lithographic technologies. The first demonstration of such a 3D device was reported by Nolte *et al.*[58] A 1 × 3 splitter was realized in fused silica by an amplified Ti:sapphire laser. The three branches split at angles of 120°, each lying on a different plane. An almost equal 32:33:35 splitting ratio was measured at 1050 nm wavelength. The input port of the splitter was observed to be slightly multimode, likely because of the above mentioned multiple writing of the common arm. In any case, the output ports showed reliable single-mode behaviour.

Directional couplers A directional coupler is a device able to transfer optical power from one waveguide to another without any intersections between the waveguides. This mechanism is due to the transmission of electromagnetic waves from one medium to another by means of the evanescent, exponentially decaying electromagnetic field outside the guiding region; this phenomenon is known as evanescent-field coupling. To exploit evanescent coupling, two or more optical waveguides must be placed very close to each other in the so-called interaction region. The amount of optical power transfer to the neighbour waveguides depends on many parameters, i.e. mode field diameter, propagation constant, separation distance and interaction length. Power transfer and splitting ratio at the directional coupler output can be finely controlled by adjusting one or more of these parameters. Femtosecond laser micromachining permits to directly inscribe interacting waveguides in the same substrate in a arbitrary 3D fashion.

The first directional coupler fabricated by femtosecond laser writing was demonstrated in a borosilicate glass by employing a high repetition rate Ti:sapphire oscillator, frequency doubled to enhance the efficiency of the nonlinear absorption process.[59] The coupler consisted of a 15 mm-long straight waveguide and a three-segment waveguide with 5 cm radius arcs at the ends and a straight 9 mm-long sector in the middle, which was parallel to the first straight waveguide with a 3.5 μm centre-to-centre distance forming the interaction region. Further, coupler fabrication was demonstrated by a 4 MHz Ti:sapphire oscillator in a commercial soda lime glass (Corning 0215).[60] A transverse writing geometry was adopted with 10 mm/s writing speed and 20 nJ pulse energy. Characterization of the device was performed at 633 nm for different interaction lengths and waveguide separations. A good agreement with coupled mode theory was obtained, but the splitting ratio was limited to 60% due to some mismatch between the two waveguides.

A directional coupler with similar design was also manufactured in silica glass by means of a low-repetition-rate amplified Ti:sapphire laser system.[61] A longitudinal writing geometry was adopted, and by exploiting pulse filamentation and simultaneous bending of the filament (by the previously induced refractive index change) the resulting curved waveguide of the coupler was uniform and single mode at 633 nm. Couplers with different coupling ratios, namely 1:1 and 1:0.5, were demonstrated by varying the interaction length L. In the same work, the 3D capability of femtosecond laser technique was exploited to fabricate for the first time a 3D directional coupler consisting of three waveguides (Fig. 9.4a). When coupled by an infrared broadband signal (700–900 nm) the coupler exhibited a wavelength-dependent splitting ratio, and this represented the first demonstration of wavelength division capability from a femtosecond laser-written device.

Previous results clearly showed the feasibility of directional coupler fabrication by femtosecond laser micromachining. A further step towards device efficiency and reliability is represented by the demonstration of low-loss directional couplers operating in the full range of telecom wavelengths.[62] The device was fabricated in borosilicate glass by means of a 1 MHz repetition rate system operated at high scan speed (7–25 mm/s). High-quality single-mode and low propagation loss waveguide fabrication was demonstrated as well as negligible bending losses. These achievements allowed an effective S-bend design of the coupler, which was butt-coupled to standard single mode fibres and exhibited impressively low insertion losses.

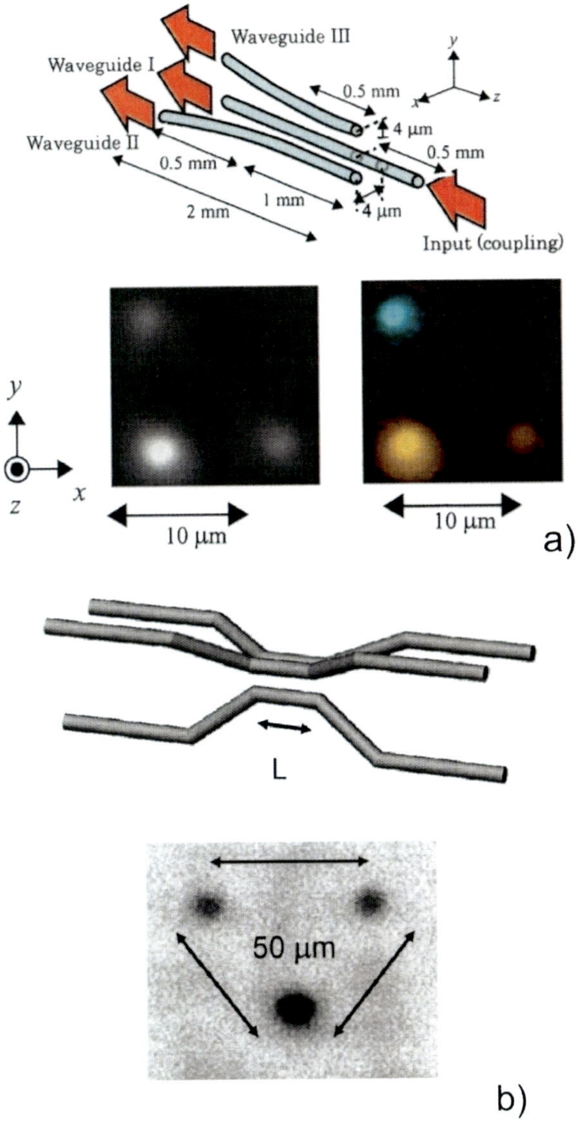

Figure 9.4 (a) Top: schematic of three curved waveguides in the interaction region. Bottom left: CCD camera image of the coupler output when excited at a wavelength of 632.8 nm. Bottom right: CCD camera image of the coupler output when excited with a broadband light source.[61] (b) Top: schematic of the symmetric three-waveguide directional coupler. Bottom: inverse greyscale CCD image of the waveguide outputs.[64]

An X design, employing straight waveguides, was also used to write a directional coupler by means of a high-repetition-rate laser in a commercial alkali-zinc-silicate glass (Schott IOG-10).[63] Such a design exploits the 3D potential of the femtosecond laser writing technique to place the coupled waveguides at different depths, thus achieving a vertical coupling. The device is obtained by writing two straight waveguides with an angle α and with a depth displacement h. Several couplers with different values of the two parameters were fabricated and several coupling ratios were achieved at telecom wavelength; wavelength division functionality was also demonstrated with a high rejection of 980 nm wavelength with respect to 1550 nm.

Another interesting example of a 3D coupler is represented by a 3 × 3 directional coupler manufactured in soda-lime glass by means of a 5.85 MHz repetition rate Ti:sapphire oscillator.[64] A 3D symmetric design was adopted with the three waveguides fabricated on the edges of an equilateral triangle (Fig. 9.4b). In a first demonstration of this device a 43%:28%:29% coupling ratio at the output was reported at 800 nm wavelength. A detailed characterization and description of the device performances for the wavelength range around 1.5 μm was reported in another paper by the same group,[65] in which they also investigated both the origin of fabrication errors and how they changed the device characteristics. It turned out that the fabricated couplers have characteristics that agree reasonably well with theory.

Exploiting the well-established results reported in the previous works, femtosecond laser-written 3D directional couplers have been applied to obtain photonic devices in several different applications. In the field of telecommunications both broadband directional couplers[66] and wavelength demultiplexers[67] were demonstrated using a 1 MHz repetition rate laser in a borosilicate glass; in both cases coupling ratio in the full telecom spectrum was controlled by properly setting the writing parameters.

Other very interesting directional couplers were fabricated in a single translation sweep by simultaneously writing two or more waveguides, controlling their relative positions on the fly.[68,69] This was achieved by employing a variable phase mask along the femtosecond writing beam path, that changed the beam profile by suitable software. These works were performed using a liquid crystal spatial light modulator (SLM) for simultaneous waveguide writing in fused silica. The 3D capability of this technology was exploited to fabricate a directional coupler in a single longitudinal sweep with input waveguides placed on a horizontal plane while output waveguides were positioned on a vertical plane (Fig. 9.5).

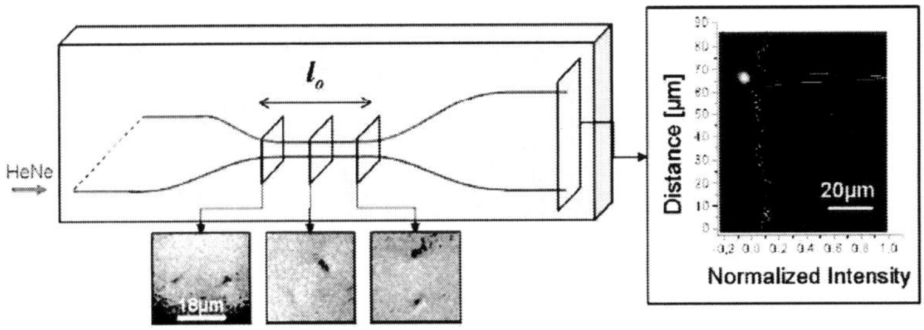

Figure 9.5 (Left) Schematic of 3D directional coupler with microscope pictures of the central region shown below the schematic and (right) near field profile at the output of the top arm under 633 nm injection in the bottom waveguide.[68]

In very recent years integrated photonics has become an attractive technology also for quantum optics. These experiments rely on the possibility to implement an integrated beam splitter, i.e. a directional coupler, which is the building block for all the interferometric elements needed in quantum information processing.

Initial results show the potentials of femtosecond laser writing in place of standard fabrication technologies.[70,71] This topic will be discussed in depth in Section 9.3.4.3.

Mach–Zehnder interferometers Waveguide splitters and directional couplers stand at the base of more complex waveguide interferometers which are commonly used for telecom as well as for sensing applications. The Mach–Zehnder interferometer (MZI) is a device used to determine the relative phase shift between two optical paths. It is usually used to measure small phase shifts in one of the two beams caused by the presence of a small sample of interest or by the change in length of one of the paths. MZIs are also used for signal modulation/routing by the electro-optic effect. The incoming laser light is split in two branches by a Y junction or a directional coupler. One branch (the so-called "sensing arm") passes close to the sample region of interest or close to an active region capable of introducing a controlled phase change. The other branch (the "reference arm") has a fixed length and is used as reference optical path. After this, the two branches are recombined by a Y junction or a directional coupler. After recombination the light intensity strongly depends on the relative phase shift introduced in the two arms, enabling detection of differences in the analysed sample or routing the light by a proper control on the active region.

One of the first demonstrations of such devices by femtosecond laser micromachining was a MZI composed of two X couplers placed back to back with a crossing angle of 2°.[60] The device was fabricated by a 4 MHz Ti:sapphire laser in soda-lime glass. The unbalanced configuration of the MZI, with a path length difference of 10 between the two arms, allowed a fixed wavelength-dependent filtering effect. The measured wavelength transfer function in the crossed interferometer arm was measured by launching a broadband signal and the results showed good agreement with the theoretically predicted wavelength dependence for a path length difference of 9.3 μm. Such a small deviation from the nominal design path length difference was attributed to small translation errors during fabrication and a slight refractive index difference between the waveguides in the two arms of the interferometer.

A very important class of devices is represented by electro-optic modulators. These devices are based on MZIs in which the relative phase shift between the two arms is induced through electro-optic phenomena in the substrate by the application of an external electric field. By combining femtosecond laser writing and thermal poling, an electro-optic modulator based on a 3D MZI was demonstrated in fused silica.[72] A 238 kHz repetition rate laser at 790 nm was employed for the fabrication and the writing speed was 100 μm/s. The device was composed of an input Y splitter and an output directional coupler, both deeply buried in the substrate in order to avoid surface and edge defects. In contrast, the two arms of the MZI were written closer to the surface (about 11 μm) in order to maximize the spatial overlap between the guided mode and the thin nonlinear layer (about 20 μm thick) that was induced by thermal poling after the interferometer was fabricated. Finally, gold electrodes were manufactured on the surface by a sputter coating and lift-off process and the spectral response of the device under different voltages was investigated (Fig. 9.6a). A 1 nm spectral shift at 1.55 μm was obtained with an applied voltage of 400 V, corresponding to an estimated effective electro-optic coefficient of 0.17 pm/V.

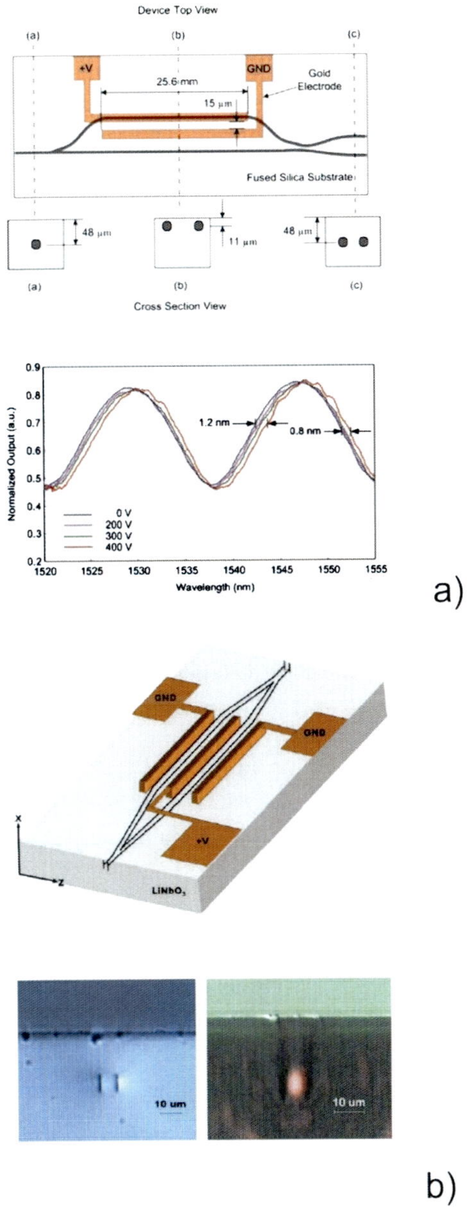

Figure 9.6 (a) Top: schematic of the 3D MZI electro-optic modulator. Bottom: spectral response at 4 different voltages.[72] (b) Top: schematic of the electro-optic modulator in LN. Bottom: microscope image and near-field intensity distribution at the exit of the MZI.[74]

In Section 9.2.2 the importance of LN in the field of electro-optics has already been discussed. At first, a low-loss MZI was fabricated in LN by means of a low-repetition-rate Ti:sapphire laser system.[73] Based on this demonstration, electro-optic modulators were also validated by different groups. By realizing optical waveguides inscribed with a femtosecond laser and by subsequently fabricating embedded microelectrodes using femtosecond laser ablation and selective electrodeless plating, an optical modulator in LN was fabricated (Fig. 9.6b) with a half-wave voltage close to 19 V at a wavelength of 632.8 nm with an interaction length of 2.6 mm.[74] The so-called "type II" optical waveguides were employed to build the MZI; in this kind of waveguides the guiding region results in between two damaged regions induced by femtosecond laser irradiation (see Section 9.2.2). The fabrication process of embedded electrodes mainly consists of a matrix-ablation technique activated by the same 1 kHz repetition rate Ti:sapphire laser source. Such a technique allows one to control the shape and the dimension of the obtained electrodes and to have a nearly uniform electric field across the waveguide.

A further improvement of such a device was also demonstrated, adding the integrated capability of frequency conversion.[75] Again the MZI was obtained by type II waveguiding structures but this time the electrodes were realized by femtosecond laser ablation of a gold layer sputtered onto the sample surface. Enhancement of the modulator extinction ratio was achieved by introducing a narrowing in the waveguide size (named dip) to suppress higher order modes. The frequency doubling was obtained by inserting a periodically poled region in the initial straight waveguide.

Another promising application of MZI is related to optical sensing. In this modality, the sensing arm passes by or directly crosses the region of interest that can be represented by a biological sample to be analysed.[76] The relative phase shift is indeed strongly correlated to the sample properties that in turn can be studied by simply detecting the intensity output signal from the MZI. Such applications are becoming very attractive in microfluidic devices and will be discussed in detail in Section 9.3.4.1.

Multi-mode interference devices A multi-mode interference (MMI) device is based on the self-imaging principle which occurs for multimode waveguides whereby an input field profile is reproduced in single or multiple images at periodic intervals along the propagation direction of the guide. The central structure of an MMI device is a waveguide designed to support a large number of modes. In order to launch light into and recover light from that multi-mode waveguide, a number of access waveguides (usually single-mode) are placed on both sides. Such devices are generally referred to as $N \times M$ MMI couplers, where N and M are the number of input and output waveguides, respectively.[77]

The MMI devices have gained wide usage in power coupling, splitting, switching and coarse wavelength-division multiplexing, because of their simple structure, large optical bandwidth, polarization insensitivity and robustness against fabrication tolerances. For all these reasons, MMI devices are attractive compared with directional couplers and Y branches, in particular for large fan-outs, in which only a single planar MMI waveguide is required, instead of cascading a large number of directional couplers or Y branches. The use of MMI couplers thus makes photonic devices more compact, which is beneficial for the dense integration of photonic circuits.[78]

The first demonstration of an MMI device by means of femtosecond laser writing was achieved in silicate glass by a 1 kHz amplified Ti:sapphire laser using a filamentation process.[79] Waveguide writing was performed in the longitudinal translation configuration and the MMI waveguide was formed by a multiscan technique, allowing precise control of both width and length of the multi-mode region (Fig. 9.7a). The MMI device was characterized by coupling light at different wavelength and observing the output mode profile (Fig. 9.7b); results were in good agreement with theoretical predictions made with a beam propagation method. Moreover, further experiments were performed to study the influence on the mode distribution of both width and length of the MMI waveguide.

A similar MMI device was fabricated in silica glass with a transverse writing geometry.[80] In this case, the MMI slab waveguide had a vertical orientation and was characterized by coupling light at different input positions.

9.3.1.2 Discrete waveguide arrays

Discrete structures play an important role in modern physics such as quantum mechanics and solid-state physics. However, many of the theoretical predictions obtained in these fields are difficult to verify experimentally. Therefore, model systems are required which exhibit similar structures in the mathematical formalism and additionally provide the possibility of experimental verification of

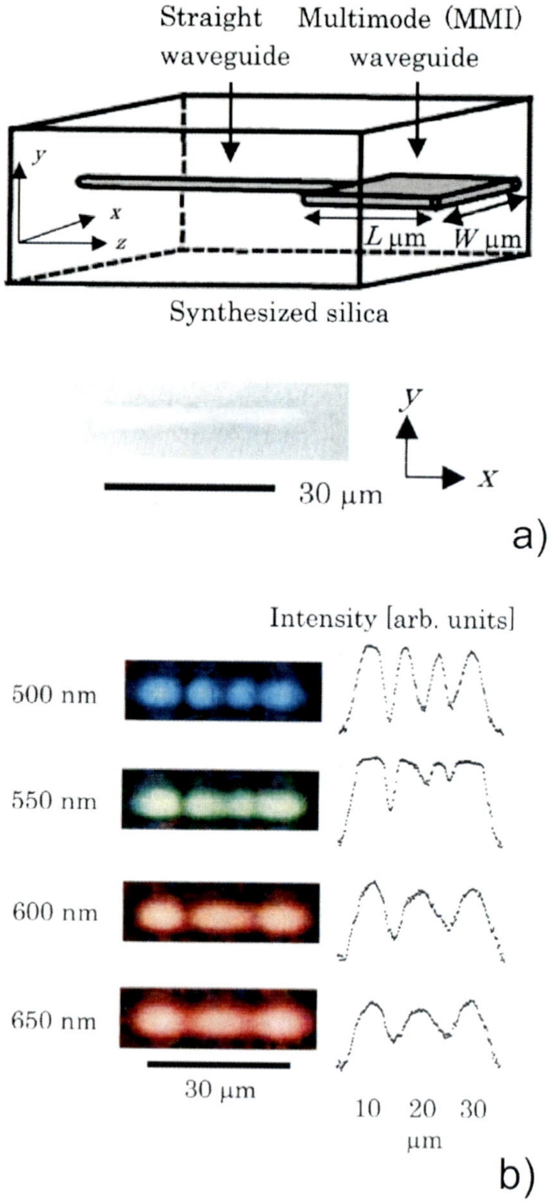

Figure 9.7 (a) Top: schematic of the MMI device. Bottom: optical image of the fabricated MMI waveguide output. (b) Near-field pattern at the MMI waveguide output for various wavelengths. The central wavelengths of the coupled light are indicated.[79]

a variety of effects. One of the most prominent systems are arrays of evanescently coupled waveguides. In contrast to homogeneous media, in a waveguide array the transverse directions are physically different from the longitudinal direction of propagation since the transverse coordinates are not continuous but discrete. This causes a variety of new unusual effects such as unique imaging and propagation properties. Moreover, classical optical waves in photonic structures offer rather unique advantages, such as the ability to directly visualize the evolution of analogous quantum mechanics problems in space rather than in time and the possibility to explore interesting regimes not yet accessible for real quantum systems. Consequently, arrays of evanescently coupled waveguides open a wide field of research, which has been investigated for decades but still provides new perspectives. In this context, the freedom offered by femtosecond laser waveguide writing can be widely exploited to study an almost infinite number of experiments owing to its unique 3D and fast prototyping capabilities.

At first, two-dimensional waveguide arrays were simply used to experimentally describe evanescent-field coupling theory in three dimensions.[81,82] A large array of waveguides were fabricated, one single waveguide of the array was excited by cw laser light at the input and the power distribution at the array output was studied, representing a discrete diffraction experiment (Fig. 9.8). Waveguides were fabricated by a Ti:sapphire femtosecond laser with wavelength of 800 nm and a focusing objective with 0.45 NA. Discrete diffraction was demonstrated in waveguide arrays with both square and hexagonal symmetry, showing a very good agreement with theoretical predictions calculated by a coupled-mode approach. Excitation wavelength and waveguide separation dependences were analysed.

A very powerful technique was introduced to arbitrarily control the coupling coefficient in different directions for different wavelengths. This method allows one to tailor the strength and the sign of the beam diffraction inside the two-dimensional waveguide array with the desired geometry.[83] It is based on the fabrication of periodically curved waveguides in an array. By tuning the period and the geometrical properties of the curved waveguides, it is possible to change the diffraction pattern at the end of the array; many examples were presented, i.e. the suppression of coupling along one or two directions, the self collimation of light beams, the different

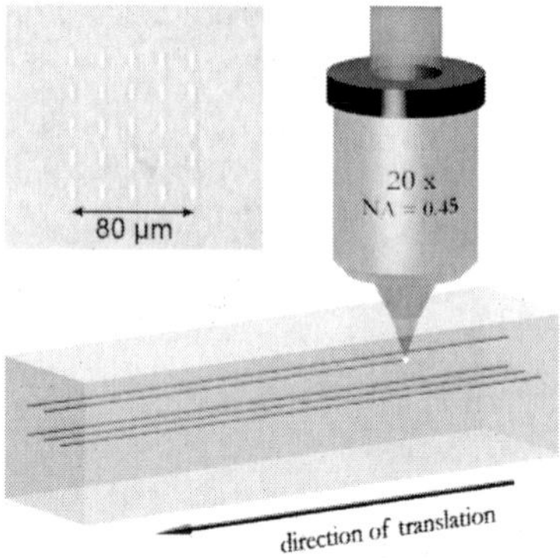

(a)

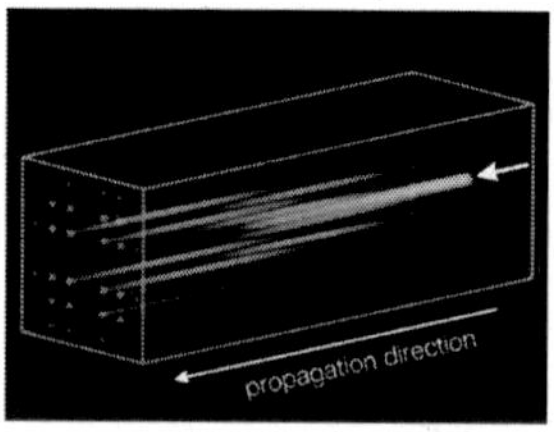

(b)

(c)

Figure 9.8 (a) Schematic of the writing of a waveguide array; inset, microscope image of the end facet of a 5 × 5 array. (b) Simulation of discrete light propagation by excitation of the centre waveguide of a 7 × 7 matrix. (c) Output intensity distributions for excitation of different waveguides of a 7 × 7 waveguide matrix; the arrow indicates the excited waveguide. Comparison between experimental results and simulations is shown.[81]

diffraction response of the same lattice with respect to excitation wavelength. Another way to control the coupling coefficient is to realize waveguides in an array with harmonically modulated refractive index along the propagation direction.[84,85] This is possible by suitably controlling the writing speed during the femtosecond laser irradiation; another important parameter is the phase shift between adjacent waveguide modulations. This technique allows one to exactly engineer the coupling properties (Fig. 9.9). In order to verify this effect by experimental measurements, a special fused silica glass with high content of OH was used. For the direct observation of the propagation pattern from above, the fluorescent light of nonbridging oxygen hole colour centres, which were formed during the writing process, was exploited.

A very interesting phenomenon is the formation of two-dimensional solitons. This is well predicted by the coupling theory, in the so-called discrete nonlinear Schrödinger equation, when k is not negligible; the value k is a measure for the waveguides effective third-order nonlinearity which is determined by the nonlinear refractive index n_2 of the material. The formation of a soliton corresponds to a discrete localization in waveguide arrays when excited with a high peak power obtained by a pulsed laser source. For low peak excitation power the nonlinear effects are negligible; instead, for high peak power soliton formation was demonstrated by observing a localization, or "trapping", of light in the excited waveguide showing very reduced or even inhibited coupling to neighbouring waveguides (Fig. 9.9).[86,87]

Discrete waveguide arrays with arbitrary 3D layouts have been recently employed to model physics problems such as discrete system response, physics of solid-state systems and quantum mechanics experiments. A problem that has been studied is the periodic motion of an electron wave packet in a crystalline lattice subjected to an external field, known as Bloch oscillations. The optical counterpart of Bloch oscillations can be realized in waveguide arrays with a transverse ramp of the refractive index.[88] Such ramp can be created by a curved waveguide array where a constant curvature induces a linear index ramp, which is equivalent to a linear potential.[89] Experiments were performed using Er/Yb-doped phosphate glass and the green upconversion fluorescence emitted by excited erbium ions was exploited to image the light propagation along the array. A further step forward was represented by the demonstration of Bloch–Zener oscillations in binary lattices, important because Bloch oscillations and Zener tunnelling are fundamental transport phenomena of electrons in periodic potentials.[90]

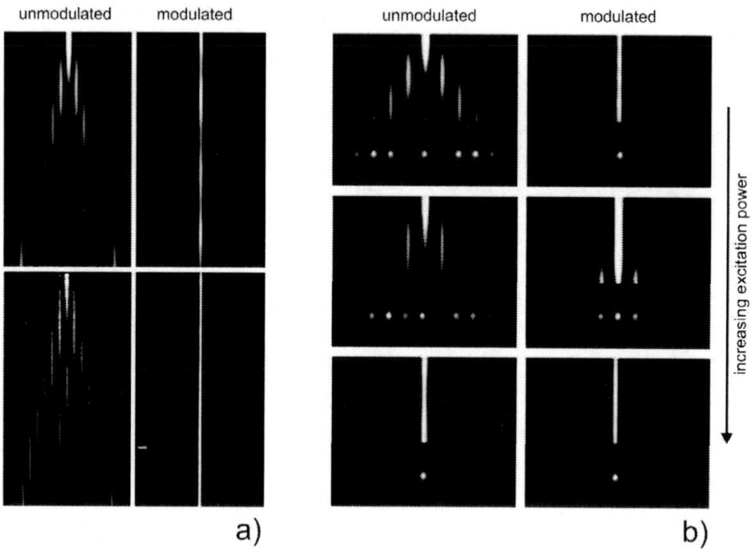

Figure 9.9 Light localization in waveguide arrays can be achieved either by tailoring the waveguide properties or by increasing the excitation power (soliton formation). (a) Theoretical (top) and experimental (bottom) intensity distributions in an array for low power excitation. Left panels show unmodulated arrays; right panels show an array with modulated refractive index such to obtain light localization in the central waveguide. (b) Left panels show an unmodulated waveguide array where localization is achieved by increasing the excitation power; right panels show modulated waveguide array responses for increasing excitation powers.[85]

The dynamic localization is an interesting phenomenon that has its origin in solid-state physics. In optical waveguide systems, the external field is mimicked by a curvature of the waveguides itself. Dynamic light localization was achieved in periodically bent waveguide arrays due to the different effective paths in the individual guides caused by the bending.[91–93]

A very interesting work was focused on the theoretical discussion and experimental demonstration of the Snell's and Fresnel's laws in discrete optical media.[94] The interface between two different substrates was realized by two waveguide arrays with different coupling coefficient. A very detailed characterization was performed to study refraction and reflection at the interface in different conditions. Again, the observation of the light propagation was achieved by imaging the fluorescent light from colour centre from above.

Waveguide arrays have also been used to simulate quantum optics phenomena. Photon number correlations in a variety of two-dimensional optical waveguide lattices were studied to characterize the quantum interference of photons.[95,96] The study reported the observation of photon bunching, meaning that two independent photons launched into the array can become path-entangled, while entangled photons can anti-bunch and become spatially separated for certain configurations. In particular, the two-dimensionality of the structures supports a variety of complex manipulations on the quantum state of light, including combinations of entangled and separable output states.

9.3.1.3 Waveguide Bragg grating devices

Femtosecond laser writing has been successfully applied to Bragg grating fabrication. The interface between media with different refractive indices creates a reflection of light (the so-called Fresnel reflection). By alternating media with different refractive indices the reflected light at each interface interferes and, at suitable conditions, this gives rise to a high portion of reflected light at a specific wavelength, controllable by designing the refractive index variation and the period. Bragg gratings are achieved by a periodic variation of the refractive index of the substrate to obtain a desired interference pattern. Therefore, they can be used as inline optical filters to block certain wavelengths or as wavelength specific reflectors.

The feasibility of femtosecond laser writing for the fabrication of high-quality waveguide Bragg gratings (WBGs) at telecom wavelengths (1.5 μm) was demonstrated. Since the latter was a remarkable step towards the development of integrated photonic devices, this section is focused on WBGs.

In one of the first works, a weak high-order WBG was fabricated in a reverse proton exchanged LN channel waveguide by means of a regeneratively amplified Ti:sapphire laser.[97] The grating was obtained by exposing the channel waveguide under the interference fringe pattern generated by a phase mask. With the same technique, WBG fabrication in borosilicate ion exchanged waveguides was also reported,[98] and a detailed investigation on optimal writing parameters demonstrated the feasibility of moderately strong Bragg gratings with reflection bandwidth (3 dB) as narrow as 0.76 nm. In these cases, femtosecond laser micromachining was employed to add Bragg grating functionality to optical waveguides fabricated with other technologies.

Another possibility is to fabricate both the waveguide and the Bragg grating by direct laser inscription. The first femtosecond laser-written optical waveguide that included a grating structure was demonstrated in fused silica by exploiting a point-by-point writing technique.[99] The process was actually a double-step writing method in which the laser pulses from a low-repetition-rate Ti:sapphire laser were scanned twice over the same glass sample region. In the first writing step the pulse energy was as high as 2 μJ, the translation speed as slow as 20 μm/s and a linear channel waveguide was obtained. In the second writing step the pulse energy was attenuated to 0.4 μJ and the sample was translated at 1.07 mm/s. Such a high writing speed partially removes the spatial overlap between consecutive pulses, giving rise to a periodic modification of the underlying waveguide, resulting in a refractive index grating. Furthermore, with the grating pitch being determined by the spatial distance between subsequent pulses, the period of the grating is simply given by the ratio between the writing speed and the repetition rate of the laser.

Point-by-point femtosecond laser writing of WBGs was further developed in a single step process for simultaneous waveguide and Bragg grating formation.[100] Such a periodic-structured waveguide exhibited both low propagation losses and inherent Bragg grating response. With the same single scan technique, ultrafast laser pulses from a low-repetition-rate Ti:sapphire laser system induced an array of refractive index voxels. By optimizing the laser processing conditions, strong first-order WBGs at telecom wavelength were demonstrated in borosilicate glass.[101] Tunability of the resonance in the 1200–1620 nm wavelength range was also reported as well as multiple wavelength and cascaded WBGs.

Similar quality WBGs were also fabricated in fused silica by means of a slightly different single-step femtosecond laser writing method employing a higher repetition rate laser system with an external acousto-optic modulator.[102] Actually, in weakly photosensitive glasses like fused silica, the single pulse technique was ineffective in generating a strong refractive index change; an external modulation allowed multiple ultrafast pulses, instead of a single pulse, to induce each index voxel, thus increasing the induced refractive index change (Fig. 9.10). The grating period is here given by the ratio between the writing speed and the intensity modulation frequency. The writing conditions were first optimized to achieve low loss continuous waveguides, i.e. without intensity modulation. Then the WBGs were fabricated by varying the duty cycle of the acousto-optic modulator from 20% to 100% and characterized both in terms of insertion loss and spectral response. The grating demonstrated narrow bandwidth (0.2 nm) and low radiation mode losses (2 dB).

Chirped WBGs of up to 20 nm bandwidth were also demonstrated in fused silica by a single-step point-by-point burst-writing method.[103,104] It is worth noting that these femtosecond laser-written WBGs presented comparable or even better performance, in terms of strength and bandwidth narrowness, with respect to WBGs fabricated by another single step technique such as continuous-wave UV writing, which requires germanium doping of the glass substrate and a more complex setup to focus the two interfering UV beams.[105]

An important application was demonstrated by using WBG as an optical sensor network inside bulk glass that directly measures environmental parameters such as temperature and strain, including their 3D distribution.[106] The reflectance peak position of the Bragg grating is sensitive to the chip temperature and the substrate strain; several WBGs were fabricated in different positions on the same fused silica glass to have a 3D map of temperature and strain with high spatial resolution. The thermo-optic and strain-optic responses of WBG devices were reported and found to be similar to traditional FBG responses in silica fibre. Further, thermal annealing revealed a high stability of both optical guiding and Bragg resonance for these devices, which is ideal for high temperature sensing environments. Such 3D grating networks open new directions for strain and temperature sensing in optical circuits, MEMS, actuators, windows and other large display or civil structures.

WBGs were also used as sensors for the measurement of refractive index of liquids. On a fused silica substrate, a microchannel was fabricated by femtosecond laser irradiation followed by chemical etching in hydrofluoric acid solution (see Chapter 10), and with the

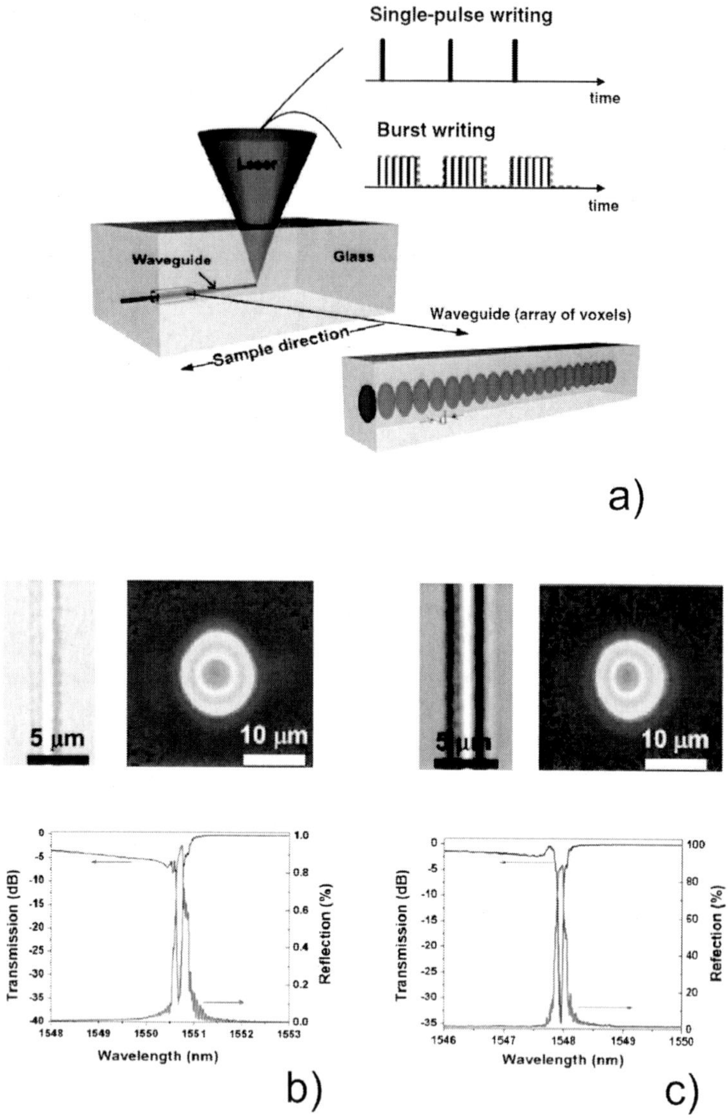

Figure 9.10 (a) Comparison between single-pulse writing and burst writing method of Bragg gratings. (b) Microscope images, 1550 nm guided mode profiles (top), transmission and reflection spectra (bottom) for a single-pulse type II WBG with grating length of 12 mm. (c) Microscope images, 1550 nm guided mode profiles (top), transmission and reflection spectra (bottom) for a burst WBG with grating length of 25 mm.[103]

same femtosecond laser source a WBG was inscribed running parallel to the microchannel.[107] Evanescent field penetration of the waveguide mode into the parallel microfluidic channel induced Bragg resonant wavelength shifts to enable refractive index characterization of the fluidic medium in the 1 to 1.452 range. Therefore, by measuring the reflection spectrum from the Bragg grating device, it is possible to obtain optical sensing in 3D optofluidic and reactor microsystems.

9.3.2 Active Devices

Femtosecond laser micromachining has been applied to several active materials suitable to demonstrate active photonic devices such as waveguide amplifiers and lasers. Typical active ions are ytterbium (Yb^{3+}) and erbium (Er^{3+}), due to their good properties suited for the telecommunication band. In particular, Er-doped and Er/Yb-doped glasses allowed the demonstration of waveguide amplifiers operating in the whole C-band (1530–1565 nm), as well as tunable lasers and single-mode or mode-locked laser operation.

The majority of active waveguide devices fabricated using the femtosecond laser direct-write technique has been reported in the following materials: Nd-doped silicate glass, Nd-doped YAG matrix, Er/Yb co-doped phosphate glass, Er-doped and Er/Yb co-doped oxyfluoride silicate glass. These samples were specifically chosen due to their target application, region of wavelength emission and ease of doping with rare-earth ions. Among these materials, the phosphate glass hosts are best suited to active device fabrication for use in telecommunications as tens of percent by weight of rare-earth ions can be held in solution offering the highest gain-per-unit length value in the C-band without detrimental effects such as ion clustering.[108]

9.3.2.1 Waveguide amplifiers

Optical amplifiers are achieved by inserting (doping) an active material inside a host passive transparent substrate. The active elements are usually atoms able to absorb photons at the pump wavelength to invert the electronic populations; then, by stimulated emission, the electrons decay by emitting photons at the signal wavelength to be amplified.

The first active device manufactured by femtosecond laser writing was an active Nd-doped waveguide. The waveguide was fabricated in a commercially available Nd-doped silicate glass rod by means of a low-repetition-rate Ti:sapphire laser.[109] According to the ON-OFF method, 1.5 dB/cm internal gain was measured at 1054 nm for an incident pump power of 350 mW at 514 nm (provided by an Ar-ion laser). Nd-doped systems are very efficient four-level active media, but unfortunately result in being unsuitable for applications at telecom wavelengths. Therefore, much effort has been devoted to femtosecond laser writing in other active glasses, with particular interest in Er-doped and Er/Yb-doped glasses allowing operation in the C-band.

A phosphate glass (Kigre QX) was found to be suitable for femtosecond laser writing with a low-repetition-rate system and allowed the first demonstration of an active waveguide exhibiting internal gain of 0.7 dB at 1557 nm.[110] The waveguide was 25 mm long and diode-pumped at 980 nm under a bi-propagating scheme (Fig. 9.11a). The single-mode guided field was approximately Gaussian but significantly large and asymmetric compared with the mode field supported by a standard single-mode fibre, resulting in high coupling losses. By exploiting an astigmatic shaping of the writing laser beam, active waveguides with a more symmetric refractive index profile were obtained in the same substrate, allowing lower coupling losses to the single-mode fibre and consequently higher gain efficiency in the whole C-band.[30] However, a real waveguide amplifier able to provide net gain was still lacking.

A significant improvement was demonstrated by exploiting a cavity-dumped Yb oscillator. In this intermediate regime the waveguide writing process exploits both the positive effects of moderately high translation speed and high pulse energy, as previously discussed in Section 9.2.1.2. The same phosphate glass was used, but the doping concentrations were optimized in order to obtain higher gain per unit length and longer devices. The cavity-dumped system led to high optical quality of the waveguides: coupling losses of 0.2 dB/facet and propagation loss of about 0.3 dB/cm. Thanks to these improvements, a net gain in femtosecond laser-written waveguides was achieved for the first time.[111] Also, as a full demonstration of the feasibility of a real device, a 37 mm-long waveguide amplifier operating in the whole C-band of optical communications was fabricated and characterized in terms of insertion losses, spectral gain, saturation and noise figure.[112] A pair of 975 nm wavelength InGaAs laser diodes in a bi-propagating pumping scheme were adopted to supply up to 460 mW incident pump power through 980/1550 nm wavelength division multiplexers (WDMs). Signal radiation at 1.5 μm was provided by a tunable laser diode. The overall insertion loss was 1.9 dB and a net gain was demonstrated in the whole C-band with a peak net gain of 7.3 dB at 1535 nm. Note that the peak internal gain per unit length turns out to be 2.5 dB/cm, and thus is comparable to the value of 3 dB/cm reported in state-of-the-art Er/Yb-doped waveguide amplifiers fabricated by ion exchange in a phosphate glass with the same doping concentration.[113]

Erbium-doped oxyfluoride silicate glass is a particularly promising host for rare earth ions because it combines the attractive spectroscopic properties of fluoride glass (similar to phosphate) with the structural compatibility and stability of silica. Femtosecond laser micromachining was applied to Er-doped oxyfluoride glass, demonstrating the capability of obtaining internal gain. The waveguide amplifier was fabricated by a Ti:sapphire laser system at 5 kHz repetition rate; however, both high propagation losses and coupling losses at 1550 nm prevented an efficient pumping.[114] Improved results were obtained by exploiting the multiscan technique, in which the desired waveguide cross section was obtained by writing many lines of modified material slightly shifted with respect to each other.[24]

The feasibility of a waveguide amplifier in this oxyfluoride silicate glass (co-doped with Yb to enhance the pumping efficiency at 980 nm) was conclusively demonstrated by a net gain in a 10 mm-long waveguide.[115] The low-repetition-rate Ti:sapphire laser was replaced by a cavity-dumped Yb:glass laser, yielding lower propagation losses, and a multiscan technique was adopted to produce waveguides with a square cross section, exhibiting good mode field matching with a standard single-mode fibre (Fig. 9.11c). A peak net gain of 0.72 dB under double-wavelength pumping was measured (Fig. 9.11b). The same cavity-dumped Yb:glass laser also allowed the fabrication of high-quality femtosecond laser-written waveguides in another promising active substrate, a Bi-doped silica glass,[116] which is known to exhibit broadband emission around 1.3 μm. At 980 nm pump wavelength a fluorescence bandwidth of 500 nm was reported.

Finally, femtosecond laser writing has also been employed to fabricate waveguide amplifiers in several other glasses.

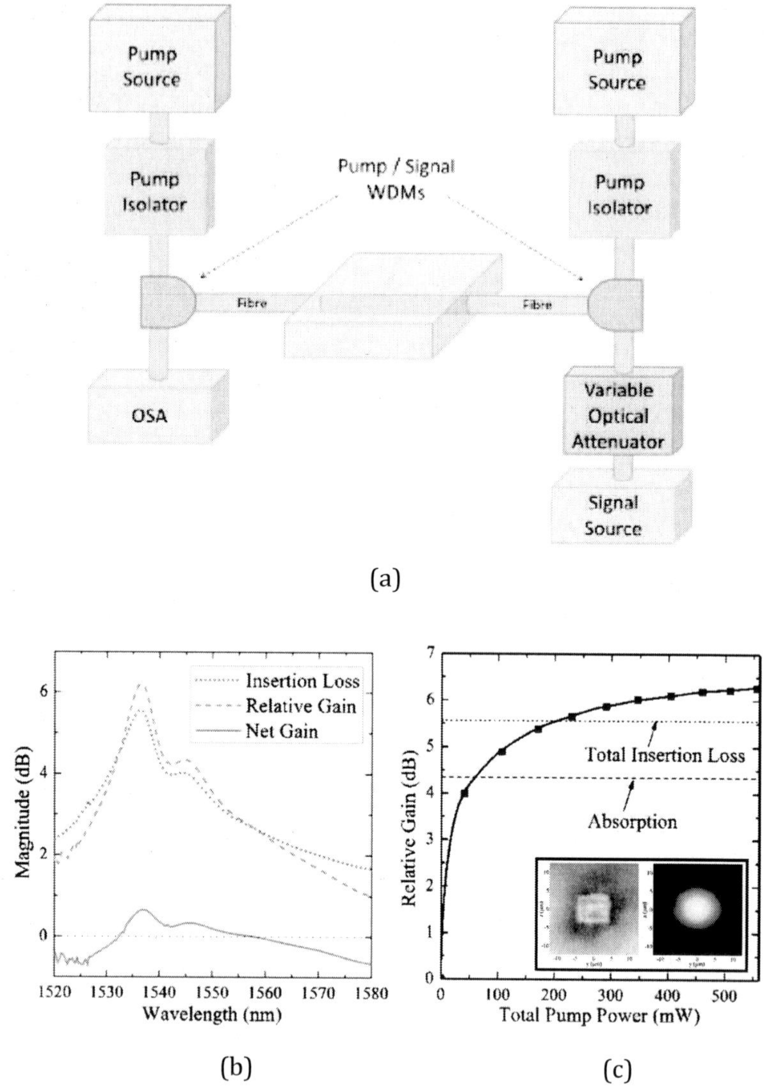

Figure 9.11 (a) Typical characterization setup for active waveguide devices; OSA—Optical Spectrum Analyser.[108] (b) Insertion loss, relative gain and net gain spectra in Er/Yb doped oxyfluoride silicate glass waveguide. (c) Relative gain vs. pump power at the peak of the gain spectrum using dual wavelength pumping; the inset shows microscope image and near-field mode image of the optimal waveguide.[115]

Micromachining with a low-repetition-rate Ti:sapphire laser was successfully applied to a new Er-doped tellurite glass modified by phosphate and aluminium.[117] Recently, by using a cavity-dumped Yb-based writing system, active waveguides exhibiting internal gain in the whole C + L band of optical communications (1530–1610 nm) have also been demonstrated in an erbium-doped phospho-tellurite glass.[118,119] An Er/Yb co-doped P_2O_5–La_2O_5-based glass was synthesized and used for producing 1.6 cm-long active optical waveguides using a low-repetition-rate Ti:Al_2O_3 femtosecond laser. The measurements provided internal gain figures comparable to the best ones reported in phosphate glasses for low-repetition-rate writing.[120]

9.3.2.2 Waveguide lasers

Waveguide lasers are based on the same principle of waveguide amplifiers, with the fundamental addition of a closed optical cavity (or resonator) for the signal wavelength. In order to achieve laser action, the amplifier must provide net gain. The cavity can be formed by applying different approaches: (i) external dielectric mirrors; (ii) external Bragg reflectors such as fibre Bragg gratings (FBGs) acting as cavity end mirrors; (iii) distributed feedback (DFB) designs achieved by the inscription, with the same femtosecond laser source, of waveguide Bragg gratings (WBGs) along the active waveguide itself, providing frequency-selective feedback distributed over the entire length of the laser medium instead of being localized at the cavity ends.[108]

Using the second approach, i.e. external Bragg reflector cavity, several waveguide lasers have been demonstrated, reaching a good quality in terms of lasing threshold power and slope efficiency. Moreover, this system allows for a simple change of the laser wavelength by simple substitution of the end gratings. The first femtosecond laser-written waveguide laser was fabricated in an Er/Yb-doped phosphate glass (Kigre QX) by means of a frequency-doubled cavity-dumped Yb:glass laser, operated at 166 kHz repetition rate.[121] The waveguides provided 2.3 dB peak net gain at 1533.8 nm under 420 mW pump power at 980 nm wavelength in a bi-propagating pumping scheme. The laser cavity was obtained by means of two FBGs with peak reflectivity at 1533.5 nm butt-coupled to the waveguide: a 99.8% reflectivity FBG at one side and a 68%

reflectivity FBG as output coupler. A maximum output power of 1.7 mW with 2% slope efficiency was reported. A significant improvement of the waveguide quality and a laser cavity optimization were responsible for higher laser performance. A maximum output power of 30 mW with 8.4% slope efficiency was obtained at 1533.5 nm and 23 mW output power with 6.6% slope efficiency was also reported at 1560 nm, thus ascertaining laser action and wavelength tunability in the whole C-band.[111]

In the previous experiment, the laser cavity was longer than 50 cm and the output coupler had a bandwidth of 0.25 nm, thus causing a spectral broadening of the source due to multimode oscillation. To achieve single-mode operation, the output coupler was replaced by a narrow bandwidth FBG and the cavity length was reduced to 5.5 cm by cutting the FBG fibres.[122] Such a compact cavity resulted in a reduced number of modes falling within the FWHM bandwidth of the FBG, thus producing a single-mode laser up to a maximum output power of about 55 mW.

A laser device was also reported in Er/Yb-doped oxyfluoride silicate glass by exploiting the high-quality waveguides previously demonstrated that exhibited net gain.[123] The laser cavity was based on FBGs and a dual-wavelength bi-propagating pumping scheme. A maximum output power of 30 μW was demonstrated in multimode operation. Recently laser action from Yb:KGW and Yb:KYW femtosecond-written channel waveguides was also demonstrated.[42] As very often happens in crystals, guiding occurred in regions surrounding the irradiated focal volume. The maximum output power achieved was 18.6 mW with a threshold of approximately 100 mW from Yb:KGW at 1023 nm in a diode-pumped compact cavity.

Another demonstration of advanced functionality in a femtosecond laser-written waveguide laser was the feasibility of passive mode-locking operation.[124] Classically, the technique of passive mode-locking employs semiconductor saturable absorber mirrors and allows the generation of picosecond and femtosecond laser pulses; recently, a new technology based on carbon nanotubes (CNTs) has emerged as an alternative for the realization of saturable absorbers with comparable performances. CNTs specially designed to have efficient saturable absorption at 1.5 μm were embedded in free-standing polyvinyl alcohol films of 50 μm thickness; by sandwiching such a film between a couple of ferrule plane connectors (FC/PC) a fibre-pigtailed CNT mode-locker was packaged, with the waveguide amplifier written in an Er/Yb-doped glass. All these components were assembled in a ring cavity configuration (Fig. 9.12a). Continuous-wave laser action and self-starting single-pulse stable mode-locking was observed just above the laser threshold (Fig. 9.12b). The duration of transform-limited sech^2 pulses was estimated to be about 1.6 ps (Fig. 9.12c); the repetition rate of the cavity was 16.7 MHz. Recently, exploiting exactly the same ring cavity configuration, a mode-locked Er-doped bismuthate glass waveguide laser was demonstrated.[125] Mode-locking was again initiated and stabilized by the use of a single wall CNT saturable absorber; the waveguide laser produced 320 fs pulses at 1.56 μm with a pulse repetition rate of 40 MHz and average output power of 1.25 mW.

An improvement in power stability could be achieved by using the third approach listed at the beginning of this section, i.e. by fabricating the Bragg gratings in the active substrate leading to a monolithic DFB laser configuration. This approach exploits the already well-established capability of fabricating WBGs by femtosecond laser inscription (see Section 9.3.1.3). Moreover, the frequency selectivity of the Bragg effect predestinates the DFB for mode selection in lasers with broadband gain media. The laser output wavelength can be tuned by altering the temperature of the device, since it causes a grating pitch change due to thermal expansion. Recently, the feasibility of such a DFB laser by femtosecond laser micromachining was demonstrated.[126] The active substrate was an Er/Yb-doped phosphate glass (Kigre QX). A 20 mm long waveguide and a superimposed Bragg grating were fabricated in a single step writing process by using a low-repetition-rate Ti:sapphire laser (Fig. 9.13a). The DFB laser device demonstrated stable single-mode operation at 1537.627 nm wavelength and delivered 0.36 mW maximum output power under 710 mW incident pump power (Fig. 9.13c) in a bi-propagating pumping scheme (Fig. 9.13b). A higher performance version of the same laser was also presented. A femtosecond-laser-written monolithic DFB waveguide laser was fabricated in Yb-doped phosphate glass.[127] The device lased at 1032.59 nm with an output power of 102 mW and a bandwidth of less than 2 pm when bidirectionally pumped at 976 nm. The pump power threshold was approximately 115 mW and the optical efficiency was over 17%.

Laser action was demonstrated also on lithium fluoride (LiF) crystal. A hologram technique was developed to encode grating

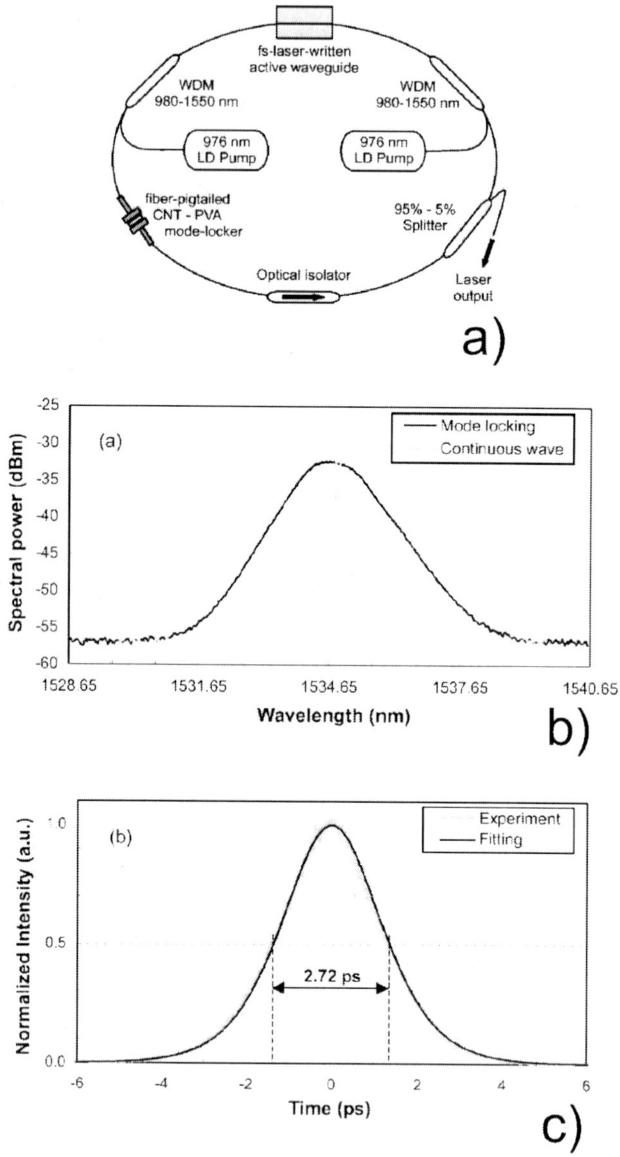

Figure 9.12 CNT-mode-locked femtosecond-written waveguide laser in an Er/Yb-doped glass. (a) Schematic setup of the ring cavity. (b) Laser output spectrum in continuous wave and mode-locked regimes. (c) Autocorrelation trace of the mode-locked pulse.[124]

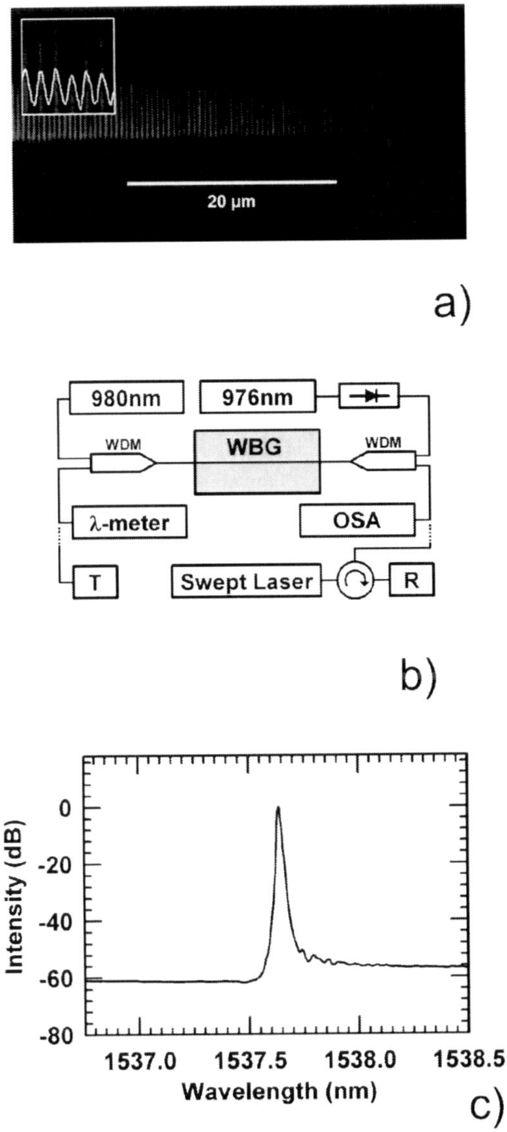

Figure 9.13 Monolithic DFB femtosecond-written waveguide laser. (a) Transmission DIC micrograph of the first-order WBG. (b) Waveguide laser pumping and characterization setup. (c) Output spectrum of the waveguide laser at maximum pump power; the ordinate axis scale is referenced to the laser peak.[126]

structures inside a LiF crystal by a single interfered femtosecond laser pulse.[128] The resulting periodic refractive index contrast was attributed to the formation of F_2 colour centres. When side-pumping the grating with 450 nm light, a DFB laser oscillator at 707 nm was created. However, the grating structure is not permanent as the colour centres can be eliminated by annealing. In addition, the linewidth was significantly broader than that normally expected in DFB lasers.

The laser devices reviewed above are all based on Er or Er/Yb laser media. The very good spectral properties of erbium make Er-based active devices the best candidates for several photonic applications. However, femtosecond laser micromachining has also been applied to neodymium (Nd) activated materials demonstrating efficient waveguide lasers at 1.06 μm due to the intrinsically very efficient four-level laser scheme offered by Nd. The first femtosecond laser-written Nd-doped waveguide laser was fabricated by means of a low-repetition-rate Ti:sapphire laser in an yttrium aluminium garnet (YAG) crystal.[129] A depressed cladding-type waveguide was fabricated by writing parallel tracks around the unmodified central volume serving as a core. The waveguide facets were coated for 1064 nm wavelength with a highly reflective dichroic coating on one side and an antireflective one on the other side; a flat mirror directly attached to the antireflective coating provided 24% output coupling. A laser threshold as low as 30 mW and conversion efficiency of 11% were demonstrated.

Recently an efficient femtosecond laser-written waveguide laser has been demonstrated in a Nd:YAG ceramic material.[130] YAG ceramics have recently attracted much interest because of the lower production cost and the possibility of higher doping concentration with respect to YAG crystals. Again, the waveguide was fabricated by exploiting the negative refractive index change induced in this ceramic material. A flat dichroic dielectric mirror was attached to one end-facet of the waveguide chip and 8% Fresnel reflection at the other end-facet provided optical feedback. This waveguide cavity delivered up to 80 mW output power in stable single-mode operation at 1064 nm. A remarkable 60% slope efficiency was demonstrated, which is comparable to state-of-the-art Nd-based waveguide lasers so far reported by other fabrication techniques.

9.3.3 Fibre-Based Devices

Fibre Bragg gratings and long-period gratings (LPG) are among the most used fibre devices. Conventional fabrication techniques make use of UV light to inscribe the grating into the fibre, exploiting photosensitivity of the material. Phase masks are commonly used for the inscription of short-period FBG, while the LPG are often produced through a point-by-point irradiation. Multiphoton absorption of pulsed light at longer wavelengths from a femtosecond laser can give some important advantages over the conventional technology.[131] First, while UV light is strongly absorbed by optical elements and even by air, the infrared beam of an ultrafast laser (or the visible-near UV wavelengths of its harmonics) can be easily manipulated. Second, it has been found that gratings inscribed by multiphoton interaction are much more stable over long time scales and can be exposed to high temperatures[132] without any degradation.

Starting from the seminal work of Kondo *et al.*[132] (1999), regarding femtosecond laser fabrication of LPG by point-by-point irradiation, several techniques and methods have been developed,[131] both for manufacturing LPG and FBG. These techniques can mainly be divided into two categories. In the first one, the femtosecond laser beam is adopted as a practical substitute of the conventional UV laser, thus mildly focused and used together with a phase mask in the case of FBG. In the second one the femtosecond pulses are tightly focused into the fibre core. Due to the small volume irradiated in a single exposure, a phase mask is not necessary and even FBG can be inscribed point-by-point.

In the following, we will concentrate on the latter category, since it is the one in which the three-dimensional capabilities of femtosecond laser processing are fully exploited.

In the first work by Kondo *et al.* femtosecond pulses from a Ti:sapphire laser, at 200 kHz repetition rate, were focused into the fibre core by a 0.46 NA objective, for a spot size in the focus of the order of 2–3 μm. Each spot was irradiated for 10 s, and by translating the fibre an LPG with 460 μm period was inscribed. The transmission spectrum showed some complex features, due to irregularities in the irradiation, anyway clear loss peaks caused by the LPG were observed, showing the validity of the technique. Interestingly, it was also found that the grating did not suffer any degradation for thermal treatment up to 500°C, whereas the conventional UV written gratings begin to degrade even at temperatures below 100°C. This thermal stability was attributed to a different origin of the index increase in the case of femtosecond pulses irradiation with respect to the UV case.

Martinez et al.[133] in 2004 used pulses from a Ti:sapphire amplified oscillator (150 fs pulse-length and 1 kHz repetition rate), with energies of hundreds of nJ, tightly focused with a 100× microscope objective, to demonstrate the first FBG, fabricated with a point-by-point technique. The fibres employed were standard telecommunication fibres and dispersion shifted fibres, with no prior procedure of photosensitization.

The tight focusing was necessary in order to obtain a spot size sufficiently small, so that each laser pulse produced a grating pitch in the fibre core. The grating period can be set by changing the translation speed of the fibre under the laser beam.

It has to be noticed that, given the high numerical aperture employed, the laser pulses modify a very small volume in the fibre, and the localization of refractive index modulation is further emphasized by the nonlinear nature of the process. This requires careful control of the fibre position under the objective. On the other hand the grating can have a cross section of only 1–2 μm, much smaller than that of the fibre core. This allows also the production of FBG which are not centred with respect to the fibre axis, and thus can be sensitive to bending of the fibre, as was indeed realized[134] by Martinez *et al.* shortly after their first work. The grating was inscribed away from the centre of the fibre, as shown in Fig. 9.14a, so that bending the fibre causes a stretching or a compression of the grating, depending on the bending direction (convex or concave). Measuring the transmission spectrum one can observe a shift in the reflection peak of the grating, depending on the variation in the period caused by the stretching or compression action.

One of the main advantages of the point-by-point approach is the possibility to fabricate gratings with a complete control of the position of each index modification that comprises the grating. Tailoring the local phase, amplitude and spacing of the grating's refractive index modulations has made possible to create chirped gratings or, in general, non-uniform gratings with complex transmission spectra. This was demonstrated experimentally very recently by Marshall *et al.*[135] Pulses of 150 fs duration and energies between 120 and 250 nJ from a Ti:sapphire laser system were focused through 0.8 NA oil-immersion objective, into the core of a standard telecommunication fibre. Modulations of the grating periodicity and phase were realized by acting on the function generator that operated as the laser's master clock. For example, sweeping linearly the repetition rate of a small amount about the 1 kHz base frequency enabled the fabrication of a chirped grating. Controlling the trigger signal of the laser amplifier, so to suppress or delay the emission of the pulses with determined periodicity, enabled also the fabrication of amplitude or phase modulated gratings (see Figs. 9.14b,c). A novel technique for amplitude modulation of the grating was also demonstrated, which is based on controlling the lateral displacement of the fibre with respect to the laser focus position, synchronizing it with the motion of the fibre. In this way the position of the grating can be continuously adjusted laterally from the centre of the fibre core (see Fig. 9.14d), causing a modulation in the overlap between the propagating mode and the grating. A modulation in the coupling is equivalent to a modulation in the amplitude of the grating. This technique was exploited to create a grating which exhibited a series of 11 narrowband resonances spaced by 50 GHz (or approximately 0.40 nm at 1545 nm).

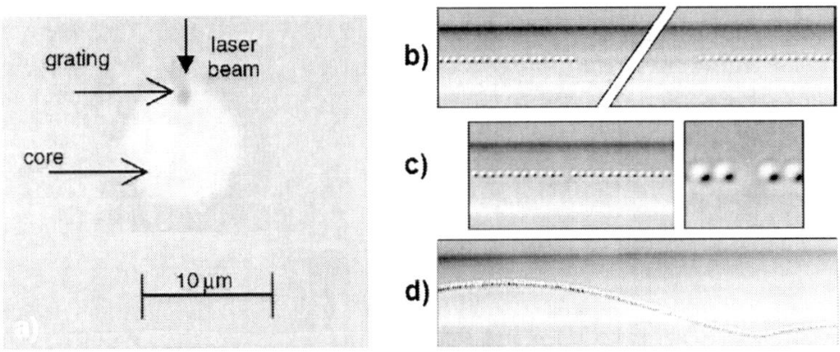

Figure 9.14 (a) Cross section of an optical fibre with FBG fabricated away from the core centre, in order to obtain a fibre bending sensor.[134] (b–d) Microscope image of a fibre in which different gratings are inscribed, respectively modulating amplitude, phase and lateral displacement.[135]

9.3.4 Integrated 3D Devices

In previous sections a wide range of integrated optical components fabricated by femtosecond laser writing have been reviewed, which may be seen as fundamental building blocks of more complex photonic circuits. In addition, the striking microfabrication capabilities of the present technique can be exploited to integrate complete devices, having as a starting point a simple plain glass substrate and as the only manufacturing tool the femtosecond laser. These devices may incorporate more than one of the building blocks presented above. A few examples will be reviewed in the following section.

9.3.4.1 Optofluidic devices

The capability of femtosecond laser to integrate in the same chip both photonic components and microfluidic channels (via irradiation followed by chemical etching), makes this technique very promising for the prototyping and fabrication of complete optofluidic devices in glass chips[136,137] (see Fig. 9.15a). It can also be applied to realize photonic components on chips where a network of microchannels has already been fabricated by other techniques, showing unprecedented postprocessing possibilities.[138] Biological applications have been successfully demonstrated, such as DNA fragments analysis[139] or optical stretching and deformation measurement of living cells.[140]

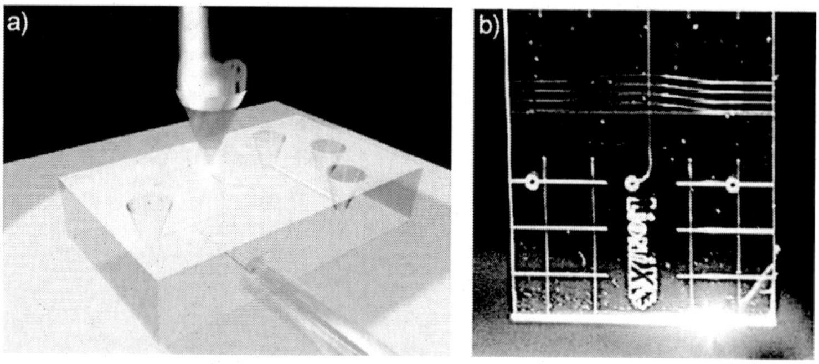

Figure 9.15 (a) Schematic of femtosecond laser waveguide writing in a microfluidic device. (b) Four Mach–Zehnder interferometers fabricated in a commercial microfluidic chip.[141]

Due to the relevance of this application field and the broad range of examples already present in the literature, the entire Chapter 10 of this book is devoted to illustrate the capabilities of the technique in realizing optofluidic devices. It is anyway worth noting here that, due to the need of interfacing optical components with microchannels realized in the bulk of the chip, the three-dimensional capabilities of femtosecond laser microfabrication are particularly interesting.

One noteworthy example is the integration of an unbalanced Mach–Zehnder interferometer[141] with one arm (the sensing arm) directly crossing a microfluidic channel, while the reference arm passes above it. Y splitters are adopted to separate and recombine the two arms. The device geometry, lying on a tilted plane, is impossible to obtain with any conventional photolithographic technology. Integration of the optical device is shown both in a commercial microfluidic chip (see Fig. 9.15b), where channels were already present, fabricated by conventional lithographic techniques, and in a home-made chip, where microchannels are also fabricated by femtosecond laser. The interferometer's waveguides are inscribed in fused silica, using the second harmonic (515 nm wavelength) of 350 fs pulses from a cavity-dumped Yb:KYW laser oscillator, emitted at 1 MHz repetition rate. 90 nJ pulses are focused through a 0.6 NA objective, while the sample is translated at 100 μm/s constant speed. The interferometer was successfully used to sense the refractive index variations of the content of the microchannel, down to 10^{-4} RIU. Perspective applications include label-free sensing in electrophoresis, where one can take further advantage of the spatial resolution in the microchannel, given by the direct-crossing geometry.

9.3.4.2 Optomechanical devices

Femtosecond laser irradiation followed by etching, which allows the production of optofluidic devices in a straightforward fashion, can also be employed to fabricate monolithic micro-optomechanical systems (MOMS). As shown in a work by Bellouard *et al.*[142] a complex flexure-based micro-mechanism can be etched out from a fused silica glass sample, maintaining a monolithic structure. Movement is allowed by solid glass connections, which are thin enough to become flexible. Fig. 9.16 shows the realized optomechanical sensor. The mechanical part is composed of a platform with a tip, which can move with respect to a stationary framework. The optical part of the sensor consists of nine waveguides, interfaced with a linear encoder on the mobile platform. Both the optical and the mechanical part were fabricated with a single irradiation process. A displacement of the tip, and hence of the linear encoder, causes a modulation in the waveguides transmission. The displacement was experimentally retrieved from the transmission measurement with a 50 nm resolution.

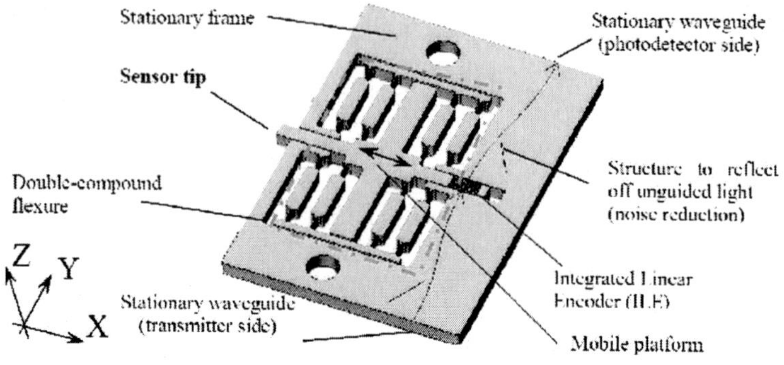

Figure 9.16 Micro-opto mechanical sensor, fabricated in monolithic fused silica substrate by femtosecond laser irradiation followed by etching.[142]

Femtosecond laser waveguide writing has also been employed to integrate the optical part in a micro-mechanical system fabricated by other techniques. Kamata *et al.*[143] showed a displacement optomechanical sensor, based on a mechanical structure formed by a series of three pieces of soda-lime glass. The central one is mounted on a suspended beam to make it sensitive to mechanical vibration, acceleration or external forces. The other two are fixed. A waveguide is fabricated by femtosecond pulses across the three pieces. Displacement of the central piece can thus be detected by measuring the change in optical transmission, giving an acceleration sensitivity down to 0.01 m/s^2.

9.3.4.3 Integrated circuits for quantum information experiments

A recently introduced field of application of femtosecond laser writing is the on-chip integration of quantum information experiments conducted with single photons. Only a few works have been published to date, but they seem to have initiated a very promising field of application, where the potentials of this waveguide fabrication technology could lead to fundamental advances in quantum optical physics. Quantum information technologies are based on encoding information on quantum systems,[144] such as trapped ions, quantum dots, superconductors or photons, and on harnessing nonclassical phenomena in order to improve the performances in several operations, e.g. increase the computational speed of certain algorithms or the measurement sensitivity and resolution in interferometry. In particular, photons can be conveniently used for these purposes because of their low decoherence; it has also been shown that universal quantum computation can be reached with linear optical networks and single photon detector.[145] A large number of works has thus been published in the past decade, exploring with bulk optics equipment the potentials of quantum optics for information processing and sensing. However, phase stability issues pose severe limitations to the development of large and complex circuital schemes. This problem has been very recently addressed and solved by Politi *et al.*[146] by the adoption of integrated optical waveguides, realized by standard lithographic techniques. This extremely stable and compact architecture allowed to perform quantum optics experiments based on nonclassical interference of photons with results comparable, where not superior, with respect to those achieved with bulk optics.

Lithography may not be an ideal solution for a rapidly developing field such as quantum optics, where testing new circuit configurations is needed with short turnaround times, more than replication on large scale of the same design. Hence, femtosecond laser writing can indeed provide a powerful alternative, as shown by Marshall *et al.*[70] In this work several directional couplers have been inscribed in high purity fused silica samples, using a 1 kHz repetition rate, 800 nm, 120 fs laser. A 40 × 0.6 NA objective for focusing was used in combination to a 520 μm slit for beam shaping, in order to obtain waveguides with practically circular cross section and a diameter of about 5 μm, able to support a single transverse mode at the 806 nm wavelength adopted for the quantum optics experiments. The mode dimensions were 6.1 μm × 6.1 μm, thus allowing excellent matching with optical fibres; insertion losses of the fabricated devices (including input and output fibre coupling) were of the order of 3 dB.

Experiments showing two- and three-photon nonclassical interference were conducted in the laser fabricated directional couplers. The two-photon experiment exploited a directional coupler with 1:1 splitting ratio and showed an interference visibility as high as 95.8%. The three-photon experiment needed a directional coupler with 1:2 splitting ratio and yielded 84% peak visibility. In the two cases, the results compares very well to those obtainable both with lithographically realized circuits and bulk optics implementations, proving the high quality of the devices produced by the femtosecond laser technology.

In these experiments the polarization of the photons was fixed and the qubits were encoded in the path of the photons. On the other hand, many quantum information processes and sources of entangled photon states are based on the polarization degree of freedom, i.e. the information is encoded in the polarization state of single photons. Hence, including the use of photon polarization in integrated quantum circuits is of essential interest for really opening the way to an integrated approach in quantum optics experiments.

In a very recent work,[71] Sansoni *et al.* showed that waveguides fabricated by femtosecond laser can be used to this purpose, demonstrating the first integrated optical chip able to support polarization encoded qubits.

A directional coupler was realized in borosilicate glass, machined by 300 fs pulses at 1030 nm from a Yb:KYW cavity-dumped oscillator. The thermal accumulation effects,[16] enabled by the 1 MHz repetition rate of the pulses, allowed to obtain an almost circular guided mode at 806 nm, without any shaping of the laser beam. Also, the writing speed as high as 40 mm/s allows fast realization of the devices, an important characteristic in perspective of a rapid prototyping employment of the femtosecond laser technique.

The most critical point when dealing with polarization encoded qubits is to provide low birefringence waveguides. A large birefringence can not only rotate the polarization of incoming photons, which can be easily compensated by waveplates, but can indeed destroy the coherence of the large bandwidth photons used in quantum experiments. Parameters in femtosecond laser waveguide writing can be optimized[71] in order to obtain a waveguide birefringence as low as 7×10^{-5}. This has to be compared with the 4×10^{-4} value commonly obtained in conventional silica-on-silicon photolithography. It has to be noticed that lithographic techniques for reducing or trimming waveguide birefringence have indeed been demonstrated, but at the expense of a much increased complexity of the fabrication process. On the contrary direct waveguide writing with femtosecond laser produces "naturally" these low birefringence waveguides, which have shown to preserve the polarization degree of the incoming beam with more than 99% accuracy.

Hong–Ou–Mandel experiments with entangled two-photon states were conducted with the integrated directional coupler and visibilities of the nonclassical interference dips and peaks higher than 90% were observed, indicating the high quality of the device produced.

9.4 Conclusions and Future Outlook

Fabrication of photonic devices by femtosecond lasers is rapidly shifting from an exotic approach to a competitive choice for producing high-quality devices at reasonable costs with unprecedented prototyping capabilities. The maskless direct writing technique allows variation and fine tuning of device parameters from one device to the other according to characterization feedback, thus shortening the optimization process and dramatically reducing the costs with respect to parallel photolithographic techniques where a new mask is required and several devices are fabricated at every run. In addition, femtosecond laser waveguide writing has unique three dimensional capabilities, which enable novel device layouts. The wealth of basic photonic components that have been demonstrated by this technology, such as couplers, interferometers, Bragg gratings, waveguide lasers and amplifiers, opens up the possibility to achieve a straightforward fabrication of quite complex microsystems that find application in several different fields. A few examples are reviewed in this chapter ranging from optofluidics to optomechanics and integrated quantum optics, however the potential and flexibility of this technology is so high that in the near future several more fields will benefit from the introduction of a femtosecond laser-written 3D photonic device.

References

1. J. W. Chan, T. Huser, S. Risbud, and D. M. Krol, Structural changes in fused silica after exposure to focused femtosecond laser pulses, *Opt. Lett.* **26,** 1726–1728 (2001).
2. K. Itoh, W. Watanabe, S. Nolte, and C. Schaffer, Ultrafast processes for bulk modification of transparent materials, *MRS Bull.* **31,** 620–625 (2006).
3. M. Ams, G. Marshall, P. Dekker, M. Dubov, V. Mezentsev, I. Bennion, and M. Withford, Investigation of ultrafast laser-photonic material interactions: Challenges for directly written glass photonics, *IEEE J. Sel. Topics Quantum Electron.* **14,** 1370–1381 (2008).

4. K. M. Davis, K. Miura, N. Sugimoto, and K. Hirao, Writing waveguides in glass with a femtosecond laser, *Opt. Lett.* **21,** 1729–1731 (1996).
5. Streltsov and N. Borrelli, Study of femtosecond-laser-written waveguides in glasses, *J. Opt. Soc. Am. B.* **19,** 2496–2504 (2002).
6. W. J. Reichman, D. M. Krol, L. Shah, F. Yoshino, A. Arai, S. M. Eaton, and P. R. Herman, A spectroscopic comparison of femtosecond-laser-modified fused silica using kilohertz and megahertz laser systems, *J. Appl. Phys.* **99,** 123112–123112-5 (2006).
7. C. Ponader, J. Schroeder, and A. Streltsov, Origin of the refractive-index increase in laser-written waveguides in glasses, *J. Appl. Phys.* **103,** 063516 (2008).
8. D. Little, M. Ams, P. Dekker, G. Marshall, and M. Withford, Mechanism of femtosecond-laser induced refractive index change in phosphate glass under a low repetition-rate regime, *J. Appl. Phys.* **108,** 033110 (2010).
9. L. Sudrie, M. Franco, B. Prade, and A. Mysyrowicz, Writing of permanent birefringent microlayers in bulk fused silica with femtosecond laser pulses, *Opt. Commun.* **171,** 279–284 (1999).
10. E. Bricchi, B. G. Klappauf, and P. G. Kazansky, Form birefringence and negative index change created by femtosecond direct writing in transparent materials, *Opt. Lett.* **29,** 119–121 (2004).
11. G. Cheng, K. Mishchik, C. Mauclair, E. Audouard, and R. Stoian, Ultrafast laser photoinscription of polarization sensitive devices in bulk silica glass, *Appl. Phys. Lett.* **80,** 219–221 (2009).
12. L. A. Fernandes, J. R. Grenier, P. R. Herman, J. S. Aitchison, and P. V. S. Marques, Femtosecond laser fabrication of birefringent directional couplers as polarization beam splitters in fused silica, *Opt. Express.* **19,** 11992–11999 (2011).
13. E. N. Glezer, M. Milosavljevic, L. Huang, R. J. Finlay, T.-H. Her, J. P. Callan, and E. Mazur, Three-dimensional optical storage inside transparent materials, *Opt. Lett.* **21,** 2023–2025 (1996).
14. M. Watanabe, S. Juodkazis, H. Sun, S. Matsuo, H. Misawa, M. Miwa, and R. Kaneko, Transmission and photoluminescence images of three-dimensional memory in vitreous silica, *Appl. Phys. Lett.* **74,** 3957 (1999).
15. S. Eaton, H. Zhang, P. Herman, F. Yoshino, L. Shah, J. Bovatsek, and A. Arai, Heat accumulation effects in femtosecond laser-written waveguides with variable repetition rate, *Opt. Express* **13,** 4708–4716 (2005).
16. S. M. Eaton, H. Zhang, M. L. Ng, J. Li, W.-J. Chen, S. Ho, and P. R. Herman, Transition from thermal diffusion to heat accumulation in high

repetition rate femtosecond laser writing of buried optical waveguides, *Opt. Express* **16,** 9443–9458 (2008).

17. C. B. Schaffer, A. Brodeur, J. F. Garcia, and E. Mazur, Micromachining bulk glass by use of femtosecond laser pulses with nanojoule energy, *Opt. Lett.* **26,** 93–95 (2001).
18. K. Minoshima, A. M. Kowalevicz, I. Hartl, E. P. Ippen, and J. G. Fujimoto, Photonic device fabrication in glass by use of nonlinear materials processing with a femtosecond laser oscillator, *Opt. Lett.* **26,** 1516–1518 (2001).
19. D. Low, H. Xie, Z. Xiong, and G. Lim, Femtosecond laser direct writing of embedded optical waveguides in aluminosilicate glass, *Appl. Phys. A: Mater.* **81,** 1633–1638 (2005).
20. G. Cerullo, R. Osellame, S. Taccheo, M. Marangoni, D. Polli, R. Ramponi, P. Laporta, and S. De Silvestri, Femtosecond micromachining of symmetric waveguides at 1.5 μm by astigmatic beam focusing, *Opt. Lett.* **27,** 1938–1940 (2002).
21. Y. Cheng, K. Sugioka, K. Midorikawa, M. Masuda, K. Toyoda, M. Kawachi, and K. Shihoyama, Control of the cross-sectional shape of a hollow microchannel embedded in photostructurable glass by use of a femtosecond laser, *Opt. Lett.* **28,** 55–57 (2003).
22. M. Ams, G. Marshall, D. Spence, and M. Withford, Slit beam shaping method for femtosecond laser direct-write fabrication of symmetric waveguides in bulk glasses, *Opt. Express* **13,** 5676–5681 (2005).
23. Y. Nasu, M. Kohtoku, and Y. Hibino, Low-loss waveguides written with a femtosecond laser for flexible interconnection in a planar light-wave circuit, *Opt. Lett.* **30,** 723–725 (2005).
24. R. Thomson, H. Bookey, N. Psaila, S. Campbell, D. Reid, S. Shen, A. Jha, and A. Kar, Internal gain from an erbium-doped oxyfluoride-silicate glass waveguide fabricated using femtosecond waveguide inscription, *IEEE Photonic. Technol. Lett.* **18,** 1515–1517 (2006).
25. D. Little, M. Ams, P. Dekker, G. Marshall, J. Dawes, and M. Withford, Femtosecond laser modification of fused silica: the effect of writing polarization on Si-O ring structure, *Opt. Express.* **16,** 20029–20037 (2008).
26. Marcinkevicius, S. Juodkazis, M. Watanabe, M. Miwa, S. Matsuo, H. Misawa, and J. Nishii, Femtosecond laser-assisted three-dimensional microfabrication in silica, *Opt. Lett.* **26,** 277–279 (2001).
27. R. Osellame, N. Chiodo, V. Maselli, A. Yin, M. Zavelani-Rossi, G. Cerullo, P. Laporta, L. Aiello, S. De Nicola, P. Ferraro, *et al.*, Optical properties

of waveguides written by a 26 MHz stretched cavity Ti: sapphire femtosecond oscillator, *Opt. Express.* **13,** 612–620 (2005).

28. V. Bhardwaj, E. Simova, P. Corkum, D. Rayner, C. Hnatovsky, R. Taylor, B. Schreder, M. Kluge, and J. Zimmer, Femtosecond laser-induced refractive index modification in multicomponent glasses, *J. Appl. Phys.* **97,** 083102 (2005).

29. J. Chan, T. Huser, S. Risbud, J. Hayden, and D. Krol, Waveguide fabrication in phosphate glasses using femtosecond laser pulses, *Appl. Phys. Lett.* **82,** 2371 (2003).

30. R. Osellame, S. Taccheo, M. Marangoni, R. Ramponi, P. Laporta, D. Polli, S. D. Silvestri, and G. Cerullo, Femtosecond writing of active optical waveguides with astigmatically shaped beams, *J. Opt. Soc. Am. B.* **20,** 1559–1567 (2003).

31. Z. Wang, K. Sugioka, and K. Midorikawa, Three-dimensional integration of microoptical components buried inside photosensitive glass by femtosecond laser direct writing, *Appl. Phys. A Mater. Sci. Process.* **89**(4), 951–955 (2007).

32. Z. Wang, K. Sugioka, Y. Hanada, and K. Midorikawa, Optical waveguide fabrication and integration with a micro-mirror inside photosensitive glass by femtosecond laser direct writing, *Appl. Phys. A Mater. Sci. Process.* **88**(4), 699–704 (2007).

33. Bettiol, S. Rao, E. Teo, J. Van Kan, and F. Watt, Fabrication of buried channel waveguides in photosensitive glass using proton beam writing, *Appl. Phys. Lett.* **88,** 171106 (2006).

34. J. Burghoff, S. Nolte, and A. Tunnermann, Origins of waveguiding in femtosecond laser-structured $LiNbO_3$, *Appl. Phys. A: Mater.* **89,** 127–132 (2007).

35. H. Bookey, R. Thomson, N. Psaila, A. Kar, N. Chiodo, R. Osellame, and G. Cerullo, Femtosecond laser inscription of low insertion loss waveguides in z-cut lithium niobate, *IEEE Photonic. Technol. Lett.* **19,** 892–894 (2007).

36. Y. Lee, N. Yu, C. Jung, B. Yu, I. Sohn, S. Choi, Y. Noh, D. Ko, W. Yang, H. Lee, *et al.*, Second-harmonic generation in periodically poled lithium niobate waveguides fabricated by femtosecond laser pulses, *Appl. Phys. Lett.* **89,** 171103 (2006).

37. R. Osellame, M. Lobino, N. Chiodo, M. Marangoni, G. Cerullo, R. Ramponi, H. Bookey, R. Thomson, N. Psaila, and A. Kar, Femtosecond laser writing of waveguides in periodically poled lithium niobate preserving the nonlinear coefficient, *Appl. Phys. Lett.* **90,** 241107 (2007).

38. J. Burghoff, C. Grebing, S. Nolte, and A. Tunnermann, Efficient frequency doubling in femtosecond laser-written waveguides in lithium niobate, *Appl. Phys. Lett.* **89,** 081108081108 (2006).

39. J. Burghoff, C. Grebing, S. Nolte, and A. Tünnermann, Waveguides in lithium niobate fabricated by focused ultrashort laser pulses, *Appl. Surf. Sci.* **253,** 7899–7902 (2007).

40. J. Thomas, M. Heinrich, J. Burghoff, S. Nolte, A. Ancona, and A. Tunnermann, Femtosecond laser-written quasi-phase-matched waveguides in lithium niobate, *Appl. Phys. Lett.* **91,** 151108–151108 (2007).

41. S. Eaton, C. Merchant, R. Iyer, A. Zilkie, A. Helmy, J. Aitchison, P. Herman, D. Kraemer, R. Miller, C. Hnatovsky, *et al.*, Raman gain from waveguides inscribed in $KgD(WO_4)_2$ by high repetition rate femtosecond laser, *Appl. Phys. Lett.* **92,** 081105–081105 (2008).

42. F. Bain, A. Lagatsky, R. Thomson, N. Psaila, N. Kuleshov, A. Kar, W. Sibbett, and C. Brown, Ultrafast laser inscribed Yb: $KgD(WO_4)_2$ and Yb: $Ky(WO_4)_2$ channel waveguide lasers., *Opt. Express.* **17,** 22417–22422 (2009).

43. S. Sowa, W. Watanabe, T. Tamaki, J. Nishii, and K. Itoh, Symmetric waveguides in poly (methyl methacrylate) fabricated by femtosecond laser pulses, *Opt. Express.* **14,** 291–297 (2006).

44. S. Sowa, W. Watanabe, J. Nishii, and K. Itoh, Filamentary cavity formation in poly (methyl methacrylate) by single femtosecond pulse, *Appl. Phys. A: Mater.* **81,** 1587–1590 (2005).

45. H. Mochizuki, W. Watanabe, Y. Ozeki, K. Itoh, K. Matsuda, and S. Hirono, Fabrication of diffractive optical elements inside polymers by femtosecond laser irradiation, *Thin Solid Films.* **518,** 714–718 (2009).

46. Y. Li, K. Yamada, T. Ishizuka, W. Watanabe, K. Itoh, and Z. Zhou, Single femtosecond pulse holography using polymethyl methacrylate, *Opt. Express.* **10,** 1173–1178 (2002).

47. P. Scully, D. Jones, and D. Jaroszynski, Femtosecond laser irradiation of polymethyl-methacrylate for refractive index gratings, *J. Opt. A: Pure Appl. Opt.* **5,** S92–S96 (2003).

48. Baum, P. Scully, W. Perrie, D. Jones, R. Issac, and D. Jaroszynski, Pulse-duration dependency of femtosecond laser refractive index modification in poly (methyl methacrylate), *Opt. Lett.* **33,** 651–653 (2008).

49. D. Kallepalli, N. Desai, and V. Soma, Fabrication and optical characterization of microstructures in poly (methylmethacrylate) and

poly (dimethylsiloxane) using femtosecond pulses for photonic and microfluidic applications, *Appl. Optics.* **49,** 2475–2489 (2010).

50. C. Mendonca, L. Cerami, T. Shih, R. Tilghman, T. Baldacchini, and E. Mazur, Femtosecond laser waveguide micromachining of PMMA films with azoaromatic chromophores, *Opt. Express.* **16,** 200–206 (2008).
51. L. Ding, R. Blackwell, J. Kunzler, and W. Knox, Femtosecond laser micromachining of waveguides in silicone-based hydrogel polymers, *Appl. Optics.* **47,** 3100–3108 (2008).
52. H. Mochizuki, W. Watanabe, R. Ezoe, T. Tamaki, Y. Ozeki, K. Itoh, M. Kasuya, K. Matsuda, and S. Hirono, Density characterization of femtosecond laser modification in polymers, *Appl. Phys. Lett.* **92,** 091120–091120 (2008).
53. S. Hirono, M. Kasuya, K. Matsuda, Y. Ozeki, K. Itoh, H. Mochizuki, and W. Watanabe, Increasing diffraction efficiency by heating phase gratings formed by femtosecond laser irradiation in poly (methyl methacrylate), *Appl. Phys. Lett.* **94,** 241122–241122 (2009).
54. D. Homoelle, S. Wielandy, A. L. Gaeta, N. F. Borrelli, and C. Smith, Infrared photosensitivity in silica glasses exposed to femtosecond laser pulses, *Opt. Lett.* **24,** 1311–1313 (1999).
55. L. Gui, B. Xu, and T. C. Chong, Microstructure in lithium niobate by use of focused femtosecond laser pulses, *IEEE Photonic. Technol. Lett.* **16,** 1337–1339 (2004).
56. J. Liu, Z. Zhang, S. Chang, C. Flueraru, and C. P. Grover, Directly writing of 1-to-n optical waveguide power splitters in fused silica glass using a femtosecond laser, *Opt. Commun.* **253,** 315–319 (2005).
57. I.-B. Sohn, M.-S. Lee, and J.-Y. Chung, Fabrication of optical splitter and passive alignment technique with a femtosecond laser, *IEEE Photonic. Technol. Lett.* **17,** 2349–2351 (2005).
58. S. Nolte, M. Will, J. Burghoff, and A. Tuennermann, Femtosecond waveguide writing: a new avenue to three-dimensional integrated optics, *Appl. Phys. A: Mater.* **77,** 109–111 (2003).
59. Streltsov and N. Borrelli, Fabrication and analysis of a directional coupler written in glass by nanojoule femtosecond laser pulses, *Opt. Lett.* **26,** 42–43 (2001).
60. K. Minoshima, A. Kowalevicz, E. Ippen, and J. Fujimoto, Fabrication of coupled mode photonic devices in glass by nonlinear femtosecond laser materials processing, *Opt. Express.* **10,** 645–652 (2002).
61. W. Watanabe, T. Asano, K. Yamada, K. Itoh, and J. Nishii, Wavelength division with three-dimensional couplers fabricated by filamentation of femtosecond laser pulses, *Opt. Lett.* **28,** 2491–2493 (2003).

62. S. Eaton, W. Chen, L. Zhang, H. Zhang, R. Iyer, J. Aitchison, and P. Herman, Telecomband directional coupler written with femtosecond fiber laser, *IEEE Photonic. Technol. Lett.* **18,** 2174–2176 (2006).

63. R. Osellame, V. Maselli, N. Chiodo, D. Polli, R. Vazquez, R. Ramponi, and G. Cerullo, Fabrication of 3D photonic devices at 1.55 μm wavelength by femtosecond Ti: Sapphire oscillator, *Electron. Lett.* **41,** 315 (2005).

64. Kowalevicz, V. Sharma, E. Ippen, J. Fujimoto, and K. Minoshima, Three dimensional photonic devices fabricated in glass by use of a femtosecond laser oscillator, *Opt. Lett.* **30,** 1060–1062 (2005).

65. K. Suzuki, V. Sharma, J. Fujimoto, E. Ippen, and Y. Nasu, Characterization of symmetric [3 × 3] directional couplers fabricated by direct writing with a femtosecond laser oscillator, *Opt. Express.* **14,** 2335–2343 (2006).

66. W.-J. Chen, S. M. Eaton, H. Zhang, and P. R. Herman, Broadband directional couplers fabricated in bulk glass with high repetition rate femtosecond laser pulses, *Opt. Express.* **16,** 11470–11480 (2008).

67. S. Eaton, W. Chen, H. Zhang, R. Iyer, J. Li, M. Ng, S. Ho, J. Aitchison, and P. Herman, Spectral loss characterization of femtosecond laser written waveguides in glass with application to demultiplexing of 1300 and 1550 nm wavelengths, *J. Lightwave Technol.* **27,** 1079–1085 (2009).

68. C. Mauclair, G. Cheng, N. Huot, E. Audouard, A. Rosenfeld, I. Hertel, and R. Stoian, Dynamic ultrafast laser spatial tailoring for parallel micromachining of photonic devices in transparent materials, *Opt. Express.* **17,** 3531–3542 (2009).

69. M. Pospiech, M. Emons, A. Steinmann, G. Palmer, R. Osellame, N. Bellini, G. Cerullo, and U. Morgner, Double waveguide couplers produced by simultaneous femtosecond writing, *Opt. Express.* **17,** 3555–3563 (2009).

70. G. Marshall, A. Politi, J. Matthews, P. Dekker, M. Ams, M. Withford, and J. O'Brien, Laser written waveguide photonic quantum circuits, *Opt. Express.* **17,** 12546–12554 (2009).

71. L. Sansoni, F. Sciarrino, G. Vallone, P. Mataloni, A. Crespi, R. Ramponi, and R. Osellame, Polarization entangled state measurement on a chip, *Phys. Rev. Lett.* **105,** 200503 (2010).

72. G. Li, K. Winick, A. Said, M. Dugan, and P. Bado, Waveguide electro-optic modulator in fused silica fabricated by femtosecond laser direct writing and thermal poling, *Opt. Lett.* **31,** 739–741 (2006).

73. C. Mendez, G. Torchia, D. Delgado, I. Arias, and L. Roso. Fabrication and characterization of Mach–Zehnder devices in $LiNbO_3$ written with femtosecond laser pulses. In *Fibres and Optical Passive Components,*

2005. Proceedings of 2005 IEEE/LEOS Workshop on, 131–134. IEEE (2005).

74. Y. Liao, J. Xu, Y. Cheng, Z. Zhou, F. He, H. Sun, J. Song, X. Wang, Z. Xu, K. Sugioka, *et al.*, Electro-optic integration of embedded electrodes and waveguides in $LiNbO_3$ using a femtosecond laser, *Opt. Lett.* **33,** 2281–2283 (2008).
75. S. Ringleb, K. Rademaker, S. Nolte, and A. Tunnermann, Monolithically integrated optical frequency converter and amplitude modulator in $LiNbO_3$ fabricated by femtosecond laser pulses, *Appl. Phys. B: Lasers Opt.* **102,** 59–63 (2011).
76. R. Heideman and P. Lambeck, Remote opto-chemical sensing with extreme sensitivity: design, fabrication and performance of a pigtailed integrated optical phase-modulated Mach–Zehnder interferometer system, *Sensor. Actuat. B Chem.* **61,** 100–127 (1999).
77. L. Soldano and E. Pennings, Optical multi-mode interference devices based on self-imaging: Principles and applications, *J. Lightwave Technol.* **13,** 615–627 (1995).
78. E. Pennings, R. Deri, A. Scherer, R. Bhat, T. Hayes, N. Andreadakis, M. Smit, L. Soldano, and R. Hawkins, Ultracompact, low-loss directional couplers on InP based on self-imaging by multimode interference, *Appl. Phys. Lett.* **59,** 1926–1928 (1991).
79. W. Watanabe and K. Itoh, Fabrication of multimode interference waveguides in glass by use of a femtosecond laser, *Opt. Lett.* **30,** 2888–2890 (2005).
80. L. Da-Yong, L. Yan, D. Yan-Ping, G. Heng-Chang, Y. Hong, and G. Qi-Huang, Transverse writing of multimode interference waveguides inside silica glass by femtosecond laser pulses, *Chinese Phys. Lett.* **25,** 2500 (2008).
81. T. Pertsch, U. Peschel, F. Lederer, J. Burghoff, M. Will, S. Nolte, and A. Tunnermann, Discrete diffraction in two-dimensional arrays of coupled waveguides in silica, *Opt. Lett.* **29,** 468–470 (2004).
82. Szameit, D. Bloemer, J. Burghoff, T. Pertsch, S. Nolte, and A. Tunnermann, Hexagonal waveguide arrays written with fs-laser pulses, *Appl. Phys. B: Lasers Opt.* **82,** 507–512 (2006).
83. Garanovich, A. Szameit, A. Sukhorukov, T. Pertsch, W. Krolikowski, S. Nolte, D. Neshev, A. Tuennermann, and Y. Kivshar, Diffraction control in periodically curved two-dimensional waveguide arrays, *Opt. Express.* **15,** 9737–9747 (2007).
84. Szameit, Y. Kartashov, M. Heinrich, F. Dreisow, R. Keil, S. Nolte, A. Tunnermann, V. Vysloukh, F. Lederer, and L. Torner, Nonlinearity-

induced broadening of resonances in dynamically modulated couplers, *Opt. Lett.* **34,** 2700–2702 (2009).

85. Szameit, Y. Kartashov, F. Dreisow, M. Heinrich, T. Pertsch, S. Nolte, A. Tunnermann, V. Vysloukh, F. Lederer, and L. Torner, Inhibition of light tunneling in waveguide arrays, *Phys. Rev. Lett.* **102,** 153901 (2009).

86. Szameit, D. Blomer, J. Burghoff, T. Schreiber, T. Pertsch, S. Nolte, A. Tunnermann, and F. Lederer, Discrete nonlinear localization in femtosecond laser written waveguides in fused silica, *Opt. Express.* **13,** 10552–10557 (2005).

87. Szameit, J. Burghoff, T. Pertsch, S. Nolte, A. Tunnermann, and F. Lederer, Two-dimensional soliton in cubic fs laser written waveguide arrays in fused silica, *Opt. Express.* **14,** 6055–6062 (2006).

88. Szameit, T. Pertsch, S. Nolte, A. Tunnermann, U. Peschel, and F. Lederer, Optical Bloch oscillations in general waveguide lattices, *J. Opt. Soc. Am. B.* **24,** 2632–2639 (2007).

89. N. Chiodo, G. Della Valle, R. Osellame, S. Longhi, G. Cerullo, R. Ramponi, P. Laporta, and U. Morgner, Imaging of Bloch oscillations in erbium-doped curved waveguide arrays, *Opt. Lett.* **31,** 1651–1653 (2006).

90. F. Dreisow, A. Szameit, M. Heinrich, T. Pertsch, S. Nolte, A. Tunnermann, and S. Longhi, Bloch-Zener oscillations in binary superlattices, *Phys. Rev. Lett.* **102,** 76802 (2009).

91. F. Dreisow, M. Heinrich, A. Szameit, S. Doering, S. Nolte, A. Tuennermann, S. Fahr, and F. Lederer, Spectral resolved dynamic localization in curved fs laser written waveguide arrays, *Opt. Express.* **16,** 3474–3483 (2008).

92. Szameit, I. Garanovich, M. Heinrich, A. Sukhorukov, F. Dreisow, T. Pertsch, S. Nolte, A. Tunnermann, and Y. Kivshar, Polychromatic dynamic localization in curved photonic lattices, *Nat. Phys.* **5,** 271–275 (2009).

93. Szameit, I. Garanovich, M. Heinrich, A. Sukhorukov, F. Dreisow, T. Pertsch, S. Nolte, A. Tunnermann, S. Longhi, and Y. Kivshar, Observation of two-dimensional dynamic localization of light, *Phys. Rev. Lett.* **104,** 223903 (2010).

94. Szameit, H. Trompeter, M. Heinrich, F. Dreisow, U. Peschel, T. Pertsch, S. Nolte, F. Lederer, and A. Tunnermann, Fresnel's laws in discrete optical media, *New J. Phys.* **10,** 103020 (2008).

95. R. Keil, A. Szameit, F. Dreisow, M. Heinrich, S. Nolte, and A. Tunnermann, Photon correlations in two-dimensional waveguide arrays and their classical estimate, *Phys. Rev. A.* **81,** 023834 (2010).

96. R. Keil, F. Dreisow, M. Heinrich, A. Tunnermann, S. Nolte, and A. Szameit, Classical characterization of biphoton correlation in waveguide lattices, *Phys. Rev. A.* **83,** 013808 (2011).

97. D. Grobnic, S. Mihailov, C. Smelser, F. Genereux, G. Baldenberger, and R. Vallee, Bragg gratings made in reverse proton exchange lithium niobate waveguides with a femtosecond IR laser and a phase mask, *IEEE Photonic. Technol. Lett.* **17,** 1453–1455 (2005).

98. D. Grobnic, S. Mihailov, C. Smelser, M. Becker, and M. Rothhardt, Femtosecond laser fabrication of Bragg gratings in borosilicate ion-exchange waveguides, *IEEE Photonic. Technol. Lett.* **18,** 1403–1405 (2006).

99. G. Marshall, M. Ams, and M. Withford, Direct laser written waveguide-Bragg gratings in bulk fused silica, *Opt. Lett.* **31,** 2690–2691 (2006).

100. H. Zhang, S. Eaton, J. Li, and P. Herman, Femtosecond laser direct writing of multi-wavelength Bragg grating waveguides in glass, *Opt. Lett.* **31,** 3495–3497 (2006).

101. H. Zhang, S. Eaton, J. Li, A. Nejadmalayeri, and P. Herman, Type ii high-strength Bragg grating waveguides photowritten with ultrashort laser pulses, *Opt. Express.* **15,** 4182–4191 (2007).

102. H. Zhang, S. M. Eaton, and P. R. Herman, Single-step writing of Bragg grating wave guides in fused silica with an externally modulated femtosecond fiber laser, *Opt. Lett.* **32,** 2559–2561 (2007).

103. P. Herman. Direct femtosecond laser writing of 3-D waveguides and gratings for optical communications. In *Optical Fiber Communication Conference.* Optical Society of America (2008).

104. H. Zhang and P. Herman. Femtosecond laser direct writing of chirped Bragg grating waveguides inside fused silica glass. In *Bragg Gratings, Photosensitivity, and Poling in Glass Waveguides.* Optical Society of America (2007).

105. G. Emmerson, S. Watts, C. Gawith, V. Albanis, M. Ibsen, R. Williams, and P. Smith, Fabrication of directly UV-written channel waveguides with simultaneously defined integral Bragg gratings, *Electron. Lett.* **38,** 1531–1532 (2002).

106. H. Zhang, S. Ho, S. Eaton, J. Li, and P. Herman, Three-dimensional optical sensing network written in fused silica glass with femtosecond laser, *Opt. Express.* **16,** 1401514023 (2008).

107. V. Maselli, J. Grenier, S. Ho, and P. Herman, Femtosecond laser written optofluidic sensor: Bragg grating waveguide evanescent probing of microfluidic channel, *Opt. Express.* **17,** 11719–11729 (2009).

108. M. Ams, G. Marshall, P. Dekker, J. Piper, and M. Withford, Ultrafast laser written active devices, *Laser Photon. Rev.* **3,** 535–544 (2009).

109. Y. Sikorski, A. Said, P. Bado, R. Maynard, C. Florea, and K. Winick, Optical waveguide amplifier in Nd-doped glass written with near-IR femtosecond laser pulses, *Electron. Lett.* **36,** 226–227 (2000).

110. R. Osellame, S. Taccheo, G. Cerullo, M. Marangoni, D. Polli, R. Ramponi, P. Laporta, and S. De Silvestri, Optical gain in Er-Yb doped waveguides fabricated by femtosecond laser pulses, *Electron. Lett.* **38,** 964–965 (2002).

111. R. Osellame, N. Chiodo, G. Della Valle, G. Cerullo, R. Ramponi, P. Laporta, A. Killi, U. Morgner, and O. Svelto, Waveguide lasers in the C-band fabricated by laser inscription with a compact femtosecond oscillator, *IEEE J. Sel. Top. Quant.* **12,** 277–285 (2006).

112. G. Della Valle, R. Osellame, N. Chiodo, S. Taccheo, G. Cerullo, P. Laporta, A. Killi, U. Morgner, M. Lederer, and D. Kopf, C-band waveguide amplifier produced by femtosecond laser writing, *Opt. Express.* **13,** 5976–5982 (2005).

113. Y. Jaouen, L. du Mouza, D. Barbier, J. Delavaux, and P. Bruno, Eight-wavelength Er-Yb doped amplifier: combiner/splitter planar integrated module, *IEEE Photonic. Technol. Lett.* **11,** 1105–1107 (1999).

114. R. Thomson, S. Campbell, I. Blewett, A. Kar, D. Reid, S. Shen, and A. Jha, Active waveguide fabrication in erbium-doped oxyfluoride silicate glass using femtosecond pulses, *Appl. Phys. Lett.* **87,** 121102 (2005).

115. N. Psaila, R. Thomson, H. Bookey, A. Kar, N. Chiodo, R. Osellame, G. Cerullo, A. Jha, and S. Shen, Er: Yb-doped oxyfluoride silicate glass waveguide amplifier fabricated using femtosecond laser inscription, *Appl. Phys. Lett.* **90,** 131102 (2007).

116. N. Psaila, R. Thomson, H. Bookey, A. Kar, N. Chiodo, R. Osellame, G. Cerullo, G. Brown, A. Jha, and S. Shen, Femtosecond laser inscription of optical waveguides in bismuth ion doped glass, *Opt. Express.* **14,** 10452–10459 (2006).

117. P. Nandi, G. Jose, C. Jayakrishnan, S. Debbarma, K. Chalapathi, K. Alti, A. Dharmadhikari, J. Dharmadhikari, and D. Mathur, Femtosecond laser written channel waveguides in tellurite glass, *Opt. Express.* **14,** 12145–12150 (2006).

118. T. Fernandez, G. Della Valle, R. Osellame, G. Jose, N. Chiodo, A. Jha, and P. Laporta, Active waveguides written by femtosecond laser irradiation in an erbium-doped phospho-tellurite glass, *Opt. Express.* **16,** 15198–15205 (2008).

119. T. Fernandez, S. Eaton, G. Della Valle, R. Vazquez, M. Irannejad, G. Jose, A. Jha, G. Cerullo, R. Osellame, and P. Laporta, Femtosecond laser written optical waveguide amplifier in phospho-tellurite glass, *Opt. Express.* **18,** 20289–20297 (2010).

120. Valles, A. Ferrer, J. Fernandez-Navarro, V. Berdejo, A. Ruiz de la Cruz, I. Ortega-Feliu, M. Rebolledo, and J. Solis, Performance of ultrafast laser written active waveguides by rigorous modeling of optical gain measurements, *Opt. Mater. Express.* **1,** 564–575 (2011).

121. S. Taccheo, G. Della Valle, R. Osellame, G. Cerullo, N. Chiodo, P. Laporta, O. Svelto, A. Killi, U. Morgner, M. Lederer, *et al.*, Er: Yb-doped waveguide laser fabricated by femtosecond laser pulses, *Opt. Lett.* **29,** 2626–2628 (2004).

122. G. Della Valle, S. Taccheo, R. Osellame, A. Festa, G. Cerullo, and P. Laporta, 1.5 μm single longitudinal mode waveguide laser fabricated by femtosecond laser writing, *Opt. Express.* **15,** 3190–3194 (2007).

123. N. Psaila, R. Thomson, H. Bookey, N. Chiodo, S. Shen, R. Osellame, G. Cerullo, A. Jha, and A. Kar, Er: Yb-doped oxyfluoride silicate glass waveguide laser fabricated using ultrafast laser inscription, *IEEE Photonic. Technol. Lett.* **20,** 126–128 (2008).

124. G. Della Valle, R. Osellame, G. Galzerano, N. Chiodo, G. Cerullo, P. Laporta, O. Svelto, U. Morgner, A. Rozhin, V. Scardaci, *et al.*, Passive mode locking by carbon nanotubes in a femtosecond laser written waveguide laser, *Appl. Phys. Lett.* **89,** 231115–231115 (2006).

125. S. Beecher, R. Thomson, N. Psaila, Z. Sun, T. Hasan, A. Rozhin, A. Ferrari, and A. Kar, 320 fs pulse generation from an ultrafast laser inscribed waveguide laser mode-locked by a nanotube saturable absorber, *Appl. Phys. Lett.* **97,** 111114 (2010).

126. G. D. Marshall, P. Dekker, M. Ams, J. A. Piper, and M. J. Withford, Directly written monolithic waveguide laser incorporating a distributed feedback waveguide-Bragg grating, *Opt. Lett.* **33,** 956–958 (2008).

127. M. Ams, P. Dekker, G. Marshall, and M. Withford, Monolithic 100 mW Yb waveguide laser fabricated using the femtosecond-laser direct-write technique, *Opt. Lett.* **34,** 247249 (2009).

128. Kawamura, M. Hirano, T. Kurobori, D. Takamizu, T. Kamiya, and H. Hosono, Femtosecond-laser-encoded distributed-feedback color center laser in lithium fluoride single crystals, *Appl. Phys. Lett.* **84,** 311–313 (2004).

129. Okhrimchuk, A. Shestakov, I. Khrushchev, and J. Mitchell, Depressed cladding, buried waveguide laser formed in a YAG: Nd3+ crystal by femtosecond laser writing, *Opt. Lett.* **30,** 2248–2250 (2005).

130. G. Torchia, A. Rodenas, A. Benayas, E. Cantelar, L. Roso, and D. Jaque, Highly efficient laser action in femtosecond-written Nd: yttrium aluminum garnet ceramic waveguides, *Appl. Phys. Lett.* **92,** 111103–111103 (2008).

131. D. Nikogosyan, Multi-photon high-excitation-energy approach to fibre grating inscription, *Meas. Sci. Technol.* **18,** R1 (2007).

132. Y. Kondo, K. Nouchi, T. Mitsuyu, M. Watanabe, P. G. Kazansky, and K. Hirao, Fabrication of long-period fiber gratings by focused irradiation of infrared femtosecond laser pulses, *Opt. Lett.* **24,** 646–648 (1999).

133. Martinez, M. Dubov, I. Khrushchev, and I. Bennion, Direct writing of fibre Bragg gratings by femtosecond laser, *Electron. Lett.* **40,** 1170–1172 (2004).

134. Martinez, Y. Lai, M. Dubov, I. Khrushchev, and I. Bennion, Vector bending sensors based on fibre Bragg gratings inscribed by infrared femtosecond laser, *Electron. Lett.* **41,** 472–474 (2005).

135. G. D. Marshall, R. J. Williams, N. Jovanovic, M. J. Steel, and M. J. Withford, Point-by-point written fiber-Bragg gratings and their application in complex grating designs, *Opt. Express.* **18,** 19844–19859 (2010).

136. Said, M. Dugan, P. Bado, Y. Bellouard, A. Scott, and J. Mabesa. Manufacturing by laser direct-write of three-dimensional devices containing optical and microfluidic networks. In *Photon Processing in Microelectronics and Photonics III,* vol. 5339, 194–204. SPIE (2004).

137. R. Osellame, V. Maselli, R. Vazquez, R. Ramponi, and G. Cerullo, Integration of optical waveguides and microfluidic channels both fabricated by femtosecond laser irradiation, *Appl. Phys. Lett.* **90,** 231118 (2007).

138. R. Martinez Vazquez, R. Osellame, D. Nolli, C. Dongre, H. H. Van den Vlekkert, R. Ramponi, M. Pollnau, and G. Cerullo, Integration of femtosecond laser written optical waveguides in lab-on-chip, *Lab Chip.* **9,** 91–96 (2009).

139. Dongre, J. van Weerd, G. A. J. Besselink, R. M. Vazquez, R. Osellame, G. Cerullo, R. van Weeghel, H. H. van den Vlekkert, H. J. W. M. Hoekstra, and M. Pollnau, Modulation-frequency encoded multi-color fluorescent DNA analysis in an optofluidic chip, *Lab Chip.* **11,** 679–683 (2010).

140. N. Bellini, K. Vishnubhatla, F. Bragheri, L. Ferrara, P. Minzioni, R. Ramponi, I. Cristiani, and R. Osellame, Femtosecond laser fabricated monolithic chip for optical trapping and stretching of single cells, *Opt. Express.* **18,** 4679–4688 (2010).

141. Crespi, Y. Gu, B. Ngamsom, H. Hoekstra, C. Dongre, M. Pollnau, R. Ramponi, H. van den Vlekkert, P. Watts, G. Cerullo, and R. Osellame,

Three-dimensional Mach–Zehnder interferometer in a microfluidic chip for spatially-resolved label-free detection, *Lab Chip.* **10,** 1167–1173 (2010).

142. Y. Bellouard, A. Said, and P. Bado, Integrating optics and micro-mechanics in a single substrate: a step toward monolithic integration in fused silica, *Opt. Express.* **13,** 66356644 (2005).

143. Kamata, M. Obara, R. Gattass, L. Cerami, and E. Mazur, Optical vibration sensor fabricated by femtosecond laser micromachining, *Appl. Phys. Lett.* **87,** 051106–051106, (2005).

144. T. Ladd, F. Jelezko, R. Laflamme, Y. Nakamura, C. Monroe, and J. OBrien, Quantum computers, *Nature.* **464,** 45–53 (2010).

145. E. Knill, R. Laflamme, and G. Milburn, A scheme for efficient quantum computation with linear optics, *Nature.* **409,** 46–52 (2001).

146. Politi, J. Matthews, M. Thompson, and J. O'Brien, Integrated quantum photonics, *IEEE J. Sel. Top. Quant.* **15,** 1673–1684 (2009).

Chapter 10

Fabrication of Microfluidic Chips and Integrated Optofluidic Devices in Glass by Femtosecond Laser Direct Writing

Ya Cheng[a] **and Koji Sugioka**[b]
[a] *Shanghai Institute of Optics and Fine Mechanics, Chinese Academy of Sciences, China*
[b] *RIKEN, Japan*
ya.cheng@siom.ac.cn, ksugioka@riken.jp

Integrated optofluidic devices or microsystems are highly functional miniaturized devices composed of microfluidic and micro-optical components synergistically combined with each other. They have found important applications in chemical sensing, biosensing, and reconfigurable photonics. This chapter describes the principles and techniques for fabricating microfluidic chips and integrated optofluidic devices in glass by femtosecond laser direct writing. It then gives a few examples of successful applications of fabricated devices. Major challenges and opportunities are also discussed.

10.1 Introduction

By integrating fluidic functions such as valving, metering, mixing,

---

Ultrafast Laser Processing: From Micro- to Nanoscale
Edited by Koji Sugioka and Ya Cheng

ISBN 978-981-4267-33-5 (Hardcover), 978-981-4303-69-9 (eBook)
www.panstanford.com

transport, and separation on a single substrate, microfluidic systems can be used to control and manipulate tiny volumes of liquids with high precision, enabling downsizing of both chemical and biological analysis [1, 2]. Further integration of optical functions on microfluidic chips has not only eliminated the need for external optical analysis systems in chemical and biosensing applications, but also led to the development of tunable and reconfigurable photonic devices [3]. Today, the most popular fabrication technology in microfluidics is soft lithography using poly(dimethylsiloxane) (PDMS) substrates [4]. Although soft lithography is rapid and cost effective, it cannot be used to directly form three-dimensional (3D) microfluidic structures such as microchannels and microchambers embedded in transparent substrates without stacking and bonding. In addition, unlike glass, PDMS is chemically incompatible with many organic solvents and compositional inhomogeneities frequently cause optical scattering. Thus, new strategies need to be developed to overcome these problems.

Femtosecond laser direct writing is the main technique currently used to modify the interior of transparent materials in a spatially selective manner. It can be used to directly form 3D hollow structures in substrates. This is possible due to the nonlinear interaction between the tightly focused femtosecond laser beam and the transparent material since the interaction is effectively confined to the vicinity of focal point where the laser intensity exceeds the threshold for multiphoton absorption (see Section 1.2.5 in Chapter 1 for more details). Femtosecond laser direct writing is particularly suitable for fabricating integrated optofluidic systems since irradiating glass with femtosecond laser pulses can alter not only its optical properties (e.g., refractive index and absorption spectrum) but also its chemical properties (e.g., acid etch rate). For instance, femtosecond laser pulses have been used to write optical waveguides in both passive and active materials by locally modifying their refractive indices [5, 6]. In combination with wet chemical etching, femtosecond laser direct writing has also been used to fabricate microfluidic structures, including microchannels and chambers [7, 8], microvalves [9], and micropumps [10]. The same technique has been extended to fabricate free-space optics such as micromirrors and micro-optical lenses in glass materials [11–13]. The unique ability of maskless femtosecond laser direct writing to simultaneously alter the chemical and optical properties in bulk glass materials opens up new avenues for fabricating a wide variety of integrated optofluidic microsystems and devices such as microfluidic lasers [14], nanoaquariums for observing living organisms [15, 16], and optofluidic sensors with various functions including refractive index monitoring [17], single-cell detection and manipulation [18, 19], and rapid screening of algae populations [20]. Although this chapter focuses on applications of femtosecond laser direct writing in optofluidics, micro-electrodes can also be integrated in microsystems along with microfluidic and micro-optical components by femtosecond-laser-induced local metallization [21, 22]. A biomicrochip integrated with a microheater for temperature control [21] and an integrated electro-optic modulator [23] fabricated based on this approach have been demonstrated to be fully functional.

This chapter describes the fabrication of microfluidic chips and integrated optofluidic devices by femtosecond laser direct writing and the application of the fabricated biochips in biological and biomedical research. We begin by giving an overview of the key technologies for fabricating microfluidic structures in glass materials. We describe the fabrication of 3D micro-optical components and integrated 3D optofluidic systems. We then give several examples of biomedical applications of microfluidic chips and optofluidic devices. We conclude with a discussion of future directions and some challenges that need to be overcome.

10.2 Fabrication of Microfluidic Structures in Glass

10.2.1 Femtosecond-Laser-Assisted Wet Chemical Etching

Photosensitive glass that can be structured by single-photon absorption of ultraviolet (UV) light was discovered by S. Donald Stookey over half a century ago [24]. However, only surface microstructures can be fabricated using single-photon absorption in UV lithography. Fabrication of 3D microstructures in photosensitive glass by laser direct writing was demonstrated in the late 1990s [25, 26]. These early investigations used nanosecond UV lasers operating at nonresonant wavelengths [25] and frequency-doubled Ti:sapphire lasers [26] to achieve spatially localized nonlinear absorption in the vicinity of the focal volume. It was subsequently discovered that irradiation of photosensitive glass by near-infrared (near-IR; e.g., ~800 nm) femtosecond laser pulses could induce similar photosensitization as UV and frequency-doubled femtosecond lasers [8], although the photosensitization mechanism at near-IR wavelengths may differ from that at UV and visible wavelengths [27]. Currently, near-IR femtosecond lasers with various pulse durations operating at different repetition rates are the dominant light source for fabricating 3D microstructures embedded in photosensitive glass.

Most current research into 3D microfabrication employing photosensitive glass uses Foturan glass, which is lithium aluminosilicate doped with trace amounts of silver and cerium. The cerium (Ce^{3+}) ion functions as a photosensitizer by absorbing a UV photon at a wavelength near 315 nm and releasing an electron to become Ce^{4+}. The silver ions doped in Foturan glass then capture some of the free electrons and become silver atoms. In the subsequent heat treatment, these silver atoms diffuse and agglomerate to form nanoclusters at ~500°C. By then raising the annealing temperature to ~600°C, the crystalline phase of lithium metasilicate grows around the silver clusters (which act as nuclei) in the amorphous glass matrix [28]. The crystalline phase of lithium metasilicate can be preferentially etched in dilute hydrofluoric (HF) acid since lithium metasilicate has a ~50 times higher etch rate than the unmodified glass matrix. UV lamps are conventionally used when performing 2D microfabrication on Foturan glass surfaces. However, irradiation by femtosecond laser pulses can confine the absorption region to an area near the focal spot (even within the glass) by using a multiphoton process. This improves the axial resolution, which is vital for achieving true 3D microprocessing [8, 29]. The photoreaction initiated by multiphoton excitation has been found to have a different mechanism from that of single-photon excitation by UV irradiation because the interaction of high-intensity femtosecond laser pulses with Foturan glass can directly produce a large amount of free electrons even without any photosensitizer. Consequently, it is not necessary to dope Foturan glass with cerium when using femtosecond lasers.

Figure 10.1 depicts the general procedure for fabricating microfluidic structures embedded in Foturan glass. It consists of the following three steps [30]: (1) formation of a latent image in Foturan glass by scanning a tightly focused femtosecond laser beam; (2) transformation of the latent image into an etchable phase by post-annealing; and (3) removal of the modified material in the affected areas by wet chemical etching in a 10% aqueous solution of HF acid in an ultrasonic bath. Unfortunately, the hollow microstructures obtained after etching usually have rough inner surfaces with a typical root-mean-square (RMS) roughness of a few tens of nanometers; however, for applications requiring smooth inner surfaces, an additional post-annealing step can be applied after wet chemical etching to reduce the RMS roughness to a few nanometers or less [11]. Although this post-annealing is performed at a slightly lower temperature (570°C) than the melting point of Foturan (659°C) to prevent deformation of the fabricated microstructures, a thin liquid layer forms on Foturan surfaces, which results in the formation of a smooth surface on the inner wall due to the surface tension of the liquid layer [12].

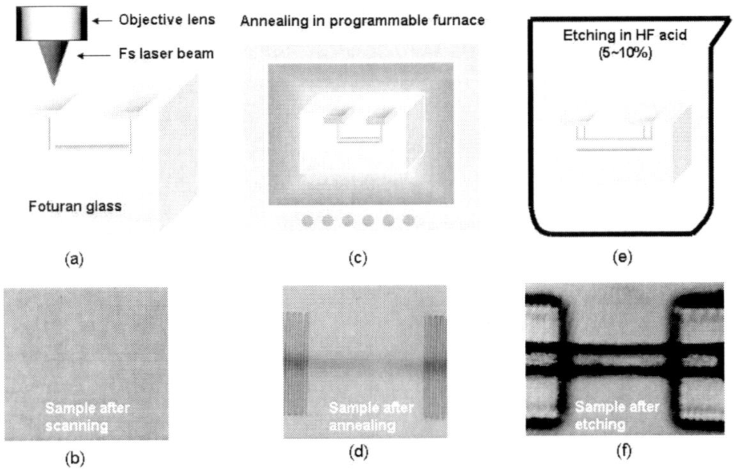

Figure 10.1 Schematic illustration of femtosecond laser 3D fabrication process. (a) Femtosecond laser direct writing for creating 3D patterns in Foturan glass; (b) inscription of latent image in sample; (c) post-annealing of sample in a programmable furnace; (d) modified areas turn brown due to the formation of silver nanoparticles; (e) etching of sample in dilute HF acid; and (f) formation of hollow structures in glass substrate after chemical etching. The final step of smoothing the internal surface by post-annealing is not depicted [30].

A variety of 3D microfluidic structures have been fabricated using the above-mentioned technique (Fig. 10.2). Figure 10.2(a) shows an X-shaped microfluidic channel formed 300 μm below the sample surface with a channel length of ~2800 μm and a width of ~45 μm [29]. The channels appear to have a nearly constant width (i.e., they have a small taper angle) as a result of using a high-intensity ultrasonic bath for wet etching. We found that it is essential to use an ultrasonic bath to fabricate homogeneous structures by chemical etching because it greatly increases the supply of fresh HF acid and reactant in the narrow channels and tiny chambers during etching. Moreover, a chemical microreactor with a 3D configuration was fabricated by arranging several microchannels and microchambers in Foturan glass (Fig. 10.2(b)) [31]. In addition to passive microfluidic networks, active components (e.g., microvalves enclosed in microfluidic chambers) can be fabricated [32]. Amazingly, regardless of how complex their geometries are, all 3D microstructures can be simultaneously formed in a glass substrate by just a single continuous laser scanning step, thereby completely eliminating the need to use multistep procedures such as stacking and bonding. This unique characteristic represents the greatest advantage of 3D femtosecond laser micromachining for fabricating 3D microfluidic structures since it greatly increases the throughput and reduces the cost.

The photoetchability of Foturan glass induced by exposure to femtosecond laser pulses can be reasonably expected as a multiphoton counterpart of the single-photon UV lithography process for microstructuring of photosensitive glass. Surprisingly, photoetchability can also be induced in nonphotosensitive glasses such as fused silica by femtosecond laser irradiation [7]. Fused-silica-based microfluidic chips have the advantages of having superior optical properties (e.g., a wide transmission range and low autofluorescence) and chemical inertness. In addition, fused silica is much easier to obtain than photosensitive glass, which can only be purchased from a very limited number of suppliers. Unfortunately, in the early stages of research, the etch selectivity induced in fused silica by femtosecond laser irradiation [7] appeared to be much lower than that in Foturan glass [8], so that it was considered that only small hollow microstructures could be fabricated in fused silica. Nevertheless, in 2004, Bellouard *et al.* formed high-aspect-ratio channels that are open on the top surface of fused silica by femtosecond-laser-assisted etching with many different channel lengths [33].

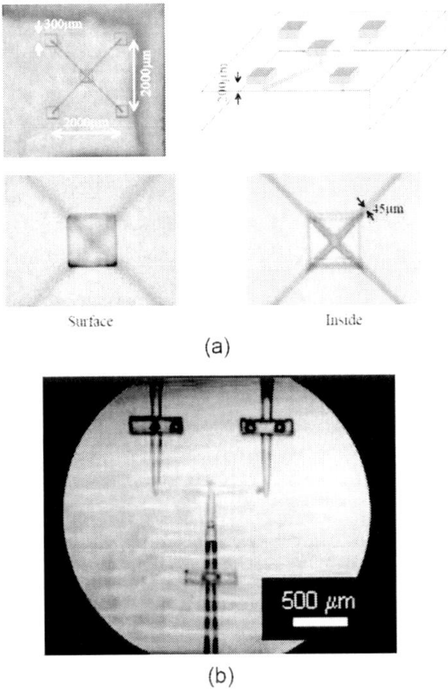

Figure 10.2 (a) A large-scale crossed-channel microstructure [29]. (b) Optical micrograph of a 3D vertical microfluidic structure fabricated inside Foturan glass by femtosecond laser direct writing followed by programmed annealing and successive etching in 10% HF solution for 75 min [31].

Hnatovsky *et al.* achieved a substantial breakthrough in 2005 [34] when they discovered that the etch rate and selectivity strongly depend on the polarization of the writing pulses. This is due to the formation of periodic nanograting-like structures that are always oriented perpendicular to the laser polarization in the laser-affected regions [35, 36]. These nanogratings consist of alternating regions with high and low etch rates. If the nanogratings are oriented perpendicular to the microchannel, the alternating regions will block the HF solution from entering the microchannel, which results in the lowest etch rate. Therefore, to achieve the highest etch rate, the nanogratings must be oriented parallel to the microchannel; this has the potential to enhance the etch rate by two orders of magnitude relative to that obtained when the nanogratings are oriented perpendicular to the microchannel. In addition, it has recently been found that circularly polarized femtosecond laser pulses are the most suitable form of irradiation when fabricating complex 3D microfluidic structures consisting of portions with different orientations because they produce uniform etch rates in the microchannel irrespective of the scanning direction [37]. With these optimizations, the etch selectivity induced by femtosecond laser irradiation in fused silica is comparable to that in photosensitive glass.

Very recently, Kiyama *et al.* found that the etch selectivity can be further enhanced by using KOH solution instead of dilute HF acid solution as the etchant [38]. Figure 10.3 shows microfluidic channels obtained by immersing fused silica samples that had been irradiated by femtosecond laser pulses in HF and KOH solutions. It clearly shows that the channels formed using KOH have much more uniform diameters. A 1-cm-long through channel was obtained using KOH etchant; it has a diameter of less than 60 μm near its ends and a much smaller diameter in the middle. These results were obtained without using an ultrasonic bath; thus, there is potential to further improve the microchannel homogeneity.

Microfluidic channels fabricated by femtosecond laser pulses generally need to satisfy certain requirements for practical applications. The first requirement is that it should be possible to control the cross-sectional aspect ratio; however, the diffraction-limited focal spot created by an objective lens is generally elongated along the optical axis. Therefore, several beam shaping techniques have been developed to tailor the focal spot shape, including the use of elliptical incident beams [39, 40], orthogonally crossed foci [41], and spatio-temporally focused femtosecond pulses [42]. Currently, slit-beam shaping to produce elliptical beams is widely used for writing microfluidic channels and optical waveguides due to its ease of implementation. In the future, spatio-temporal focusing may be further investigated due to its unique ability to create a 3D-homogeneous spherical focal volume using only a single objective lens.

The second requirement is that the taper angle of the microchannel must be minimized as inhomogeneity in the diameter will induce variations in the fluid flow velocity. Although the taper angle can be straightforwardly reduced by enhancing the etch selectivity, the problem cannot be completely solved by this approach as the etch selectivity cannot be increased without limit [38]. One approach to produce homogeneous microchannels is to pre-compensate the taper by suitably wobbling the glass sample during laser irradiation [43]; however, this strategy can only be used when fabricating wide

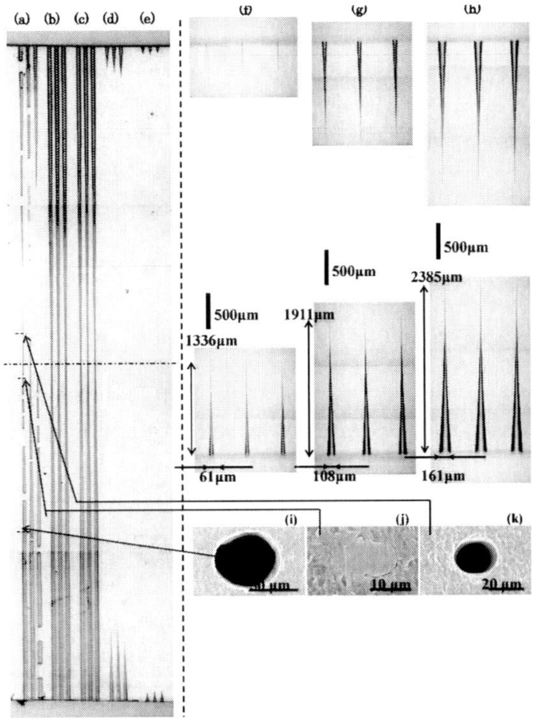

Figure 10.3 Comparison of etching profiles in a 9.2-mm-long silica substrate (9.2 × 10 × 0.5 mm) produced by etching by 10 M (35.8%) aqueous KOH (20 mL) and 2.0% aqueous HF (20 mL) observed by reflectance optical microscopy. (a–e) Channels fabricated after soaking for 60 h in KOH at 80°C after irradiation by femtosecond pulse trains (interval: 0.1 mm) focused by a 40× objective lens (NA: 0.65) at a depth of 10 mm below the surface for pulse energies of (a) 500, (b) 400, (c) 300, (d) 200, and (e) 100 nJ. (f–h) Two images of each set of etching fronts, which represent the time evolution of channel formation in aqueous HF solution at ambient temperature. A specimen that had been irradiated by pulses with energies of 360 nJ through a 40× objective lens (NA: 0.65) was submerged in the solution for (f) 24, (g) 48, and (h) 72 h. Pulse trains were irradiated from top to bottom with the electric vector of the laser beam oriented parallel to the scanning direction. Laser writing was performed three times at each laser pulse energy to confirm reproducibility. Rows (i–k) show cross-sectional Field-Emission Scanning Electron Microscope (FE-SEM) images of the channel on the far left in (a), showing open holes (i and k) and a hole filled with precipitates (j). Reprinted with permission from [38]. Copyright 2009 American Chemical Society.

channels and the channel length is limited to a few millimeters. To overcome these problems, glass drawing can be applied to a glass sample containing a tapered microfluidic channel [44]. As shown in Fig. 10.4, the microchannel undergoes dramatic changes during drawing. The microchannel cross section becomes perfectly symmetrical and the channel diameter becomes uniform along the entire channel length. This technique allows thin microfluidic channels to be produced with extremely long lengths. A 12 mm-long microfluidic channel with a diameter of only ~10 μm was fabricated, as shown in Fig. 10.4. An additional benefit of using glass drawing is that the RMS roughness of the inner surface of the microchannel is improved from ~50 to ~0.3 nm. This excellent surface smoothness will certainly be beneficial for optical and optofluidic applications.

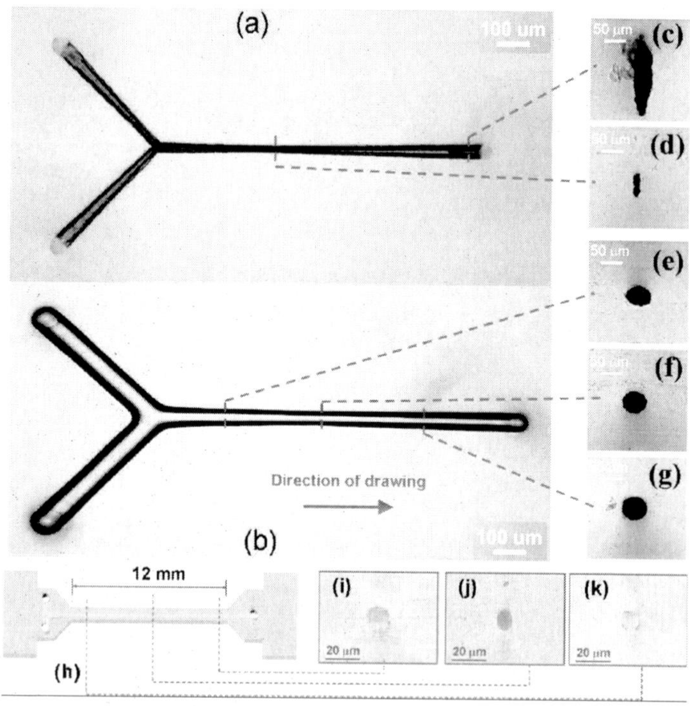

Figure 10.4 Optical micrographs of Y-branched channels. Top view of the channels (a) before and (b) after drawing, and (c–g) cross-sectional views of the channels at the locations indicated by the dashed lines. (h) Schematic drawing of a 12-mm-long channel. (i–k) Cross sections at the locations indicated by the dashed lines [44].

In addition to glass materials, embedded microfluidic channels have also been fabricated in sapphire by femtosecond laser irradiation and subsequent wet chemical etching using either HF or KOH as the etchant [45, 46]. A significant advantage of using sapphire instead of glass for microfluidic applications is that the contrast ratio in etch selectivity between unmodified areas and areas modified by femtosecond laser irradiation of sapphire ($\sim 10^4$:1) is much higher than that of glass ($\sim 10^2$:1). Consequently, microfluidic channels with lengths of ~1 mm and diameters of ~1 μm have been directly formed in sapphire with aspect ratios (i.e., the ratio between the length and diameter) as high as ~1,000. On the other hand, sapphire has a very low etch rate and its etched surfaces are rough.

10.2.2 Liquid-Assisted Femtosecond Laser 3D Drilling

3D microfluidic structures in glass materials can also be fabricated by femtosecond laser 3D drilling in distilled water from the rear surface of the glass; this process is frequently referred to as liquid-assisted femtosecond laser drilling [47]. The water introduced into the microchannel can help remove ablation debris, greatly enhancing the drilling length relative to that achievable by drilling in ambient air. 3D microdrilling is realized by translating the focal spot in the sample in the transverse direction. Compared with fabricating microfluidic channels by femtosecond-laser-assisted chemical etching, this technique is easier to implement and is more environmentally friendly. More importantly, since it does not rely on generating etch selectivity in materials by femtosecond laser irradiation, it can be applied to any material that is transparent to the writing pulses [48]. However, even with the assistance of water, debris generated by femtosecond laser ablation still clogs the microchannel when the depth/length of the channel reaches several hundreds of micrometers, restricting the size of fabricated microstructures [49]. Employing an ultrasonic bath can enhance the debris removal rate, allowing the channel length to be extended to nearly 1 mm [50]. Nevertheless, this length scale is still too short for many microfluidic applications.

On the other hand, liquid-assisted femtosecond laser drilling can be used to fabricate smaller diameter microchannels than chemical etching. The reason for this has not been systematically investigated, but it is probably at least partially due to the fact that chemical etching affects not only the laser-modified region but also unmodified regions because of the limited etch selectivity. Water-assisted femtosecond laser 3D drilling has been used to produce the smallest diameters achieved to date for microfluidic channels fabricated by femtosecond laser irradiation, as shown in Fig. 10.5 [51]. These subsurface nanochannels, which have diameters of only ~700 nm and can have arbitrary geometries, were fabricated using low-energy femtosecond laser pulses tightly focused by a high-numerical-aperture (NA) objective lens. The channel lengths are again limited to several hundred micrometers because debris cannot be expelled from longer channels by bubbles created in water by laser irradiation [51].

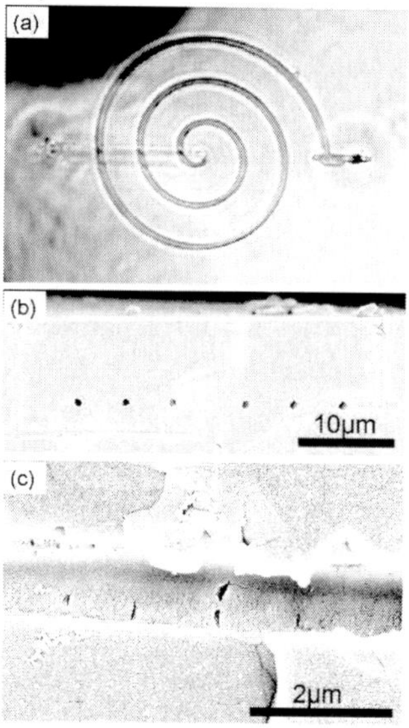

Figure 10.5 A spiral channel demonstrates the ability to produce long channels in a small space for separation, hydrodynamic flow resistance, or to allow mixing. (a) Transmitted light microscopy image of the spiral. The spiral was machined using pulses with energies of 18 nJ to produce a channel diameter of 900 nm and a length of 143 μm. (b) SEM image of cross section of the spiral revealing that it does not contain debris. Scale bar: 10 μm. (c) Close up of a segment of the spiral showing the roughness of the inner channel surface. Scale bar: 2 μm. [51]. Courtesy of the authors.

The fundamental limit on the length of microfluidic channels fabricated by femtosecond laser micromachining has recently been exceeded by using water-assisted drilling on a porous glass [52]. Figure 10.6 shows a schematic view of the experimental setup (Fig. 10.6(a)) and a flow diagram of the fabrication process (Fig. 10.6(b)). The main fabrication process involves two steps: (1) direct formation of hollow microchannels in a porous glass substrate immersed in water by femtosecond laser ablation and (2) post-annealing at ~1150°C to consolidate the porous glass substrate. Unlike glass drawing, which can only fabricate 1D microfluidic structures with unlimited lengths, the new technique permits 3D microfluidic channels to be fabricated inside glass with arbitrary lengths and arbitrary geometries. To demonstrate this, Fig. 10.6(b) shows a square-wave channel with a total length of ~14 mm and a diameter of ~64 μm produced ~250 μm below the glass surface.

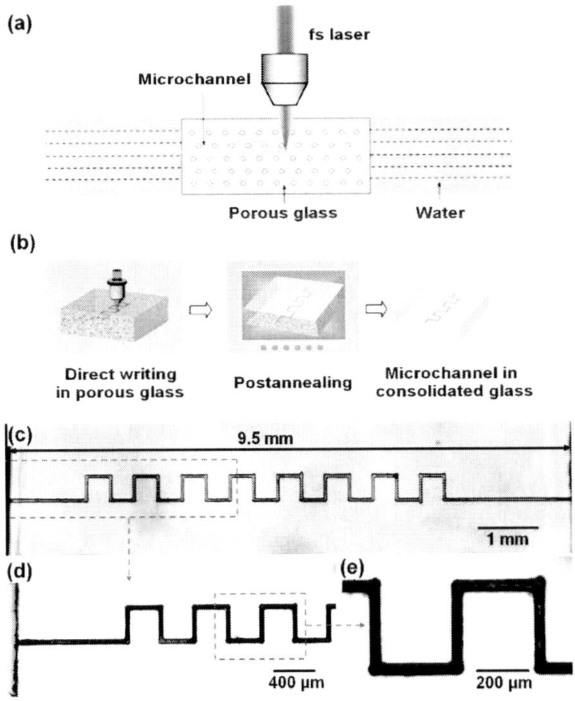

Figure 10.6 (a) Schematic diagram of experimental setup and (b) flow diagram of fabrication process. (c) Top-view micrograph of a 1.6-cm-long microchannel embedded in porous glass before post-annealing. Close-up views of post-annealed microchannel filled with red ink for (d) low and (e) high magnifications [52].

10.3 Fabrication of Micro-Optical Components and Optofluidic Integration in Glass

10.3.1 Free-Space Micro-Optical Components

Since both waveguide-based and free-space micro-optical components have found important applications in optical sensing, it is very important to integrate them with microfluidic components. Femtosecond laser direct writing is a powerful tool for constructing 3D photonic circuits based on waveguides in glass due to its unique ability to induce localized refractive index changes in the vicinity of the focal volume [5, 53]. Real-time tunable waveguides can be realized in glass by filling microchannels with liquids having suitable refractive indices as the core materials [54]. Since chapters 8 and 9 discuss fabrication of optical waveguides by femtosecond laser pulses, we focus on fabrication of free-space micro-optical components in glass here.

The procedures used to fabricate microfluidic components in Foturan can be easily extended to fabricate micro-optical components. To fabricate optical components such as mirrors and lenses in Foturan, a tightly focused femtosecond laser beam is first scanned across a 2D surface to define the contours of the optical component. The same heat treatment and chemical etching steps are then applied to remove material from the laser-affected regions. However, for optical applications, extra care must be taken to achieve smooth surfaces with nanoscale RMS surface roughnesses. The inner walls of embedded 3D structures that are encased by glass cannot be mechanically accessed and thus it is not possible to apply surface polishing. The only technique that can currently be used to smooth inner walls is post-annealing glass samples at a suitable temperature, which can produce a reflow layer on the surface. Micro-optical components such as micromirrors and micro-beamsplitters [11] and microlenses [12, 13] have been fabricated in Foturan in this manner, opening the way for monolithic integration of fluidics and optics in single glass substrates.

The latest progress in this area is the replacement of Foturan by fused silica because the latter has several superior optical properties (e.g., a wide transmission range and low autofluorescence), which are highly desirable for chip-based optical sensing applications [55].

The excellent material uniformity of fused silica also results in the formation of smooth, regular surfaces over a large area, whereas for Foturan, irregular ripple-like structures can randomly appear that can deform the waveform and scatter light [12].

Figures 10.7(a) and (b) show a micro-optical lens fabricated on a fused silica chip that offers a high optical resolution approaching the diffraction limit [56]. Layer-by-layer annular scanning is used to form the micro-optical lens and the surface smoothness was improved by oxyhydrogen (OH) flame polishing. The full width of the focal spot at $1/e^2$ of the peak intensity is ~4 μm at 633 nm (the He–Ne laser wavelength), which is close to the theoretically calculated value. Indeed, the microlens can be used for two-photon fluorescence microscopy with a lateral resolution of ~1.7 μm and an axial resolution of ~12 μm. Figure 10.7 shows two-photon fluorescence micrographs obtained with the microlens (Fig. 10.7(e))

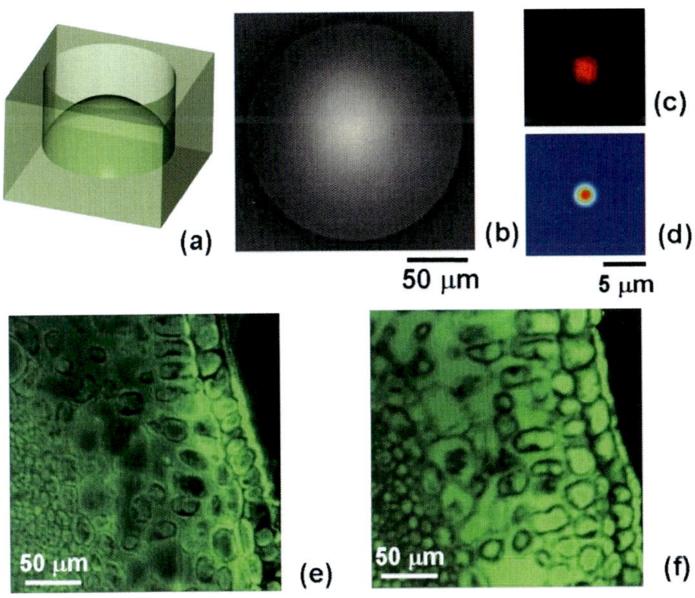

Figure 10.7 (a) Schematic and (b) optical micrograph of microlens fabricated by femtosecond laser micromachining. (c) Measured focal spot and (d) calculated point-spread function of the microlens in (b). Two-photon fluorescence images of the leaf tissue acquired using (e) the microlens and (f) an objective lens (5 × magnification) [55, 56].

and a conventional 5× objective lens (Fig. 10.7(f)), providing clear evidence that the microlenses have a comparable optical resolution to that of a commercial objective lens. In this example, the surface of the microlens was smoothed by OH flame polishing because the microlens was on the top of the fused silica substrate; however, the inner surfaces of embedded structures can be smoothed by post-annealing the fused silica sample at 1150°C, as we demonstrated very recently [57].

10.3.2 Integrated Optofluidic Systems

Due to the compatibility of the above-mentioned techniques for fabricating both microfluidic and micro-optical components in glass, optofluidic integration is straightforward by femtosecond laser micromachining. A typical example of an integrated optofluidic system currently under intensive development is the miniaturization of flow-cytometer type systems by simultaneously fabricating microfluidic channels and waveguides in glass using femtosecond laser pulses [17, 18, 20, 58–63]. Some of these innovative microdevices have already been employed in biomedical applications, which are described in the next section.

In addition to optical waveguides, free-space optics can be incorporated into microfluidic networks fabricated in glass to produce compact and robust microdevices [14, 64, 65]. Figure 10.8 shows a 3D microfluidic dye laser embedded in glass that has an optical ring cavity composed of four 45° micromirrors perpendicular to the substrate, a horizontal microfluidic chamber that is ~400 μm below the glass surface, and a microfluidic through channel that passes through the center of the microchamber. Lasing is observed when the microchamber is filled with a gain medium of rhodamine 6G (Rh6G) and then pumped by a frequency-doubled neodymium-doped yttrium aluminum garnet (Nd:YAG) laser [14].

Figure 10.9 shows another 3D integrated optofluidic system in which a 6 mm-long optical waveguide is connected to a microfluidic chamber with dimensions of 1.0 mm × 1.0 mm × 1.0 mm, and two microlenses with a radius of curvature of 0.75 mm are separately arranged on the left side of the microchamber for fluorescence measurements and on the opposite side from the optical waveguide across the microchamber for absorption measurements. Experimental demonstration of optical biosensing using the integrated microchip revealed that fluorescence analysis and absorption measurements of liquid samples can be performed with efficiencies enhanced by factors of 8 and 3, respectively, due to the micro-optical lens [64].

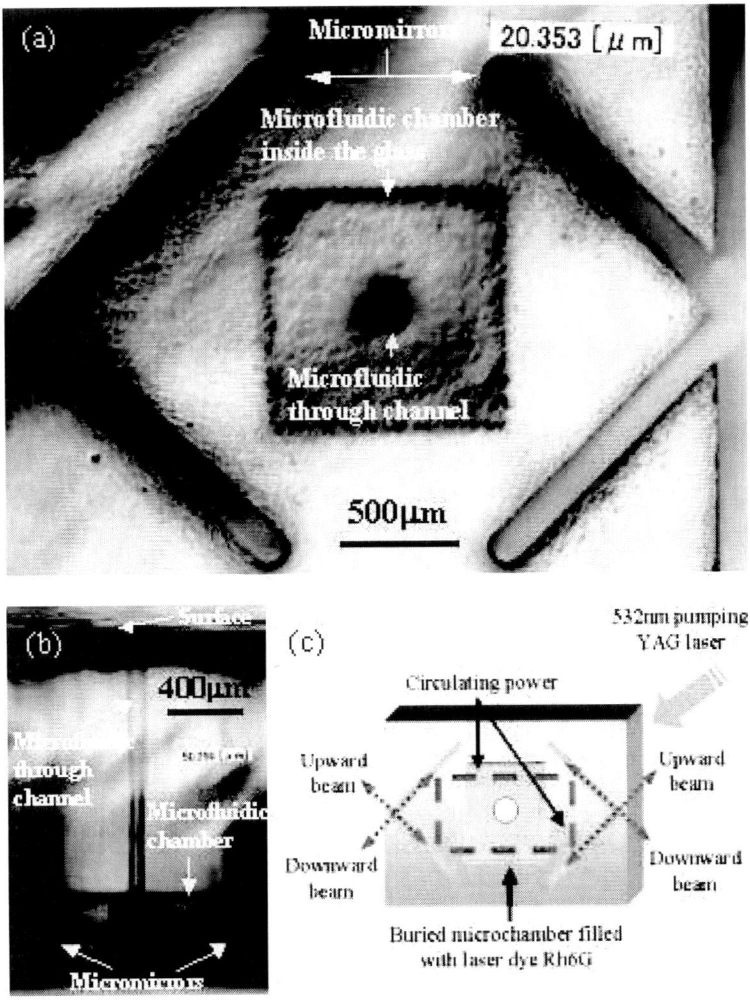

Figure 10.8 (a) Top-view optical micrograph of microfluidic laser; (b) side view optical micrograph of microfluidic chamber and through channel; and (c) illustration of the light path in microfluidic laser [14].

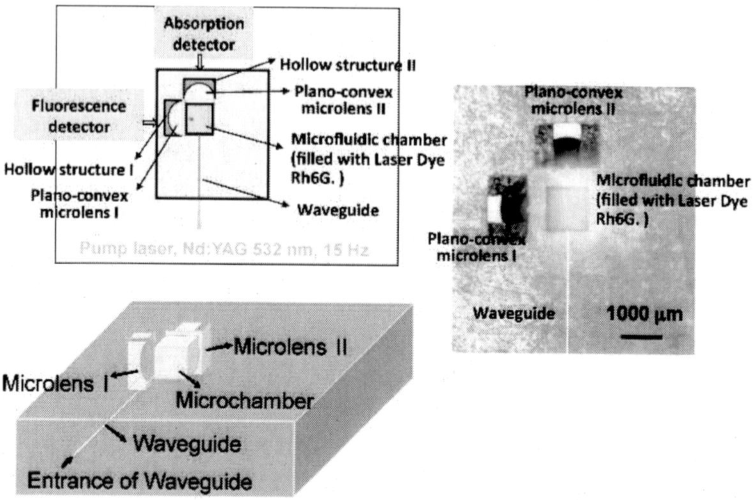

Figure 10.9 2D (upper left) and 3D (lower left) schematic configurations of a biophotonic microchip in which micro-optic components, such as micro-optical planoconvex lenses and an optical waveguide, are integrated with a microfluidic chamber in a single glass chip. Optical microscope image of top view of a fabricated microchip is shown on the right-hand side [64].

10.4 Applications of Microfluidic and Integrated Optofluidic Chips in Biomedical Research

10.4.1 Dynamic Observation of Living Cells Using Nanoaquariums

Despite their brief history, microfluidic chips and integrated optofluidic devices fabricated by femtosecond laser micromachining are attracting tremendous interest for biological and biomedical research. This field is currently undergoing rapid growth, as evidenced by the increasing number of papers published in the last few years [15–20, 58–63, 66, 67]. Three-dimensional microfluidic chips fabricated by femtosecond laser micromachining can provide a platform for observing microorganisms in a confined environment; thus, they have been termed nanoaquariums [15, 66]. Figures 10.10(a) and (b) show a nanoaquarium fabricated for observing the motion of *Euglena gracilis*. It is composed of a 1 mm-long channel

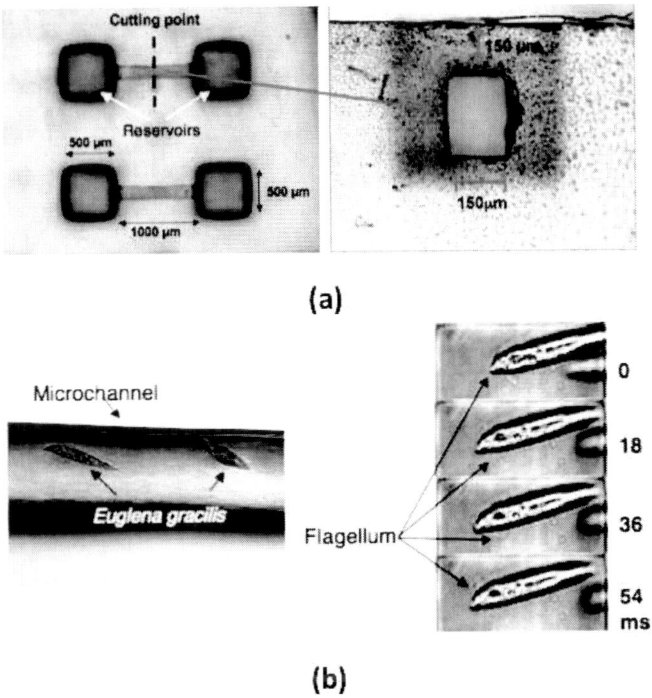

Figure 10.10 (a) Nanoaquarium fabricated for observing the motion of *Euglena gracilis*. (b) Microscope images of *Euglena gracilis* swimming in a microchannel [15].

with a cross section of 150 μm × 150 μm embedded 150 μm below the glass surface and two open reservoirs with dimensions of 500 μm × 500 μm connected to both ends of the channel for introducing *Euglena gracilis* in water [15]. Figure 10.10(b) shows microscope images of *Euglena gracilis* swimming in the microchannel. Such dynamic observation is impossible using conventional cover glass slips that only confine microorganisms in a 2D space since living organisms will quickly escape from the field of view of a microscope. In addition, the vertical reservoirs at the two ends of the microchannel allow *Euglena gracilis* to be observed from either its front or rear side, which again is impossible by conventional methods. Due to the flexibility of femtosecond laser direct writing, a variety of nanoaquariums with different geometries and functionalities can be easily fabricated for various applications, including determining the information transmission process in *Pleurosira laevis*, observing the high-speed motion of *Cryptomonas*, and attaching *Phormidium* to seedling roots to promote the growth of *Komatsuna* [66, 67].

The functionalities of the nanoaquariums have recently been greatly enhanced by incorporating integrated optical waveguides or optical attenuators, both fabricated by femtosecond laser direct writing in Foturan glass after the formation of a hollow microfluidic channel [16]. In this case, the nanoaquarium was used to elucidate the gliding mechanism of *Phormidium*, which is a genus of filamentous cyanobacteria that forms endosymbiotic associations with seedling roots to accelerate the growth of vegetable seedlings. For this purpose, two optical waveguides intersected by a microfluidic channel were integrated into the nanoaquarium to allow the pH change of the water containing the seedling root or the CO_2 in the microchannel to be quantitatively determined by absorption spectroscopy. The results suggest that CO_2 secreted from the seedling root is the attractant that induces the gliding movement of *Phormidium*. It was also found that the gliding movement of *Phormidium* is sensitive to the illumination conditions (e.g., the intensity and spectral characteristics of the illumination light). Therefore, an elegant method has been devised to investigate the effect of illumination, namely to control the optical transmission of visible light by integrating optical attenuators around a portion of the microfluidic channel in the nanoaquarium, as shown in Fig. 10.11(a). The optical attenuators were formed by irradiating the glass sample with a femtosecond laser beam after the microfluidic channel had been fabricated and it was then annealed again (but not etched). After annealing the microchip, the laser-irradiated regions turned brown due to the formation of lithium metasilicate crystallites, as can be seen in Fig. 10.11(b), which act as optical attenuators. The optical transmission can be tuned by stacking different numbers of layers of attenuators. Figure 10.11(b) shows that *Phormidium* glided to the root in the microfluidic channels when five or fewer layers of optical attenuators were used, but that it remained at the entrance of the channel even after 2 h when six or more layers of optical attenuators were used. This indicates that there a certain threshold intensity for the white light illumination needs to be exceeded before *Phormidium* glides to a seedling root. Further investigation reveals that only light in the red wavelength region (i.e., in the spectral range 640–700 nm) effectively promotes *Phormidium* gliding. These results are important for developing ways to accelerate vegetable seedling growth.

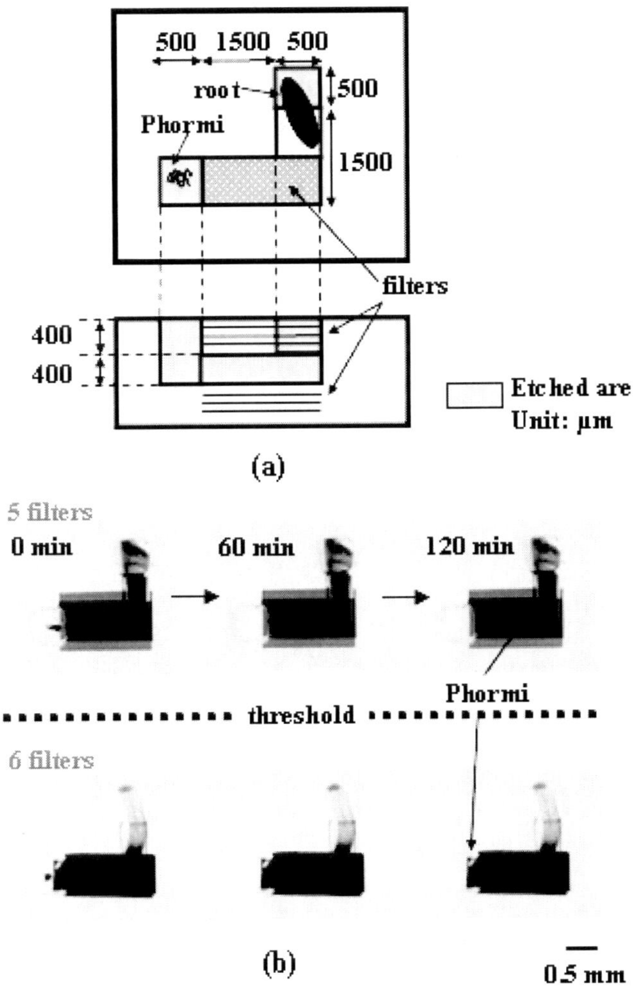

Figure 10.11 Integrated microchip design and investigation of white light intensity required to initiate *Phormidium* gliding. (a) Structural diagrams of microchip with integrated optical attenuators. The same numbers of layers of optical attenuators were formed above and below the microfluidic channel to control the white light intensity in the microfluidic channel. (b) Sequential microscope images of the gliding movement in the microchip. *Phormidium* glided to the seedling root in the microfluidic channel covered with up to five layers of optical attenuators, but it did not glide to the root when there were six layers of optical attenuators, even after 2 h [16].

10.4.2 Optical Sensing with Integrated Optofluidic Chips

At present, most integrated optofluidic chips fabricated by femtosecond laser direct writing consist of integrated micro-optical waveguides and microfluidic channels because the techniques for fabricating both structures are already mature. Micro optical waveguides perpendicular to the microfluidic channels offer various functions and can function as illumination sources and scattering/fluorescence light collectors [17, 18, 20, 58–62]. In addition, they can be used to construct a Mach–Zehnder interferometer (MZI) for spatially selective refractive index sensing [17], as illustrated in Fig. 10.12(a). Figure 10.12(b) shows a microscope image of the microfluidic channel and the two arms of the MZI. The two arms are distributed in a tilted plane such that the reference arm is constructed above the microchannel and the sensing arm and the microchannel intersect at right angles. Since the MZI and the microchannel are not coplanar, such 3D microdevice cannot be fabricated by conventional lithographic techniques due to their intrinsically planar nature. The microchannel was filled with glucose solution as a test sample and the sensitivity of the microdevice in Fig. 10.12 was determined to be 10^{-4} refractive index units (RIU), which corresponds to a detection limit of 4 mM [17]. Further improvement of the sensitivity and detection limit is anticipated by reducing temperature fluctuations in the microfluidic channel, which is a major cause of fluctuations in the refractive index of the analyte in the current experiments.

10.5 Conclusions

We have demonstrated that both microfluidic chips and integrated optofluidic devices can be fabricated in glass materials by femtosecond laser micromachining. To date, this is still the only approach that can be used to fabricate hollow structures with nearly arbitrary 3D geometries buried in a variety of glass materials. Technical advances over the last few years have enabled the fabrication of long channels with uniform diameters and micro-optical components that have performances approaching those of bulk optics. In particular, various microfluidic chips and optofluidic devices fabricated by this technique have been increasingly used in biological and biomedical research.

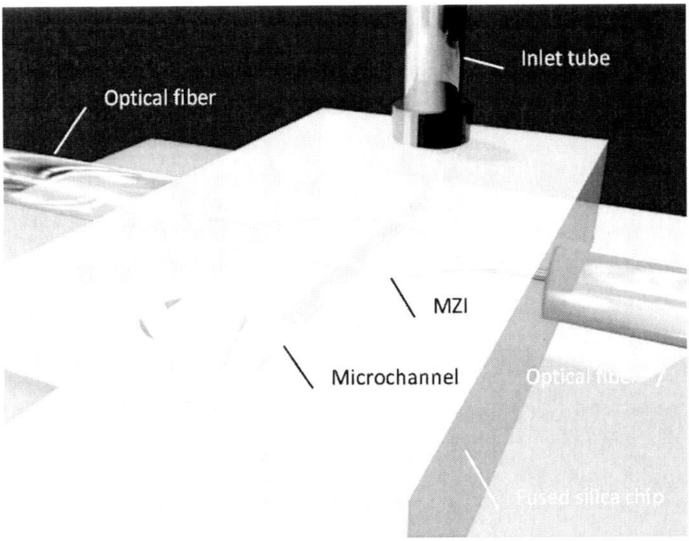

(a)

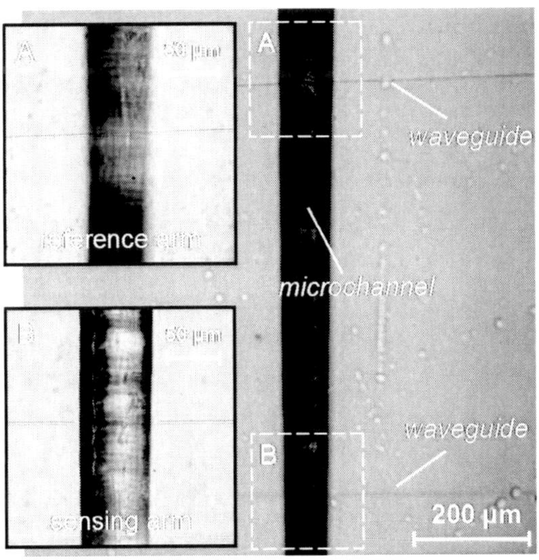

(b)

Figure 10.12 (a) Schematic diagram of femtosecond-laser-fabricated microfluidic channel and integrated MZI. Because of the tilted geometry, the sensing arm intersects the channel orthogonally, whereas the reference arm passes over it. (b) Microscope image showing the two arms of the MZI crossing the microfluidic channel [17]. Courtesy of the authors.

The major disadvantage of fabricating microfluidic and integrated optofluidic structures using femtosecond laser micromachining is the high cost of the amplified femtosecond laser system. In addition, femtosecond laser direct writing is a serial processing technique, making it inherently less efficient than projection lithography. To solve these problems, novel femtosecond lasers need to be developed that are inexpensive, have high average powers and repetition rates to permit high-throughput 3D micromachining. Nevertheless, the fabrication time is acceptable for optofluidic systems that mainly consist of 1D structures such as microchannels and waveguides. Fabrication of microfluidic and integrated optofluidic systems by femtosecond laser micromachining is still in its infancy; the future development of this technique will lead to innovative devices that cannot be realized by any existing technique.

Acknowledgments

The authors would like to sincerely thank Dr. Y. Hanada, Dr. H. Kawano, Mr. M. Masuda, and Dr. K. Midorikawa of RIKEN, and Dr. F. He, Dr. Y. Liao, Dr. L. Qiao, and Prof. Z. Xu of SIOM for their contributions.

References

1. Manz, A., Graber, N., and Widmer, H. M. (1990). Miniaturized total chemical analysis systems: A novel concept for chemical sensing, *Sens. Actuators, B***1**, 244.
2. Whitesides, G. M. (2006). The origins and the future of microfluidics, *Nature*, **442**, 368.
3. Monat, C., Domachuk, P., and Eggleton, B. J. (2007). Integrated optofluidics: A new river of light, *Nat. Photon.*, **1**, 106.
4. McDonald, J. C. and Whitesides, G. M. (2002). Poly(dimethylsiloxane) as a material for fabricating microfluidic devices, *Acc. Chem. Res.*, **35**, 491.
5. Davis, K. M., Miura, K., Sugimoto, N., and Hirao, K. (1996). Writing waveguides in glass with a femtosecond laser, *Opt. Lett.*, **21**, 1729.
6. Ams, M., Dekker, P., Marshall, G. D., and Withford, M. J. (2009) Monolithic 100 mW Yb waveguide laser fabricated using the femtosecond-laser direct-write technique, *Opt. Lett.*, **34**, 247.

7. Marcinkevičius, A., Juodkazis, S., Watanabe, M., Miwa, M., Matsuo, S., Misawa, H., and Nishii, J. (2001). Femtosecond laser-assisted three-dimensional microfabrication in silica, *Opt. Lett.*, **26**, 277.

8. Masuda, M., Sugioka, K., Cheng, Y., Aoki, N., Kawachi, M., Shihoyama, K., Toyoda, K., Helvajian, H., and Midorikawa, K. (2003). 3-D microstructuring inside photosensitive glass by femtosecond laser excitation, *Appl. Phys.*, *A***76**, 857.

9. Masuda, M., Sugioka, K., Cheng, Y., Hongo, T., Shihoyama, K., Takai, H., Miyaloto, I., and Midorikawa, K. (2004). Direct fabrication of freely movable microplate inside photosensitive glass by femtosecond laser for lab-on-chip application, *Appl. Phys, A***78**, 1029.

10. Matsuo, S., Kiyama, S., Shichijo, Y., Tomita, T., Hashimoto, S., Hosokawa, Y., and Masuhara, H. (2008). Laser microfabrication and rotation of ship-in-a-bottle optical rotators, *Appl. Phys. Lett.*, **93**, 051107.

11. Cheng, Y., Sugioka, K., Midorikawa, K., Masuda, M., Toyoda, K., Kawachi, M., and Shihoyama, K. (2003). Three-dimensional micro-optical components embedded in photosensitive glass by a femtosecond laser, *Opt. Lett.*, **28**, 1144.

12. Cheng, Y., Tsai, H. L., Sugioka, K., and Midorikawa, K. (2006). Fabrication of 3D micro-optical lenses in photosensitive glass using femtosecond laser micromachining, *Appl. Phys.*, *A***85**, 11.

13. Wang, Z., Sugioka, K., and Midorikawa, K. (2007). Three dimensional integration of micro-optical components buried inside photosensitive glass by femtosecond laser direct writing, *Appl. Phys., A***89**, 951.

14. Cheng, Y., Sugioka, K., and Midorikawa, K. (2004). Microfluidic laser embedded in glass by three-dimensional femtosecond laser microprocessing, *Opt. Lett.*, **29**, 2007.

15. Hanada, Y., Sugioka, K., Kawano, H., Ishikawa, I. S., Miyawaki, A., and Midorikawa, K. (2008). Nano-aquarium for dynamic observation of living cells fabricated by femtosecond laser direct writing of photostructurable glass, *Biomed. Microdevices*, **10**, 403.

16. Hanada, Y., Sugioka, K., Ishikawa, I. S., Kawano, H., Miyawaki, A., and Midorikawa, K. (2011). 3D microfluidic chips with integrated functional microelements fabricated by a femtosecond laser for studying the gliding mechanism of cyanobacteria, *Lab Chip*, **11**, 2109.

17. Crespi, A., Gu, Y., Ngamsom, B., Hoekstra, H. J. W. M., Dongre, C., Pollnau, M., Ramponi, R., van den Vlekkert, H. H., Watts, P., Cerullo, G., and Osellame, R. (2010). Three-dimensional Mach–Zehnder interferometer in a microfluidic chip for spatially-resolved label-free detection, *Lab Chip*, **10**, 1167.

18. Kim, M., Hwang, D. J., Jeon, H., Hiromatsu, K., and Grigoropoulos, C. P. (2009). Single cell detection using a glass-based optofluidic device fabricated by femtosecond laser pulses, *Lab Chip*, **9**, 311.
19. Bragheri, F., Ferrara, L., Bellini, N., Vishnubhatla, K. C., Minzioni1, P., Ramponi, R., Osellame, R., and Cristiani, I. (2010). Optofluidic chip for single cell trapping and stretching fabricated by a femtosecond laser, *J. Biophoton.*, **3**, 234.
20. Schaap, A., Bellouard, Y., and Rohrlack, T. (2011). Optofluidic lab-on-a-chip for rapid algae population screening, *Biomed. Opt. Express*, **2**, 658.
21. Hanada, Y., Sugioka, K., and Midorikawa, K. (2008). Selective metallization of photostructurable glass by femtosecond laser direct writing for biochip application, *Appl. Phys.*, *A***90**, 603.
22. Xu, J., Liao, Y., Zeng, H., Zhou, Z., Sun, H., Song, J., Wang, X., Cheng, Y., Xu, Z., Sugioka, K., and Midorikawa, K. (2007). Selective metallization on insulator surfaces with femtosecond laser pulses, *Opt. Express,* **15**, 12743.
23. Liao, Y., Xu, J., Cheng, Y., Zhou, Z., He, F., Sun, H., Song, J., Wang, X., Xu, Z., Sugioka, K., and Midorikawa, K. (2008). Electro-optic integration of embedded electrodes and waveguides in $LiNbO_3$ using a femtosecond laser, *Opt. Lett.*, **33**, 2281.
24. Stookey, S. D. (July 18, 1950). Photosensitive gold glass and method of making it, *U.S. Pat.*, No. 2, 515, 937.
25. Hansen, W. W., Janson, S. W., and Helvajian, H. (1997). Direct-write UV-laser microfabrication of 3D structures in lithium-aluminosilicate glass, *Proc. SPIE,* **2991**, 104.
26. Kondo, Y., Qiu, J. R., Mitsuyu, T., Hirao, K., and Yoko, T. (1999). Three-dimensional microdrilling of glass by multiphoton process and chemical etching, *J. Jpn. Appl. Phys.*, **38**, L1146.
27. Hongo, T., Sugioka, K., Niino, H., Cheng, Y., Masuda, M., Miyamoto, I., Takai, H., and Midorikawa, K. (2005). Investigation of photoreaction mechanism of photosensitive glass by femtosecond laser, *J. Appl. Phys.*, **97**, 063517.
28. Fuqua, P., Janson, S. W., Hansen, W. W., and Helvajian, H. (1999). Fabrication of true 3D microstructures in glass/ceramic materials by pulsed UV laser volumetric exposure techniques, *Proc. SPIE*, **3618**, 213.
29. Cheng, Y., Sugioka, K., Masuda, M., Toyoda, K., Kawachi, M., Shihoyama, K., and Midorikawa, K. (2003). 3D microstructuring inside Foturan glass by femtosecond laser, *RIKEN Rev.*, **50**, 101.

30. Cheng, Y., Sugioka, K., Midorikawa, K., and Xu, Z. (2006). Integrating 3D photonics and microfluidic using ultrashort laser pulses, *SPIE Newsroom,* http://spie.org/x8513.xml.

31. Sugioka, K., Masuda, M., Hongo, T., Cheng, Y., Shihoyama, K., and Midorikawa, K. (2004). Three-dimensional microfluidic structure embedded in photostructurable glass by femtosecond laser for lab-on-chip application, *Appl. Phys.*, *A***78**, 815.

32. Sugioka, K., Cheng, Y., and Midorikawa, K. (2005). Three-dimensional micromachining of glass using femtosecond laser for lab-on-a-chip device manufacture, *Appl. Phys.*, *A***81**, 1.

33. Bellouard, Y., Said, A., Dugan, M., and Bado, P. (2004). Fabrication of high-aspect ratio, micro-fluidic channels and tunnels using femtosecond laser pulses and chemical etching, *Opt. Express*, **12**, 2120.

34. Hnatovsky, C., Taylor, R. S., Simova, E., Bhardwaj, V. R., Rayner, D. M., and Corkum, P. B. (2005). Polarization-selective etching in femtosecond laser-assisted microfluidic channel fabrication in fused silica, *Opt. Lett.*, **30**, 1867.

35. Shimotsuma, Y., Kazansky, P. G., Qiu, J., and Hirao, K. (2003). Self-organized nanogratings in glass irradiated by ultrashort light pulses, *Phys. Rev. Lett.*, **91**, 247405.

36. Bhardwaj, V. R., Simova, E., Rajeev, P. P., Hnatovsky, C., Taylor, R. S., Rayner, D. M., and Corkum, P. B. (2006). Optically produced arrays of nano-planes inside fused silica, *Phys. Rev. Lett.*, **96**, 057404.

37. Yu, X., Liao, Y., He, F., Zeng, B., Cheng, Y., Xu, Z., Sugioka, K., and Midorikawa, K. (2011). Tuning etch selectivity of fused silica irradiated by femtosecond laser pulses by controlling polarization of the writing pulses, *J. Appl. Phys.*, **109**, 053114.

38. Kiyama, S., Matsuo, S., Hashimoto, S., and Morihira, Y. (2009). Examination of etching agent and etching mechanism on femtosecond laser microfabrication of channels inside vitreous silica substrates, *J. Phys. Chem.*, *C***113**, 11560.

39. Cheng, Y., Sugioka, K., Midorikawa, K., Masuda, M., Toyoda, K., Kawachi, M., and Shihoyama, K. (2003). Control of the cross-sectional shape of a hollow microchannel embedded in photostructurable glass by use of a femtosecond laser, *Opt. Lett.*, **28**, 55.

40. Osellame, R., Taccheo, S., Marangoni, M., Ramponi, R., Laporta, P., Polli, D., Silvestri, S. D., and Cerullo, G. (2003). Femtosecond writing of active optical waveguides with astigmatically shaped beams, *J. Opt. Soc. Am.*, *B***20**, 1559.

41. Sugioka, K., Cheng, Y., Midorikawa, K., Takase, F., and Takai, H. (2006). Femtosecond laser microprocessing with three-dimensionally isotropic spatial resolution using crossed-beam irradiation, *Opt. Lett.*, **31**, 208.

42. He, F., Xu, H., Cheng, Y., Ni, J., Xiong, H., Xu, Z., Sugioka, K., and Midorikawa, K. (2010). Fabrication of microfluidic channels with a circular cross section using spatiotemporally focused femtosecond laser pulses, *Opt. Lett.*, **35**, 1106.

43. Vishnubhatla, K. C., Bellini, N., Ramponi, R., Cerullo, G., and Osellame, R. (2009). Shape control of microchannels fabricated in fused silica by femtosecond laser irradiation and chemical etching, *Opt. Express*, **17**, 8685.

44. He, F., Cheng, Y., Xu, Z., Liao, Y., Xu, J., Sun, H., Wang, C., Zhou, Z., Sugioka, K., Midorikawa, K., Xu, Y., and Chen, X. (2010). Direct fabrication of homogeneous microfluidic channels embedded in fused silica using a femtosecond laser, *Opt. Lett.*, **35**, 282.

45. Wortmann, D., Gottmann, J., Brandt, N., and Horn-Solle, H. (2008). Micro- and nanostructures inside sapphire by fs-laser irradiation and selective etching, *Opt. Express*, **16**, 1517.

46. Mazuli, M., Juodkazis, S., Ebisui, T., Matsuo, S., and Misawa, H. (2007). Structural characterization of shock-affected sapphire, *Appl. Phys.*, *A***86**, 197–200.

47. Li, Y., Itoh, K., Watanabe, W., Yamada, K., Kuroda, D., Nishii, J., and Jiang, Y. (2001). Three-dimensional hole drilling of silica glass from the rear surface with femtosecond laser pulses, *Opt. Lett.*, **26**, 1912.

48. Kim, T. N., Campbell, K., Groisman, A., Kleinfeld, D., and Schaffer C. B. (2005). Femtosecond laser-drilled capillary integrated into a microfluidic device, *Appl. Phys. Lett.*, **86**, 201106.

49. An, R., Li, Y., Dou, Y., Yang, H., and Gong, Q. H. (2005). Simultaneous multi-microhole drilling of soda-lime glass by water-assisted ablation with femtosecond laser pulses, *Opt. Express*, **13**, 1855.

50. Hwang, D. J., Choi, T. Y., and Grigoropoulos, C. P. (2004). Liquid-assisted femtosecond laser drilling of straight and three-dimensional microchannels in glass, *Appl. Phys.*, *A***79**, 605.

51. Ke, K., E. F. Hasselbrink, Jr., and Hunt, A. J. (2005). Rapidly prototyped three-dimensional nanofluidic channel networks in glass substrates, *Anal. Chem.*, **77**, 5083.

52. Liao, Y., Ju, Y., Zhang, L., He, F., Zhang, Q., Shen, Y., Chen, D., Cheng, Y., Xu, Z., Sugioka, K., and Midorikawa, K. (2010). Three-dimensional

microfluidic channel with arbitrary length and configuration fabricated inside glass by femtosecond laser direct writing, *Opt. Lett.*, **35**, 3225.

53. Wang, Z., Sugioka, K., Hanada, Y., and Midorikawa, K. (2007). Optical waveguide fabrication and integration with a micro-mirror inside photosensitive glass by femtosecond laser direct writing, *Appl. Phys.*, *A***88**, 699.

54. Sun, H., He, F., Zhou, Z., Cheng, Y., Xu, Z., Sugioka, K., and Midorikawa, K. (2007). Fabrication of microfluidic optical waveguides on glass chips with femtosecond laser pulses, *Opt. Lett.*, **32**, 1536.

55. He, F., Cheng, Y., Qiao, L., Wang, C., Xu, Z., Sugioka, K., Midorikawa, K., and Wu, J. (2010). Two-photon fluorescence excitation with a microlens fabricated on the fused silica chip by femtosecond laser micromachining, *Appl. Phys. Lett.*, **96**, 041108.

56. Qiao, L., He, F., Wang, C., Liao, Y., Cheng, Y., Sugioka, K., Midorikawa, K., Pan, C., and Wei, X. B. (2011). Fabrication of a micro-optical lens using femtosecond laser 3D micromachining for two-photon imaging of bio-tissues, *Opt. Commun.*, **284**, 2988.

57. He, F., Lin, J., and Cheng, Y. (2011). Fabrication of hollow optical waveguides in fused silica by three-dimensional femtosecond laser micromachining, *Appl. Phys.*, *B***105**, 379.

58. Bellouard, Y., Said, A. A., Dugan, M., and Bado, P. Monolithic three-dimensional integration of micro-fluidic channels and optical waveguides in fused silica, in *Proceedings of Materials Research Society Fall Meeting Symposium A*, Vol. **782** (Materials Research Society, 2003), 63–68.

59. Cheng, Y., Sugioka, K., and Midorikawa, K. (2005). Freestanding optical fibers fabricated in a glass chip using femtosecond laser micromachining for lab-on-a-chip application, *Opt. Express*, **13**, 7225.

60. Osellame, R., Maselli, V., Vazquez, R. M., Ramponi, R., and Cerullo, G. (2007). Integration of optical waveguides and microfluidic channels both fabricated by femtosecond laser irradiation, *Appl. Phys. Lett.*, **90**, 231118.

61. Vazquez, R. M., Osellame, R., Nolli, D., Dongre, C., Vlekkert, van den H. H, Ramponi, R., Pollnau, M., and Cerullo, G. (2009). Integration of femtosecond laser written optical waveguides in a lab-on-chip, *Lab Chip*, **9**, 91.

62. Applegate, R. W., Squier, J., Vestad, T., Oakey, J., Marr, D. W. M., Bado, P., Dugan, M. A., and Said, A. A. (2006). Microfluidic sorting system based on optical waveguide integration and diode laser bar trapping, *Lab Chip*, **6**, 422.

63. Maselli, V., Grenier, J. R., Ho, S., and Herman, P. R. (2009). Femtosecond laser written optofluidic sensor: Bragg grating waveguide evanescent probing of microfluidic channel, *Opt. Express*, **17**, 11719.
64. Wang, Z., Sugioka, K., and Midorikawa, K. (2008). Fabrication of integrated microchip for optical sensing by femtosecond laser direct writing of Foturan glass, *Appl. Phys.*, *A***93**, 225.
65. Qiao, L., He, F., Wang, C., Cheng, Y., Sugioka, K., and Midorikawa, K. (2011). A microfluidic chip integrated with a micro-optical lens fabricated by femtosecond laser micromachining, *Appl. Phys.*, *A***102**, 179.
66. Hanada, Y., Sugioka, K., Kawano, H., Ishikawa, I. S., Miyawaki, A., and Midorikawa, K. (2009). Nano-aquarium with microfluidic structures for dynamic analysis of *Cryptomonas* and *Phormidium* fabricated by femtosecond laser direct writing of photostructurable glass, *Appl. Surf. Sci.*, **255**, 9893.
67. Sugioka, K., Hanada, Y., and Midorikawa, K. (2010). Three-dimensional femtosecond laser micromachining of photosensitive glass for biomicrochips, *Laser Photon. Rev.*, **4**, 386.

Chapter 11

Fabrication of 3D Functional Microdevices by Two-Photon Photopolymerization

Dong Wu,[a] Xiao-Feng Lin,[a] Qi-Dai Chen,[a] Hong Xia,[a] Yong-Lai Zhang,[a] and Hong-Bo Sun[a,b]
[a]*State Key Laboratory on Integrated Optoelectronics, College of Electronic Science and Engineering, Jilin University, 2699 Qianjin Street, Changchun 130012, China*
[b]*College of Physics, Jilin University, 2699 Qianjin Street, Changchun 130012, China*
hbsun@jlu.edu.cn

With the trend of device miniaturization and integration, the realization of three-dimensional controllable microstructures and functional devices was highly desirable because of their wide applications in micro-optical systems, micromachines, lab-on-a-ship, and integrated MEMS (Micro Electro Mechanical Systems) systems. However, conventional microfabrication technology is not suitable for preparing true three-dimensional (3D) microstructures. Femtosecond laser direct writing, as a kind of advanced 3D microfabrication technique, could realize arbitrary 3D microstructures by point-to-point scanning. In this chapter we will

Ultrafast Laser Processing: From Micro- to Nanoscale
Edited by Koji Sugioka and Ya Cheng

ISBN 978-981-4267-33-5 (Hardcover), 978-981-4303-69-9 (eBook)
www.panstanford.com

introduce the recent progress of three-dimensional functional devices by means of two-photon-induced photopolymerization, mainly on high-precision micro-optical devices, functional micromachines, and 3D microfluidic devices.

11.1 Introduction

The fabrication of various 3D micronanostructures and functional devices in a controlled fashion is of great importance to fluidic [1–5], photonics [6–10], sensors [11–14] and tissue engineering [15–17], which is driving the progress of microfabrication techniques such as electron beam [18–21], focused ion-beam [22–25], ultraviolet photolithography [26, 27] and X-ray lithography [28–29]. Photolithography has been the dominant microfabrication technique in the past fifty years. The electronic components have been fabricated in batches, huge leaps, high speed, and low cost. The feature sizes have been also pushed below a half pitch of 32 nm by recent improvements in the resolution of 193 nm immersion lithography. This achievement is a technological marvel; however, one of the important issues is that photolithography is essentially a planar technique. To build up the 3D microstructures, more complex layer by layer processes were required. Moreover, the control currently available in the vertical dimension does not begin to approach what can be achieved in the other two dimensions. To fabricate 3D controllable microstructures, new techniques such as ink-base writing [30–32], layer-by-layer assemblies [33], and laser interference lithography [34–39], were developed. Ink direct writing utilized a nozzle to deliver inks composed of polymers, colloids, or polyelectrolytes. Some 3D lattice-type scaffolds have been demonstrated with this technique and the resolution could reach below 1 μm. However, the inks are usually not rigid enough to fabricate arbitrary microstructures. Layer-by-layer assemblies could create 3D microstructures by stacking planar patterns. However, this method not only required multistep complex process but also suffered from the fabrication of microstructures with geometric limitation. Multibeam interference lithography is a promising approach to rapidly fabricate large-area and periodic 3D microstructures on the micrometer and nanometer scale. Interference lithography (IL) transferred a periodic light intensity pattern formed by the interference of multiple beams of light into the resin [40–42].

However, they are inherently restricted to locally periodic patterns and therefore are not capable of arbitrary patterning in three dimensions. In the past few years, femtosecond laser direct writing via two-photon polymerization [43–52] has attracted great attention because of its unique 3D microfabrication ability, which emerged as a new technology in 1997 [53]. The resins are not polymerized by absorbing one UV photon, but by simultaneously absorbing two photons at longer wavelength, usually the red-infrared spectral region. Compared to conventional UV single-photon absorption, the two-photon polymerization exhibited many advantages. Due to the longer wavelength, the common polymers have negligible linear absorption and the light could deeply penetrate into the materials. In addition, the TPP rate is proportional to the square of the laser intensity, so high-precision (<50 nm) could be realized under the laser threshold effect, with no restraints on the diffraction limits of the light beam. Owing to the above features, the two-photon polymerization laser microfabrication exhibited intrinsic high-precision 3D fabrication capability by point-to-point scanning. Therefore, TPP has been successfully applied to the fabrication of various complex 3D photonic, micromechanical, and microfluidic devices. Here, we will discuss the basic principles of two-photon polymerization and some important parameters, such as the spatial resolution and surface roughness. Then, we will discuss the recent progress of high-precision 2D and 3D micro-optical devices and their excellent optical properties. In the third part, we will introduce functional micromachines containing light-driven and magnetic-driven microcomponents. At last, we will present the microfluidic devices and biological application of two-photon polymerization.

11.2 Femtosecond Laser Micronanofabrication Based on Two-Photon Photopolymerization

11.2.1 Chemical Processes

Two-photon polymerization (TPP) was a powerful microfabrication technology for various 3D complex microstructures. Its principle is two-photon-induced polymerization, which is one of the most important types of photochemical reactions that have been used for laser microfabrication. Two-photon photopolymerization is to induce the conversion of small unsaturated molecules in the liquid state to solid macromolecules thorough light-induced polymerization reactions, which is the process that occurs through simultaneous absorption of photons via virtual states in a medium. After laser irradiation, the polymer material showed a significant phase transition from liquid to solid. The unpolymerized resin is easily removed by the solvent during the developing process so that 3D structures were obtained. Although it was theoretically predicted by Maria Göppert-Mayer in 1931 [54], the requirement of a high peak-power laser delayed the first experimental evidence of this nonlinear phenomenon for 30 years. For two-photon polymerization, the basic components of the liquid polymer material are monomers and oligomers (or prepolymer). Upon light excitation, the monomers or oligomers was solidified by two means: polymerization and cross-linking [55–57]. An important feature of polymerization is the chain reaction by which macromolecules are created; while cross-linking deals with the formation of cross-links with chemical bonds. Photopolymerization happened via chain reactions (Eq. 11.1).

$$M \xrightarrow{M} M_2 \xrightarrow{M} M_3 \cdots M_{n-1} \xrightarrow{M} M_n \tag{11.1}$$

Here M is the monomer or oligomer unit, and M_n, the macromolecule containing n monomer units. In addition, for practical photopolymer systems, photoinitiators and photosensitizers are needed because the quantum yield of general monomers and oligomers is low. To enhance the initiating efficiency, some low-weight molecules that are more sensitive to light irradiation are added. Photoinitiators form radicals by absorbing one or two photons. The production of active species attack monomers or oligomers, which are the most important step in photopolymerization. Generally, the initiation step is showed as follows:

$$I \xrightarrow{h\nu} I^* \longrightarrow R\cdot \tag{11.2}$$

where I is photoinitiator, radical ($R^\cdot$) and I^* are both intermediate states of the photoinitiator after absorbing a photon. The polymerization process is described by the following equation:

$$R\cdot + M \rightarrow RM\cdot \xrightarrow{M} RMM\cdots\cdot \rightarrow RM_n \tag{11.3}$$

The photoproduced radicals react with monomers or oligomers, producing monomer radicals, which combine with new monomers, and so on; so the monomer radicals expand in a chain reaction, until two radicals meet with each other. Therefore the polymerization process consists of several steps: (i) photoinitiation, (ii) chain propagation, and (iii) termination.

11.2.2 The Spatial Resolution

High fabrication resolution is pursuit in various microfabrication technologies. For example, X ray, electron beam lithography and focused ion beam microfabrication methods have reached the order of 10 nm precision. Two-photon polymerization laser direct writing could realize various complex 3D microstructures by point-to-point scanning, so great attention has been paid for enhancing the resolution to several nanometers. Due to its nonlinear absorption, the TPP have shown the feature widths below the diffraction limit of conventional one-photon process. In 2001, Kuebler *et al.* [58] have reported on the fabrication of woodpile structures with line widths of 200 nm using 730 nm laser excitation by optimizing two-photon initiators and exposure conditions, as shown in Fig. 11.1. They adopted a kind of large two-photon absorption cross-sections photoinitiation (D-π-D chromophores), which D is an electron electron-donor and π is a biphenyl, stilbene, or bis(styryl)benzene

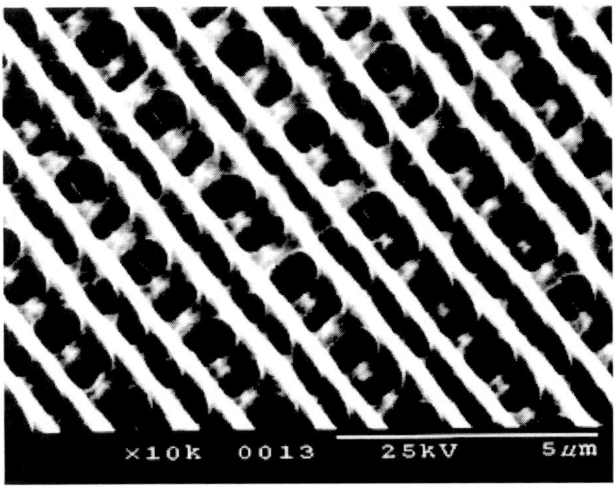

Figure 11.1 SEM image of a photonic bandgap crystal fabricated by TPIP. Excitation wavelength: 730 nm; laser power: 0.45 mW; scan speed: 50 μm/s. The average periodicity is 1 μm and the line width 0.2 μm.

conjugated bridge. A three-dimensional test structure was prepared to evaluate how the intensity and the exposure times affect the shape and dimensions of the polymerization voxel. Experimentally, as the intensity and the exposure times increased, the dimensions of the polymerization voxel became big. In addition, the exposure conditions also affect the degree of cross-linking and the mechanical properties of the structure. The smallest voxel could reach as small as ~200 nm in width and ~700 nm in length.

To realize high-precision functional microdevices, Kawata *et al.* [59, 60] try to design voxels of small size by choosing an appropriate laser pulse energy and exposure time. They believed that the diffraction limit imposed by Rayleigh criteria was simply a measure of focal-spot size, and did not put any restraint on the actual size of solidified voxels. Only the resin at the region with energy above the two-photon absorption (TPA) threshold was solidified. By optimal parameters, the resolution reached as small as 120 nm (Fig. 11.2(c)). A 10 μm-long, 7 μm-high bull, about the size of a red blood cell, was fabricated; the smallest model animals ever made artificially. Moreover, the lateral spatial resolution could be further reduced to around 100 nm (Fig. 11.2(h)) by intentionally introducing radical quenchers in the resin.

In 2007, Duan *et al.* [61] reported a highly sensitive and efficient photoinitiator (anthracene derivative (9,10-bis-pentyloxy-2,7-bis[2-(4-dimethylamino-phenyl)-vinyl]anthracene (BPDPA)]. The curable resin containing 0.18 mol % BPDPA exhibited a low polymerization threshold of 0.64 mW and the lateral spatial resolution in two-photon induced polymerization was improved to 80 nm at a laser power of 0.80 mW and the linear scan speed of 50 μm/s. This also indicates that the threshold power and the exposure time played an important role in improving the resolution. To further enhance the fabrication resolution, Perry *et al.* [62] designed a radical-initiated, cross-linkable acrylate resin of a "donor-p-donor" chromophore for effective excitation at a shorter wavelength and adopted a 50 nm femtosecond pulse to excite the resin. Then, they demonstrated woodpile-type photonic crystal microstructures under 0.6 μW and 0.45 μW from a Ti:Sapphire regenerative amplifier (Spitfire, Spectra-Physics). At a scan speed of 60 μm/s with the lateral spacing of 0.5 μm, the average line widths are about 75 nm and 63 nm (Fig. 11.3), respectively. Transmission spectra showed that these PC structures exhibited sharp photonic bandgap.

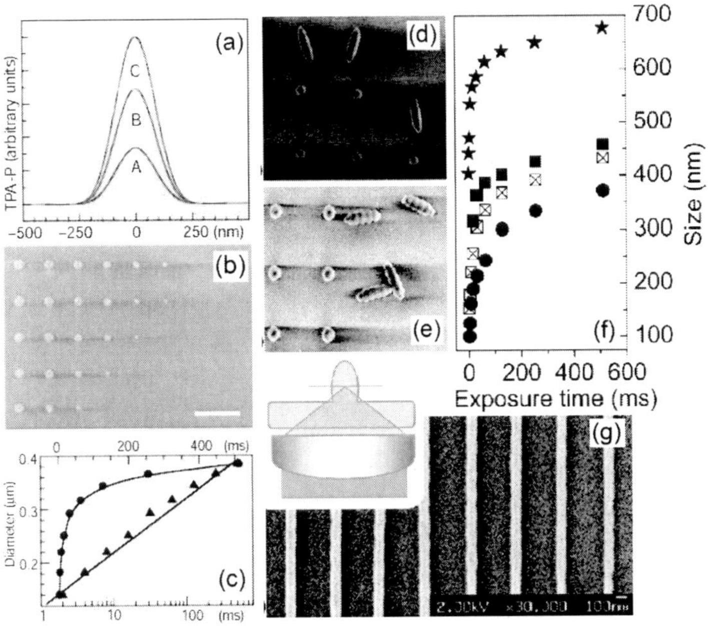

Figure 11.2 Spatial resolution of two-photon polymerization after the function of radical quenchers. (a) Achievement of subdiffraction-limit resolution. (b) Scanning electron micrograph of voxels formed at different exposure times and laser-pulse energies. (c) Dependence of lateral spatial resolution on exposure time (d) and (e) voxels produced in thin resin film of different thicknesses; (f) the exposure duration dependent lateral resolution under different resin compositions. (g) The 100 nm-width line achieved with optimized quencher concentration.

SU-8 is a high-aspect-ratio epoxy-based negative photoresist tailored for the fabrication of micromechanical devices using optical lithography in the blue–ultraviolet spectral region. Recently, it has been used for the preparation of high-precision three-dimensional photonic crystals (Figs. 11.4(d–f)) at telecommunications wavelength [63] by TPP due to its high transmittance for light from the visible to the near-infrared wavelengths, low polymer volume shrinkage, good mechanical properties (Young's modulus, $E \sim$ 4–5 GPa and biaxial modulus of elasticity ~5.2 GPa), and high thermal stability (the degradation temperature ~380°C). It has been reported that 30 nm-resolution [64] could be realized by two-photon polymerization at a much lower laser pulse energy (0.3 nJ). As shown in the Figs.

11.4(a–c), the SU-8 nanorods were obtained, suspended in air between the thicker counterparts.

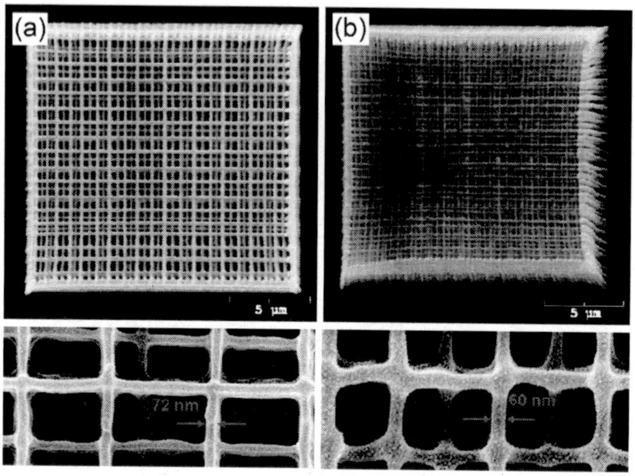

Figure 11.3 SEM overview images of woodpile-type PC structures fabricated with 520 nm excitation at (a) 0.60 µW and at (b) 0.45 µW using the DABP-triacrylate resin.

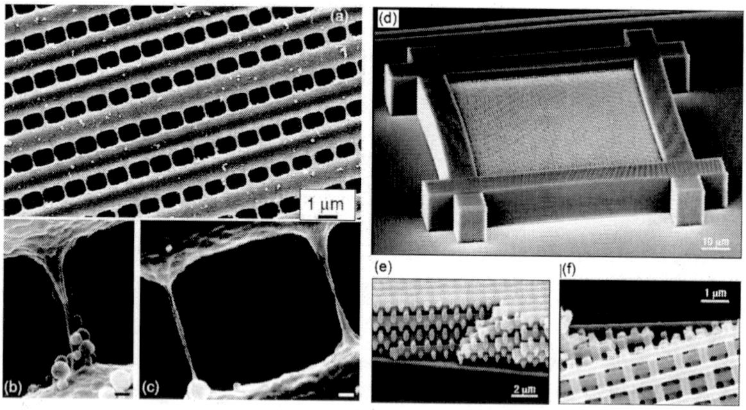

Figure 11.4 (a–c) SEM images of nanorods recorded in SU-8. Scale bars are 100 nm. Three-dimensional photonic crystals fabricated by DLW. (d) Layer-by-layer structure with 40 layers and a massive wall that prevents bending and reduces distortions due to polymer shrinkage during polymerization, completely fabricated by DLW. (e) and (f) Side and top view of a different broken sample with 12 layers, illustrating the sample quality obtained with the DLW process.

SCR500 is a commercial resin from JSR, Japan, which is widely used in two-photon microfabrication. By controlling the laser power and laser focus scan speed, sub-20 nm feature size without adding a radical quencher was demonstrated by Li *et al.* [65]. They prepared polymerized fibers at the scan speed limit (600 μm/s) when the spacing between two supports were 600 nm. Due to the fast scan speed, the feature size (~23 nm) was realized, as shown in Fig. 11.5(a). Moreover, based on repolymerization between two structures close to each other, the feature size was further reduced to ~18 nm (Fig. 11.5(b)), the smallest line ever reported. Such high fabrication resolution is beneficial to the preparation of various micro-optical, micromechanical, and biological microstructures.

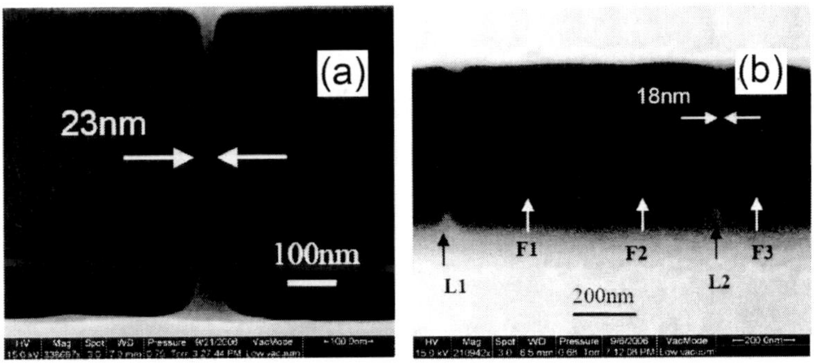

Figure 11.5 (a) Feature sizes with 23 nm. (b) Feature sizes of lines fabricated between two closely transposed supports. The pulse power was 35 mW, the scan speed was 700 μm/s, and the spacing was 600 nm.

11.2.3 Surface Roughness

Surface roughness is another important factor for laser microfabrication since it significantly affects the functions of nanomechanical devices and micro-optical devices. Especially, for photonic elements like waveguides, mirrors, and switches, rough surface will lead to serious light loss due to coarseness-induced scattering. Sun *et al.* [60] first investigated the roughness factor of a flat surface (12 μm × 12 μm) by TPP. They systemically studied the relationship between the roughness factor and the scanning space, and found that the scanning space seriously affect the surface quality.

In order to make quantitative characterization, atomic force microscope (AFM) was used to analyze the sample surface. According to the AFM measurement, the surface roughness was 9 nm, 18 nm, 38 nm, and 76 nm for 100 nm, 200 nm, 300 nm, 400 nm spacing, respectively (Figs. 11.6(a–d)). As the scanning space became big, the roughness factor significantly increased. Because liquid resins among adjacent voxels (Fig. 11.6(i)) is difficult to remove during the developing process due to the surface Gibbs free energy that is minimized by the negative curvature, the roughness propagated to the polymerized surface becomes negligible compared with those caused by other factors when the overlapping degree is high enough, or equivalently the pitch is sufficiently small. The roughness

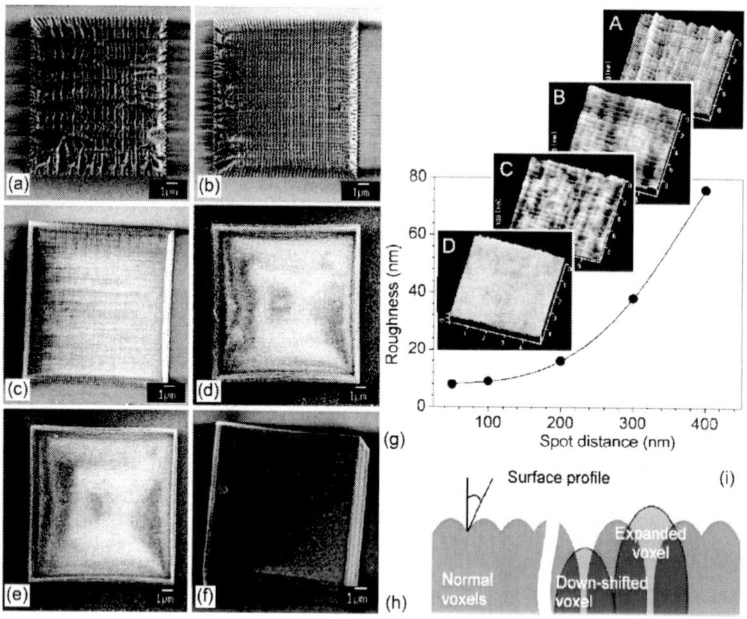

Figure 11.6 Cube structures for testing the surface roughness under different voxel distance or depicting pitch: (a) 400 nm, (b) 300 nm, (c) 200 nm, (d) 100 nm (e) 50 nm, (f) tilted view of the 50 nm cube. (g) The surface roughness. (h) Illustrative model to show the voxel-induced roughness cannot propagate to the sample surface. (i) Therefore ambient and material factors should be responsible for the surface unflatness.

of a flat surface spin-coated with the same resin is about 1 nm while the fabricated surfaces prepared by different laser wavelengths, exposure conditions, and focusing were about 4–11 nm. The surface roughness is insensitive to the variation of these parameters. The surface roughness may be caused by the fluctuation of the voxel sizes when the scanning step is sufficiently small.

Although the 10 nm surface roughness could satisfy the most common need of micro-optical and micromechanical devices, it is still indispensible to further improve the surface quality for high-precision laser microfabrication. Wu *et al.* [66] then took many measures to enhance the surface quality containing flat surface and curved ones. For example, they believed that the voxel fluctuation was associated with the quadratic dependence of photopolymerization rate on the laser light intensity. The slight variation of laser intensity will cause significant surface protrusion. So, to solve this problem, they stabilized the laser pulse energy by maintaining the local temperature within 23 ± 0.2°C. Moreover, they build an output feedback system to *in situ* monitor and adjust the laser power so that the pulse energy fluctuation was lowered to less than 0.5%. By this way, the surface roughness defined by the root of mean square on 10 × 10 μm planar squares were less than 2.5 nm at the 100 nm voxel distance when self-smoothing effect plays an essential role in the overall surface roughness (Fig. 11.7(a–d)).

For 3D curved microstructures, such as typical microlens structures, the slicing layer thickness is another important factor for high surface smoothness. The typical lens surface was scanned with equal-height slicing by which the vertical step is constant h = 100 nm. This method is simple, but it usually caused the significant solenoid traces (Figs. 11.7(e) and (g)) due to the lateral offset of voxels location, which is beyond the capability that is able to be flattened by the self smoothing effect. In order to improve the large inward shift in voxels layer by layer, an equal-arc scanning strategy was proposed, by which the distance between adjacent voxels keeps constant and the layer height $\Delta h = \Delta L \times \sin\theta$ becomes a variant, depending on the arc length L and the sloping angle θ. By this novel scanning method, high-quality curved surfaces were attained (Figs. 11.7(f) and (h)). Such high surface quality could satisfy various needs of high-precision micro-optical, micromechanical, microsensing and biological applications.

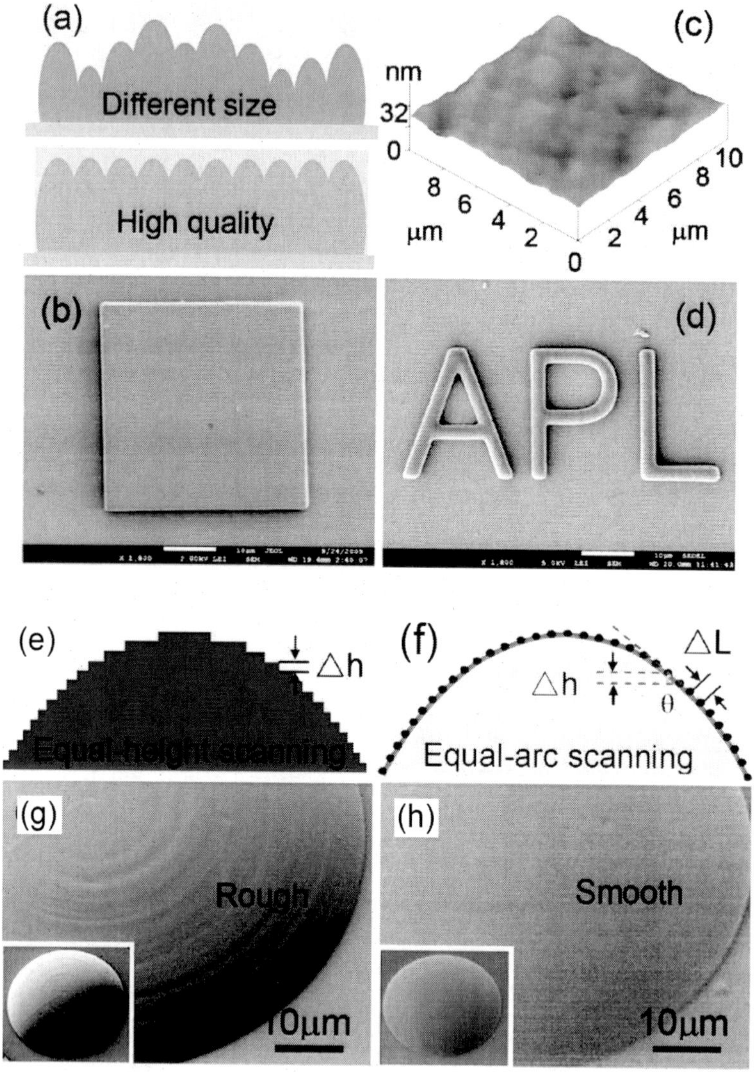

Figure 11.7 High surface smoothness realized by high reproducibility control of voxels. (a) Self-smoothing effect arising from surface tension. (b) A 10 × 10 μm^2 square fabricated by this improved femtosecond technology. (c) The AFM image showing high surface smoothness. (d) Three microletters "APL" with high surface quality. (e) Equal-height scanning mode. (f) Equal-arc scanning mode. (g) and (h) SEM images of single microlens prepared by the two scanning methods.

11.2.4 Materials Functionalization

In many practical applications, various optical, electric or mechanical functions were highly desirable. However, the common polymer, such as SCR 500, SU8, and NOA 61, did not exhibit these properties. So, people began to design functional materials by doping special substance into the common polymer and demonstrated novel microdevices by femtosecond laser-induced two-photon polymerization.

An important work is reported by Duan *et al.* [67], who reported the *in situ* synthesis of semiconductor-polymer luminescent nanocomposites and realized 3D multicolor microstructures by TPP. By tuning the cross-linking density of polymer, they controlled the size of *in situ* synthesized CdS nanoparticles, which show unique tunable light emission properties arising from quantum size effects because the size of the CdS nanoparticles within the polymer matrix is confined by the space available within the cross-linking polymer network. The resins consisted of precursors of CdS NPs (cadmium methacrylates), monomers (Methacrylic acid and methyl methacrylate), oligomers (dipentaerythritol hexaacrylate), photoinitiator (benzyl (Aldrich), 1 wt %) and photosensitizer (2-benzyl-2-(dimethylamino)-4′-morpholinobutyrophenone (Aldrich), 1 wt %). Fluorescence spectra measurement showed that the colors of the nanocomposite resin could be tuned from green to blue by increasing the oligomer content from 0 to 48.7 wt%. The blue-shift of the maximum wavelength of the main emission band of the nanocomposites from 528 to 446 nm is almost linear. A typical TEM image and electron diffraction pattern of CdS NPs encapsulated in polymer networks shows the presence of clearly visible lattice fringes. The 0.33 nm fringe spacing can be assigned to the (111) plane of face-centered cubic CdS. The CdS NPs synthesized within the polymer matrix appear to be a face-centered cubic structure. The decrease in the average size of *in situ* synthesized NPs from about 6 to 2.6 nm upon increasing the cross-linker content from 0 to 48.7 wt% has also been confirmed, which is in good agreement with the calculated results from absorption onsets. Figures 11.8(c) and (d) showed the size distribution of CdS NPs synthesized with different amounts of the cross-linker; these results clearly show that highly cross-linking polymeric networks can narrow the size distributions of CdS NPs.

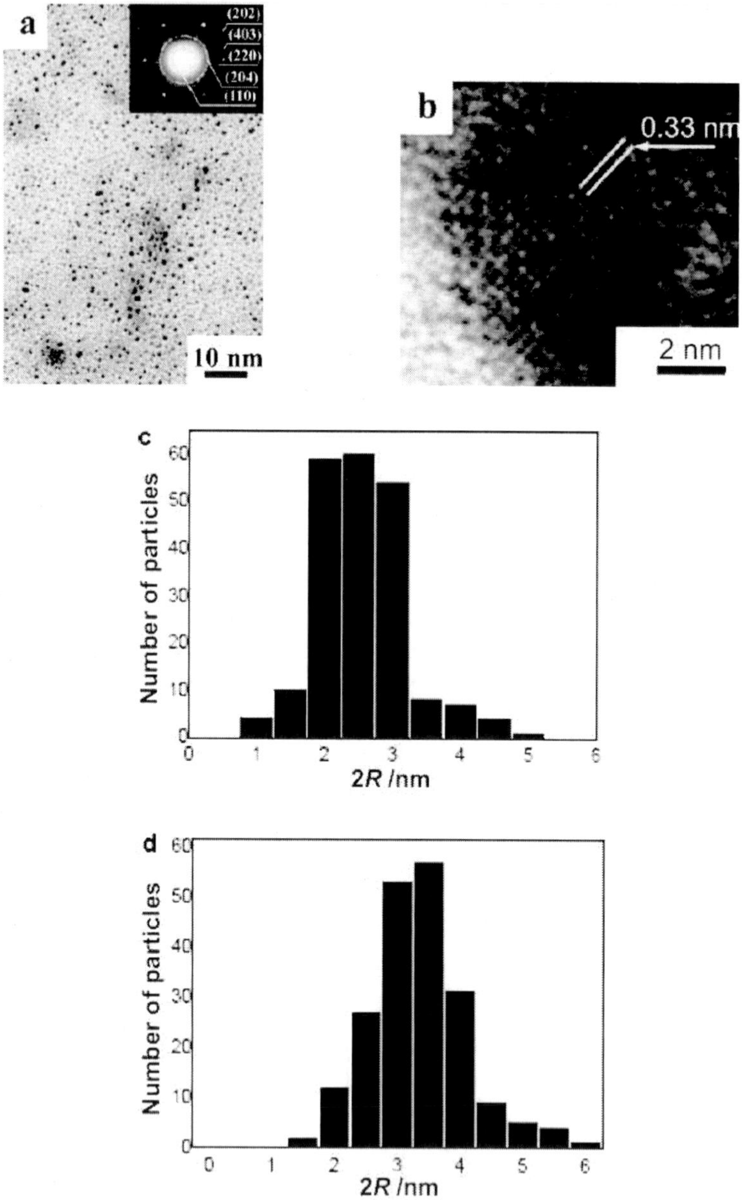

Figure 11.8 (a) TEM and (b) typical high-resolution TEM images of CdS–polymer nanocomposites (resin F). Corresponding histograms showing the size distribution of CdS NPs for (c) resin F and (d) resin E.

Furthermore, multicolor microstructures were successfully fabricated by the combination of TPP and the *in situ* synthesized CdS NPs within 3D microstructured polymer matrices, as shown in Fig. 11.9. Fluorescent microscopy measurement showed that the microbull emitted blue light. For comparison, a 3D microbull was fabricated the resin without a cross-link, which showed green fluorescence (central FM image in Fig. 11.9(a)). Simultaneously, they also demonstrated a cyan light emission feature after the *in situ* synthesis of CdS NPs using a resin with 48.7 wt% cross-linker content for 3D microbull fabrication (right FM image in Fig. 11.9(a)). A 15 μm-size multicolor lizard has been fabricated, as shown in Fig. 11.9(b). Such structures will play an important role in functional MEMS/NEMS and integrated optical devices.

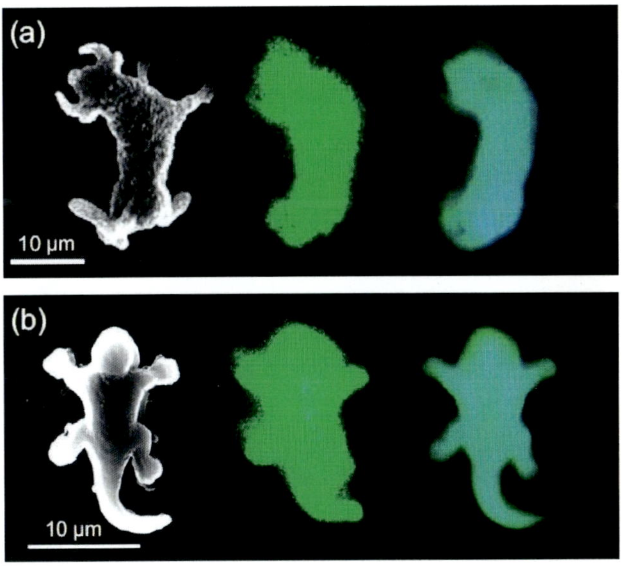

Figure 11.9 Images of scanning electron microscopy (left) and fluorescence microscopy of a 3D microbull (a) and a 3D microlizard (b) fabricated from resin A (centre) and resin F (right).

Besides luminescent polymer materials, Sun *et al.* [68] recently reported a kind of functional micromachines made of a common resin doped by surface-modified Fe_3O_4 nanoparticles, which can be driven by magnet remotely. The Fe_3O_4 nanoparticles were synthesized by a traditional Massart method. First, $FeCl_3{\cdot}6H_2O$ (5.38 g) and

$FeCl_2 \cdot 4H_2O$ (1.98 g) was dissolved by deionized water (200 mL) under N_2 gas protection for about 0.5 h. Then they quickly charged Ammonia aqueous solution (25%, 7 mL) into the solution with vigorous stirring, and the color of solution changed to black sharply. The solution was stirred for an additional 3 h at room temperature. After washed with deionized (DI) water for 3 times, the reaction product was collected by a ferromagnet. To better dope the ferric oxide particle cores into a common resin, they chemically modified the surface of such powder with surfactant oleic acid (OA) (Fig. 11.10). While the particles were stirring in DI water (200 mL), it was necessary to drop-wise add OA (1.2 g) into the solution to ensure the uniformity of reaction, and then the stirring lasted 30 min at 80°C under N_2 gas protection. Finally, they got OA–ferric oxide particles after washed three times with hexane to remove unreacted OAs.

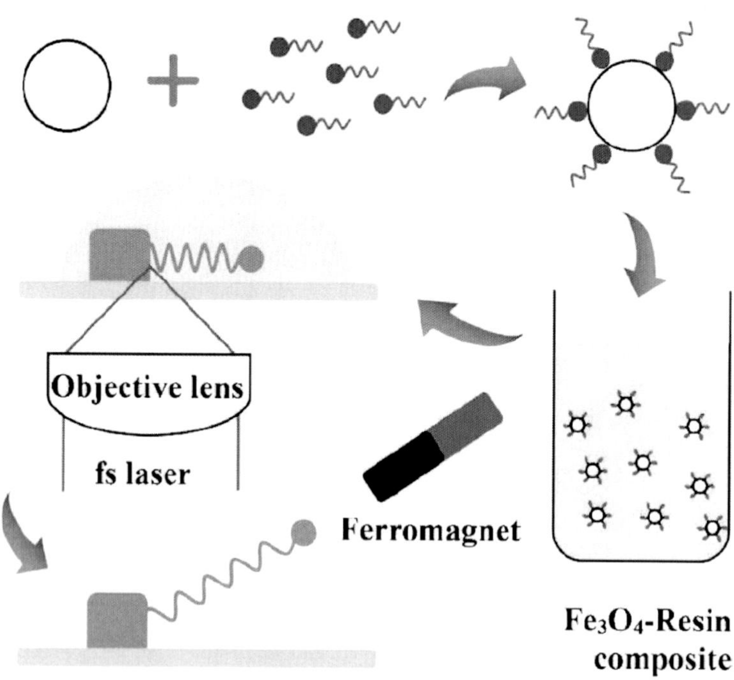

Figure 11.10 Experimental scheme for preparation of surfactant oleic acid OA–Fe_3O_4 magnetic nanoparticles, its doping to photopolymerizable resin, femtosecond laser fabrication of microdevices, and magnetic driving of the devices.

It was demonstrated that the nanoparticles were monodispersed with diameter of about 10~20 nm by TEM. For the purpose of fabricating magnetic micronanomachines, they mixed such magnetic nanoparticles with a kind of homemade photopolymerizable resin consisting of 37.14 wt. % methyl acrylate as monomer, 60.00 wt. % pentaerythritol triacrylate as cross-linker, 1.66 wt. % benzyl as photoinitiator, and 1.20 wt. % 2-benzyl-2-(dimethylamino)-1-(4-morpholinophenyl) butan-1-one as photosensitizer. The resin with magnetic nanoparticles was still visible as expected, because the OA modified around the surface of Fe_3O_4 could prevent them from aggregation essentially. Also, the optimized doping rate of magnetic nanoparticles in the resin mixture was about 2.40 wt % according to the measurement of thermogravimetric analysis. Based on such magnetic photopolymerizable resin, various functional micromachines were fabricated by pinpoint laser writing.

11.3 Micro-Optical Components Macros

11.3.1 Planar Optical Devices

In recent years, micro-optics has gained increasing attention in many fields, such as micro imaging, micro focus, and beam shaping and so on. Femtosecond laser-induced TPP microfabrication does not need mask and could exhibit high-precision 3D controllability, so it has been used for the preparation of various planar and three-dimensional micro-optical devices. Among various micro-optical devices, Fresnel zone plates, as a kind of focusing and imaging component, have played an important role in laser shaping and X ray imaging. Huang *et al.* [69] first reported amplitude-type Fresnel zone plates by TPP of the resin SCR500. They optimized the fabrication parameter by using the annular scanning mode with the lateral step pace and prepared a series of Fresnel lens with different scanning steps of 0.1 μm, 0.2 μm, 0.3 μm, and 0.4 μm. They found that the scanning steps seriously affected the quality of Fresnel lens. The best Fresnel lens, 17 μm in diameter, is shown in Fig. 11.11(a). The total fabrication takes about 8 min. The Fresnel zone plate (FZP) showed excellent focusing ability, which well agreed with the theoretical simulation.

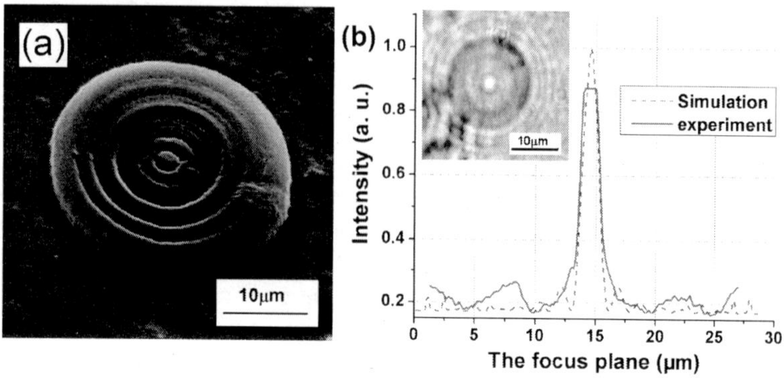

Figure 11.11 (a) Microfresnel zone plate. (b) The simulated and experimental diffractive intensity distribution of micro Fresnel lens.

Diffractive efficiency, defined as the ratio of the power collected to the primary focal spot to the total incidence light, was an important parameter for evaluating the optical property of Fresnel zone plates. Compared with amplitude-type FZP, the phase-type FZP whose thickness was determined by the light wavelength (λ) and the material refractive index (n) showed high diffractive efficiency. The formula is $d = \lambda/N(n - 1)$. Sun *et al.* [70] reported phase-type Fresnel zone plates by precise control of the thickness of every zone. For the two-level, four-level, and eight-level lens, the thickness of the zone is 475 nm, 238 nm, and 119 nm in order to induce the phase changes of π, $\pi/2$, and $\pi/4$, respectively. The theoretical diffractive efficiency is 40.5%, 81%, and 95.1%. Figure 11.12 shows the SEM images and the focal spots: (a) for the four-level and (d) for the eight-level structures. The measured diffraction efficiencies reached as much as 30.5%, 54%, and 68%, which are comparable with the ones of FZPs prepared by other technologies such as planar lithography. In addition, Sun *et al.* also designed and fabricated reflective FZPs by electroless plating on the polymer FZPs. A two-level phase lens was used as the template of electroless plating, on which a 50 nm silver film was formed. To further enhance the diffractive efficiency, Li *et al.* [71] reported high-quality phase-type Fresnel zone plates (FZPs) produced by femtosecond laser TPP of the resin SU-8 with optical experimental parameters. The diffractive efficiency was enhanced to 35.8% for the two-level phase lens to 67% for the four-level phase lens and 73.9% for the eight-level phase lens. Moreover, the lens imaging functions "JLU" and "吉大" (English and Chinese abbreviations, respectively, for Jilin University) were also demonstrated. Fresnel zone plate arrays (FZPAs), as a kind of an important integrated micro-optical device, have attracted great attention. However, the fill factor of present FZPAs by femtosecond technology is a little low, which leads to serious light loss and low signal-to-noise. Recently, Niu *et al.* [72] demonstrated 100% fill-factor square and hexagonal FZPAs by femtosecond laser TPP of the resin SU-8. Their excellent optical focusing and imaging 'F' were realized, which showed the high uniformity and high fidelity of these FZPAs.

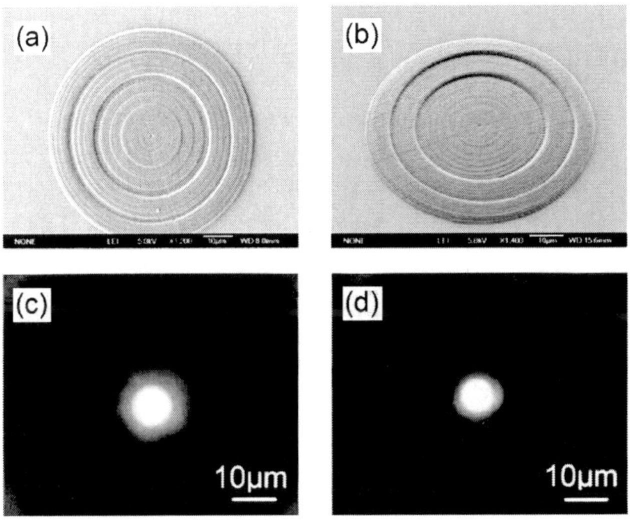

Figure 11.12 Multistep diffractive phase lenses. (a) and (b) are from four level and eight-level structures, and (c) and (d) are are photographs of foci for the two lenses.

The above progress about Fresnel zone plates focused on the enhancement of the diffractive efficiency. As we know, the FZP showed color dispersion, which is not beneficial for the application of high-precision micro-optical systems. Fractal zone plates (FraZP) [73] are constructed with a one-dimensional binary function with a regular Cantor fractal profile, followed by a rotation of the transformed function around one of its extremes. It could be considered a modified FZP with some missing zones and showed multifoci along the optical axis. The FraZP could reduce chromatic aberration under white light illumination because it showed multiple foci with internal fractal properties along the optical axis and provides an extended depth of field. From the imaging "JLU" (Fig. 11.13), it could be obviously seen that the FraZP could improve the color dispersion. An effective measure to enhance the diffractive efficiency of the FZP is to design a multilevel phase-type lens. Likewise, to further enhance the diffractive efficiency, they realized a four-level fractal phase lens, whose diffraction efficiency reached as high as 37.6%.

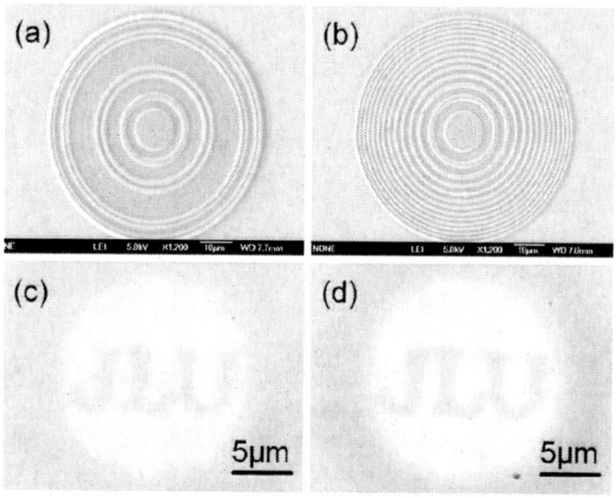

Figure 11.13 Focusing and imaging characteristics of the FraZP (N = 2, S = 3) and FZP. (a) and (b) are SEM images of the FraZP and FZP. (c) and (d) are the images of the letters JLU by the FraZP and FZP.

Besides the FZP and FraZP, Dammann gratings, as beam splitters and coherent signal generators, have attracted great attention. Lin *et al.* [74] first adopted TPP microfabrication to realize high efficiency Dammann gratings (Fig. 11.14). Femtosecond laser pulses (800 nm wavelength, 120 fs pulse width, mode locked at 82 MHz, Tsunami, Spectra-Physics) were tightly focused by a high numerical aperture (NA = 1.40) oil immersion objective lens. The focal spot was scanned laterally by manipulating a two-galvano-mirror set and along the optical axis by a piezo stage. A He:Ne laser beam with the wavelength of 632.8 nm was used as the incident light. The Dammann gratings generated an array of 2 × 2, 3 × 3, 4 × 4, 5 × 5, and 6 × 6 spots with theoretical diffractive efficiencies of 81.06%, 66.42%, 70.63%, 77.38%, and 84.52%, respectively. Most important, even an input signal that is not coherent (noncoherent white light) could produce the coherent patterns when they passed the Dammann grating.

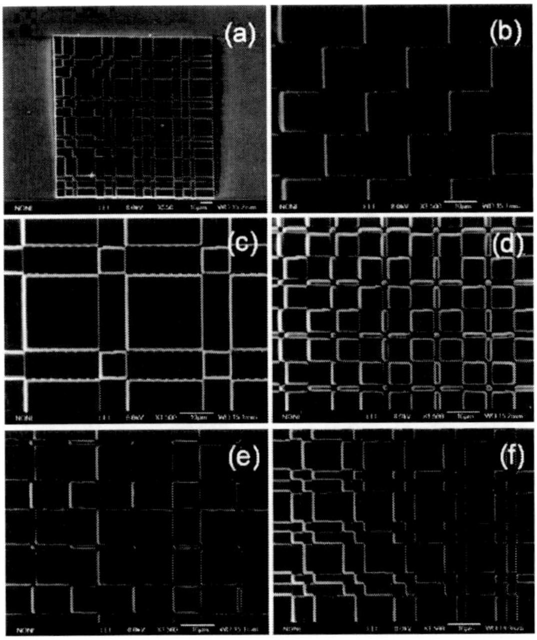

Figure 11.14 Dammann gratings fabricated by two-photon photopolymerization. (a) Top-view SEM image and locally magnified views of the Dammann gratings that generate (b–f) 2 × 2, 3 × 3, 4 × 4, 5 × 5, and 6 × 6 spots.

11.3.2 Three-Dimensional Complex Photonic Structures

Femtosecond laser-induced TPP adopted point-to-point scanning mode to realize various 3D complex microstructures, which were usually difficult for conventional planar lithography. Photonic crystals have attracted great attention because of its unique optical properties. These periodically structured dielectric materials could represent the optical analogue of semiconductor crystals, and provide a novel platform for the realization of integrated photonics. In 1999, Misawa *et al.* [75] first realized three-dimensional photonic crystals by TPP of resin. 20-layer photonic structures with different in-plane rod spacing (d = 1.2 μm, 1.3 μm, and 1.4 μm) were realized. The fabricated areas of the samples were all 40 μm × 40 μm. These 3D photonic crystals showed excellent photonic bandgap. The transmittance dips under normal incidence occur at wave numbers of 2553, 2507, and 2454 cm^{-1}, respectively. This is the first time when 3D photonic structures by TPP have been demonstrated. Misawa *et al.* also demonstrated diamond-lattice photonic crystal structures consisting of around 500 nm-diameter (111) rods and 580 nm-diameter photonic atoms (Figs. 11.15(a) and (b)) [76]. To improve the quality of 3D PC, shape compensation were proposed and demonstrated. With optimal parameters, nearly perfect photonic lattices were obtained (Figs. 11.15(c) and (d)), from which single-period power rejection of approximately 35% was achieved.

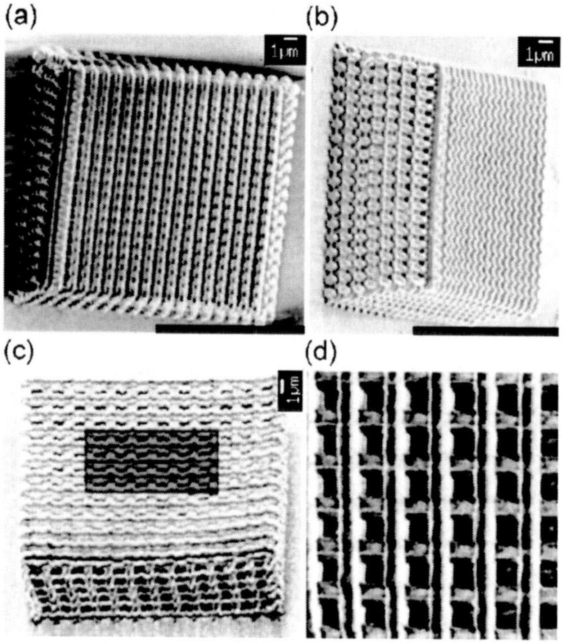

Figure 11.15 Precompensation of logpile PhCs. All structures were designed with a horizontal cross section of 20 μm × 20 μm. (a) No precompensation, PhC top area shrunk to 17 μm × 17 μm. (c) A nearly cubic PhC structure after precompensation according to calculated deformation. (b) and (d) are from the same samples as (a) and (b), respectively. Notice the highlight region in (c) was clockwise rotated by 90° and shown in (d). Their shape difference is caused by different view angle.

The aforementioned works focused on the photonic bandgap at 3.9 μm wavelength by layer-by-layer or woodpile photonic crystals. Importantly, fundamental stop bands or photonic bandgaps at telecommunication wavelengths have not been demonstrated. Markus *et al.* [63] reported the fabrication and detailed characterization of high-quality large-scale f.c.c. layer-by-layer structures, with fundamental stop bands ranging from 1.3 to 1.7 μm by TPP of the resin SU-8. Three kinds of different 3D photonic crystals with in-plane rod spacings of 1.0 μm, 0.9 μm, and 0.8 μm were obtained. The transmission spectra for three different samples with an estimated SU-8 filling fraction of about 35% were measured. The corresponding fundamental stop bands are clearly visible as pronounced dips at 1.7 μm, 1.5 μm, and 1.3 μm. Besides the so-called woodpile structures in a layer-by-layer manner, a kind of novel spiral architecture [77], though not amenable to layer-by-layer fabrication, has shown promising photonic bandgap properties. By femtosecond laser direct writing, Misawa *et al.* realized three-dimensional spiral architecture photonic crystals with the designed parameters a = 1.8 μm, L = 2.7 μm, and c = 3.04 μm (L = 1.5a, c = 1.69a) , as shown in Fig. 11.16. The dimensions of the structure are about 48 μm × 48 μm × 30 μm. In addition, to realize 3D photonic crystals with shorter wavelength, they adopted a fabrication scheme [78] to construct spiral structures such that the spiral advancing direction (spiral axis) is horizontal or perpendicular to the beam focusing direction. Although this modification does not affect the voxel volume, its different orientation helps avoid excessive overlap within the spirals. Using this approach, the spiral pitch could be reduced as small as 0.88 μm, and their optical characterization revealed stop gaps with sizeable attenuation of about 15% in the spectral interval of 0.88 μm, which is very approximate to the visible wavelength light.

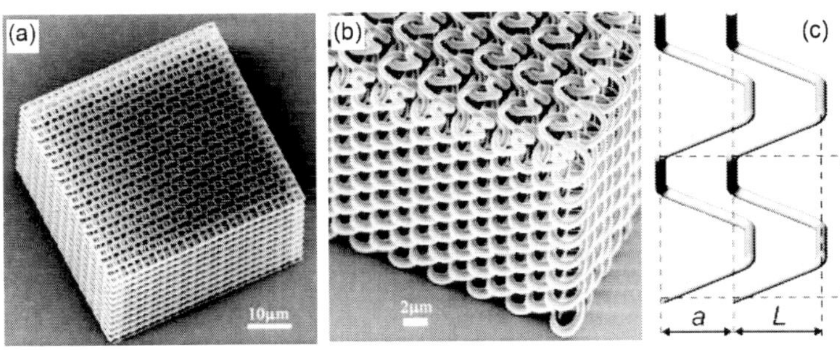

Figure 11.16 (a) SEM image of spiral architecture photonic crystals. (b) The circular spiral sample with 180° phase shift between the adjacent spirals. (c) Individual spirals separated from the structure.

Michael *et al.* [79] fabricated a type of high-quality chiral polymeric 3D spiral photonic crystals with polarization stop bands via direct laser writing. Both of the left-handed and right-handed spiral 3D PC were prepared. The sample parameters are: in-plane lattice constant 1.3 μm, pitch 1.3 μm, spiral diameter 0.78 μm, volume filling fraction 34.7%, lateral diameter of the spiral arms 380 nm. Shown in Fig. 11.17 are the measured transmittance spectra both left-handed and right handed dielectric circular spirals and for both left-handed and right-handed incident circular polarization of light. The transmittance is below 5% for one circular incident polarization and above 95% for the other—for just eight lattice constants along the propagation direction at wavelengths around 1.8 μm. The experimental data are compared with scattering-matrix calculations for the actual finite structures, leading to good agreement. In addition, Michael *et al.* proposed and experimentally realized a bi-chiral dielectric photonic crystal (Fig. 11.18) for the first time [80]. The artificial structures give access to all four bi-chiral combinations, whereas only two are realized in nature. Moreover, Michael *et al.* showed that the ones missing in nature are the ones exhibiting the strongest effects.

Besides the bandgap properties, Gu *et al.* [81] demonstrated the wavelength-dependent super-refraction phenomena in superprisms based on polymer woodpile structures. They fabricated 3D photonic crystal by femtosecond laser TPP. The bandgap was in agreement with the calculated value, where the PBG could be found at a wavelength range of 1.19 μm to 1.23 μm for rod distance 700 nm. To visualize the pathways of the light propagating inside the crystal, the top surface of the crystal was imaged onto a charge-coupled device (CCD) camera. A small amount of the propagating light scattered by imperfections in the PC structure is imaged onto the CCD. So, the images of the crystal and the propagating laser light could be recorded simultaneously (Fig. 11.19). As the wavelength is tuned from 1000 nm (Fig. 11.19(a)) to 1020 nm (Fig. 11.19(c)), the light inside the crystal changes its direction of propagation by more than 10°. This result shows the great potential of the low-refractive-index three-dimensional (3D) crystals to be directly used as functional micro-optical devices in the near-infrared (NIR) wavelength regime.

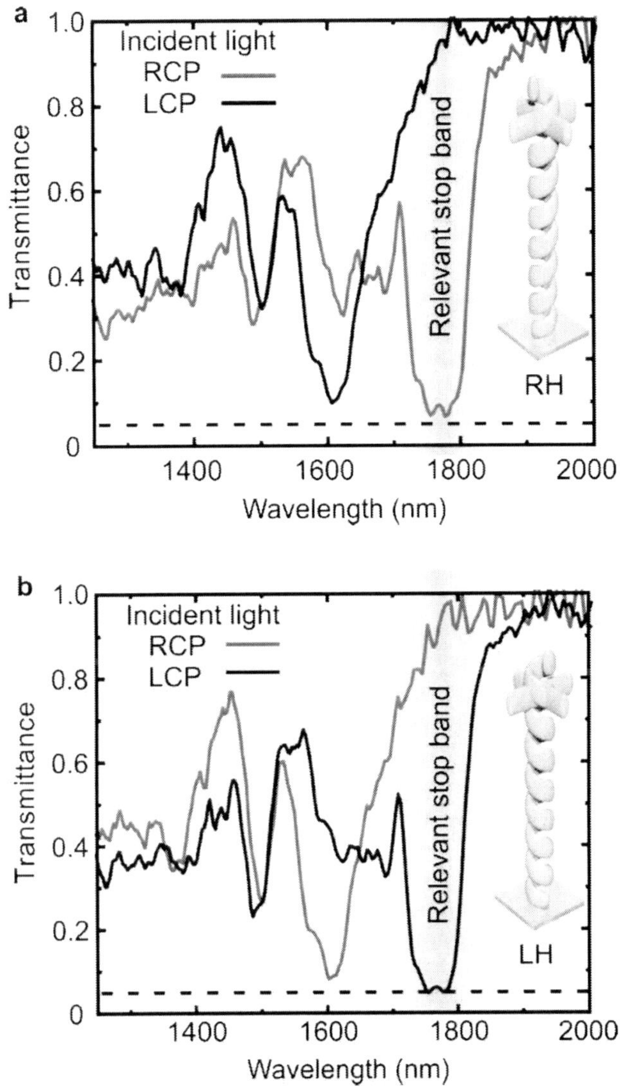

Figure 11.17 Measured transmittance spectra of 3D spiral photonic crystals. (a) Right-handed (RH) dielectric spirals. The transmittance spectrum for right-handed circular (RCP; left-handed LCP) polarization of the incident light is shown in red (LCP: blue). The dashed horizontal lines correspond to a transmittance level of 5%. (b) Same as (a), but for left-handed (LH) dielectric spirals.

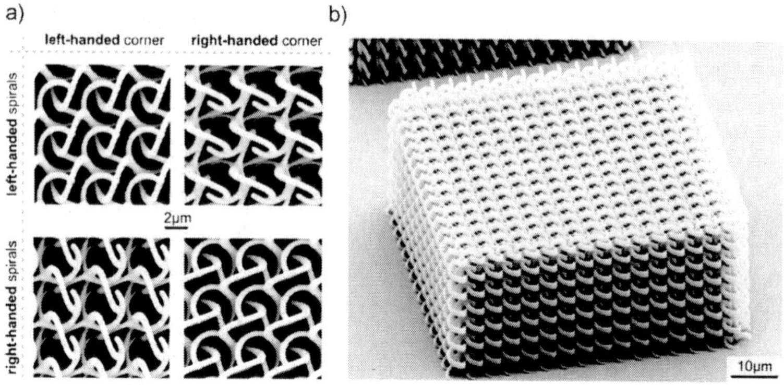

Figure 11.18 Electron microscopy images of fabricated bi-chiral photonic crystals. (a) Top views on left/left, right/left, left/right, and right/right-handed photonic crystals. (b) Oblique view on a right/left-handed polymer structure made by direct laser writing.

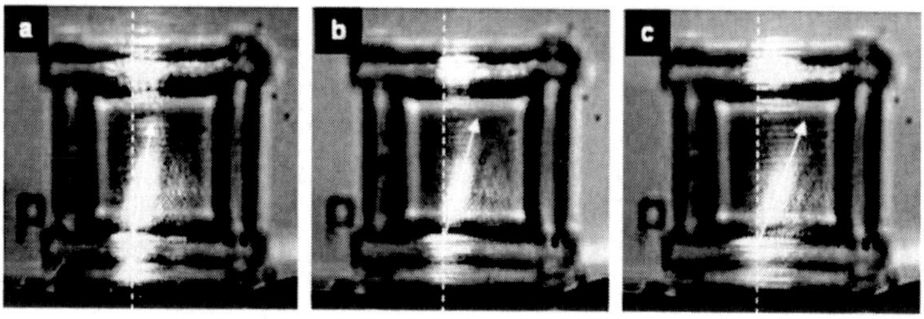

Figure 11.19 Microscope images showing the scattered light of the propagating beams inside the woodpile structures for different wavelengths: (a) 1000 nm, (b) 1010 nm, and (c) 1020 nm. The angle of incidence was kept constant at 4°. The bright spots at the lower and upper edges of the crystals are due to light that did not propagate inside the PC and hence indicate the direction of the incident light.

11.4 Functional Micromachines

11.4.1 Light-Driven Microdevices

In integrated systems, a variety of devices, such as optical, sensing, micromachines and electronics, were needed for multifunctional application. Among this, functional micromachines have played an important role and thus have been paid great attention.

Microrotor is a powerful instrument to measure the property of microscopic systems, e.g., determining the torsional elasticity of nanoparticles, DNA molecules, and proteins, and it can also be used as a tool of nanotechnology in the production and manipulation of microscopic objects. Pal *et al.* [82] first adopted TPP technology to build microscopic rotors (Fig. 11.20) in the resin Norland NOA 63. Moreover, they could control the shape of the microrotors, e.g., helix, sprinkler, and propeller, with rotational symmetry of various folds. The microrotors were manipulated by laser tweezers and the dynamics of the rotation are evaluated. The microrotor rotated at $w = 7\ s^{-1}$ with the viscous drag torque of 3.6×10^{-17} Nm, which well agreed with the theoretical value. Besides single rotor, the two-photon polymerization method also could construct more complicated micromechanical systems. For example, Pal *et al.* demonstrate a system where light driven rotors are used to drive cogwheels. Both the rotor and the cogwheels are produced by one-step TPP method. The cogwheels are rotating on axes fixed to the glass cover of the sample compartment; the wheels cannot leave the axes. The rotor is manipulated by the laser tweezers. The rotor and the cogwheel

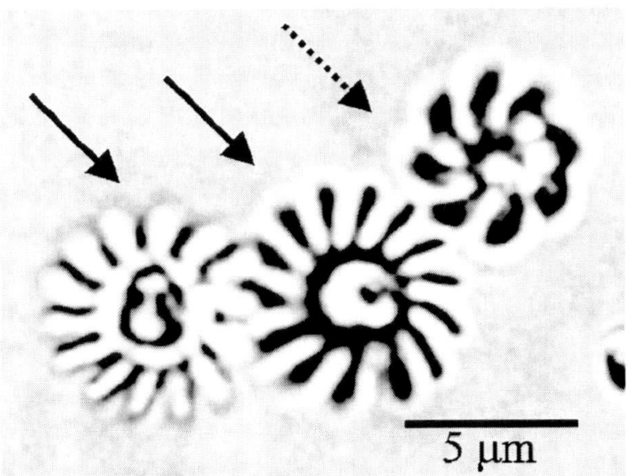

Figure 11.20 Complex micromachine built by the two-photon technique: two engaged cogwheels are rotated by a light driven rotor. The solid arrows point to the cogwheels rotating on axes fixed to the glass surface. The rotor is held and rotated by the laser tweezers and the rotating propeller drives the system.

are engaged by appropriately positioning the rotor with the laser tweezers. As the rotor turns the cogwheels, the machine works. The example shows the availability of TPP method for the preparation and manipulation of complex functional micromachines and their integrated systems.

Tweezers are useful for grabbing or probing nanoscale objects in chemical and biological applications. However, since present tweezers are based on electrostatic force, they are not ideal for aqueous solution work such as manipulation of cells, microbes, and single molecules. Maruo *et al.* [83] reported the microtweezers with 1.8 μm-length and 250 nm-diameter probe tips by two-photon microstereolithograpy. The probe tips were fabricated by point-by-point exposure with 30 mW laser power. Each exposure lasted 100 ms, and the entire fabrication process was completed in only 6 min. Then, the microtweezers were driven by the optical system with a Ti:sapphire laser operated in cw mode. Moreover, Maruo *et al.* also designed and fabricated a microneedle, as shown in Fig. 11.21. Since the probe tip of the needle is attached to the slotted arm, this microneedle can provide translational motion and rotational motion

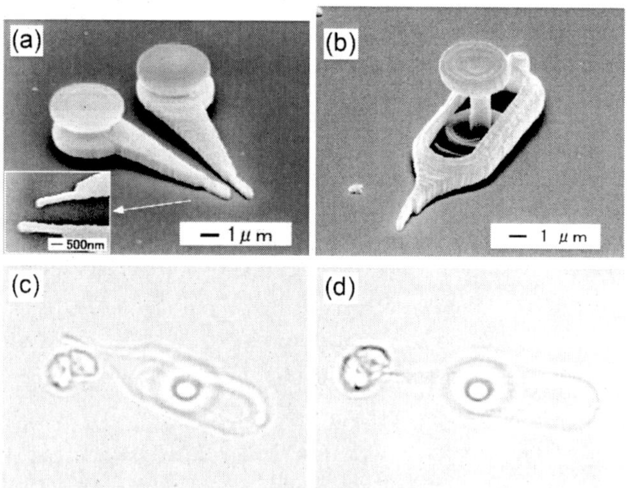

Figure 11.21 (a) and (b) SEM image of the microtweezers and microneedles with submicron probe tips fabricated by two-photon microstereolithography. (c) Pushing a micro-object with an arm of the microneedle. (d) Pricking a micro-object with a submicron probe tip of the microneedle.

under the laser tweezers. A dot is attached to the end of the slotted arm for laser manipulation. By trapping the dot with a focused laser beam, the microneedle can be manipulated with the desired motion. Maruo *et al.* also demonstrated that the microneedle was useful for manipulating a micro-object in the liquid. Figure 11.21 illustrates the mechanical stimulation of a micro-object by the microneedle. A tiny particle of dust floating in the liquid was pushed with the arm of the microneedle or pricked with its probe tip. The probe tip was sufficiently strong to prick the micro-object without any deformation of the tip. The optically controlled functional micromachines may be useful for bionanotechnology applications such as nanosurgery systems for living cells and nanoanalysis systems for single molecules.

Micropump is a kind of component widely used in microfluidic devices. In the past two decades, a variety of pressure-driven micropumps have been prepared by conventional microfabrication techniques. Most of these micropumps are diaphragm types, which was drove by piezoelectric actuators or pneumatic actuation for fluid transport. To obtain adequate displacement of the membrane, the diameter of the membrane must range from 100 to 500 μm. As a result, the overall dimensions of the micropumps are much larger than the microchannels. This hinders the miniaturization and integration of microfluidic components. Maruo *et al.* [84] reported an entirely optically driven micropump, in which two lobed rotors are incorporated into a microchannel by TPP method. The length of the major axis of each rotor is only 9 μm and the thickness of the rotor is 2.5 μm (Fig. 11.22(a)). The lobed rotors are individually confined to a microchannel by their own shaft, thus preventing the rotors from moving unless continuously irradiated with light. This allows the micropump to be easily integrated into a practical biochip. The lobed rotors are then incorporated into a microchannel whose inlet and outlet are in 5 μm wide and 7 μm high (Fig. 11.22(b)). Both the rotors and the microchannel are fabricated using 3D two-photon microfabrication system. This micropump is driven by means of radiation pressure generated by a focusing laser beam. When the laser beam is focused on the side of the rotor, the net radiation pressure applied to the rotor points toward the focus, allowing the rotor to be controlled by changing the trajectory of the scanning laser beam.

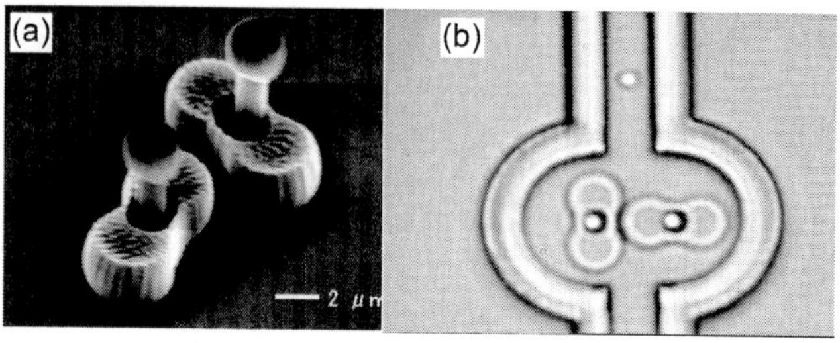

Figure 11.22 (a) SEM image of a prototype of the lobed micropump. (b) The optically driven micropump.

To enhance the rotation performance of the micropump, Maruo *et al.* developed an optically driven micropump that employs viscous drag exerted on a spinning microrotor with left- and right-handed spiral blades on its rotational axis [85]. It was demonstrated that the twin spiral microrotor provides a higher rotation speed than a single spiral microrotor. The rotation speed reached 560 rpm at a laser power of 500 mW. The twin spiral microrotor was also applied to a viscous micropump with a U-shaped microchannel. To pump fluid, the twin spiral microrotor located at the corner of the U-shaped microchannel was rotated by focusing a laser beam. The viscous micropump with a twin spiral microrotor can be used for pumping fluid and the transportation of micro-objects. Figure 11.23 shows a sequence of images taken at 2 s intervals showing a tracer particle being pumped through the channel when the microrotor was rotating at a speed of 300 rpm. Moreover, by connecting multiple viscous micropumps with a crooked microchannel, they constructed a tandem micropump consisting of multiple twin spiral microrotors.

11.4.2 Magnetic-Driven Micromachines

Light-driven is an effective method for manipulating micromachines, but it suffers from high laser intensity, which is unsuitable for biological and chemical applications. The magnetic force drive technique, which has been widely applied in macroscopical instruments, would be an ideal method for remote control due to its simple, safe and noncontact properties. However, up to now, magnetic force is still not properly applied to remote control of micronanomachines for accomplishing desired task. Sun *et al.* [86] reported controllable magnetic micromachines realized by femtosecond laser-induced TPP of the resin doped by magnetic nanoparticles. For example, magnetic microsprings and microturbines were realized.

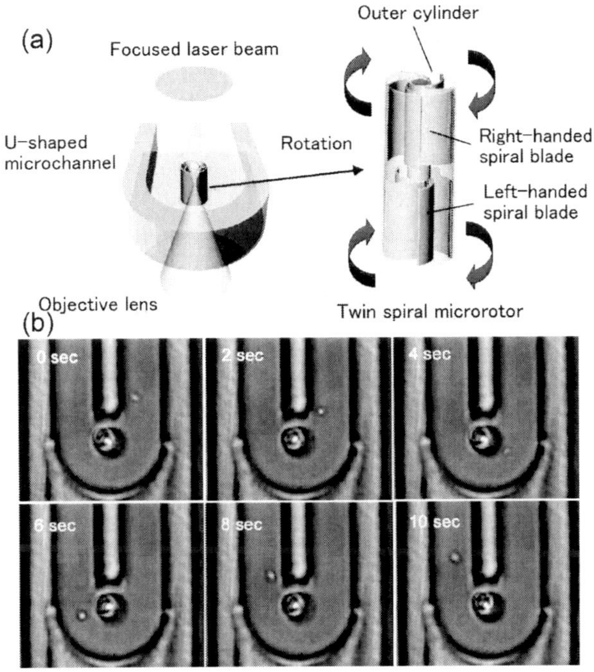

Figure 11.23 (a) Optically driven viscous micropump using a twin spiral microrotor. (b) Particle transportation in a viscous micropump using a twin spiral microrotor.

As illustrated in the schematic model (Fig. 11.24(a)), one end of the spring was fixed to a polymerized cubic anchor attached to the substrate, and the other end is a sphere shape used as a moving part. The spring structure was immersed into a solvent and placed under an optical microscope. The MPS-Fe_3O_4 nanoparticles as magnetic dipoles were homogeneously incorporated in both driven part (bead) and spring structure, thus the microspring could be driven under the external magnetic field gradient. Figures 11.24(b–f) show a sequence of frames of this movement process. The original length of the microspring is 60 μm (Fig. 11.24(b)). During its elongation, length values of 81 μm (Fig. 11.24(c)), and 86 μm (maximum, Fig. 11.24(d)) were observed. It is of interest to note that different cycles of the microspring exhibit different elongation ratio along the axis direction. In addition, different slope angle of the microspring was demonstrated. The fabrication and remote control of micronanomachines are not only limited to microspring, a microturbine was also developed for remote control. As shown in Fig. 11.25, a collarjoint microturbine was successfully created according to the pre-designed model (Fig. 11.25(a)). The microturbine was about 35 μm in diameter with a central axletree and three blades. For magnetic control, a piece of ferromagnet was placed on a vortical device around the microturbine. The microturbine rotated with an average rate of 3 revolutions per second. The development

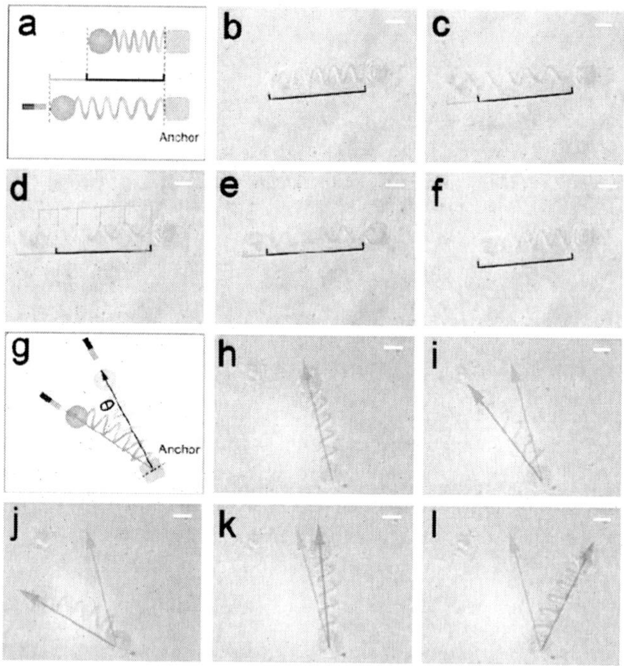

Figure 11.24 Remote control of the micro-spring in acetone. (a–f) Elongation movement: (a) Scheme model for elongation of the micro-spring. (b) Original length of the microspring. (c–e) Micro-spring with various elongations; (f) recovery of the micro-spring; (g–l) sway movement: (g) scheme model for sway movement of the micro-spring; (h–l) micro-spring with different slope angles. Scale bar, 10 μm. A piece of magnet with surface magnetic field strength of 1800 Gs is used for remote control of the micro-spring.

of microturbine represents the availability of remote control of micronanomachines, and it is believed that the microturbine would be of considerable interest in microfluidic field.

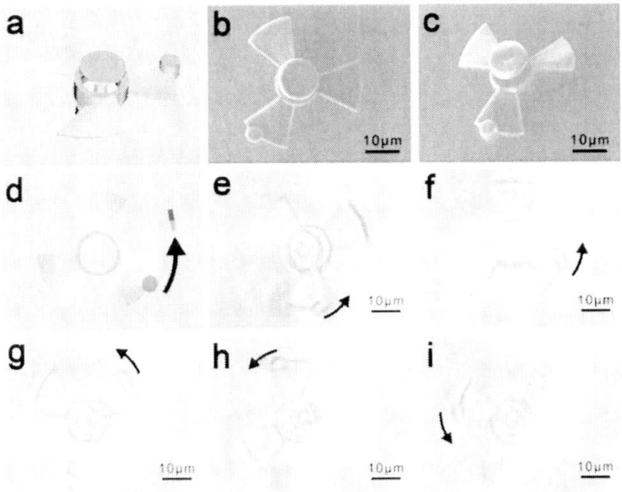

Figure 11.25 Remote control of the micro-turbine in acetone. (a) model of the micro-turbine. (b,c) SEM images of the micro-turbine. (d) Top view scheme model for circumgyratetion. (e–i) optical microscopy images of the microturbine in a circumgyratetion cycle. For remote control and observation of the micro-turbine, a piece of ferromagnet was placed on a vertical device around the objective lens.

11.5 Other Applications

11.5.1 Functional Microfluidic Components

In the past two decades, microfluidic systems have been intensively investigated due to their broad applications in chemistry, physics, biology, and medicine. Compared with macroscopic settings, microfluidic devices exhibited many distinct advantages, such as small reagents requirement, short reaction time, portability, low cost, low consumption of power, and high integrity, which strongly stimulate the development of new techniques for the fabrication of microfluidic devices. However, until now, it is still difficult to precisely prepare microfluidic chips with complex 3D topological microstructures in a designable, flexible, and controllable fashion. Moreover, it remains challenging for local mending, modification, functionalization and integration of existing microfluidic devices for special applications. Femtosecond laser TPP is ideal due to its many advantages such as nanometer spatial resolution, three-dimensional prototyping capability, and the diversity of usable materials.

Wu *et al.* [87] first adopted femtosecond laser nanoshell scanning to rapidly realize high-precision 3D microfluidic devices. For example, a complex microfluidic device proposed by Jamil *et al.* was reproduced, which was designed as a disk structure of 100 μm diameter and 15 μm height (Fig. 11.26(b)). A 3D filtration network with six 2 μm-sized bores was designed for the separation function of impurity and cells with different sizes (Fig. 11.26(d)). The version with stand-alone shells directly defining the device with walls (Figs. 11.26(a) and (c)) and that with buried channels created with the aid of additional ultraviolet exposure are both capable of functioning as designed. In the latter case, The TPP fabrication time is dramatically reduced from 20 h to less than 30 min.

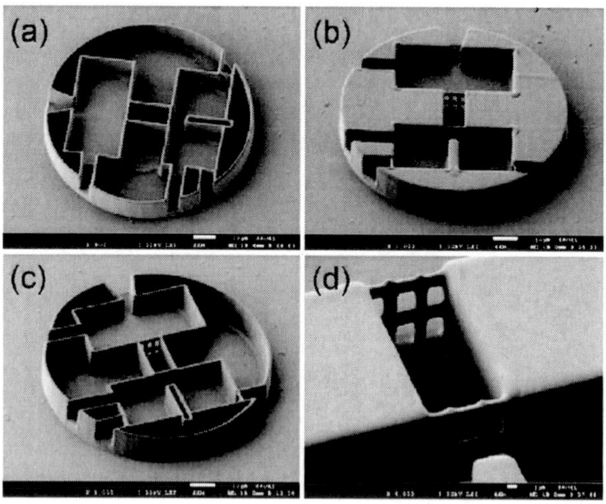

Figure 11.26 A conceptual microfluidic device fabricated by TPP of SU-8. (a) Overall SEM view of the 3D microfluidic systems with 100 μm diameter disk and 15 μm height. Here, the internal portion of the background volume is solidified by additional ultraviolet exposure. (b) Two infiltration networks with six 2 μm-sized bores. (c), (d) SEM images of the same microfluidic devices but consisting of stand-alone shell as walls.

To further enhance the fabrication efficiency, Wang *et al.* [88] demonstrate a novel post-embellishment strategy for appending extra functions on traditional microfluidic channels via femtosecond laser micronanofabrications. Additional photopolymer functional parts were integrated into existing microchannels prepared by conventional photolithography. As a representative example, they realized microsieves with different pore sizes and shapes (Figs. 11.27(a) and (b)) inside the glass microchannel, including straight edge (edges of square and triangle), cambered edge (round), acute angle (triangle), and right angle (square). A series of sieve wafers with 6, 5.5, 5, 4 and 3.5 μm pores were fabricated for sieving particles with

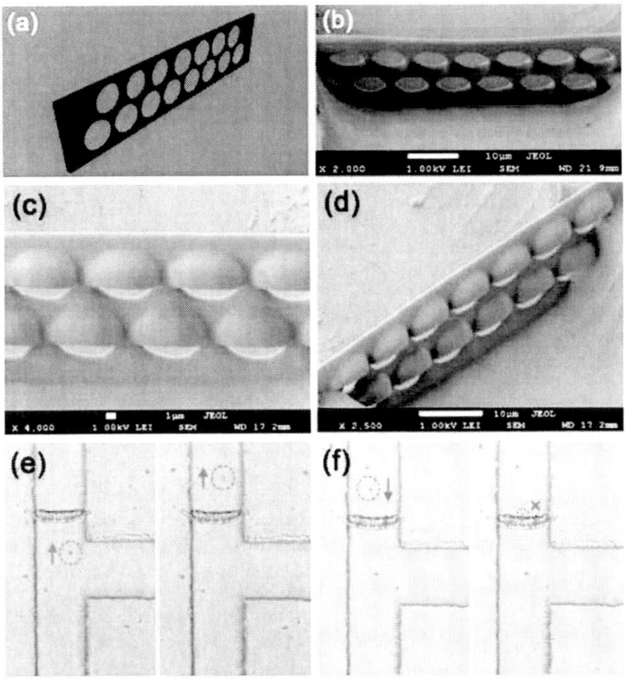

Figure 11.27 (a) 3D structures designed by 3D max. Pores in it with different round shapes (b). Side view (c, d) of the valve with fish scale-like structures. (e) If the particles come from the non-eave direction, they would meet the eaves after they passed the pores. (f)If the particles come from the eaves direction, they meet the eaves before the wall, and they lose their horizontal momentum when they bang against the eaves, and would be blocked by the valve.

sizes from 2.5 to 5.5 μm. In addition, they also showed a fish scale-like structure (Figs. 11.27(c) and (d)), which was designed as a one-way valve (Figs. 11.27(e) and (f)) for microfluidic chips. This work showed that femtosecond laser micro/nanofabrications could be used for embellishing microfluidic devices for chip functionalization and integration, which will greatly promote the applications of LOC devices.

For integrated microfluidic devices, various mixing structures are needed to increase the mixing efficiency. Generally, they were classified into two types of active type and passive type according to the presence of an external power source for the operation of a micromixer. Compared to the active type mixer, the passive type is very simple and cost effective because the designed microstructure induced the mixing of fluids, which did not rely on a complicated movable part or external power source. Yang *et al.* [89] reported a neo-conceptive three-dimensional crossing manifold micromixer (Fig. 11.28(a)) integrated into microchannel by sequential processes of photolithography and two-photon polymerization. First, a microchannel pattern was fabricated by a conventional photolithography process. The negative photoresist (SU8-3035, MicroChem) was spin-coated on a cover glass at 2000 rpm for 30 s. A prebaking process was then performed at 95°C for 10 min. After UV exposure of 200 mJ/cm^2 on the photomask with the microchannel pattern, it was postbaked at 95°C for 10 min for the cross-linking process. The pattern is developed using propylene glycol methyl ether acetate and rinsed using isopropyl alcohol. Finally, a microchannel with a 50 μm height and width was obtained. After fabrication of the microstructures in the channel, the microfluidic system was sealed by bonding with polydimethylsiloxane (PDMS) slabs with two inlet and one outlet holes. A mixing ratio of 93.9% (Fig. 11.28(b)) was demonstrated from the standard deviation, which was a little higher than that of the simulation result (90%). The mixing ratio could be further enhanced by an optimized design that considered the mixing ratio and pressure drop.

11.5.2 Biological Applications

Recent researches have paid great attention to the intriguing differences between cell migration in two and three dimensions because the study of cell migration are important for understanding

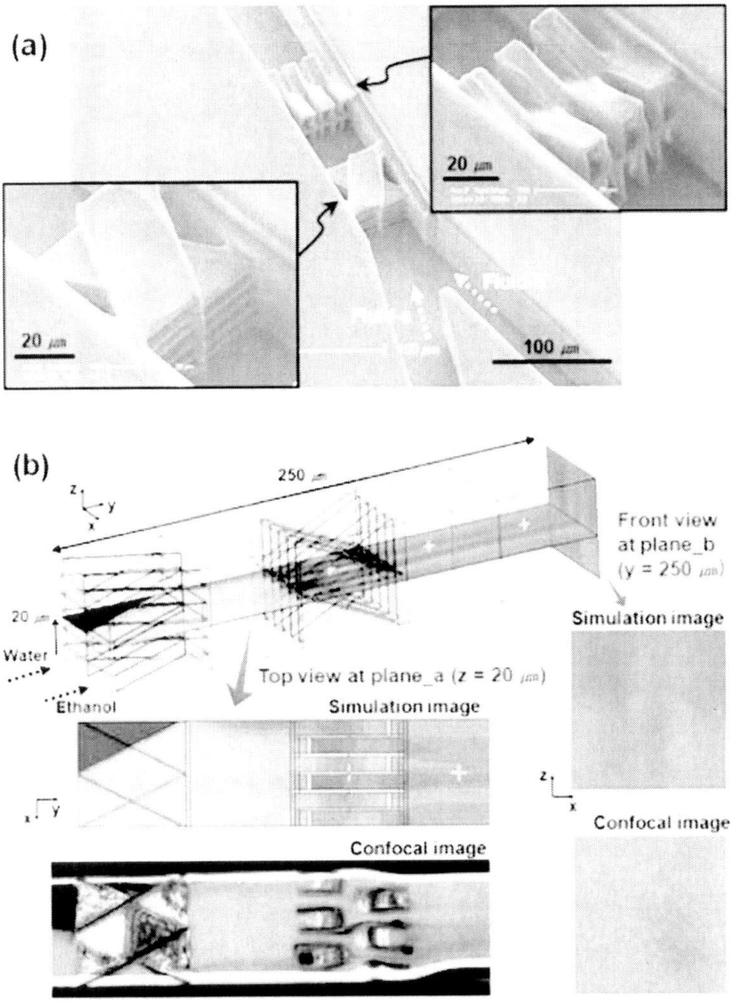

Figure 11.28 (a) SEM images of the H/V-CMM fabricated in the microchannel. A dead zone of fluids is observed in the triangle region between the manifolds and the side wall of the channel. It is formed by the accumulated dose between the structures. However, the dead zone is not significantly large, and it would be resolved by reducing the laser power. (b) Grayscale images of simulation and experimental results. Ethanol with rhodamine B (0.1 wt%) and water flow from left to right; the white color is the fluorescence of the dye. (c) The number of pixel vs. the distribution of brightness of the confocal image at plane_b.

a variety of physiological and pathological processes such as embryonic development, cancer metastasis, blood vessel formation and remodeling, tissue regeneration, immune surveillance, and inflammation. The mechanisms and regulation of cell migration in 2D cell-culture models have been extensively studied, where migration is predominantly a function of adhesion and de-adhesion events. Mazur *et al.* [90] prepared 3D cell scaffolds by TPP. The architectural parameters of the 3D matrix like pore size, shape, porosity and permeability was precisely controlled to the submicrometer scale. The human fibrosarcoma cell line HT1080 was used to study cell migration in the scaffolds. They seed the fluorescently-labeled cells onto the scaffolds at a density of 500,000 cells mL^{-1}. Figure 11.29(a) shows a top-view bright-field image of cells taken 5 h after seeding inside a 52 μm pore-sized scaffold. Initially, the cells are located only on the cover slip, but begin to move in 3D after a few hours. The cells readily move in three dimensions inside the scaffold, as seen

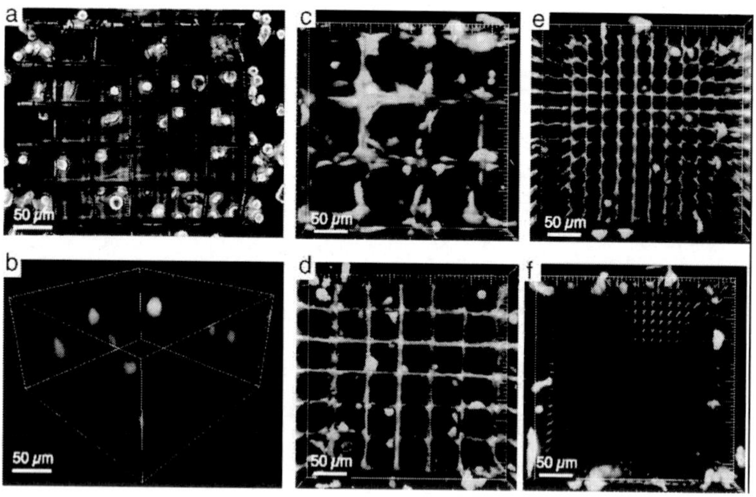

Figure 11.29 (a) Brightfield optical image showing top view of cells inside a 52 μm pore-sized scaffold 5 h after of seeding cells. (b) Fluorescence image showing isometric 3D rendered view of HT1080 cells inside a scaffold 24 h after seeding cells (c–f) Top view overlay of fluorescence and differential interference contrast images of cells in 110, 52, 25 and 12 μm pore-sized scaffolds showing non-uniformity in the distribution of cells in different matrices owing to variations in physical obstruction.

from a 3D rendered image of GFP labeled cells (Fig. 11.29(b)). The cells attach and move predominantly along the beams in the 110 μm scaffold (Fig. 11.29(c)), while they occupy the entire pore in the smaller scaffolds. Also, the cells are more uniformly dispersed inside the 52 μm scaffold than inside the 25 μm and 12 μm scaffold (Figs. 11.29(d–f)), respectively.

Figure 11.30(a) depicts a typical track of the motion of a cell in a 25 μm scaffold. They demonstrated that the overall mean speed

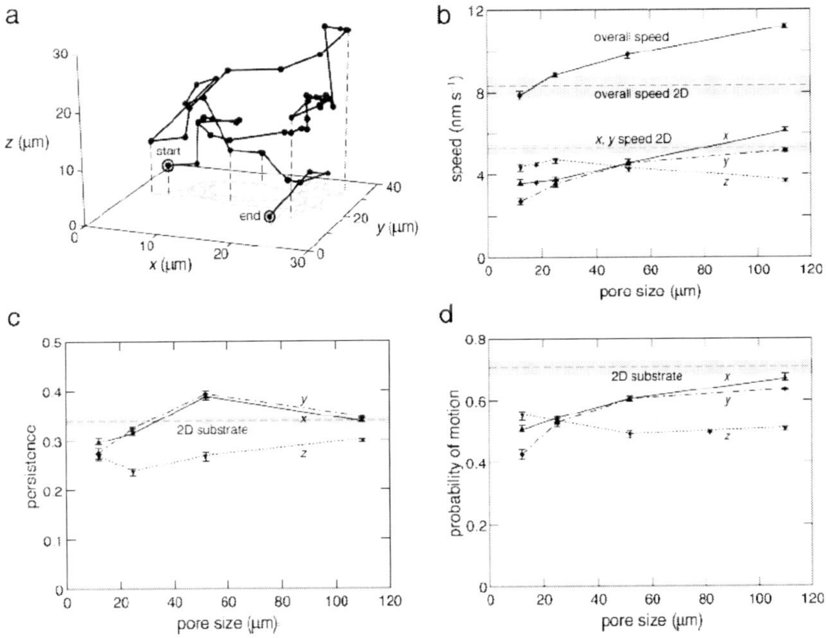

Figure 11.30 (a) Typical track of a cell in a 25-μm scaffold. (b) Overall mean speed and directional speed, (c) persistence and (d) probability of movement as a function of pore size. Data points: 3D matrices; dashed lines: values obtained on 2D substrates. The error bars represent the standard error around the mean value. Statistical significance was evaluated for difference in values between adjacent pore-sized scaffolds. It was also evaluated between 2D and 110 μm pore-sized scaffold. The difference between adjacent data points has p values of less than 0.05 for all pairs except those marked by an asterisk.

of the cells is higher in the scaffold than it is on 2D substrate (Fig. 11.30b), but the lateral speed and persistence in the x and y directions are the same (Fig. 11.30(c)). In contrast, Fig. 11.30(d) shows that the probability of movement in the x and y directions is lower in the scaffold than it is on 2D substrate, where cells encounter fewer obstructions. They also showed that as the pore size increases, the overall mean speed increases, but the persistence and probability of movement remain the same.

11.6 Outlook

11.6.1 Novel 3D Optical Devices and Integrated Optical Circuits

With the improvement of micro/nanofabrication technology, micro-optical devices [91–97] are developing towards functionality, miniaturization and integration. 3D optical devices are crucial to realize novel functions, such as 3D photonic crystal [98], 3D metamaterials [99]. Researchers have demonstrated that two-photon polymerization could realize high-quality 3D devices with high-performance optical functions. As more and more novel optical devices have been developed, TPP will play an important role in this field. In addition, femtosecond laser TPP has the instinct ability to integrate various 3D optical components into a chip, so it may be used for the preparation of integrated optical circuits in future.

11.6.2 Microfluidic Devices with Functional 3D Components for Biological Application

The ability to fabricate various 3D microstructures [100–105] on the microfluidic channels is fundamental in the study of biological Microsystems. Although recent advances on controllable microdevices by TPP have been made, for example, microsieves and micromixer have been integrated into microfluidic systems for filtering and mixing functions, it is still attractive to fabricate other 3D functional components into microchannels for broader applications in chemical sensing, cell engineering, and so on.

11.6.3 Electrical-Optical Complex Micromechanical Systems

The ability to create novel micromachines and conductive microstructures [106–111] by TPP is a recent development. Magnetic functional micromachines and microinductors, which will be useful in cell culture and nanosicence, have been demonstrated. How to assemble these separate microdevices into multifunctional microsystems remains challenging, which also will be an important research area in future.

11.7 Conclusions

The work discussed in this chapter only touches the recent progress of TPP for the preparation of micro-optical, micromachines, and microfluidic devices due to the unique advantages of this technique. Thanks to the great efforts by research groups worldwide, the basic "toolbox" of this technique is nearly complete. Efficient TPP microfabrication systems are now commercially available. Although some additional developmental work is needed, this technology is mature enough, and therefore the focus is now shifting to broader applications such as biology, sensor, and medical diagnosis. We believe that multiphoton microfabrication has the potential to become an important tool in industry as well.

References

1. Lee, S. A., Chung, S. E., Park, W., Lee, S. H., and Kwon, S. (2009) Three-dimensional fabrication of heterogeneous microstructures using soft membrane deformation and optofluidic maskless lithography, *Lab Chip*, **9**, 1670–1675.
2. Liu, X., Wang, Q., Qin, J., and Lin, B. (2009) A facile "liquid-molding" method to fabricate PDMS microdevices with 3-dimensional channel topography, *Lab Chip*, **9**, 1200–1205.
3. Cheng, Y., Sugioka, K., and Midorikawa, K. (2005) Microfabrication of 3D hollow structures embedded in glass by femtosecond laser for Lab-on-a-chip applications, *Appl. Surf. Sci.*, **248**, 172–176.
4. Sugioka, K., Masuda, M., Hongo, T., Cheng, Ya., Shihoyama, K., and Midorikawa, K. (2004) Three-dimensional microfluidic structure

embedded in photostructurable glass by femtosecond laser for lab-on-chip applications, *Appl. Phys. A,* **79**, 815–817.

5. Zhou, W. H., Kuebler, S. M., Braun, K. L., Yu, T., Cammack, J. K., Ober, C. K., Perry, J. W., and Marder, S. R. (2002) An efficient two-photon-generated photoacid applied to positive-tone 3D microfabrication, *Science,* **296**, 1106–1109.
6. Fleming, J. G., Lin, S. Y., Kady, I. E. Biswas, R., and Ho. K. M. (2002) All-metallic three-dimensional photonic crystals with a large infrared bandgap, *Nature,* **417**, 52–55.
7. Qi, M. H., Lidorikis, E., Rakich, P. T., Johnson, S. G., Joannopoulos, J. D., Lppen, E. P., and Smith, H. I. (2004) A three-dimensional optical photonic crystal with designed point defects, *Nature,* **429**, 538–542.
8. Jeong, K. H., Kim, J., and Lee, L. P. (2006) Biologically inspired artificial compound eyes, *Science,* **312**, 557–561.
9. Meisel, D. C., Diem, M., Deubel, M., Willard, F. P., Linden, S., Gerthsen, D., Busch, K., and Wegener, M. (2006) Shrinkage precompensation of holographic three-dimensional photonic-crystal templates, *Adv. Mater.,* **18**, 2964–2968.
10. Hermatschweiler, M., Ledermann, A., Ozin, G. A., Wegener, M., and Freymann, G. V. (2007) Fabrication of silicon inverse woodpile photonic crystals, *Adv. Funct. Mater.,* **17**, 2273–2277.
11. Li, H., Wang, J., Yang, L., and Song, Y. (2008) Superoleophilic and superhydrophobic inverse opals for oil sensors, *Adv. Funct. Mater.,* **18**, 3258–3264.
12. Wu, J., Day, D., and Gu, M. (2008) A microfluidic refractive index sensor based on an integrated three-dimensional photonic crystal, *Appl. Phys. Lett.,* **92**, 071108.
13. Gu, F. X., Zhang, L., Yin, X., and Tong L. (2008) Polymer single-nanowire optical sensors, *Nano Lett.,* **8**, 2757–2761.
14. Gu, Z. Z., Horie, R., Kubo, S., Yamada, Y., Fujishima, A., and Sato, O.(2002) Fabrication of a metal-coated three-dimensionally ordered macroporous film and its application as a refractive index sensor, *Angew. Chem. Int. Ed.,* **41**, 1154–1156.
15. Cho, E. C., Kim, J. W., Nievew, A. F. N., And Weitz, D. A. (2008) Highly responsive hydrogel scaffolds formed by three-dimensional organization of microgel nanoparticles, *Nano Lett.,* **8**, 168–172.
16. Freed, L. E., Engelmayr, G. C., Borenstein, J. T., Moutos, F. T., and Guilak, F. (2009) Advanced material strategies for tissue engineering scaffolds, *Adv. Mater.,* **21**, 1–9.

17. III, R. A. B., Shepherd, R. F., Hanson, J. N., Nuzzo, R. G., Wiltzius, P., and Lewis, J. A. (2009) Direct-write assembly of 3D hydrogel scaffolds for guided cell growth, *Adv. Mater.*, **21**, 2407–2410.

18. Koller, D. M., Hohenaus, A., Ditlbachera, H., Gallera, N., Baudriona, A. L., Reila, F., Schausbergera, S., Aussenegga, F. R., Leitnera, A., and Krenna, J. R. (2008) Three-dimensional SU-8 sub-micrometer structuring by electron beam lithography, *Microelectron. Eng.*, **85**, 1639–1641.

19. Liu, N., Guo, H., Fu, L., Kaiser, S., Schweizer, H., and Giessen, H. (2008) Three-dimensional photonic metamaterials at optical frequencies, *Nat. Mater.*, **7**, 31–37.

20. Ishiia, Y., and Taniguchi, J. (2007) Fabrication of three-dimensional nanoimprint mold using inorganic resist in low accelerating voltage electron beam lithography, *Microelectron. Eng.*, **84**, 912–915.

21. Subramania, G., and Lin, S. Y. (2004) Fabrication of three-dimensional photonic crystal with alignment based on electron beam lithography, *Appl. Phys. Lett.*, **85**, 5037–5039.

22. Valentine, J., Zhang, S., Zentgraf, T., Avila, E. U., Genov, D. A., Bartal, G., and Zhang, X. (2008) Three-dimensional optical metamaterial with a negative refractive index, *Nature*, **455**, 376–379.

23. Tjerkstra, R. W., Sergerink, F. B., Kelly, J. J., and Vos, W. L. (2008) Fabrication of three-dimensional nanostructures by focused ion beam milling, *J. Vac. Sci. Technol. B.*, **26**, 973–977.

24. Jun, K., Joo, J., and Jacobson, J. M. (2009) Focused ion beam-assisted bending of silicon nanowires for complex three dimensional structures, *J. Vac. Sci. Technol. B.*, **26**, 3043–3047.

25. Jeon, J., Floresca, H. C., and Kim, M. J. (2010) Fabrication of complex three-dimensional nanostructures using focused ion beam and nanomanipulation, *J. Vac. Sci. Technol. B.*, **28(3)**, 549–553.

26. Lim, C. T., Low, H. Y., Ng, J., Liu, W. T., and Zhang, Y. (2009) Fabrication of three-dimensional hemispherical structures using photolithography, *Microfluid. Nanofluid.*, **7**, 721–726.

27. Meier, R. C., Badilita, V., Brunne, J., Wallrabe, U., and Korvink, J. G. (2011) Complex three-dimensional high aspect ratio microfluidic network manufactured in combined PerMX dry-resist and SU-8 technology, *Biomicrofluidics.*, **5**, 034111.

28. Casse B. D. F., Moser, H. O., Lee, J. W., Bahou, M., Inglis, S., and Jian, L. K. (2007) Towards three-dimensional and multilayer rod-split-ring metamaterial structures by means of deep x-ray lithography, *Appl. Phys. Lett.*, **90**, 254106.

29. Awazu, K., Wang, X., Fujimaki, M., Kuriyama, T., Sai, A., Ohki, Y., and Imai, H. (2005) Fabrication of two- and three-dimensional photonic crystals of titania with submicrometer resolution by deep x-ray lithography, *J. Vac. Sci. Technol. B.*, **23**, 934–939.

30. Barry, R. A., Shepherd, R. F., Hanson, J. N., Nuzzo, R. G., Wiltzius, P., and Lewis, J. A. (2009) Direct-write assembly of 3D hydrogel scaffolds for guided cell growth, *Adv. Mater.*, **21**, 2407–2410.

31. Gratson, G. M., Santamaria, G., Lousse, V., Xu, M., Fan, S., Lewis, J. A., and Braun, P. V. (2006) Direct-write assembly of three-dimensional photonic crystals: Conversion of polymer scaffolds to silicon hollow-woodpile structures, *Adv. Mater.*, **18**, 461–465.

32. Gratson, G. M., Xu, M., and Lewis, J. A. (2004) Direct writing of threedimensional webs, *Nature*, **428**, 386.

33. Zhao, X. M., Xia, Y., and Whitesides, G. M. (1996) Fabrication of three-dimensional micro-structures: Microtransfer molding, *Adv. Mater.*, **8**, 837–840.

34. Ullal, C. K., Maldovan, M., Thomas, E. L., Chen, G., Han, Y. J., and Yang, S. (2004) Photonic crystals through holographic lithography: Simple cubic, diamond-like, and gyroid-like structures, *Appl. Phys. Lett.*, **84**, 5434–5436.

35. Mao, W. D., Liang, G. Q., Pu, Y. Y., Wang, H. Z., and Zeng, Z. (2007) Complicated three-dimensional photonic crystals fabricated by holographic lithography, *Appl. Phys. Lett.*, **91**, 261911.

36. Jang, J. H., Dendukuri, D., Hatton, T. A., Thomas, E. L., and Doyle, P. (2007) A route to three-dimensional structures in a microfluidic device: Stop-flow interference lithography, *Angew. Chem. Int. Ed.*, **46**, 9027–9031.

37. Maldovan, M., Ullal, C. K., Jang, J. H., and Thomas, E. L. (2007) Sub-micrometer scale periodic porous cellular structures: Microframes prepared by holographic interference lithography, *Adv. Mater.*, **19**, 3809–3813.

38. Ramanan, V., Nelson, E., Brzezinski, A., Braun, P. V., and Wiltzius, P. (2008) Three dimensional silicon-air photonic crystals with controlled defects using interference lithography, *Appl. Phys. Lett.*, **92**, 173304.

39. Campbell, M., Sharp, D. N., Harrison, M. T., Denning. R. G., and Turberfield, A. J. (2000) Fabrication of photonic crystals for the visible spectrum by holographic lithography, *Nature,* **404**, 53–56.

40. Shoji, S., and Kawata, S. (2000) Photofabrication of three-dimensional photonic crystals by multibeam laser interference into a photopolymerizable resin, *Appl. Phys. Lett.*, **76**, 2668–2670.

41. Sun, H. B., Nakamura, A., Shoji, S., Duan, X. M., and Kawata, S. (2003) Three-dimensional nanonetwork assembled in a photopolymerized rod array, *Adv. Mater.*, **15**, 2011–2014.

42. Jang, J. H., Ullal, C. K., Kooi, S. E., Koh, C. Y., and Thomas, E. L. (2007) Shape control of multivalent 3D colloidal particles via interference Lithography, *Nano Lett.*, **7**, 647–651.

43. Sun, H. B., Tanaka, T., Takada, K., and Kawata, S. (2001) Two-photon photopolymerization and diagnosis of three-dimensional microstructures containing fluorescent dyes, *Appl. Phys. Lett.*, **79**, 1411–1413.

44. Sun, H. B., Tanaka, T., and Kawata, S. (2002) Three-dimensional focal spots related to two-photon excitation, *Appl. Phys. Lett.*, **80**, 3673–3675.

45. Deubel, M., Wegener, M., Kaso, A., and John, S. (2004) Direct laser writing and characterization of Slanted Pore Photonic Crystals, *Appl. Phys. Lett.*, **85**, 1895–1897.

46. Serbin, J., Ovsianikov, A., and Chichkov, B. (2004) Fabrication of woodpile structures by two-photon polymerization and investigation of their optical properties, *Opt. Express,* **12**, 5221–5228.

47. Teh, W. H., Durig, U., Salis, G., Harbers, R., Drechsler, U., Mahrt, R. F., Smith, C. G., and Guntherodt, H. J. (2004) SU-8 for real three-dimensional subdiffraction-limit two-photon microfabrication, *Appl. Phys. Lett.*, **84**, 4095–4097.

48. Coenjarts, C. A., and Ober, C. K. (2004) Two-photon three-dimensional microfabrication of poly(dimethylsiloxane) elastomers, *Chem. Mater.*, **16**, 5556–5558.

49. Kato, J. K., Takeyasu, N., Adachi, Y., Sun, H. B., and Kawata, S. (2005) Multiple-spot parallel processing for laser micronanofabrication, *Appl. Phys. Lett.*, **86**, 044102.

50. Park, S. H., Lee, S. H., Yang, D. Y., Kong, H. J., and Lee, K. S. (2005) Subregional slicing method to increase three-dimensional nanofabrication efficiency in two-photon polymerization, *Appl. Phys. Lett.*, **87**, 154108.

51. Nakanishi, S., Shoji, S., Kawata, S., and Sun, H. B. (2007) Giant elasticity of photopolymer nanowires, *Appl. Phys. Lett.*, **91**, 063112.

52. Kaneko, K., Yamamoto, K., Kawata, S., Xia, H., Song, J. F., and Sun, H. B. (2008) Metal-nanoshelled three-dimensional photonic lattices, *Opt. Lett.*, **33**, 1999–2001.

53. Maruo, S., Nakamura, O., and Kawata, S. (1997) Three-dimensional microfabrication with two-photon-absorbed photopolymerization, *Opt. Lett.*, **22**, 132–134.

54. Göppert-Mayer, M. (1931) Elementray actions with two quantum leaps, *Ann. Phys.*, **9**, 273–295.

55. Odian, G. (1991) *Principles of Polymerization*, 3rd Edn. Wiley, New York.

56. Reiser, A. (1989) *Photoreactive Polymers: The Science and Technology of Resists.* Wiley, New York.

57. Fouassier, J. P. (1993) (Eds) *Radiation Curing in Polymer Science and Technology*. Elsevier, London.

58. Kuebler, S. M., Rumi, M., Watanabe, T., Braun, K., Cumpston, B. H., Heikal, A. A., Erskine, L. L, Thayumanavan, S., Barlow, S., Marder, S. R., and Perry, J. W. (2001) Optimizing two-photon initiators and exposure conditions for three-dimensional lithographic microfabrication, *J. Photopolym. Sci. Technol.*, **14**, 657–668.

59. Kawata, S., Sun, H. B., Tanaka, T., and Takada, K. (2001) Finer features for functional microdevices, *Nature*, **412**, 697–698.

60. Takada, K., Sun, H. B., and Kawata, S. (2005) Improved spatial resolution and surface roughness in photopolymerization based laser nanowriting, *Appl. Phys. Lett.*, **86**, 071122.

61. Xing, J. F., Dong, X., Chen, W., and Duan, X. M. (2007) Improving spatial resolution of two-photon microfabrication by using photoinitiator with high initiating efficiency, *Appl. Phys. Lett.*, **90**, 131106.

62. Hashe, W., Chen, V. W., Hales, J. M., Dong, W., Barlow, S., Marder, S. R., and Perry, J. W. (2007) 65 nm feature sizes using visible wavelength 3-D multiphoton lithography, *Opt. Express*, **15**, 3427–3436.

63. Deubel, M., Freymann, G., Wegener, M., Pereira, S., Busch, K., and Soukoulis, M. (2004) Direct laser writing of three-dimensional photonic-crystal templates for telecommunications, *Nat. Mater.*, **3**, 444–447.

64. Juodkazis, S., Mizeikis, V., Seet, K. K., Miwa, M., and Misawa, H. (2005) Two-photon lithography of nanorods in SU-8 photoresist, *Nanotechnology*, **16**, 846–849.

65. Tan, D., Li, Y., Qi, F., yang, H., Gong, Q., Dong, X., and Duan, X. M. (2007) Reduction in feature size of two-photon polymerization using SCR500, *Appl. Phys. Lett.*, **90**, 071106.

66. Wu, D., Wu, S. Z., Niu, L. G., Chen, Q. D., Wang, R., Song, J. F., Fang, H. H., and Sun, H. B. (2010) High numerical aperture microlens arrays of close packing, *Appl. Phys. Lett.*, **97**, 071106.

67. Sun, Z., Dong, X. Z., Chen, W. Q., Nakanishi, S., Duan, X. M., and Kawata, S. (2008) Multicolor polymer nanocomposites: *In situ* synthesis and fabrication of 3D microstructures, *Adv. Mater.*, **20**, 914–919.

68. Wang, J., Xia, H., Xu, B. B., Niu, L. G., Wu, D., Chen, Q. D., and Sun, H. B. (2009) Remote manipulation of micronanomachines containing magnetic nanoparticles, *Opt. Lett.*, **34**, 581–583.

69. Guo, R., Xiao, S., Zhai, X., Li, J., Xia, A., and Huang, W. H. (2006) Micro lens fabrication by means of femtosecond two photon photopolymerization, *Opt. Express*, **14**, 810–816.

70. Chen, Q., Wu, D., Niu, L. G., Wang, J., Lin, X. F., Xia, H., and Sun, H. B. (2007) Phase lenses and mirrors created by laser micronanofabrication via two-photon photopolymerization, *Appl. Phys. Lett.*, **91**, 171105.

71. Li, Y., Yu, Y., Guo, L., Wu, S., Chen, C., Niu, L., Li, A., and Yang, H. (2010) High efficiency multilevel phase-type Fresnel zone plates produced by two-photon polymerization of SU-8, *J. Opt.*, **12**, 035203.

72. Niu, L. G., Wang, D., Jiang, T., Wu, S. Z., Li, A., and Song, J. F. (2011) High fill-factor multilevel Fresnel zone plate arrays by femtosecond laser direct writing, *Opt. Comm.*, **284**, 777–781.

73. Wu, D., Niu, L. G., Chen, Q. D., Wang, R., and Sun, H. B. (2008) High efficiency multilevel phase-type fractal zone plates, *Opt. Lett.*, **33**, 2913–2915.

74. Chen, Q. D., Lin, X. F., Niu, L. G., Wu, D., Wang, W. Q., and Sun, H. B. (2008) Dammann gratings as integratable micro-optical elements created by laser micronanofabrication via two-photon photopolymerization, *Opt. Lett.*, **33**, 2559–2561.

75. Sun, H. B., Matsuo, S., and Misawa, H. (1999) Three-dimensional photonic crystal structures achieved with two-photon-absorption photopolymerization of resin, *Appl. Phys. Lett.*, **74**, 786–788.

76. Sun, H. B., Suwa, T., Takada, K., Zaccaria, R. P., Kim, M. S., Lee, K. S., and Kawata, S. (2004) Shape precompensation in two-photon laser nanowriting of photonic lattices, *Appl. Phys. Lett.*, **85**, 3708–3710.

77. Seet, K. K., Mizeikis, V., Matsuo, S., Juodkazis, S., and Misawa, H. (2005) Three-dimensional spiral-architecture photonic crystals obtained by direct laser writing, *Adv. Mater.*, **17**, 541–545.

78. Seet, K. K., Mizeikis, V., Juodkazis, S., and Misawa, H. (2006) Three-dimensional horizontal circular spiral photonic crystals with stop gaps below 1 um, *Appl. Phys. Lett.*, **88**, 221101.

79. Thiel, M., Decker, M., Deubel, M., Wegener, M., Linden, S., and Freymann, G. V. (2007) Polarization stop bands in chiral polymeric three-dimensional photonic crystals, *Adv. Mater.*, **19**, 207–210.

80. Thiel, M., Rill, M. S., Freymann, G. V., and Wegener, M. (2009) Three-dimensional Bi-Chiral photonic crystals, *Adv. Mater.*, **21**, 1–3.

81. Serbin, J., and Gu, M. (2006) Experimental evidence for superprism effects in three-dimensional polymer photonic crystals, *Adv. Mater.*, **18**, 221–224.

82. Galajda, P., and Ormos, P. (2001) Complex micromachines produced and driven by light, *Appl. Phys. Lett.*, **78**, 249–251.

83. Maruo, S., Ikuta, K., and Korogi, H. (2003) Submicron manipulation tools driven by light in a liquid, *Appl. Phys. Lett.*, **82**, 133–135.

84. Maruo, S., and Inoue, H. (2006) Optically driven micropump produced by three-dimensional two-photon microfabrication, *Appl. Phys. Lett.*, **89**, 144101.

85. Maruo, S., Takaura, A., and Saito, Y. (2009) Optically driven micropump with a twin spiral microrotor, *Opt. Express*, **17**, 18525.

86. Xia, H., Wang, J., Tian, Y., Chen, Q. D., Du, X. B., Zhang, Y. L., He, Y., and Sun, H. B. (2010) Ferrofluids for fabrication of remotely controllable micro-nanomachines by two-photon polymerization, *Adv. Mater.*, **22**, 3204–3207.

87. Wu, D., Chen, Q. D., Niu, L. G., Wang, J. N., Wang, J., Wang, R., Xia, H., and Sun, H. B. (2009) Femtosecond laser rapid prototyping of nanoshells and suspending components towards microfluidic devices, *Lab Chip*, **9**, 2391–2394.

88. Wang, J., He, Y., Xia, H., Niu, L. G., Zhang, R., Chen, Q. D., Zhang, R. L., Li, Y. F., Zeng, S. J., Qin, J. H., Lin, B. C., and Sun, H. B. (2010) Embellishment of microfluidic devices via femtosecond laser micronanofabrication for chip functionalization, *Lab Chip*, **10**, 1993–1996.

89. Lim, T. W., Son, Y., Jeong, Y. J., Yang, D. Y., Kong, H. J., Lee, K. S., and Kim, D. P. (2011) Three-dimensionally crossing manifold micro-mixer for fast mixing in a short channel length, *Lab Chip*, **11**, 100–103.

90. Tayalia, P., Mendonca, C. R., Baldacchini, T., Mooney, D. J., and Mazur, E. (2008) 3D cell-migration studies using two-photon engineered polymer scaffolds, *Adv. Mater.*, **20**, 4494–4498.

91. Watanabe W., Asano, T., Yamada, K., and Itoh, K. (2003) Wavelength division with three-dimensional couplers fabricated by filamentation of femtosecond laser pulses, *Opt. Lett.*, **28**, 2491–2493.

92. Huang, Y. Y., Paloczi, G. T., Poon, J. K.S., and Yariv, A. (2004) Bottom-up soft-lithographic fabrication of three-dimensional multilayer polymer integrated optical microdevices, *Appl. Phys. Lett.*, **85**, 3005–3007.

93. Pan, C. T. (2004) Design and fabrication of sub-micrometer eight-level bi-focal diffraction optical elements, *J. Micromech. Microeng.*, **14**, 471–479.

94. Huang, Y. Y., Paloczi, G. T., Yariv, A., Zhang, C., and Dalton, L. R. (2004) Fabrication and replication of polymer integrated optical devices using electron-beam lithography and soft lithography, *J. Phys. Chem. B*, **108**, 8606–8613.

95. Xu, Q. F., Sradley, B., Pradhan, S., and Lipson, M. (2005) Micrometre-scale silicon electro-optic modulator, *Nature*, **435**, 325–327.

96. Tong, L. M., Lou, J. Y., Gattass, R. R., He, S. L., Chen, X. W., Liu, L., and Mazur, E. (2005) Assembly of silica nanowires on silica aerogels for microphotonic devices, *Nano Lett.*, **5**, 259–262.

97. Cheng, Q. D., Lin, X. F., Niu, L. G., Wu, D., Wang, W. Q., and Sun, H. B. (2008) Dammann gratings as integratable micro-optical elements created by laser micronanofabrication via two-photon photopolymerization, *Opt. Lett.* **33**, 2559–2561.

98. Fujita, M., Takahashi, S., Tanaka, Y., Asano, T., and Noda, S. (2005) Simultaneous inhibition and redistribution of spontaneous light emission in photonic crystals, *Science*, **308**, 1296–1298.

99. Rill, M. S., Plet, C., Thiel, M., Staude, I., Freymann, G. V., Linden, S., and Wegener, M. (2008) Photonic metamaterials by direct laser writing and silver chemical vapour deposition, *Nat. Mater.*, **7**, 543–546.

100. Leclerc, E., Sakai, Y., and Fujii, T. (2003) Cell culture in 3-dimensional microfluidic structure of PDMS (polydimethylsiloxane), *Biomed. Microdev.*, **5(2)**, 109–114.

101. Hofmann, O., Voirin, G., Niedermann, P., and Manz, A. (2002) Three-dimensional microfluidic confinement for efficient sample delivery to biosensor surfaces.application to immunoassays on planar optical waveguides, *Anal. Chem.*, **74**, 5243–5250.

102. Jeon, N. L., Chiu, D. T., Wargo, C. J., Wu, H. K., Choi, I. S., Anderson, J. R., and Whitesides, G. M. (2002) Design and fabrication of integrated passive valves and pumps for flexible polymer 3-dimensional microfluidic systems, *Biomed. Microdev.*, **4(2)**, 117–121.

103. Therriault, D., White, S. R., and Lewis, J. A. (2003) Chaotic mixing in three-dimensional microvascular networks fabricated by direct-write assembly, *Nat. Mater.*, **2**, 265–271.

104. Jeon, S., Malyarchuk, V., White, J. O., and Rogers, J. A. (2005) Optically fabricated three dimensional nanofluidic mixers for microfluidic devices, *Nano Lett.*, **5**, 1351–1356.

105. Bettinger, C. J., Weinberg, E. J., Kulig, K. M., Vacanti, J. P., Wang, Y. D., Borenstein, J. T., and Langer, R. (2006) Three-dimensional microfluidic tissue-engineering scaffolds using a flexible biodegradable polymer, *Adv. Mater.*, **18**, 165–169.

106. Xu, B. B., Xia, H., Niu, L. G., Zhang, Y. L., Sun, K., Chen, Q. D., Xu, Y., Lv, Z. Q., Li, Z. H., Misawa, H., and Sun, H. B. (2010) Flexible nanowiring of metal on nonplanar substrates by femtosecond-laser-induced electroless plating, *Small*, **6**, 1762–1766.

107. Maruo, S., and Saeki, T. (2008) Femtosecond laser direct writing of metallic microstructures by photoreduction of silver nitrate in a polymer matrix, *Opt. Lett.*, **16**, 1174–1179.

108. Formanek, F., Takeyasu, N., and Tanaka, T. (2006) Three-dimensional fabrication of metallic nanostructures over large areas by two-photon polymerization, *Opt. Express*, **14**, 800–809.

109. Cao, Y. Y., Takeyasu, N., Tanaka, T., Duan, X. M., and Kawata, S. (2009) 3D metallic nanostructure fabrication by surfactant-assisted multiphoton-induced reduction, *Small*, **5**, 1144–1148.

110. Tanaka, T., Ishikawa, A., and Kawata, S. (2006) Two-photon-induced reduction of metal ions for fabricating three-dimensional electrically conductive metallic microstructure, *Appl. Phys. Lett.*, **88**, 081107.

111. Ishikawa, A., Tanaka, T., and Kawata, S. (2006) Improvement in the reduction of silver ions in aqueous solution using two-photon sensitive dye, *Appl. Phys. Lett.*, **88**, 113102.

Chapter 12

Ultrafast Laser Processing: From Micro- to Nanoscale Industrial Applications

Friedrich Dausinger and Steffen Sommer
Dausinger + Giesen GmbH, Rotebühlstraße 87 70178 Stuttgart, Germany
dausinger@dausinger-giesen.de

Since many years, the advantages of ultrafast laser processing have been demonstrated by numerous research groups all over the world. Indisputably the achievable precision increases with shortening of the pulse duration by reducing the heat load. Additionally non-linear effects could be harnessed to tailor absorptivity and beam guidance in transparent materials. Until recently the acceptance of ultrafast technology for industrial applications remained quite poor, however. High costs and complexity together with still insufficient power and reliability of the available lasers distracted production engineers from using them. The recent advent of picosecond lasers with nearly two orders of magnitude higher power at the same price level as older ones emitting pulses of about 100 fs brought the breakthrough. Nowadays these systems are used in the automobile, electronic and solar industries for

- structuring of surfaces

Ultrafast Laser Processing: From Micro- to Nanoscale
Edited by Koji Sugioka and Ya Cheng

ISBN 978-981-4267-33-5 (Hardcover), 978-981-4303-69-9 (eBook)
www.panstanford.com

- milling and drilling
- separating

Structuring of surfaces is usable for a multitude of applications:

- producing tribological patterns on machine components
- patterning of embossing tools
- engraving printing rolls
- functionalizing of electro-optic devices such as photovoltaic or displaying panels

Milling is used for microtools in competition with electrical discharge machining (EDM). Drilling is of interest for fuel injection nozzles. Separating of brittle devices with lasers is on the way to replace cutting or breaking with mechanical tools. Examples are glasses for smartphones and electronic devices. This chapter will give an overview of realized industrial applications or those whose realization is expected in the near future.

12.1 Introduction

The main benefits of the use of ultrashort laser pulses for micromachining are the high achievable quality and the large variety of materials. The high quality is a result of the very short interaction time from absorbing the pulses, the heating inside the material and finally the thermal removal mainly in vaporized phase. For metals these phenomena occur within several ten picoseconds (ps) [1]. For non-metals this can occur faster. This short interaction time disables the transfer of energy within the target by heat conduction, so the thermal effects, typically known from the machining with nanosecond pulses, are reduced dramatically and the ablation results show high precision.

This is valid for the ablation with energy densities about up to two times higher than the ablation threshold. For higher energy densities the deposited energy is so high that the processes to vaporize the material can last longer and so relevant heat conduction can appear. The calculation in Fig. 12.1 shows the resulting melt layer thickness dependent on the irradiated pulse duration and the energy density for aluminium. This is resulting in melting layers, moved out of the interaction zone by the pressure of vaporization and accumulated as burr beside the structures. For fragile materials, such as ceramics, the pressure of volume expansion resulting from volume heating can be too high and cracking appears. For high power densities further the formation of plasma occurs, which by absorbing and scattering affects the interaction by strongly enlarging the interaction in space and time.

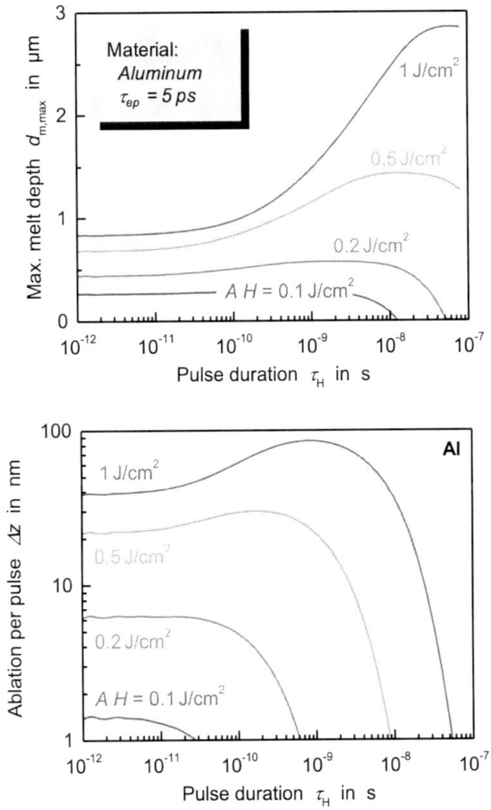

Figure 12.1 Calculation of melt layer thickness and ablation volume for different pulse duration regimes and energy densities [2].

On the other hand, the ablated volume grows with energy density [2]. This results in the classical conflict between precision and productivity, which cannot be solved but diminished by technical measures. A compromise has to be found between the highest tolerable energy density giving an efficient ablation process and the still acceptable quality loss. At a given limit of the energy density, that has to be found for every single application, a further increase of processing productivity only can be reached by a lateral or temporal splitting of the laser irradiation (see Fig. 12.2) [3].

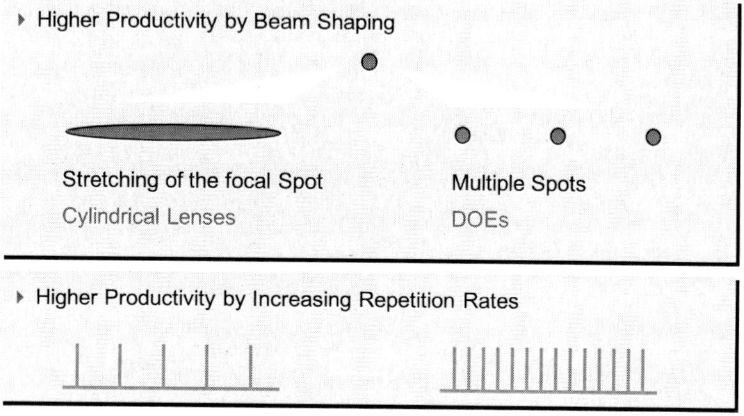

Figure 12.2 Possible strategies to improve average power by keeping constant the energy density.

Precondition for industrial applications is the availability of stable, reliable, powerful and affordable laser systems. Significant developments took place for ps-laser sources during the past years. The average output power raised from 1 W up to more than 50 W and the price per Watt dropped from 300 k$ down to 4 k$ per Watt. The stability of the systems reached a very high level and even the efficiency of the laser systems improved.

12.2 Structuring of Surfaces

The ablation of surfaces is a very successful application. In competition with mechanical milling and electrical discharge machining (EDM) the laser processing enables machining of hard and isolating materials, there is no wear of tools, a high precision is achieved and a high flexibility is provided.

12.2.1 Tribological Patterns

Friction and wear happens in every machine and costs energy and lifetime. A very smooth surface reduces the friction, but if there not enough lubricant left, this cannot remain distributed on the surface and wear appears. Therefore, especially within combustion engines, a defined roughness of the surface is state of the art ensuring lubricant distribution along the piston wall. By structuring small grooves into the surface as demonstrated in Fig. 12.3, the distribution of the lubricant can be designed by the engineers and the roughness of the complete surface can be reduced. This enables the reduction of friction losses in the engine. The main benefit, however, is the reduction of the amount of lubricant that is necessary to inject into the engine. This reduces the particle emission significantly while lifetime is prolonged. This technology has been implemented by various global car manufacturers in serial production and racing sports for many years [4].

Figure 12.3 Grooves within the cylinder wall of a combustion engine. The well-defined patterning reduces friction and wear and enables a reduction of the particle emission. Courtesy of Gehring GmbH.

Other automotive components are also structured with lasers. Some automotive suppliers ablate grooves into the cone of injection needles. Robert Bosch published the structuring of sealing faces within a high-pressure valve to avoid leakage of fuel (see Fig. 12.4) [5].

Not only the reduction of wear by laser-ablated structures enhances the properties within combustion engines, but the lifetime of deep drawing tools can be prolonged significantly. For certain applications, it was presented that the laser patterning of the surface can increase the lifetime from 0.5 million parts up to 50 million parts [6].

Figure 12.4 Groove around a high pressure coupling to avoid discharge of fuel. Courtesy of Robert Bosch.

Further applications in different mechanical components are thinkable and in development. Especially, bearings with long lifetime and reduced accessibility, i.e. in offshore wind power station, can achieve higher performances.

12.2.2 Embossing and Moulding Tools

There are many applications in replication of precise geometries. Starting with signets, a high demand for injection die moulding tools exists. Especially, for electronics parts and exclusive toys, the need in precision increased during the past years. Therefore the efficiency of the state-of-the-art technologies, milling and spark erosion drops down. The 2-1/2-dimensional ablation with lasers increased the processing productivity due to powerful lasers available and with the ultrashort laser pulses the quality increased. Furthermore, the materials, machinable with lasers include hard metals and ceramics. With very high effort a surface roughness about 0.2 μm is possible at removal rates of up to 50 mm^3/s. This is sufficient to ablate the moulds for housings of mobile phones, connector housings and more in short production time and in a cost-effective manner. Embossing tools for mints and forming punches are machined with lasers as well. Figure 12.5 shows ablated geometries in different materials.

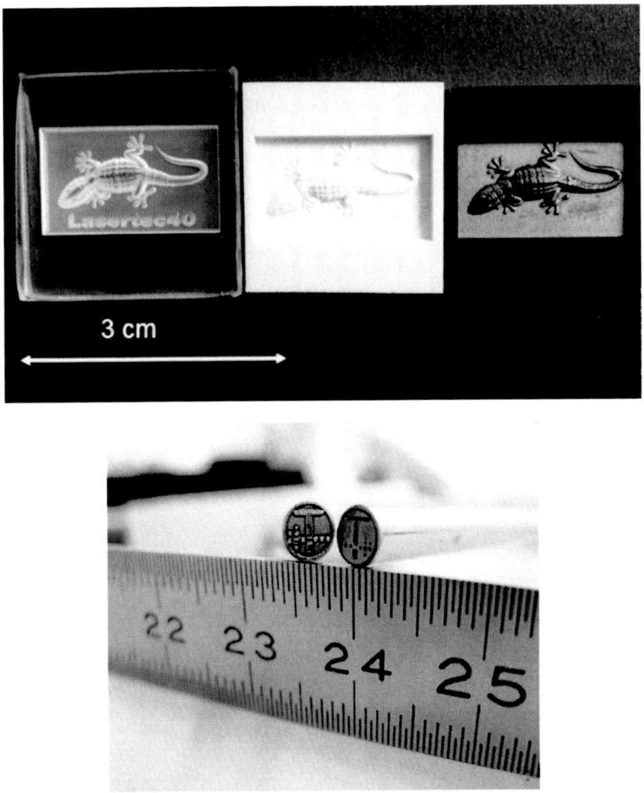

Figure 12.5 Three-dimensional machining within glass, on mammoth tusk and on tungsten carbide (top)—Courtesy of Lasertec. Machining of signets in stainless steel (bottom)—Courtesy of TGSW.

12.2.3 Printing and Embossing Rolls

To get a higher throughput in replication, the tools can be realized as rollers. Laser-engraved rollers are suitable for high precision and for a high quantity of replicas. For more than 100,000 high-quality replicas, intaglio printing is used. The rollers typically have a copper layer as information carrier. This is covered with a chromium layer to get a higher lifetime. The artwork is normally done with diamond needles, but also laser engraving for high-end applications is available.

Especially the embossing rollers with designs for servicing, paperhanging, automotive interior, lamination and even design effects in metal sheets need high-precision tools over large areas of up to several square meters. High-precision engraving to emboss optical functional elements for back light units is demonstrated in Fig. 12.6. Due to the large dimensions and very large volume that has to be removed, this application needs very high volume ablation rates to achieve short enough manufacturing times. At the same time, minor changes in irradiation conditions affect the artwork directly. This application needs a very high stability of laser and handling within very small tolerances at high average power [7, 8].

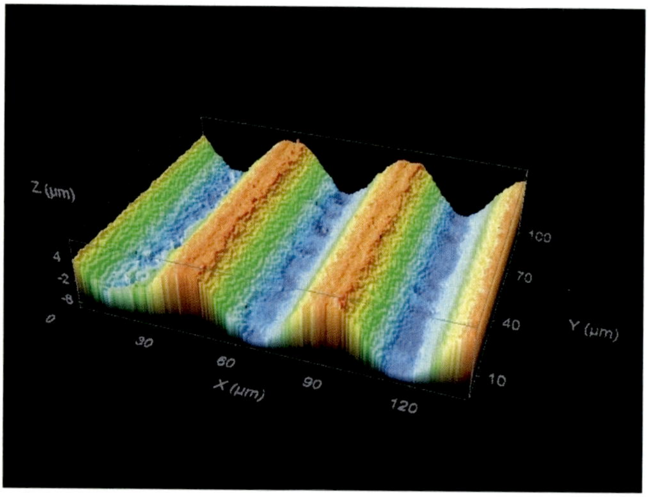

Figure 12.6 High-precision surface ablation for embossing display components. At a structuring depth of about 10 μm a surface roughness better than 0.5 μm could be achieved. Courtesy of Wetzel and TGSW.

12.2.4 Functionalization of Opto-Electronics

A very large growth of the opto-electronics market could be observed in recent years, especially for flat and large-scale applications such as photovoltaic modules and flat-screen displays. Therefore, the functionality results in a composition of several thin material films. These have to be ablated locally. The deposition of these layers is mainly done by sputtering. To ensure precise frontiers at the rim area, after the sputtering often an edge isolation process with lasers is necessary. For solar modules a series connection has to be realized to achieve higher voltage levels. This is done by scribing after the different sputtering processes. Earlier, the scribing was done with diamond needles, but laser scribing has become more and more relevant. Compared with diamond needles, state-of-the-art laser scribing systems enable smaller grooves and a smaller distance between the grooves. This reduces the non-active area (dead zone) and enables higher module efficiency. Due to the wear of the needles, maintenance time and, hence, total costs for the laser processes are lower.

Actually laser ablation with nanosecond (ns) pulses is sufficient for most scribing processes, but the reduction of layer thicknesses and different material combinations need a higher ablation quality without cracking (see Fig. 12.7). In addition, there are possibilities to enhance the quality of thin-film ablation by beam formation to squared beam profiles as shown in Fig. 12.8. New cell designs substitute glass substrates with plastics. Therefore ps pulses are necessary to ablate without thermal destruction of the substrates. In this case, the laser processing competes with the increasing quality of direct printing technologies.

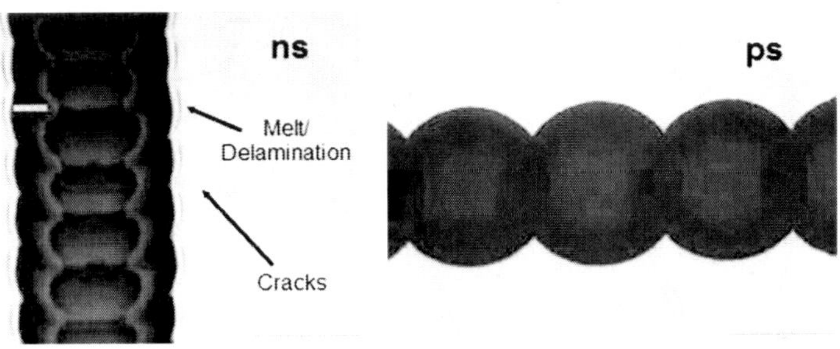

Figure 12.7 Ablation of thin metallic layers on glass. Comparison of thermal effects for pulse durations of several nanoseconds and picoseconds. Courtesy of Trumpf.

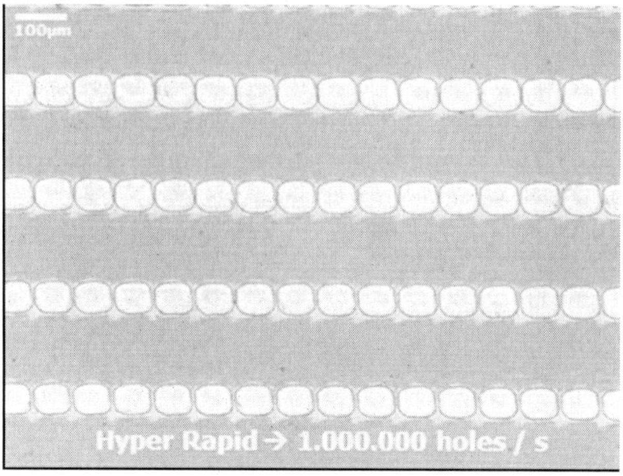

Figure 12.8 Removal of SiN-layer with a thickness of 70 nm at a possible feed rate of up to 50 m/s. Courtesy of Lumera Laser.

For different solar cells there is a shunt fault between front and rearside trough the edge after different coating processes. Therefore laser edge isolation is implemented with ns pulses in serial production. The resistance can be further enhanced up to four times by ablating with ps pulses in adequate processing times due to recent research results.

For the thin film cell design and the crystalline silicon wafer-based cells, there are many laser processes, but most of them can be done with ns or even longer pulse durations [9]. The use of ultrashort laser pulses in production of crystalline wafer-based cells can be necessary for edge isolation and generation of high absorptive surfaces. In thin film production the precision of the ultrashort pulses is necessary for ablation processes on flexible substrates and to reduce the distance between the patterning lines.

12.3 High-Precision Drilling

Laser drilling of injection components shown in Fig. 12.9 was one of the driving applications for the development of the technology of ultrashort laser pulses. Due to the short interaction time between laser radiation and the material, it is possible to drill with high aspect ratio and high roundness without post-processing.

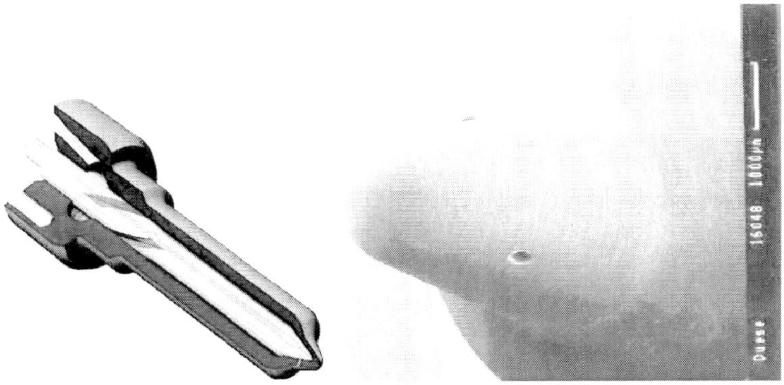

Figure 12.9 Laser drilling of injection nozzles. Courtesy of GFH.

The use of special techniques such as helical drilling requiring adequate beam handling systems is indispensable for achieving the wanted high geometric quality. Some models allow adjusting the inclination angle of the incoming beam (see Fig. 12.10). This enables to compensate losses at the inner wall and to drill holes with perfect cylindricity or holes with negative conicity (smaller entrance than outlet diameter of the borehole). Additionally the processing time is reduced due to the inclination angle. Such high-precision technology enables high aspect ratios, needs several seconds per hole and is suitable for injection applications, cooling holes and for high-end filter technologies in automotive, aerospace and chemical industries.

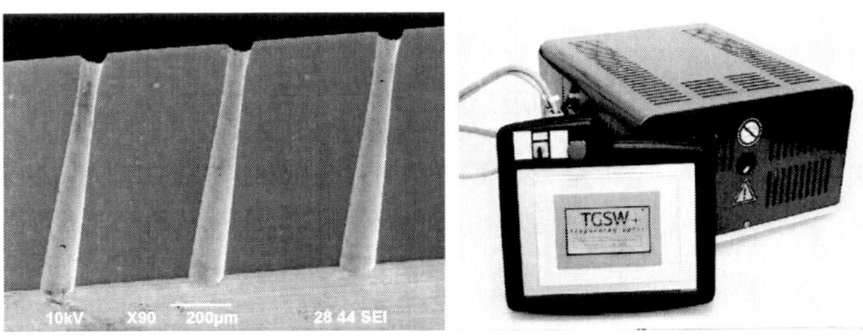

Figure 12.10 Precision laser drilling of negative conical holes with a helical drilling optic (right picture). Courtesy of TGSW.

Low-aspect-ratio holes can be processed with scanners. The applications are mainly in silicon and printed circuit boards (PCB) for three-dimensional packaging. In drilling these boards, the dimensions of the holes have to be reduced further. This raises the costs for mechanical drilling. With laser drilling it is now possible to produce up to 1000 holes per second nearly without wear and maintenance time for the tool, compared to the mechanical process. Figure 12.11 shows such holes machined with picosecond laser pulses.

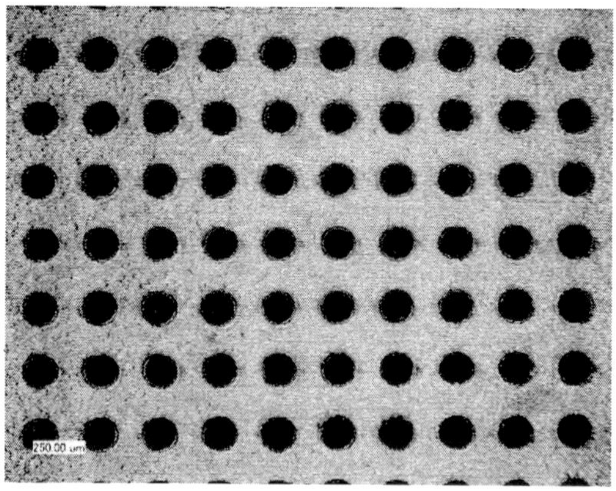

Figure 12.11 Laser drilling of PCB with up to 1000 holes per second. Courtesy of Schmoll and Lumera Laser.

12.4 Separation

The consumer industry needs to improve their products by reducing size, weight and power consumption and increasing robustness. At the same time production costs have to be lowered. This leads to a reduction of the size of parts and often the use of more brittle materials. Conventional machining technologies get slower and more expensive or even cannot reach the required accuracy. These circumstances foster the use of lasers for separation.

12.4.1 Dielectrics

Precision machining of glass becomes more and more interesting, because of its use as front panels in smartphones and as substrate of optoelectronic devices such as flat-panel televisions, diodes and solar cells. The low thermal influence of ps laser pulses enables to cut complex geometries without cracks. This high quality at sufficient machining speed is demonstrated in Fig. 12.12.

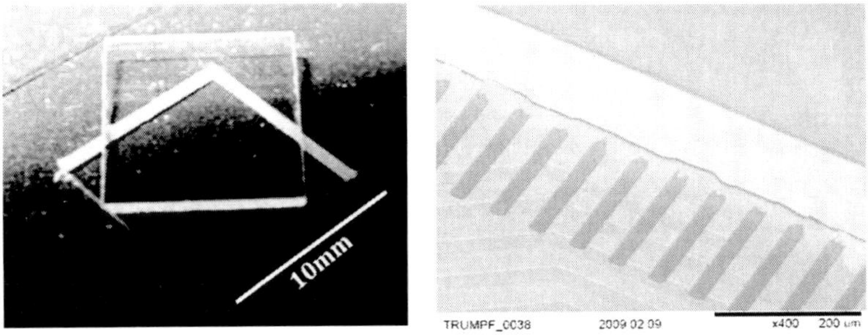

Figure 12.12 (Left) Form cutting of glass, courtesy of Dausinger + Giesen. (Right) Cutting of glasses for plat-panel displays, courtesy of Trumpf.

The high-precision machining by lasers even enables to produce microchannels in glass materials (see Fig. 12.13) that can be used for microsystems, for example, to control small dimensional chemical processes. Further applications such as inside glass marking and welding of glass sheets are in development.

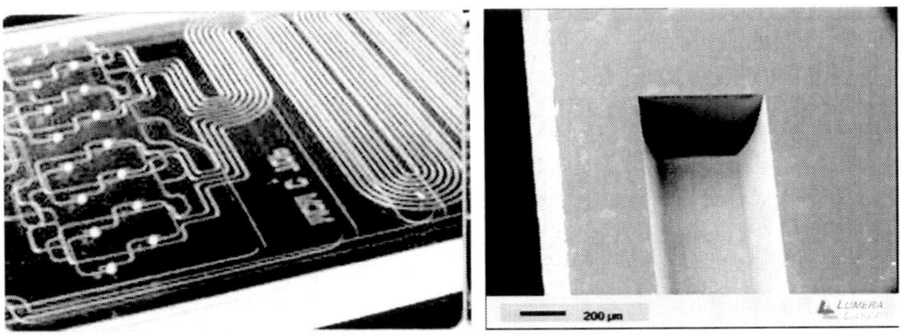

Figure 12.13 Engraving in fused silica for microfluidic chips. Courtesy of Lumera Laser.

12.4.2 Semiconductors

In the semiconductor industry the main base material is silicon. This is quite brittle, and the reduction of the material thickness causes problems in conventional processing. By processing with ps pulses, the thermal impact leading to cracks can be reduced significantly (see Fig. 12.14). This technology further allows smaller cut grooves. In cutting of wavers with small chips, it is possible to place more elements into one wafer, and this enhances production output and saves material and processing time. Besides the cutting of the chips (dies), even ablation processes within the front end are realized with ps technologies for 3D-integration, organic electronics, light-emitting diodes (LEDs) and microelectromechanical systems (MEMSs).

Figure 12.14 Cutting of silicon. Courtesy of Trumpf.

Ultrashort laser pulses can machine all kinds of materials used within microelectronic devices from metals (copper, gold, tungsten, etc.) to ceramics (SiO_2, SiC, Al_2O_3, etc.). New materials for nanoelectronics, so-called low- and high-k materials, are also machinable, even if they are very brittle and hard.

12.4.3 Medical Devices

Laser technology is well established in the production of medical devices, due to its low impact on the base material, for example, in welding pacemakers and marking of surgical instruments. Cutting of stents has been also done with different kinds of laser technologies for many years. The laser can process very thin and fragile products with very low forces.

Compared with cw-laser processing, the use of ps pulses enables considerable saving of production time and costs as shown in Figs. 12.15 and 12.16.

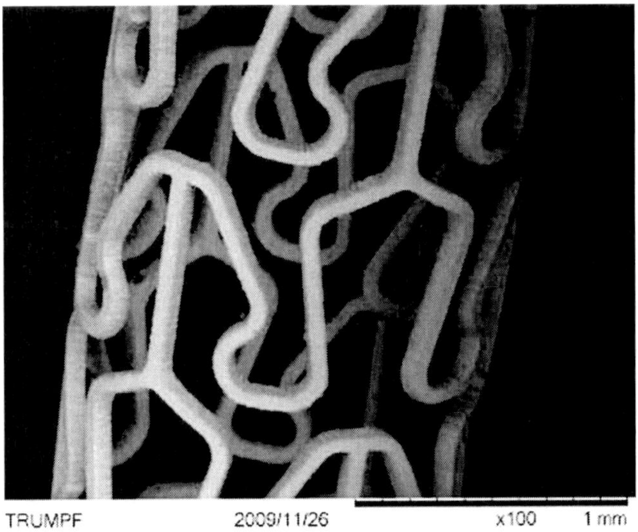

Figure 12.15 Nitinol-stent without electro polishing after picosecond processing. Courtesy of Trumpf.

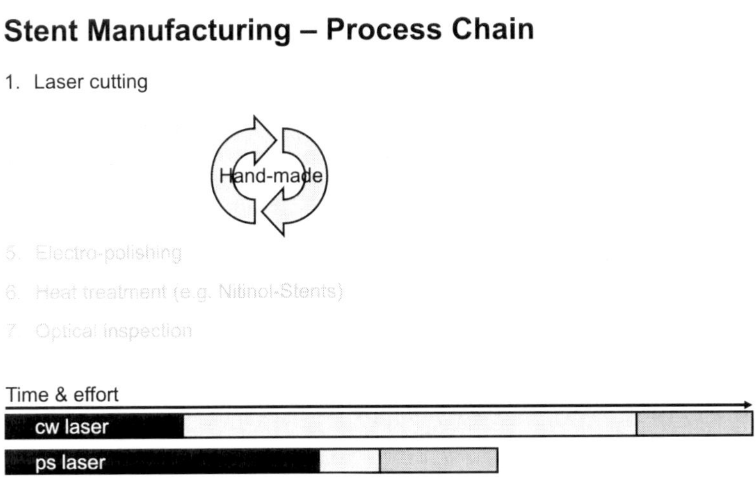

Figure 12.16 Production steps for laser cutting of stents. Comparison of the time consumption for each step for cw- and ps-laser processing. Courtesy of Trumpf.

12.5 Outlook

The main advantages of processing with ultrashort laser pulses, such as high precision, high quality and the possibility to machine nearly every material, are on the expense of low machining speed and high cost of investments. The problem of reliability of the complex laser sources appears to have been solved. For instance, Lumera claims that with more than 500 established systems, millions of operating hours have been achieved, so far. All the mentioned applications show that the advantages can predominate. The number of productive industrial solutions is increasing. Figure 12.17 shows two of different available laser sources with picosecond pulses for industrial production. Higher average power is necessary to achieve

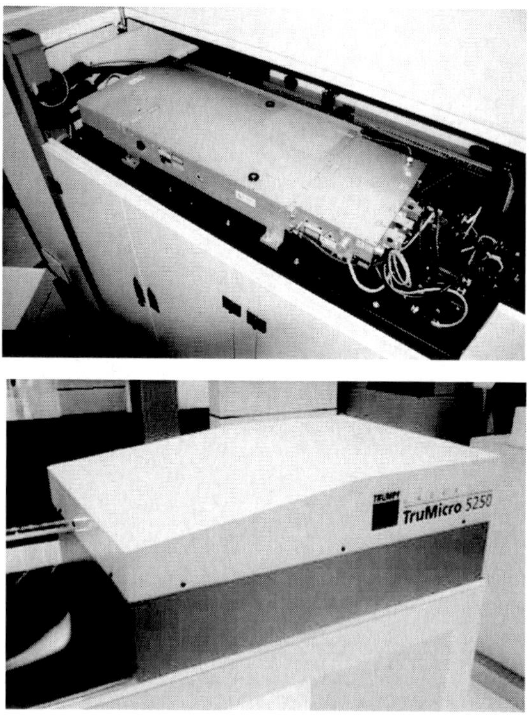

Figure 12.17 High-power picosecond laser sources at industrial grade. A Lumera HyperRapid integrated into a 3D-Micromac machining station (top) and a Trumpf TruMicro 5000 Series (bottom).

higher productivity. However, powerful system technology for part and beam handling has to be developed, as well, to enable full usage of higher power. In the coming years, the prices of laser systems will have to be reduced further or the average power will have to be increased at the same price level. Components such as optics, scanners and beam-forming solutions will have to be designed for special applications.

The advent of powerful ps lasers on the market brought a strong breakthrough in industrial application of ultrafast lasers. Femtosecond (fs) lasers are still not available at the power level of 50 Watt or more and are regarded as expensive, complicated and unstable systems. The advantages of fs pulses, i.e. mainly higher ablation efficiency and use of non-linear effects for controlled increase of absorptivity in transparent materials, are not used industrially, so far. The future availability of fs lasers with average power and reliability comparable to ps ones would certainly enable a multitude of new industrial applications even if their price is higher.

References

1. Limpert, J., Nolte, S., Tünnermann, A. (2004). Ultrashort pulse fiber lasers and amplifiers. In: Dausinger, F., Lichtner, F., Lubatschowski, H. (eds.): *Femtosecond Technology for Technical and Medical Applications. Topics Applied Physics 96* (Springer-Verlag Berlin Heidelberg, Germany).
2. Breitling, D., Ruf, A., Dausinger, F. (2004). Fundamental aspects in machining of metals with short and ultrashort laser pulses. In: Herman, P.R., Fieret, J., Piqué, A., Okada, T., Bachmann, F.G., Hoving, W., Washio, K., Xu, X., Dubowski, J.J., Geohegan, D.B., Träger, F. (eds.): *Photon Processing in Microelectronics and Photonics III* (Bellingham, WA): SPIE, (*Proc. SPIE*, Vol. 5339). 49–63.
3. Sommer, S., Kraus, M., Dausinger, F. (2006). Improvement of productivity in structuring metals with ultrashort laser pulses. In: Japan Laser Processing Society (ed.): *7th International Symposium on Laser Precision Microfabrication LPM 2006* (Kyoto, Japan). Online Proceedings of LAMP, Paper #06–11.
4. Abeln, T., Klink, U. (2003). Laseroberflächenstrukturierung—Verbesserung der tribologischen Eigenschaften. In: Hügel, H., Dausinger, F., Müller, M. (eds.): *Stuttgarter Lasertage SLT '03*

(Forschungsgesellschaft für Strahlwerkzeuge (FGSW) (Stuttgart, Germany), S. 107–110 (in German).

5. Robert Bosch GmbH (2008). Herausragendes neues Produktionsverfahren—Bosch-Forscher mit Innovationspreis ausgezeichnet. Pressrelease PI 6323, September 2008.
6. Dumitru, G., Romano, V (2008). Boosting the lifetime of deep-drawing tools by laser structuring. In: Graf, T., Kern. S. (ed.): *Proceedings of the Stuttgart Laser Technology Forum '08 (SLT'08)* (Stuttgart, Germany).
7. Lehmann, B. (2004). Lasergravur für die Druck- und Prägetechnologie. In: forum photonics BW (ed.): *Lasertechnologie in der Anwendung: Mikrostrukturieren, Mikroschweißen* (Stuttgart). Oberkochen: Photonics BW. (in German).
8. Hennig, G., Selbmann, K.-H., Mattheus, S., Kecke, R., Brüning, S. (2006). Laser precision micro fabrication in the printing industry, *Journal of Laser Micro/ Nanoengineering*, **1** (2), 2006.
9. Booth, H. (2010). Laser processing in industrial solar module manufacturing. In: Japan Laser Processing Society (ed.): *11th International Symposium on Laser Precision Microfabrication LPM 2010* (Stuttgart, Germany). Online Proceedings of LAMP 2010, Paper #10–02.

Index

The Culture and Technology of African Iron Production

EDITED BY PETER R. SCHMIDT

UNIVERSITY PRESS OF FLORIDA

GAINESVILLE · TALLAHASSEE · TAMPA · BOCA RATON

PENSACOLA · ORLANDO · MIAMI · JACKSONVILLE

Printed in the United States of America on acid-free paper ∞

Design by Louise OFarrell

01 00 99 98 97 96 6 5 4 3 2 1

LIBRARY OF CONGRESS CATALOGING-IN-PUBLICATION DATA
The culture and technology of African iron production / edited by Peter R. Schmidt.
p. cm.
Includes bibliographical references (p.) and index.
ISBN 0-8130-1384-4 (alk. paper)
1. Iron age—Africa. 2. Iron-works—Africa—History. 3. Africa—Antiquities.
I. Schmidt, Peter R. (Peter Ridgway). 1942- .
GN780. 4. A1C85 1996
960′.1—dc20 95-31930

Chapter 8, "Complex Iron Smelting and Prehistoric Culture in Tanzania" by Peter R. Schmidt and Donald H. Avery, originally appeared in *Science* 201 (September 22, 1978):1085–89. Reprinted by permission of the American Association for the Advancement of Science.

An earlier version of chapter 9, "Actualistic Models for Interpretation of Two Early Iron Age Industrial Sites in Northwest Tanzania" by Peter R. Schmidt and S. Terry Childs, appeared as "Innovation and Industry during the Early Iron Age in East Africa: The KM2 and KM3 Sites of Northwest Tanzania" in *African Archaeological Review* 3 (1985):53–94. Reprinted by permission of Cambridge University Press.

Chapter 10, "Use of Preheated Air in Primitive Furnaces: Comments on Views of Avery and Schmidt" by J. E. Rehder and "The Use of Preheated Air in Ancient and Recent African Iron Smelting Furnaces: A Reply to Rehder" by Donald H. Avery and Peter R. Schmidt, originally appeared in the *Journal of Field Archaeology* 13 (1986):351–57. Reprinted by permission of the authors and the trustees of Boston University.

The University Press of Florida is the scholarly publishing agency for the State University System of Florida, comprised of Florida A & M University, Florida Atlantic University, Florida International University, Florida State University, University of Central Florida, University of Florida, University of North Florida, University of South Florida, and University of West Florida.

University Press of Florida, 15 Northwest 15th Street, Gainesville, FL 32611

This book is dedicated to the memory of my father, Paul Marcus Schmidt: historian, amateur etymologist, businessman, and passionate reader of history.

Contents

List of Figures and Maps

Figures

Maps

List of Tables

Preface

This book provides a new way of viewing the history of iron technology in Africa. The perception of African iron technology as backward compared to that of Europe is an artifact of Western thought. This book replaces this perspective with a recuperated history based on fine-grained ethnographic studies and informed technological research. This is a story which should appeal to historians, anthropologists, economists, and scientists who are concerned about the historical and cultural setting in which technological innovation and invention occur.

It is also a story that weaves together myth, ritual, history, and science to more accurately capture Africa's complex world of technological accomplishment. In Africa ritual and technology are integral parts of the fabric of industrial production, and any attempt to understand African ironworking must incorporate both domains. This book departs from the common homogeneous representations of African iron technology precisely because it views the symbolic representations of iron working as part of the technological experience. The two cannot be separated. Thus both the student of ritual and the student of the history of science will find common ground in what follows.

Archaeological investigations conducted in western Tanzania in the 1970s revealed evidence for a complex and highly advanced iron technology several thousand years ago. These remarkable finds were elucidated by ethnoarchaeological studies of iron smelting that showed that recent peoples seemed to be using innovative techniques of iron production that were very similar but not identical to those of the past. In short, these results suggested that the accepted history of African iron technology had obscured and even erased evidence of such technological success.

After our initial ethnoarchaeological studies we turned to the further excavation of archaeological remains, particularly those that could throw further light on the history of iron technology in the region. These data provided some stunning correspondences between the ancient record and the ethnographic models, especially in regard to the possibility of preheating in iron smelting. As excavations continued into 1978 and 1980 the scope of our inquiry turned to other iron smelting technologies, most notably the Barongo iron smelters who live south of Lake Victoria. Working among the Barongo in 1979, 1980, and 1984, I developed a deeper appreciation for and understanding of the finely layered world of transformational ritual and symbolism attached to iron smelting activities. Many of those insights are shared in the chapters that follow. This book has developed from those studies as well as a film, *The Tree of Iron*, about the remarkable congruence of myth, ritual, and technology observed in western Tanzania.

While I was living in Dar es Salaam between 1985 and 1987, I shared these research results at length with N. J. Karoma and with M. M. Mulokozi, both Haya with personal and deep intellectual interests in Haya cultural history. These conversations helped to open more clearly some of the deeper meanings of Haya symbolic meaning as they are expressed in iron production and oral traditions. During that time, only after some seventeen years of study, did some of the essential connections among myth, ritual, and technology become clear. This is a poignant testimony to the fact that such studies demand a long exposure to the cultural context in which a technology is set. This is also a cautionary tale about using information from studies conducted by an investigator who stays for only abbreviated periods and consequently lacks a systemic understanding of the cultural setting.

While the Dar es Salaam sojourn was essential in my personal development of a deeper understanding of the cultural meanings and beliefs associated with iron working, it was important to expand understanding by bringing together others engaged in similar studies. Two symposia at the African Studies meetings in 1988 were organized to examine advances in the technical understanding of African iron working while also speaking to the cultural context in which technology obtains meaning.

Many chapters in this volume come directly from those symposia. The most recent and important research on symbolic and technical studies about African ironworking is represented in the chapters by Childs, David and Robertson, Childs and Dewey, Killick, Schmidt, de Maret and Thiry, and Goucher and Herbert. All of these researchers have been engaged for several decades in the rigorous documentation of African iron technology. Some, such as David, Killick, Goucher, Herbert, and Schmidt, have done this through what is often called "salvage ethnography": recovering the knowledge of African iron smelting that is dying with the passage of old men who have held this valuable and detailed knowledge. Their deaths mark the death of complex technological knowledge. These researchers have helped to recover some critical pieces in this important knowledge system, pieces without which a history of African iron technology could not be written. As important as this collection of studies is, it was necessary to go further and to include the work of Randi Barndon, a young Norwegian archaeologist who has fully documented the ritual and technological practices of the Fipa ironworkers in southwest Tanzania.

One of the key issues that arose at the 1988 symposia was the critique of the preheating hypothesis that was presented in David Killick's paper. Recounting mostly previously published criticisms, Killick argued that the Schmidt and Avery hypothesis for preheating in ancient iron working had not been proved. Killick's chapter presented a vexing problem. It could not stand alone in this book, for alone it would leave the false impression of presenting a tight case against an important scientific explanation for African technological invention. And, it would lack sufficient context, namely the considerable prior debate on the issue with detailed arguments in support of and in opposition to the preheating hypothesis.

To solve this problem and to highlight the importance of the debate to African history and archaeology, we decided to present the debate from the time of the first publication of the 1978 Schmidt and Avery article in *Science*. The upshot of this is the presentation of a coherent set of chapters that present the full range of issues that are contested. Three chapters are newly written (chapters 11, 12, and 13), one is rewritten (chapter 9), and several are reprinted from technical journals (chapters 8 and 10). Together they lay out the arguments in a sequenced dialogue for and against the preheating hypothesis. We feel confident that this will provide a basic reference for the evidence and opinions that pertain to this important historical and scientific question.

Many people in Tanzania have contributed to my work and given me personal assistance and friendship. I am particularly indebted to N. J. Karoma, M. M. Mulokozi, and Fred Kaijage, all of whom are located at the University of Dar es Salaam and have always offered critical advice tempered with patience. Their assistance and many discussions have been important in unpacking the symbolic meaning of iron smelting songs and rituals in western Tanzania. Of the many people who have helped in the field, my thanks go to the smelters and people of Nyungwe village in Kiziba and the people of Butengo in Biharamulo, Kagera Region. This is partly their story. Within

Kagera region, the cultural officer and regional development director were always helpful and cooperative. At the national level, the Department of Antiquities and the then National Scientific Research Council both provided permits and various forms of assistance. Alphonce Feliciani, Hussein Abdalla, and a whole host of archaeological supervisors and assistants made the archaeological field work possible and successful. Numerous people in Bukoba and Dar es Salaam have helped in a multitude of ways, with special thanks going to Shiraz and Zinat Fazal, Martin Wamara, G. Tibakweitira, Hussein Dhihrani, and Omukama Lukamba of Kiamutwara.

My thanks also goes to a number of colleagues who have shared my life and thoughts, including Peter O'Neill, Frank Muhly, Winnie Lambrecht, Imogene Lim, and Terry Childs. Don Avery's companionship certainly kept us focused on things technological as well as on the absurd. During the 1970s and 1980s Terry Childs gave devoted effort in the field and the laboratory. We see the culmination of her hard work in the number of important contributions she makes to this book. Many thanks go to Patricia Miller, who assisted in a wide variety of editorial tasks.

This research would not have been possible without the support of the National Science Foundation, the National Endowment for the Humanities, and the Social Science Research Council, all of which have provided generous funds for the initial research and the later analysis and writing up of these materials. My wife, Jane, has provided professional assistance as an archaeologist as well as steadfast encouragement for all phases of the enterprise. My children, raised partly in the field, know only too well that this is only one step in the ongoing quest to write a new history of African technology.

References

Childs, S. T., and D. Killick. 1993. "Indigenous African Metallurgy: Nature and Culture." *Annual Review of Anthropology* 22:317–37.

Herbert, E. W. 1993. *Iron, Gender, and Power: Rituals of Transformation in African Societies.* Bloomington and Indianapolis: Indiana University Press.

Schmidt, P. R., and D. H. Avery. 1978. "Complex Iron Smelting and Prehistoric Culture in Tanzania." *Science* 201:1085–89.

Schmidt, P. R., and S. T. Childs. 1985. "Innovation and Industry during the Early Iron Age in East Africa: The KM2 and KM3 Sites of Northwest Tanzania." *African Archaeological Review* 3:53–94.

The Culture and Technology of African Iron Production

1 Cultural Representations of African Iron Production

Peter R. Schmidt

In 1978, *Time* magazine carried an article, based on a report in *Science*, about a complex and innovative iron technology that had recently been documented in East Africa by two American researchers. Wire services and other publications around the world soon picked up this news and widely publicized what was described as a discovery that "will help to change scholarly and popular ideas that technological sophistication developed in Europe but not in Africa."[1]

Reactions to the 1978 announcements illuminate why iron production in Africa has long been a widely followed but seriously misunderstood part of Africa's history. Most perceived this new evidence to be extraordinary because it contradicted views that African technological history lacked any achievement worthy of scientific interest. Since 1978 many important studies have freshly considered some conventional ideas about African iron production and how it has been presented in the writing of African history (for example, Schmidt and Avery 1978; Goucher 1981; Goucher and Herbert, chapter 3, this volume; David et al. 1989; Rowlands and Warnier 1993; Herbert 1993). These studies challenge constructions about the technical characteristics of iron production that have been presented primarily by Western observers and scholars as well as revise representations about the meaning and significance of the ritual behavior that surrounds ironworking.

Out of these revisions and challenges has arisen controversy. This book looks at some of these new ideas and delves into important debates surrounding new interpretations of iron production in Africa. These debates affect far more than the few researchers they directly engage. We have purposefully brought the heat of debate to this book because it exposes the issues to greater scrutiny and helps others to understand the implications of these new ideas for the construction of an alternative African history (Schmidt and Patterson 1995).

The title of this book captures the two elements making up the cultural representations of iron production: the ritual (beliefs and ideology) and the technical. The first part of this book takes up the first cultural representation—the presentation of ironworking, particularly iron smelting, as a highly ritualized process. This emphasis has mostly divorced the symbolism, ritual, and taboo attached to iron production from other, sometimes parallel, belief systems, such as local healing, that help to contextualize meanings (Schmidt 1983a; Barndon, chapter 4, this volume; Herbert 1993). The second cultural representation holds that the technical processes of iron smelting throughout the continent are based on principles similar to those developed in ancient and primitive European iron smelting. This construction arises out of Western perceptions rather than out of close inquiry into African systems of technology and economy. As such it is a cultural projection of European values that masks enormous technological variability and ingenuity in Africa.

The essays that follow will also help to erode the idea that ritual and technological process are necessarily separate domains. To be sure, it is not easy to understand the relationship between the two; it is especially difficult for most Westerners to comprehend, for they generally believe that science and technology are unrelated to ideology. Several essays in this book, especially Schmidt's study of the Barongo smelters (chapter 5) and Childs's and Dewey's view of Luba and Shona forging (chapter 7), will also illustrate and help to explain how the technical process is inextricably interwoven with the myth, ritual, and healing practices that culturally situate and confer meaning to iron-producing technologies.

Transformed Representations

Iron smelting and iron fabrication, such as smithing and wire drawing, attracted the attention of early travelers, missionaries, colonial administrators, and anthropologists who were fascinated by the ritual mystifications and other "exotic" behavior associated with all aspects of the production cycle. Over the past 150 years, a diverse literature of varying ethnographic value and questionable historical value has developed out of these observations. One result has been that many of these early observers, lacking technical training and captivated by the blood sacrifice and other magical applications, placed a consistent emphasis on ritual characteristics at the expense of technical understanding. This emphasis went hand in hand with the notion that there was nothing of scientific value to learn from African technologies.

Enlightenment thought about progress and civilization also deeply influenced the way Europeans observed African iron production and how they represented what they thought they saw. Translated through European values, African iron production became a primitive expression of technological life that was seen to represent a survival of European iron production as it would have been several thousand years ago (Schmidt and Patterson in press). Thus, the historical value of iron production in Africa became its usefulness in helping to explain the characteristics of ancient technological history in Europe.

The equation of ancient European technological history with recent African technological history denies any sense of historical development for iron technology in Africa. It assumes that technology in Africa in the nineteenth and early twentieth centuries remained unchanged throughout its history. This synchronic representation has also been amplified by a strong descriptive emphasis on ritual, a characteristic that appears timeless and therefore obscures change and development. Even a recent and comprehensive attempt to understand the panoply of beliefs surrounding African iron production fails to overcome the overwhelming sense of synchronicity and timelessness that has come to characterize the history of these complex systems of meaning (Herbert 1993).

When European scholarship has tried to make sense out of the mass of information on African iron production, it has used diffusionist ideas, an approach that reduces variation by emphasizing similarities. A diffusionist emphasis completes the picture of an undynamic, ahistorical African iron technology by representing it with descriptions of traits such as furnace shapes and bellows types—categories that invite comparison often without consideration of historical relationships (see Cline 1937; Kense 1981, 1985).

The perspectives discussed thus far are transformational representations of African iron production that have had two primary implications. First, African iron production has been historically homogenized—that is, its many different expressions over time have been represented as technologically similar.

Second, technological histories that hold important stories of innovation and invention have been erased and replaced by representations that focus on ritual practice and beliefs surrounding iron production. Both of these perspectives arise from European cultural perceptions and historical experience rather than from African historical experience or cultural meanings of iron production.

The first comprehensive survey of evidence on African iron production was published more than fifty years ago as part of a larger study of metallurgy in Africa (Cline 1937). For its time, Cline's monograph was a well-informed consideration of some African technological processes; it has been used for decades as the only authoritative synthesis of observations on African iron production. Because the process was seen as fundamentally unchanging and similar throughout the continent, regardless of significantly different outward appearances, the assumptions of this study have been widely incorporated into texts and have come to color our general thinking. An example of some of these widely accepted, oversimplified observations is the interpretation that "Negroes usually make wrought iron rather than cast iron or steel. Sometimes, however, Negroes produce steel, either by accident or design" (Cline 1937, 30).

While Cline was interested in the steel making documented by several early observers, such as Bellamy (1904) and Stanley (1921), much of his speculation concentrated on whether or not cast iron was produced by African iron smelters. His discussion of many accounts that allege to document the production of high-carbon cast iron concludes with his assertion that such accounts are defective because they lack complete descriptions of all the technological steps that would have been required to produce a malleable steel from cast iron. Thus, Cline provides a tantalizing suggestion of possible technological sophistication that is overcome by the realistic assessment that such accounts were incomplete and insufficient to make technological characterizations. Cline accurately assessed the insufficient evidence for production and working of cast iron, but his interpretation of the accidental nature of steel production set an unfortunate emphasis that came to penetrate thinking about African iron production.

Randomness as a Normative View

In the face of such uncertainty about the technological capacity and characteristics of African iron production, historians understandably came to accept the conclusion that African iron production was a hit-or-miss process, lacking clear procedures or methods that ensured a predictable final product. Thus unfolds an important insight into a central feature of the culture of iron production in Africa—the European depiction of iron production as an accidental, random process, not governed by consistent principles of production.

This Eurocentric cultural projection has also been reinforced by diffusionist methodologies that use descriptions of the formal traits of iron smelting and forging and then fit these traits into generalized typological schemes. The shapes and sizes of smelting furnaces as well as bellows types have been popular foci for such European typologies based on superficial characteristics. This orientation has persisted into the last quarter of this century. The distribution of high-shaft furnaces as well as pit or bowl furnaces, for example, continued to fascinate archaeologists until as recently as the mid-1980s (Haaland and Shinnie 1985). Trait lists, reminiscent of the culture-area approach in North American archaeology during the early part of this century, are rendered in geographical form, but maps of such traits show clearly that there is a "bewildering distribution of furnace types across much of sub-Saharan Africa . . . [that is] difficult to explain" (Kense 1985). This sense of random distribution amplifies the idea that technological innovation or accomplishment is also a random or at best an irregular process.

Chapter 2, by Pierre de Maret and Genvieve Thiry, illustrates the deeply ingrained concerns for trait distributions in African archaeology and linguistic studies that continue to inform scholarship. De Maret is an eminent Belgian archaeologist who has worked for more than two decades in central and west-central Africa on questions pertaining to the development of Iron Age societies. He and Thiry, his student, bring together some of the major issues, such as linguistic evidence for early ironworking, that continue to challenge and inform much of the archaeological scholarship that has predominated in Europe about the African Iron Age. While their chapter differs from those that follow, it poignantly demonstrates that some trait studies, especially those that pertain to linguistic relationships, offer hope of understanding how such variability developed historically on the African cultural landscape, but the studies also spread confusion about more essential ideas of African technological development.

The idea of randomness with its complex configuration of formal attributes has been further reinforced by a type-comparative approach used by both historians and archaeologists. The format Cline used has been taken up by a number of later scholars; this eclectic approach focuses on prominent features such as furnace shape that may have more stylistic meaning than the functional significance that scholars might believe most important (Williams 1974; Sutton 1985; Pole 1985; Kense 1985). The result is a hodgepodge of comparative descriptions; for example, tall shaft furnaces of West Africa are compared, on the basis of outward morphology, to similar furnaces in Malawi. The same comparative studies often contrast the treatment of ores and labor, but do not relate them to technical processes, chemical results, or to ritual processes in a rigorous and systematic way. This results in a morass of detail that confuses the reader, fails to sort out the essential and critical similarities and differences, and defies the most basic canons of comparative ethnology.

A fundamental failing of such an approach is its use of superficial or surface characteristics. Another profound difficulty is that furnace form is often confused with furnace function, a confusion arising out of the false assumption that form marks similarity in internal process. Over the last fifteen years, investigators have increasingly turned to functional categories, such as natural-draft, convection furnaces to describe the tall-shaft furnaces that do not require bellows or slag-tapping and non–slag-tapping to describe the way slag is treated in a variety of other furnace forms, including bowl furnaces, low-shaft furnaces (up to 1.5 m), and tall-shaft furnaces (de Maret and Thiry, chapter 2, this volume). Finally, an eclectic comparative treatment is often based on incomplete, truncated, and imperfectly informed accounts. This observation equally applies to those accounts that have attempted to make sense out of the plethora of symbolic meanings associated with iron production (for example, Herbert 1993). This book remedies the eclectic comparative approach by presenting in-depth case studies of several societies and by using ethnographic comparison only when it helps to elicit cross-cultural patterns and important similarities in technological processes and ritual meaning.

ORIGINS, DIFFUSION, AND HISTORICAL REPRESENTATIONS

The construction of a Eurocentric interpretation of African iron production that features irregularity and randomness has come to be accepted as an integral part of the culture of iron production in Africa. Archaeologists have not helped much in trying to sort out some of the speculation and randomness that characterizes their own treatment of iron production. In fact, traditional archaeological thought about the antiquity of various types of iron smelting techniques as well as speculations about the origins of iron production in Africa have simply added confusion.

As de Maret and Thiry observe (chapter 2, this volume), questions of origins and the quest to find the earliest dates for iron production still drive the research agendas of many archaeologists. Moreover, speculation on the question of origins consistently overwhelms historical accuracy or scientific evidence. One prominent example of this proclivity is the idea that the earliest iron-smelting furnaces in Africa were the open-bowl type (Kense 1985). This interpretation tends to be closely associated with, if not derived from, the idea that furnaces lacking a superstructure are simple and more primitive (see Sutton 1985). This intuitive speculation about the greater antiquity of one type of furnace is not substantiated by any archaeological evidence.[2] Of the several hundred iron-smelting furnaces excavated on the continent, only a few have been adequately described. Among those described and dated, not one can be definitely attributed as an open hearth furnace dating to the Early Iron Age.

A distinct archaeological problem arises in trying to sort out the characteristics of early furnace types. There is ambiguity about furnace form introduced by both the purposeful and natural destruction of ancient furnace superstructures. Even if a bowl-shaped depression once used for the reduction of iron is defined archaeologically, there can be no certainty that it was not surmounted in ancient times by a furnace superstructure that has since weathered away or been physically removed. Furthermore, the idea that open-hearth, bowl furnaces are simple and not as capable of producing iron or steel has been disproven by an archaeological experiment (Lim and Schmidt 1982) in which a well-consolidated carbon steel was produced in such a furnace. This came as no surprise to those who conducted the experiment, yet it importantly contradicts the popular image of this type of furnace as distinctly less complicated and more primitive than other smelting devices.

Ideas about this particular type of African iron smelting technology typify the ignorance and facile assumptions that generally surround the subject of African iron production. The extent to which such ideas have gained acceptance is illustrated by discussions of early iron production in reputable and important textbooks on African history, to wit: "Most of Africa used the shaft furnace, perhaps of Graeco-Roman origin. This was a much more efficient device than the bowl furnace. A shaft furnace would yield about five kg of iron in one smelting, while a bowl furnace would yield only one kg at best" (Curtin et al. 1978, 21–22). There is simply no scientific evidence to support such a generalization, which conveys the false impression that efficiency studies and quantification of smelting yields have been conducted. Such ideas are not based on a scientific understanding of technological processes and of how technology fits into the cultural system in which it operates, but rather derive from descriptions by amateur observers that emphasize superficial characteristics. These poorly informed surface descriptions are used to fit technologies into arbitrary categories that assign technological significance and historical relationships. This book is designed to replace such ill-informed speculation with substantive observations and a more rigorous consideration of how iron technologies fit into the historical picture of several regions.

If attempts to categorize and typologize African iron technology have led to confusion and to speculation that does not stand the test of scientific and historical scrutiny, then the multitude of diffusionist arguments about the origins of iron production in Africa fit the same pattern. Debates over the origins of ironworking in Africa have mostly stressed the theme of external origins, but they also incorporate what was once considered highly improbable—the invention of iron production in Africa. Thought about indigenous invention has varied from unsubstantiated assertions (Diop 1968; Lhote 1952) to more reasoned considerations of early copper smelting as a pyrotechnology that gave rise to iron smelting (Schmidt 1983b). Though the dating of copper smelting in West Africa has now been convincingly fixed to the second millennium B.C. (Killick et al. 1988) and the prior conditions for iron smelting clearly established, arguments that iron smelting developed as a byproduct of using iron oxide as a flux in copper smelting remain based on analogy to developmental sequences elsewhere, such as the Near East. To be sure, it is a powerful argument based on known developmental processes of pyrotechnology in the Mediterranean; nonetheless, there are no archaeological data yet that confirm the validity of this hypothesis for Africa (Childs and Killick 1993).

The hypothesis for indigenous development awaits substantiation, but it is at least testable, with a universe of possible appropriate data that can be investigated, especially in the Agades region of Mali. In this respect the hypothesis for indigenous invention is currently the most viable among the multitude of diffusionist hypotheses. Diffusionist ideas about the origins of ironworking have proliferated to satisfy almost every conceivable possibility. These speculations have ranged from the notion that iron working derived from East Asia via Madagascar (Trigger 1969) to the idea that diffusion occurred up the Nile to Meroe and subsequently to both East and West Africa (Arkell 1961). One of the few specifically strong counterdiffusionist interpretations to develop out of modern archaeology in Africa is the observation that the diffusion of iron from Meroe across North Africa to Taruga, Nigeria, does not explain the rise of iron smelting at Taruga in the mid-first century B.C. (Tylecote 1975a, 1975b; Schmidt 1981, 1983b). Tylecote convincingly shows that the shaft furnace of Meroe dates to the early part of this era and is unrelated to the earlier furnaces of Taruga. But there remain many questions about ambiguous mid–first millennium B.C. traces of iron working at Meroe (Schmidt 1981, 1983b). Until definitive investigations of the earlier evidence occur, it will not be possible to make a more comprehensive commentary about the role of Meroe and the evolution of iron production at that location.

It is both fascinating and instructive that there have been so many explanations offered for the origins of iron production in Africa yet none of these constructs are supported by material evidence. One possible reason that European scholars have engaged in such exuberant speculation may be the plethora of technological styles in Africa. At the end of the nineteenth century there were hundreds if not thousands of different iron-production systems active on the continent. Such variation was accompanied by the need to explain it, allowing many competing hypotheses for external origin to exist simultaneously. Given the thesis that African technologies resembled ancient European iron technologies, then the origins of African iron production logically had to be sought in ancient Europe or at the very least in the southern Mediterranean. Thus grew and prospered the dominant idea that iron production came to Africa via the North African littoral.

The speculation that knowledge of iron manufacturing came from Phoenician settlements in North Africa was first put forward by the French scholar Raymond Mauny (1952). Although Mauny did not provide any archaeological evidence that substantiated the presence of iron smelting in or near cities such as Carthage, the continued absence of scientific evidence did not deter later generations of archaeologists and metallurgists from repeating and expanding this idea (see Tylecote 1975b; van der Merwe 1980; van der Merwe and Avery 1982). Though a more recent archaeological text (Phillipson 1985) takes the more responsible view that no evidence supports this diffusionist idea, the citation of the probability that iron metallurgy arrived in North Africa with the Phoenicians in the eighth century B.C. (van der Merwe 1980) leaves the impression that this is the likely source. The idea of Phoenician origins has become so entrenched that it is commonly repeated in introductory texts, some of which provide their own elaborations, such as the Phoenician settlement of Cerne as a secondary diffusion point in West Africa about 400 B.C. (Curtin et al. 1978). This phenomenon demonstrates the reification of a speculation to the extent that it becomes a concrete, materially based fact in African history. This transformational process typifies most of our "knowledge" about Africa's technological history and how it has been consistently erased over the last century of scholarship.

The preceding discussion illustrates one aspect of a cultural characterization that has arisen around ironworking in Africa. This process dwells on the external attributes of iron-production techniques and attempts to fit them into typological categories that make sense to the Western mind, thus reducing extraordinary variation to a few common-sense types. This homogenizes African iron production and encourages misleading ideas when they appear under the guise of legitimate Western scholarship, for example that "in spite of considerable local differences in furnace types and other techniques, the basic process of iron smelting was similar everywhere until the fourteenth century A.D." (Curtin et al. 1978, 21). Other reductionist conclusions easily follow, including technological generalizations: "Before that time, and long afterward in most of Africa, no furnaces were hot enough to produce molten iron as they did molten copper" (Curtin et al. 1978, 21). Such statements have grown out of the poorly informed and severely underdeveloped science of African iron production. In sum, this portrayal of technological culture is the product of Western cultural projections that obscure variation and complexity, substituting homogeneous depictions that erase local invention and history.

This book provides an alternative view to speculation, typological rendering, and homogenous characterizations. The essays in the second section provide a view that helps to restore indigenous technological culture in Africa, a view that seeks to comprehend the particular genius of scientific experiment inherent in each technological tradition. This approach accepts that each technological system is embedded in a larger cultural system that is the distinct product of a particular historical experience. This approach also accounts for a more inclusive treatment of technological knowledge, experience, beliefs, symbols, and rituals of various iron-producing systems—a bundle of characteristics that make up technological culture. While this method recognizes and gives importance to local particularities, it also incorporates a cross-cultural perspective that acknowledges the important similarities, especially in beliefs and symbolic meanings, that prevail among so many iron-producing cultures in Africa.

Systems of Symbolic Meaning

The ritual, taboo, and metaphorical representations associated with iron production, especially iron smelting, have usually been treated as if the technological process were divorced from a cultural matrix. For the most part this treatment has focused on the reproductive symbolism associated with iron smelting and iron smithing. Of particular interest are representations of the furnace as the womb of a fecund woman who gives birth to an iron child. A panoply of associated symbolisms, such as the blow pipes as phalluses, bellows as testicles, and slag as a birth placenta, are found scattered in different cultures along with more metonymic treatments, such as furnaces with breasts and furnaces structured to accommodate parturition through a birth canal (Schmidt 1988; Childs 1991; Herbert 1993). Parallel to but sometimes independent of the metonymic treatments of furnaces, in which the appearance of the transformed container (for example, a furnace with breasts) takes on attributes of a fecund woman, reproductive values are also attached to iron production by taboos, songs, medicines, and fertility rituals. Among the most prominent of these beliefs are taboos about sexual abstinence among smelters, taboos banning menstruating women, and fertility rituals—in which Herbert (1993) tries to find some common thread.

Herbert's difficulty in making sense of such variability among cultures in expression of these beliefs, sometimes with significant differences in neighboring cultures, is in good part related to the fragmentary manner in which African iron production has been observed. As noted above, such accounts have often come from casual observers, such as colonial administrators (for example, de Rosemond 1943) and travelers, who were short-term visitors with little awareness of how such ritual expressions could relate to the larger culture. Alternatively, such observations are sometimes derived from missionary accounts (for example, Lechaptois 1913), which may provide some good details but which may also omit beliefs that conflict with Christian teachings. Whatever the case, ritual has most often been treated as a phenomenon unrelated to the cultural milieu in which the technology is practiced. This book contains several case studies that provide cultural context and several key intersecting attributes common to these different cultures that will allow new insights and important connections between these cultures without overwhelming the reader in a mass of comparative data.

An emphasis on reproductive rituals and taboos has biased our understanding of the culture of African iron production. Cultural characterizations have subsumed what are often finely nuanced ritual expressions drawn from myth, agricultural ritual, marriage ritual, general taboos, rainmaking, healing practices, and oral traditions about legendary figures associated with ironworking. Recent research among ironworkers in Africa is changing this picture of the ritual and symbolism associated with iron production, replacing it with understandings of ritual that link such expressions to the broader cultural context in which the technology is being practiced (Schmidt 1983a; Schmidt in press).

The chapter by Candice Goucher and Eugenia Herbert is an important departure in this direction. Goucher, an anthropologist who has worked in Togo since the late 1970s, and Herbert, a historian who has specialized in the history of African technology, have together conducted an ethnoarchaeological study of iron smelting among the Bassari of central Togo. Their study, also summarized in an excellent film (Saltman, Goucher, and Herbert, 1986), attempts to find a deeper social meaning for the production of iron. Such an inquiry explores the domain of fertility symbolism and how it informs and transforms iron smelting from a physical, chemical process into a culturally represented biological one. The behavior they describe among the Bassari smelters shows the transformational qualities of ritual, transformations that parallel those that occur within the furnace. The transformations are not, as Rowlands and Warnier (1993) contend, transformations from one natural state to another but rather are from a natural to a cultural state: iron ore is natural matter, but iron is assuredly cultural. Similarly, the rituals applied to the Bassari smelting furnace transform it from natural matter to cultural matter, such as when a Bassari woman wears a marriage belt of beads. Rowlands and Warnier helpfully illustrate that the logic of such transformations was perceived to be more associatively than mechanically causative (1993, 513).

Goucher and Herbert see such associative symbolism as closely intertwined with gender relations within Bassari culture. For instance, they link the taboos prohibiting the smelters from sexual intercourse during the smelt to ideologies about adultery. They suggest that the Bassari belief provides a powerful metaphor that, when extended to the female furnace, forbids males who are "husbands" of the furnace from committing adultery with other women. Randi Barndon makes a similar point about the deleterious effects of an "adulterous" smelter (one who violates the sexual abstinence taboo) in her discussion about taboos among the Fipa iron smelters of southwestern Tanzania. Goucher and Herbert go on to say, however, that in addition to assuring a successful smelt (because of conservation of energy that otherwise would be expended in sex), taboos that prevent the presence and participation of fertile women in smelting are aimed at control over wives and offspring.

These materialist interpretations, especially those pertaining to energy conservation, may well have some legitimacy. But it is not as clear why sexual taboos are aimed at power over wives and offspring among the Bassari. We know from an extensive literature that male power over women extends into most productive domains except agriculture, where women dominate. One possible explanation that Goucher and Herbert do not consider may be configured as follows: because women control agricultural production and because agricultural production and hence human reproduction depends so heavily on iron production, male intrusion into this reproductive domain is alien and must be protected against female attempts to reclaim it. While Bassari sexual taboos in iron smelting strongly complement ideas about adultery among the Bassari, we cannot claim that this simply marks control over wives and offspring given the widespread distribution of such beliefs in other cultures without specific adultery connections. Since sexual abstinence beliefs are almost a universal expression (with some important exceptions) among African iron-smelting cultures, it is important to develop a deeper understanding of the variation of such beliefs and their living application to smelting. For example, my chapter on the Barongo iron smelters south of Lake Victoria illustrates that a focused study of variation in sexual taboos opens deeper understandings of the reproductive paradigm.

Randi Barndon's study of Fipa iron smelting in southwestern Tanzania is the first study of the Fipa process based on extended ethnographic field observations. Barndon, a Norwegian anthropologist, overcomes a legacy of eighty years of partial and secondhand accounts, missionary reports, and experimental studies in her contextualized study of Fipa smelting. Though the earlier work of John Wembah-Rashid (1973) with Fipa iron smelting is excellent, it was done in an experimental setting—Fipa iron smelters using sometimes strange materials in the alien setting of the village museum located in Dar es Salaam. Barndon's study captures some critical attributes of Fipa ritual applications, particularly the application of a red clay on the heads of the iron smelters. This is a key observation. Barndon notes that the red clay on the smelters' foreheads mimics a Fipa wedding ritual in which the material is applied to the bride's forehead to symbolize fertility.

This practice points out the poorly understood state of liminality in the status of the iron smelter, who is both midwife and lover to the furnace (Muhly and Schmidt 1988; Schmidt 1988). The Bahaya iron smelters, several hundred miles north of the Fipa, shared a similar liminality in which the smelter dressed in the raffia skirt of a female—a role normally reserved for a special ritual official in charge of agricultural fertility and blessing women's hoes (Schmidt 1988). Another Fipa belief, that two senior smelters act as midwife-confessors for adultery after a refining smelt has failed, further reinforces the idea of smelter as midwife and lover. These examples illustrate that important dynamics of symbolism in iron smelting have been overlooked because the evidence has not been based on extended and informed field studies but rather on piecemeal accounts that stress only what the Western investigator has expected to see or has considered acceptable. Particularly in multiethnic settings, such as among the Barongo, taboo is a fluid phenomenon, accommodated to many different cultural sensibilities and reflective, too, of marriage patterns in which wives are taken from other ethnic groups.

Perhaps the most potent materialist argument to be made on behalf of female prohibition is one posited by a smelting informant in Buhaya (Schmidt 1988; Schmidt in press), also mentioned by Goucher and Herbert as a possible interpretation of a belief related to social realities. The practice of iron smelting in traditional African cultures conferred an extraordinary amount of wealth and power. The social groups that controlled the technical secrets to iron smelting assured themselves of a monopoly over wealth and power as long as that knowledge was kept safe within the lineage or clan. In situations where exogamous marriage prevails, such as among the Haya and the Bassari, it was essential that potential spies, namely wives from other social groups, not be allowed to share and convey such knowledge to their lineages.

Interpretation along these lines has a certain attraction, yet it is important to know more about how completely and effectively control over craft activity and over specialized processing was divided by lineage and clan. It is also necessary to account for the widespread distribution of such beliefs, especially among iron-smelting groups in West Africa, such as the Mande, that are endogamous (McNaughton 1988). In the case of endogamous groups, it might be suggested that gender-based prohibition does not fit with a materialist explanation because prohibitions against sexual contact with women outside the endogamous group already exist. Finally, such exclusion of women occurs even in cultural settings where access to smelting is not limited by membership in a social group.

Among the common beliefs that surround iron smelting, the taboo applied to menstruating women is one of the most prominent. Within the "orthodox" reproductive paradigm, this taboo asserts that women in menses are in a phase of their reproductive cycle that is temporarily sterile. A menstruating woman is dangerous to a smelt because she threatens to abort the birth of the fetus should she come in contact with the fecund furnace, and, in many cultures, the charcoal, tuyeres, and other associated materials. Thus, menstruation holds the danger of a failure to conceive, a value that sets up a dichotomy between sterility (menstruation) and conception in the reproductive paradigm (see Herbert 1993, 72–73, 85–88).

My work has examined questions of iron production and its cultural settings in Africa for twenty-five years, and it brings some important insights into the adequacy of interpretations of menstruation taboos. My chapter on the Barongo iron smelters captures an important variation on the attitudes about menstruation there: the Barongo have no menstruation taboo and in fact they actively integrate menstrual symbolism into the fertility rites that they perform with their furnaces. These findings, based on repeated observations of Barongo smelting, suggest that the incorporation of menstruation is a richer and more dynamic symbolic rendering of the female reproductive cycle, which runs from the purification of the womb during menses, to fertilization, through gestation, and finally to parturition. The Barongo symbolic armature for reproduction overcomes the shortcomings of an incomplete and dichotomized sterile/fertile representation abstracted thus far from incomplete and biased accounts.

In her recent book addressing the reproductive paradigm in iron smelting, Herbert (1993) tries to account for negative evidence of this kind. She rests her argument on the observation that there has been a truncation or limitation of the sexual-abstinence taboo to an elite order of master smelters among several prominent and productive iron-smelting groups during the nineteenth century. She hypothesizes that this shrinkage of the application of the taboo is the result of expanded production—that is, accelerated economic activity modifies the belief system and some of its negative strictures, or "economies of scale seemed to go hand in hand with a diminution of ritualization" (Herbert 1993, 122). To support this idea she cites the Babungo example in the Ndop Plains of Cameroon (Fowler 1990), where "sexual taboos were consolidated in his [celibate smelter] person as a very rational response to the economic takeoff" of the nineteenth century (Herbert 1993, 122). Such an argument depends, however, upon complex and repetitive oral traditions that explicitly document beliefs from an earlier period—a difficult condition to meet.

A most interesting point in the Babungo case is the open recruitment of apprentices into iron smelting, a condition that also prevailed among the Barongo iron smelters of the twentieth century. While the Barongo do not hold taboos on menstruating women and while they restrict abstinence taboos to the chief smelters, there is no way to date this development or to tie it to increased economic production. Herbert's speculative argument is hinged to a desired but unsubstantiated relationship between increased production and what she thinks are truncated taboos; it is not clear if sufficient data exist to disprove or confirm this hypothesis. But it is hardly clear that such variability should be viewed as a truncation of an ideal template.

It is important to closely scrutinize such speculative positions to see if alternative ideas are appropriate. In the Barongo case there is no loss of ritual elaboration vis-à-vis other iron-smelting cultures in Africa. On the contrary, reproductive symbolism completely informs the Barongo process in ways that are much more richly elaborated than in most other known belief systems. It appears that the Western investigator, so mesmerized by reproductive symbolism in African iron smelting, sees and hears only that which comfortably fits the established paradigm. Exceptions are filtered out or explained away as products of change (as some may well be). It is important to listen closely to the exceptions and to substantively attach them to other clearly documented changes or to specific innovations.

It has been suggested by Helge Kjekshus (1977), in his analysis of iron production in Tanzania during the colonial era, that the ritualist school of research has come to dominate representation of ironworking. Kjekshus focuses primarily on the published accounts of ironworking, especially those about the Fipa (Wyckaert 1914; Robert 1949; Wise 1958), which he sees as creating a misplaced emphasis on supernatural influences, the smelter as magician, and communication with ancestral spirits. Kjekshus contends that the early traveler accounts have more credibility because they treat ritual as having secondary importance, "at the most . . . a type of trade union secrecy aimed at maintaining monopoly over an important economic enterprise" (Kjekshus 1977, 91). Although Kjekshus is correct that there was a preponderant emphasis on ritual attributes at the expense of technological studies during the colonial era, some of his collateral assumptions merit closer scrutiny. He also agrees with Bishop Lechaptois (1913) that culture change had caused the Fipa to lose their chemical and technical knowledge (de-development) and that this was replaced by the "rise of the magical element as a substitute explanation of the metallurgical processes" (Kjekshus 1977, 91).

The Kjekshus hypothesis that ritual replaces technical knowledge seems odd, given the only two decades of hardly intensive contact with European culture in southwestern Tanzania. There is no evidence to suggest that Fipa technical processes had experienced erosion, but there is abundant evidence that missionaries and others were extraordinarily taken by the ritual elements. If Fipa ritual marks a mental de-development in technical process, then what is the evidence for increased ritual expressions and for declining technical standards? There is none.

Recently, Sidney Lemelle (1992) has taken up Kjekshus's hypothesis to suggest that increased ritual expressions mark indigenous resistance to colonial domination and commoditization of the market. In Lemelle's view, there was an emphasis on the ideological component over the material component because the "dawa represented a symbol of knowledge (codified in the ritual) and a symbol of resistance to colonial rule and economic decline" (Lemelle 1992, 173).[3] He further suggests that evidence for the erosion of Fipa ritual knowledge between 1913 and 1958 is related not to their forgetting the ritual, but to their active "dis-remembering," a form of selective memory that "enabled them to separate themselves from the commodization and Christianizing of their ritual" (Lemelle 1992, 175). In this instance, the case for an unsubstantiated increase in ritual (as resistance) is replaced ipso facto by a diminution of ritual expression (also seen as resistance). Lemelle clearly cannot have it both ways. His contradictory positions fail not only because they are opposites but because he fails to indicate precisely what changes occurred over time.

There is a more fundamental failure in the approaches of Kjekshus and Lemelle. Both ignore at considerable risk the complex interaction of ritual and technology, preferring to see the two domains as conceptually and practically separated. How can we overcome this imperfect view and arrive at a synthetic understanding of ritual and technology? There is no exact answer to this question, but there are angles of view that allow the interplay to be seen more clearly. The Barongo case study in chapter 5, for example, provides a perspective based on the concept of bricolage, of the fabrication of ritual and technical approaches together in the experimental exploration of more productive pathways. Viewing the smelter as bricoleur or fabricator provides the opportunity to understand how ritual and medicine come together with technology in an experimental process that resembles healers who search for ways to cure patients or scientists who intuit a variety of pathways and follow multiple working hypotheses until they find one that works.

The reproductive symbolism that infuses meaning into iron smelting is often found in a wide variety of different expressions in the smithy (see Herbert 1993), but the products of smithing have not received much attention. Terry Childs and William Dewey consider in chapter 7 how iron goods help to structure social, political, and economic relations in ways that are derived from cosmology and worldview. The connections and differences between the furnace, forge, and final products are brought to the surface by Childs, an archaeologist who began her career with research into Bahaya iron smelting, and Dewey, an art historian who has studied Shona iron smelting in Zimbabwe. Their work helps us to understand better how products of the forge often take on symbolic meanings related to political power and authority, spiritual power and authority, and social and economic status. The objects that open up insights into the social and political meaning of forged iron are the ancient and recent iron axes found among the Luba of Zaire and the Shona of Zimbabwe.

Childs and Dewey document the remarkable continuity of axes as objects of prestige and authority over the last millennium in Zaire by using burial evidence from the Upemba Depression and a study of contemporary axes among the Luba; evidence for such strong continuities is not as strong in Zimbabwe, yet there are treatments of axes that suggest that they are intimately tied into systems of meaning that derive from the human life cycle.

They see the forging processes of hammering, annealing, folding, and welding as conceptually and symbolically related to the life stages marked by transformational rites. It is also clear that Luba and Shona smiths had very explicit ideas about fabrication techniques for utilitarian axes as opposed to those for ceremonial and ritual axes. Moreover, the form and the decorative techniques applied to Luba ceremonial axes, for example, identify and legitimize the power of royals, legitimize the spiritual power of spirit mediums, and enhance the separateness and power of leaders of the Bambudje secret society. The application of female cicatrix markings on Shona axes presents anthropomorphic and gender qualities that tie in closely with the scarification rite applied to young women at reproductive age. The panoply of multiple symbolic meanings in both cultures shows the rich field of such meanings that are linked not only to the cycle in reproductive life but also to conferral of power and authority, retention of legitimacy, and passage of authority to new generations. This study shows that the core symbols of the reproductive cycle found in iron smelting are reworked and translated into a symbolic system that captures the entire cycle of human reproduction through the decoration and ritual use of iron objects that structure complex relations relating to power and authority.

Chapters 1 through 7, the first section of this book, illustrate that some distance still remains before the significance of differences and similarities of beliefs surrounding ironworking will be understood. While these contributions make up considerable ground in our need to control deeper ethnographic understandings of the cultural settings for iron production, cultural change has been so significant that it is today difficult to find groups of iron smelters who even remember the old days, let alone who are able to smelt. The result is that belief systems and technical knowledge are seriously abridged. While salvage or "emergency ethnography" (Barndon, chapter 4, this volume) significantly helps to recover essential knowledge, cultural change has been proceeding apace for more than a century in most regions where ironworking once prospered. Whatever we can learn from living experts is inevitably influenced by religious conversion, imperfect memories, and fragmentary identities—all of which bias any interpretations and conclusions that we may reach. Cognizant of these problems, the authors in this book take an important step by providing the best available in-depth understandings of cultural settings and by introducing the long-lost element: technological process.

Bringing Technology into African Iron Production

The culture of African iron production suffered from representation of African iron smelting that either ignored African technological processes or equated them with ancient European approaches and solutions. It is necessary to overcome the disabling paradigm that there is nothing to be learned from African iron technology by presenting technological and archaeological evidence for the development of considerable African innovation and invention. Chapters 8 through 13, the second section of this book, form part of a larger project to understand Haya iron production in its historical and cultural milieu (Schmidt in press). Too often iron production is presented either as a set of beliefs (for example, Herbert 1993) without consideration of the interwoven technological steps or as a strictly technological process (Rostoker and Bronson 1990). This section and the Barongo chapter in the first section overcome this deficiency by demonstrating that a holistic approach can bring history and technology into the picture to present a better understanding of development, change, and innovation—all of which also dispose of the idea of a homogeneous technology.

The essays in the second section of this book explore some of the controversy attracted by recent studies of the technological culture among the Haya people of northwestern Tanzania, along the western shore of Lake Victoria. In chapter 8, Donald Avery and I revisit important technological observations that help to overcome and reverse false generalities and speculations that have heretofore come to dominate discourse about African iron technology. This article was originally published in 1978 in *Science* and was widely summarized in newspapers and popular magazines such as *Time*. Knowledge of this important publication remains distant to most historians and to many other social scientists. A historian and archaeologist, I initially worked with Avery on Bahaya and Barongo iron smelting between 1976 and 1979. Avery is a metallurgist and engineer and has studied experimental bloomeries and African smelting for several decades, most recently in Malawi during the 1980s. This chapter recaps the significance of ethnoarchaeological studies of Haya iron technology, particularly the remarkable determination with which Haya iron smelters employed the technique of preheating that leads to a hot air blast in the furnace, more efficient smelting, and a distinctive process of iron production.

These findings captured much popular interest. It is instructive to note, however, that many archaeologists and historians have continued with their speculations and homogeneous views as if the Haya findings do not exist. In a recent review of African mining and smelting of metals in Africa it was observed that the preheating hypothesis has failed to gain acceptance (Miller and van der Merwe 1994). Moreover, rarely are scientific arguments linked to the failure to acknowledge and address preheating in most scientific and historic literature (for an example see Miller and van der Merwe 1994). Curiously representations of the central argument often result in distorted or inaccurate conclusions. For example, Miller and van der Merwe conclude their recent review with the observation that in Africa processes of bloomery iron production were extremely varied, "producing products ranging from nearly pure iron through carbon steel to cast iron" (Miller and van der Merwe 1994, 31). We agree. Our work, summed up in chapter 13 by Childs, is testimony to a substantive understanding of how one technology produces such variability. But the Miller and van der Merwe acknowledgement of technological variability does not support or logically lead to their following conclusion: "This mitigates claims for the early local development of metallurgically sophisticated smelting techniques of special historical significance. In particular, claims for the indigenous development in East Africa of an early and sophisticated iron smelting technology involving pre-heated air have not won general support" (Miller and van der Merwe 1994, 31–32). It is absurd to claim that variation mitigates claims for innovative local development. In fact detailed observations of such variation are the most important part of developing new insights into why and how local innovations and inventions arose. And amongst these insights is the issue of preheating, which is accepted by other authorities such as Rostoker and Bronson (1990, 26, 80). It is just as dangerous to mask invention by the assertion of variability as it was in the past to mask invention by the assumption of homogeneity.

Why such important scientific observations have not been incorporated into mainstream historical and archaeological writing is also explained by the fact that such ideas challenge the conventional orthodoxy held by those who control the production and dissemination of knowledge about Africa. But there is another, more practical, reason, and that is simply that these arguments have been published in specialized scientific journals that are not read by social scientists. Consequently, this book also seeks to exhume the most important parts of that literature and some of the reaction to it from the remote reaches of esoteric journals and to provide the wider hearing that such important arguments deserve.

The next step in documenting the development and antiquity of the preheating process was the publication of extensive and detailed archaeological evidence for preheating (Schmidt and Childs 1985; Schmidt and Childs, chapter 9, this volume). Chapter 9 is a modified and abbreviated version of our rigorous confirmation of the use of preheating in African antiquity. It sets out precisely the archaeological evidence that supports the idea that the invention of preheating occurred in the early first millennium A.D. Of particular interest was the recovery of a long furnace tuyere that displayed all the attributes that we would expect in a preheating blowpipe, based on our observations of recent smelting. This "smoking gun" as well as a wide range of collateral evidence affirm that innovation and experimentation with different technological approaches occurred in the period between 200 and 400 A.D.

Chapter 9 also presents the methods we used to obtain archaeological evidence, our interpretations of technological characteristics of the ancient technology, and details on analysis and dating of materials found in two important sites, the KM2 and KM3 sites on the western shore of Lake Victoria. The part of the chapter referred to in the title as actualistic studies addresses the use of ethnoarchaeological observations made during Haya iron smelting. Part of that inquiry focused on the residues and physical/chemical changes that resulted from all phases of iron smelting. Observations such as the types of materials used to build the furnace, the thermal transformations that occurred in clays, and the variety of slags produced during a smelt are used to construct a set of models of material attributes and consequences of iron smelting under preheating conditions. Actualistic models for each of these physical domains are then used to assess similarities and differences in the archaeological record. This comparative process helps to isolate anomalies or differences between the actualistic models and the archaeological data that point to technological change through time. It also provides a powerful tool for sorting out the significance of technological characteristics that otherwise would remain submerged and obscure. As an approach that enables a higher level of confidence for interpretation of ancient technological process, it demonstrates the value of ethnoarchaeology to historical studies of African technology. Because the process uses recent phenomena to interpret the past, it also carries with it the inherent danger of projecting the present into the past. This theoretical concern is addressed by the careful scrutiny of the archaeological evidence for exceptions, differences, and peculiarities that contradict the actualistic models. This precaution reflects the belief that differences are often more significant than similarity, which in the case of this approach has the capacity to erase ancient history and replace it with the present.

Shortly after the publication of the archaeological evidence several counterarguments questioned the preheating hypothesis and in fact attempted to deny the innovation of preheating among ancient African populations. The most informed of these challenges has been mounted by J. E. Rehder, a metallurgist who has worked for decades in Canadian steel production. Rehder's

1986 critique opened the debate,[4] and it is reprinted here in the first part of chapter 10.

Because Rehder's critique raises some important technical points about what he considers to be negligible heat transfer in the tuyeres and normal temperatures in a charcoal fueled furnace, it was necessary for a response to be directed specifically to the technical issues raised. Thus, our response to Rehder in the second half of chapter 10 closely examines all the technical arguments used in our previous publications, including a 1979 article published in the *Journal of Metals*, as well as a recap of the overlooked archaeological evidence in support of the hypothesis.

One short section of our response to Rehder is a mathematical model that clearly sets out and further refines all scientific reasoning about the technical aspects of preheating using principles common to thermodynamics. We have attempted to make the text as clear as possible and keep it free of jargon, but the technical arguments are critical to our position. As they are intended as a complete exegesis of the scientific thinking and support on behalf of a hypothesis of merit and importance to African history, we hope that they will not be seen as mystifications.

As work progressed on further documentation, another challenge to the preheating hypothesis arose from within archaeological circles. David Killick, who conducted his Ph.D. research in Malawi during the early 1980s and who has subsequently helped Nicholas David in his study of Mafa iron smelting in northern Cameroon (David et al. 1989), delivered a paper critical of the preheating hypothesis in the 1988 symposium at the African Studies Association meetings; chapter 11 is based on that paper. Killick reprises many earlier criticisms of Rehder but also questions the application of our mathematical model and focuses much attention on the issue of blast zone or oxidation temperatures in Haya furnaces as compared to experimental furnaces observed in European laboratories.

Our response to Killick, chapter 12, by necessity covers some familiar ground, as it appears that our response to Rehder has not been understood. Our observations are intended to clarify a debate that has clearly moved toward obfuscation of the central issues as well as the selective use of evidence from experimental, laboratory conditions that in no way approximate the Haya furnace conditions. We believe that a serious problem has arisen when something as fundamental and as important as the effect of preheating on high-temperature-reducing conditions—leading to more efficient extraction of iron from slag—is ignored. Emphasis is given instead to oxidizing zone temperatures, which we believe is a red herring.

Since the debate has gathered momentum, it is essential to provide more detailed evidence for the thermodynamic and chemical processes that occur inside the replicated Haya furnaces as well as those of antiquity. Collaborating with me since 1976, Childs undertook the study of the material remains. One of the few archaeometallurgists working on African materials, Childs has extensively analyzed materials that are produced under preheating conditions. Chapter 13 represents our efforts during the late 1980s and early 1990s to marshal laboratory evidence that can be used to characterize the technological culture of the region west of Lake Victoria. Drawing on my historical and ethnoarchaeological work to contextualize the study, Childs goes on to show that laboratory analyses can provide strong commentaries on what type of product was valued; what kinds of technical problems were encountered and overcome (for example, the conditions under which innovation arose); and how the technology utilized resources appropriate to the environmental conditions as well as the political and social conditions that prevailed in the past. It is possible then to reconfigure knowledge of iron technology by using laboratory analyses to rewrite a history of the culture of technology in the ancient and recent African past.

As the debate gathers energy with the publication of this book, we hope that readers will keep in mind that change does not come easily to domains in which history has been erased, altered, and suppressed. The preheating hypothesis, as Killick puts it, has stimulated further inquiry into African iron technology. While it is impossible to say if the hypothesis in its initial form will stand the test of scrutiny over the coming decades, it certainly has yet to be disproved, and if anything the evidence is even more persuasive that preheating constituted an essential part of the technological repertoire of ancient African iron smelters. The construction of African history has changed as a consequence of this scientific research. The past seventeen years have represented an important and exciting venture into the past of African technological culture. The host of other observations made in the process attest to a rich and finely nuanced history of African innovation and invention, a legacy that will endure.

Notes

1 *Time*, September 25, 1978, 80; quoted from Schmidt and Avery 1978.

2 More recently Herbert, in her survey of beliefs surrounding iron smelting, refers to the Ekonda of Zaire who "use a bowl furnace with no shaft at all . . . in the absence of a *true* furnace" (emphasis mine). While Herbert goes on to contextualize the values surrounding the Ekonda process, her choice of words reveals a deeper attitude toward what is truly a furnace and what is not (Herbert 1993, 63–64).

3 There are other problems with Lemelle's argument. For example, once he argues for increased ritual expressions, he immediately reverses his position to suggest that Christianity leads to erosion of ritual expressions. He does not recognize that the arguments cannot coexist.

4 M. K. H. Eggert, a German archaeologist who works in the Zaire river basin, has presented a polemic that does not, in our estimation, raise valid points (Eggert 1987). It is also marred by misconceptions about iron production processes. Because it is flawed as a critique, we have chosen not to address it.

References

Arkell, A. 1961. "The Valley of the Nile." In *The Dawn of African History*, ed. R. Oliver, 7–21. London: Oxford University Press.

Avery, D. H., and P. R. Schmidt. 1979. "A Metallurgical Study of the Iron Bloomery, Particularly as Practiced in Buhaya." *Journal of Metals* 31:14–20.

———. 1986. "The Use of Preheated Air in Ancient and Recent African Iron Smelting Furnaces: A Reply to Rehder." *Journal of Field Archaeology* 13:354–57.

Barndon, R. 1992. "Traditional Iron Working among the Fipa: An Ethnoarchaeological Study from Southwestern Tanzania." Master's thesis, University of Bergen.

Bellamy, C. B. 1904. "A West African Smelting House." *Journal of the Iron and Steel Institute* 66:99–126.

Childs, S. T. 1991. "Style, Technology, and Iron Smelting Furnaces in Bantu-Speaking Africa." *Journal of Anthropological Archaeology* 10:332–59.

Childs, S. T., and D. Killick. 1993. "Indigenous African Metallurgy: Nature and Culture." *Annual Review of Anthropology* 22:317–37.

Cline, W. 1937. *Mining and Metallurgy in Negro Africa*. Menasha, Wis.: George Banta Publishing Co.

Curtin, P., S. Feierman, L. Thompson, and J. Vansina. 1978. *African History*. London and New York: Longmans.

David, N., J. R. Hiemann, D. Killick, and M. Wayman. 1989. "Between Bloomery and Blast Furnace: Mafa Iron Smelting Technology in North Cameroon." *African Archaeological Review* 7:183–208.

de Rosemond, C. C. 1943. "Iron Smelting in the Kahama District." *Tanganyika Notes and Records* 16:79–84.

Diop, L. 1968. "Traditional Iron Working and the Iron Age in Africa." *Bulletin de l'IFAN* (B) 30:10–38.

Eggert, M. K. H. 1987. "On the Alleged Complexity of Early and Recent Iron Smelting in Africa: Further Comments on the Preheating Hypothesis." *Journal of Field Archaeology* 14:377–81.

Fowler, I. 1990. "Babungo: A Study of Iron Production, Trade, and Power in a Nineteenth-Century Ndop Plain Chiefdom (Cameroons)." Ph.D. diss., University of London.

Goucher, C. 1981. "Iron Is Iron 'Til It Rust: Trade and Ecology in the Decline of West African Iron-Smelting." *Journal of African History* 22:179–89.

Haaland, R., and P. Shinnie. 1985. *African Iron Working: Ancient and Traditional.* Oslo: Norwegian University Press.

Herbert, E. W. 1993. *Iron, Gender, and Power: Rituals of Transformation in African Societies.* Bloomington and Indianapolis: Indiana University Press.

Kense, F. J. 1981. "Daboya: A Gonja Frontier." Ph.D. diss., University of Calgary.

——. 1983. "Traditional African Iron-Working." African Occasional Papers, No. 1. Calgary: University of Calgary.

——. 1985. "The Initial Diffusion of Iron to Africa." In *African Iron Working: Ancient and Traditional*, ed. R. Haaland and P. Shinnie, 11–27. Oslo: Norwegian University Press.

Killick, D. J. 1990. "Technology in Its Social Setting: The Ironworkers of Kasungu, Malawi, 1860–1940." Ph.D. diss., Yale University.

Killick, D. J., N. J. Van der Merwe, R. B. Gordon, and D. Grébénart. 1988. "Reassessment of the Evidence for Early Metallurgy in Niger, West Africa." *Journal of Archaeological Science* 15:367–94.

Kjekshus, H. 1977. *Ecology Control and Economic Development in East African History.* London: Heinemann.

Lechaptois, A. 1913. *Aux rives du Tanganyika: etude ethnographique.* Algiers: Maison-Carree.

Lemelle, S. J. 1992. "Ritual, Resistance, and Social Reproduction: A Cultural Economy of Iron-Smelting in Colonial Tanzania, 1890–1975." *Journal of Historical Sociology* 5:161–82.

Lhote, H. 1952. "La connaissance du fer en Afrique occidentale." *Encyclopedie mensuelle d'outremer* 1:269–72. Paris.

Lim, I., and P. Schmidt. 1982. *Beating the Bellows.* Film. Providence: Foundation for African Prehistory and Archaeology.

Maret, P. de. 1985. "The Smith's Myth and the Origin of Leadership in Central Africa." In *African Iron Working: Ancient and Traditional*, ed. R. Haaland and P. Shinnie, 73–87. Oslo: Norwegian University Press.

Mauny, R. 1952. Essai sur l'historie des metaux en Afrique occidentale. *Bulletin de l'IFAN* (B) 14:545–95.

———. 1961. *Tableau géographique de l'Ouest Africain au moyen age, d'après les sources écrites, la tradition et l'archéologie.* Memoires de l'IFAN, No. 61. Dakar: IFAN.

McNaughton, P. R. 1988. *The Mande Blacksmiths: Knowledge, Power, and Art in West Africa.* Bloomington and Indianapolis: Indiana University Press.

Miller, Duncan E., and Nikolaas J. van der Merwe. 1994. "Early Metal Working in Sub-Saharan Africa: A Review of Recent Research." *Journal of African History* 35:1–36.

Muhly, F., and P. Schmidt. 1988. Script for *The Tree of Iron* (film). Gainesville, Fla.: Foundation for African Prehistory and Archaeology.

Phillipson, D. 1985. *African Archaeology.* New York: Cambridge University Press.

Pole, L. M. 1985. "Furnace Design and the Smelting Operation: A Survey of Written Reports of Iron Smelting in West Africa." In *African Iron Working: Ancient and Traditional,* ed. R. Haaland and P. Shinnie, 142–63. Oslo: Norwegian University Press.

Rehder, J. E. 1986. "Use of Preheated Air in Primitive Furnaces: Comment on Views of Avery and Schmidt." *Journal of Field Archaeology* 13:351–53.

Robert, R. P. J. M. 1949. *Croyance et customes magico-religieuses des Fipa païens.* Tabora: Tanganyika Mission Press.

Rostoker, W., and B. Bronson. 1990. *Pre-Industrial Iron: Its Technology and Ethnology.* Archaeomaterials Monograph, No. 1. Philadelphia.

Rowlands, M., and J.-P. Warnier. 1993. "The Magical Production of Iron in the Cameroon Grassfields." In *The Archaeology of Africa: Food, Metals, and Towns,* ed. T. Shaw, B. Andah, P. Sinclair, and A. Okpoko, 512–50. London: Unwin Hyman.

Saltman, C., C. Goucher, and E. Herbert. 1986. *The Blooms of Banjeli: Technology and Gender in African Ironmaking.* Film. Watertown, Mass.: Documentary Educational Resources.

Schmidt, P. R. 1981. *The Origins of Iron Smelting in Africa: A Complex Technology in Tanzania.* Research Papers in Anthropology, No. 1. Providence: Brown University.

———. 1983a. "Cultural Meaning and History in African Myth." *International Journal of Oral History* 4:165–83.

———. 1983b. "Further Evidence for an Advanced Prehistoric Iron Technology in Africa." *Journal of Field Archaeology* 10:421–34.

———.1988. "Reproductive Symbolism in Iron Smelting among the Barongo." Paper presented at the thirty-eighth annual meeting of the African Studies Association, Chicago.

———. In press. *Symbolic and Material Views of Iron Technology in East Africa.* Bloomington: Indiana University Press.

Schmidt, P. R., and D. H. Avery. 1978. "Complex Iron Smelting and Prehistoric Culture in Tanzania." *Science* 201:1085–89.

Schmidt, P. R., and S. T. Childs. 1985. "Innovation and Industry During the Early Iron Age in East Africa: The KM2 and KM3 Sites of Northwest Tanzania." *African Archaeological Review* 3:53–94.

Schmidt, P. R., and T. Patterson. In press. "From Constructing to Making Alternative Histories." In *Making Alternative Histories,* ed. P. Schmidt and T. Patterson. Santa Fe: SAR Press.

Stanley, G. G. 1921. "Some Products of Native Iron Smelting." *South African Journal of Science* 28:131–34.

Sutton, J. 1985. "Temporal and Spatial Variability in African Iron Furnaces." In *African Iron Working: Ancient and Traditional,* ed. R. Haaland and P. Shinnie, 164–91. Oslo: Norwegian University Press.

Trigger, B. 1969. "The Myth of Meroe and the African Iron Age." *African Historical Studies* 2:23–50.

Tylecote, R. 1975a. "Iron Smelting at Taruga, Nigeria." *Journal of the Historical Metallurgical Society* 9:49–56.

——. 1975b. "The Origin of Iron Smelting in Africa." *West African Journal of Archaeology* 5:1–9.

van der Merwe, N. 1980. "The Advent of Iron in Africa." In *The Coming of the Age of Iron,* ed. T. Wertime and J. Muhly, 463–506. New Haven: Yale University Press.

van der Merwe, N., and D. H. Avery. 1982. "Pathways to Steel." *American Scientist* 70:146–55.

Wembah-Rashid, J. A. R. 1973. *Iron Working in Ufipa.* Dar es Salaam: National Museum of Tanzania.

Williams, D. 1974. *Icon and Image.* London: Allen Lane.

Wise, R. 1958. "Some Rituals of Iron-Making." *Tanganyika Notes and Records* 51:232–38.

Wyckaert, R. P. 1914. "Forgerons poies et forgerons chretiéns au Tanganyika." *Anthropos* 9:371–80.

2 How Old Is the Iron Age in Central Africa?

Pierre de Maret & G. Thiry

Explicit or not, the rather naive and simple question that is the title of this chapter shapes most research undertaken so far on African metallurgy, whether from ethnographic, linguistic, or archaeological points of view. Contrary to popular belief, simple questions do not necessarily get simple answers, especially not those posed in this area. As our knowledge of the African past improves, the answers become more elusive.

In the late 1960s, African history seemed very simple, and most scholars held a single understanding that African metallurgy originated in the Near East and came to Africa via the Nile Valley and Meroe. The so-called Birmingham of Africa was interpreted as the center of a large diffusionist process toward the west and south of the continent. The Nok culture, furnaces, and dates testified to the knowledge of metallurgy in Nigeria at a time that could be linked to Bantu expansion. Guthrie (1967–71) demonstrated that the Bantu had a knowledge of metallurgy at the proto-Bantu linguistic level, and the prevalent paradigm then became that the Bantu expanded because of their knowledge of iron. The use of iron spread as a result of the Bantu. In fact, the linguistic dates do not support Guthrie's certitudes, and the simplistic diffusionist schema has been challenged by new field research.

So where do we stand now? In archaeology we have moved from the Iron Age of Certitude to the Age of Incertitudes, but some major trends seem to be emerging from the research of the past five years. One is the nature of the transition period from the Late Stone Age to what we could call the Metal

Age. There is a mounting body of dates during the crucial last four millennia B.C. that suggests there was indeed some sort of neolithic, at least in the etymological meaning of the term. Villagers were probably involved in a vegeculture involving palm trees and *Carnarium* schweinfurthii. In the Cameroon grassfields, a series of dates place the appearance of bifacially retouched stone tools of the axe/hoe type, sometimes waisted and partially polished, around the fifth millennium B.C. (Maret et al. 1987). In Yaoundé, Cameroon, there is evidence from the second millennium B.C. (Maret 1982, 1985), and the Tajunu megaliths in the Central African Republic are dated by N. David (1982) to the first half of the last millennium B.C.

Many excavations have occurred in Gabon (Clist 1989), and around the estuary of the Gabon River there seem to be two neolithic traditions. One of these, on the Denis River, includes flat-bottomed ceramics with channeled rims and chert and flint debitage dated between 3000 and 1000 B.C. The other tradition is more recent, probably between 600 and 200 B.C., in sites like Ndjole or Okala, and is characterized by very large pits, similar to those of that same period commonly found between Cameroon and Zaire.

Elsewhere, in the Congo near the mouth of the Kwilu River at the Loango coast at Tchissanga West, Denbow (1990) dated to around the sixth century B.C. a coarsely tempered, grooved, channel-decorated pottery associated with quartzite and chert lithic debris. Further south, in Lower Zaire, there is a sixteenth century B.C. date for a pit with a previously unknown pottery group associated with polished stone tools (Gosselain 1988, 50–56). At the same site of Sakuzi, there are two more recent pottery groups associated with polished stone tools: the Sakuzi group, dated to the last five centuries B.C., and the Ngovo group, dated to the last three centuries B.C. (Maret 1986). In some places, populations using pottery and polished- and flaked-stone tools continued until much later. For example, on the Ubangi River the Batalimo/Maluba tradition dates from the first to the fourth century A.D. (Eggert 1987).

A recent survey conducted on Bioko Island of Equatorial Guinea confirms that the earliest occupation level so far, the Carboneras, began around the sixth century A.D. Contrary to popular belief, iron first appeared in Central Africa among villagers and must have diffused slowly among groups that had settled there previously. For a long time a mosaic pattern prevailed with various groups exploiting ecosystems at various technological stages, prefiguring the complex situation of iron metallurgy that we see from ethnographic observations.

Iron is almost everywhere in Central Africa in the first three, if not five, centuries A.D. Dates are becoming increasingly older where systematic excavation programs are in progress. In the Yaoundé area, at Obobogo and Ndindan, for example, iron appeared at the latest by the fourth century B.C. (Maret 1985). In the center of Gabon, two C^{14} dates for the same furnace at Otumbi go back to 690 ± 70 B.C. (Beta 14.834) and to 450 ± 50 B.C. (Gif-7130) (Peyrot and Oslisly 1987). At Moanda, in the southeast, iron smelting seems to have started around 400–350 B.C. (Schmidt et al. 1985). In the north, near Oyem, two pits with iron slag were dated to c. 330–270 B.C. (Clist 1987b). All these results agree with the Taruga dates in Nigeria, where Nok metallurgy is dated to the sixth century B.C., if not earlier (Calvocoressi and David 1979).

Moving east to the interlacustrine area, there are, altogether, fourteen dates in the last seven centuries B.C. for Rwanda, Burundi, and Buhaya in Tanzania (van Grunderbeek et al. 1983; Schmidt and Childs 1985; Clist 1987a). It is important to keep in mind, however, the C^{14} fluctuations between 2520 and 2160 B.P. (570–210 B.C.) (Stuiver and Pearson 1986). When calibrated, dates between 2520–2400 B.P. (570–450 B.C.) cannot be resolved to better than anywhere in the 780–420 B.C. range, and dates between 2260–2160 B.P. (310–210 B.C.) spread in a 380–200 B.C. interval, making it very difficult to establish a precise chronology of the spread of iron technology through Africa during this critical period.

Given the spectrum of dates, it is reasonable to assume that by the fifth if not by the seventh century B.C., iron was smelted in the corner of the Gulf of Guinea and in the interlacustrine area. But more perplexing are a number of older dates reaching as far back as the middle of the second millennium B.C. for iron metallurgy. At Obobogo, Cameroon, there are four dates between the tenth and seventeenth centuries B.C.: 950 ± 100 B.C. (Hv 0582); 1120 ± 95 B.C. (Hv 0583); 1105 ± 110 B.C. (Hv 10833); and 1675 ± 165 B.C. (Hv 11046) (Maret 1982, 1985). In Zaire, one date at Gombe, 1575 ± 35 B.C. (GrN 7279), is associated with slag and pottery (Maret and Nsuka 1977).

In the interlacustrine area, eight dates fall in the same range, several on charcoal from furnaces: Muganza I.l (1665 ± 205 B.C.) (Hv 12126) and Kabacusi I (865 ± 165 B.C.) (Hv 12123) in Rwanda; Rwiyanje I (1230 ± 155 B.C.) (Hv 11144), Rwiyanje I (905 ± 285 B.C.) (Hv 12130), and Mubuga V (1210 ± 135 B.C.) (Hv 1141) in Burundi (van Grunderbeek et al. 1983); and Katuruka N894 (1250 ± 120 B.C.), N897 (1080 ± 110 B.C.), and N899 (1470 ± 120 B.C.) in Tanzania (Schmidt 1978).[1]

Except for van Grunderbeek et al. (1983), we have disregarded those dates as resulting from contamination, problems with a specific radiocarbon laboratory, or strong inconsistency with other results. Outside Central Africa, we see the same problem in Niger: After examining the "Cuivre I" of Grébénart that turned out to be inconclusive, except for furnace 8 with a radiocarbon date of 1710 ± 110 B.C., Killick et al. (1988) suggest the date should be viewed with caution because it may reflect the use of old wood or fossil charcoal. This may be a valid explanation under arid conditions, but as far as equatorial Africa is concerned, usually only trees less than 30 cm/diam and younger than fifty years are used to produce charcoal.[2] One should keep in mind that most of those trees rot in less than four years.

The more appropriate question really is how young a date associated with iron metallurgy should be acceptable. It seems the answer is probably that the date should postdate the invention of iron metallurgy outside of Africa, but isn't that being prejudiced? Only more dates will lead to more answers, but the chronological frontier is clearly moving. The archaeological data at this point suggest that iron metallurgy demonstrates an impressive continuity and is at least 2,500 years old in Africa. It appeared roughly at the same time in the corner of the Gulf of Guinea (Nigeria/Cameroon/Gabon) and in the interlacustrine area, but this distribution may reflect nothing more than the pattern of systematic archaeological research carried out in Central Africa.

But the vacuum in the central basin should be noted. At this stage, and contrary to Vansina (1984), Schmidt et al. (1985) and others, in light of these dates and their degree of uncertainty and overlapping, it is impossible to make any diffusionist hypothesis that assumes iron spread from east to west or vice versa. So far, the data favor only a north-south diffusion of iron technology, metallurgy attested thus far only in the first century in Lower Zaire at Sakuzi, with the Kay Ladio group, and the third century A.D. around Lumbashi (with copper).

Technology and Ethnography

In a general survey of the various aspects of smelting technologies throughout Central Africa (Thiry 1987), based on secondhand or bibliographical data, there are as many different typologies as writers. Therefore, it is necessary to choose a schema that encompasses the great morphological variety of furnaces and that is based on technological criteria applied to archaeologically identifiable characteristics, such as the presence or absence of slag-tapping, which can include bowl furnaces, shaft furnaces, and dome furnaces. The reconstruction of other technological characteristics is a difficult task because of factors not always detectable from archaeological evidence only, such as the original chemical composition of ore in some cases and/or the presence of flux.

There is an absence of any straightforward or obvious path of diffusion of any one type of iron-smelting furnace. There are overlapping characteristics that can be used to attempt to devise typologies that account for both formal and technological attributes. (For instance, in Zaire, Madi furnaces show a small chimneylike wall and slag flow; slag flow is also present in the funnel-shaped pit furnaces of the Songo, in Tanzania; in Zambia, the Tetela dig a bowl furnace so deep that it is used like a shaft; and the Sundi, in Zaire, cover the bowl.) Dome furnaces are found only among the Cewa of Zambia, the Shona of Zimbabwe, the Okefu of Zaire, and the Labwor of Uganda. Shaft furnaces are widespread in East and Central Africa by the Early Iron Age, but they are also found in the forest, where we find them together with bowl (open hearth) furnaces. Slag-tapping furnaces are present in the eastern part of Central Africa and in East Africa and extend to the southwest, whereas non–slag-tapping furnaces follow a vertical direction roughly in the middle of Central Africa. In sum, these rather perplexing distributions afford no clear idea of how or why such variability arose.

This typology reveals morphological similarities between zones, but it avoids the establishment of relations or interactions between groups using similar techniques unless they are archaeologically or historically confirmed. Such connections are not confirmed in the example of the very tall convection furnaces (2 to 4 m) often used without bellows, present in West Africa (Togo, Niger, and Senegal) as well as between Lakes Tanganyika and Nyasa in East Africa (Sutton 1985). They are not documented so far in the equatorial forest.

The East African Early Iron Age seems to be a unique case of apparently close concordance between a ceramic tradition and a coherent technological tradition, that is, a close association between Urewe pottery and non–slag-tapping shaft furnaces. In contrast, in the Central African Republic there was a cultural tradition, based on common symbolic conceptions, that runs through the Gbaya and Manza peoples, who use permanent slag-tapping shaft furnaces that differ in details and in some aspects of the smelting process. These variations could relate to ecological constraints such as ore composition, for example, but the regional variation in soil does not explain the technological variations (Monino 1983).

Economic, political, social, and symbolic considerations, such as population density, economic needs, availability of people and time, and ritual practices obviously affect the development or selection of a given technological form. This phenomenon holds true for copperworking. In the copper belt, where primarily malachite is worked, metallurgical traditions and furnaces differ greatly. The technological tradition chosen depends on cultural and economic options. Consequently, this warns us that a chronological framework based solely on functional efficiency cannot be applied to furnaces.

Symbolism is also found in furnace construction as evidenced by small pits or depressions for medicines and small pots found under the furnaces in Kivu (eastern Zaire), East Africa, and among the Cewa, Sukuma, and Barongo (Schmidt, chapter 5, this volume); in Congo at the site of Moubiri (eleventh to fourteenth century) (Manima-Moubouha 1987); and in Gabon among the Fang, and at the Mbam site (eleventh to fifteenth century). According to Schmidt and Childs (1985), such symbolism represents a Bantu trait; but we note the presence of a similar practice among the Madi, a non-Bantu group from northeastern Zaire. According to van Noten (1985), however, this practice was employed for technological reasons there.

The multiplicity of technical factors associated with many human factors makes any attempt to argue for distribution, diffusion, or chronology very difficult. More data are needed if we are to gain any meaningful insights into the relationship of furnace form to technological capacity, let alone the historical relationships that may pertain. While archaeological remains are available, traditional practices of metallurgy and their human components are dying—pointing to the urgency to record ethnographic accounts of African iron production.

Linguistics

Contrary to Guthrie's opinion (1967–71), today there is general agreement that the proto-Bantu were not metallurgists. Guthrie proposed proto-Bantu reconstructions of nouns for *iron, forge, hammer,* and *bellows.* In the 1970s, with more primary data on hand, two scholars reexamined Guthrie's comparative lexis on larger scale (Maret and Nsuka 1977), and confirmed that some widespread words for *hoe, axe,* or *spear* were very probably proto-Bantu, because modern words based on these roots with the same meaning are found throughout the Bantu-speaking area. This certainly does not hold for terms of a more technical nature, however. For instance, some fifty different stems for *iron* and more than twenty for *blacksmith* exist in the western and eastern Bantu together. Such a striking diversity of meanings and their distribution in separate geographical blocks throughout the Bantu-speaking world frustrate any attempt to reconstruct technical metallurgical vocabulary to the proto-Bantu level.

Although often phonologically correct, Guthrie's reconstructions for more specifically iron-related terms are too semantically variable to stand as firm proto-Bantu roots. The distribution of the stem *-gèdà*, first reconstructed as *iron* by Guthrie (CS 800), provides a good example of the semantic skewing often associated with the comparative Bantu metallurgical lexis. Reflexes of that stem where they are documented show various meanings, such as *knife* in Kongo or *needle* or *pin* in Shi. The meaning *iron* is attested in an area restricted to the linguistic zones A and L (Maret and Nsuka 1977).

Another root, *-tádè*, reconstructed as *stone, iron ore,* and *iron (wire)* by Guthrie (CS 1642, 1643, 1644), means *iron ore* in East Africa but conserves the meaning *stone* in the northwest. It may have undergone a semantic shift[3] in the logical direction of a specialization accompanying the development of smelting iron from iron stone. A similar argument can be advanced in the case of *-túd-* (to forge) because it also means to beat, to strike, or to grind, a more general and earlier sense (Maret and Nsuka 1977).

J.-M. Hombert (1979) shows regular correspondence between the proto-grassfields Bantu stem *-cúkà* (hoe). Reflexes of some of the protograssfields Bantu roots for *axe* and *hoe* are restricted in distribution to the eastern Bantu area, and others are found only to the south, confirming the existence of the eastward and southward Bantu migratory routes. But again, similarly distributed reflexes do not occur in words more closely related to metallurgical work. Protograssfields Bantu *-sa* (iron ore)[4] has no ancestor with the same meaning in general proto-Bantu.

In conclusion, Guthrie's reconstructions can only be established for items that are not specifically related to iron technology per se (Hombert 1979). These nouns indicate only that proto-Bantu speakers used hoes, axes, spears, and arrows. Guthrie's linguistic evidence says nothing about the material from which those tools were made. A more specialized metallurgical vocabulary cannot yet be reconstructed for proto-Bantu; instead, the picture emerges of a lexicon connected with a stone technology, having evolved and diversified in relation to new metallurgical techniques. Thus the conclusion that proto-Bantu–speaking communities did not use metal until a more recent stage in their expansion seems unavoidable.

According to lexicostatistical analysis, proto-Bantu dialects differentiated as speakers moved southward and eastward. This view is confirmed by the distribution of items associated with tools and weapons. Many dialects restricted to the linguistic zone J in the East African Great Lakes region are semantically distinct but may be connected to proto-Bantu reconstructions with reflexes present in the northwest source area (Hombert 1979).

Proto-Bantu speakers may, then, have acquired metallurgy on the way to the Great Lakes region from a yet unknown source (or sources), or they may have acquired it after settling in the Great Lakes region if the Urewe people did not speak a Bantu language. However, Schoenbrun (1990) strongly suggests a correlation between Urewe potters and Bantu speakers. In that case, Urewe potters as Bantu-speaking people could have acquired iron-smelting knowledge while settled in the Great Lakes region, given the close archaeological association of Urewe pottery and iron-smelting evidence. Such a scenario would be affirmed if there were a larger amount of borrowings than actually have been identified so far.[5]

In a survey of the metallurgical vocabulary in zone J (Great Lakes), Thiry (1989) found that reflexes of technical words for metallurgical practice, with underlying proto-Bantu forms, were relatively few or had undergone a semantic shift. Some of them show a great diversity of meanings, even in closely related languages. Newly reconstructed items in zone J (ibid.) concern mostly finished products. The relative homogeneity of stems for finished iron products surely reflects their circulation in regional exchange networks.

The words meaning "to forge," including the newly identified stem *-péec-* (ibid.),[6] are too numerous to support the hypothesis that the entire body of metallurgical expertise was adopted as a whole with a complete and bounded terminology. Moreover, we know that metallurgy was introduced in southern Africa by Bantu speakers coming from the north, but the stem *-péec-*, probably a lexical innovation proper to zone J, seems to be absent from the southern Bantu languages. Southward Bantu expansion through the forest is not documented archaeologically, a deficiency that also prevails for the proto-Bantu homeland at the Nigeria-Cameroon border, although some work there (Asombang and Schmidt 1990) suggests cultural affinities between the homeland and the Yaoundé area in mid–first millennium B.C. Archaeological work is needed along all the possible expansion roads to clarify the modalities of the Bantu migrations.

Bantu historical linguistics, formerly especially concerned with lexicostatistics and relative chronologies, now turns more and more to comparison of technical and specialized lexicons. Only through a systematic collaboration of linguists, archaeologists, and ethnographers may we achieve a better understanding of the history of metallurgy in Africa. The top priority in this matter is to assemble interdisciplinary teams to do fieldwork before some of the most significant data are irretrievably lost.

Notes

1 *Editor's note:* A close reading of Schmidt (1978) shows that date number N-899 (1470 ± 120 B.C.) explains the three early dates at the Rugomora Mahe site in Katuruka. It is explicitly stated that this date is linked to a horizon earlier than the Iron Age and that it appears to represent a forest fire episode. The other two dates are apparently derived from the same phenomenon (Schmidt 1978, 179, 192).

2 *Editor's note:* This is an interpretative position, not an authoritative generalization. Certainly where forest clearance occurred in gallery forests as part of a clearance process for agriculture, very large and mature trees were harvested and used for smelting charcoal.

3 We translate "semantic shift" as the notion of *glissement sémantique*, which is a progressive change of meaning, as has been shown by Grégoire (1976) and Bastin (1985).

4 The item *-cá* (point) (CS 237) may be represented by the form *-sá* (small iron blade for digging) in Lunyoro, Luhyia, and Ecejiita. But a link with Hombert's reconstruction *-sa* (iron ore), which Schoenbrun suggested to me in a personal communication, should be closely verified for semantic reasons. Moreover, Hombert does not give any of the reflexes upon which his reconstruction is based.

5 This seems to be confirmed by Schoenbrun (1990, 268), who hypothesizes that Central Sudanic-speaking people would have played "an important role in introducing metallurgy to their then forest-dwelling proto-Eastern Bantu speaking neighbors."

6 Guthrie has *-pék-* (CS 1467) (rub [fire] with firestick), a PB-B item of zones G, N, and P expanding southward, in zone S, under a skewed form *-péc-*. It may be possible to connect these forms with *-péec-* because their respective meanings constitute a complementary distribution. Moreover, Guthrie's reconstruction is based on reflexes occurring in languages that no longer use vowel quantity, except for P21 (Yao), but some written sources for Yao do not transcribe it, or, possibly Yao borrowed the word from an adjacent language. Yet we need a short causative—X—to explain the palatalization of stem final k. The question remains open.

References

Asombang, R., and P. Schmidt. 1990. "Rock Shelters and the Greater History of the Bamenda Grassfields, Cameroon." *Nyame Akuma* 34:5–10.

Bastin, Y. 1985. "Les Relations Sémantiques dans les langues Bantoues." *Académie Royale des Sciences d'Outre-Mer.* Mémoires, No. 48.

Calvocoressi, D., and N. David. 1979. "A New Survey of Radiocarbon and Thermoluminescence Dates for West Africa." *Journal of African History* 20:1–29.

Clist, B. 1987a. "A Critical Reappraisal of the Chronological Framework of the Early Urewe Iron Age Industry." *Muntu* 6:35–62.

——. 1987b. "Travaux Archéologiques Récents en République du Gabon." *Nsi* 1:9–12.

——. 1989. "Archaeology in Gabon, 1886–1988." *African Archaeological Review* 8:59–85.

David, N. 1982. "Tazunu: Megalithic Monuments of Central Africa." *Azania* 12:43–77.

Denbow, J. 1990. "Congo to Kalahari: Data and Hypotheses about the Political Economy of the Western Stream of the Early Iron Age." *African Archaeological Review* 8:139–76.

Eggert, M. K. H. 1987. "Imbonga and Batalimo: Ceramic Evidence for Early Settlement of the Equatorial Rain Forest." *African Archaeological Review* 5:129–45.

Gosselain, O. 1988. *Sakuzi: Fouille d'un Premier Village du Néolithique et de l'âge des Métaux au Zaire.* Mémoire de licence, Universite Libre de Bruxelles.

Grégoire, C. 1976. "Le Champ Sémantique du Thème Bantou Banja." *African Languages* 2:1–13.

Guthrie, M. 1967–1971. *Comparative Bantu.* Farnborough: Gregg Press Ltd.

Hombert, J.-M. 1979. "Early Bantu Population Movements and Iron Metallurgy: The Linguistic Evidence." In *Fifth Annual Meeting of the Berkeley Linguistic Society,* 465–70.

Killick, D., N. van der Merwe, R. B. Gordon, and D. Grébénart. 1988. "Reassessment of the Evidence for Early Metallurgy in Niger, West Africa." *Journal of Archaeological Science* 15:367–94.

Manima-Moubouha, A. 1987. "A Propos des Recherches Archéologiques sur la Métallurgie du Fer et du Cuivre en R. P. du Congo." *Nsi* 1:3–5.

Maret, P. de. 1982. "New Survey of Archaeological Research and Dates for West-Central and North-Central Africa." *Journal of African History* 23:1–15.

——. 1985. "Recent Archaeological Research and Dates from Central Africa." *Journal of African History* 26:129–48.

——. 1986. "The Ngovo Group: An Industry with Polished Stone Tools and Pottery in Lower Zaire." *African Archaeological Review* 4:103–33.

Maret, P. de, B. Clist, and W. van Neer. 1987. "Résultats des Premières Fouilles dans les Abris de Shum Laka et d'Abeke au Nord-ouest du Cameroun." *L'Anthropologie* 91:559–84.

Maret, P. de, and F. Nsuka. 1977. "History of Bantu Metallurgy: Some Linguistic Aspects." *History in Africa* 4:43–65.
Maret, P. de, F. van Noten, and D. Cahen. 1977. "Radiocarbon Dates from West Central Africa: A Synthesis." *Journal of African History* 18:481–505.
Monino, Y. 1983. "Accoucher du Fer: La Métallurgie Gbaya (Centra-Afrique)." In *Mémoire de la Société des Africanistes*, ed. N. Echard, 281–309. Paris.
Pearson, G., and M. Stuiver. 1986. "High-Precision Calibration of the Radiocarbon Time Scale, 500–2500 BC." *Radiocarbon* 28:839–62.
Peyrot, B., and R. Oslisly. 1987. "Paléoenvironnement et Archéologie au Gabon: 1985–1986." *Nsi* 1:13–15.
Schmidt, P. 1978. *Historical Archaeology: A Structural Approach in an African Culture.* Westport, Conn.: Greenwood Press.
Schmidt, P., and S. T. Childs. 1985. "Innovation and Industry during the Early Iron Age in East Africa: The KM2 and KM3 Sites of Northwest Tanzania." *African Archaeological Review* 3:53–94.
Schmidt, P., L. Digombe, M. Locko, and V. Mouleingui. 1985. "Newly Dated Iron Age Sites in Gabon." *Nyame Akuma* 26:15–8.
Schoenbrun, D. 1990. "Early History in Eastern Africa's Great Lakes Region: Linguistic, Ecological, and Archaeological Approaches, ca. 500 B.C. to A.D. 1000." Ph.D. diss., University of California, Los Angeles.
Stuiver, M., and G. Pearson. 1986. "High-Calibration of the Radio-Carbon Time Scale, A.D. 1950–500 B.C." *Radiocarbon* 28:805–38.
Sutton, J. E. G. 1985. "Temporal and Spatial Variability in African Iron Furnaces." In *African Iron Working, Ancient and Traditional*, ed. R. Haaland and P. Shinnie, 164–91. Oslo: Norwegian University Press.
Thiry, G. 1987. *Les Techniques de Réduction des Métaux en Afrique Centrale.* Mémoire de licence en Histoire de l'Art et Archéologie, Université Libre de Bruxelles.
——. 1989. *Contribution de la Linguistique Comparative à l'Histoire de la Métallurgie dans la Région Interlacustre.* Mémoire de licence en Linguistique Africaine, Université Libre de Bruxelles.
van Grunderbeek, M.-C., E. Roche, and H. Doutrelepont. 1983. *Le Premier Age du Fer au Rwanda et au Burundi: Archéologie et Environnement.* Butare: Institut National de Recherche Scientifique.
van Noten, F. 1985. "Ancient and Modern Iron Smelting in Central Africa: Zaire, Rwanda and Burundi." In *African Iron Working, Ancient and Traditional*, ed. R. Haaland and P. Shinnie, 102–20. Oslo: Norwegian University Press.
Vansina, J. 1984. "Western Bantu Expansion." *Journal of African History* 25:131–49.

3 The Blooms of Banjeli: Technology and Gender in West African Iron Making

Candice L. Goucher & Eugenia W. Herbert

In the Bassar region of northwestern Togo, near the Ghana border, people have been making iron since at least the twelfth century. At the height of the industry's operations in the nineteenth century, production levels reached a semi-industrial scale. Today, heaps of slag—the by-product of the smelting process—and the remains of thousands of abandoned furnaces dominate the local landscape. When the Germans arrived in 1885, they were struck by the size and extent of smelting operations in Banjeli: to one writer, the noise of the blacksmiths' hammers and the fires from hundreds of furnaces burning through the night recalled his native Ruhr. Just before the turn of the twentieth century, a German official (von Doering 1895) estimated the number of working furnaces in the area to be in the hundreds.[1] The region's technology thus constitutes one of the largest industrial systems known in precolonial West Africa.[2]

By the nineteenth century, the scale of the industry had led to a high degree of regional specialization. Individual villages specialized in different stages and metallurgical processes: smelting, refining, smithing, mining, and charcoal making. Archaeological survey and excavations conducted by Philip de Barros (1986) have established the antiquity of the industry to about the twelfth century, demonstrating some variations in the levels of production over time. It is quite likely that Early Iron Age levels may also exist but with little chance of recovery given the degree of later environmental degradation. Written descriptions of the industry date from the German period and oral histories collected by Goucher, de Barros, Kuevi, Cornevin, and others do little to extend historical coverage beyond the nineteenth-century limits. Bassari iron was traditionally traded either in the form of blooms (the raw iron smelted in a characteristic horseshoe shape) or of discs for hoe blades. Trade networks recorded by German observers extended over wide areas: to Mango to the north, and west to Dagomba, east to the interior of Dahomey, and south to Kete Krachi, Atakpame, and even to the coast. So plentiful was iron locally that it was sold daily at the Bassar market, its greatest consumers probably the justly famous Kabye blacksmiths. As elsewhere in Africa, iron circulated not only as a raw material to be worked by local blacksmiths but also as a currency.

Regular smelting of local ores declined rapidly in the early years of this century in the face of competition from cheaper scrap metal imported from abroad. The decline came not only at a time when scrap was becoming increasingly abundant and when colonial controls prohibited the exploitation of resources, but also when the cost of producing iron locally had increased to prohibitively high levels, partly as a result of the depletion of hardwood fuels. As elsewhere in Africa, the industry revived somewhat during World War II when imported metal was scarce; indeed, a local missionary claimed that some fifty furnaces were still operating in Banjeli as late as the early 1950s (Hahn 1991, 11n. 5). For some thirty years prior to our reconstruction in 1985, therefore, the Bassari had not smelted iron on a regular basis, although there were several attempts during this period to carry out demonstration smelts in existing furnaces. Our reconstruction was followed almost four years later by that of Hans Hahn.

The earliest written records of Bassari ironworking date from the German occupation in the early 1890s. These records are supplemented by a remarkable film of iron smelting in Banjeli made in 1914 by a German big-game hunter and filmmaker, Hans Schomburgk.[3] Recent archaeological studies have also attempted to reconstruct the evolution of the industry, and laboratory analysis of slags and furnace materials from excavated sites provides technical data from different sources (for example, see two UCLA dissertations, Goucher 1984 and de Barros 1985).

Despite the relative abundance of documentation, however, much of the knowledge of iron smelting survives only in the oral traditions of specialist families and villages, especially in the case of the rituals and medicines that accompany the technical procedures and that constitute an integral part of them. Although iron has not been regularly smelted since the 1950s and not at all in some areas since the 1920s, a few of the elders, now in their seventies and eighties, observed or assisted smelters in their youth. As they die, their knowledge will disappear completely because young men are not learning the craft; already, they have lost the skills that come with constant practice.

During the brief resumption of smelting in World War II, Bérard (1951) collected some of these traditions. Subsequently, Robert Cornevin (1962), a colonial administrator, and most recently Bruno Martinelli (1982) and others obtained more extensive information through interviews conducted in the early 1980s. None of them, however, seems actually to have watched iron smelting in progress. In fact, none of the writers on Bassari ironworking at any period before our reconstruction appears to have seen a furnace being built or the rituals associated with this and with the initial smelt, so that only secondhand information exists on this phase of the process. This fact is not surprising since during the heyday of the industry furnaces could be used over and over for a period of years, with relatively minor repairs, unless rains were extraordinarily intense (as in 1987). In addition, there is evidence that in the past this operation took place outside the village, in seclusion, so that it would not have been readily visible to travelers or officials. Nevertheless, any attempt to understand the rituals surrounding ironworking would be inadequate if it did not include the building of the furnace and the practices that embed the technology into a social matrix.

In January 1985, two historians and a filmmaker went to Banjeli, in the heart of the Bassari iron country, to investigate traditional methods of iron smelting. We all were women, a circumstance that will require some comment in its own right. We undertook to have an entirely new furnace constructed in the village of Banjeli, earlier one of the foremost centers of Bassari iron smelting. The site chosen was adjacent to an older furnace, and smelting was carried out in both. We filmed the entire process (Saltman, Goucher, and Herbert 1986), relying on two interpreters from the village to help us gather information and improve our vocabulary.[4]

Although we could compare the techniques we observed with earlier accounts and with Schomburgk's film, the only extended comparison of rituals we could make was with the accounts provided by Martinelli's informants; no older information appears to exist. The differences between what we observed and what Martinelli's informants described may result simply from variation not only between villages but even between different smelting clans, which tended to keep secrets within the family group. Furthermore, the rituals performed in Hahn's subsequent reconstruction in Banjeli also differed in some details, albeit not in major respects. In any case, the two reconstructions and Martinelli's information represent the outer limits of oral memory in the area. A reconstruction, arranged and paid for by outsiders, will always be an exercise in artificiality, and one can never be sure what may have been unintentionally forgotten or deliberately withheld. The same is true of information obtained through interviews. In addition, there are no comprehensive ethnographies of the Bassari to aid us and no dictionary of the language.

These limitations of the data must always be kept in mind. Nevertheless, a comparison of Bassari techniques, rituals, and beliefs with those of other African ironworkers helps to buttress our interpretations, for the understanding of technology presented here is by no means unique to the Bassari: behind particular practices lie patterns of greater universality.

Furnace Construction and Rituals

The project had of necessity to be submitted to the chief of Banjeli at the time, the late Sertchi Magnibo, himself a member of a smelting clan. Once he approved and money matters were negotiated, he designated the master smelter, Belam Diyambo, and his three helpers and agreed to the other arrangements. He was consulted at each stage of the process. When asked if the fact that we were all women would present a problem, he said that it would not because we would be observers, rather than participants, an explanation that puzzled us. Nevertheless, we were under the same sexual interdictions as the men taking part; that is, we were forbidden to have sexual relations until the smelts were completed. In a sense this gave us the status of provisional males, since women were otherwise excluded from the smelting site.

The site selected for the new furnace (*kpai*; pl. *m'pampu*) lay within the compound of the chief, not far from a furnace already standing. It appears that clusters of furnaces were often built in the bush, within walking distance of the village, as at Nangbani M'pampu, or Belemele. That may not always have been the case, however, because our new furnace was to be constructed on the site of a preexisting one, and the 1914 film shows furnaces in close proximity to a cluster of houses.

Bassari iron making, like iron making elsewhere in Africa, was family-based and exclusively a male occupation. A man became a smelter by birth into a smelting lineage and then by lengthy apprenticeship under a master smelter. The training consisted of learning the practical details of the craft plus the rituals and medicines that were essential to the process. Formerly, the entire team building and operating a furnace would have been smelters; the head of the team would customarily have been the oldest member of the family using the furnace. Mining and charcoal making were similarly family affairs.

Once the site was chosen and cleared it was marked off with a leafy branch, *lienaboul*, which protected against bad spirits. But before construction could begin, the master smelter performed rituals on the site itself. Kneeling beside a small pile of fragments from old furnaces, he offered libations of *chakpa* (millet beer) to drive away evil spirits and to bless the enterprise. He first called upon Ounimbote, the high god, then on his ancestors (his grandfather and great-grandfather), then on various spirits, pouring a drop of beer on the ground with each name:

> "Grandfather Tangma, this is your drink. Receive it.
> Father of my grandfather, Fandi, I have heard your name,
> but I do not know you. This is your drink. Receive it.
> The spirits are called Banja, this is your drink."

Closing, he repeated three times:

> "All the ancestors whose names I have forgotten, this is your drink. All the spirits whom I have forgotten, this is your drink. We want the fire of the furnace to be a real fire, not just smoke. You all come and build this furnace. It is not we who build it."

The enterprise would involve not just spirits but also the past owners and metallurgists of preexisting furnaces on this site. Indeed, bits of old furnace were incorporated into the new one. Like any living being, the furnace would have a genealogical connection with earlier ones.

The master smelter and his helpers then made a shallow circular trench, using a string as compass and marking it with small stones and bits of old furnace. This would provide the base of the new furnace, with the dimensions taken from the neighboring furnace. Next, the master smelter performed a ceremony to attract customers from near and far. Standing in the middle of the circle and casting sand mixed with medicines in each direction, he repeated: "All buyers, come and buy this iron."

In a small calabash he mixed a potion made from shaved bark (*n'yoo*), used to protect the future furnace, and daubed the hands and feet of all the participants with it, ourselves included, to keep away bad spirits. Henceforth, no one was allowed on the site who had not been purified with this, and once purified, everyone had to refrain from sexual relations until the smelt was completed. It was a sacred male domain, taboo for women of childbearing age.

Next the smelter placed the medicines with the pebbles in the shallow trench outlining the base of the furnace. The vegetable medicines (a melon-like fruit, *lepotadjur*, and the dried leaves and wood of the *boufobo* tree) had been gathered by the medicine man (*inoxodjal*), who was not one of the smelters, going out naked in the middle of the night to the surrounding mountains. To these plants, he added ground porcupine skin and small chunks of hippopotamus hide. All of these ingredients were intended to strengthen and to protect the furnace from contact with those who were having sexual intercourse and were added to each tier of the walls during construction—at the point where the bloom would fall at the end of the smelt, at the level where the "stomach" of the furnace met its "chest," and near the top of the furnace. Once the furnace was completed, medicines were also smeared on the interior furnace surface and shelf.

The incorporation of the living world of plants, trees, and animals into the furnace reflects a holistic view of the unity of creation and a complex understanding of correspondence between domains of living matter. As the furnace was transformed into a living being, it would assume some of the characteristics of the materials embedded into it. The leaves may also be similar to those used medicinally in comparable contexts, for example, those given to new mothers to increase the flow of milk.

The fact that the fetisher went out to gather plants entirely nude is also relevant: nudity is often associated with key undertakings in Africa, representing unmediated contact between man and the powers and spirits that control the world. Bassari smelters were required to work bare-chested, wearing only a leather loin cloth and a hide apron; elsewhere they often were required to work entirely naked (see Martinelli 1982, 50; Herbert 1993, 93–94).

The furnace was constructed of clay mixed with grass (*n'some*), formed into flattened loaves and placed vertically next to each other in courses, an extremely common building technique in West Africa. The exterior and interior were then smoothed over. The completed furnace was between 2.5 and 3 m high, cylindrical in shape but narrowing toward the top. At Nangbani, the remains of furnaces show them to have been much larger, which may be related to the lower quality of iron ores in that region and the need for higher temperatures to reduce them. On the inside perimeter a shelf was built just above the air intake holes (tewels) to hold the heavy charge of ore and charcoal and to separate it from the layer of sand and broken bits of furnace below until the metal had been reduced. This can also be observed in old furnaces in the region but has never been noted in literature.

Once construction of the new furnace was underway, a celebratory meal was prescribed. Because menstruating or sexually active women were considered to pose a grave danger to the furnace and to the future smelt, it was prepared and brought to the furnace site by Yamimbe Sakare, a widow well past childbearing age who was also the only woman present during the palavers with the chief. She was assisted by two girls whom she had kept with her the previous night to make sure they were "safe." They had also been purified with the protective liquid. We were invited to share this meal with the master smelter and his fellow workers.

As the furnace took shape, it was regarded more and more as a human being. The bosh or midportion was referred to as *apo* (belly); the stack is *debend* (chest), and the top is *dyerekan* (*se*) *kem* (head), with a mouth through which the charge was fed. The interior with its charge was known as *m'poyo* (the lung), and when smelting had begun, *mfomfotem* (respiration) was said to be taking place—the furnace was breathing. The master smelter was regarded as the "father" of the furnace; he and his assistants would also become its husbands as the furnace became more explicitly a female being who could deliver the iron bloom from her womb.

The Banjeli furnace relies on natural draft rather than the forced air produced by bellows. Air enters through holes bored evenly around the lower wall of the furnace, with a cluster of these holes in its walled-up mouth. In this case there were nine holes, although as many as twelve holes have been observed in archaeological contexts (Goucher 1984, 99). During smelting, the hot gases inside the furnace rise, drawing a constant supply of cooler air from the outside. The holes are fitted with plugs (*mpulu*) fashioned from wooden poles covered with wet clay: the poles can be inserted or withdrawn to control the volume of induced air, and hence the heat of the furnace, during different phases of the smelting process. Although the natural draft furnace is by no means as fuel efficient as the bellows furnace, it requires less manpower, undoubtedly an important consideration given the scale of Bassari smelting: operating bellows is very laborious work, whereas the induced air furnace can be operated by a single smelter who adjusts the flow of air according to the progress of the smelt.

The last element to be added to the furnace was a low, curving retaining wall built to the left of the furnace mouth to hold back slag that might run off. At the same time, the area immediately in front of the furnace was dampened and pounded smooth to give it a hard surface, and the interior furnace floor was built up with broken pieces of old furnace and sand gathered from a nearby river bed that provided a bed on which the bloom and slag would fall. Although the smelters made much of the special qualities of the sand, its main purpose seems to have been to facilitate removal of the slag without damage to the furnace floor.

The construction of the furnace was completed in five days. The weather was very hot and dry at this time of the year, but the thick clay would nonetheless take some time to dry thoroughly. To speed up the process, dried grass was burned inside the furnace. It may not, however, have been sufficiently dry to operate properly—at least one informant suggested that the first smelt in a new furnace was rarely expected to be successful.

Rituals of the First Smelt in a New Furnace

The furnace was loaded with ore, charcoal, and wood and lit from above, but before smelting proper could begin, further rituals were necessary to enable the furnace to "breathe." The master smelter painted a circle of white ash mixed with water (*lurekpai*) around the "chest" of the furnace, adding a loop above its entrance. Then a sprig of a plant (*budu*) was placed in the center of the loop, held in place by red clay. The circle represents the "person" of the furnace, with two arms and mouth. Such circles are also found on furnaces in northern Ghana and more generally serve as forms of symbolic protection.

Yamimbe Sakare arrived with a special meal of *fufu* and mudfish. The master smelter placed a gourd of food in front of the entrance to the furnace. Holding a handful of food, he addressed the furnace: "Give birth to good iron; give birth to good iron," then smeared some of the fufu and mudfish on the sprig in the ashen loop. He called upon the furnace to be fertile as the tree that produced the leaf, to bring forth "bracelet iron"—iron of the very best quality.

The master smelter then beat a young boy with a switch until he cried, brought him around to the front of the furnace, and left him with a bowl of food before joining the others with the meal (the only event that we were excluded from, along with Yamimbe Sakare). The child may represent a surrogate sacrifice, intended to drive away evil spirits and invoke the aid of ancestors and spirits. The cries of the child may also correspond to those of a newborn, carrying through the analogies of smelting and childbirth. They signal that respiration has begun; the furnace breathes. The human identity of the furnace also emphasizes the independence of the natural draft system in which no smelters with bellows are required to breathe on its behalf.

The Smelting Process

Smelting is the process by which iron ore is heated in a reducing (oxygen-free) atmosphere to the point where the raw metal separates from the rocky matrix in which it is found and metallic iron is released from other compounds found in the ore. The chemical reaction occurring is customarily expressed as:

$$Fe_2O_3 \rightarrow 2Fe + 3CO_2$$

Achieving this result requires a careful knowledge of the ores and fuels at hand and an ability gained only through long experience to monitor what is going on inside the furnace through sound and sight: the sounds of the slag dropping and the colors of the gases formed at various stages of the operation. The smelter is constantly listening, peering inside the draft holes, and tapping slag (*tukur*) through them, all the while controlling the temperature by means of the plugs according to his sense of what is happening.

The heat should ideally build gradually to about 1200–1300°C to separate metal from slag and produce the horseshoe-shaped "bloom" (*nkuyeke*) of raw (unrefined) iron characteristic of Bassari smelting. The characteristic shape is thus a product of the arrangement of the shelf and air openings. The slag consists primarily of silica in the ore which combines with iron oxide. This melts at a lower temperature than iron and runs off as molten fayalite:

$$2FeO + SiO_2 \rightarrow Fe_2SiO_4$$

Iron ore (*aspetan*) in the form of ferric oxide (hematite) was abundant in the Bassar region. According to Hupfeld (1899), the richest ores were found in the mountains and valleys near Banjeli: they contained as much as 69 percent iron (Goucher 1984). Ore was mined especially by women and children and possibly by slaves as well. Miners usually, they dug it out of surface pits but sometimes also out of underground shafts. In 1914 Meg Gehrts (1915), travelling with Schomburgk, described the women miners: "It was pitiful to see some of them, with babes at their breasts . . . digging out the ore with a curious horseshaped tool. Besides being a hard occupation, it is also a dangerous one. Only a day or so before our visit, one of the miners had been killed owing to a shaft falling in" (246).

Yamimbe Sakare, who came of a smelting family and who had mined as a child, located the ore and directed its preparation: it was broken up with a stone and sorted to winnow out impurities. Ideally, it was mined the same day it would be used and kept "moist" under a covering of green leaves. It was loaded into the furnace in a small calabash, used only for this purpose, which served to measure the amount of ore added to the charge.

CHARCOAL MAKING

In the past, women also prepared the charcoal (*nko*). Entire villages such as Dimouri specialized in the craft (Goucher 1984, 65). In 1985, however, the chief assigned the work to men because they wanted to earn the wages. Ntube Dichiyou, a woman over seventy, recounted how as a little girl she had helped her mother make charcoal for her father, who was a smelter: "As a young girl I cut down trees with my mother. On the second day, we burned them in the pits, and on the third day, we carried the charcoal and left it at my father's furnace."

Not all types of wood could be used to produce charcoal for smelting, which differs from that used for smithing or for cooking. Ideally, the wood should be hard and dense, with high alkali and silica content, slow-burning and capable of achieving high temperatures. The preferred species here (Goucher 1981, 179–89), as across the savanna regions from the western Sudan to Malawi and southern Africa, was the *oussidja* (*Burkea africana*). This tree grows slowly, however, so that charcoal makers had to range farther and farther to find choice trees: Ntube Dichiyou described having to go as far away as Kabou (20 km) for suitable wood. In the past wood was even imported from the Asante region of Ghana to the southwest and by Hausa caravans from the north (Goucher 1984, 25). Where the exhaustion of charcoal woods in England in the seventeenth and eighteenth centuries led to the adoption of new fuels, namely coke and coal, West Africans, with no suitable substitute, were forced to rely on inferior wood fuel and ultimately on imported metals.

Inputs and outputs indicate the economic impact of smelting activities. Charcoal was measured in standard basketfuls, alternating with ore and sticks of green wood in loading the furnace. The total charge consisted of ore (54.75 kg), charcoal (183.5 kg), and wood (15 kg). The resulting products were slag (29 kg), iron bloom (14 kg), and excess charcoal (5 kg). The charcoal-making efforts yielded a wood to charcoal ratio of twelve to one, a rather inefficient rate of recovery, no doubt due in part to the inexperience of the charcoal makers. In the past a single furnace charge would have been expected to produce a bloom of 25 to 30 kg, or approximately twice the size achieved in the reconstruction.[5]

The African iron industry paid a high environmental price for the levels of production it achieved: each furnace consumed as many as three trees or 180 kg of charcoal per smelt. A single furnace could therefore consume between one hundred and three hundred trees annually. If the Germans correctly estimated that the overall production of the region exceeded four hundred tons of iron per year, the smelters would have required some one hundred tons of charcoal or a minimum of 18,000 trees. Despite the inevitability of rising fuel costs, however, locally produced iron was widely believed to make stronger and more durable hoes and axes than metal from abroad, a belief common to many African peoples.

The smelt lasted about forty-five hours, during most of which the master smelter stayed close to the furnace, constantly checking the progress of the smelt and tapping the small amounts of slag that began flowing on the second day. He was also concerned to guard it against contact with people who might not be abstaining from sexual relations. As time passed he listened for the thud of the bloom falling to the base of the furnace, which, like the fetus dropping in the womb, would signal that the smelt was complete. Once this happened, his assistants would be released from the prohibition against sexual activity, but he would have to wait until the bloom was extracted from the furnace. Once the bloom dropped, too, women could freely approach the furnace; their presence was no longer a danger.

The smelt in the new furnace, like that carried out in the old one, was only partially successful: instead of the traditional horseshoe of raw iron, the bloom came out as an amorphous mass containing a great deal of slag, charcoal, unreduced ore, and furnace clay still adhering to it. A number of theories were offered to explain the disappointing result. The smelter himself concluded that the taboos against sexual relations must have been violated. The chief suggested that too much sand at the base of the furnace may have lowered the temperature so that the ore was not fully reduced. It is possible that oral memory had erred as to the exact proportions of ore, charcoal, and wood, or the furnace may still have retained too much moisture, which could have adversely affected the temperatures achieved and thus the separation process. There was little understanding of the use of the furnace plugs to regulate temperature.

More fundamentally, it is obvious that without constant practice smelters cannot be expected to maintain their skills at their earlier level, to sharpen the almost intuitive sense of what is happening inside the furnace, and to adapt to the process. Like cooking, smelting is both a science and an art. Then, too, some of the rituals have also become hazy through lack of usage. Even the prohibition against sexual relations seemed to be regarded with ambivalence, at least by some of the participants. Although certainly the subject of a great deal of joking, it was difficult to know how seriously it was taken. And what about the presence of three foreign women, a possibly unsettling condition and yet one that was never put forth to explain the failure of the smelt? When we returned seven years later and showed the film, viewers who had not participated in the 1985 smelt were unanimous in their belief that a violation of sexual taboos was responsible for failure of the smelt; no one mentioned our presence.

While the local iron industry still flourished, blooms would have been sold in the market to neighboring smiths. Some would have been taken to Bidjabe, which specialized in further refining the metal before it was worked into hoes, axes, knives, spears, bracelets, and the like. These blooms would have been broken up in enclosed in clay crucibles (*lokokol*), which would then be heated in small furnaces using goatskin bellows. We were not able to document the refining process because there was no one able to repair the village bellows, a sad testimony to the disappearance of the craft. Indeed, it was difficult enough to persuade the village blacksmiths to forge a small piece of iron bloom because they were unaccustomed and unprepared to handle the medium-carbon steel we had produced.

Technology and Gender

Technologies of radical transformation, of which iron making is a prime example, involve not only complex skills but a necessary framework of explanation. In preindustrial societies a constant point of reference is the human, the live world of the body and of social relationships. Because these are genderized, gender becomes a means by which the world is experienced and categorized. The analog of human processes becomes the bridge to understanding the natural world.

To the Bassari, the furnace becomes a living being, created by the metallurgist both through his control of "natural" forces and his ability to summon "supernatural" powers, his own ancestors and the spirits whose assistance is deemed indispensable to all major enterprises. The furnace is not overtly female; the words used to designate its parts—mouth, chest, lungs, stomach—are generically human, just as when air is drawn through the draft holes and the furnace is said to breathe. What not only brings it to life but also gives it a sexual and social identity are the ritual acts and invocations of the Bassari smelter. He is initially the "father" of the furnace. Later, he and his assistants become symbolic husbands. It should be noted that our informants were evasive when asked about the sexual aspects of the furnace and smelting. It was only when we obtained a fuller translation from nonparticipant Bassari of the prayers and combined these with the visible rituals that we could confirm our interpretation. Some of the rituals that would have made this more evident have been lost: Martinelli's informants (1982, 45), for example, told him that when the foundation of a new furnace was first outlined with pebbles, the head of the smelting lineage would pour out millet flour mixed with water, a symbol here as elsewhere of fertility, while he invoked the high god and the spirits to make the enterprise fruitful. During the 1985 reenactment only millet beer was used for these libations.

Why the exclusion of women, and especially menstruating women, from the smelting site and the prohibition against sexual relations by those participating in the operation? As one aged informant put it, "If a smelter slept with a woman while the smelt was going on, he would not return to the furnace until it was finished. Otherwise, he would risk his life and the success of the smelt." As we have seen, protection against women and against those having sexual relations was one of the main purposes of the medicines used and of the invocations of ancestral and other spirits. In the past, women were not even supposed to look at the flames spewing out of the furnace.

Such prohibitions and exclusions are extremely common, not only in Africa but in other parts of the world, as Mircea Eliade (1978) has shown. These restrictions often affect other areas of activity that are considered dangerous or problematic, and that are equally dependent on the control of natural forces: hunting, warfare, rites of passage. These reasons are undoubtedly complex and multiple and are on the one hand so much a matter of custom that they do not require explanation; on the other hand, they are linked with bodies of esoteric knowledge that are not meant to be iterated. Hence, any broad interpretation must be partly inferential and based on a larger body of information from peoples who share the same exclusions and prohibitions (Herbert 1993).

On the most down-to-earth level, the presence of women always represents a distraction and a temptation to men, best avoided by keeping them at a distance. It is a common assumption that sexual activity simply diminishes one's energies for other tasks, energies that are especially vital to difficult or dangerous undertakings. More specifically, the view that sexual activity on the part of anyone coming into contact with the furnace can ruin the smelt seems to be intimately related to the notion that the furnace is analogous not simply to a pregnant woman but to a pregnant *wife*, impregnated by the smelter-husband and gestating a fetus of raw iron in its womb. Concepts of adultery and socially sanctioned sexual rights to women can be invoked to explain these taboos: like the detailed regulation of marital sexual relations during pregnancy, the regulations concerning who has access to the furnace are aimed simultaneously at insuring a successful outcome and asserting control over wives and offspring. They guard the furnace against her adultery as well as her husband's, an adultery that could cause the smelt to abort just as a woman's adultery is commonly believed to cause a difficult labor or a deformed or stillborn baby. Indeed, failure to observe sexual taboos is the most common explanation for the failure of a smelt given in virtually all the reports available from all parts of the subcontinent. (Witchcraft is the second most popular explanation.) This was as true for our reconstructions at Banjeli as for others in Zambia (Chaplin 1961; Armor 1987). Similarly, women's menstrual blood represents failure to conceive and could cause a similar failure on the part of the furnace. It may also be that sexual activity and menstrual blood are thought to be dangerous because they are both hot, and control of heat and fire are essential in the smelting process. Such beliefs have been documented elsewhere (for example, see Richards 1956), but we were not able to corroborate them among the Bassari. Beliefs concerning menstrual blood are just beginning to receive the attention they deserve. Above all, we should be wary about assuming they correspond to Western notions of filth and "pollution" (Herbert 1993, 85–86).

Whatever its metaphysical justification, the exclusion of women from all but the preparatory stages of the smelting process also serves practical purposes. It bars them from access to economic wealth and thus from political power. It might be argued, too, that the exclusion of wives and daughters helps maintain proprietary and industrial secrets, functioning much like a patent system for the production of valuable technical information from loss to outside groups via exogamous marriage or kinship ties. The same purpose would have been accomplished by the exclusion of slaves, reported by early observers. Ironically, such discrimination may unintentionally have also protected both groups from the hazards of industrial pollution, including exposure to the fumes, gases, and heavy metals released from the tops of furnaces into the surrounding air.

Nevertheless, it is hard to escape the conclusion that smelting represents a symbolic usurpation of female reproductive capacities. The Bassari smelter reenacts the drama of copulation, gestation, and birth—most graphically exemplified by the phallic function of the airhole plugs (mpulu). On the most fundamental level, smelting taboos represent the profound ambivalence men feel toward women: fears of pollution and danger are at the same time an acknowledgment of men's sexual dependence on women and of their envy of women's procreative powers. It should be noted that the exclusion of women at Banjeli did not apply to older women, such as Yamimbe Sakare, since they were no longer sexually active nor menstruating—in sum, no longer essentially female and hence no longer a "danger" to men. Iron smelting reflects simultaneously both positive and negative views of femaleness—creativity and danger. By enforcing the exclusion of women from smelting, the male smelter reaffirms his control over productivity and reproductivity.

Conclusion

The Bassar region plays a critical role in the history of African iron technology. Metal production here began in the twelfth century, if not earlier, and continued into the middle of the twentieth, reaching protoindustrial levels on the eve of German colonial conquest. Quite apart from the scale and longevity of the region's industry, iron production here is among the best recorded of the continent's metallurgical systems. As appears to hold for other West African smelting regions, the environmental and human factors were complex and interrelated. The Bassari adaptation of natural draft furnaces to the scale of production constituted a significant technological innovation but with a major trade-off. Because bellows were not used for smelting, demands for male labor were kept to a minimum. At the same time, the inordinate fuel consumption of these furnaces could only be met by expanding charcoal production through an increase in female labor. Presumably, more women were acquired thanks to the wealth of iron produced and traded. Thus the technology was efficient in terms of male labor inputs but not necessarily in terms of female. Finally, the insatiable appetite of the natural draft furnaces for charcoal pushed the capacity of the environment almost to its limits.

Here as elsewhere in Africa, technology must also be understood in a cosmological context. The technology at Banjeli is embedded in belief systems about social relationships and the natural world, above all in the interplay of gender and age. As we have seen, smelting rituals endowed the furnace with both a gender and a genealogy; ancestors and spirits were part of the unseen work force. Because the furnaces could be used over and over, rituals were mostly concentrated on the inauguration of a new furnace, although behavioral restrictions were imposed on each smelt. Far from impeding the expansion of the industry, therefore, ritual requirements could be adapted to its intensification.

The particular vocabulary of technical and ritual practice at Banjeli is unique. In essentials, however, it corresponds to paradigms found throughout sub-Saharan ironworking. Sadly, little of the richness of this culture of metallurgy was recorded by earlier observers, and here, as everywhere in Africa, the smelting industry has now disappeared virtually beyond recall.

Notes

This essay is based on fieldwork in Togo in 1982, 1985, and 1992. We are grateful to the Ministère de l'Education Nationale et de la Recherche Scientifique for research authorization. We would also like to thank Dovi Kuevi, Philip de Barros, John Treacy, Morgan Kulla, and Johanna Kowitz for their assistance in Togo. Dave Killick has offered many valuable suggestions and comments on the work. Goucher's field and archival research was partially supported in 1982 by a grant from the Friends of Archaeology at UCLA and in 1985 by grants from Portland State University and the Oregon Committee for the Humanities. Herbert's research was funded by a grant from the William H. Donner Foundation to the Mount Holyoke Project for Gender in Context. Her return to Togo in 1992 to show *The Blooms of Banjeli* and record responses to it was supported by a faculty grant from Mount Holyoke College. Carlyn Saltman filmed the reconstructions and is largely responsible for editing the resulting video.

Most of all, we thank the people of the Bassar region for their participation and generous hospitality.

1 This number is clearly supported by surveys around Nangbani and Banjeli alone.
2 Comparable but slightly less extensive industrial sites have been investigated in several other parts of West Africa (Haaland 1980; Warnier and Fowler 1979; Célis 1991).
3 We gained access to Schomburgk's film, located at the Institut für den Wissenschaftlichen film in Göttingen, only after our fieldwork was completed. Excerpts have been included in *The Blooms of Banjeli: Technology and Gender in West African Ironmaking* with permission of Schomburgk's widow.
4 We subsequently showed the raw footage to three Bassari in the United States who had not been involved in the project and who did not know any of the people who were. We asked them to provide translations for portions of the soundtrack that we had not been able to have translated at the time and to suggest interpretations for the rituals.
5 Calculations are based on 1982 and 1985 interviews and experimental work in Dimouri and Banjeli and on information recorded by the Germans. See Goucher (1984:65–74) for previous estimates. For his 1988 reconstruction, Hahn gives the ratio of charcoal to iron ore as ten to one in terms of volume, four to one by weight (1991:55). Finished blooms were collected by Hupfeld and can be found in the Berlin Museum für Völkerkunde.

References

Armor, Murray. 1987. Personal communication.

Bérard, M. 1951. "La Métallurgie Bassari au Togo Français." Première Conférence Internationale des Africanistes de l'Ouest (Dakar 1945). Paris: Librairie d'Amérique et d'Orient.

Célis, Georges. 1991. Eisenhütten in Afrika/Les fonderies africaines du fer. Frankfurt am Main: Museum für Völkerkunde.

Chaplin, J. H. 1961. "Notes on Traditional Smelting in Northern Rhodesia." *South African Archaeological Bulletin* 16:58.

Cornevin, Robert. 1962. *Les Bassari du Nord Togo*. Paris: Berger-Levrault.

de Barros, Philip. 1985. "The Bassar: Large-scale Iron Producers of the West African Savanna." Ph.D. diss., University of California at Los Angeles.

———. 1986. "Bassar: A Quantified, Chronologically Controlled, Regional Approach to a Traditional Iron Production Centre in West Africa." *Africa* 56:148–73.

Eliade, Mircea. 1978. *The Forge and the Crucible*. Chicago: University of Chicago Press.

Gehrts, M. 1915. *A Camera Actress in the Wilds of Togoland*. London: Seeley, Service.

Goucher, Candice L. 1981. "Iron Is Iron 'Til It Rust: Trade and Ecology in the Decline of West African Iron-Smelting." *Journal of African History* 22:179–89.

———. 1984. "The Iron Industry of Bassar, Togo: An Interdisciplinary Investigation of African Technological History." Ph.D. diss., University of California at Los Angeles.

Haaland, Randi. 1980. "Man's Role in the Changing Habitat of Mema . . ." *Norwegian Archaeological Review* 13:31–46.

Hahn, Hans P. 1991. "Eisentechniken in Nord Togo." Unpublished manuscript.

Herbert, Eugenia W. 1993. *Iron, Gender and Power. Rituals of Transformation in African Societies*. Bloomington: Indiana University Press.

Hupfeld, F. 1899. "Die Eisenindustrie in Togo." *Mitteilungen aus den deutschen Schutzgebieten* 12:178.

Martinelli, Bruno. 1982. *Metallurgistes Bassar: Techniques et Formation Sociale*. Lomé: Université du Bénin.

Richards, Audrey. 1956. *Chisungu: A Girls' Initiation Ceremony among the Bemba of Northern Rhodesia*. London: Faber.

Saltman, Carlyn, Candice Goucher, and Eugenia Herbert. 1986. *The Blooms of Banjeli: Technology and Gender in West African Ironmaking*. Film. Watertown, Mass.: Documentary Educational Resources.

von Doering, Erich. 1895. "Reiseberichte von Premierlieutenant v. Doering aus den Jahren 1893 bis 1895." *Mitteilungen aus den deutschen Schutzgebieten* 8:231–76.

Warnier, Jean-Pierre, and Ian Fowler. 1979. "A Nineteenth-Century Ruhr in Central Africa." *Africa* 49:329–51.

4 Fipa Ironworking and Its Technological Style

Randi Barndon

This study of traditional ironworking was carried out among the Fipa, who inhabit the Rukwa region in southwest Tanzania (map 4.1). Fieldwork in Ufipa fell under the heading of "urgent ethnoarchaeology," as the knowledge of the craft is now rapidly disappearing. Consequently, fieldwork carried out in 1990 and 1991 attempted to establish a meaningful context of Fipa iron smelting within a cultural framework while it was still possible to observe and document.

The first iron production observed in 1990 included the entire process from initial building of smelting furnaces to final forging of the iron produced. The two smelts observed in 1991 were conducted by reusing a furnace first built and initiated in 1990 (figure 4.1) and an old furnace dating to sixty years ago (figure 4.2). Old abandoned smelting sites were surveyed and documented and numerous interviews with the ironworkers and their relatives were also conducted.

Introduction

Among researchers of African traditional metallurgy, the tall Fipa iron smelting furnace is a known ethnographic example (Wise 1958; Wembah-Rashid 1969). The furnace has often been cited in comparative studies of technologies and rituals of iron making (Cline 1937; Eliade 1978; Kense 1983; Collett 1985; Sutton 1985, 1990; van der Merwe and Avery 1987; Childs 1991; Herbert 1993).

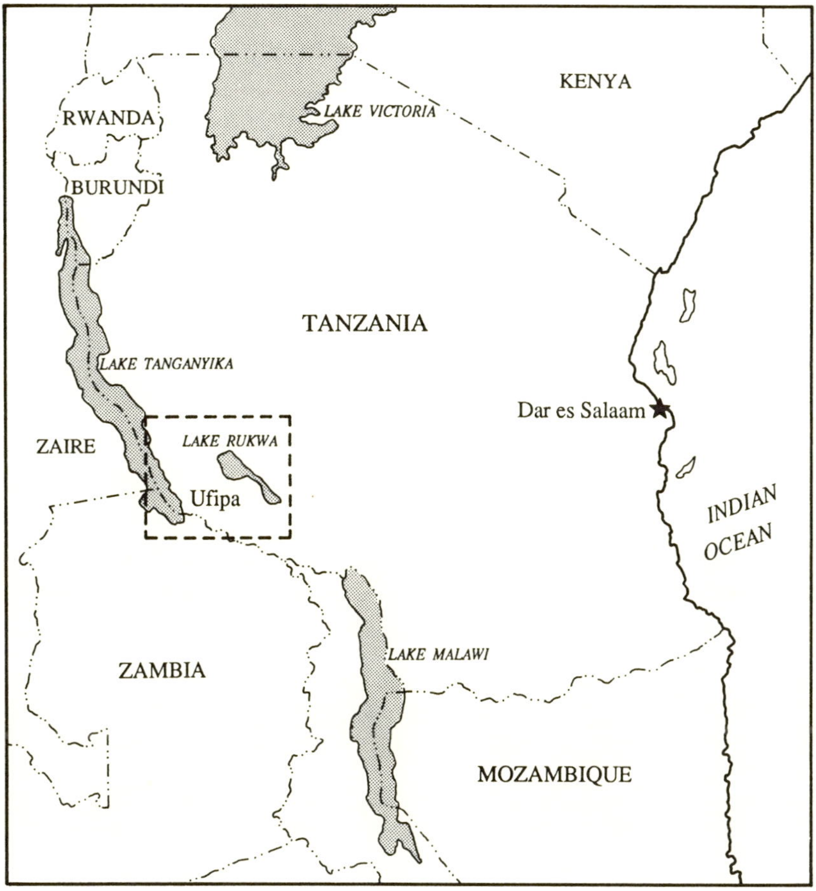

MAP 4.1. The geographical location of Ufipa country within southwestern Tanzania.

These comparisons have been based primarily on documentation made by explorers and missionaries from the late nineteenth and early twentieth centuries, and these works have emphasized the tall chimney furnace, which constitutes only one stage in Fipa ironworking. This study, however, will examine Fipa ironworking within a cultural framework rather than describing technological aspects as they are conceived in Western chemical terms. As pointed out by Collett (1985, 184), iron smelting may be conceptualized in cultural terms, thereby examining the origins of ideas about smelting rather than the origins of technologies.

Magic and rituals associated with iron smelting are common among sub-

Saharan Bantu-speaking metallurgists (van der Merwe and Avery 1987, 143), and African iron smelting is surrounded by fertility symbolism and sexual taboos that often exclude women (Maret 1985). This essay will attempt to demonstrate how the magic of iron smelting, fertility symbolism, and sexual taboos are related to a Fipa system of thought. A relational view involving symbolism (the use of material objects, metaphors, and analogies as part of the technological behavior) (Lechtman 1977; Ingersoll 1987) rather than a cause/effect explanation will be employed (Ingersoll 1987, 7).

Technology as Cultural Behavior

Studying technological behavior within a cultural framework has not been approached systematically within ethnoarchaeology or traditional archaeology; as a result, the relation between technology and culture has remained rather vague (Gosselain 1992, 549). Lechtman (1977), Childs (1991), and more recently Gosselain (1992) and Pfaffenberger (1992) have indicated how cultural variables in material culture production and technological behavior can be approached in archaeological and ethnoarchaeological studies.

FIGURE 4.1. *Left*, a tall *iluungu* furnace built in 1990 in Ufipa; *right*, the same furnace being used again in a 1991 smelt. This is the primary furnace in a multistep reduction process. Note the scaffolding attached to the exterior.

FIGURE 4.2.
A sixty-year-old *iluungu* furnace that was reused for one of the 1991 smelts.

Lechtman combines technological aspects, such as the practical process of metal production, with more ritual aspects, symbolism, and ideological modeling (1977, 15). Therefore, technological behavior is characterized by the many elements that make up technological activities, for example, by technical modes of operation, attitudes toward materials, some specific organization of labor, and ritual practices and observances—elements that are unified nonrandomly in a complex of formal relationships (ibid. 6). Together, these elements form a technological style. The use of this concept prevents a polarization between technology and symbolism. The technological style exhibited by a specific iron-smelting tradition results from cultural options and constraints, functional and symbolically structuring principles in the way people produce iron, think about the process, and explain and distribute the knowledge of iron smelting in local terms. As Childs explains (1991, 332): "A theory of technological style requires that we direct our attention to the iron smelters and the operations they perform." Kent (1984) points out, however, that the end products of material culture production and use represent the archaeological reality. Thus, ethnoarchaeological studies should not only focus on technological behavior and operations per se but combine studies of these activities with the end products produced.

A distinction between the everyday practices and the specifically goal-oriented and seasonal production of material culture is needed when considering technological behavior as cultural. Material culture in everyday practices involves the use of material items in households. Spatial organization in the household is part of the general flow of life (Moore 1986), in contrast to specific-purpose technologies and manufacturing. More specific artifact manufacturing takes place apart from the general flow of life, often as part-time occupations on a seasonal basis. Therefore, the participants will extend their consciousness of material culture manufacturing, and the technological activities and behavior might evoke cultural codes and messages absent from or more implicit in everyday life. Furthermore, a specific manufacturing context is also used by the participants to communicate cultural norms and values. Ultimately, then, systems of thought (which are a part of iron smelting as a specific material-culture production) are materially manifested on the smelting sites.

Ethnographic Background

The Fipa inhabit a high mountainous and today almost treeless area in southwest Tanzania. The plateau is situated between 1500 and 2000 m above sea level, and vegetation consists of open grassland with scrub. Woodlands still occur in the northern and southern part of the region. The Ufipa plateau has numerous iron-production sites, with remains of abandoned iron furnaces (*iluungu*) connected to voluminous slag heaps, indicating the extension of the craft until recent times. Deforestation in Ufipa might be partly explained by the extensive charcoal production needed for iron smelting in the past (Haaland 1980; Goucher 1981; Schmidt 1994). Apart from surplus production during especially active smelting times, however, local iron products were mostly distributed on a inter- and intravillage level, connected to an internal and local trade network of transactions in commodities such as cattle, cloth, and salt for iron hoes.

The Fipa ironworking process involves three separate stages. The reduction of a limonite ore (*inyiimbo*) and primary extraction of iron is conducted in a 2- to 3-m tall chimney furnace (iluungu). The furnace operates with natural draft through ten tuyere inlets (*incheelu*) evenly distributed around the furnace base (see figure 4.3). The main entrance/exit and tuyere inlet faces west and is called in Fipa the "mother door" (*palinyina*), and the door

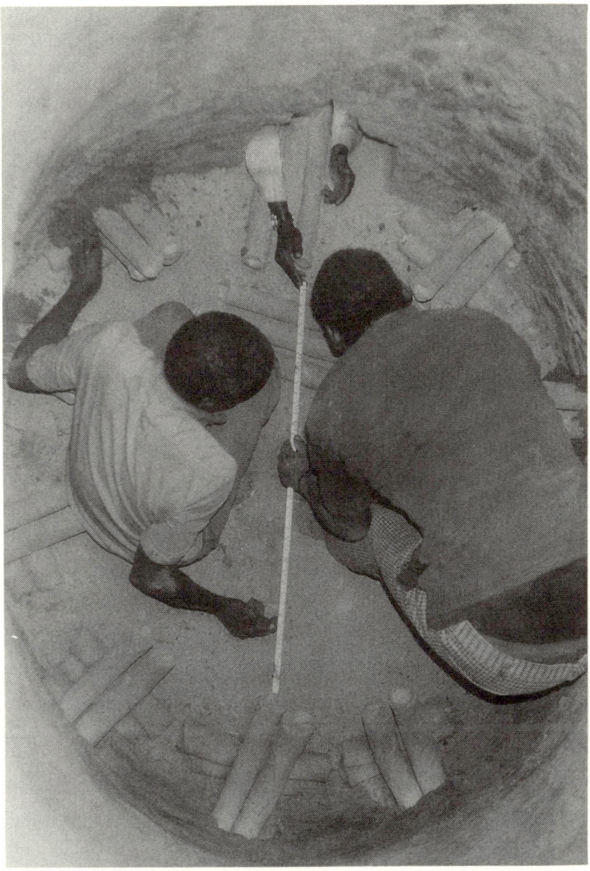

FIGURE 4.3. Positioning the tuyeres in 1991 before the first smelt.

facing east is named "the father door" (*palisi*). Above the eastern tuyere inlet is a small peephole called a star (*ntaanda*), which the smiths use to control the furnace temperatures and ore reduction. The furnace is charged with layers of charcoal, green logs of wood, and iron ore.

A secondary furnace (*icinteengwe*) is needed to remove slag and to consolidate the bloom. This furnace is 40 cm tall, charged with charcoal, and driven by three bagbellows inserted into tuyeres (see figure 4.4). The furnace has a rake hole, also called the palinyina, for slag removal in the front. Forging of the produced iron is carried out over open fire using two bagbellows. Similar ironworking technologies are known in adjacent regions, including among the Nyika (Brock 1965), Lungu (Chaplin 1961), and Phoka and Chewa in

Malawi (Davidson and Mosley 1988). It has been suggested that the tradition of iron smelting among these people should be called the Tanganyika-Nyasa corridor tradition, as both technological aspects and cultural traits are similar (Barndon 1992).

Iron smelting was practiced in Ufipa during the dry season from mid-June until the end of August. The Fipa smelters were traditionally part-time specialists who, like the rest of the population, provided for their own subsistence. Because iron objects, especially hoes, were considered things of value, blacksmiths were able to accumulate more wealth than the rest of the population. In the active smelting days, all the iron produced was owned by the master smith, who distributed iron as tribute (*ilaango*) to the royal chiefs (*mweene*) of Lyangilele and Nkansi and among his assistants. Master smiths were highly respected because of their knowledge of transforming ore into iron; master smiths often became village headmen. Unworked iron for a hoe or forged iron hoes were used as general currency (*ifyuuma*) and constituted an important part of bride wealth.

FIGURE 4.4. *Left,* a smaller *icinteengwe* refining furnace built by the Fipa in 1991. This furnace used a forced air blast. There are two clay extensions called the "inside of the thighs"; *right,* tap slag running between the thighs of the *icinteengwe* furnace.

Within a group of approximately ten to fifteen ironworkers (*asiluungu*) there was a hierarchical allocation of titles reflecting authority and skills. The master smelter (*umwaami*) was recognized as the owner of a basket (*intangala*) containing magical substances (*ifingila*, "things that enter") as well as of the chimney furnace and the iron produced. The status of a master smith was strongly associated with possession of the basket. The use of the ifingila, in addition to medicinal barks (*ingailo*) and a cockerel sacrifice, were considered necessary to build a furnace and conduct a successful smelt. The contents in the basket are believed to help the smelt turn out successfully; iron smelting without the use of magical substances is unattainable. The title *umwaami* was traditionally also assigned to the leader of an elephant hunt. The master smith's assistant (*unnsole*) represented the leader of the crew whenever needed. The assistant could then allocate tasks and arrange the organization of the work. Two elder men (*aluumba*) were the master smith's closest assistants and shared the knowledge and secrets of iron smelting with the master smelter and his younger assistant.

Rituals of Fipa Iron Smelting

Among the Fipa, as in general, rituals are separate from the general flow of life, much like iron smelting. Rituals differ from everyday communication and signification and are carried out for specific purposes (Bloch 1992, 46). Therefore, participants in rites and in iron smelting do something as well as say something (Richards 1956, 112). Rituals of iron making among the Fipa have been approached in this way, although not all actions and events in the process of iron making are rituals or are symbolically communicative. The use of magical substances during the iron-smelting process might be compared to curative rituals or hunting cults (Turner 1967), in this case not assuring a successful hunt but assisting a successful smelt. Smelting magic is considered an important and integrated part of ironworking technology, a part of the formula and technological behavior employed to produce forgeable iron. The actual content in the intangala, items such as pieces of skin, nails, shells, or bark, are not curative but instead possess the quality ascribed the idioms—the animal or stone the item represents. Thus the transferred qualities of these magic items will procure benefits and prevent harm during the smelting. Explaining why a piece of an elephant's ear is used, the smith said, "Even when the elephant sleeps, his ears are moving. We apply a piece of the ear since we want the fire inside the furnace to burn continuously." The master smelter is free to select from the basket the items he finds purposeful or necessary for the specific smelt. The basket is inherited over several generations and items are never removed, only added. Nobody knows the exact content. The knowledge of the basket's contents, like the entire process of iron smelting as it is both traditional and gradually adjusting to individual needs, is therefore an example of "culture in the making" (Obeyesekere 1981).

RITUALS AT THE FOUNDATION OF THE FURNACE (*ULUFULA*)

When smelting iron, the smiths move out of the village and settle in the bush; they therefore also move out of everyday life, something that creates a new social, symbolic, and spatial dimension in regard to technology and material culture production. They are told by the master smelter that they must remain faithful to the craft and refrain from sexual intercourse as long as they remain at the smelting site. Preparations for a smelt start when the contents of the intangala basket is poured out on a black he-goat skin called "the winnowing basket of the magical substances" (*uluupe uluua intangala*) and the smiths each color their foreheads with a red powder (*unnkolo*). The red powder is normally used in wedding ceremonies by the bride (*nawiinga*) and her relatives to symbolize fertility; during iron smelting it will mark those who are professional iron smelters.

Before the actual building of the furnace, some magical substances are cut into small pieces and placed evenly around the furnace base. A boy and girl who have not reached puberty and who are therefore "pure" perform the task. Assisted by the master smelter, they sacrifice a white cockerel (*nhkoko*). The cockerel's blood is sprinkled over the magical substances and the bark, and its head is placed in a black horn together with some bones of a lion. The cockerel's mouth points toward the east. It is called the furnace caretaker (*umpakasi iluungu*) and the smiths expect that it will mature and crawl like a hen during the smelting. The pieces of lion bones and medicinal bark will make the cockerel strong and fierce. When the furnace has been raised, one of the cockerel's wings is placed above the main tewel inlet and called the "caretaker for the mother door" (*umpakasi pamweela palinyina*). The building of the furnace can then start and after one day's hard work it is left to dry in the sun.

KANAWIINGA RITUAL

When a new furnace has been constructed, it is decorated with white flour and praised as a marriageable woman (nawiinga). Both the smiths and their wives dance around the furnace while they sing that the adulterous women of the village should come and see this beautiful bride, who is the one who will help them (*imwiila*). The furnace has now symbolically matured into a marriageable woman and the smiths are as her husbands celebrating the wedding. The furnace shape itself is not symbolically linked to women, but the songs, the ritual ceremony, the decorations on the furnace, and the linguistic terms given the furnace and the tuyere inlets are all relationally linked to a wedding.

When the chimney furnace is ready for charging, the master smelter will enter through the main door (palinyina) to arrange the tuyeres, ore, wood, and charcoal from the inside (see figure 4.3). The furnace is filled with fuel handed down to the master smelter in baskets from a scaffold, and the master smelter descends from the top after putting the furnace to fire. Entering the furnace, the master smith is symbolically entering the woman's womb, something reflected in his movements and the songs. When the smelting has started, a feast occurs, with songs and local brew, and the smiths sleep around the furnace during the night. The smelting of the ore and cooling of the furnace take approximately twenty-four to thirty hours.

ICINTEENGWE, "THE ONE WHO REVEALS THE SECRET"

The smaller furnace (*icinteengwe*) (figure 4.4) is given the assistance of special medicinal bark and pieces of animal bones, which are placed within the combustion chamber. In secrecy the master smelter, his assistant, and the two elder men will add some special bark with a white juice (*itoole*). Two elongated clay lumps are formed and placed outside the rake hole (figure 4.4). They are called the "inside of women's thighs" (*impaamba*). As discussed above, the chimney furnace represents a bride and a symbolic intercourse gives the furnace iron ore. The secondary furnace represents a mother giving birth and the iron is born inside the furnace chamber. During the process of smelting, the furnace will reveal the quality of the iron, secrets, and maybe unconfessed adultery. If the sexual taboos laid upon the smiths have not been followed, the iron will be impure (*untare*), and one of the smiths will be blamed for the unsuccessful smelt and accused of adultery (*insoni*).

The analogical linking between iron smelting and childbirth is common in sub-Saharan iron-smelting traditions (Herbert 1993). Among the Fipa this relational thinking might be explained by the fact that adultery is generally believed to affect fertility and viability. If a woman in delivery has problems and there is a threat of miscarriage, two midwives (*amaloombwe*) will demand that she confess adultery and reveal the name of the father. The two older men assisting the master smelter are analogous to the two midwives. During the technological process they are given special ritual authority and knowledge, and they arrange songs and rituals. The same secrets of adultery are revealed during iron smelting as in childbirth and the iron produced is metaphorically linked to a newborn child. The quality of the iron produced reflects the life/death dichotomy stressed during childbirth. In the general Fipa system of thought, sickness (*inciila*) caused by contact with adulterous people is viewed as one of the most dangerous diseases, and its invisible pains might lead to death. If one of the smiths admits that he has been unfaithful, he will be called "one of those who commit adultery" (*wachiluungu*). The term refers to a Fipa myth in which two Twachi (Twa) women came to Ufipa and cheated the Milansi chief into dividing his territory with them (Willis 1981). If the smiths can prove that they have been faithful during the smelt, they are told to look among their wives for the adulterous one. The informants claimed that in reality this rarely occurred, however, and the pure iron (*ululu*) from the small furnace was brought home to the village to be forged into iron hoes.

Socially Created Space and Technological Style

Space in itself has no meaning, but when socially constructed the meaning of space structures the way people use, manipulate, and interact within it (Moore 1986). The Fipa traditionally performed the process of iron smelting in the bush, including the preliminary work with ore collecting and crushing, charcoal preparations, and furnace building and charging. The ritual events in Fipa iron smelting create a specific cosmos that can be related to other spheres of Fipa social and cultural reality. It has been demonstrated that during iron production the smiths insert social and cosmological ideas into the technological activities in which they participate. At production sites still found in Ufipa it is possible to trace certain dimensional and spatial rules connected to the organization of the ironworking. The spatial organization of the sites is inevitably geographically significant in regard to raw materials available and technologically or functionally important in regard to the chemical processes that are needed to produce iron. But the dimensional aspects are also socially significant, as a socially created space reflects symbolic and stylistic features of the Fipa mode of thought employed in iron making. The ritual performances and dimensional features of furnaces and smelting sites are partly determined by the way that the Fipa conceptualize iron technology. Figure 4.5 illustrates spatial organization of a production site used during one of the smelts demonstrated in 1991. The same spatial arrangements were documented at old, abandoned smelting sites within a wider area of Ufipa, sites known and unknown to the informants (Barndon 1992). The dimensional features at the iron-smelting sites are relationally linked to procurement and fertility symbolism in the Fipa system of thought. The tuyere inlets on the chimney furnace face east and west, the cockerel faces east, and the slag removal and rake hole face west on both furnaces. These traits are not, technically speaking, necessary to produce iron, but within the system of thought applied they have meaning.

The Fipa secondary blast furnace receives anthropomorphic attributes of a woman, a tradition throughout much of Bantu-speaking Africa. It is often suggested that this system of thought can be related to concepts of heat and fire as a transformative agent (Eliade 1978; de Heusch 1982; Collett 1985; Herbert 1993). Among the Mashona, two elongated clay masses were often extended from the central rake hole at the furnace base, suggesting open human legs at the base of the belly (Childs and Dewey, chapter 7, this volume). More common is the modeling of female breasts on the furnace walls, such as among the Bena (Sutton 1985). As recorded by van der Merwe and Avery, East African ironworkers commonly prohibit women from the smelting site and from engaging in sexual intercourse during the smelting process (van der Merwe and Avery 1987, 143; Muhly and Schmidt 1988). The idea of sexual taboo is a cross-cultural phenomenon, possibly related to the fact that where metallurgical knowledge is highly respected it is commonly carried out in secrecy (Schmidt, chapter 1, this volume).

The use of magic, sexual taboos, and secrecy controls transmission of metallurgical knowledge. Female exclusion among the Fipa is ideal but not applied in many behavioral domains. For example, women participate in various parts of the work, such as preparing soil for furnace building, carrying water and charcoal to the smelting site, and celebrating a new chimney furnace. The smiths are more particular about preventing the women from having any contact with the iron ore or the smelted iron, however. Women, especially pregnant ones, are seen as a threat to the purity of the iron and to the ore's ability

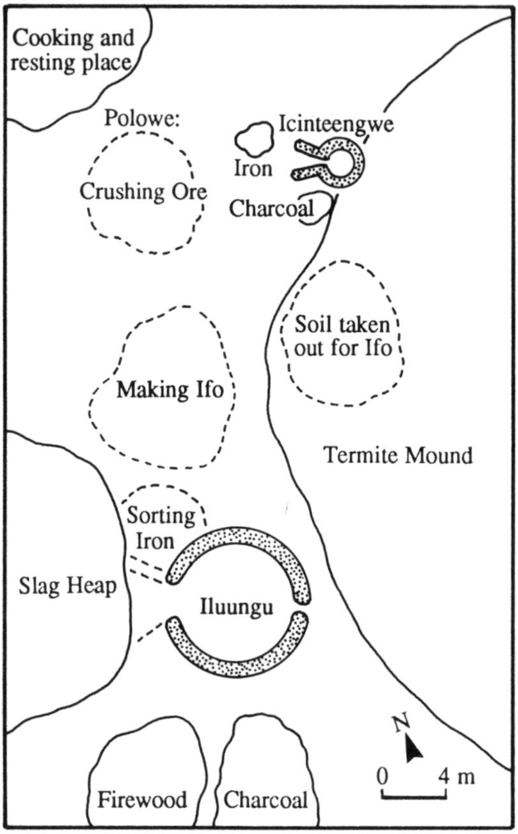

FIGURE 4.5.
The structure of the Fipa work space during iron smelting. This configuration, arranged according to cosmological principles, shows great continuity over the last century.

to produce iron. Women's contact with men and iron will weaken the smiths' strength, which they need for the iron smelting. The general prohibition against sexual intercourse during the smelting process is related to a mode of thought in which the smiths associate the furnace with a woman.

Within this system of relational thinking it is contradictory not to remain faithful to the furnace (woman). Female exclusion is further rendered meaningful because the Fipa smiths themselves are symbolically representing women when they color their foreheads with red powder as women do during wedding ceremonies, a liminal status that is also recognized among the Haya smelters of northwest Tanzania (Muhly and Schmidt 1988). The smiths are the husbands of a marriageable woman (the furnace) and exclude their wives from the iron-smelting process for this reason. In all actions and events during the iron smelting, the smiths use concepts and cosmological ideas from real life to prevent harm and provide forgeable iron.

Conclusions

African iron-smelting furnaces provide both a technical means to produce iron and a medium to express a variety of cultural messages. The cultural messages expressed in Fipa ironworking are related to norms of proper moral behavior in social relations. Adultery is feared because it can cause death in childbirth and impure iron. Childs (1991) has addressed factors that cause stylistic variation in African iron-smelting furnaces, and she relates the furnace decorations to other material items such as pots and grain bins. The spatial organization of the production sites themselves is part of these stylistic variations and should be addressed more cross-culturally and systematically.

Concepts of iron smelting will yield new dimensions to our knowledge of traditional iron technology if approached within a cultural framework. The present study has documented the material and spatial manifestation of the production sites as incorporated into iron smelting as a technological style. Repetitive technological behavior exhibits a specific style, but which ecological possibilities, social factors, and cultural processes are involved? If the organization of space at production sites can be recognized as "products of people's ideas" (Barth 1961), it is possible to establish new dimensions in the study of traditional ironworking. These new dimensions will be related to questions of cultural continuity or change in a certain region in regard to the iron-smelting technology practiced. Identifying ironworking within a cultural framework that exhibits a technological style will assist in interpreting how and why material culture is organized spatially in a specific way.

Note

I would like to thank Professor Randi Haaland, who first introduced me to African archaeology and ethnoarchaeology at the University of Bergen and who later introduced me to Tanzania and the Fipa. She generously included me in her project (NORAD Tanzania 064). My research was funded through a scholarship from NORAD and the Center for Development Studies at the University of Bergen. Permission to conduct fieldwork in Ufipa was granted by a research clearance from the Tanzania

Commission of Science and Technology. I want to especially acknowledge my local contact while in Tanzania, Dr. Seth I. Nyagava, coordinator at the archaeology unit, University of Dar es Salaam, and my fellow student Bilham J. Kimathi, who assisted in all possible ways and helped reduce the language barriers while in Ufipa. Finally, I owe all the people in Ufipa great thanks for their generosity and openness toward us.

References

Barndon, R. 1992. "Traditional Ironworking among the Fipa: An Ethnoarchaeological Study from Southwestern Tanzania." Master's thesis, University of Bergen.

Barth, F. 1961. "Diffusjon: Et tema i studie av kulturell prosesser." In *Kultur og Diffusjon,* ed. A. M. Klausen. Universitetets Etnografiske Museums Bulletin, No. 10.

Bloch, M. 1992. *Prey into Hunter: The Politics of Religious Experience.* Cambridge and New York: Cambridge University Press.

Brock, B. P. 1965. "Ironworking among the Nyiha of Southwestern Tanzania." In *Tanzania before 1900,* ed. A. D. Roberts. Nairobi: East African Publishing House.

Chaplin, J. H. 1961. "Notes on Traditional Iron Smelting in Northern Rhodesia." *South African Archaeological Bulletin* 16:53–60.

Childs, S. T. 1991. "Style, Technology, and Iron Smelting Furnaces in Bantu-Speaking Africa." *Journal of Anthropological Archaeology* 10:332–59.

Cline, W. B. 1937. *Mining and Metallurgy in Negro Africa.* Menasha, Wisconsin: George Banta Publishing Co.

Collett, D. P. 1985. "The Spread of Early Iron-Producing Communities in Eastern and Southern Africa. Ph.D. diss., Cambridge University.

Davidson, S., and P. N. Mosley. 1988. "Iron-Smelting in the Upper North Rukuru Basin of Northern Malawi." *Azania* 23:57–100.

de Heusch, L. 1982. *Rois nés d'un Cœur de Vache, Mythes et Rites Bantous* Vol. 2. Paris.

Derricourt, R. 1980. *People of the Lakes: Archaeological Studies in Northern Zambia.* Zambian Papers, No. 13. Manchester: Manchester University Press.

Eliade, M., ed. 1978. *The Forge and the Crucible: The Origins and Structures of Alchemy.* Chicago: University of Chicago Press.

Gosselain, O. P. 1992. "Technology and Style: Potters and Pottery Making among Bafia of Cameroon." *Man* 27:559–86.

Goucher, C. L. 1981. "Iron Is Iron 'Til It Rust: Trade and Ecology in the Decline of West African Iron-smelting." *Journal of African History* 22:179–89.

Haaland, R. 1980. "Man's Role in the Changing Habitat of Mema during the Old Kingdom of Ghana." *Norwegian Archaeological Review* 13:1–46.

——. 1985. "Iron Production: Its Socio-Cultural Context and Ecological Implications." In *African Iron Working, Ancient and Traditional,* ed. R. Haaland and P. Shinnie. Oslo: Norwegian University Press.

Herbert, E. 1993. *Iron, Gender, and Power: Rituals of Transformation in African Societies.* Bloomington and Indianapolis: Indiana University Press.

Ingersoll, D. W. 1987. Introduction. In *Mirror and Metaphor: Material and Social Constructions of Reality*, ed. D. W. Ingersoll and G. Bronitsky. Lanham, Md.: University Press of America.

Kense, F. 1983. *Traditional African Iron Working.* Alberta: University of Calgary Press.

Kent, S. 1984. *Analyzing Activity Areas: An Ethnoarchaeological Study of the Use of Space.* Albuquerque: University of New Mexico Press.

Lechtman, H. 1977. "Style in Technology: Some Early Thoughts." In *Material Culture: Style, Organization and Dynamics of Technology*, ed. H. Lechtman and R. Merill. St. Paul: West Publishing Co.

Maret, P. de. 1985. "The Smith's Myth and the Origin of Leadership in Central Africa." In *African Iron Working, Ancient and Traditional*, ed. R. Haaland and P. Shinnie. Oslo: Norwegian University Press.

Moore, H. L. 1986. *Space, Text, and Gender: An Anthropological Study of the Marakwet of Kenya.* Cambridge and New York: Cambridge University Press.

Muhly, F., and P. R. Schmidt. 1988. Script for *The Tree of Iron.* Gainesville: Foundation for African Prehistory and Archaeology.

Obeyesekere, G. 1981. *Medusa's Hair.* Chicago: University of Chicago Press.

Pfaffenberger, B. 1992. "Social Anthropology of Technology." *Annual Review of Anthropology* 21:491–516.

Richards, A. I. 1956. *Chisungu: A Girl's Initiation Ceremony among the Bemba Of Northern Rhodesia.* London: Faber and Faber.

Schmidt, P. R. 1994. "Historical Ecology and Landscape Transformation in Eastern Equatorial Africa." In *Historical Archaeology: Cultural Knowledge and Changing Landscapes*, ed. Carole Crumley. Santa Fe: SAR Press.

Sutton, J. E. G. 1985. "Temporal and Spatial Variability in African Iron Furnaces." In *African Iron Working, Ancient and Traditional*, ed. R. Haaland and P. Shinnie, 168–91. Oslo: Norwegian University Press.

———. 1990. *A Thousand Years of East Africa.* Nairobi: British Institute in East Africa.

Turner, V. 1967. *The Forest of Symbols.* Ithaca: Cornell University Press.

van der Merwe, N. J., and D. H. Avery. 1987. "Science and Magic in African Technology: Traditional Iron Smelting in Malawi." *Africa* 57:143–72.

Wembah-Rashid, J. A. R. 1969. "Iron Workers in Ufipa." *Bulletin of the International Committee of Urgent Anthropological Research* 37:425–52.

Willis, R. 1981. *A State in the Making.* Bloomington: Indiana University Press.

Wise, R. 1958. "Iron Smelting in Ufipa." *Tanganyika Notes and Records* 51:106–11.

5 Reconfiguring the Barongo: Reproductive Symbolism and Reproduction among a Work Association of Iron Smelters

Peter R. Schmidt

"Every sacrifice seems to activate the process of reproduction."

—*de Heusch 1985*

"At 3:30 Rwamakoga and Masaka bring out the remaining bellows and seal them with cow manure. This is done in a matter of five minutes. Then the last bellows is bound (the bamboo strips lining the outside of the clay tube are tied tightly around the tube) by Kikundi and Munubwa. As I look around the compound and observe the children over by the main house, I see one of eight years carrying a baby and at the same time combing and braiding the hair of a friend while she struggles to keep her balance with the baby on her back. Behind her is a neighbor woman who has just finished helping to pound cassava in a mortar. She bends over, her infant on her back, and inspects her long, pendulous breasts—picking off something, brushing them roughly, and then binding them with the same cloth that holds her child to her back. The flies softly buzz in the background while Mbuzehose makes sisal string on his knee and Masaka's wife sorts beans, throwing them into an aluminum pan. A rooster struts between the two groups, chasing small chicks out of his way while Avery reads his book on the history of World War II."

—*Field notes, among the Barongo, June 26, 1979*

This is a story about bricolage, or fabrication from whatever is at hand, among the Barongo ironworkers south of Lake Victoria. The Barongo are a group of smelters and smiths who first came to the attention of the Western world through travelers' descriptions of their workshops. Over the years, the Barongo have come to be characterized by European descriptions about their social organization, beliefs, and practices, such as their use of an extensive ritual pharmacopoeia (de Rosemond 1943) and their successful participation in a regional trade system (Kjekshus 1977). While these observations do in fact apply to the history of the Barongo, alone they do not reveal the historical importance of the Barongo or the significance of the Barongo system of production to building a larger understanding of iron production in Africa.

The Barongo ironworkers fabricate iron, they fabricate ritual, they fabricate processes of production, and they fabricate reproduction. As bricoleurs, they participate simultaneously in a number of transformational domains—myth, ritual, healing, and iron smelting. A profoundly innovative spirit of bricolage prevails in Barongo iron culture. The fluidity of the Barongo experience—from smelt to smelt—bears witness to a significant departure from notions of formulaic approaches to iron smelting. The Barongo are bricoleurs who, every new day that they are smelting, try different permutations, different innovations in their attempt to find a viable (re)productive pathway.

Claude Lévi-Strauss captures the inspiration of the bricoleur when he notes that to make a "decision as to what to put in each place . . . depends on the possibility of putting a different element there instead, so that each choice which is made will involve a complete reorganization of the structure, which will never be the same as one vaguely imagined nor as some other which might have been preferred to it" (Lévi-Strauss 1966, 19). Lévi-Strauss, of course, draws a complex metaphor between the technological bricoleur and the myth maker, a relationship that takes on a metonymical quality with the Barongo bricoleur, who simultaneously participates in the contiguous domains of technology and ritual.

I conducted fieldwork among the Barongo on three separate occasions during 1979, 1980, and 1984. During these periods, the Barongo performed eight different smelting operations, each of which differed significantly from the others. While it is not possible to set out the details of each smelting operation, this essay will focus on the bricolage that is manifest within the repertoire of the Barongo. Any reconstructive study of this sort creates tensions and responses that are alien to the normal lives of these smelters, who in 1979 had not practiced their craft in twenty-eight years. In this respect a critic can claim that whatever innovation or bricolage occurred simply resulted from shredded memory and lost skills. There is some validity to this observation, and without doubt a goodly portion of the scrabbling and searching for solutions described here derives directly from the passage of time and the practice of an economic activity that has fallen into disuse. But there are also important insights to be gained from decisions to use specific combinations of ritual procedure or technological sequence. As the Barongo bricoleurs dipped into their reservoir of ritual and technological process, they continued to fabricate distinctive solutions, many of which provide invaluable understandings of technological experimentation and ritual meanings.

Who are the Barongo? Although this seems a simple question, its answer has been obscured over the years. Much misunderstanding of Barongo identity certainly has penetrated the written accounts of these ironworkers. One of the earliest accounts is by de Rosemond, a British district officer who published a short vignette in *Tanganyika Notes and Records* (1943). His report was stimulated by an account, published six years earlier, by Greig (1937), of the iron-smelting rituals of the Fipa. De Rosemond structured his account of the Barongo of Kahama District similarly: the first section of his report is devoted to beliefs surrounding ironworking. He ventures the "opinion" that Barongo iron smelters had "been initiated into the knowledge of the secret *dawa* [*medicine* in Kiswahili; *omubazi* in Kizinza] for smelting" (1943, 79). Only those initiated into the secret of the dawa may smelt; others are relegated to forging activities. Finally, he claims that the name *Warongo* applied generally to all of the people living south of Lake Victoria. All of these claims deserve examination because they impinge on our understanding of Barongo culture and identity.

The term *Barongo* certainly included both iron smelters and forgers. There is no evidence to suggest that all iron smelters experienced a formal initiation into the secrets of the rituals applied to iron smelting. To the contrary, the Barongo incorporated many children and young male adults who acted as apprentices, sometimes for many years. As time passed, some of these young people would also serve simultaneous apprenticeships in the ritual processes. Two of the smelters who participated in the smelts conducted between 1979 and 1984 were ritual specialists, but the other smelters were no less Barongo than these specialists. As one seventy-six-year-old smelter remarked in 1979, "The Barongo are those who pumped the bellows." De Rosemond understood correctly that the smelts were conducted by elders who possessed complete ritual knowledge, but he erred in thinking that the other participants were not also identified as Warongo.

It is also apparent that de Rosemond's claim that all people south of Lake Victoria came to be called by the same name completely missed the nature of Barongo identity. There are many different groups living in the quadrant that runs from Emin Pasha Sound to the southwest corner of Lake Victoria, south to the Malagarazi River, and east to Kahama—the Basubi, the Bazinza, the Baha, the Basumbwa, the Basukuma, and the Banyamwezi, among others (map 5.1). Confusion over Barongo identity has continued in more recent attempts to "construct" who they were. For example, Lemelle (1992, 177) questions informant testimony obtained by Nchotti (1975) that the Barongo were an open group, admitting anyone who was willing to pay the imposed fee. He mistakenly assumes that this openness to membership somehow marked the Barongo as different from groups in other ironworking regions, "where iron smelting was enshrouded with mystery" (Lemelle 1992, 177). In the case of the Barongo, ritual elaboration remained rich until very recently, suggesting that open membership is not related to less ritual activity.

The testimonies gathered by Nchotti (1975) in Geita district about Barongo identity as a work group are precisely the Barongo voices for which Lemelle is searching. But like many studies that are limited to one geographical area, confined to one ethnic group, Nchotti concluded that the Barongo were made up only of those claiming only Bazinza ethnicity. Lemelle (1992, 171) is not able to reconcile this dual identity—that is, an iron worker who is both Murongo and Muzinza, and further confuses the question of Barongo identity by referring to them as a "discrete people" for reasons of expediency. This leads him to err again when he concludes that the Barongo "were a set of lineages of Zinza origin which took on a separate and distinct ethnic identity" (Lemelle 1992, 171). This is a rather interesting position: Lemelle makes the Barongo into a separate ethnic group, a process that is identical to colonial creations of ethnicity that he faults in which "tribes like the 'Warongo' were in many instances colonial constructs" (1992, 171).

In this instance emic definitions provide the greatest clarity. As one smelter remarked, the Barongo are not a tribe (*kabila* in Kiswahili) but are a functional group that shares the same language. Similar descriptions from others confirmed the idea that the Barongo are an industrial and trade guild or work association with a common lexicon, *kirongo*. Kirongo is a technical vocabu-

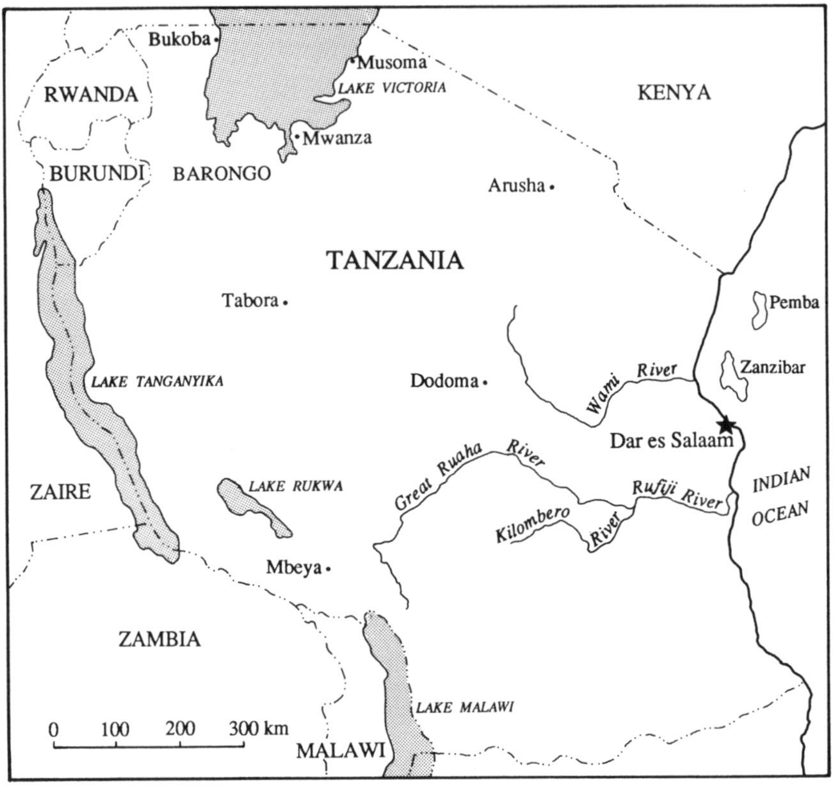

MAP 5.1. The Barongo iron workers are situated to the southwest of Lake Victoria.

lary to which the smelters and forgers refer when they identify themselves as Barongo; one is a Murongo (sing.) if one speaks kirongo. It is not a language per se but a set of technical and specialized terms applied to the ritual and technical processes that encompass iron smelting.

The Barongo who lived between 1979 and 1984 in Butengo village, near the border of Biharamulo and Geita Districts, came from many different origins. Butengo itself was then a new village, the product of villagization—the Tanzanian resettlement scheme of the mid-1970s. Most Barongo had previously resided in the area around Bwanga, a crossroads community a dozen miles to the west along the road between Geita and Biharamulo. While the Barongo were known to be concentrated in the Geita District during the late nineteenth and early twentieth centuries, the eastern part of Biharamulo District (contiguous to Geita) was also a primary locale because of its plentiful resources.

The Barongo of Butengo traced their ethnic origins to a number of groups or subgroups within central, western Tanzania: the Bazinza, the Basumbwa, the Basubi, the Baha, and the Bahaya. The Basubi of Biharamulo District origin seem to have constituted the largest number among the Butengo smelters. Some smelters also attest to multiethnic backgrounds, for example as children of marriages between Basubi and Basumbwa. This quality of plural origins within the group as well as mixed marriages is quintessentially Barongo. For the most part Barongo smelters/smiths followed their fathers into the trade, but some came into iron smelting and forging through apprenticeship outside the family.

One smelter in the Butengo group gained status as a Murongo at eighteen years of age. His admission took a distinctive route. First he participated in the salt trade to the east in Sengerema, where he would buy a pan of salt and later trade it at home for fifty-five used hoes, which he would reforge into twenty-five new hoes. He would trade them at two hoes for one goat, and then return to Buyovu (an area near Karutage Hill, the iron-ore source), where the goats would be sold at six to eight shillings each. He was trained as a Murongo by a friend of his father's and smelted for three years starting in 1938. He left iron smelting, he said, because he detested and feared the molten slag flying out of the iron bloom while it was being forged. He remembers that there were eight men in his work group at Karutage,[1] including those who fetched water and cooked. As many as fourteen *emitindi* (sing. *kitindi*) or smelting houses operated simultaneously at Karutage during this period of the early World War II era. He emphasized strongly that he was a Murongo only insofar as his ironwork was concerned.

Beyond the esoteric language they spoke, other characteristics separated the Barongo from other populations and contributed to their mystification. During the smelting season, they resided and worked deep in the forest, also hunting for game. They were viewed with suspicion and seen as somewhat dangerous, marked by mysterious rituals, special powers, and their nakedness during smelting.

Barongo emic commentary about who they are resembles but is not identical to the Western institution of a craft association: a group with different grades of skilled workers, such as apprentices, master smelters, and smiths/smelters; open membership unrestricted by family, lineage, or ethnic group; participation in the control of trade and preservation of the craft only for members of the Barongo.

The Barongo, as the collective members of a multiethnic work association, seem to have arisen south of Lake Victoria during the era of slave trading in that region in the mid-nineteenth century. The region was subject to the deprivations of slavers as well as of other groups, such as the Ngoni, who harassed Ha and Zinza territory as early as the 1850s, and the Baganda, who actively engaged in local disputes in the 1880s (Roberts 1969). Internal disputes and dislocations within the region occurred during the 1880s in Buha, while both the Mirambo and the Ngoni raided Buzinza during the 1870s and 1880s. Buganda actively interceded in the affairs of Buzinza and Busubi (to the west) during the 1880s and helped to conquer the Uyovu chiefdom in Busumbwa (immediately to the south of Buzinza proper) (Roberts 1969). Slave caravans in the trade passed through Busubi, within Barongo territory; consequently, the products of the Barongo smelters were easily traded for goods flowing to and from the coast and converted to considerable profits. The disruptions and dispersals of many different craft groups at the edge of the caravan trade created the conditions for the rise of the Barongo. Many Barongo appear to be a product of those times, essentially gatherings of different displaced iron smelters from various ethnic groups in the region, working under secret conditions within the deep forest, and producing goods used in the caravan trade. At the same time, other Barongo groups, particularly those of Geita, appear to have been made up of ironworkers who also claimed identity as Bazinza—a claim consistent with the plural quality of Barongo identity.

The Barongo attest to the earlier presence of other ironworkers in the region sometime before the twentieth century. Although they did not trace any direct ancestry to the Bayovu, when the Barongo ironworkers traveled south to Karutage Hill to mine iron ore they constantly referred to the Bayovu (of Uyovu chiefdom), a group of ironworkers who once lived there. In the area surrounding Karutage Hill, Barongo ritual incorporates prayers and special offerings to Bayovu spirits that must be appeased before iron ore is mined in the area.

Trade

The market history of the Murongo recounted earlier indicates a fair degree of prosperity among the Barongo who produced iron and then participated in regional trade. While it is difficult to reconstruct precise configurations of the trade networks in which Barongo goods circulated before the twentieth century, it is highly probable that many Barongo hoes ended up in Uganda and in the Tabora market, where they were traded for slaves and used in trade as caravans moved to the coast. After the collapse of the caravan trade at the end of the nineteenth century, Barongo production seems to have continued to satisfy significant regional needs.

The testimony of smelters who were active in production and trade during the 1920s, 1930s, and 1940s provides important insights into which goods were marketed and how much profit the Barongo made. While Barongo smiths made a variety of iron goods, such as axes, spears, adzes, knives, and specialized agricultural tools for the local markets, their preeminent product for trade was the heart-shaped hoe. There are clear reasons why the Barongo process continued longer than, for instance, iron production in Buhaya to the northwest or in other areas of eastern Africa. First, the Barongo serviced a zone where transport was difficult, roads were few, and scrap steel and manufactured goods of European origin were difficult to obtain. Second, the Barongo worked in an area that was rich in iron ores and rich in forest resources—predominantly thickly wooded *miombo* woodland and a relatively light population density. Finally, because the Barongo process existed during the 1930s, it was easy for them to respond to colonial government directives to increase production during World War II, when the number of European-manufactured iron goods dropped off precipitously. Production among the Barongo increased significantly as demand throughout the region for locally produced goods increased during the war years.

The most detailed testimony about trade comes from the smelters themselves, particularly from one who was born in 1903. When this man was active in smelting and trading the Barongo hoes during the 1920s, he produced up to thirty hoes a month over a ten-month period. This level of productivity normally occurred in a work group of six to eight men, who would cooperatively mine the iron ore, cut wood, make charcoal, and conduct the smelts. It is also apparent that many young men—some not more than ten years of age—worked in the Barongo industry. One smelter noted that his father and grandfather would beat him on his arms to toughen him so that he would not tire during the long and grueling work.

The profits seen by the Barongo can be measured by the experience of the smelter who worked the trading circuit during the 1920s. This trade was mostly directed primarily toward Burundi, where deforestation was virtually complete by this time. The smelter would carry fifteen large hoes on his head from Buzinza. Once in Burundi he would trade the hoes for goats and cattle, ten hoes fetching a cow, five a bull, and one a goat. An emphasis on goats evidently lent itself to a quicker turnover. Upon his return to Bwanga, he would sell each goat for a minimum of three hoes, thus turning a 200 percent profit for a round trip to Burundi. If he traveled with his goods to Sukumaland to the east, then he could purchase a goat for two hoes, a 50 percent return. According to this account, the trade in hoes to Burundi and Sukumaland proceeded actively during the smelting period, with various smelters rotating responsibility. This particular man traveled to Burundi every month and to Sukumaland every other month. This trading cycle was based on his personal yield of three smelts a month, or approximately fifteen hoes.

It is clear that the distance traveled in trade yielded different profits but that longer trade also cut down on the number of hoes that could be produced per smelter. The head smelter during our observations, Masaka, claimed that seven smelts were held each month and that thirty hoes would be produced in a month per smelter. This scale of production, which occurred during the 1940s, did not depend on longer distance trade to regions such as Burundi. Instead most trade seems to have been localized, with most hoes flowing to the east into Sukumaland.

How many hoes could be produced from each smelt? Testimony varies widely, but a fairly consistent number is between five and twelve, with some testimony claiming as many as sixteen to twenty. De Rosemond (1943) relates information that as many as twenty hoes resulted from one smelt. It is important to keep in mind that children and apprentices would not have been provided a full share. Average production likely lies somewhere in between these figures. As each smelter was allocated the products of a smelt on a rotational basis, approximately thirty-six smelts (at five hoes per smelt) to twelve smelts (at fifteen hoes per smelt) would have been required to obtain sufficient bloom for the 180 hoes a work group produced in a month. It is indeed possible that each smelter would have been allocated the product of seven smelts per month and that production could have exceeded thirty smelts per month. Our observations show that it is technically possible to conduct two smelts in a long day, given the ease with which the Barongo furnace is constructed and dismantled.

The level of production near Karutage in the 1940s suggests that the approximately fourteen work groups in residence were turning out 25,000 hoes per year from approximately 40,000 to 50,000 kilos of bloom. This is not an improbable picture; during the mid-nineteenth century as many as 150,000 hoes were sold annually at Tabora for the caravan trade (Kjekshus 1977, 90). The full scale and impact of the industry can be better imagined when it is realized that many groups of Barongo were scattered throughout the miombo woodlands of Buzinza, Busumbwa, and Busubi.

In the twentieth century, then, the Barongo appear to have prospered in meeting regional demands, particularly during the war years, when demand for iron goods increased significantly. The rise in demand is read in the lower prices for bulls or oxen during this period—only three hoes. Lemelle (1992) paints a considerably different picture of Barongo fortunes during the early twentieth century. He argues that the discovery of gold in Geita occurred in the hills used by the Barongo to mine iron ore[2] and that the gold-mining company, in collusion with the colonial government, prevented the Barongo from exploiting those iron-ore mines. This led, he thinks, to the collapse of iron smelting among the Barongo in the 1930s. We encountered no testimony from smelters about such events; these poorly documented reports need to be confirmed by independent archival research and field research within Geita District.

Without question the most important observation about Barongo fortunes in the 1930s is the extensive revival of the industry stimulated by a severely diminished supply caused by the war effort. The prosperity issuing from the war experience continued into the postwar era. With one exception, all the Barongo smelters with whom we worked were deeply involved in this phase of production. Even when commercial iron goods began to reappear on the market, the Barongo continued to satisfy local and regional needs. They had so successfully positioned themselves in the marketplace that it was only by threat of force that the colonial government was able to suppress Barongo industry in the early 1950s. As one smelter recounts, people were directed to buy the *mamba jembe*, or the crocodile brand name hoe from local shops; smelters were threatened with jail if they continued to smelt. As a result, hoards of hoes were buried in the forest near Karutage so that violations of the decree would not be discovered (Kulahira, in Butengo, June 26, 1979).

The Barongo Study Introduced

The Barongo as iron smelters and smiths in 1979, 1980, and 1984 provide important insights into technological bricolage. Though they had not worked iron on a regular basis since 1951, they readily and eagerly took up the challenge to return to their old ways. Part of their willingness rested squarely in their nostalgia for the past. As a people recently displaced by villagization, alienated from their productive homesteads, and longing to return to their old ways, they gladly accepted the proposition to smelt again. These people regularly walked long distances back to their old homesteads to work the soil and to pay tribute to their ancestors. The idea of iron smelting, incorporating as it did the return to the sites that they habituated during their youth, closely meshed with their need to connect with the land of their past.

One of the first tasks the Barongo undertook was to mine the clay to produce the clay blowpipes or tuyeres called *nkelo*. Rather than return some distance to the clay source used during the 1940s, the head smelter took his workers to a swamp a mile east of Butengo village. The clay was loaded into *milindo* (pl.) baskets and carried to the village, where a pit that was 75 cm deep and wide was excavated slightly to the west of the head smelter's house. They lined the pit with banana leaves and then began the first ritual initiating the smelting cycle: a black powder, *rukago* (or *orukago*, sing., *enkago*, pl.), was sprinkled on the bottom of the pit by the head smelter. Within the region, rukago is a protective medicine used to overcome witchcraft or other malevolent forces (Cory 1949; Bjerke 1981). The head smelter said a short prayer in which he identified this as his work and enjoined that no stranger come to spoil it—followed by the sign of the cross. The syncretic nature of the twentieth-century Barongo rite is seen in the Christian-type prayer combined with the use of a traditional transformational device.

Once the smelters had placed the clay in the pit, they pounded strips of omukoma bark in a mortar and soaked them in water, creating an opaque fluid that they then sprinkled and poured on the clay. This substance, which eventually covers the clay, acts as an electrolyte to bind the clay together. The clay matured in this pit for a five-day period, after which it was removed and formed into thick cylindrical strips for molding. In the first case, however, the clay was contaminated by rain and the smelters added a dry sandy clay to it by pounding and mixing it in mortars. The clay was then molded around a stick called *mulotero* (or *omulotero*, sing., a distinctly kirongo word), which was first lubricated by the slippery bark and then covered in ash. As the smelters worked the clay around each mulotero, they continually dipped into a pot holding a solution from the omukoma bark, wetting the clay with this substance to make it easier to work (figure 5.1).

FIGURE 5.1. Forming a tuyere, using ash as a lubricant on the stick around which the tuyere is formed.

Tuyeres produced in this manner are 60 to 70 cm long with internal diameters at the distal mouth of 3.0–4.5 cm. The tuyeres were allowed to dry in the sun, but when they cracked rapidly the smelters applied ash to shield them from direct sun and to draw out the moisture. The smelters were then confronted with the problem of insufficient numbers of wooden, double-drum bellows, *mizuba* (pl.). Like the bellows of the Haya to the north, these are two-drum bellows carved out of one piece of wood. The smelters explained that when faced with such shortages in the past, they commonly made bellows of clay (*vichunga*, pl.). Thus the same day that tuyeres were made, work began on the manufacture of bellows; bowls were made with coils, much like a pot, and then attached to the still-plastic clay of the tuyere.[3] Three days later, the

FIGURE 5.2. Special clay bellows were made by attaching tuyeres to a clay bowl; the longer section was then reinforced with bamboo strips and sealed with cow manure.

tubes of the bellows were covered with bamboo strips that were shoved into the wall of the bellows bowl. Banana bast was slipped beneath the bamboo splints, which were bound to the tube to reinforce the fragile clay and then covered with cow manure to provide a seal against any cracks that should develop (figure 5.2). While bellows and tuyere manufacturing was continued by the older smelters, who confidently shaped them, the other smelters began to cut poles and wood to construct the *kitindi,* or smelting house.

Harvesting the wood for the smelting house took the better part of two days, with the smelters ranging up to a kilometer from the smelting site, which was located a quarter kilometer across and south of the Geita Road. Many different species were used, but finding the correct size trees was difficult in an area that has been cut back severely over the last five years for charcoal production for the urban marketplace. Construction of the 5.5 m/diam structure took only a half day, with the upright forked posts positioned in a circle around a mostly flat anthill. A horizontal member, an *embazi,* ran through the upright forked posts, and approximately sixty-five rafters were tied to this post with rope made of bark. A small lean-to was built to the eastern side of the kitindi to store charcoal and bellows. The binding of the rafters to more horizontal members and the tying of grass to the rafters were completed during part of the second day (figure 5.3). Meanwhile, many of the initial batch of tuyeres made from the specially mixed clay had broken and cracked, an event that precipitated much discussion about an alternative locale remembered as the former source of clay for smelting tuyeres. The felling of a large ironwood tree for charcoal also occurred during this period, as did the difficult cutting of the dense wood.

Nine days before actual smelting began the smelters wanted to travel to Karutage Hill, where they had mined iron ore (*embale*) and smelted iron during their youth. Because they have been resettled and the site is about 25 miles distant, the smelters agreed to visit the site only if transported there. At Karutage there is a 2 to 3 km walk to the mining site, where a banded ironstone is obtained from an eroded rubble at about 60 cm depth. While walking through this wooded savanna we encountered several heaps of tuyeres, fired termite earth, and slag, the waste heaps of smelters who worked the area several decades earlier. Most interesting from an archaeological point of view was the relative paucity of slag on these sites in view of the testimonies about the length and volume of smelting. One of the smelters in the group pointed

FIGURE 5.3. A *kitindi* smelting house built in 1979 by Barongo iron smelters.

out places where smelting groups practiced their craft during the pre–World War II period when he personally worked there. The reason for the low archaeological visibility of slag relates directly to the type of technological process used in Barongo iron smelting, as will be discussed later.

Before mining at this site, Misema Forest at Karutage, south of Bwanga, Biharamulo, the head smelter, praised the spirits of the Bayovu, asking these ancestors of former ironworkers in the area to grant them peace in using the iron ore.[4] The iron ore was transported out of the forest by suspending it in containers slung on a pole carried over the shoulders of two smelters (figure 5.4). In the same area, small termite mounds that stand approximately 1 m high were broken off and shaped into *empiga*, the flattish ovals that serve as the major component in the furnace wall, highly refractory and with pores and interstitial spaces that are natural insulators.

Taboos

The head smelter related that since the day that he placed the rukago medicine into the pit in which the tuyere clay was maturing, he had not had sexual intercourse with his wife. We had learned through our prior experience with the Bahaya smelters that, if asked direct questions about taboos and specific beliefs surrounding iron smelting, smelters often give idealized responses to such queries. With the Barongo we waited for spontaneous commentaries or ritual performances to understand the application of taboos. Thus we learned that the taboo requiring a period of sexual abstinence begins among the Barongo at the first step in the technological process, which also happens to intersect with the first application of ritual medicinal elements.

During the day that construction of the kitindi was completed, the head smelter began to discuss the other steps in preparation for the actual smelting. Among other issues, he explained that on the day before the first smelt occurs, the furnace pit is excavated and ritually consecrated. On that day he and his chief assistant would go out to obtain special medicine. For the following five days, neither man could have sexual intercourse. Should either of these two violate the five-day taboo, the smelter warned, then the smelt would be spoiled and no iron obtained. In addition, anyone who placed bad medicine in the furnace would become ill or die. The head smelter emphasized that all others are free to have sexual intercourse.

This explanation about a taboo on sexual activity (applied to the head

FIGURE 5.4. Barongo smelters transporting iron ore from the Karutage mining site.

smelter and his assistant) meshes with what is known from virtually hundreds of other iron smelting cultures (see Herbert 1993), including the Fipa (Barndon, chapter 4, this volume), who draw out the metaphor that violation of the taboo will cause an abortion of the fetus, as adultery is seen to cause miscarriage. But the absence of a prohibition of sexual intercourse for other smelters is a significant departure from the almost-universal application of this belief among iron smelters in the interlacustrine area, elsewhere in Bantu-speaking Africa, and in much of the continent (for example, see chapter 13, this volume). In some respects, Barongo proscriptions about sexual intercourse are closer to iron smithing, in which reproductive symbolism also prevails, but it does so in ways that differ significantly from smelting (Herbert 1993, 118).

Barongo beliefs are also distinctive in their treatment of women and the smelting process. Women can touch tuyeres and iron ore and can enter the smelting house without violating social taboos. They normally do not enter the house and observe the work; in the past they stayed away out of fear, said the head smelter, so that they would not be faced with a situation in which the men are naked during their work. (When the men wore skins, they would remove them during work so they would not be spoiled.)

There are many other instances in Africa where women mine the ore, carry the ore, and help in a number of other tasks. There are also several notable examples, such as the Kwanyama of southern Angola, in which the women pumped the bellows and helped to conduct the smelting (Estermann 1976,

148; Herbert 1993, 29). The explicit incorporation and tolerance for women bellows-pumpers among the Barongo is a significant sign that the Barongo belief system and technological process depart from normative representation.

Moreover, the Barongo smelting ritual lacks the common taboo against menstruating women, a belief that is seen by some scholars as a metaphorical contradiction to fecundity: women in menses are at the height of their sterility, and any contact with the furnace will sterilize the smelting process. However, the Barongo provide an important lesson about the erroneousness of such an interpretation. Menstruation is but one of a number of transformative processes in human reproduction; it is an essential component of the female reproductive cycle, and its exclusion in other smelting rituals in Africa has presented an odd contradiction that has confounded most observers (Herbert 1993, 67). Never, during three separate periods of studying the Barongo, was there any reference to the belief that menstruating women have a destructive effect on the fecundity of the furnace. It is also noteworthy that de Rosemond (1943) makes no mention of such a taboo; given the popular interest in such matters at that time, the absence of such an observation is important. After several seasons of work with the smelters, specific questions on the matter elicited only denials that menstruation in the presence of the furnace was a concern. In view of the fact that much of the ritual surrounding the Barongo process pertains to human reproduction and the representation of the furnace as a fecund bride ready to give birth to a child/bloom, it might be expected that a taboo against menstruation—the height of infertility (Muhly and Schmidt 1988; Herbert 1993)—would also constitute an important part of the symbolic repertoire. The absence of the menstruation taboo among the Barongo opens a significant window of understanding in the construction of other systems of reproductive symbolism associated with iron smelting. It also provides important understanding of how the Barongo bricoleur has fabricated a system of meaning in ethnically diverse circumstances that still fits within a reproductive paradigm yet also accommodates variation and self-expression. This question will be addressed in a later section in which the Barongo practice of ritual bricolage is discussed.

Before Smelting

More than a week before the first smelt was scheduled, the head smelter organized a short expedition to an area about two km southwest of Bwanga where he had once smelted iron from ore that had been carried from Karutage Hill.

About 400 m into a cotton field and on the top of a small knoll we found pieces of slag, tuyere, and burned termite earth. This site was immediately excavated to obtain porous-looking slag called *muselo* (expensive, higher value; *msilo* in de Rosemond 1943). Fifty-five kg of muselo were harvested from this site, along with thirty-six pieces (26 kg) of forging slag, initially called *entoto* (semifluid form, muddy), obtained from a contiguous forging site. The head smelter explained that these weathered pieces of slag would later be re-smelted when they were mixed with the iron ore. In some cases we could observe well-oxidized pieces of iron in the slag matrix. The smelters excavated a piece of a smaller dolerite hammer called an *ensule*, used in the final shaping of tools. Discovery of this fragment elicited a long discussion about the need to arrange for the *muhimbo* hammer, the large granite hammer used to express slag from the metal in the first stage of smithing.

Within Butengo village there were a variety of tasks underway, including the sorting of slags to be smelted and the working of skins by applying butter and rubbing them with a piece of forging slag and an iron tool (*linyukiso*). They also had to carve the *mpimpi*, or bellows sticks, and to weave the string to tie the goatskins to the bellows. During these tasks the smelters discussed the need to obtain a black goat to sacrifice at the beginning of the first smelt but could not agree on where to purchase it; the matter was allowed to linger.

Charcoal burning began several days before the smelt. Heaps of wood cut from the large ironwood tree were burned in the early evening. Before the fire was lit, the head smelter assembled a medicine bundle that contained the roots and leaves of one plant, the leaves of two other plants, and two additional roots, plus "sticks" about 15 cm long made of a broken empiga. This concoction was placed under the wood, he explained, to deal with anyone who might have polluted or brought bad medicine to this place since these activities had begun. It would also cause harm to anyone who came with ill intent, a function similar to the rukago medicine.

The charcoal burning began with an open distillation process. The open fire was tended for nearly four hours, with occasional additions, until the bottom of the pile was well burned. Then the smelters threw earth onto the fire from four piles they had prepared earlier. The smothering of the fire took fifteen minutes, after which the pile was left to complete the distillation process under reducing conditions. Two days later the charcoal was sifted from the heap of earth and sorted—the larger pieces were set aside for the interior of the furnace.

The day before the smelting began, the head smelter and the chief ritual assistant/specialist foraged for the ingredients for the special medicinal concoction placed in the bottom of the furnace before the start of the smelt. They visited two different farms to find the herbs and tubers required; there was also a close inspection of trees within the 3 or 4 km zone back from the Geita-Biharamulo Road—an area of once-lush woodland forest that has been cut and sold off for charcoal production in new village enterprises. Secondary regrowths on stumps were carefully inspected to find the several different species of trees required. The ingredients were laid out on a wooden tray in sequential order, with the first and fifth items, pieces of trees, bracketing three tubers; the next two tubers and one root were also fit between the tree samples, with the final ingredients being a piece of tree *(kakuba)* with red dye and a herbaceous plant (which belonged to one of the previously collected tubers) placed on top of the entire collection (figure 5.5). This special medicinal package was carried clockwise around the kitindi four times, beginning at the northern door.

The head smelter then began to chop the outline for the furnace pit into the almost-flat termite mound on which the kitindi had been sited. Another smelter then cut a sloped bowl with walls at 45° angles. After he had reached 40 cm deep, the head smelter measured and then began to cut deeper in the center with a *mukobero,* which resembles the heavy cleaver used to chop banana plant roots. With this he constructed the *nyaruwengi,* the medicinal hole in the bottom of the furnace. Interestingly, the piece of *kakuba* wood with its symbolic menstrual blood was placed into the ritual pit first, a sequencing that captures the reproductive cycle—the womb is cleansed before it is fertile and ready to reproduce.[5] It was difficult to see the order in which the other medicinal elements were placed into the furnace or to elicit further commentary from the smelters. However, one of the tubers (*muhingura*) had been identified as a local medicine used to cure infertility. We later learned that the leaves of the muhingura plant were placed beneath the piece of bleeding kakuba tree once the latter had been put into place. It must be noted that the Barongo integration of elements symbolic of menstrual blood is not unique; the Wapangwa, who live in southern Tanzania, also place a plant with red flowers symbolic of menstrual blood into a ritual medicine hole in the bottom of their furnaces (Stirniman 1976, 277).

After all medicinal ingredients were in place, both Masaka and Munubwa (the chief assistant) removed their pants and sat down on the earth that had been pulverized from termite earth and heaped up in a ridge around the

FIGURE 5.5. *Above,* Barongo smelters collecting medicinal ingredients for the furnace; *below*, some of the tubers, herbs, bark, and wood used in the medicinal fertility package that is placed in the base of the furnace pit.

nyombe (furnace pit). They sat opposite each other, back to back, and moved 180° counterclockwise, pushing the earth into the pit with their genitalia and buttocks (see figure 5.6), an activity also associated with "contempt for bad things" among the Bazinza (Bjerke 1981, 197). Once they had covered the medicine, they put their pants back on and tamped down the earth with a meter-long pole. After the bottom was smoothed to fit the contour, Masaka took out two small banana-bast packages of powdery rukago medicine. He sprinkled one across the pit in several directions very quickly. He then took the second, larger package and sprinkled its contents slowly over the pit. He moved across it about six times, east to west. He did not utter a prayer aloud, but later related that he had been saying it silently. He explained that the purpose of the buried medicine was to bring a heavy child. The other medicine, rukago, was intended to make people fear this place and to keep anyone from tampering with it.

There was clearly a relatively broad range of symbolic expression available to Barongo iron smelters in the region when it came to the bricolage of furnace consecration. In a photo essay on Barongo smelting in Geita District, Jesper Kirknaes (1980) illustrates the application of rukago medicine through the medium of a prepubescent boy, approximately five years old. His head wrapped in a goat skin, blinded like a newborn, the naked youngster applied the special medicine to the perimeter of the furnace pit, and was then pulled—or birthed—as a heavy child from the womb.

As this intensely ritualized process unfolded, the pushing of the earth into the furnace by the genitalia of the head smelter and the chief ritual specialist was both an act of contempt for evil influences and a symbolic intercourse and fertilization of the furnace womb. This is the second potent step of a ritual cycle that transforms the furnace into a fecund bride. The ritual process is metonymic in its effect, for by human coupling and the placement of symbolic menstrual blood the furnace becomes a container of reproductive materials. This is the deepest order of transformation, beyond metaphor, which merely compares, whereas metonymy identifies one order of reality as another.

The head smelter lined the nyombe with *muyenzi* leaves, which would stay in place during the smelt, and then he filled the pit with charcoal. The other smelters then quickly smashed more termite earth to build another small mound around the furnace in which the empiga blocks for the furnace wall would be set during furnace construction the following day. The smelters brought the required charcoal, iron ore, slag, and empiga and stored them in the lean-to that had been built against the east side of the smelting house.

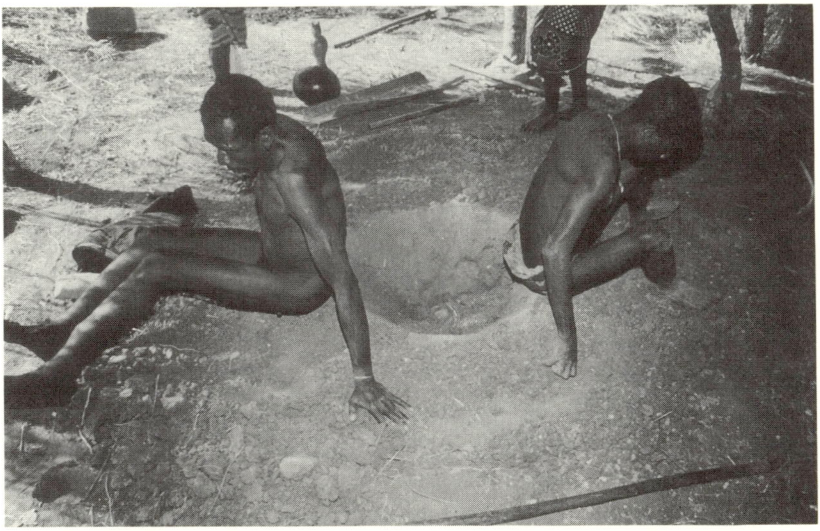

FIGURE 5.6. Barongo iron smelters ritually impregnating an iron smelting furnace.

The First Day of Smelting

Just after daybreak on the morning of the first smelt, the head smelter brought a spear and a pot to the area to the west of his house. There he prayed to the spirits of his father and grandfather and left the spear and pot. This ritual established contact with the ancestors, which was necessary before he could begin smelting.

The head smelter dug a small, angled trench on the south side of the furnace pit. Into this he placed leaves of the muhingura plant, an essential component in the medicinal package placed into the furnace bottom, under the piece of kakuba tree, which had been the first element into the pit the day before. This particular medicinal bundle was about 10 cm/diam and was buried. The head smelter announced that its purpose was to ensure that the bellows blowing would be sufficient to produce iron.

Construction of the furnace then started, with five empiga blocks set into the circular mound. Lengths of broken tuyeres were then wedged between the empiga blocks in horizontal rows to form a louvered wall between each set of empiga (figure 5.7). Newer tuyeres were broken and used when older ones proved insufficient in number. Once this superstructure, approximately 50–60 cm high, was in place, the tuyeres were then inserted inside the furnace after being rubbed at their ends with *maputo* bark, a reddish substance said to give the tuyeres further strength.[6] It seems that this ritual practice had potent metaphorical value—that is, it represented the stiffening of the phallus to withstand the heat and not droop (melt) and fail. While the ethnic groups from which the Barongo are drawn do not practice male circumcision, nearby peoples to the east employ a similar practice when conducting circumcision rituals: before circumcision the penis is smeared with a red clay to prepare it for the flowing of blood (Roles 1966a, 584). Such an interpretation affirms the symbolic value of the tuyere as phallus, but it also introduces yet another variation on the treatment of menstruation, which in ritual circumcision men appropriate, and so appropriate women's reproductive power (de Heusch 1985, 142–43). As Barndon notes about the Fipa (chapter 4, this volume), the meanings of smelting rituals may be unlocked by reference to the larger context of medicinal and ritual practices, sometimes within the region if not within the group itself.

Once the empiga were shored up by granite blocks, iron ore was loaded into the furnace. After stacking the ore against the insides of each empiga to approximately a 12 cm depth, large blocks of charcoal were placed in the furnace at the ends of the tuyeres (approximating ground level). Two to three pieces of old forging slag (initially called *nkamba*) were then brought and placed directly on top of each pile of iron ore, followed by large pieces of charcoal up to 15 cm below the top of the tuyere/empiga wall (see figure 5.7). The furnace was then loaded to its rim with smaller charcoal. At this juncture, another ritual package, *munala* (the contents of which are secret), was tied to the rafters above the furnace in the part of the smelting house called *nserere*; attached to this was a piece of entoto, a type of smelting slag. This medicine prevents the hot gases from igniting the roof of the kitindi. A hole was then hollowed out of the charcoal in the top of the furnace, and larger pieces of smelting slag, previously prepared and sorted to the north of the smelting house, were carefully stacked up by the head smelter, who was closely observed by one of the elders in his eighties. While the smelters prepared and positioned their bellows around the furnace, the head smelter brought more slag for the furnace. Finally, after nearly three hours of preparation, they stuffed live coals down a tuyere and began bellowing. Once the furnace began to smoke, the chief ritual assistant/specialist brought another medicine to the furnace, a piece of *kalinza* tuber (cut from the piece used in the previous day's ritual, pierced by a piece of *mukozanyana* wood). The *kalinza* was stuck on top of the slag at the top of the furnace, moved to accommodate yet another

FIGURE 5.7. Once the *empiga* (the wall sections made from small termite mounds) are in place, louvres are made with pieces of tuyeres and fitted between the empiga, tuyeres are inserted, large charcoal lumps are placed in the center, and iron ore and sometimes forging slag (here seen as the larger lumps) are used in the initial charge of the furnace.

charge of slag, and then was replaced. Fifteen minutes after smelting began, the fire burst from the top of the furnace; all the smelters stopped pumping their bellows and said "Kasinge," which means "thank you." More slag, which in this instance was being used as a flux, was added to the top after twenty minutes, after which the medicine bundle had disappeared.

Half an hour after smelting began, the wife of the head smelter entered the smelting house with a pot of beer. Each smelter visited the pot, sucked beer through a straw, spat it on the ground, took another sip, walked to the furnace and spat it on the furnace. The symbolic quality of beer in this culture is clearly identified with fertility, and the ritual mimics a local nuptial rite in which bride and groom spit beer or milk on each other. In this instance the fecund bride is the furnace, while each smelter is the groom. It is not clear why the beer was spit on the ground, but this ritual detail differed in the 1984 repetition of the fertility ritual. The integration of a smelter's wife into fertility rites applied to the iron-smelting furnace is also known among the Chokwe of Angola (Martins 1966; Herbert 1993), as is an offering of beer. A menstruation taboo also is not mentioned among the Chokwe, and furthermore, the classification and reprocessing of slag (and its use as a flux) plays a central role in the Chokwe process (Redinha 1953, 137).

INTERMEDIATE REFLECTIONS

This description pertains to the first time that the Barongo smelted iron in my presence and in that of Don Avery. What was observed in this instance differed from events at later smelts in 1979, 1980, and 1984. The differences derive from several possibilities. The initial episode in 1979 was conducted under conditions in which we were not well known to the Barongo smelters, having resided for only six weeks in their neighborhood. While they were willing to share some of the ritual processes with us, they were quite secretive about other aspects (such as the arrangement of specific meanings of items that went into the furnace pit). During later episodes they were more familiar with us and had a fuller understanding of our desire to know about all associated rituals.

Thus, the ritual cycle took on greater variation if not complexity, but the meaning of some ritual materials remained secret or opaque. Also, various reasons were put forward during 1979 about why a full ritual cycle had not been possible; for example, there were no black goats available in the vicinity and no one wanted to travel to a distant market to obtain one. In this instance, the critical element of blood sacrifice was omitted during the first and second smelts in 1979 for reasons of time and trouble.

THE SMELT CONTINUES

Approximately an hour after the smelt began, the head smelter brought small grass balls of medicine, each about 2 cm/diam, containing the leaves of the *musinira* tree, which were pounded and mixed with ash. In an instance of bricolage, he pushed the mixture down the tuyeres to help prevent them from melting, another ritual process incorporating white ash. This was a variation on a theme drawn from a ritual associated with fertility in the region.

TECHNOLOGICAL INSIGHTS

Soon thereafter, more slag and charcoal were added; this time, the pieces of slag were larger (6 cm^2) and, according to our measurements, made a vertical column of approximately 60 cm deep—well within the blast zone. As the first smelt proceeded after the beer rite, we found that the furnace was operating at temperatures of 1400–1460°C, more than sufficient for a bloomery process. This process used mostly iron-rich slags and slags that contained partially reduced iron ore. The iron ore within the furnace was being presmelted, later to be separated from the huge matrix excavated from the furnace and then reprocessed through the furnace in the subsequent smelt. The Barongo process, while different from a European blast furnace that used slag as feed stock (Rostoker and Bronson 1990), nonetheless depends extensively on the use of slags from previous smelts and from smithing. These are iron-rich slags high in wustite (FeO) from the forge and various glassy fayalite-wustite smelting slags with melting temperatures below 1250°C. Consequently, the Barongo furnace obtains a fluid, dripping slag in the furnace stack above the blast zone. As the slag drips down and over the large pieces of charcoal, flakes of iron in the slag solution catch on the charcoal and begin to sinter or agglomerate into a typical sponge iron boom. This appears quite different from Rostoker's (Rostoker and Bronson 1990, 94) rather idealized scenario where sintering of the iron takes place in the bottom of the slag pool. A continued charge of lower temperature slags maintains a flow through the furnace while the FeO and already sintered iron from previous smelts provide a constant flow in which further sintering occurs.

In the first smelt, the head smelter began to complain of the poor quality of the tuyeres after an hour of smelting. He said that they were not adequate, that they were slumping, and that they were not passing the air. He then admitted that he did not mine the clay at the same site he had used in the past but had compromised and used a site with clay he thought would be adequate and that was much closer and convenient to the village.

Slag charges continued at two hours (2 cm^2 pieces) and again at three hours; after the three-hour charge, six and a half pieces of forging slag were placed on top, immediately followed by more charcoal and then the last of the peanut-size pieces of slag. At this time, the tuyeres were conducting air directly up against the slag, a reoxidizing condition that prevailed for some twenty minutes until more charcoal entered the blast zone and its temperatures returned to 1480–1540°C. After five hours of smelting, when the charcoal level had reached the top of the furnace, the head smelter initiated different types of bellows rhythms. After six and a half hours, the smelt concluded, the tuyeres were immediately removed, and earth was shoveled and thrown onto the furnace to smother the fire.

The following morning, the *izimu*, or slag mass, was removed from the furnace pit by prying it out with a stick (figure 5.8). Bits of partially reduced iron ore and charcoal were sifted from the detritus, and 12.36 kg were recovered with the remaining 9 kg incorporated into the 50 x 60 x 40 cm high slag mass. A large sledgehammer was used to break apart the mass, and the slag and iron (*ruoma*) were sorted into separate piles. Three types of slag were recognized at this time: the iron-rich muselo, the well-consolidated and dense nkamba, and the glassy and fluid-looking entoto (also the term applied to forging slags).[7] Only 0.63 kg of acceptable iron was recovered, while other smaller pieces of iron were discarded into the pile of muselo. A spontaneous discussion arose about the tuyeres, and the head smelter maintained that the newness of the smelting pit was to blame rather than the clay. He went to the pit and used a stick to push dirt into the interstitial spaces among the medicines.

Further Smelting

For the second smelt different leaves were used on the pit floor. *Mutwarachengo* leaves would change the color of the slag, we were told, from gray to metallic black with a sparkle. The smelting furnace was configured in slightly smaller dimensions for the second smelt and only ten forging slags were placed in the bottom with the iron ore. This time there were 50 kg of muselo available for smelting—some of it from the previous smelt. As the embers were placed in the furnace, the head smelter announced that he had not washed since the charcoal was burned five days before and that he would not wash until the smelting was finished, a practice also documented by de Rosemond thirty-six years earlier (1943, 81). This prohibition, within the context of a symbolic reproductive cycle, seems to bear strong affinities to a puberty rite of passage among the Nyakusa (southern Tanzania), where girls reaching menstruation are secluded and not allowed to wash for a month—again suggesting that the complete female cycle is the structuring principle. It is also possible that principles applied to ritual specialists in this region of Tanzania are manifest in the washing prohibition (Bjerke 1981).

After nearly two hours of smelting, the head smelter brought a tray with salt and *omusesene* root (dug at Karutage); at each tuyere he took a bit of each, spat on it, and shoved it to the end of the tuyere. He explained that this action was intended to improve the yield of the smelt. Next he returned with *musengati* root cut into pieces, wet them in his mouth, dipped them in salt,

FIGURE 5.8.
The *izimu* or slag mass (containing the bloom on the top) being pulled from the furnace pit.

and placed them in each tuyere, enjoining the bellows pumpers to increase their tempo. He said that this process would open the tuyeres. When finished, he dumped the remaining root and salt on top of the furnace charge. This smelt used iron ore in the charge, with the second and final charge coming about 2.3 hours into the smelt, followed immediately by large blocks of smelting slag.

The second smelt was plagued by severe problems with the tuyeres, which caused much consternation. Some had only 1 cm/diam openings, and the blast zone of the furnace was operating at only 1100–1200°C as a consequence. The smelters twisted the tuyeres in an attempt to free them from the slag mass, but results were mixed. Between the fourth and fifth hours, large charges of muselo slag were then followed by 8 kg of forging slag recovered from the first smelt. There were vigorous attempts to keep the tuyeres clear and exhortations to pump harder. After five hours, an interesting interlude occurred. Drawn to the site out of curiosity, a pubescent girl was invited into the smelting house. When she refused, one smelter left his bellows and physically pulled her into the smelting house. He thrust the bellows sticks into her hands, and told her to pump the bellows. This she did for a minute or so while the smelters roared with laughter. There was no concern evidenced by the smelters over the menstrual condition of this young woman. Toward the end of a seven-hour smelt the head smelter publicly complained about the poor performance of the tuyeres.

Exegesis of Ritual

One way to gain a fuller understanding of the symbolic meanings of the otherwise opaque Barongo ritual vignettes is to examine rituals that occur during major points of transition in the life cycle in nearby cultures. Nyakyusa rites appear to provide excellent entree to the meanings of the Barongo rituals using roots and salt placed in the tuyeres. For example, immediately after first menstruation, "A pungent tasting root is pushed through the tip of a funnel and salt is poured into [it]. The girl takes the tip of the root in her mouth and pulls it inward with her teeth, thus causing the salt to trickle into her mouth" (Wilson 1957, 87). Monica Wilson argues that the form of delivery of the medicine symbolizes intercourse, and that the medicine is a prophylactic against the pain of intercourse and menstruation. A local woman further confirmed that "it is the pain of periods we symbolize in the sharpness of the [root] and salt" (Wilson 1957, 102).

This rite seems to affirm that both impregnation and menstruation are symbolically captured in the Barongo ritual, but it is also important to recognize that another important Nyakyusa ritual upon first menstruation is the burial of the bark of a tree under the hearth. While this medicine is identified as the blood of the lineage, its ritual association with menstruation is the key element. The analogy with Barongo burial of a bark (symbolic of menstrual blood) beneath the furnace hearth needs no further comment. Most important for linking the symbolism of menstruation to iron smelting is the Nyakyusa conceptualization of iron and menstrual blood: "As soon as a girl grows up people begin to think about her bearing children. They say: 'The molten iron is discharged, the blood for bearing children has come.' When they grind and brew beer in her honour, they do it in honour of her 'iron' [that is, her menstrual blood]" (Wilson 1957, 104).

The meaning of some Barongo rituals, then, can be understood from Wilson's excellent documentation of Nyakyusa puberty rites before Christianity had obliterated them. We feel confident that the Barongo use of a symbolic armature that brings menstruation into the reproductive cycle is consistent with other beliefs associated with female reproduction. The varied Barongo rituals form a suite of interrelated meanings that unify all stages of the reproductive process.

RESUMPTION OF SMELTING

The third day began with an assessment of the results and further refinements of terminology applied to slags. Usage now indicated that the term *nkamba* is also applied to the forging slags, which in this instance were found loose and free above the slag mass.[8] The yield of 0.56 kg of bloom disappointed the smelters, who were clearly accustomed to much better results. They did not accept sponge iron that smelters from other cultures might (such as the smelters of Babungo in Cameroon, who extracted each small piece of iron by cold hammering the slag). They considered much of the sinter marginal, to be processed again and consolidated in the subsequent smelt. Their judgments seemed related to the difficulty of welding the fine and unconsolidated bloom that they encountered in the top of the slag mass; indeed, it is difficult to keep such fine pieces together in the hearth of the forge because they tend to be carried away and incorporated into the slag at the bottom of the hearth. It also became apparent at this point that the problem of poor tuyere clay without sufficient refractories meant that temperatures high enough to drain slag away from the sinter could not be maintained within the interior of the smelting furnace.

THE BARONGO BRICOLEURS AT WORK

The third smelt was marked by the absence of leaves on the bottom of the pit because the pit had dried, we were told. This smelt again captured the Barongo bricoleur at work. First, the head smelter brought a different type of medicine for the slags—three roots tied together with bark. This combination was rubbed on the muselo slag so that it would produce iron. It was also rubbed on a new category of slag separated from muselo: *migomba* (not expensive, less value) which comprises fused and cracked pieces of iron ore. During this smelt only iron ore that had been through the previous day's smelt was used—fresh iron ore was not placed in the base of the furnace. New medicinal cures were employed, new categories of slag were designated and treated, and different resources and procedures were brought to this bricolage. After the furnace was completely charged and the top stuffed with muselo slag, the head smelter and his chief ritual assistant fetched a goat, which was brought to the kitindi while the smelters were eating. Once the ember was placed in a tuyere, just inside the northern wall the goat's throat was cut by the head smelter, who with two assistants then rushed the goat around the furnace three times while its arterial blood jetted upon the slag at the top of the furnace stack as well as upon all of the furnace equipment, including tuyeres (see figure 5.9).

The third smelt was also distinctive in its large cone of muselo slag. Because of this thick cap, flame did not break through until an hour after smelting started. The new migomba slag was used after 1.6 hours, and another enormous muselo charge 35 cm high was applied after 2.75 hours. Monitoring of the tuyeres was much more regular than in previous smelts. Another mostly muselo slag charge with the remaining migomba slag occurred at three hours; we saw the slag quickly settle into the furnace, indicating a greater slag fluidity, which agrees with the steady 1500°C temperatures in the blast zone that we monitored with an optical pyrometer. Though the smelt appeared to proceed better, one smelter opined that the smelts had gone poorly because of the chief assistant—he was plagued by bad luck and had a reputation for bringing bad luck to things he touched. This smelt lasted for seven hours and experienced one broken tuyere and two completely clogged tuyeres. The next day's inspection revealed that, despite the complications with tuyeres, 1.45 kg of workable iron was obtained, sufficient to forge a tool.

The Smelting Technology

At this juncture it is important to turn to some of the highlights of the technological process that was observed in Barongo iron smelting. The Barongo process is not distinctive in its reprocessing of iron sinter and slag. As we have noted, the Chokwe of Angola, to cite one example, also differentially value their slags and are careful in how they are reprocessed (Redinha 1953, 137).

FIGURE 5.9. Ritual sacrifice with blood covering the Barongo furnace.

The Barongo furnace represents a three-step process inside each smelting operation: (1) the preparation and partial reduction of iron ore in the bottom of the furnace; (2) the production of large amounts of iron-rich slags, some with significant iron sinter; and (3) the production of sponge-iron blooms. Some technologies separate each of these steps and combine others. For example, the Haya prepare ore and produce a final well-consolidated bloom in their smelting furnaces. The Fipa produce an iron-rich sinter and then consolidate it in a refining furnace, which results in a consolidated bloom. The Barongo process is a continuous one that combines all these steps into one activity.

The analysis of Barongo products reveals some important insights into the technology. Metallographic examination of one of the pieces of iron sinter set aside as muselo (slag is our word) after the first smelt shows characteristics that we would identify as a sponge iron bloom (figures 5.10 and 5.11) that incorporates an iron-carbon alloy (steel). Metallographic examination revealed that the steel structure showed signs of rapid air cooling and decarburization suggesting that the steel was altered when the bloom was removed from the furnace pit. However, the thin, continuous sheets of iron that grow from

FIGURE 5.10. A photomicrograph of a bloom from a piece of *muselo* slag. This shows why the muselo slag is an integral part of the smelting process: It is iron-rich; in fact, it is a heterogeneous iron-carbon alloy (steel) with approximately 0.7 carbon on the left side. The right side shows a Widmanstatten structure or array of iron plates, with all of the plates crystalographically aligned. The structure suggests air cooling; 50 times magnification, 2 percent nital etch. Photo by S. Terry Childs.

FIGURE 5.11. (*facing*) A close-up (magnification 500 times) of the area in the right side of figure 5.10. It shows a Widmanstatten structure with ferrite grains (white) interspersed between grains of pearlite (mottled). This is not a piece of forged steel, but rather a piece of iron within an iron-rich slag set aside for further processing, which means that the cooling of the steel—resulting in a Widmanstatten structure—occurred when the bloom was removed from the smelting pit; 2 percent nital etch. Photo by S. Terry Childs.

platelets at higher temperatures were poorly consolidated, so that when this material was cold worked—to remove the slag—it fell apart.[9] As a result, it is difficult to consolidate the iron in the forge, so it goes into the muselo pile for further consolidation in the furnace. In the Barongo furnace, the practice was to add slag with high amounts of iron sinter, which contributes two important features to the smelt: (1) it provides iron nuclei that quickly pass through the furnace within fluid slag until reaching the large pieces of charcoal, where they await further passage of fluid slag; and (2) the slag charges provide a fluid slag that then fluxes the iron ore charges and forms a fluid iron-rich slag that drips down and over the iron plates developed in the previous smelt. As the

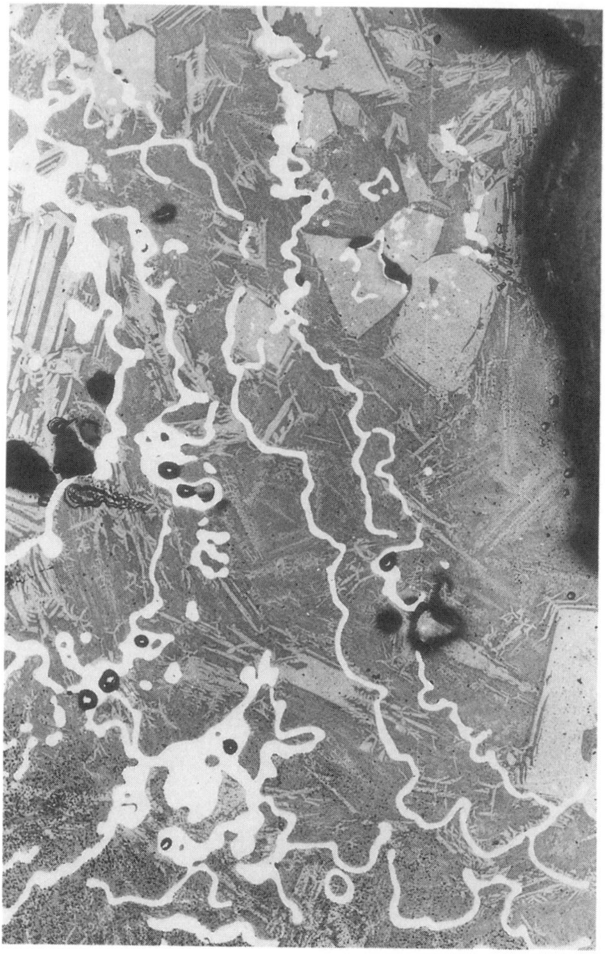

FIGURE 5.12. A photomicrograph illustrating sheets of ferrite—pure iron—forming in a fayalitic slag, or a process where there is secondary growth of iron plates consuming an earlier fine precipitate in the slag. Photo by Donald H. Avery.

fluid slag passes over these older agglomerations and encounters the highest temperatures, the sheets and nodules of iron dissolved in the slag precipitate onto the stationary massive pieces, leading to the formation of a continuous iron network (figure 5.12) and the eventual consolidation of an iron sponge (Avery et al., 1988). As wustite (FeO) in the slag comes into contact with the charcoal, iron will also form along the boundary with the charcoal (carbon). The remaining slag, high in silica content, flows to the bottom of the furnace bowl.

Examination of the slags shows that the growth of iron platelets and nodules occurred within the context of the free-running slag. In this case the slag acted as a solvent that dissolved the iron particles and then transported them by convection to the earlier deposited platelets (which derive from the muselo), which then further grew as iron was precipitated from the slag flowing over them. Thus, the fluid slag is the key for understanding how iron is deposited and consolidated in the Barongo furnace. If, however, too much high-silica slag is placed in the short furnace in the initial charges, the slag quickly runs through the 50–60 cm stack to the bottom, fills the pit, and clogs the tuyeres.

Forging

Forging of the sponge bloom obtained in the third smelt occurred ten days later. A Barongo forge consists of a large granite anvil at ground level, with a 60 cm deep contiguous pit in which the smith stands when striking the anvil. Thus, the anvil is a platform upon which the smith works, first with a large granite hammer, a muhimbo, and later with a ensule hammer for final shaping work. The hearth is situated about a meter from the anvil and is a modest 22 cm/diam pit filled with charcoal and outfitted with a single tuyere and a wooden double-drum bellows. The muhimbo hammer for this forging episode had been made several weeks before by the chief assistant, who shaped it from a rough piece of granite by pecking with a small iron hammer (figure 5.13). This resulted in a 16.7 kg hammer that was 33 cm long and had a 150 cm^2 working surface. It is easy to understand from the weight of the hammer that Barongo forging is very demanding work (see David and Robertson, chapter 6, this volume).

The sponge bloom was first cold worked, with the smiths attempting to mechanically consolidate the many pieces so that they would to stay together in the hearth. They also separated pieces that they considered to be muselo and in need of further consolidation in the smelting furnace. The sponge mass was set on top of the charcoal in the hearth, where it sat for about ten minutes in temperatures of 1250°C. It was removed with wooden tongs and placed on the anvil, where another smith beat it six times with the heavy hammer. Many pieces of incandescent fluid slag flew from the slag, scattering spectators (figure 5.14). When it was returned to the hearth, several remaining pieces of sponge were added; it rested in the hearth for another ten minutes, after which it was returned to the anvil for eleven whacks.

This is a process of fluid and mechanical removal of slag from the iron sponge, which in the Barongo case was done with tremendous force with a heavy, large hammer used at maximum mechanical advantage. This method of treating sponge bloom is widely known in Africa, but the Barongo variation is testimony to the fairly large amount of slag incorporated into the iron sponge. A third treatment resulted in thirty-six compressions and the return to a forge now operating at 1300–1350°C, higher temperatures to deal with slag that was deeply captured in the more consolidated mass. It was clear after the third treatment that the amount of iron was sufficient for an axe but not for a hoe. Over the next two hours, the large hammer was used to bring the axe to its basic shape. As a ensule hammer was not available, a steel hammer was used over the next hour to bring the axe to its final shape. It was cold-edge worked, with the axe tip chiseled off to eliminate cracks and to bevel it. The

FIGURE 5.13. A *muhimbo*, granite forge hammer being made by pecking with a steel hammer.

FIGURE 5.14. A Murongo smith wielding the 17 kg *muhimbo* hammer at the forge, in the early stages of working a sponge bloom.

axe was then finished by burning a horn on its surface while it was still hot (figure 5.15).

Because the Barongo smiths processed sponge iron that was sometimes poorly consolidated, some of the iron sinter was lost with slag into the hearth of the forge; the slag in the hearth was highly magnetic. But the hearth also operates at high temperatures in oxidizing conditions, which reoxidizes much of the iron sinter to FeO or wustite. In any case, it is easy to understand why the Barongo smelters eagerly reprocess forging slags in their furnaces. They also had a discriminating eye toward the sponge iron that did not merit forging because it was too poorly consolidated to survive the hearth once the slag matrix melted; they separated this material for further processing as muselo.

Metallographic analysis of a section from the tip of the forged Barongo axe shows a heterogeneous carbon content, ranging from pure ferrite to about 0.7 percent carbon (figure 5.16), but also containing quite a large number of slag inclusions. These thin pockets of slag are commonly found in such products, and as figure 5.17 indicates, these mostly wustite (FeO) slag stringers are roughly analogous to a laminated wrought iron. The tip of the axe shows excellent carburization (figure 5.18a) indicating that despite the heavily oxidizing conditions of the forge the smiths were able to retain carbon in the product by incorporating layers of higher-carbon steel—an effective way to produce a carbon steel tool. The butt end of the axe shows a steel structure (figure 5.18b) that documents the rapid cooling of the steel in that region—where hardness matters more than brittleness. The Barango smiths did use quenching, both by immersion in water and by light sprinkling, perhaps ac-

FIGURE 5.15. An axe forged from sponge bloom produced by the Barongo smiths. Photo by S. Terry Childs.

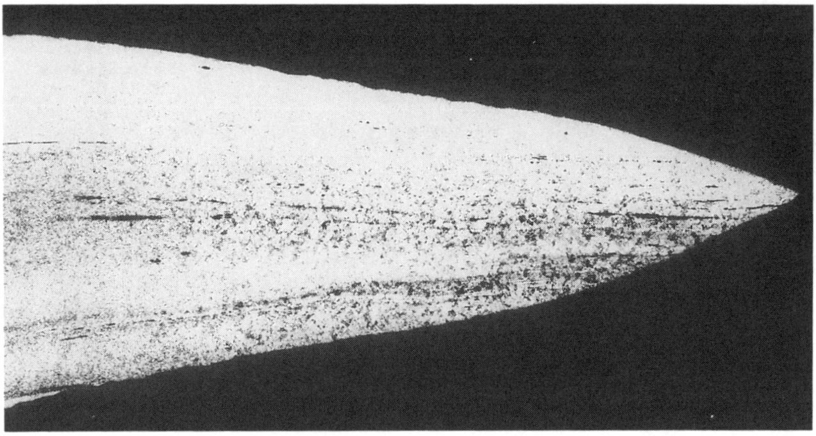

FIGURE 5.16. A photomicrograph of a cross section of the tip of the axe illustrated in figure 5.15; the section shows an annealed structure with a very heterogeneous carbon content, ranging to almost pure ferrite at the bottom edge to 0.6–0.8 percent carbon at the upper edge. Although the iron bloom was submitted to heavy pressure from the granite hammer, significant amounts of slag remain behind as slag stringers, the elongated dark strips; 125 percent magnification, 2 percent nital etch. Photo by S. Terry Childs.

counting for this structure; the rapid cooling along *one* edge (figure 5.18a) is likely linked to quenching. In short, the Barongo smiths are master craftsmen capable of manipulating an unfamiliar and difficult raw product and producing an excellent high-quality finished product.

RETURN TO SMELTING

A break was taken for three weeks after the third smelt, but upon return to work the smelters/bricoleurs openly reflected on their search for the best solution. As they discussed their impressions they also diagnosed their progress and assessed their choices. The head smelter made a point of explaining his choices of slag as he sorted the slag stored in the furnace pit. He explained that the muselo with iron sinter is not selected for forging because it would not hold up but would disappear into the slag in the forge hearth. As a result, it must go back into the furnace for further consolidation.[10] As he sorted slag,

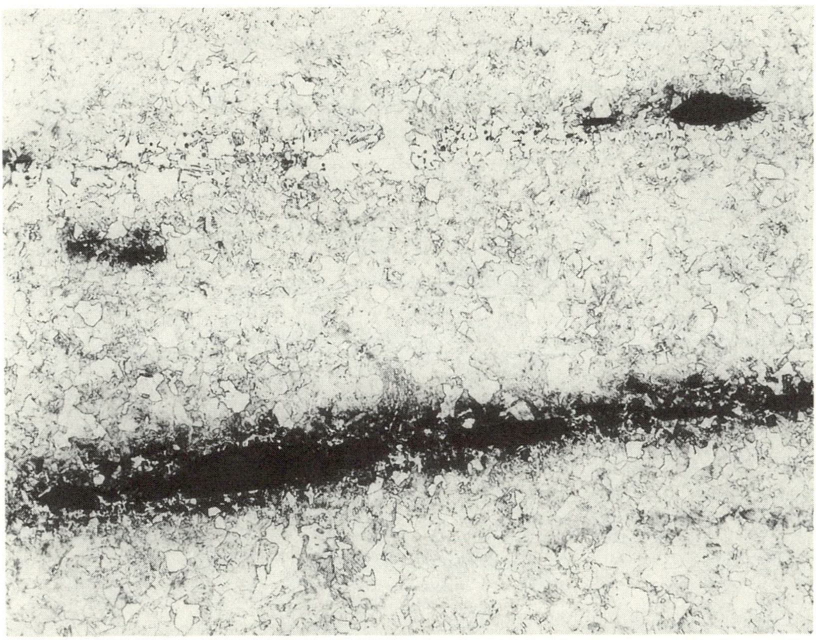

FIGURE 5.17. A photomicrograph showing a close-up view (200 times magnification) of the mostly wustite slag stringers in the axe, creating a structure analogous to a laminated wrought iron; 2 percent nital etch. Photo by S. Terry Childs.

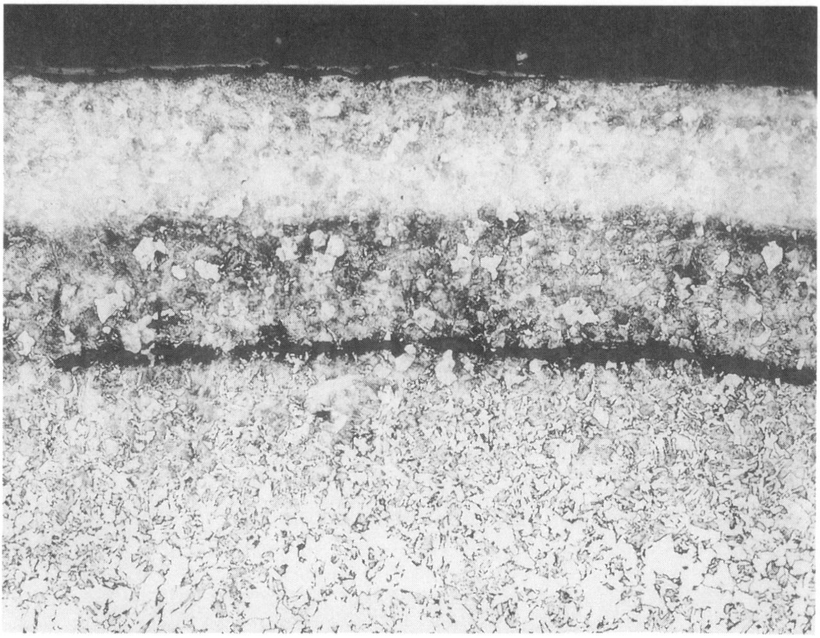

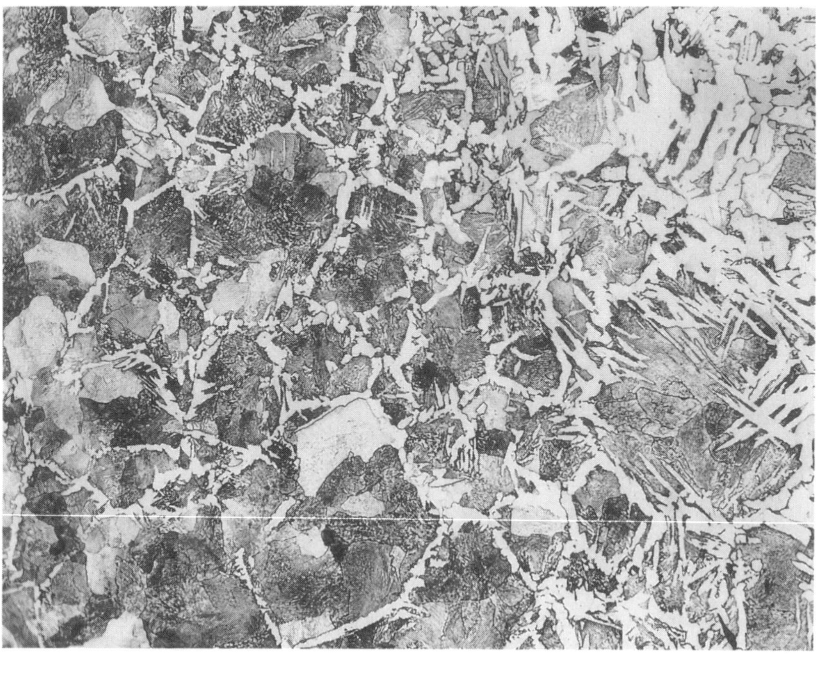

the smelter took out those pieces with obvious iron sinter and placed them into a small pit, calling them *muselo gwa insingiro*. The glassy slags were placed in a pit and called *muselo gwa mbabiro* (expensive, for the future), a bubbly, light-density cinder with much charcoal. He also placed the dense, massive slag (sometimes with large bubbles) that incorporated iron ore into the migomba pile. Finally, the dense and massive slag found on the edges and bottom (previously called *nkamba*) was instead called *entoto* and was placed with the forging slag.[11] This time the head bricoleur smashed eight of the nkamba forging slags into small pieces and mixed them with muselo insingiro and migomba for the charge. He also explained that the top of the slag mass is called *musakamira* (the section with muselo insingiro) and that he intended to mix this with the entoto from the bottom for the charge—a deep insight into the continuous experimental bricolage. Virtually all slag was reused. It became apparent that the Barongo process leaves behind little waste product, leading to an archaeological signature on the landscape that is much less obvious than that of other iron technologies. This fact serves as a caution against using the slag volumes to estimate levels of iron production, or even slag sampling, to characterize past technologies without a deeper knowledge of how slag may have been used.

When the fourth smelt occurred in late July 1979, 53.3 kg of slag were prepared for reprocessing. Unsmelted iron ore was placed in the furnace this time, perhaps because of the need to build a supply of unoxidized slag with good sinter. The furnace was charged with muselo insingiro first, followed by muselo mbabiro; large pieces (15 x 15 x 10 cm) totaling 16 kg were placed into the first charge. This time the head smelter used *muhingira* root and salt (dipping the chewed and wetted tuber into salt), and pushed this into the tuyere, saying, "Nakurunga," or cook well. The addition of a fertility drug into the womb and the use of a cooking metaphor bring together two powerful trans-

FIGURE 5.18. (*facing*) *Above*, a photomicrograph showing carburization of the axe along the edge of the blade where carbon content is 0.6–0.7 percent. This is a deliberate surface weld, marked by a discrete change in composition. The area of lower carbon below is segregated by slag at the weld. The opaque, light color along the edge may result from elemental segregation; 200 times magnification, 2 percent nital etch. Photo by S. Terry Childs. *Below*, a photomicrograph at 200 times magnification of a section taken from the pointed butt of the axe; it shows that fairly rapid cooling—indeed, the quenching employed by the Barongo smiths—occurred to create a Widmanstatten structure with ferrite along grain boundaries in the pearlite (*center, left*); 2 percent nital etch. Photo by S. Terry Childs.

formational tropes, reproduction and cooking. De Heusch's (1982) exegesis of salt symbolism among the Kuba sheds some light on this observation. He argues that salt is symbolic of the conjugal union and also as a symbol of cooking is a mediating term between man and woman (e.g., male tuyere and female furnace). David Collett (1985, 1993) has pointed out the appropriateness of the cooking analogy, and David Killick's observation that the Chewa words for *iron* (cold-cooked porridge) and *slag* (half-cooked porridge) seem to suggest that the direct intercession of the Barongo bricoleur as cook has grounding elsewhere in iron-smelting cultures (Killick 1990; Herbert 1993).

On this occasion, previously smelted iron ore was used as a separate charge and another large mixed charge of iron sinter and forging slag was added after 2.25 hours. Only half an hour after this large charge, the tuyeres began to plug, having taken on large amounts of slag. While the head smelter ordered vigorous pumping, it was clear that the smelt was finished. He tried to free tuyeres and continued to add charcoal, but without results. The smelt was abandoned after 4.75 hours, with the head smelter asserting that the new clay used for the tuyeres came from the old, traditional source and hence could not be the problem. He believed that a witch had done something harmful to the furnace and announced that he would consult a diviner.

The next morning's inspection of the izimu showed that one tuyere was clearly forming a bloom; it appeared similar to the well-formed blooms in the Haya process. The furnace was constructed in thirty minutes and the head smelter brought a brown powder (a rukago) that he obtained from a diviner; he sprinkled it on the west wall, the pit, and the east wall, repeating the process for the north and south walls and placing a special stick in the bottom center of the pit. He started the smelt with a large charge of muselo insingiro (as the day before) and capped it with muselo mbabiro. While the blast zone of the furnace operated at 1500°C within an hour, he followed this with another large charge of iron sinter. Within fifty minutes, the tuyeres were plugged with slag (up to 12 cm deep, we later learned) and the smelt had to be stopped. Despite a spirit of the best bricolage, the Barongo had yet to get the process right, in fact fading from the near success of smelt three.

Smelting among the Barongo did not resume until July 1980. During a trip to mine ore at Karutage, the head smelter cut a root from the *mususemi* shrub to use in a rite that appeases the spirits of the Bayovu ancestors. New tuyeres, made from traditional clay obtained at Buseresere, were manufactured. With these changes in place it was announced that new medicines were not required in or next to the furnace, and blood sacrifice was also not required, because the medicine from the previous season was still potent. The first smelt started with the rubbing of *mupala* root on the tuyeres and with the burial of a special medicine made of two sticks plus rukago medicine near the site of the goat sacrifice.[12] Masaka informed us that these medicines would renew the power of the medicines in the base of the pit. The medicine for the top of the furnace was also employed again.

The muselo on hand weighed 27.68 kg, but most of it had been mined near Bwanga and was therefore weathered and reoxidized. The first charge this time was with raw iron ore (also placed in the bottom of the furnace, but without forging slags), which was immediately followed by ten pieces of forging slag. On top of this concoction was placed a charge of muselo. Once this charge was in place, a different medicine bundle, *nsamba* (an aphrodisiac), was placed on top of the slag. Two objects were inside: an iron needle, used to sew baskets and remove jiggers, and a piece of copper wire (as well as five roots, one of which had been used in a previous smelt).[13] Immediately after flames broke through the surface, Masaka removed the iron needle and copper wire from the bundle, which was returned to the slag heap. Muselo was heaped again on the furnace after an hour, and again at 1.5 and 1.6 hours. A charge of dense entoto slag was added at two hours and it was clear shortly thereafter that the tuyeres were beginning to clog. The head smelter did not respond to signs that the smelt was finished but continued to try to clear the tuyeres. He added more charcoal over the next hour, trying to get the "water" out of the entoto. Meanwhile, the chief assistant asserted that the tuyeres were stopped up because of the *maji* (water) from the entoto. When the furnace was opened, it was discovered that the tuyeres were mostly full of slag. The smelters said that this phenomenon resulted from *musigwa*, or their ancestors' spirits, which brought misfortune. The smelters agreed that they must consult a *mufumu* in order to discern the reason and to learn how to purge the evil forces.

The next day's dismembering of the izimu surprisingly showed a yield of 0.667 kg of acceptable bloom, a quantity that suggested that if the smelt had continued uninterrupted the yield would have been significantly greater. Attention was now focused on migomba slag, which included a significant amount of partially reduced ore. As in the first smelt of 1979, muyenzi leaves lined the pit. Of the slag charges readied for this smelt, 2.3 kg was *migomba*, 17.12 kg was muselo, and 13 kg was entoto; it is important to note that the muselo included all of the recovered bloom. Once the furnace bottom was charged with iron ore, most of which had been presmelted (but not with forging slags on the bottom) and filled with charcoal, the head smelter sprinkled a white powder, which he would not discuss, over the top of the furnace. He then sang a chant that had been given to him by an unknown mufumu healer and charged the furnace with muselo. This time he used large pieces along with the pieces of sponge bloom, but he let them sink down to the rim level before slowly adding more light pieces of iron-rich sinter.

Fifteen minutes later, the head smelter placed a large charge of migomba (the partially reduced iron ore and slag) with pieces of presmelted iron ore on its periphery. He followed this by placing wustite-rich muselo in the middle, again surrounding it with presmelted iron ore and then covering it all with charcoal. This was followed by 4 kg of entoto and the last small pieces of muselo twenty minutes later. Masaka then took a mususemi root, stripped its bark, crushed some rock salt, wet the bark in his mouth, dipped it in the salt, and stuck it down the tuyere, saying, "Nakurungira tata" (He has fed you well). He repeated this fertility rite in alternate tuyeres in a clockwise fashion until all the tuyeres were treated with the prescription of the diviner. Masaka explained that this was a gift to satisfy the ancestral spirits and result in a successful smelt. His actions brought into prominence the cooking metaphor within the context of reproduction, suggesting that the ancestors would need a well-cooked reproductive product from the furnace.

The technique of furnace charges took a significant turn with the new ritual/healing approaches, as Masaka continued to lay slags in a circular configuration around different central charges: entoto were then added in a circle, with muselo scraps added in the middle. Similarly, the last of the roasted iron ore was added in a circular form over the entoto (separated by a layer of charcoal), followed by more entoto in the center and soon afterward by a few more muselo scraps for the center. When the first tuyere check occurred at the two-hour point, all were clear. The slag charges and charcoal fell steadily into the furnace. Spirits were high. My wife suddenly was asked to help pump the bellows, which she reluctantly did, accompanied by the whooping of thanks to the cooperative spirits. After only four hours of smelting, the furnace was sealed.

The yield from this smelt was 4.0 kg of well-consolidated sponge bloom, which was concentrated in the top center of the izimu mass. Mbuzehose remarked that it was sufficient for at least one large hoe, while another smelter said it was easily enough for one and a half large hoes, suggesting that this smelt reached a level of productivity that was minimally 30 percent of what was produced in the late 1940s. It was achieved with 45.5 kg of charcoal.

Notably, in this series of seven smelts the Barongo never recovered their levels of former production. But their bricolage shows a remarkable conjunction of ritual and technological fabrication that is richly interwoven and that leads to empirical solutions, illustrating that any attempt to analytically separate the two domains is meaningless and misses the point: experimentation is both ideological and procedural. It is just a matter of eventually bringing all the right pieces together to make an engine that works. Western scientists, if honest in apprehending the scientific method, will admit to beliefs, intuitions, and even superstitions that consciously and unconsciously interpenetrate, influence, and guide the bricolage that we often witness in scientific investigation and discovery.

There is one final piece to this remarkable smelting story that needs to fit into the picture: the two smelts of 1984, arranged so that the process could be recorded in a documentary film. It is informative that objections about price and acquisition of a goat were put aside in 1984, with the strong insistence that a goat be obtained and that the sacrifice be conducted. Let us turn for a moment to review precisely what happened during the first smelt of 1984.

Once the smelters had conducted the ritual procedures to impregnate and protect the furnace womb, they added a variation to their procedures on the morning of the first smelt. A hole was excavated close to the north side of the kitindi at the spot where the goat would be sacrificed. There a package of *vuma* (a root) medicine was buried, and the furnace was constructed and filled. When the smelt had started, the head smelter's wife brought beer to the smelting house, where it was placed above the buried vuma. She was then joined by one of the smelters, who, like the woman, wore on his head a protective charm (*kizingo*) made of cowry shells. Together both took beer from the tightly woven basket and spat it three times on the furnace as they stood between different tuyeres. The beer was then passed to two other senior smelters, who spat it on the furnace at their tuyere stations.[14] After this ritual, the head smelter and his assistant opened the artery of a goat at the same ritual station; with arterial blood spurting across the slag charge on top of the furnace and splattering on the furnace walls and bellows, the goat was taken once around the furnace counterclockwise (figure 5.9). Compared to the three circuits during the third smelt of 1979, nearly as much blood was spilled from a larger wound. The goat was removed to the north side of the smelting house where it was then butchered, the liver immediately roasted, and the meat prepared for boiling.

Again, a woman played an integral role in one of the most important transforming rituals, during which the furnace was again paired with smelters as a fecund bride.[15] The woman's menstrual state was of no consequence, but her reproductive capacity was essential in a ritual in which she substituted, as the bride, for the furnace. This is consistent with the symbolic armature in which the Barongo operate, as evidenced by the smelters' invitation to the young pubescent girl to join them in the smelting house at the end of the second smelt, and to my wife during the second smelt of 1980. It was no accident that each woman involved was obviously in the flower of her reproductive life. The Barongo perspective is not an isolated case. The Chewa iron smelters of Chulu in northern Malawi, just across the Tanzania border, incorporate a younger, fecund woman into the fertility rite that initiates the furnace (van der Merwe and Avery 1987). A nursing mother is accompanied by her child and, much like the Murongo woman, she eats the porridge symbolic of fertility. While the observers in the Chewa report assert that the woman is sexually inaccessible because she is nursing, this interpretation fades in importance to the wider practice of reproductive women playing key roles in the fertility rituals of iron smelting.

The role of blood sacrifice in the Barongo smelt is distinctive in its attention to the full saturation of the slags stacked on the top of the furnace. Given what we already know about the representation of menstruation and periodicity of reproduction in the Barongo reproductive symbol system, this ritual bears much closer exegesis. The Barongo industrial intent is to induce slag to flow through the furnace, and in this ritual context—given what we already understand of metonymic menstrual characteristics of the furnace—it appears that the sacrificial blood is a purifying agent that cleanses the womb and prepares it for reproduction. The Barongo rites contain two reproductive cycles, the first captured in menstrual cleansing and impregnation during the pit ritual, the second captured in menstrual cleansing during the sacrifice and the later impregnation of the furnace, when the smelt is underway, by the injection of fertility objects into the phallic tuyere to make it stronger in the reproductive process.

This interpretation meshes with the other components of Barongo belief and ritual practice. It may not be an interpretation, though, that will excite those wedded to the idea that blood sacrifice cannot be tied to menstruation. However, evidence suggests that running slags are represented as menses; among the Ekondo of Zaire, slag is compared to menses, which prompts Eugenia Herbert to remark that "the correspondence between slag and menstruation is apt: both are the waste products in the creative process" (Herbert 1993, 64). Herbert is right, but for the wrong reasons. Slag is a waste product only in the minds of Western observers. As the Barongo brilliantly demonstrate (along with others, such as the Babungo in Cameroon and the Chokwe of Angola), the slags are enormously valuable (muselo is derived from *okusela*, which is expensive and highly valued) and are the primary component in successful smelting. It is critical in the case of the Barongo process that they begin to flow through the furnace womb so that proper fertilization and then gestation will occur.

The ritual structure is indeed a homology for reproduction, a model much more complete than the contradictory opposition of sterility. The Barongo dialectical resolution of this heretofore baffling contradiction compels us to set aside the Western orthodoxy that blood sacrifice is not associated with menstruation (for example, Herbert 1993, 86). Luc de Heusch opens some space for us in this discourse, for he elegantly constructs a dialectic from Dogon myth in which "Blood from throat slitting, called 'menses of the earth,' contains the life of the world" (Griaule and Dieterien 1965, 23435, cited in de Heusch 1985, 138). "Sacrifice is, then, a beneficent cosmic menstruation" (de Heusch 1985, 137). The Barongo and Wapangwa integrate menstruation into the ritual reproductive cycle, which may also be the case among many other ironworking cultures that scheduled production by phases of the moon (Herbert 1993). As a system of thought that does not construct menstruation as polluting, there are no associated taboos that address the prohibition of menstruating females.

These observations lead to several important questions. Are other descriptions of iron smelting based on insufficient observational data and truncated understandings of ritual behavior so that the full and dynamic range of beliefs have not been revealed? Is it possible that many of the interpretations about the genderization of iron smelting are a product of abbreviated knowledge and that ideas about male appropriation of female reproduction obscure a much more richly textured phenomenon? The Barongo ritual world compels us to reexamine such issues.

Unraveling Bricolage: Thoughts on Variation

It is helpful to turn for a moment to several other examples of African iron-smelting groups that supposedly incorporate contradictions to the reproductive paradigm in their omission of beliefs and other normative rituals in their iron smelting activities (Herbert 1993, 122-25). The Njanja of Zimbabwe are characterized as a highly productive group in the late nineteenth century that welcomed apprentices from neighboring groups and freely employed women in the smelting process (Herbert 1993; Shimmin 1893). Some resemblance to the Barongo is clear. Moreover, the Koni of the Ivory Coast observed no rituals or taboos during any of their smelting-related activities, and the Busanga of northern Ghana did not adhere to taboos about women's participation (Eckert 1974, 174; Pole 1975, 34; Herbert 1993, 124–25). Where do these contradictions and variations lead?

First, it is important to realize that most reports of African iron smelting are woefully abbreviated accounts that do not begin to capture internal variability—how one smelt differs from the next as well as how one village or lineage in the same region differs from another. It has been implicitly assumed that iron smelting is a uniform process, a point made in the introductory chapter of this book. The Barongo resoundingly put this false notion to rest. Moreover, confronting the assumption of uniformity with contradictions or departures from the idealized norm leads to explanations that these variations are strictly productive phenomena (Pole 1975; Herbert 1993)—that is, taboos have been abandoned to facilitate increased production. Again the Barongo have taught us that ritual and technological bricolage is the modus operandi in a world that always presents challenges. It is essential to come to more informed understandings of variation and why it appears as such a dynamic part of the productive cycle. For example, the residual presence of sexual taboos among the Njanja, documented by William Dewey (1990), and contradicting their nineteenth-century inclusion of women, may result from observations of different smelting groups or local normalization of smelting using well-known tropes—that is, possibly a way of restoring or creating a legitimate past.

There is another important point about ritual variation in iron smelting. The Barongo case shows that transformational rituals focused on sexuality and human reproduction are much more complete and powerful precisely because they incorporate menstruation and women. The periodicity of female reproduction is fully appropriated but not at the exclusion of male rites of sexual passage. Menstruation is not antithetical to fecundity in the Barongo reproductive cosmology; rather, it is causatively linked to fecundity, which cannot prevail without menstruation—clearly much more than the omission of a menstruation taboo. Symbolic menstrual blood placed in the womb of the furnace and blood sacrifice representing renewal of the womb are powerful reiterative images that disclose a well-integrated ritual cosmology. Given this rich logic (which I suspect prevails in other African iron-smelting cultures but has been obscured by Western obsessions), it follows that women are welcome to pump the bellows. And it follows that the wife of the head smelter (regardless of her menstrual status) is an integral part of the fertility rite that initiates the furnace. The elaborateness of Barongo reproductive symbolism creates strong suspicions that investigators may be missing important variations in other iron-smelting cultures.

The Barongo maintain a finely textured reproductive paradigm, but they do so as bricoleurs who innovate across ethnic boundaries, into economic opportunities, and into new and unpredictable resources. Iron smelting is a risky business, always subject to failure. Mines give out, clay sources dry up, people are relocated because of civil disruptions: all of these challenges test the skill and innovative powers of the bricoleur. Likewise, the amalgamation of many different ethnic groups into a common work force presents significant creative challenges to the ritual bricoleur and to the smelter. The symbolic reservoir into which these bricoleurs dip seems to draw extensively on ritual transformation from local rites de passage as well as the local pharmacopoeia.

The challenges met by the Barongo bricoleurs did not occur only during the nineteenth and twentieth centuries. As other chapters in this book show, expanded iron production was a cyclical phenomenon related to resource availability, trade opportunities, population, and technological skills. We know from controlled archaeological evidence in Bahaya (150 km to the northwest of the Barongo) that Early Iron Age ritual expressions pertaining to reproductive symbolism in iron-smelting furnaces waxed and waned through time (Schmidt and Childs 1985; Schmidt and Childs, chapter 9, this volume). Reasons for such variability 1,500 to 2,000 years ago are not immediately apparent, but they may be related to the different ritual identities of the varied, perhaps competing social groups practicing iron smelting in the same region. Ritual variability has also been observed in twentieth-century Haya iron smelting, when restricted resources seem to have played an important role rather than increased production.

Perhaps Herbert's most important observation is that we "should probably be more suspicious of information that is *too* consistent than of that containing anomalies" (Herbert 1993, 121). Recent archaeological theory (Gould, 1980; Gould and Watson, 1982; Binford 1987) in the practice of ethnoarchaeology makes a similar point. Inquiries should focus on the anomalies, not the similarities. The fallacy of studies of iron-smelting rituals is one of seeking similarity and erasing contradiction.

Notes

Author's note: I wish to thank the smelters and people of Butengo village in Biharamula District. Thanks also go to the cultural offices of Kagera Region and Biharamulo District as well as to the then National Scientific Research Council for research permission. Research and analysis was supported by the National Science Foundation (BNS78-07855) and the National Endowment for the Humanities (RO-10570-83).

1 A low and broad hill where a banded iron ore was mined. It is located 1 km north of the Kahama Road, where it intersects with the old road running north to Bwanga.

2 This observation is geologically perplexing; gold would be expected in alluvial contexts or *in situ* within igneous formations, not within the banded ironstone deposits.

3 Much larger and deeper bellows were possible when clay was used; the diameter averaged about 30 cm and the depth averaged about 12.5 cm. The clay bellows were disposable and were abandoned when the season's smelting was finished. Because they were not fired, however, we would not expect to find any archaeological traces of these tools.

4 Before Masaka began to excavate the iron ore, he grabbed two stalks of grass, held them at chest level, and faced Karutage Hill. He then intoned a prayer to the spirits of ancestors who once made iron here. He asked the spirits of the Bayovu, who are the *wakubwa* of this hill, for good luck in getting embale and in obtaining iron without the furnace breaking or failing. If they were to carry off the iron ore without begging permission, ill luck would befall the smelt because the Bayovu would consider that they had stolen it.

5 Recently Alan Roberts (1993) has argued that among the Tabwa, west of Lake Tanganyika, the color red symbolizes violent, bloody transformations, and that red is the untamed, "bloody" power of women. While this is not explicitly parallel, it

points to allied symbolism in which the redness of women is transformative and creative. Among the Zinza, medicines often take their names from plants from which they are derived. It is noteworthy that none of the Barongo medicines are mentioned by Bjerke (1981) as commonly used by a shaman in dealing with spirit afflictions; the one exception is the generic *rukago*, the protective powder. The medicines collected were as follows: a piece of *musisi* tree; leaves from the kalinza, a tuber; a root from a *mugereka*, a tuber, 30 cm long and 6 cm wide, like a long carrot; a *muhingura*, a small tuber; a stump from an *ihondogolo*, a small tree (using the piece of musisi to measure the length; it is placed on the opposite side from the musisi and encloses the tubers); *mutazia*, a tuber with round leaves; small roots from a *zuguma*, an herb; *kafrungusha*, a beet-shaped tuber without leaves; *masaguye*, a sprout of a tree, measured with the musisi piece; a 15 cm/diam piece of a *kakuba* tree, which has a red dye in the inner bark that stains; and leaves of the *muhingura*, broken off at root level.

6 It is unlikely that the maputo bark contains any chemical that could significantly assist the refractory qualities of the tuyere; the high temperatures would quickly burn off the substance.

7 It is important to note that these categories were provisional and were later changed and considerably refined.

8 The forging slags were placed in the furnace toward the end of the fifth hour.

9 I thank Don Avery for joining me in the field and for sharing his insights into this process. I am also indebted to Terry Childs for her help with the metallographic analysis and photomicrographs. Much of the caption detail has also come from discussions with Childs and Avery.

10 This assessment agreed exactly with our independent one.

11 Note that forging slag was previously called *entoto*, but both are used in the furnace and thus carry interchangable names.

12 The sticks were stripped of bark and were called *kabula* and *mushishi*; the rukago was last used during the maturing of the tuyere clay in 1979.

13 *Vuma* was buried previously near the outside of the smelting house. The other roots are *musombwe*, *mwama*, *mogonile*, and *musazi*. The significance of the needle is not clear. However, Bjerke (1981, 193) mentions a needle as part of a medicinal bundle attached to a figurine used in an exorcism ritual conducted by a Zinza shaman.

14 Note that beer was not spat on the ground this time, and the rite actively involved smelters, not just surrogates.

15 Among the Bazinza, spitting beer, with a woman participating, is considered propitiating the spirits—reinforced by the Kizingo, used by spirit medium when communicating with ancestral spirits (Bjerke 1981).

References

Avery, D. H., N. J. van der Merwe, and S. Saitowitz. 1988. "The Metallurgy of the Bloomery in Africa." In *The Beginning of the Use of Metals and Alloys*, ed. R. Madden, 261–82. Cambridge: Massachusetts Institute of Technology Press.

Binford, L. R. 1987. "Researching Ambiguity: Frames of Reference and Site Structure." In *Method and Theory for Activity Area Research: An Ethnoarchaeological Approach*, ed. S. Kent, 449–572. New York: Columbia University Press.

Bjerke, S. 1981. *Religion and Misfortune: The Bacwezi Complex and Other Spirit Cults of the Zinza of Northwestern Tanzania*. Oslo: Universitetsforlaget.

Collett, D. 1985. "The Spread of Early Iron-Producing Communities in Eastern and Southern Africa." Ph.D. diss., Cambridge University.

——. 1993. "Metaphors and Representations Associated with Precolonial Iron-Smelting in Eastern and Southern Africa." In *African Archaeology: Food, Iron, and Towns*, ed. T. Shaw, B. Andah, P. Sinclair, and A. Okpoko, 499–511. London: Unwin Hyman.

Cory, H. 1949. "The Ingredients of Magic Medicine." *Africa* 17:13–32.

de Heusch, L. 1982. *The Drunken King, or, The Origin of the State*. Bloomington: Indiana University Press.

——. 1985. *Sacrifice in Africa: A Structuralist Approach*. Bloomington: Indiana University Press.

de Rosemond, C. C. 1943. "Iron Smelting in the Kahama District." *Tanganyika Notes and Records* 16:79–84.

Dewey, W. 1990. *Weapons for the Ancestors*. Film. Iowa City: University of Iowa.

Eckert, H. E. 1974. "Les fondeurs de Koni." *Géographie: Ann. de l'Université d'Abidjan* 6:169–89.

Estermann, C. 1976. "The Non Bantu People." In *Ethnography of Southwestern Angola*, ed. G. D. Gibson, 1–49. New York: Africana Publishing.

Fowler, I. 1990. "Babungo: A Study of Iron Production, Trade, and Power in a Nineteenth-Century Ndop Plain Chiefdom (Cameroons)." Ph.D. diss., University of London.

Gould, R. A. 1980. *Living Archaeology*. Cambridge: Cambridge University Press.

Gould, R. A., and P. J. Watson. 1982. "A Dialogue on the Meaning of Use of Analogy in Archaeological Reasoning." *Journal of Anthropological Archaeology* 1:355–81.

Greig, R. C. H. 1937. "Iron Smelting in Fipa." *Tanganyika Notes and Records* 4:77–81.

Griaule, M., and G. Dieterien. 1965. "Le renard pâle," *Le mythe cosmogonique*. Paris: Institut d'Ethnologie.

Herbert, E. W. 1993. *Iron, Gender, and Power: Rituals of Transformation in African Societies*. Bloomington and Indianapolis: Indiana University Press.

Killick, D. J. 1990. "Technology in Its Social Setting: The Ironworkers of Kasungu, Malawi, 1860–1940." Ph.D. diss., Yale University.

Kirknaes, Jesper. 1980. *Jernsmeden—sadan arbejder warongo-smedene i Tanzania.* Holte, Denmark: Skarv.

Kjekshus, H. 1977. *Ecology Control and Economic Development in East African History.* London: Heinemann.

Lemelle, S. J. 1992. "Ritual, Resistance, and Social Reproduction: A Cultural Economy of Iron-Smelting in Colonial Tanzania, 1890–1975." *Journal of Historical Sociology* 5:161–82.

Lévi-Strauss, C. 1966. *The Savage Mind.* Chicago: University of Chicago Press.

Martins, J. V. 1966. *A Idade dos métais na Lunda.* Lisbon.

Muhly, F., and P. Schmidt. 1988. Script for *The Tree of Iron.* Gainesville, Fla.: Foundation for African Prehistory and Archaeology.

Nchotti, E. 1975. "Some Aspects of the Iron Industry of Geita, 1850–1950." Master's thesis, University of Dar es Salaam.

Pole, L. 1975. "Iron-Working Apparatus and Techniques: Upper Region of Ghana." *West African Journal of Archaeology* 5:11–39.

Redinha, J. 1953. *Campanha etnografica ao Tchiboco.* No. 1. Lisbon.

Roberts, A. 1969. "Political Change in the Nineteenth Century." In *A History of Tanzania,* ed. I. Kimambo and A. Temu, 57–84. Nairobi: East African Publishing House.

———. 1993. "Iron, Master of Them All." Paper presented at the Fifth Stanley Conference on African Art, University of Iowa, Iowa City, March 3–6.

Roles, N. C. 1966a. "Tribal Surgery in East Africa During the Nineteenth Century: Part 1, Ritual Operations." *East African Medical Journal* 43:579–94.

———. 1966b. "Tribal Surgery in East Africa During the Nineteenth Century: Part 2, Therapeutic Surgery." *East African Medical Journal* 44:17–30.

Rostoker, W., and B. Bronson. 1990. *Pre-Industrial Iron: Its Technology and Ethnology.* Archaeomaterials Monograph, No. 1. Philadelphia.

Schmidt, P., and S. T. Childs. 1985. "Innovation and Industry During the Early Iron Age in East Africa: The KM2 and KM3 Sites of Northwest Tanzania." *African Archaeological Review* 3:53–94.

Shimmin, I. 1893. "Journey to Gambisa's." In *The Story of Mashonaland,* ed. F. W. McDonald. London: Wesleyan Mission House.

Stirniman, H. 1976. *Existenzgundlagen und traditionelles Handwerk de Pengwa von SW.-Tansania.* Freiburn: Schweiz.

van der Merwe, N., and D. Avery. 1987. "Science and Magic in African Technology: Traditional Iron Smelting in Malawi." *Africa* 57:143–72.

Wilson, M. 1957. *Rituals of Kinship among the Nyakyusa.* London: Oxford University Press.

6 Competition and Change in Two Traditional African Iron Industries

Nicholas David & Ian Robertson

Abstract

Muslim and Montagnard societies of the Mandara region of North Cameroon have sharply differentiated traditions of ironworking, each of which is undergoing rapid but contrasting development.

The two sets of smiths have reacted very differently to changing access to raw materials and other factors associated with modernization during the colonial and independence periods. Muslim blacksmiths have shown remarkable technical and managerial initiative in coping with new challenges and opportunities. Montagnard smiths, having first lost a near monopoly of the supply of bloomery iron as a consequence of the availability of modern steel and scrap, are now ceding their share of the market for finished tools to Muslim competitors. In this chapter we consider and, to an extent, explain these opposed trajectories in the context of the economic, social, and political history of the region. We conclude with suggestions for the interpretation of such technological successions in the archaeological record and predictions regarding the future of ironworkers and potters in the Mandara region.

This chapter offers a preliminary report and interpretation of two competing traditional technologies and of the spread of one at the expense of the other. Archaeologists frequently observe evidence of such processes in the archaeological record, and indeed technological successions are *par excellence* the phenomena in terms of which entities such as cultures are delimited and defined. This case study also has implications for those involved in rural animation and development in that it demonstrates that a holistic approach is required to make sense of technological change.

The Mandara highlands of North Cameroon are occupied for the most part by small scale societies practicing traditional religions and ranging in complexity from unhierarchized communities to petty chiefdoms (maps 6.1 and 6.2). We refer to such societies as Montagnard in contrast to the plains-dwelling Muslim societies, primarily the inheritors of the precolonial Wandala state and of various Fulbe (Fulani) chiefdoms, the latter formerly (and often only formally) "feudal" dependencies of the Sokoto caliphate. In precolonial times, the highlands were regarded by the Muslims as a reservoir of slaves, who were obtained by raiding or by trade with those Montagnard groups either willing or pressured to act as middlemen. The Fulbe, raiding from the west and east, were not entirely predatory, engaging, for example, in the exchange of cattle for grain and other Montagnard products (Mouchet 1944, 58), but there was little sustained and patterned interaction. The Wandala, on the other hand, settled on the plains around the northern end of the mountain chain, maintained close relationships at least with the Montagnard Mora (Muraha), among whom they were on occasion constrained to seek refuge while escaping more powerful invaders (MacEachern 1991; Forkl 1984). In general, however, not only Muslim-Montagnard but also intra-Montagnard relationships were strained. Competition for farmland and wives among the hill peoples contributed to an endemic state of hostility that pitted settlement against settlement but nonetheless did not preclude peaceful economic interaction or even interdependence between Muslims and Montagnards.

This paper addresses only the sharply contrasting Wandala and Montagnard traditions of ironworking, and not that of the Fulbe. The latter have long obtained their iron and their ironworkers from various sources and do not represent a distinct, coherent metallurgical tradition. Although there are now Fulbe wholesalers of iron stock, their ironworking has had no significant impact on this area.

In the central Mandara highlands (Lembezat 1950)—the area around Mokolo that is the southern part of our research area—the smiths among the Kapsiki (van Beek 1987), the Mafa (the largest linguistic entity within the improperly grouped "Matakam" cluster) (Martin 1970), and many of the

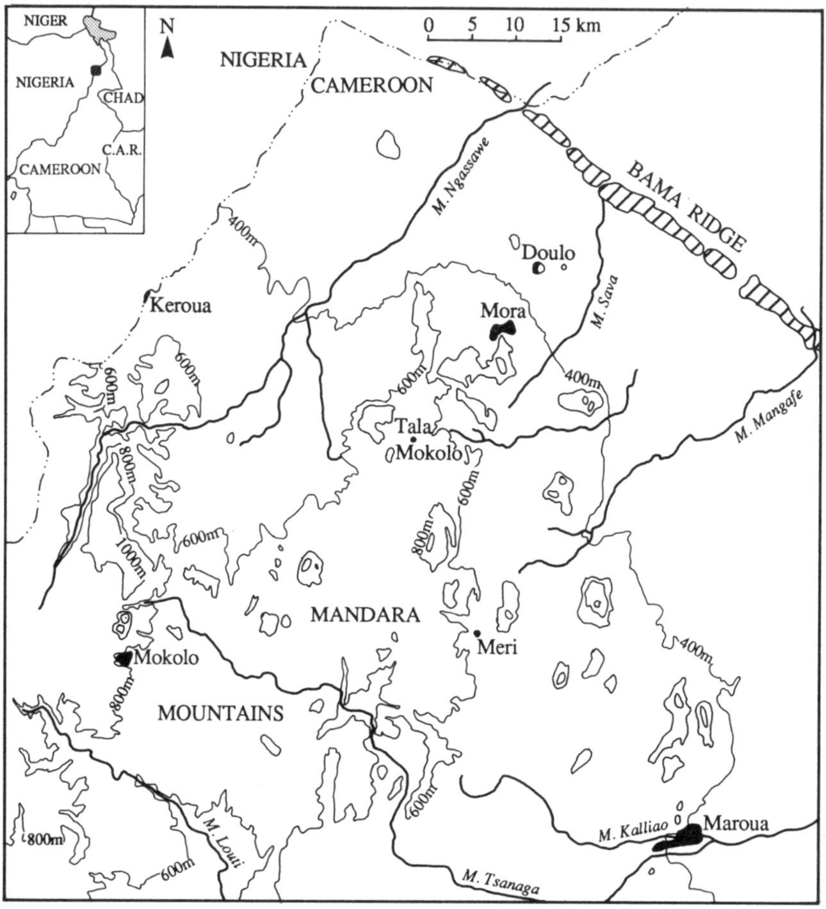

MAP 6.1. The northern Mandara highlands and adjacent plains, showing topography, rivers (river = mayo in Fulfulde), and selected towns and settlements.

other Montagnard societies form an endogamous caste (Podlewski 1966; Genest 1974; Sterner and David 1991). As commonly elsewhere in West Africa, this caste is regarded with considerable ambivalence (see, for example, McNaughton 1988) (map 6.2). Its members traditionally perform a variety of transforming functions, often including divination, the disposal of the dead, and the smelting, fining, and forging of iron; the women hold a monopoly on potting and specialist midwifery. Both sexes commonly act as curers, women specializing in pediatrics, while senior male smiths may perform important ritual duties on behalf of society at large. In the northern part of the range,

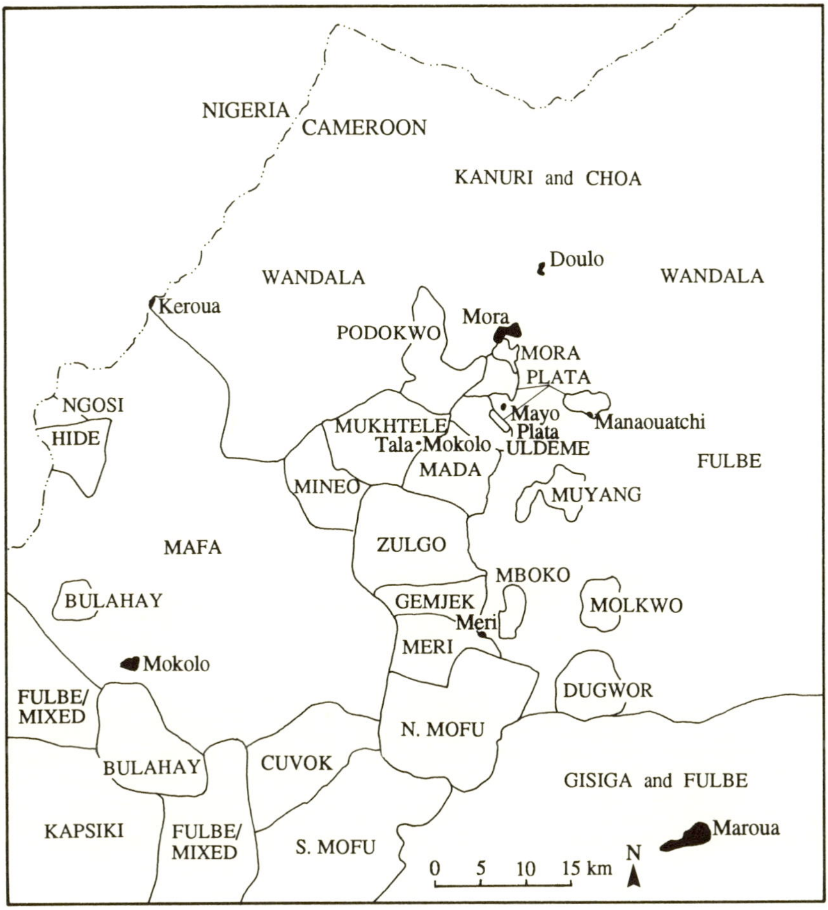

MAP 6.2. Ethnolinguistic groups in the Mandara highlands with selected towns and markets.

the smiths of the Uldeme (de Colombel 1986) and related Montagnard groups are not casted. While they belong only to certain major lineages, neither are they subject to proscriptions (although there are other lineages whose members it is considered dangerous to marry), nor do they customarily fulfill other of the southern smiths' traditional functions.

Among the Wandala, ironworking is hierarchically organized. There are two major production centers: Manaouatchi, where Robertson concentrated his work, is the senior; the other is Keroua, on the Nigerian border. Manaouatchi, a town of 300–400 inhabitants, contains the largest concentration of

Muslim smiths in the Mandara region. Situated 12 km southeast of the Wandala capital of Mora and 10 km east of the major Uldeme market at Mayo Plata, this town has been a focus of Wandala ironworking since the reign of May Aji Bukar (1731–53) (Seignobos 1986, 34; Mohammadou 1982, 11). The hereditary chief of the Wandala blacksmiths resides here, and there are at present about thirty active blacksmiths. Not all Muslim smiths enjoy the benefits that a guild town of this sort affords. Many craftsmen, perhaps born and raised in Manaouatchi, now work in other small towns and villages, serving the more immediate needs of their local clienteles. The smiths of Manaouatchi regard their town as the headquarters of the Wandala iron trade and the forges in the second level of the hierarchy as branch establishments.

According to Forkl (1984, 11–12) Wandala blacksmiths are, along with butchers, "Méprisés et placés au niveau social le plus inférieur, on n'entra presque jamais, jusqu'au temps de l'indépendence du Cameroun [1960], par mariages dans leurs familles." Robertson found no evidence that they are presently held in low regard, however. On the contrary, they are respected and important members of their communities. Instances of marriage between blacksmiths and the daughters of Muslim leaders not only corroborate this interpretation but suggest that it holds for more than one generation into the past. The discrepancy, Forkl suggests, probably results from lingering conservative attitudes in the Wandala capital (Forkl, personal communication).

Wandala smiths are best described as members of a guild with its own rules and responsibilities, governed by its own chief, and into which members are recruited through a rather formal apprenticeship. Although apprentices are most commonly sought among the Wandala smithing community, some are recruited elsewhere. In at least two recent instances, Montagnard apprentices have been taken on and, given the requisite talent and conversion to Islam, may eventually become full-fledged blacksmiths. The degree to which this may have occurred in the past is unclear; a current shortage of suitable apprentices results largely from a greatly expanded field of career options in the modern world. Members of the guild claim that none of the qualified smiths working in Manaouatchi are of non-Muslim or non-Wandala origin.

Up to the 1940s and early 1950s, most of the iron consumed in the Mandara region was locally smelted by Montagnards using a unique type of bellows-driven, downdraft furnace and high-quality magnetite ore eroded from local granites (David, Heimann, Killick, and Wayman 1989). Because the ore originates in the mountains, Montagnards had a natural logistic advantage.

Although Wandala ironworkers have told us that they sometimes traded with the Montagnards for ore that they smelted themselves, their production was, it seems, quite insufficient to supply all Wandala forges. Major Dixon Denham (in Bovill 1966, 3:356–57), visiting the region in 1823, was told that the area around Kerawa was the main source of iron supply. In describing a Wandala forge, he mentions that "large masses of the iron, as nature produces it, were lying about; and they appeared to me as so many rusty earthy masses." Almost certainly what he saw were blooms that had been carried down from the hills. The Wandala smiths fired and forged them into various goods for local distribution and into "hinges, small bars and a sort of hoe used to weed the corn, and send them for sale to the Bornou towns." The rich iron ore of the Mandara region and the iron trade north to the southern margins of Lake Chad were first reported by Lorenzo Anania in the mid-sixteenth century (Lange and Berthoud 1972, 343, 351; Hunwick 1971, 210). The trade may well be as old as the Iron Age itself and is still of importance today.

Not all Montagnards produced iron; certain villages and groups with access to concentrated ores specialized in smelting. Among these were the Sukur (25 km west of Mokolo and just across the border in Nigeria), the Mabas, Hide, and several Mafa villages in the south, and the Uldeme, Molkwo, and Podokwo (de Colombel 1986; Cuignet 1968; Lembezat 1950) further north. We do not know how the prices of ore and blooms were set, but we presume that they could be manipulated to some limited extent by Montagnards and that by such adjustments they could and would have resisted any attempt by Muslim smiths to obtain a share of their market in finished tools.

Starting before World War II and increasing rapidly in the late 1940s, imported iron stock, vehicles, and other machinery from the industrialized West offered new sources of metal that by the early 1950s had put almost all the Montagnard smelters out of business. Scrap replaced bloomery iron and steel, and the flow of metal was reversed. The larger towns, all except for Mokolo situated on the plains, became entrepôts of iron stock. For the first time the Muslim smiths found themselves closer to the source of supply than their Montagnard counterparts. In spite of Montagnards' occasional attempts to revive their smelting and so free themselves of a debilitating dual dependence on Muslims and Western steel (see Sassoon 1964 for an example from Sukur), they have been unsuccessful. The Muslims now have a near monopoly of iron supply; we suspect that in Mokolo, the seat of the prefecture that covers most traditional Mafa territory, Mafa converts to Islam control it.

The Wandala smiths were quick to exploit the newly favourable terms of access, benefitting both directly and from substantial markups on stock sold to Montagnards.

Montagnard smiths have on the whole persisted in their traditional roles; Western industrial products in particular have made few inroads. Nevertheless, changes have occurred; the reversal of the flow of iron and the establishment of the modern market system are perhaps the two main factors that have placed Montagnard products in direct competition with those of Muslim smiths. While in the south, largely due to the presence of Mokolo as a secondary entrepôt, the hill peoples still obtain the majority of their iron goods from blacksmiths of their own group, this is less and less the case in the north, where an increasing number of items, particularly those associated with agriculture, are produced in the forges of Manaouatchi.

Wandala smiths stressed forging over smelting even prior to the general availability of industrial scrap metal. They have since shown remarkable technical and managerial initiative in responding to the increased opportunities offered by improved access to raw materials and by a more open national society. The potential for increased sales accompanying market centralization has been realized through the adoption of improved smithing and marketing techniques. Individual smiths vary their products according to their age and abilities and to changing patterns in the supply and demand for apprentices, assistants, and metal.

Approximately half the local markets, including Mayo Plata, have a significantly non-Muslim, ethnically diverse clientele. Shoppers are offered a wide range of products, some specially targeted toward specific ethnic groups. The development of distinct ethnic markets and the movement of hill peoples onto the plains have also created a need for new modes of product distribution. Wide sales territories are now serviced from Manaouatchi by two specialist smiths who act as dealers, distributing finished products to the various markets. They are the middlemen for most of the Manaouatchi blacksmiths. Anyone accepted as a smith can expect to have his products purchased by one of these two, who will then take them to the appropriate markets, which may be quite distant. Although blacksmiths could realize a higher unit profit by selling the goods in person, their time is better spent in the forge, doing what they do best.

We will exemplify the organizational and technical advantages of the Wandala smiths by contrasting aspects of their profession with those of their

Montagnard counterparts. First are the simple facts of their concentration in Manaouatchi and their membership in the brotherhood of Islam, which offer numerous benefits, facilitating their ability to act as a unit under the guidance of a guild chief, to obtain the best possible prices for raw materials, and to plan production in such a way as to maximize output of a full range of goods while minimizing internal competition. The Montagnard smiths, on the other hand, live in widely dispersed compounds, and have only limited regard for social ties that are not kin-based. Smiths of other clans and lineages, even if of the same community or linguistic group, may be regarded with a degree of suspicion that precludes major cooperative endeavor.

Many of the Wandala smiths approach full-time employment as blacksmiths; they specialize individually in the production of a limited range of types and do so to sell to specific groups of clients. Most Montagnard smiths both engage in a number of other specialist activities and until recently were prepared to reproduce the complete tool kit to order. Although markets located in the hills are a colonial introduction, smiths did forge tools for sale with the assistance of their unmarried sons, whom they trained in this manner. Whenever, depending on the political situation, Montagnards dared to attend, they would sell some of their wares at Mora and other Wandala markets at the feet of the mountains.

Due also to their more primitive forging equipment, the Montagnards are less efficient blacksmiths than the Muslims. Heavy forging is conducted with massive (about 6 kg), unhafted boulder hammers, wielded either by an assistant or, especially when a tool is being made to order, by the client (figure 6.1). The smith directs the blows and does finer forging with small cobbles or an iron bar hammer. In Uldeme forges the bellows are worked by a second assistant; the placement of this apparatus is such that its operator is effectively prevented from contributing further to the forging (figure 6.2). Although the layout of some Montagnard forges allows the bellows operator to double as striker, the physical effort involved cannot be maintained for very long. A third worker, often a lad, is therefore needed for sustained production.

Wandala smiths normally work in pairs, and always do so when manufacturing the better quality of hoes (figure 6.3). The apprentice or mate blows the bellows and both he and the smith collaborate on the forging. Their roles are well defined by custom and practice, and they are generally expert. Because Muslim smiths are colleagues, as Montagnard smiths are not, and because particular smiths tend to specialize in certain products, technical innovations

FIGURE 6.1. Heavy forging with a boulder hammer in a Mafa forge.

seem to occur and spread more easily. An example is the addition of Western-style sledgehammers to the traditional repertoire of forging equipment. Within the memory of practising smiths, heavy forge work employed the same boulder hammers still used today in the hills. The energy expended by the Wandala striker is much less than that of his Montagnard competitor, as is the time in which a hoe can be forged to shape.

Tool production is further streamlined through highly standardized and efficient ways of cutting up stock into tool blanks. The number of hoes, for example, that can be expected from a given truck wheel rim can be estimated quickly and exactly, and the amount of steel wasted is minimal. By such means they are capable of manufacturing twenty quality hoes a day; a tradi-

FIGURE 6.2. An Uldeme forge, showing the position of the bellows man.

tional Montagnard forge with a greater input of labor and charcoal can produce only about twelve.

Muslim smiths, small businessmen animated by the profit motive and having ready access to market trends and demands through their ironmonger colleagues, also appear far more willing to innovate in other ways. Robertson (1992) has described how the Muslims have developed a range of hoes targeted at various ethnic markets. These products are differentiated on the one hand functionally, in part to supply tools adapted to the stony soils of the terraced hillsides, and on the other stylistically, in order, or so the Wandala suppose, to appeal to differing ethnic tastes. The forms of these various types are maintained through a set of sheet metal templates that, largely as a teaching aid for new apprentices, promotes standardization. The Wandala smiths have developed a form of ethnic packaging (figure 6.4).

We strongly suspect that historical, cultural, and economic factors have facilitated the penetration of Montagnard markets by Wandala blacksmith middlemen. At the time of the mid-eighteenth century conversion of the Wandala elite to Islam, their smiths fled from Doulo, where they had been established, to Bogo, 77 km east of Mora. They were then persuaded to return by the Sultan, Aji Bukar. However, they settled neither in Doulo nor in the

FIGURE 6.3. Smithing in a Wandala forge. Note the sledgehammer.

capital, but at a discreet distance in Manaouatchi (Seignobos 1986, 34). This tradition betrays a tension between the Wandala state apparatus and the smiths that sets the latter somewhat apart from the bulk of the population. We hesitate to speak of submerged ethnicity, but ironworkers whose profession was profoundly embedded in an autochthonous ideology cannot have converted wholeheartedly and without retention of an important residue of their former beliefs. It would appear that one by-product of their liminality and ambiguous status was an ability to continue to operate on both sides of the major ethnic divide.

In precolonial times, before the decline of the traditional smelting industry, ironworkers of diverse ethnic origins interacted and were to an extent exempt from the imminent threats of violence and enslavement that inhibited travel. From the Wandala viewpoint, iron was too critical a commodity to justify either killing or alienating those with the knowledge and the skills to obtain and work it. Similarly, the Montagnards could not afford to lose access to certain necessities—salt and dried fish are the most often mentioned. Most interaction between Montagnard and Wandala ironworkers was, we believe, occasioned by Wandala going to the mountains either to purchase ore or bloom or to carry out a smelt. If this is so, the Muslims, favored in any case by the colonial administration, would have been preadapted through knowledge and experience to expand their sales into the territories opened up by pacification.

Montagnard responses to Muslim initiatives vary. By 1940 the Mukulehe, a Podokwo subgroup, were obtaining most of their iron tools from Wandala smiths, probably those of the Keroua-Assigassia center (Lembezat 1952, 162). Several Montagnard groups no longer have any practicing smiths to serve them. In the northern part of the area, the Podokwo, Mukhtele, and Mada do not produce their own iron tools but supply themselves at the Mayo Plata, Mora, Tala Mokolo, and lesser markets. Indeed, perhaps because of their relative ease of access to Mora and Mokolo and of the importance of the Mayo Plata market, the Uldeme smiths appear to be the only northern Montagnard ironworkers still in full production. But their twenty-odd forges are unable to satisfy local demand, and even their home market has been invaded by goods

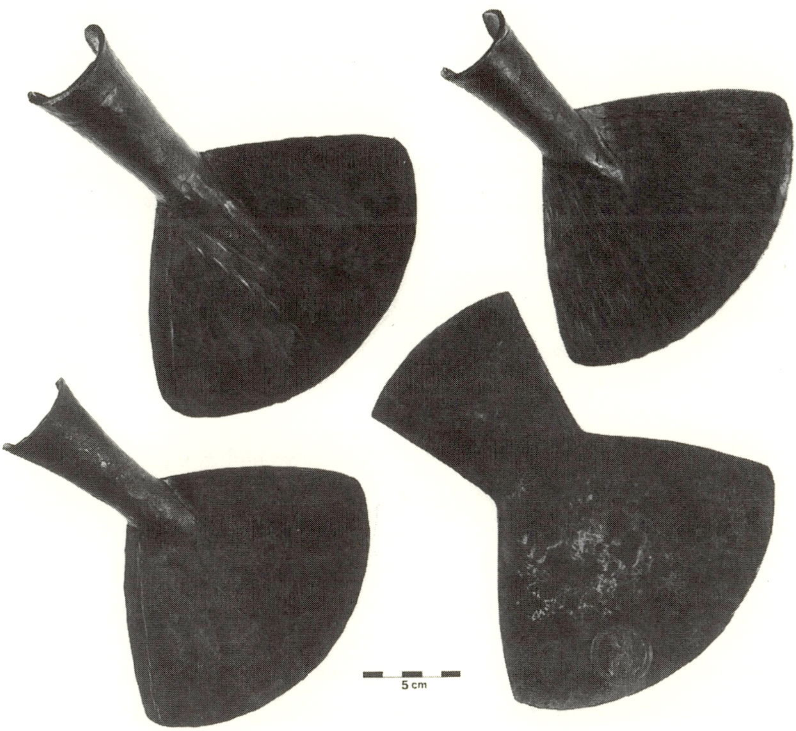

FIGURE 6.4. Three named Wandala hoe types and their common template.

from Manaouatchi. Some 20 km further south, the Gemjek obtain their iron tools either from the weekly market at Meri, now in large part supplied from Manaouatchi, or at their own market to which some Mafa smiths, who we suppose can obtain iron relatively cheaply from Mokolo, bring their wares. A little further from the sources of supply, there is still a smith among the southern Zulgo, but he works irregularly and serves mainly to sharpen and repair tools manufactured elsewhere.

The acceptance of a dependence upon others, even upon Muslims, for basic tools thus constitutes one Montagnard response to changing circumstances. Some of their smiths are attempting to retain market share by investing in new technology. A sledgehammer allows a smith to forge tools with only his wife or a child to assist in blowing the bellows, a significant reduction in labor costs (figure 6.5). The actual cost of the sledge is not great, perhaps the equivalent to the gross returns of two days smithing, but it requires at least one relatively expensive day trip into Maroua, the provincial capital and uncertain source of supply. Thus, some Montagnard smiths are modifying their traditional practices, for the most part slowly, with a view to increasing productivity and thus reducing labor costs.

In conclusion, the foregoing sketch suggests some generalizations that, albeit near truisms, may be of some use to prehistorians confronted with technological successions in the archaeological record. First, context is crucial. In this case the factors that set the stage for the competitive success of Muslim technology are economic (the reversal in the tide of iron), sociological (the liminal status of the Wandala smiths), and political (the colonial pacification). More proximately, the organization of production—centralized on the plains and fragmented in the hills—plays a critical role, one that appears to repeat a pre-Wandala pattern on the plain detectable archaeologically at the site of Mehe (David and MacEachern 1988). Next, the two technologies are in genuine economic competition; neither is being imposed, nor is the market distorted by state intervention. In such situations we may predict that whereas the artisans of the expanding system will innovate by expanding both the stylistic and functional ranges of their products in the hope of attracting a wider clientele, the craftsmen of the threatened group are likely to be forced to make changes in their technology, usually toward that of their rivals, in order better to compete. (This is a slight oversimplification; the Uldeme smiths are indeed expanding their repertoire of hoes, largely to fill niches vacated by smiths of neighbouring Montagnard groups.) Conversely, were one society dominating

FIGURE 6.5.
A Mafa smith who often works alone with his wife on the bellows, using a small sledgehammer for heavy forging.

the other and imposing its technology, we would expect that, along with the disappearance of one school of craftsmanship, there would be substantial stylistic continuity in the products of the other.

Finally, let us look to the future. Muslim blacksmiths efficiently and cheaply satisfy a demand that is, if anything, likely to increase in the next decades. As they control recruitment into the profession, their immediate future as specialist craftsmen is probably secure. Some of the casted southern Montagnard smiths, those centered on Mokolo, will survive through increased professionalism and by accepting technological change. The roles of these smiths' womenfolk have changed little and appear in the foreseeable future likely to change much less than those of the men. Pots are still produced in quantity and very few women give birth in hospitals. This should translate into an improvement in the status of women, although less so than in the north, where Montagnard women potters, in marked contrast to male smiths, now monopolize pottery supply to most Wandala markets.

Whereas in precolonial and colonial times some division of labor occurred among Montagnard smiths, the independence period has seen the loss of the smelting specialization and, locally, of other prerogatives—of burial as a result of the spread of the influence of Christian missions and of some healing with the increased availability of Western health care. As the task of directing traditional funerals is closely bound up with ritual practices that are unacceptable to the growing numbers of Christian and Muslim converts, there is little possibility of adjusting this speciality to new circumstances, but this is not the case with either divination or curing, especially of ailments with an important psychosomatic component. Peaceful conditions enabling people and information to move about much more freely have allowed certain experts in these skills to establish reputations that transcend their own settlements and even ethnic groups. But the number of such consultants will remain limited. Underemployment of smiths is already a reality. Some casted Montagnard smiths are turning to full-time farming, an activity denied to them in the past, and more will do so in the future. Over time such persons, no longer practicing any of the specialities reserved for smiths, become scarcely distinguishable from their neighbors. This is the likely fate of the vast majority of the members of the smithing lineages in the northern mountains. Even in those societies where smiths are casted, their children, in spite of remarkably persistent cultural attitudes toward them (Sterner and David 1991), will eventually lose their distinct status. Meanwhile it remains to be seen whether the remaining specialists, diviners and healers, will retain one foot in the forge and thereby their connection to the potent set of symbols associated with ironworking, or whether, as increasing division of labour might seem to render inevitable, there will be further disassociation of once interrelated roles.

Notes

Authors' note: The research on which the paper is based was carried out under authorizations from the Ministry of Higher Education and Scientific Research of Cameroon, and within the context of a Protocole d'Accord de Coopération between the University of Calgary and the Office de la Recherche Scientifique et Technique Outre-Mer of France. It was supported by Canadian Social Science and Humanities Research Council grants and an SSHRC Leave Fellowship awarded to Nicholas David.

A draft was presented at the Thirty-First Annual Meeting of the African Studies Association (Chicago, October 1988). N. David thanks Peter Schmidt, who made possible his attendance. We also thank Scott MacEachern for information on the northern Montagnard groups, Stephen Peitchinis for an economist's view of our materials, and Christian Seignobos for copies of several important papers in advance of publication.

References

Bovill, E. W. 1966. *Missions to the Niger.* Cambridge: Cambridge University Press.

Cuignet, M. 1968. "Les Mada—Contribution à l'Étude des Populations du Nord-Cameroun." *Bulletin de l'IFAN*, Série B, 30:1062–1146.

David, N., R. Heimann, D. Killick, and M. Wayman. 1989. "Between Bloomery and Blast Furnace: Mafa Iron-smelting Technology in North Cameroon." *African Archaeological Review* 7:183–208.

David, N., and A. S. MacEachern. 1988. "The Mandara Archaeological Project: Preliminary Results of the 1984 Season." In *Le Milieu et les Hommes. Recherches Comparatives et Historiques dans le Bassin du Lac Tchad,* ed. D. Barreteau and H. Tourneux, 51–80. Paris: ORSTOM.

de Colombel, Véronique. 1986 . *Phonologie Quantitative et Synthétique avec Application à l'Ouldémé, Langue Tchadique du Nord-Cameroun.* Langues et Cultures Africaines, No. 7. Paris: Société d'Études Linguistiques et Anthropologiques de France.

Forkl, H. 1984. "Recherches Ethno-Historiques chez le Peuple de la Langue Mandara (17 Février–10 Novembre 1984)." Report to the Ministry of Higher Education and Scientific Research, Yaoundé, Cameroon.

Genest, Serge. 1974. "Savoir Traditionnel chez les Forgerons Mafa." *Revue Canadienne des Études Africaines* 8:495–516.

Hunwick, J. 1971. "Songhay, Bornu and Hausaland in the Sixteenth Century." In *History of West Africa,* ed. J. F. A. Ajayi and M. Crowder, 1:202–39. London: Longman.

Lange, Dierck, and Silvie Berthoud. 1972. "L'Intérieur de l'Afrique Occidentale d'après Giovanni Lorenzo Anania (XVIe siècle)." *Cahiers d'Histoire Mondiale* 14:298–351.

Lembezat, Bertrand. 1950. *Kirdi—Les Populations Paiennes du Nord-Cameroun.* Mémoire de l'Institut Français d'Afrique Noire. Yaoundé: I. F. N. Centre du Cameroun.

———. 1952. *Mukulehe—un Clan Montagnard du Nord-Cameroun.* Paris: Éditions Berger-Levrault.

MacEachern, A. S. 1991. "Les Gens de Ngolele: An Examination of Prehistoric Ethnic Relations in the Northern Mandara Mountains." In *Du Politique à l'Économique: Études Historiques dans le Bassin du Lac Tchad,* ed. J. Boutrais, 165–92. Paris: ORSTOM.

Martin, J.-Y. 1970. *Les Matakam du Cameroun. Essai sur la Dynamique d'une Société Pré-industrielle.* Mémoire ORSTOM, No. 41. Paris: ORSTOM.

McNaughton, Patrick R. 1988. *The Mande Blacksmiths.* Bloomington: Indiana University Press.

Mohammadou, E. 1982. *Le Royaume du Wandala ou Mandara au XIXe siècle.* African Languages and Ethnography, No. 14. Tokyo: ILCAA.

Mouchet, J. 1944. "Rites agraires et classes d'âge." *Études camerounaises* 6:51–61.

Podlewski, A. M. 1966. *Les Forgerons Mafa: Description et Évolution d'un Groupe Endogame.* Cahiers ORSTOM, sér. Sciences Humaines 3, No. 1. Paris: ORSTOM.

Robertson, I. G. 1992. "Hoes and Metal Templates in Northern Cameroon." In *An African Commitment: Papers in Honour of Peter Lewis Shinnie,* ed. J. Sterner and N. David, 231–40. Calgary: Calgary University Press.

Sassoon, H. 1964. "Iron-smelting in the Hill Village of Sukur, North-eastern Nigeria." *Man* 64:174–78.

Seignobos, C. 1986. "Les Zumaya ou l'Ethnie Prohibée." Expanded version of a paper presented at the IIIe Colloque Internationale Méga-Tchad, Paris, September 1986.

Sterner, J., and N. David. 1991. "Gender and Caste in the Mandara Highlands: North-eastern Nigeria and Northern Cameroon." *Ethnology* 30:355–69.

van Beek, Walter E. A. 1987. *The Kapsiki of the Mandara Hills.* Prospect Heights: Waveland Press.

7 Forging Symbolic Meaning in Zaire and Zimbabwe

S. Terry Childs & William J. Dewey

Iron smithing has received relatively little attention in the recent literature on prehistoric African ironworking, particularly in contrast to iron smelting. Africanist archaeologists, for example, are increasingly examining the interrelationships among ancient techniques of iron smelting, social relationships, cultural ideology, and power negotiations (see Schmidt 1983; Collett 1985; Killick 1990; Childs 1991b; Herbert 1993; Childs and Killick 1993), yet similar studies focused on iron smithing remain rare (Maret 1980, 1985a; Childs 1991c). In particular, there has been little investigation of the specific smithing techniques and processes (figure 7.1) by which indistinct metal pieces are rendered into recognizable objects and are given function and meaning in prehistoric societies.

This chapter has two primary objectives. The first is to evaluate how iron smithing and object use involve patterned decision-making that can embody, be activated by, or manipulate cultural values and concerns. The second is to compare how iron objects were conceived, made, and used in two areas of central Africa in order to examine these objects' similarities and differences in light of regional histories and sociocultural configurations.

The specific focus of this inquiry is the production, use, and meanings of iron axes in precolonial Zaire and Zimbabwe. Axes are particularly useful for our objectives because they exemplify how decisions made during production and use influence and are influenced by sociopolitical relations and cultural concerns based in worldview and cosmology (Pfaffenburger 1988, 1992;

Lechtman 1977). Because all social behavior, including technology, is structured by beliefs and values that explain human existence and impose order on society, these ideas help to pattern traditions of object production and use. Thus, smiths from different societies often manufacture axes differently. They may choose to hammer out a lump of metal and cut out a form, fold and weld the metal into shape, pile weld different pieces of metal (some may be pieces of recycled objects) together, or cast an axe from liquid metal. Another result of the interplay between technology and culture is the wide variety of forms, functions, and meanings of axes across Africa (Widstrand 1958).

The comparative approach provides some insight into the differing relationships between iron smithing and the precolonial societies in which they were practiced by focusing on two areas of central Africa, the Upemba Depression of southeastern Zaire (where the current occupants, the Luba, and their predecessors have resided for some 1500 years) and Zimbabwe (where the Shona peoples of today and their antecedents have lived for a comparable period of time) (map 7.1). The rich archaeological and ethnographic evidence from these areas provides unique foundations on which to study the long history of metal production and use within contexts of varying degrees of sociopolitical complexity.

The first section of this chapter compares the two culture areas in several ways. It provides brief accounts of the cultural histories of the two areas as they have been reconstructed to date and then examines the iron-production techniques used in each area, how axes were used in different societies over time, and the multiple levels of meaning they exhibited and acquired. The results show some remarkable continuities in the form and decoration of axes used in each area along with some provocative similarities in the manufacture, functions, and meanings of axes between the two areas. Yet there are also some interesting differences in how the primary functions and meanings of axes changed in Zaire and Zimbabwe over time. The second section offers some speculations that provide potentially fruitful avenues for future research.

Comparing Axe Production and Use in Zaire and Zimbabwe: Historical Backgrounds

The culture history of iron-using peoples in the Upemba Depression of southeastern Zaire is reconstructed from excavations at five burial sites located along several lakeshores (Nenquin 1963; Hiernaux et al. 1971; Maret 1985a, 1992).

Some ceramic evidence indicates that the area was first occupied by farmers and fishermen around the sixth century A.D. The earliest evidence of metal use, however, comes from a grave dating to approximately 600–700 A.D. (Maret 1982, 1992).

By the Ancient Kisalian period, beginning in the late eighth century, the local people used well-made iron tools (axes, hoes, and spears) and pottery for food production and other activities. Some individuals were buried with iron jewelry and distinctively shaped iron objects for symbolic and ritual purposes. Recovered from several opulent graves were iron axes of unusual shape and decoration that were nonetheless noticeably similar to recent Luba ceremo-

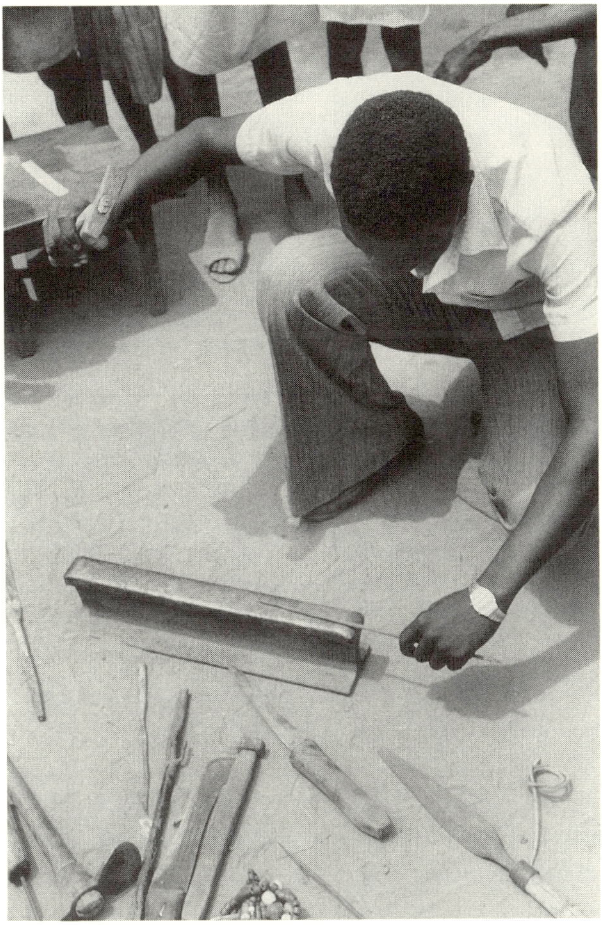

FIGURE 7.1. A Luba blacksmith, 1988. Photo by William J. Dewey.

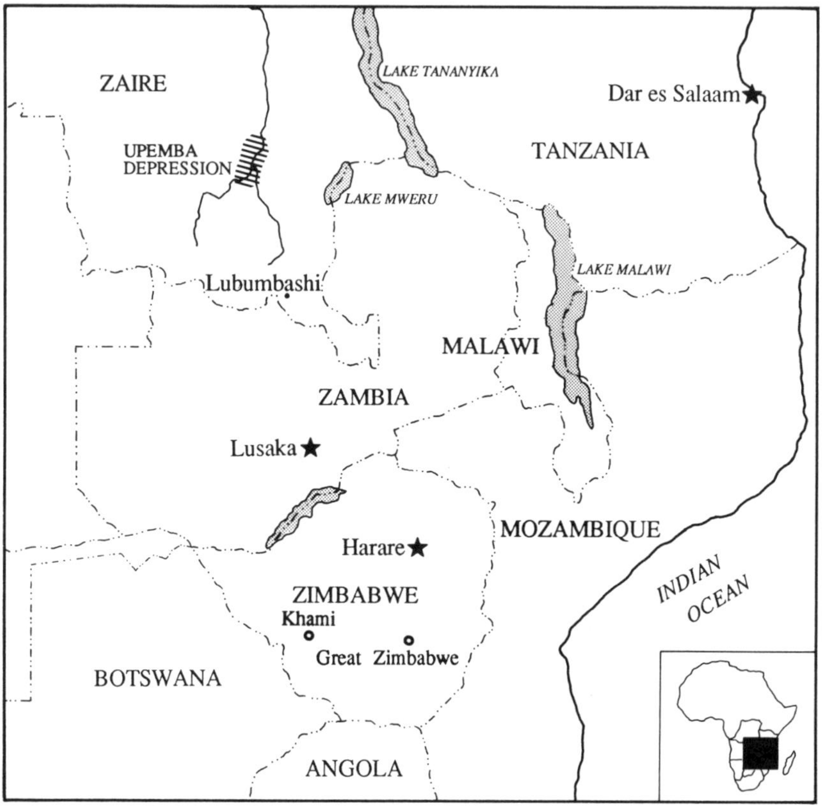

MAP 7.1. The locations of the Upemba Depression in Zaire and the principal archaeological sites in Zimbabwe.

nial axes (Maret 1979, 1985b, 1992). Small amounts of copper also began to be imported into the Upemba during the Ancient Kisalian period, probably from the Katanga copperbelt to the south, and were primarily used for jewelry. In addition, a few of the Ancient Kisalian burials contained a disproportionately large number and variety of metal, ceramic, and other goods suggesting some degree of sociopolitical differentiation and ranking (Maret 1977, 1979; Bisson 1976).

Some changes in the decoration and shapes of ceramic vessels, a noticeable increase in the quantity and types of copper objects, and a complex funerary practice associated with the graves help mark the advent of the Classic Kisalian period in the tenth century. Another highlight is the sharp increase in the number of people buried during this period, suggesting an increase in population (Maret 1982). Fishing was an important component of the local economy, as indicated by many iron and copper fishhooks and harpoons in the graves. The fine craftsmanship of various ceramic, iron, copper, and ivory objects suggests that they were manufactured by specialists, sometimes only for funerary purposes (Maret 1979, 1982). Finally, the inequities in the quantities and types of objects placed in different burials, particularly the amount of copper ornaments and the occasional ceremonial axes and bells,[1] indicate that the society became increasingly hierarchical through the thirteenth century and may have become an early state. Some of the richest burials were of children, indicating that rank and status were inherited (Maret 1982).

Changes in burial ritual and pottery manufacture mark the beginning of the Kabambian period in the fourteenth century. These changes coincided with a decrease in the amount and variety of grave goods, particularly such objects as ceremonial axes.[2] Although the number of copper objects also decreased during this period, special-purpose currency was introduced in the form of cast copper crosses, possibly for bride-wealth payments. Over time, the crosses became smaller and more standardized and were used as general-purpose currency for trade and tribute (Bisson 1975, 1982; Maret 1981).

The Luba achieved a level of sociopolitical organization consistent with mature state formation by at least the eighteenth century, although possibly long before (Maret 1979; Reefe 1981). Luba society was and still is hierarchical, is led by a sacral king (the *Mulopwe*) and once included numerous small neighboring chiefdoms as tributaries. Today, Luba royalty, a hierarchy of title-holders, the secret *Bambudye* society, and spirit mediums (*Bwana Vidye*) still play important social and political roles (Nooter 1991).

Iron has been critically important to the Luba economy over the centuries as an item of trade, for daily subsistence use, and for a variety of ritual and symbolic purposes. Control over access to rich iron ore deposits provided an important source of economic and political power for some Luba leaders (Reefe 1981). Today, iron smelting has died out, but numerous local smiths make tools of scrap iron, and people commonly carry axes and knives for daily use.

Current understanding of the complex culture history of iron-using peoples in Zimbabwe derives principally from excavations at occupation rather than burial sites. The earliest Iron Age sites of the Gokomere, Ziwa, and Zhizo occupations date between 200–900 A.D. and were small farming villages with pole-and-*daga* houses. The occupants domesticated animals and made iron and copper objects, as well as distinctive pottery (Huffman 1978; Hall 1987). Increasingly hierarchical societies, such as Gumanye and Leopard's Kopje, developed during the tenth century. These cattle-herding people are associated with the commencement of "Shona" culture (Huffman 1978, 1986; Beach 1984).

A distinctly hierarchical culture emerged in the Shashi-Limpopo region during the early twelfth century A.D. (Huffman 1982). The first cultural center was Mapungubwe, which is distinguished from earlier period sites, such as K2, by profound changes in architecture (such as houses made of solid, red-daga walls and stone walls to demarcate important areas), settlement layout (elite residences located at higher elevations than commoners), and technology (such as ceramics and gold) (Huffman 1986). The elite at Mapungubwe also controlled trade networks to the Indian Ocean. As Mapungubwe collapsed in the thirteenth century, another center garnered power and took control of the trade networks. Building began to the north at the new elite center of Great Zimbabwe in the thirteenth century (Huffman and Vogel 1991). Now renowned for its magnificent stone architecture, Great Zimbabwe housed a stratified society led by a sacral king and a minority elite who traded gold, ivory, iron, and exotic skins with the coast, exacted payments from tributary chiefdoms, and kept large herds of cattle as their bases of wealth. Furthermore, the elite legitimized their power base through extensive rituals and use of numerous symbols (including specially shaped axes and spears) directed to the ancestral spirits (Garlake 1973, 1982; Huffman 1986).

Great Zimbabwe began to decline in the middle of the fifteenth century (see also Collett et al. 1992), probably partly as a result of the overuse of resources and of a shift in power relations as other Shona states such as Torwa and Mutapa arose. Repeated episodes of fusion and fission between smaller chiefdoms and states, related to each other through trade and tribute payment, characterized the last few hundred years of Zimbabwe history prior to colonization (Beach 1980).

Shona society seems to have become less hierarchical after the demise of the far-reaching centralized state system. Today, traditional chiefs have almost no political power, although spirit mediums and diviners remain very active religious leaders. Just as in the Luba case, iron was a critical component of Shona society for centuries, and iron implements played key roles in ritual, everyday uses, and trade. Iron smelting died out in the early 1900s largely as a result of colonial attempts to regulate the manufacture of weapons and to market European iron.[3] The relatively few smiths who still forge iron do so with scrap iron, primarily making objects to serve religious or ritual demands (Dewey 1986, 1991b).

Techniques of Iron Production

Very little is known about the ancient smelting techniques in either area through either excavations or metallographic analyses, yet iron is found in many of the Upemban graves in Zaire and at many of the early occupation sites in Zimbabwe. Some aspects of ancient smithing practices have, therefore, been reconstructed. Our recent laboratory analyses reveal that the metal produced from the eighth century onward in Zaire and between the thirteenth and seventeenth centuries in Zimbabwe ranged from malleable wrought iron to high-grade steel (Childs 1991a, 1991c). The analyses also show that many ancient smiths were highly skilled; they understood the properties of the various metals produced and knew how to adapt them for particular functions.

FIGURE 7.2.
A ceremonial axe from the Kamilamba site, dating to approximately A.D. 900. Scale in centimeters. Photo by Pierre de Maret.

The predecessors of the Luba were adept at pressure welding—hammering together different pieces of metal at high temperatures—and used this forging technique to form many objects (Childs 1991c). To distinguish axes for daily use from those for ritual and symbolic functions, the ancient Upemban smiths formed different blade shapes, decorated the blades, and ornamented the handles (Maret 1985a, plate 43; Connah 1987, figure 9.5) (figure 7.2). The smiths at the royal centers of Great Zimbabwe (thirteenth to sixteenth centuries), and Khami and Danangombe of the Torwa state (fifteenth to seventeenth centuries) were also excellent welders. They, however, do not seem to have used as many folds and layers of metal to make an object as the ancient smiths of Zaire.[4] They, like the Upemban smiths, distinguished prestige and ritual axes from everyday axes by forming different blade shapes and decorating some axe handles with intricately woven copper wire (figure 7.3).

There is much more information about iron smelting and smithing among the nineteenth and twentieth century Luba and Shona than there is about their predecessors. The more recent information highlights some interesting cultural and technical similarities between the two groups that need to be

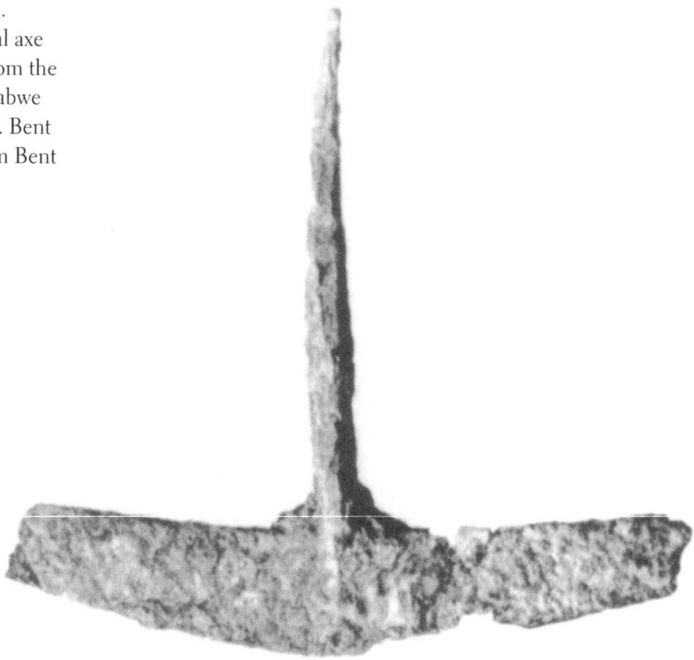

FIGURE 7.3. A ceremonial axe excavated from the Great Zimbabwe ruins by J. T. Bent in 1891. From Bent 1892, 182.

explored further. The missionary William Burton described Luba iron smelting in the 1920s: "The furnace itself is shaped roughly like a woman, and often even clay breasts are added, while the actual extraction of the molten metal takes place in such a manner as to represent a birth" (1961, 119).

Working among the Shona in the late 1800s, J. Theodore Bent observed that a furnace "is made of clay, and is another instance of the design being taken from the human form, for it is made to represent a seated woman; the head is the chimney, decorated in some cases with eyes, nose and mouth, resting on shoulders; the legs are stretched out and form the sides of the furnace, and to complete the picture they decorate the front with breasts and the tattoo decorations usually found on female stomachs" (1892, 46; see Dewey 1991b and Childs 1991b for other accounts of Shona furnaces).

Conceptual linkages between furnace and fertile woman and between iron smelting and human birth are indicated by the explicit imagery of the Luba and Shona furnaces as well as by various taboos and vocabulary associated with the furnaces and smelting. These furnace designs are quite rare in Africa yet occur in an as-yet unexplained diagonal swath from Zimbabwe in the southeast up through southern Zaire and across into northwestern Angola. Other examples include the Chokwe in Zaire and Angola (Baumann 1935, cited in Maret 1980; Herbert 1993) and the Luchazi in Angola (Housden and Armor 1959).

Although analysis of the smelting imagery among the Luba is just beginning, it appears that the metaphors of Shona smelting (figure 7.4) were not just based on human reproduction but also reproduced Shona kinship and social structure. Married ironworkers followed widespread taboos against sleeping with their regular wives during smelting because this "adultery" against the furnace as their true wife could cause the death of the expected child/iron. Furthermore, Shona furnaces were sometimes built with a representation of a woman's waist belt, known as a *mutimwi* (Dewey 1990). These belts not only had and still have an erotic connotation,[5] but they are used symbolically to mark important stages in a Shona woman's life, such as first menstruation, marriage, childbirth, and death. The belts also "strengthen her sexual power and . . . safeguard her fertility" (Aquina 1968). It can be interpreted that when a husband pays bride wealth, some of which often is used to buy beads for a new belt, he temporarily acquires the use of his bride's fertility from her family to produce children for his own. The smelters were symbolically doing the same, acquiring the fertility of the furnace to produce iron offspring.

Traditional Luba and Shona forging involved extensive hammering, folding, and welding to create distinctive axe shapes. The Luba and Shona formed the axes used for daily utilitarian tasks into similarly shaped, elongated triangles. These forms are (and presumably were in the past as well) attached through the shafts of the wooden handles and are used only for chopping. The smiths made different forms of ritual and symbolic axes, such as the Luba *kibiki* and *kasolowa* axes with flaring ends (figures 7.5 and 7.6). These Luba axes sometimes have shapes similar to the most elaborate ones from the ancient burials. The Shona *gano*-type axe blade was and still is crescent-shaped, with a narrow central tang, and the *tsomho*-type axe has a narrow blade and a curved tang (figure 7.7). The wooden handles of these axes were often highly decorated with metal attachments or, in the case of the Luba, with carved figurative representations.

Analysis to date of the production and decoration of the Luba axes and those of their predecessors takes one step further the linkages between furnace and wife or fertile woman and between smelting and human birthing. The evidence suggests that the transformative actions of repeated hammering, annealing, folding, and welding at the forge were metaphorically linked to the growth and maturation of individual members of society and to the reproduction of society in general. This is partly indicated by some of the

FIGURE 7.4. Shona iron-smelting furnaces in the Charter district, c. 1920s. From NADA 1926, 52; photo by J. W. Posselt.

FIGURE 7.5. A Luba ceremonial axe. Courtesy of the Linden Museum, Stuttgart.

vocabulary associated with iron smithing. The verb *kusunga* (van Avermaet and Mbuya 1954, 649–50), for example, means both to fabricate a weapon and to carry out a ritual associated with a rite of passage. These rites include announcing a woman's pregnancy, the death of a spouse, the installation of a new chief, the end of circumcision ceremonies, and the purification of a new diviner. It is significant that most rites occur after a purification ritual when the person is adorned with ornaments. The manufacture of weapons involves analogous transformations, refinements, and adornments and, hence, the use of the same verb.

Various embellishments anthropomorphized the Luba ritual axes (figures 7.5 and 7.6), thus visually symbolizing the social transformations that some individuals underwent and the resulting power they acquired. First, many handles of ritual axes were elaborately carved with a high-ranking (signified by her coiffure and scarification) female figurine at one end. Second, Luba infor-

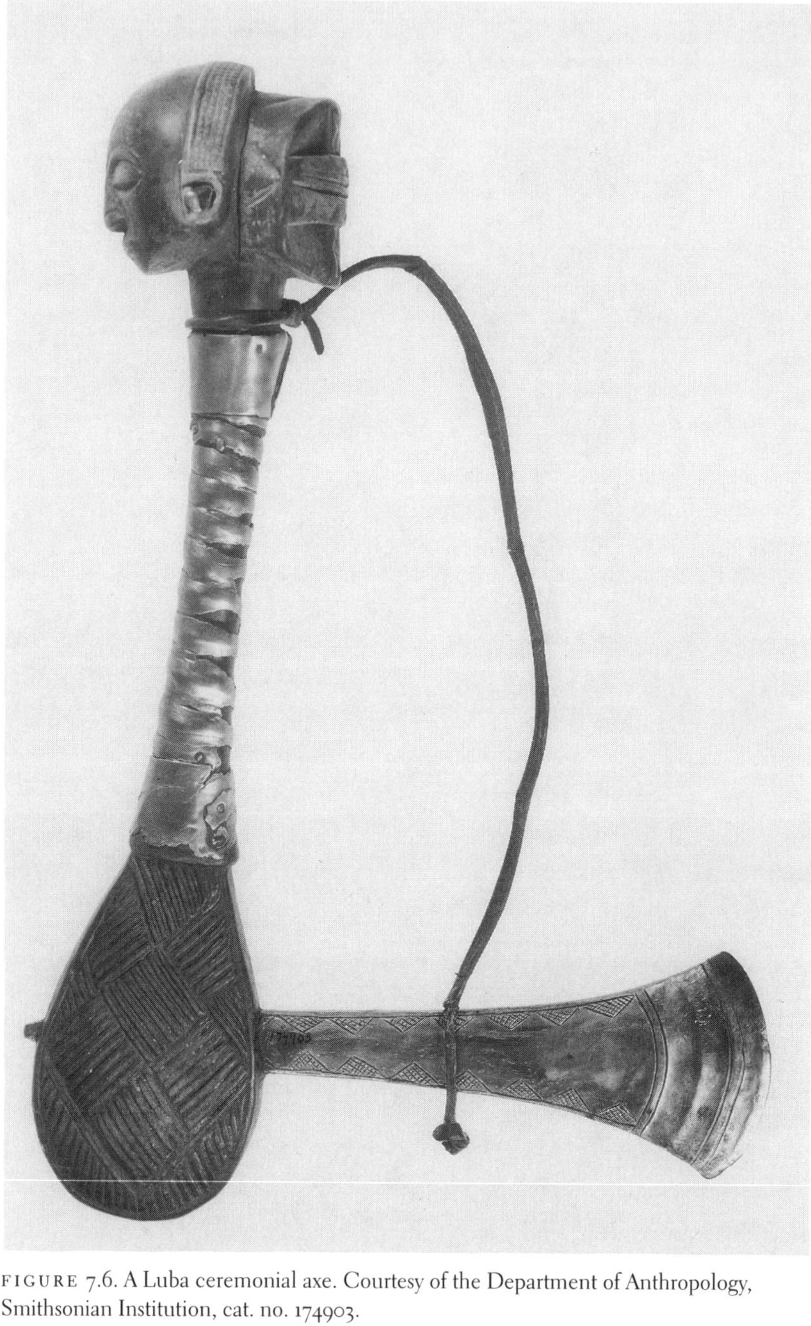

FIGURE 7.6. A Luba ceremonial axe. Courtesy of the Department of Anthropology, Smithsonian Institution, cat. no. 174903.

mants consistently called the engravings on the axe blades *ntapo*, or female scarification. Cicatrix marks beautify a girl at the threshold of adulthood as she becomes a wife, a productive mother, and a contributor to society (Kitenge-Ya 1976). Certain Luba scarification patterns, such as one known as *nkaka*, are said to represent pangolin scales. These, in turn, are metaphors for strength and durability, both of women and political power (Nooter 1990, 44–45). Furthermore, spiraled copper ribbon on many axe handles may represent the Luba waist belt that symbolizes the life of a person and generational continuity.[6] Finally, iron or copper nails[7] ornament the handle or were inserted into the hair of the figurative sculptures on some axe handles. It is believed that these nails symbolically fasten spirit power to the axe, whether divination spirits or kingly sacral powers (Nooter 1991, 259; Dewey 1991a; Dewey and Childs in press). The antiquity of the association between pins or nails and coiffure is now well established because of the recent excavation of a Kisalian-period burial at Sanga. The individual wore an elaborate hair decoration of copper and iron nails in a form similar to those on the axes (figure 7.7).

The ideological foundation of Shona axe production and decoration is less developed than for the Luba. It is important to note, however, that the Shona similarly engraved the axe blades used for symbolic and ritual purposes with *nyora*, or cicatrization decorations. Here, too, Shona girls received such markings as they became women. The Shona smiths also decorated the handles with copper, a prestige metal, although it was usually in the form of plaited wire, in contrast to the ribbon used by the Luba. Whether these Shona embellishments have a deeper symbolic meaning is a topic for future investigation. It is worth noting that plaited wire ornamentation now appears on ritual items of the religious and political elite (Dewey 1991b).

Axe Functions and the Social Context of Their Use

The Luba, Shona, and their ancestors recognized two principal types of axes. They used elongated triangular axes without decoration for daily, utilitarian functions, such as felling trees and cutting wood. The second type was specially shaped and decorated for ritual and symbolic use relating to social position as well as to political and religious power. Categorizing ritual axes can be problematic, because they often had more than one function. In one context an axe may have had a practical function as a weapon or an instrument of justice, but in another context it was held as a political symbol. Although

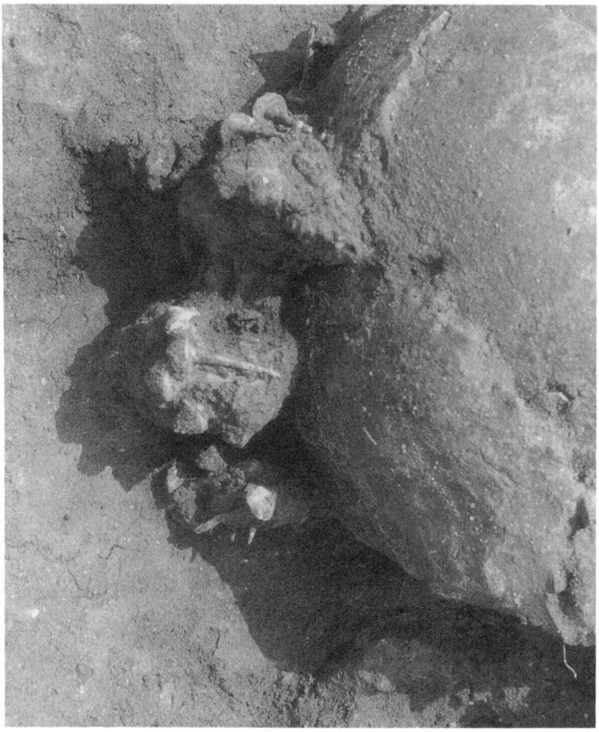

FIGURE 7.7. Burial no. 189 at Sanga, showing the hair ornament of copper and iron nails at the back of the skull. Photo by S. Terry Childs.

these specially shaped axes were not utilitarian per se, their meanings sometimes depended on their ability to cut, chop, or kill.

A number of ceremonial axes have been recovered from the prehistoric burials in the Upemba, but the social functions of the axes can only be hypothesized. Notably, they almost always occur in high-status burials along with numerous other grave goods (Maret 1982). Laboratory analyses indicate that the axes were also usually well made and could be used for utilitarian purposes (Childs 1991a). Furthermore, excavations of a child burial, probably dating between the tenth and thirteenth centuries, revealed that the child held a miniaturized ceremonial axe. A Luba elder, Papa Michelle of Sanga village, identified the axe without knowing its age. He claimed it was a *mutole*, the axe of a chief's son from a Luba subgroup. Some of the ancient axes, then, like their modern counterparts, may have been linked to political power and prestige. The importance of the spiritual and religious basis of that power is a matter for future consideration.

Today the utilitarian axe is ubiquitous in daily life among the Luba and can be carried by anyone. There are, however, clear-cut dictates concerning the cultural contexts in which elaborate, specially shaped axes can be used. First and foremost, the Luba king, local chiefs, and titled elders are entitled to carry axes such as the kibiki or kasolowa at all times. The ceremonial axe embodies the power that the royals and their chiefs acquired from the royal ancestors during their investiture. The axe also legitimizes their social position as sanctioned by the ancestors. A second group, the spirit mediums known as Bwana Vidye, are the chiefs of the ancestral world when they are in trance and communicating with spirits. They carry and brandish the ceremonial axe at this time, because it embodies the spiritual power necessary for their social role. Finally, the leaders of the secret Bambudye society, which guards and controls the ideological secrets of the Luba political system, are entitled to carry ceremonial axes when they are in session and dancing, a time when they are in communication with the ancestors. The modern axes, therefore, are strongly linked to both politics and religion. Some manipulation of their religious significance by the political elite, perhaps in the recent past, is suggested by the careful control over the times when the Bwana Vidye can use these axes (figure 7.8).

The history of ritual axe use in Zimbabwe is incomplete. No axes from the Early Iron Age sites have been adequately described, although perusal through publications of major Zimbabwe sites, such as Great Zimbabwe, Khami and Danangombe, reveals that both ritual and utilitarian axes were used at royal centers. It remains unclear whether the ritual axes functioned in either or both political and religious domains at those sites, yet most of the early ritual axes examined in the laboratory were not well made[8] and would have functioned better for display than for cutting or chopping.

Within the memory of living Shona, the uses of ritual axes are more varied, have changed more in function and meaning (figure 7.9), and are less dictated by sociopolitical rules than for the Luba. The same form of axe is known to have been used as a weapon of war, to hunt elephants, as a symbol of chiefly authority, and as a symbol of the ancestors used by many Shona, especially spirit mediums (Dewey 1985). Whereas some chiefs still keep and display ritual axes to symbolize their rank and status,[9] its religious function is most important in modern life. In particular, axes are used during spirit possession dances for two types of spirits, the *vadzimu*, ancestors who represent the collective ideals of Shona society and are the essence of morality, and the *shave*, alien and wandering spirits (Dewey 1991b; Bourdillon 1976). Most informants associate the gano-type axe with the vadzimu and the differently shaped tsomho axe with shave spirits (figure 7.10). These axes are used as conduits for spirit communication (the gano-type axe is often used on a day-to-day basis as a direct focus for prayers to the ancestors) and are symbolic of a vital social role that the spirits play.

The Multiple Meanings of Axes

The production and uses of axes clearly involved multiple levels of meaning for the Luba, Shona, and their predecessors, and many of these meanings were quite similar for both groups. The symbolic meanings given to iron as it

FIGURE 7.8. A Luba spirit medium/diviner (*Bwana Vidye*) with ceremonial axe over his shoulder, 1988. Photo by William J. Dewey.

FIGURE 7.9. Thomas Baines's drawing of a man and his weapons in the Wedza area: "Mashona at Umtigesi's place, July 20th 1870." By permission of the British Museum (Natural History), London.

was made into objects, however, differed from those associated with using a finished product. Religious and political concerns helped to shape the meanings of the ritual axes as they were formed and used in both culture areas over time, although the evidence suggests that different emphases evolved in each area.

Fundamental to the religious and political potency of the axes was how the metal was conceptualized as it was smelted and then forged. In both areas, the iron bloom was the product of an explicit birthing process in which a new, potentially productive element was brought into society. The smith, perhaps like a parent under the guidance of the ancestors, took that bloom and shaped it into a productive item with many potential uses, from agriculture to politics. In fact, the Luba smith, or *sendwe,* is seen as an expert and someone with special knowledge (van Avermaet and Mbuya 1954, 591). Several verbs relating to forging and the smith, *kufula* (to beat) and *kubunda* (to forge) (van Avermaet and Mbuya 1954, 95, 164), are also strongly associated with the creative actions of the Luba deities or spirits.

The way Luba ritual—as opposed to utilitarian—axes were made reproduced the Luba hierarchy of social relations: the elite, ritual axe versus the more common everyday axe. The production and decoration process was a metaphor for the life cycle of the elite, who underwent numerous rites of passage, "refinements," and accumulated prestige goods. The use of iron itself is now and may once have been a metaphor for the strength and durability of a chief's power (Nooter 1990, 43; 1991, 166).[10] The decorative metaphors used to equate the elite with the transformations of a girl to a fertile woman were significant, particularly given that the ultimate responsibility of the political elite was to encourage productivity and perpetuate society.

There are suggestions of similar linkages between Shona smithing, the human form, and the human life cycle, but these connections need to be

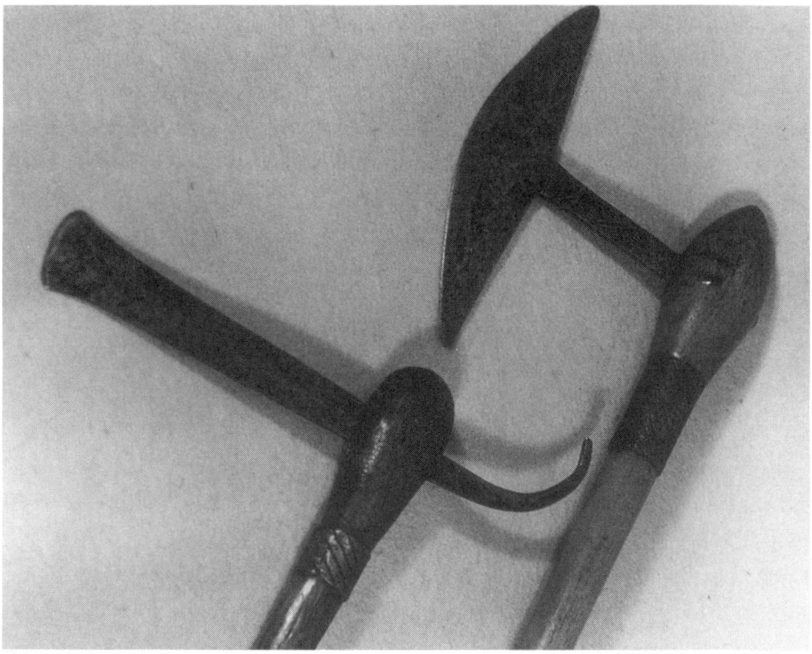

FIGURE 7.10. Shona axes (left, *tsomho;* right, *gano*). By permission of the South African Museum, Cape Town. *Left*: Acc. no. 1723 (handle 48 cm, blade 26 cm), wood, iron, brass wire. Purchased from Rev. J. O'Neil, St. Trias Hill, 1913. *Right:* Acc. no. 1813 (handle 69 cm, blade 17 x 18 cm), wood, iron, brass wire. Purchased from Rev. M. A. Fleischer, St. Trias Hill, 1914. Both are labeled "Manyika." They are likely from either the Makoni or Mutasa areas, which are both near the mission.

explored more fully in the future. The inclusion of nyora (female cicatrix) markings on the tang of the axe, for example, anthropomorphized and genderized what was otherwise a very abstract symbol of power, since the Shona did not traditionally carve figurative representations on their axe handles in the manner of the Luba. Again, girls only received nyora when they came of age as potentially reproductive women.

An interesting case of the Shona gano-type axe in recent events provides an excellent example of how it takes on many dimensions of meaning during use. At a meeting in 1962 (figure 7.11), the old warriors of the first Chimurenga (the 1896–97 war against the white settlers) passed a gano-type axe to the new warriors of Zimbabwe, who were to lead the second Chimurenga (the 1966–80 war against the white settler regime) toward national independence. The choice of an axe as a vehicle with which to transfer authority was not accidental. Most obviously, the axe was a symbol of war. On a deeper level, it expressed aspirations to regain the political power that the former Shona rulers once held. Most importantly, it symbolized the ancestors, who took an interest in current political events and gave their moral sanction. As a personal heirloom passed down through the generations, the axe provided a potent symbol of real ancestors and embodied considerable power (Dewey 1985).

Future Avenues of Research

The above discussion comparing the production, use, and meaning of iron axes over time in southeastern Zaire and Zimbabwe focuses on the interrelation between technology and culture, suggests some provocative technical and cultural similarities between the areas, and shows where comparative information is lacking. This section suggests three major avenues for future interdisciplinary study: (1) more complete historical reconstructions; (2) a closer look into the intraregional dynamics of axe uses and meanings; and (3) more detailed insights on the movement of people and ideas. These are topics offered both to fill in existing lacunae in our knowledge and to explore new directions.

Historical Perspectives

The historical contexts in which we have examined the production, uses, and meanings of axes in Zaire and Zimbabwe contain significant gaps, beginning with the earliest periods of iron use. Presently, it is generally assumed that the

FIGURE 7.11. A veteran of the 1896–97 war handing a ceremonial axe to the leaders of the nationalist movement in 1962. Photo from the Zimbabwe National Archives, Harare.

first iron axes made were for utilitarian purposes and ritual axes evolved from them. Did this really happen, and why? Possible explanations for such an evolutionary scheme include (1) increasing value placed on the axe based on its contributions to food production and related socioeconomic endeavors, as well as to the construction of political centers (Nooter 1991, 155); and (2) manipulation of the object by ironworkers to create a symbol of their acquired wealth and prestige. These possibilities highlight the need to consider the economic, political, and religious motivations that may have influenced changes in object form and function. It is important, therefore, to establish what kinds of axes were first produced in each area and attempt to determine if the production of ritual axes actually coincided with greater sociopolitical complexity as is assumed here.

Similarly, present historical reconstructions between the period in which both axe types were used by an emerging hierarchical society and the ethnographic present of the Luba and Shona need further investigation. The data presented above show some remarkable continuities in ritual axe forms for each area as well as some important similarities between the areas in the relationship between ritual axes, politics, and religion. Some significant differences in how axe functions and meanings changed are also indicated for each area, particularly the associations with a political and religious elite, and these must be explored more thoroughly. It is possible that the material manifestations of political centralization were similar during the early periods of Luba and Shona society, perhaps because of some common ancestry, past cultural linkage, or information transfer that linked technological information and social patterning. The meanings and uses of items such as axes, however, changed because of different social and political histories.

Some of the evidence presented above leads to the speculation that because the centralized authority of the Luba state was maintained until quite recently, the political elite regulated the social contexts in which ritual axes were used. This may have been done to ensure that their primary meanings were related to political authority. Furthermore, many of the meanings associated with axe use have remained relatively stable because the Luba state remained active and powerful into the colonial era and still commands attention today.

In contrast, the centralized political authority associated with Great Zimbabwe and its successors, Torwa and Mutapa, collapsed centuries ago. This left numerous small chiefdoms to fight for autonomy, along with a less regulated citizenry. Authority and guidance that came directly from all the ancestors may have gained precedence over that of the royal ancestors through a centralized king. This change may have elevated the power and prestige of Shona spirit mediums and placed greater emphasis on the religious qualities of the ritual axes, as is currently the case. Modern political events also must be considered because traditional Shona chiefs became largely discredited and unpopular with local populations in the 1960s and 1970s, owing to their collaboration with the ruling white government. The spirit mediums, on the other hand, became noted for their opposition to the white regime and their collaboration with guerrillas who fought against the government (Fry 1976; Lan 1985).

INTRAREGIONAL DYNAMICS OF AXE FORMS, USES, AND MEANINGS

Another focus for future study is the degree to which the forms, meanings, and uses of ritual axes varied in each area over time. For example, some significant differences exist in the shapes of the ritual axes excavated from relatively contemporaneous graves in the Upemba Depression in Zaire. Because the axes come from different cemeteries along different lakes (see Maret 1985a, 1992 for discussions of the site locations and details of the burials) yet come from similarly rich graves, it is possible that the individual owners of these axes held similar leadership roles among different, localized groups. We might expect that as a central authority arose and strengthened, axe shapes and meanings became more standardized to conform with royal dictates. Careful ethnographic and art historical research that documents the dimensions along which Luba ritual axes vary today and examines the significance of these dimensions can be used to set up criteria with which to evaluate the nature of axe variation in the past.

Another important example of the need to better understand intraregional variability of ritual axes relates to the history of the two Shona ritual axe types, the gano and tsomho, which now coexist in many Shona areas. Many present-day Shona insist that the gano axe, with a crescent-shaped blade and a narrow central tang, is associated with the vadzimu, or ancestors who embody Shona ideals, and is primarily used by men. The tsomho axe, with a curved tang, is linked to outsider or alien spirits, called shave, and is associated with women. The origins of these axes remain unclear, although gano-like axes are found at Great Zimbabwe–period sites (Bent 1892; Caton-Thompson 1931). Some archaeological evidence has arisen that provides insights of historical and ideological importance into this dilemma and underscores the need for more archaeological and art historical research.

Concurrent with the hierarchical Great Zimbabwe society and its prominent elite centers in the south was an independent Musengezi tradition where scholars believe that Musengezi people continued to live under the political domination of people from Great Zimbabwe. Provocatively, an oddly shaped iron object that resembles a tsomho axe was found at a Musengezi tradition site (figure 7.12). This evidence suggests that the two axe traditions now identified with the Shona may have once originated among independent culture groups, the Zimbabwe and Musengezi. It seems that as the Zimbabwe-related people incorporated the Musengezi people of northeastern Zimbabwe, they adopted the foreign tsomho axe. Interestingly, this axe type was given a distinct ritual function associated with the spirits of a forever alien group.

Interregional Dynamics

Finally, research into the variable production processes, uses, and meanings of axes at an interregional level must be continued because it can contribute

FIGURE 7.12. An iron axe blade excavated from Wazi Hill (15.2 cm) (Musengezi tradition). Photo by William J. Dewey.

new data with which to examine movements of people and ideas. Recent attempts to trace Late Iron Age groups along several different cultural dimensions, such as pottery types and settlement patterns (Huffman 1989), may benefit from the input of patterns of metal object style, usage, and manufacturing technique.

This chapter sought to begin this comparative approach by outlining several similarities in axe production, meaning, and use between the far distant neighbors in southeastern Zaire and Zimbabwe. These include the similar practices of anthropomorphizing and genderizing Luba and Shona furnaces, the common use of some metaphors linking ironworking and human life, the importance of ritual axes in politics, and the strong association of iron axes with ancestral spirits. Do these similarities suggest cultural connections in the distant past? The greatest differences between the two areas lie in the distinctively different axe shapes and embellishments used by the Luba and their predecessors as opposed to those of the Shona and their predecessors. Future stylistic, symbolic, and historical linguistic analyses of the furnaces and ritual axes must incorporate the surrounding and intermediate culture groups. The resulting patterns may yield some provocative insights into past cultural connections, mechanisms for the transfer of technological and ideological knowledge, and a more complete range of roles that iron production and use played in the rise of African sociopolitical complexity.

Notes

This research was generously supported by the J. Paul Getty Trust, which recognized the potential of our interdisciplinary work incorporating archaeology, art history, ethnography, history, and materials science. We would also like to thank the governments of Zaire and Zimbabwe for research clearance; Muya wa Bitanko, the director of the Musée de Lubumbashi, Zaire, and his staff; the people of Sanga village, Zaire; the National Museums and Monuments of Zimbabwe; the Archaeology Unit of the History Department, University of Zimbabwe; the staff at the Queen Victoria Museum, Harare, Zimbabwe; and Pierre de Maret, who helped the project in innumerable ways. The laboratory work was conducted at the Center for Materials Research in Archaeology and Ethnology at the Massachusetts Institute of Technology where we were both postdoctoral fellows.

1 Bells are well-known symbols of political status in central Africa (Vansina 1969).
2 Some ceremonial items, however, remained in use. A large ceremonial knife, for example, was found in association with a utilitarian-like axe, a hoe, and other iron objects in a grave at the Malemba-Nkulu site (Maret 1985a, 1992).
3 Similar influences seem to have affected the demise of iron smelting among the Luba, although they are poorly documented to date.
4 The study sample subjected to metallographic analysis from Zimbabwe is much smaller than that from Zaire. Our observations to date, therefore, are preliminary.
5 A belt is usually hidden under a woman's outer garments and is manipulated by her husband during lovemaking.
6 This speculation is inspired by Allen Roberts's work among the nearby Tabwa (1980: 96–101, 488).
7 Note that some of these nails have the same shape and name as the blacksmith's anvil/hammer.
8 The welds of several axes were incomplete and the metal used was soft wrought iron (Childs 1991a).
9 The form of the chief's axe is the same as that used for other purposes, but the former sometimes has more elaborate wirework embellishing the axe handle.
10 Several informants (Mutombo Mukulu, September 21, 1988; Lunda wa Lunda September 22, 1988) told us that iron symbolized strength and durability and that many present-day chiefs wear a single iron bracelet to gain those powers.

References

Aquina, Sister Mary (A. K. H. Weinrich). 1968. "Mutimwi—A Note on the Waist Belt of the Karanga." *nada* 11:5.

Beach, David N. 1980. *The Shona and Zimbabwe 900–1850*. Gweru, Zimbabwe: Mambo Press.

——. 1984. *Zimbabwe before 1900*. Gweru, Zimbabwe: Mambo Press.
Bent, J. Theodore. 1892. *The Ruined Cities of Mashonaland*. London: Longmans, Green.
Bisson, Michael. 1975. "Copper Currency in Central Africa: The Archaeological Evidence." *World Archaeology* 6:276–92.
——. 1976. "The Prehistoric Coppermines of Zambia." Ph.D. diss., University of California, Santa Barbara.
——. 1982. "Trade and Tribute: Archaeological Evidence for the Origin of States in South Central Africa." *Cahiers d'Études Africaines* 87–88:343–61.
Bourdillon, Michael. 1976. *The Shona Peoples*. Gweru, Zimbabwe: Mambo Press.
Burton, William. 1961. *Luba Religion and Magic in Custom and Belief*. Annales, Musée Royal de l'Afrique Centrale, Sciences Humaines, No. 35. Tervuren, Belgium.
Caton-Thompson, Gertrude. 1931. *The Zimbabwe Culture*. Oxford: Clarendon Press.
Childs, S. Terry. 1991a. "Iron as Utility or Expression: Reforging Function in Africa." In *Metals in Society: Theory Beyond Analysis*, ed. R. Ehrenreich, 57–67. MASCA Research Papers in Science and Archaeology, No. 8:2. Philadelphia.
——. 1991b. "Technology, Style, and Iron Smelting Furnaces in Bantu-Speaking Africa." *Journal of Anthropological Archaeology* 10:332–59.
——. 1991c. "Transformations: Iron and Copper Production in Central Africa." In *Recent Trends in Archaeo-metallurgical Research*, ed. P. Glumac, 33–46. MASCA Research Papers in Science and Archaeology, No. 8:1. Philadelphia.
Childs, S. Terry, and David Killick. 1993. "Indigenous African Metallurgy: Nature and Culture." *Annual Review of Anthropology* 22:317–37.
Collett, David. 1985. "The Spread of Early Iron-Producing Communities in Eastern and Southern Africa." Ph.D. diss., Cambridge University.
Collett, David, A. E. Vines, and E. G. Hughes. 1992. "The Chronology of the Valley Enclosures: Implications for the Interpretation of Great Zimbabwe." *African Archaeological Review* 10: 139–61.
Connah, Graham. 1987. *African Civilizations, Precolonial Cities, and States in Tropical Africa: An Archaeological Perspective*. Cambridge: Cambridge University Press.
Dewey, William. 1985. "Shona Ritual Axes." *Insight: The National Gallery of Zimbabwe Journal* July:1–5.
——. 1986. "Shona Male and Female Artistry." *African Arts* 19: 64–67, 84.
——. 1990. *Weapons for the Ancestors*. Film. University of Iowa Video Production Unit.
——. 1991a. "Blacksmiths and Kings: Forging Symbolic Meaning in Luba Metal Arts." Paper presented at the annual meeting of the College Arts Association, Washington, D.C.
——. 1991b. "Pleasing the Ancestors: The Traditional Art of the Shona People of Zimbabwe." Ph.D. diss., Indiana University.
Dewey, William, and S. Terry Childs. In press. "Memory and the Forging of Iron and Kings." In *Memory: Visual Representations and Luba History*, ed. Mary Nooter Roberts and Allen Roberts. New York: Museum of African Art.

Fry, Peter. 1976. *Spirits of Protest.* Cambridge: Cambridge University Press.

Garlake, Peter. 1973. *Great Zimbabwe.* New York: Stein and Day.

——. 1982. *Life at Great Zimbabwe.* Gweru, Zimbabwe: Mambo Press.

Hall, Martin. 1987. *The Changing Past: Farmers, Kings, and Traders in Southern Africa, 200–1860.* Cape Town: Creda Press.

Herbert, Eugenia. 1993. *Iron, Gender, and Power: Rituals of Transformation in African Societies.* Bloomington: Indiana University Press.

Hiernaux, J., E. de Longress, and J. De Buyst. 1971. *Fouilles Archéologiques dans la Vallée du Haut-Lualaba. I. Sanga 1958.* Annales, Musée Royal de l'Afrique Centrale, Sciences Humaines, No. 73. Tervuren, Belgium.

Housden, J., and M. Armor. 1959. "Indigenous Iron Smelters." *Northern Rhodesia Journal* 4:135–38.

Huffman, Thomas. 1978. "Origins of the Leopards Kopje: An Eleventh Century Difaquane." *Arnoldia* 8:1–23.

——. 1982. "Archaeology and Ethnohistory of the African Iron Age." *Annual Review of Anthropology* 11:133–50.

——. 1986. "Iron Age Settlement Patterns and the Origins of Class Distinction in Southern Africa." In *Advances in World Archaeology,* ed. F. Wendorf and A. Close, 5:291–338. New York: Academic Press.

——. 1989. "Ceramics, Settlements, and Later Iron Age Migrations." *African Archaeological Review* 7:155–82.

Huffman, Thomas, and J. C. Vogel. 1991. "The Chronology of Great Zimbabwe." *South African Archaeological Bulletin* 46:61–70.

Killick, David. 1990. "Technology in Its Social Setting: Bloomery Iron-Smelting at Kasungu, Malawi, 1860–1940." Ph.D. diss., Yale University.

Kitenge-Ya. 1976. "Les Ntapo: Essai d'Interpretation Generale des Tatouages dans la Monde Afro-Bantou." *Problemes Sociaux Zairois* 114–15:159–69.

Lan, David. 1985. *Guns and Rain: Guerrillas and Spirit Mediums in Zimbabwe.* Berkeley and Los Angeles: University of California Press.

Lechtman, Heather. 1977. "Style in Technology—Some Early Thoughts." In *Material Culture—Styles, Organization and Dynamics of Technology,* ed. R. Merrill and H. Lechtman, 3–20. St. Paul, Minn.: West Publishing.

Maret, Pierre de. 1977. "Sanga: New Excavations, More Data, and Some Related Problems." *Journal of African History* 18:321–37.

——. 1979. "Luba Roots: The First Complete Iron Age Sequence in Zaire." *Current Anthropology* 20:233–35.

——. 1980. "Ceux Qui Jouent avec le Feu: La Place du Forgeron en Afrique Centrale." *Africa* 50:263–79.

——. 1981. "L'évolution Monetaire du Shaba Central entre 7e et le 18e Siècle." *African Economic History* 10:117–49.

——. 1982. "The Iron Age in the West and South." In *The Archaeology of Central Africa,* by F. van Noten, 77–96. Graz: Akademische Druck.

———. 1985a. *Fouilles Archéologiques dans la Vallée du Haut-Lualaba, Zaire. Sanga et Katongo, 1974.* 2 vols. Annales, Musée Royal de l'Afrique Centrale, Sciences Humaines, No. 120. Tervuren, Belgium.

———. 1985b. "The Smith's Myth and the Origin of Leadership in Central Africa." In *African Iron Working*, ed. R. Haaland and P. Shinnie, 73–87. Oslo: Norwegian University Press.

———. 1992. *Fouilles Archéologiques dans la Vallée du Haut-Lualaba, Zaire. III. Kamilamba, Kikulu, et Malemba-Nkulu, 1975.* Annales, Musée Royal de l'Afrique Centrale, Sciences Humaines, No. 131. Tervurven, Belgium.

Nenquin, Jacques. 1963. *Excavations at Sanga, 1957. The Prehistoric Necropolis.* Annales, Musée Royal de l'Afrique Centrale, Sciences Humaines, No. 45. Tervuren, Belgium.

Nooter, Mary (Polly). 1990. "Secret Signs in Luba Sculptural Narrative: A Discourse on Power." In *Iowa Studies in African Art*, ed. Christopher Roy, 3:35–60.

———. 1991. "Luba Art and Polity: Creating Power in a Central African Kingdom." Ph.D. diss., Columbia University.

Pfaffenburger, Bryan. 1988. "Fetished Objects and Humanised Nature: Towards an Anthropology of Technology." *Man* 23:236–52.

———. 1992. "Social Anthropology of Technology." *Annual Review of Anthropology* 21:491–516.

Reefe, Thomas. 1981. *The Rainbow and the Kings.* Berkeley and Los Angeles: University of California Press.

Roberts, Allen. 1980. "Heroic Beast, Beastly Heroes: Principles of Cosmology and Chiefship among the Lakeside BaTabwa of Zaire." Ph.D. diss., University of Chicago.

Schmidt, Peter. 1983. "An Alternative to a Strictly Materialistic Perspective: A Review of Historical Archaeology, Ethnoarchaeology, and Symbolic Approaches in African Archaeology." *American Antiquity* 48:62–79.

Soper, Robert, and Gilbert Pwiti. 1988. "Preliminary Report on Excavations at Wazi Hill, Centenary, Northern Zimbabwe." *Zimbabwe Prehistory* 20:16–20.

van Avermaet, E., and J. Mbuya. 1954. *Dictionnaire Kiluba-Francais.* Annales, Musée Royal de l'Afrique Centrale, Linguistique, No. 7. Tervuren, Belgium.

Vansina, Jan. 1969. "The Bells of Kings." *Journal of African History* 10:187–97.

Widstrand, Carl. 1958. *African Axes.* Studia Ethnographica upsaliensia, No. 15. Uppsala: Almquist and Wiksell Boktryckeri.

8 Complex Iron Smelting and Prehistoric Culture in Tanzania

Peter R. Schmidt & Donald H. Avery

Western scientists and students of history have long understood the iron bloomery process through evidence available from European archeology. Ethnographic, technological, and archaeological research into the technological life of the Haya of northwestern Tanzania show that these people and their forebears 1500 to 2000 years ago practiced a highly advanced iron smelting technology based on preheating principles and, as a result, produced carbon steel. This sophisticated technology may have evolved as an adaptation to overexploited forest resources. These discoveries are significant for the history of Africa and the history of metallurgy.[1]

One of the primary problems in the prehistory of Africa is the question of how iron metallurgy developed in sub-Saharan Africa. Since 1937 when W. Cline wrote *Mining and Metallurgy in Negro Africa,* there has been no systematic attempt to understand the technological complexities and developmental history of African iron metallurgy. Cline's study is a continentwide ethnological survey which prehistorians, historians of technology, and metallurgists continue to use as their main reference source. These scholars either subscribe to or passively accept the assumption that the sponge iron bloomery process, well known from European archaeological contexts, also describes African iron smelting. This view persists despite the great variability known to exist in African iron-smelting processes. Yet no one has tried to determine whether or not the European bloomery model applies to iron smelting in Africa. This is a particular problem that we address in this article: whether or not the bloomery process as known from European archeology is similar to iron-smelting technology in Africa.

It is also apparent that some authorities have overlooked important variables in African smelting furnaces, and consequently have failed to see possible explanations for the production of high-grade steel. For example, when Tylecote (1965) looks at the evidence from Oyo, southwestern Nigeria, reported by Bellamy (1904), he fails to attach importance to the production of a high-carbon (1.67 percent) steel in the induced-draught shaft furnace with tuyeres (blowpipes) inserted inside the furnace. Tylecote attaches little significance to the possibility that preheating occurred in the Oyo furnace: "At first sight it would seem that the long tuyeres are designed for efficient preheating of the ingoing air, but a calculation shows that the degree of preheat would only be 10°C" (1965, 340). We show here that these statements are based on unrealistic laboratory and theoretical models and, in fact, do little justice to the technological processes that occur in authentic furnaces such as those operated by the Haya who live in the Kagera Region (formerly West Lake Region) of Tanzania (map 8.1).

Our current perspective is limited to one culture, but we expect that it can be extended to other areas of the continent as we gain more archeological and ethnographic information about iron metallurgy. In Africa, there are several areas where traditional iron bloomery smelters actively practice their craft, or remember how it is practiced. Among the Haya there is an active blacksmithing tradition in which scrap iron is used, but there are no active contemporary bloomery smelters. However, some of these smiths and other old men did smelt iron in the traditional way during their youth fifty to sixty years ago; they are still alive, able to smelt, and, in many cases, eager to relive the experience. The knowledge held by such old men in Kagera Region, Tanzania, is profound in its implications for the history of iron smelting and for the prehistory of complex African civilizations. But this knowledge is threatened every day by the passage of time, by death, and by age-related infirmities occurring in this quickly shrinking group of expert smelters.

The Haya are a Bantu-speaking agricultural people who live along the western shore of Lake Victoria. The Haya live in densely populated villages while practicing some cattle herding and a banana- and bean-based agriculture. Coffee and tea are grown as cash crops. In precolonial times, the culture area contained six small kingdoms, three of which were ruled by different branches of the Bahinda clan.

MAP 8.1.
The traditional kingdoms of the interlacustrine region.

Archeological verification of oral traditions in one kingdom, Kyamutwara, indicate that oral traditions accurately document the history of some ancient religious and historical sites seized by the more recent Bahinda dynasty (Schmidt 1975, 1978). Archeological investigations at the Rugomora Mahe site confirmed the accuracy of traditions about iron production there and dated the iron production to 2400 to 2550 years ago. The later occupation of the site by the royal dynasty was dated by the C^{14} method to 250 to 300 years ago (Schmidt 1975, 1978; Sutton 1972), confirming dates suggested by oral, royal genealogies. Archeological analysis of this site included examination of technological materials, principally slag and fragments of tuyeres.

As part of our study of the ancient technology, we attempted to discover the flow temperatures of slag that had been formed in Early Iron Age furnaces. Test results showed a temperature of 1350°–1400°C. On the basis of European archeological evidence, this is within the highest temperature limits obtained during most experiments performed in bloomeries. In addition, the physical properties of the tuyere fragments suggested that they had come from inside the furnace. From these archeological data we developed the hypothesis that preheating of the air blast had been a prominent feature of African Early Iron Age smelting. We reasoned that higher combustion temperatures achieved by this preheating would have formed slag at temperatures similar to or higher than those indicated by our heat tests on prehistoric slag. Preheating would permit the attainment of much higher furnace temperatures and better fuel economy than was obtainable in cold blast European bloomeries.

This hypothesis was subsequently tested by an experiment that showed that twentieth-century metallurgists and anthropologists, given sufficient money, instrumentation, and effort, could preheat air and make some iron in a reconstructed furnace. But we felt that, in order to obtain reliable and authentic results, we had to observe traditional iron smelters in Africa. We knew from previous research, especially from interviews with old men who had smelted, that the Haya had placed their tuyeres inside their furnaces, thereby creating a situation that would have led to a hot air blast. As Tylecote observes (1965), enormous gains were made in nineteenth-century blast furnaces with low degrees of preheat (100°C) brought about by "inefficient [rough] pipe stoves." Although Cline (1937) mentions several furnace types with tuyeres inside furnaces as part of his broad ethnological consideration of iron metallurgy, these examples, along with Oyo, have never been recognized as possible preheating. We presumed that if a similar technological phenomenon occurred during prehistoric times in Kagera Region, then there was an excellent possibility that an efficient and complex technology prevailed there during the Early Iron Age.

The implications of our studies suggested that ethnographic research on iron smelting and iron forging, including quantitative technological observations, was critical to testing the hypothesis as well as to understanding the "bloomery" process itself and its associated cultural behaviors. Before we formed our preheating hypothesis, we had assumed—like Maddin, Muhly, and Wheeler (1977, 122)—that "the highest temperature that could be reached in a primitive smelt appears to have been about 1,200°." That assumption is accompanied by the idea that "smelting iron ore at that temperature yields not a puddle of metal but a spongy mass mixed with iron oxide and iron silicate." These were apparently valid assumptions until our research among the Haya demonstrated at least two unique and important characteristics in this advanced iron technology.

The first demonstrated example of preheating in an African smelting furnace is the Haya process. As oral information had indicated, the Haya do place tuyeres inside their furnaces (figure 8.1). The furnace is what we call a shaft-bowl type, with a forced-air draught. A cone-shaped shaft 140 cm high (figure 8.2) and constructed with old, refractory slag and mud made from the earth of a termite mound is built over a bowl 50 to 60 cm deep and lined with termite mud (figure 8.3). Swamp grass is burned in the bowl until the bowl is filled with the charred swamp reeds. The charred reeds provide a charcoal bed of filamentary alkaline-coated carbon fiber that is readily wetted and infiltrated by the molten iron slag. Eight drum bellows covered with goatskins are used to force air into tuyeres, 50–60 cm in length, which are inserted inside the furnace. The combustion zone just beyond the tuyere develops very high temperatures. The high-temperature products of combustion divide, part passing up through the furnace shaft and part back-flowing along the tuyère: this heats the tuyere and the air blast passing through it. Eventually, external tuyere surfaces are wet with slag and dissolve in it at about 1250°–1300°C. This

FIGURE 8.1. Smelters and assistants place tuyeres inside the furnace to variable depths. Tuyeres inserted inside the furnace cause preheating of the air blast and result in high temperatures in the combustion zone, as well as in the interior reduction zone. The eight tuyeres are placed between large blocks of refractory slag, which are used as the foundation for the furnace chimney.

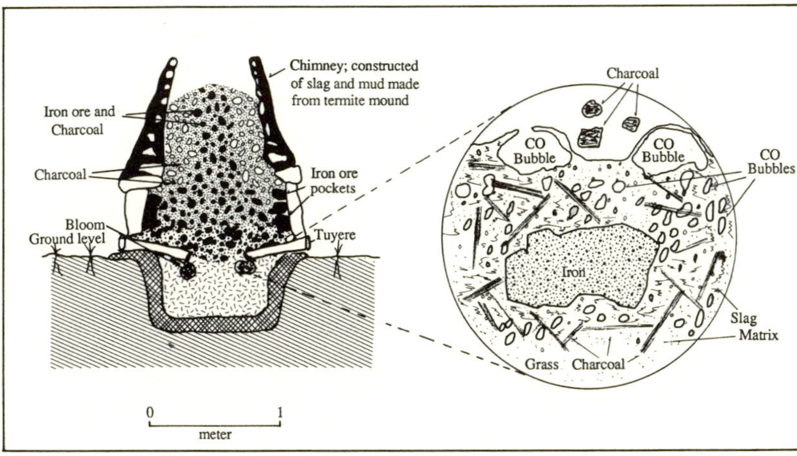

FIGURE 8.2. A schematic profile of a Haya iron smelting furnace; this elaborates on the profile published in the 1978 *Science* article. It adds an inset that shows a summary of the reduction process inside the furnace, in particularly illustrating the explosion of CO bubbles—a carbon boil—on the top of the slag mass. Note that the tuyeres are inside the furnace and therefore are conduits that preheat the air passing through them. Iron ore pockets are added inside the foundation blocks to prepare and partially sinter iron ore for the next smelt.

means the external temperature of the tuyere clay is in excess of 1250°C. Slag penetrated approximately one-third through the tuyere wall.

The air temperatures inside the tuyere were measured by passing a ceramic-insulated thermocouple down the tuyere. Figure 8.4 shows—*within one tuyere at 2.5 cm intervals*—typical curves at different times and tuyere lengths.[2] The rapid rise in the last 10 cm probably represents significant radiation heating of the thermocouple; however, the preheating temperature of the air is clearly in the range of 600°C or more.[3] The preheating efficiency of the tuyere (as reflected in the combustion temperature *as well as the temperatures in the more interior reduction zone*) is related to its length (figure 8.4). As the tuyere shrinks in size, the temperatures in the combustion zone—*and the internal reduction zone*—decrease. Combustion zone temperatures were found to vary, but in many cases were above the melting point of a Pt-PtRh (10 percent rhodium) thermocouple, which melts at 1820°C. The Zelechovise I type furnace (Northern Moravia, eighth century A.D.) tested by Pleiner (1969, 458) in Czechoslovakia during 1964 reached a maximum temperature of 1450°C in its small underground shaft. During tests of an experimental second century

FIGURE 8.3. A smelt under way in Nyungwe village in the Bukoba area of Kagera Region in Tanzania. Eight double-drum bellows are operated for up to eight hours in a process that yields a carbon-steel bloom, a process that is distinctive in its maintenance of high temperature yet reducing conditions through the use of preheating.

Roman shaft furnace, Tylecote et al. (1971) recorded a combustion zone temperature as high as 1600°C. We feel confident that the high combustion zone temperature in the Haya furnace is caused by preheating of the air blast. The high temperatures achieved through preheating are critical to the Haya process, *especially in terms of the higher than expected temperatures in the reducing zones*. Given the other unique characteristic of the process—the crystalline formation of the bloom discussed below—the attainment of high temperatures means that iron will be produced more efficiently.

The other remarkable characteristic that makes this process of great interest to science and history is that, in the Haya process, the bloom does not form by the sintering of fine solid particles. Throughout Europe and in many areas of the world, the bloomery process is known as one in which reduction takes place in a CO atmosphere while the bloom forms by sintering of fine solid particles in an iron sponge and by liquation of a fluid slag. But in the Haya smelting furnace of Kagera Region, the bloom is precipitated as large crystals growing in a fayalite-wustite slag: the molten slag undergoes a carbon boil in much the same way that an open-hearth furnace has a carbon boil. The carbon boil is caused by an intimate and large contact area with solid carbon (the charred swamp reeds) throughout the molten slag mass. The critical variables in the process appear to be the roasting of the iron ore, the use of charred swamp reeds in the furnace bowl, and the high temperatures involved.

"Roasting" of the ore is an important initial step, for then carbon is introduced into the ore. The ore is "roasted" in a reducing atmosphere in a pit with wet wood and limited O_2 access. This reduces the ore, with the ore providing the O_2 for combustion:

$$2Fe_3O_4 + 8CO \rightarrow 6Fe + 8CO_2$$

After roasting the ore cools off in a CO_2 atmosphere and is reoxidized with carbon deposited:

$$6Fe + 3CO^2 \rightarrow 6FeO + 3C$$

This and the use of charred swamp grass is important: when the iron ore melts above the tuyeres, it forms a fayalite-wustite slag with carbon inclusions from the roasting. It then infiltrates and interacts with the fibrous, carbonaceous interface provided by the burned swamp grass. This, in turn, provides an extremely high carbon-slag contact area. The solid carbon reacts with the slag to form CO bubbles that coalesce and rise to the slag surface, forming large bubbles 5 cm/diam. As the carbon boil removes oxygen from the slag, the slag becomes supersaturated with iron. Iron crystals then precipitate and grow. As the liquid slag changes its composition and loses wustite (FeO), it becomes increasingly refractory, and its melting point rises. Thus, a high temperature allows further and more efficient precipitation of iron crystals.

Metallographic examination of these iron precipitates shows a planar growth interface (figure 8.5a). Here we see the massive iron (white area) growing into the fayalite-wustite slag (gray area). The growing interface seems always to be pure (carbon-free) iron. The crystal perfection is very good with large straight Neuman bands and large well-formed crystals, as illustrated in the micrographs of a traditional bloom (figures 8.5a, 8.5b, and 8.5c). As the iron grows, some slag and carbon are entrapped. The entrapped carbon then locally carburizes the iron surrounding it, producing a highly variable local carbon content (figure 8.5d). If this local carburization were to continue beyond 2.4 percent carbon, then the cast iron droplets would form.[4] But this level of carburization is not typical in the Haya process. Our analyses show that the bloom obtained from six smelts in Kagera Region during 1976 have a

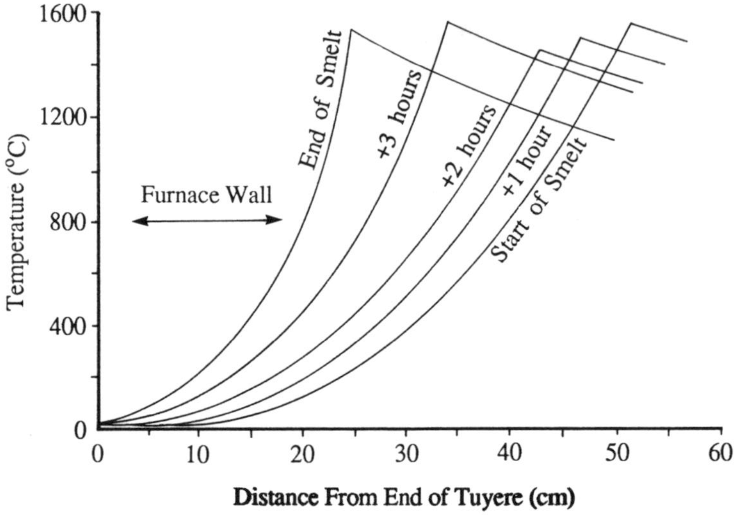

FIGURE 8.4. Temperature profiles along one tuyere (measured at 2.5 cm intervals), and through the blast zone. As smelting progresses, the tuyeres burn off, the preheating length decreases, and furnace temperatures drop. As the Haya smelters experimented and adopted more refractory tuyere clays, tuyere length was better maintained and preheating was applied more effectively for longer periods, resulting in higher yields.

carbon content ranging from pure iron to eutectic and hypereutectic microcompositions. Further sampling and testing will determine whether the eutectic compositions predominate.

The assemblage of Early Iron Age artifacts is not complex, being limited to two braceletlike objects from both the Rugomora Mahe and KM2 sites, and one possible knife blade fragment from each site. Often the iron is severely oxidized, and therefore we await better archeological evidence for the cultural and technological aspects of prehistoric iron implements.[5]

We find in traditional Africa an Iron Age technological process that was exceedingly complex and, in historical and relative terms, also advanced. There are indications from cultures near the Haya that similar types of furnaces were used in Uganda (Tylecote 1965). Ethnographic data, although incomplete, indicate that many other neighboring cultures practiced an entirely different kind of technology. However, we suggest that this level of complexity was not limited to East Africa. The probable use of preheating in Yoruba furnaces (Bellamy 1904) and the high-carbon steel produced in them suggests that this and other African smelting technologies may have technological characteristics that are highly advanced and technologically sophisticated according to contemporary historical and scientific values. The insertion of multiple tuyeres in large (4 m high) shaft-type furnaces such as those among the Wafipa people of southwestern Tanzania (Greig 1937; Wembah-Rashid 1973) also suggests preheating and the possible production of steel by a process different from that observed in Kagera Region.

Our ethnographic study of recent smelting has also provided material evidence that the technological process revealed in 1977 archeological discoveries in the Kagera Region has remained unchanged during the last 1500 to 2000 years.[6] Archeological excavations at the KM2 site, an Early Iron Age industrial locale near the Lake Victoria shore (1°28'30"S, 31°44'45"E), have yielded excellent data which demonstrate that similar furnace form and preheating were employed during the Early Iron Age. Furthermore, archeological surveys tentatively show that Early Iron Age peoples lived in large villages where they practiced iron production. It has been suggested (Schmidt 1975) that large settlements may be linked to localized high population densities caused by an industry "which demanded significant manpower." Our recent observations of the diversity of skills required for preparation of materials, operation of the furnace, and processing of the iron bloom show an organized, highly cooperative labor force. We suggest that the recently observed exploitative behavior is a reasonable, general model for the Early Iron Age. This does not mean that we believe the two are the same, but that the demands of the prehistoric industry were very similar to those of the recent past. It is abundantly clear that the industry is labor intensive, particularly during the direct production phase, during which a minimum of one dozen craftsmen are engaged in the cooperative enterprise. It is also possible that an economically complex and technologically advanced culture with high population densities may be linked to the evolution of politically centralized states in Kagera Region and in neighboring areas such as Rwanda.

Further archeological research is required to discover why such a complex technology grew up along the western shore of Lake Victoria. One possible hypothesis is that the heavy exploitation of forests (for charcoal and for agricultural purposes) may have triggered the development of an efficient, fuel-economizing technology. The widespread distribution of Early Iron Age industrial sites in Kagera, the manner of charcoal production and its 10 to 1 weight ratio of wood to charcoal, and the need for 500 lbs of charcoal all suggest that, if smelting were widely practiced in prehistoric times, then the

FIGURE 8.5

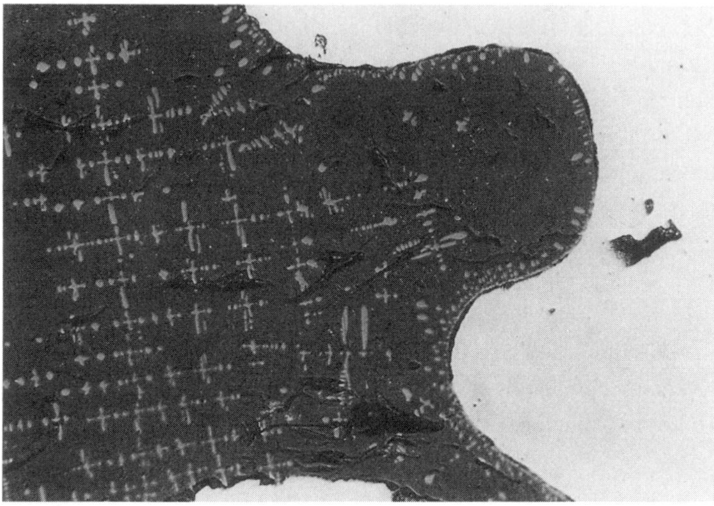

A: Bloom from 1976 smelt, showing a growth interface between iron (white) and slag (gray) (unetched, magnified twenty times).

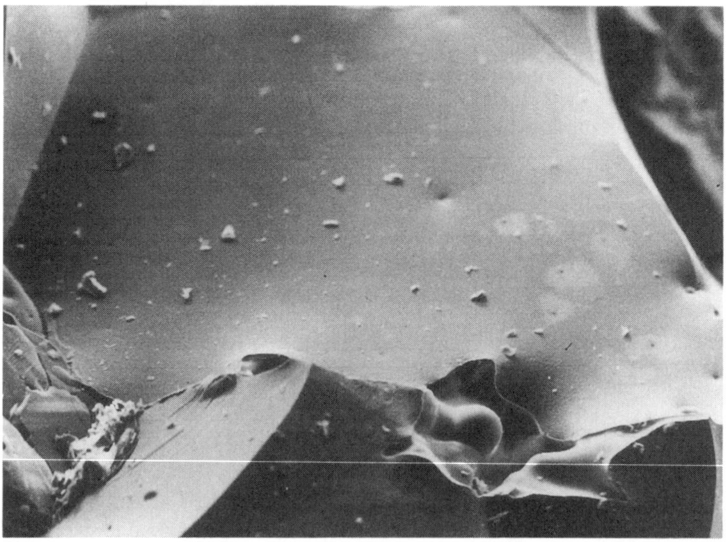

B: A scanning electron micrograph view of intercrystalline fracture in traditional bloom.

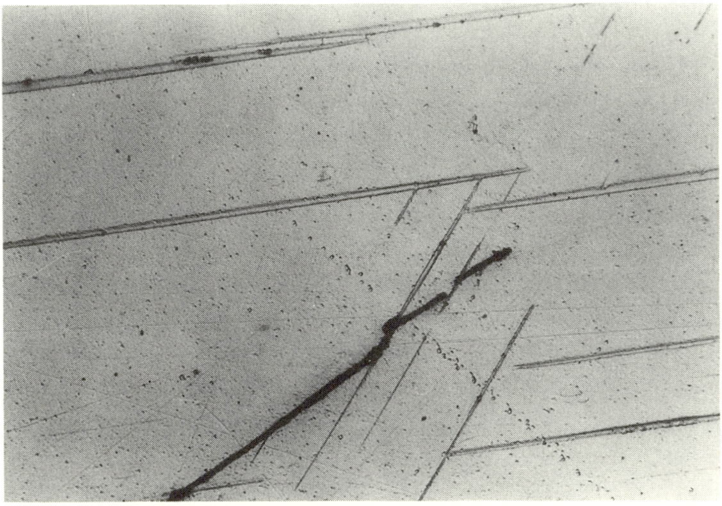

C: A photomicrograph of same bloom illustrated in *B*, showing long undistorted Neuman bands (2 percent nital, magnified 18 times).

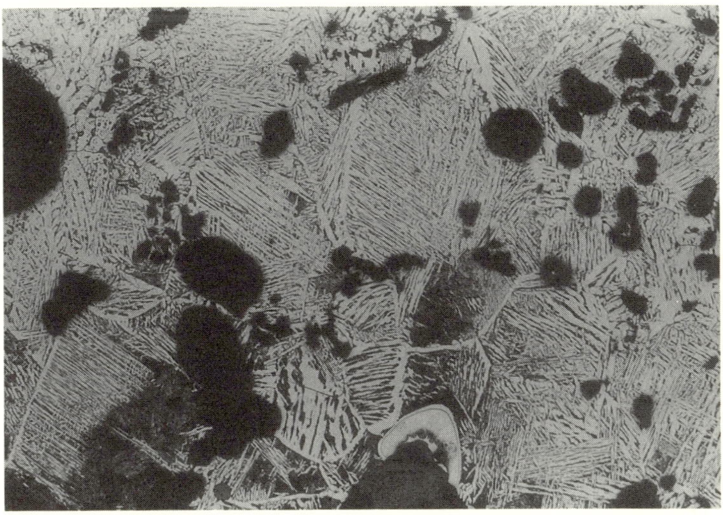

D: An etched micrograph of bloom from 1976 smelt, showing wide local range of carbon content in what averages out as a medium carbon steel (magnified 18 times).

impact of the technological system must have been severe. The decline of productivity may be linked to an overexploited forest resource base: the local smelters may have developed the fuel-efficient preheated furnace to adapt to that depleted resource.

Preliminary information on prehistoric vegetation obtained through study of prehistoric pollen indicates possible widespread forest clearance during prehistoric times. The enormous drop in productivity since the first colonial government in 1890 is attributed by the smelters to the availability of cheap, imported iron tools and spring steel and to the greater economic rewards of coffee farming.

One of the more profound implications of the Tanzania discoveries is that we are now able to say that a technologically superior iron-smelting process developed in Africa more than 1500 years ago. This knowledge will help to change scholarly and popular ideas that technological sophistication developed in Europe but not in Africa. In that respect the ramifications are significant for the history of Africa and her people.

Notes

1 We thank the United Republic of Tanzania, especially the University of Dar es Salaam and the Antiquities Unit for permission to conduct research in Kagera Region. A feasibility study of field research was assisted by a grant (P.S.) from the Wenner Gren Foundation for Anthropological Research. Field research and analyses were supported by NSF grant BSN 75–19611.
2 We have introduced few changes in the original *Science* text. Several minor changes have been made to clarify scientific questions and our argument about the effect of preheating on higher reduction temperatures. These changes are marked in the text by italics.
3 See chapters 11 and 12, this volume, for a modification of this figure.
4 See chapter 13, this volume, for new observations on how cast iron droplets do in fact form above and back of the tuyere mouths.
5 More evidence for artifacts is now available. Some of the further evidence is reported in chapters 9 and 13, this volume.
6 There is now more complete evidence for technological change over time. See Schmidt 1994.

References

Bellamy, C. V. 1904. "A West African Smelting House." *Journal of Iron Steel Industry* 66:99–126.

Cline, W. 1937. *Mining and Metallurgy in Negro Africa.* Banta: Menasha, Wisconsin.

Greig, R. C. H. 1937. "Iron Smelting in Fipa." *Tanganyika Notes and Records* 4:77–81.

Maddin, R., J. Muhly, and T. S. Wheeler. 1977. "How the Iron Age Began." *Scientific American* 237(4):122–31.

Pleiner, R. 1969. "Experimental Smelting of Steel in Early Medieval Furnaces." *Pomatky Archeologicky* 60:458–86.

Schmidt, P. R. 1975. "A New Look at Interpretations of the Early Iron Age in East Africa." *History in Africa* 2:127–36.

———. 1978. *Historical Archeology: A Structural Approach in an African Culture.* Westport: Greenwood.

———. 1994. "Historical Ecology and Landscape Transformation in Eastern Equatorial Africa." In *Historical Ecology: Cultural Knowledge and Changing Landscapes,* ed. Carole Crumley, 99–125. Santa Fe: SAR Press.

Sutton, J. E. G. 1972. "New Radiocarbon Dates for Eastern and Southern Africa." *Journal of African History* 13:1–24.

Tylecote, R. F. 1965. "Iron Smelting in Pre-Industrial Communities." *Journal of Iron Steel Institute* 203:340–48.

Tylecote, R. F., J. N. Austin, and A. E. Wraith. 1971. "The Mechanism of Bloomery Process in Shaft Furnaces." *Journal of Iron Steel Institute* 204:342–63.

Wembah-Rashid, J. A. R. 1973. *Iron Working in Ufipa.* Dar es Salaam: National Museum of Tanzania.

9 Actualistic Models for Interpretation of Two Early Iron Age Industrial Sites in Northwest Tanzania

Peter R. Schmidt & S. Terry Childs

One of the paradoxes in the archaeology of Iron Age Africa is the absence of well-dated and well-documented evidence for Early Iron Age technology. The paradox arises out of investigations of the Iron Age over the last three decades that have focused almost exclusively on the construction of elaborate diffusionist schemes that seek to explain the distribution of Early Iron Age cultures. This preoccupation with diffusion has been accompanied by assumptions about Early Iron Age technology that have inhibited investigations into the technological attributes of Early Iron Age life in Africa. The strength of such assumptions in Western scholarship has inhibited archaeologists of Iron Age Africa from studying iron technology as something distinctly African, as a technological experience that drew on ecological adaptations and innovative discoveries altogether different from those of Europe or Asia.

There is now a body of archaeological evidence and ethnoarchaeological studies of iron technology that emphatically suggests that iron technology in Africa developed along an evolutionary path independent of the Eurasian tradition (Schmidt and Avery 1978; Schmidt 1981, 1983a). Knowledge of iron production may ultimately derive from Europe or Asia or from multiple sources (Childs and Killick 1993), but it is now clear that there arose in Africa a tradition of iron technology separate and unconnected with what we know from outside the continent.

The most important observation on the evolution of iron technology to develop during our archaeological and technological studies in northwest Tanzania is the technological significance of the placement of clay tuyeres inside the iron smelting furnace (Schmidt and Avery 1978; Schmidt 1981, 10–11, 17–18; 1983a, 421). Systematic observation of iron smelting in an experimental and ethnographic context has shown that when a clay tuyere is placed inside the smelting furnace, the air blast passing through the clay pipes is preheated: that is, the air entering the blast zone is heated to nearly 200–400°C above ambient air temperature. The result is a reducing temperature that is significantly beyond the range achievable in prehistoric European furnaces. The genesis of the preheating hypothesis lies in the archaeological materials excavated at the Rugomora Mahe site in Katuruka (Schmidt 1978, 235–43); heat tests on slag suggested temperatures between 1300–1400°C were obtained during Early Iron Age smelting. This rather elementary but effective observation, when joined with the further observation that 39 percent of the tuyere fragments showed characteristics that affirmed their alteration in a slagging, vitrifying, and reducing environment, led to the hypothesis for preheating (Schmidt and Avery 1978, 1086). This hypothesis was subsequently tested in an experimental furnace (Childs and Schmidt 1985), and confirmed during a series of nine smelts performed by Haya iron workers during 1976, 1979, and 1984.

The implications of the preheating tests are significant for how we think about and study iron technology in Africa, for they swing the focus of studies of iron technology to a recognition that Africa had its own scientific inventiveness, and they compel us to understand how and why African iron producers embarked on this important developmental path. The application of preheating principles to iron smelting in nineteenth-century Europe, for example, had a revolutionary effect by increasing productivity multifold over the cold blast technologies. Yet, the archaeological evidence from northwest Tanzania suggested that a preheated technology had developed in Africa nearly two millennia before. It was important that we go to the archaeological record to obtain physical proof for the antiquity and the developmental context of this technique in Africa. Before excavations in Tanzania to obtain more definitive material evidence, we needed to understand the material signatures for preheating in the archaeological record. This required actualistic studies of the material transforms that occur under conditions of preheating. These actualistic studies resulted in a set of material observations that enhance our confidence in functional inferences for Early Iron Age technology.

Observations of tuyeres used in the recent smelting show that as much as 40 cm of a 55–60 cm tuyere used in a preheated furnace may display evidence of slagging, vitrification, and reduction—all of which are diagnostic markers of preheating. Hence, a critical component of our archaeological analysis of tuyeres from the KM2 and KM3 sites is an examination of tuyere sections and fragments in order to document the presence or absence of preheating. We want to encourage other investigators to use similar analyses, so that we can come to a better understanding of the environmental and sociological contexts in which preheating developed in the prehistoric past. Our excavations at the KM2 site in 1977 and at the KM3 site during 1978–79 were explicitly designed to obtain a more thorough understanding of Early Iron Age technology in general and to retrieve evidence for the antiquity of the complex preheated iron technology of northwest Tanzania.

Ethnoarchaeological Observations

Before we discuss the furnace structures of KM2 and KM3, it is appropriate to review the origins of the technological, processual, and formal criteria that we employ in this assessment. A suitable first step is to consider several aspects of the actualistic model derived from Schmidt's ethnotechnology of the Haya process. The recent Haya furnaces provide some basic insights into the significance of several categories of evidence. First, it is apparent from the structure of the furnace pit and the slag remains in the replicated furnaces that the Early Iron Age smelting furnaces are non–slag-tapping. In this respect they closely resemble the historic Haya process. This means that there is no structural provision to drain the slag from the smelting furnace itself; instead, the slag drains away from the bloom to the bottom of the furnace pit. In order to accomplish this in a shaft-bowl furnace, there must be some medium that allows the slag to drain unimpeded from the blast zone to the base of the furnace. In the case of the recent Haya furnaces, the slag drains through and around the porous charred reeds in the furnace pit. In the recent past massive skulls of slag were formed in the furnace pit. But it also results in the formation of small slag prills, often as beads that come to rest on the furnace floor. These pieces of flow slag form as slag drips down through the grass matrix; when frozen, they sometimes may be observed as larger lumps of slag that are flat on the bottom, glassy-looking, smooth, and dense. Small beads (3–6 mm) and larger twisted prills are found on the floor of the furnace and are, therefore, excellent diagnostic markers for smelting when found in an archaeological context, particularly on the floor of a suspected furnace pit.

Slag is a good indicator of the materials used in the furnace environment. The slag flows around the carbonaceous materials that are used in the base of the furnace as well as in the furnace blast zone. In the recent furnaces this means that most of the slag, having ended up in the base of the furnace, shows the imprints of the charred reeds that are used to fill the pit. Slag casts of reeds are common and the imprints are easily recognized. Slag found toward the top of the furnace pit close to the blast zone is often imprinted with small pieces of charcoal. Observations of Barongo smelting in Biharamulo district south of Lake Victoria provide further information about slag formation (Schmidt, chapter 5, this volume). One helpful observation is that the Barongo fill their furnace pits with gross pieces of charcoal, thus creating a matrix with large interstitial space capable of receiving large amounts of fluid slag.

Another fundamental observation of the Haya process that helps to sort out the function of archaeological features is the presence of a reddish ring of burned earth slightly larger than the circumference of the furnace pit. This is an oxidation isotherm that marks the placement of the furnace wall (usually the center of the wall) and penetrates only 10–20 cm deep from the ground surface on which the furnace chimney sits. The earth inside this isotherm shows signs of reduction if the burning activity inside the pit was conducted under reducing conditions; thus its inner portion marks the boundary between oxidizing and reducing atmospheres.

The pits of the smelting furnaces are not used exclusively for smelting purposes. In the Haya process the pit is first used to roast iron ore in preparation for the first smelt. This prepares the ore by driving off water, partially sintering the ore, and impregnating it with carbon (Schmidt and Avery 1978, 1088). It is altogether possible that furnace pits were used periodically to roast ore between smelts or that pits abandoned for smelting purposes were subsequently used for roasting purposes. The relevance of these observations will become apparent as we discuss the technological features of KM2 and KM3.

The furnace pits constructed by the Haya elders of Nyungwe village are lined with a mud made from the earth of termite mounds. As a consequence, the liner is a highly siliceous material that is more refractory than normal soil. This attribute results because termites select quartz sand grains from the subsoil to build their mounds. Thus, we would expect to find a significantly higher proportion of sand in liners if they were made with termite earth. Furnace pits, once excavated and lined, are used a number of times. This leads to the gradual destruction of the liner as slag fuses with parts of the furnace wall. When slag is later pried out of the furnace, pieces of the liner are removed with it. Smelting continues without repair to the liner. One of the common attributes of the Early Iron Age furnaces at KM2 is the lining applied to the furnace bowl. Oftentimes only small, remnant pieces of the liner are found on the floor or walls of the furnace, suggesting that sections of the liners were removed when slag to which they were fused was removed from the furnace. Liners are not always made of earth. The Barongo smelters south of Lake Victoria use special leaves to line the furnace pit (Schmidt, chapter 5, this volume). This results in the imprints of the lining material only on the base of the slag mass. It is conceivable that a variety of other materials would also function in a similar manner.

The ethnoarchaeology of the Haya and Barongo technologies provides the most complete inferential models for our interpretation of Early Iron Age technological processes, while the most complete and important report on Early Iron Age technology thus far focuses on the high plateau environment of Rwanda and Burundi (van Grunderbeek, Roche, and Doutrelepont 1983a, 1983b). Van Grunderbeek has continued the work begun by Hiernaux and Maquet (1957, 1960) and continued by van Noten (1979) on the Urewe ware complex and its associated iron technology. Because the Rwanda and Burundi excavations are described in some detail and because they are found in a region contiguous to northwestern Tanzania, we will examine this evidence to discern possible affinities with the technology that we have been studying in Tanzania. Van Grunderbeek and others have concentrated their research in the Butare area in south central Rwanda, where the greatest concentrations of Early Iron Age smelting furnaces have been located, and in the Gitega area of central Burundi.

The most apparent difference between the Rwanda/Burundi deposits and those of Tanzania is that iron-smelting furnaces there are found as isolated units, devoid of associated stratigraphic cultural materials (van Grunderbeek et al. 1983a, 7), mostly as a consequence of agricultural activity and erosion. The northwest Tanzania furnaces often appear in relief, with furnace bricks standing above the subsoil, but we have always found that the surrounding matrix holds cultural material.

The Sites and Their Dating

KM2

Before its destruction the KM2 site was located at an altitude of 14 m above Lake Victoria on a knoll 500 m west of the Kemondo Bay port facility (map 9.1). It was discovered in early 1977 by a Tanzanian member of our survey team as he walked over the newly exposed road surface that had been opened as part of the new Kemondo Bay port access road. Circles of slag and circles of burned brick had been exposed during removal of the top 30–45 cm of topsoil by heavy earth-moving equipment. All spoil had been pushed to the roadside to form two long rows of earth (figure 9.1). The size of the pit features and their contents resembled the Early Iron Age smelting furnaces that had been excavated during 1970 by Schmidt at the Rugomora Mahe site in Katuruka, just across Kemondo Bay and within sight of the KM2 site (Schmidt 1975, 1978). Earth removal had proceeded to the sterile subsoil in much of the cleared zone, but deposits in the areas where excavation Blocks A and B were subsequently opened remained as deep as 20 to 30 cm overlying the cultural features.

The excavation strategy at the KM2 site addressed the need for recovery of stratified remains, while at the same time it balanced the need to obtain the best possible horizontal definitions of structures along with their careful stratigraphic excavation. As the agreed time for excavations drew to a close, we cleared the overburden north of Block B in order to rescue the known cultural evidence in that area, a tactic that we accompanied with shovel scraping of already disturbed zones north of Block B and near Block A (figure 9.2). That proved to be a fortuitous decision, for many industrial features were exposed and excavated north of Block B, leading to the definition of the most complex Early Iron Age industrial site, other than Taruga in Nigeria, thus far excavated in Africa.

Because of the complexity of the site, its many radiocarbon dates, and the antiquity of the technology, we want to provide a more detailed review of the KM2 (and KM3) industrial features. Our comments on both the KM2 and KM3 sites focus on the industrial structures and their associated artifacts. The furnace bowls excavated on these sites were filled with a variety of materials that usually included slag, pieces of partly reduced iron ore, fired bricks used to build the furnace superstructure, charcoal, pieces of tuyeres, and sometimes domestic pottery. Similar fill is found in the iron-smelting furnaces of Rwanda and Burundi (van Noten 1979, 64–67; van Grunderbeek et al. 1983a, 7). Test

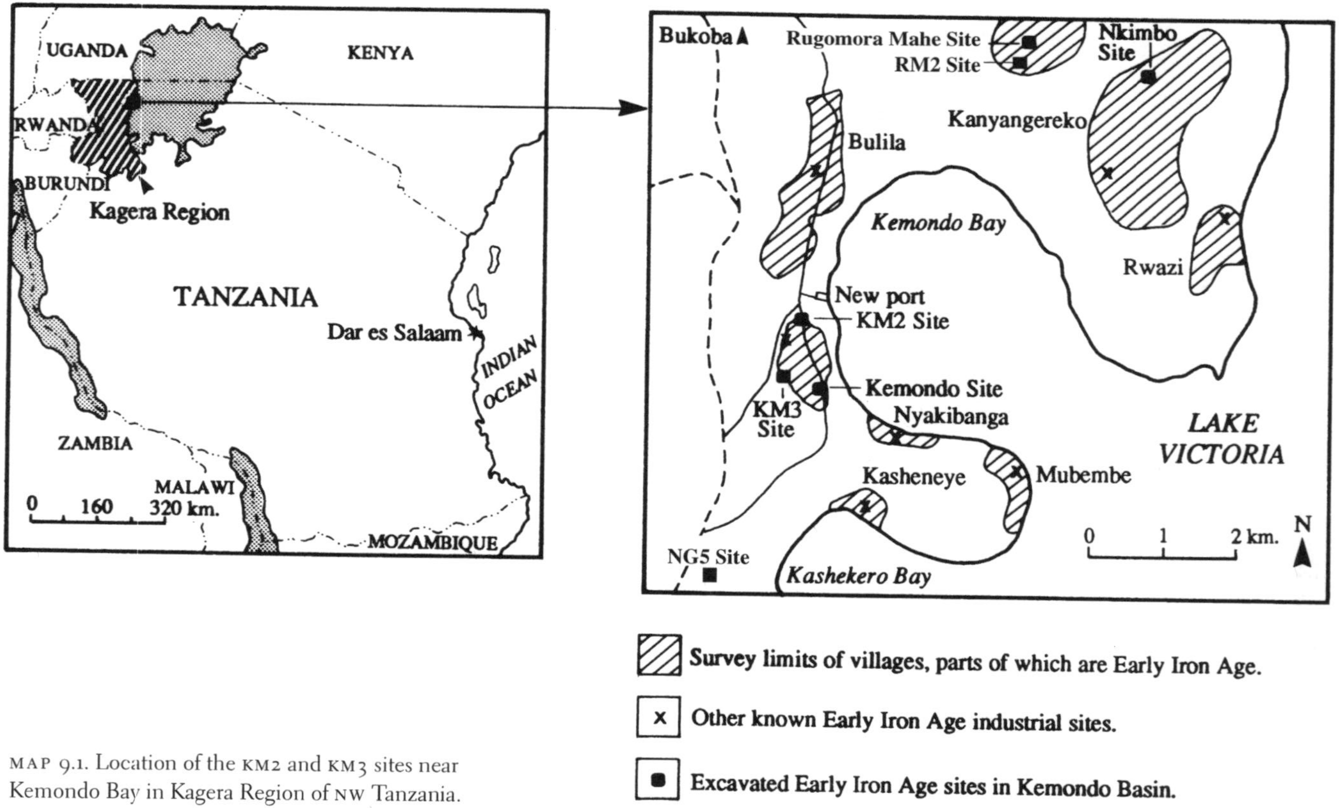

MAP 9.1. Location of the KM2 and KM3 sites near Kemondo Bay in Kagera Region of NW Tanzania.

FIGURE 9.1. Looking NE across the Block B extension of the KM2 site after completion of excavations. Each person is positioned near the base of an Early Iron Age smelting furnace. A large industrial refuse pit (feature 8) is right center and another furnace is left center. Photo by Peter R. Schmidt.

excavations were also opened in the western part of the site, where only minor evidence was discovered. Other 2 x 2 m tests and block excavations in areas of feature concentration were opened to recover remains that were observable on the transformed surface (figure 9.1). Survey east of the road spoil indicated that the site continued in this direction.

KM3

The KM3 site was discovered during our village survey by a Tanzanian surveyor who observed the bricks of a furnace in a main path bisecting the site (figure 9.3). Situated 1 km south of the KM2 site, the KM3 site sits at an altitude ranging between 75 and 85 m above Lake Victoria. Immediately to the south of the site there is a steep hill filled with sandstone scree. The northern part of the site was under intense agricultural development when investigations began in 1978, and as a consequence a zone 30 m immediately north of the arbitrary site boundaries depicted in figure 9.3 was open only to cursory examination. Recent and continuous deep agricultural development in this zone had removed and pulverized at least three furnace structures with bricks. Had we obtained permission to excavate that area, no intact structures would have been recovered.

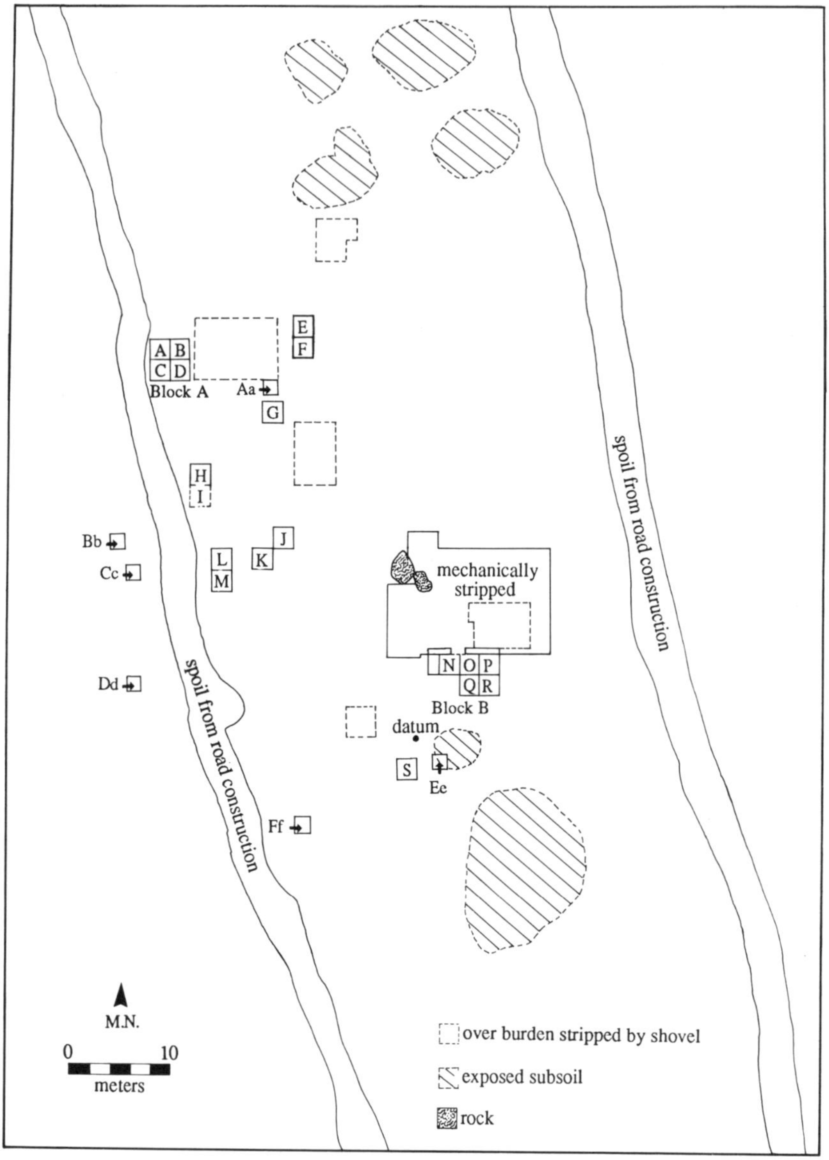

FIGURE 9.2. An excavation plan at the KM2 site.

We returned to excavate at the KM3 site because it offered the possibility of technological evidence that would be closely related to KM2 in time and space and, we reasoned, the best context for recovery of further evidence for the antiquity and development of preheating in African iron smelting. Our research strategy for the site was to focus stratigraphic excavations on zones of intense industrial utilization and to established their possible relationships. To determine those zones we employed diagnostic techniques that included controlled surface collections, dousing rod surveys, and a proton magnetometer survey. All provided distinctive and mutually complementary information, with the proton magnetometer isolating magnetic anomalies within sensitive zones delimited by density contours of the surface scatters. The northern part of the site had the most significant densities of industrial artifacts dating to the Early Iron Age. It was precisely in this area that we concentrated many excavation units to form Block A with a trench extension to the south (figure 9.3). The remaining block excavations and test trenches were placed according to similar criteria and proton magnetometer results. For example, Block C was located over a set of furnaces (not yet excavated) and a large brick and ceramic feature that was located by proton magnetometer. Block B was located on the edge of a sensitive zone marked by a partially exposed smelting furnace.

All excavated furnaces have been radiocarbon dated, with three having been dated twice. As predicted, the northern sector had the highest density of well-preserved industrial features, as four of the seven furnaces were located in that area. The second major temporal and spatial cluster was found in Block B, where two furnaces were located. Today the KM3 site sits on the border of Kigarama village partly in a rocky zone that does not lend itself to hoe agriculture. As today, in the past the southern area would have been marginal compared to the much finer soils immediately to the north. Survey to the west of the site grid indicated a strong likelihood of remains there, while to the north and the east dense settlement and agriculture with old coffee trees and deep ground cover inhibit inspection until some future date.

KM2 RADIOCARBON DATES

The radiocarbon dates from the KM2 site are all derived from wood charcoal that has been recovered from sealed contexts within the prehistoric pit features excavated on the site. Ten of the fourteen samples were excavated from iron-smelting furnaces, with three furnaces (7, 9, and 14) dated twice. The

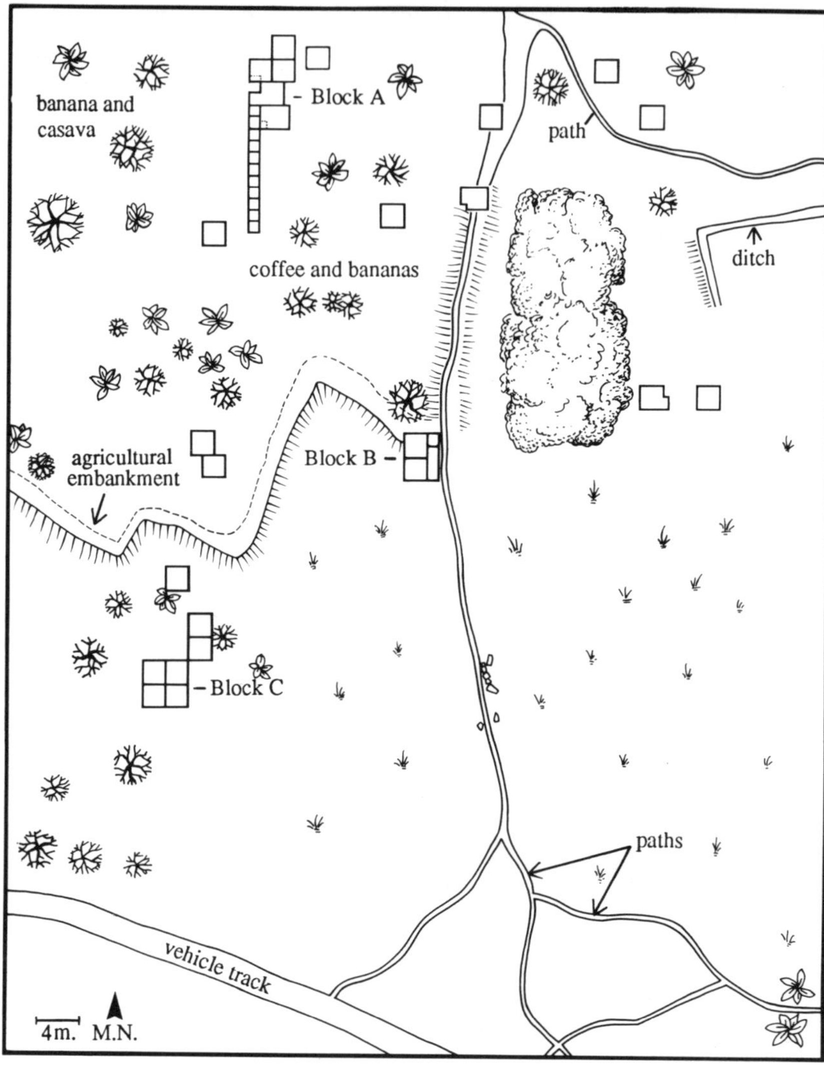

FIGURE 9.3. An excavation plan at the KM3 site.

remaining four samples were excavated from a house floor (feature 12); a large industrial refuse pit (two samples from feature 8); and a small slag refuse pit, feature 29 (table 9.1 and figures 9.4 and 9.5). With the exception of the house floor, the charcoal samples were obtained from beneath furnace bricks or slag in the deeper stratigraphic zones of the pit features. This means that there is an unusually high degree of contextual integrity for the samples. Nonetheless,

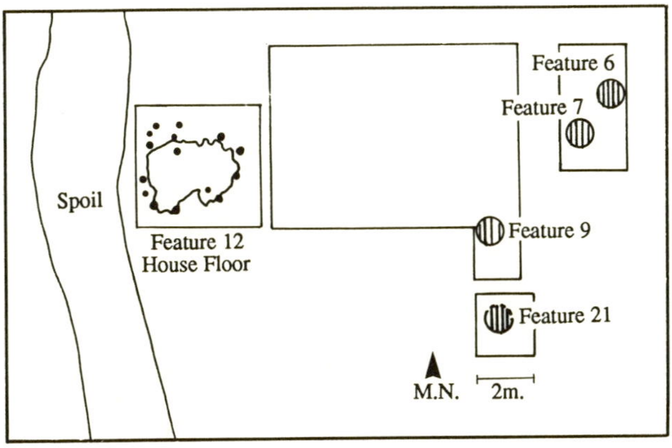

FIGURE 9.4.
A plan view of the location of four iron smelting furnaces in the NW sector of the KM2 site near Block A.

these pit features were not always free of disturbance. When the furnace pit was filled with massive concretions of brick and slag, then disturbances such as rodent activity were not present. But when the fill was of a less dense variety, then such features were more susceptible to rodent and agricultural disturbances.

The recent history of agricultural activity on the site shows that it was used for fallow-type agriculture, with the planting of groundnuts and cassava. Thus, it is probable that the furnace pits suffered alteration by hoe agriculture, with one of the more notable adverse effects being the truncation of the furnace pits. The results would have been a more shallow profile and an archaeological definition that does not represent the prehistoric dimensions.

Most of the radiocarbon dates fall in the early first millennium A.D. If we consult the calibration curves for the A.D. era calculated by Stuiver (1982) using Douglas fir and Sequoia dendrochronological ages, then these dates as a group will be younger. The decadal tables provided by Klein et al. (1982) are based on similar and additional data and are used here to figure the calibrated ranges at the 95 percent level of confidence in table 9.1, as well as to reckon the corrected dates. The list of dates in tables 9.1 and 9.2 is the definitive report at this time.

The tight spatial clustering of furnaces in the northwest portion of the site (figure 9.4) forms a distinct and separate group. The C^{14} dates from this cluster, though, are not as closely related as the morphology of the structures. Feature 6, dated to 1590 ± 60 B.P. (360 A.D. ± 60, corrected to a calendrical date of 412 A.D. ± 83), appears to be younger than what we might expect from this group, for this furnace has a shape similar to numbers 9 and 7 and also

TABLE 9.1. Radiocarbon dates from the KM2 site

Date no.	*Feature no.*	*Function*	*Lab. no.*
1	6B	Smelting furnace	Beta-2245
2	7C	Smelting furnace	RL-1010
3	7C	Smelting furnace	Beta-10,490
4	8B	Refuse pit	RL-1011
5	8A	Refuse pit	SMU-1436
6	9B	Smelting furnace	Beta-10,491
7	9D	Smelting furnace	RL-1012
8	12D	House floor	RL-1013
9	14A	Smelting furnace	Beta-10,492
10	14B	Smelting furnace	Beta-2244
11	16C	Smelting furnace	Beta-2247
12	18B	Smelting furnace	RL-1014
13	29B	Slag refuse pit	Beta-2246
14	34B	Smelting furnace	RL-1015

[a]8A C^{13} correction to C^{14} age: 1815 ± 42 bp, used in calibration

has a small hole in the furnace floor. Yet feature 6 did not have whole furnace bricks, as did numbers 7 and 9, along with the large refuse pit (no. 8). Feature 9 has been dated twice, 1860 ± 50 B.P. (90 A.D. ± 50) and 1930 ± 120 B.P. (20 A.D. ± 120); if we calculate a weighted average (Long and Rippeteau 1974) for these two dates, then feature 9 dates to 1870 ± 46 B.P. (80 A.D. ± 46, corrected to 113 A.D. ± 59). The feature 9 dates agree closely with those from the refuse pit, no. 8: 1870 ± 120 B.P. (80 A.D. ± 120), and 1826 ± 42 B.P. (124 A.D. ± 42), which result in a weighted average of 1831 ± 40 B.P. (119 A.D. ± 40, corrected to 138 A.D. ± 59). There is a difference between the uncorrected mean dates of only 0.45 of their standard deviations. When corrected the difference between both dates narrows, so that when the t-test (Long and Rippeteau 1974) is applied, we find that there is a greater than 80 percent probability that the dates are contemporaneous.

year half-life	ad/bc	Date bp 5568 Range at 95% confidence level	
		UNCALIBRATED	CALIBRATED
1590 ± 60	ad 360 ± 60	ad 240-480	AD 245-590
1820 ± 110	ad 130 ± 110	90 bc-ad 350	10 BC-AD 415
1450 ± 60	ad 500 ± 60	ad 380-610	AD 445-635
1870 ± 120	ad 80 ± 120	160 bc-ad 310	145 BC-AD 340
1826 ± 42[a]	ad 124 ± 42	ad 40-208	AD 20-255
1860 ± 50	ad 90 ± 50	10bc-ad 190	AD 1-240
1930 ± 120	ad 20 ± 120	220 bc-ad 260	165 BC-AD 240
2120 ± 120	170 ± 120 bc	410 bc-ad 70	400 BC-AD 40
1410 ± 60	ad 540 ± 60	ad 410-660	AD 560-655
1480 ± 60	ad 470 ± 60	ad 350-590	AD 425-625
1620 ± 60	ad 330 ± 60	ad 220-450	AD 235-580
1440 ± 110	ad 510 ± 110	ad 290-730	AD 415-770
1580 ± 60	ad 370 ± 60	ad 250-490	AD 250-590
1670 ± 120	ad 280 ± 120	ad 40-520	AD 65-585

Feature 7 has also been dated twice, 1820 ± 110 B.P. (130 A.D. ± 110, corrected to 203 A.D. ± 106) and 1450 ± 60 B.P. (500 A.D. ± 60, corrected to 540 A.D. ± 48). The difference between these two dates is greater than two standard deviations; the second date can be rejected on statistical grounds as deviant from the other dates derived from the furnaces east of Block A. There is a possibility there was contamination during storage or transport of the sample that yielded the younger date. The larger standard deviation of date 2 (feature 7: 1820 ± 110 B.P.) lowers the probability of its contemporaneity with feature 8. Outside of the question of contemporaneity, it is clearly the case that dates 2 and 4–8 constitute a grouping that marks a distinct period of early industrial activity on KM2.

The argument that the date from the house floor (feature 12) also belongs to this group is not strong. If we compare it to the older of the feature 9 dates (no. 7), calibrate both, and apply the t-test, then there is less than a 20 percent chance that they are contemporary. This separation in time suggests that the house floor belongs to an earlier separate occupation on the site inasmuch as it is dated to 2120 ± 120 B.P. (170 B.C. ± 120, corrected to a calendrical date of 180 B.C. ± 110). However, the presence of fragments of furnace brick and slag in the top stratum suggests that we should not exclude the possibility that the house structure may have coexisted in space and time with the industrial activities found immediately to the east on the site.

The second major cluster of dates that point to the third period of site use (figure 9.5) is associated with two shallow furnace pits and with one furnace that has vertical walls, feature 6 in the northwest part of the site (figure 9.4).

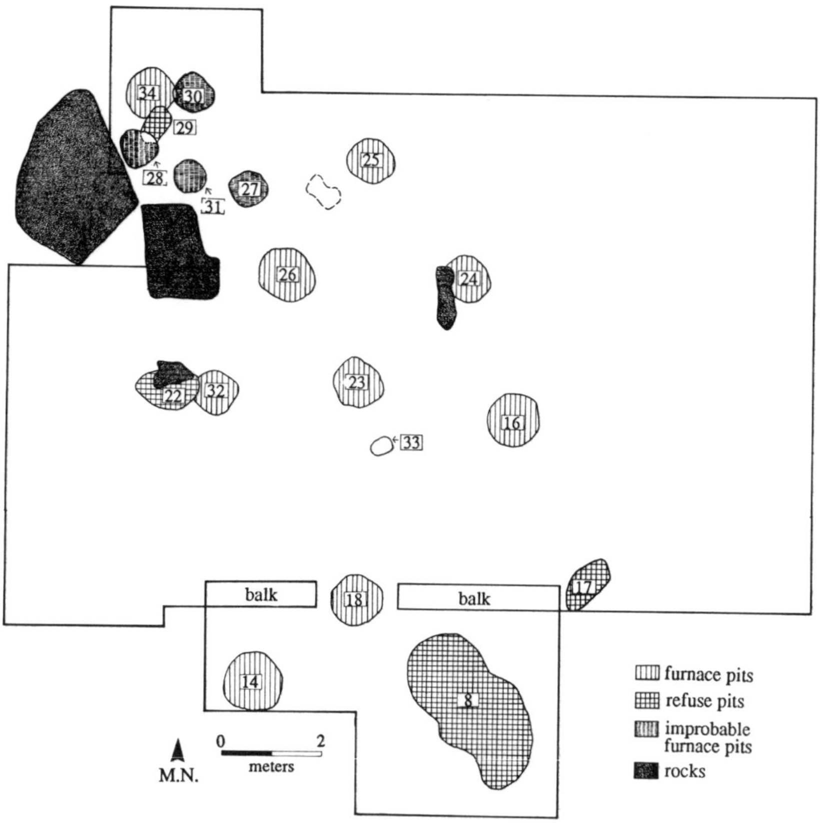

FIGURE 9.5. A plan view of the Early Iron Age industrial features of Block B and its extension at the KM2 site.

The two shallow furnace pits in Block B are features 16 and 34, dated to 1620 ± 60 B.P. (330 A.D. ± 60, corrected to 408 A.D. ± 86) and 1670 ± 120 B.P. (280 A.D. ± 120, corrected to 325 A.D. ± 130), a difference of only .28 of their uncorrected standard deviations. There is no significant difference in age between the dates for feature 16 and feature 6, which is dated to 1590 ± 60 B.P. (360 A.D. ± 60, corrected to c. 418 A.D. ± 86). While feature 6 belongs to this group temporally and resembles the others in its lack of large, complete bricks, it also differs from them; the steep, vertical sides of the furnace pit differ along with the presence of a small hole excavated in the floor of the furnace. These attributes (see table 9.3) more closely resemble the furnaces among which feature 6 was found. This temporal cluster also includes a refuse pit filled with slag, feature 29, that cuts through the floor of feature 34 (see figure 9.12b). This pit has been dated to 1580 ± 120 B.P. (370 A.D. ± 60, corrected to c. 420 A.D. ± 95). The archaeological observation that feature 29 is younger than no. 34 is confirmed by the C^{14} dating, for the calibrated date from feature 29 is approximately ninety-five years younger than the date for feature 34.

Two shallow dish-shaped furnace pits constitute the fourth and final temporal grouping of smelting pits, both of which lacked large bricks as pit fill. The first example, feature 14, has been C^{14} dated to 1410 ± 60 B.P. (540 A.D. ± 60) and to 1480 ± 60 B.P. (470 A.D. ± 60), with a mean date of 1445 ± 42 B.P. (505 A.D. ± 42, calendrically corrected to 540 A.D. ± 48). Feature 18, which is similar in morphology to 14 and also its close neighbor, is dated to 1440 ± 110 B.P. (510 A.D. ± 110, corrected to 593 A.D. ± 89), so that statistically there is no significant difference between the two dates after calendrical calibration.

This discussion of the KM2 C^{14} dates shows that there are four distinct use periods on the site. These periods are easier to conceptualize if we use the calibrated dates at one standard deviation: the first is an occupation period that spans the third and second centuries B.C. The second is an industrial period that covers the first two centuries A.D., followed by a hiatus of one century, between 200–300 A.D. The third period falls in the fourth and fifth centuries A.D., and the fourth follows from the beginning of the sixth century into the early seventh century (figure 9.6).

KM3 RADIOCARBON DATES

The variety of structural features dated at the KM3 site is more restricted than at KM2. All ten dates are derived from furnace pits, with three furnaces, num-

TABLE 9.2. Radiocarbon dates from the KM3 site

Date no.	Feature no.	Function	Lab. no.
1	1C	Smelting furnace	Beta-2243
2	1B	Smelting furnace	SMU-1433[a]
3	3C	Smelting furnace	Beta-2241
4	6B, C	Smelting furnace	Beta-2242
5	7B	Smelting furnace	Beta-2239
6	8B	Smelting furnace	Beta-10,493
7	8D	Smelting furnace	RL-1260
8	9C	Smelting furnace	Beta-2240
9	9C	Smelting furnace	SMU-1434[a]
10	10	Iron smelting refuse pit	Beta-10,494

[a]corrected for fractionation

bers 1, 8, and 9, dated twice. There are three temporal periods represented at the KM3 site. The first period agrees with the earliest industrial period at the KM2 site and is manifest in two furnaces located in Block B in the center of the site (table 9.2 and figure 9.7). The oldest furnace is no. 8, dated to 1920 ± 120 B.P. (30 A.D., corrected to a calendrical date of 42 A.D. ± 91).[2] Partially destroyed on its northern side by hoe agriculture, this furnace contained critical evidence for the manner in which ancient iron was produced (Schmidt 1981, 1983a). It was also one of the furnaces that exhibited some signs of disturbance. It contained one pottery sherd decorated with a carved cylinder, a roulette technique that dates to the fourteenth century A.D. elsewhere in the region. This intrusive evidence and some rodent burrowing on the south edge indicate possible contamination of the second sample dated to 2520 ± 100 B.P. (570 B.C. ± 100, corrected to 617 B.C. ± 106). The sample was also very small, thus increasing the probability for an error in calibration. Based on the other reasonable date from the furnace and the remainder of the dates from Block B and the entire site, this date must be considered too early.

This example of dating aberration illustrates well the wide variability that

year half-life	ad/bc	Date bp 5568 Range at 95% confidence level	
		UNCALIBRATED	CALIBRATED
1660 ± 70	ad 290 ± 70	ad 150-430	AD 225-565
1717 ± 37	ad 233 ± 37	ad 159-307	AD 70-440
1560 ± 80	ad 390 ± 80	ad 230-550	AD 235-620
1620 ± 70	ad 330 ± 70	ad 190-470	AD 235-580
1830 ± 100	ad 120 ± 100	80 bc-ad 320	15 BC-AD 405
2520 ± 100	570 ± 100 bc	770-370 bc	830-405 BC
1920 ± 120	ad 30 ± 120	210 bc-ad 270	165 BC-AD 240
1500 ± 60	ad 450 ± 60	ad 330-570	AD 415-615
1684 ± 55	ad 266 ± 55	ad 156-376	AD 100-550
1740 ± 70	ad 210 ± 70	ad 70-350	AD 60-425

can occur in dating wood charcoal from the early era of the Iron Age. It is possible, for example, that date no. 6 from furnace 8 is derived from charcoal of a mature forest tree, a reasonable likelihood in an area where industrial groups were beginning to exploit mature forest stands. Alternatively, there is a remote possibility that the younger date has been influenced by charcoal from furnace 7, which slightly intersected furnace 8 on its southeast border and has been dated to 1830 ± 100 B.P. (120 A.D. ± 100 corrected to 180 A.D. ± 98), a temporal relationship confirmed by the sequential archaeological evidence. But there is no archaeological evidence to suggest that this could have happened. Van Grunderbeek et al. (1983a, 44, 48–49) found a similar phenomenon in Rwanda, where one furnace at Gasiza I has been dated to 810 B.C., and in Burundi at Rwiyange I, where a furnace was dated to 1450 B.C. and 1020 B.C., and at Mirama III, dated to 570–90 B.C. ± 85 (van Grunderbeek 1992). Van Grunderbeek accepts this series of dates as apparently valid and constructs an argument on behalf of an origin for iron technology in the region dating to the first half of the first millennium B.C. (see also van Grunderbeek 1992 and Clist 1987). Her argument is buttressed by the Rugomora

TABLE 9.3. Furnace attributes at the KM2 site

Feature no.	*Mean diameter (cm)*	*Burned earth*	*Liner*	*Bottom hole*	*Charcoal logs*	*Depth (cm)*	*Shape*
FIRST INDUSTRIAL PERIOD							
9	101	Bottom	Side	Center	Sides	23	v.w. bowl[a]
7	98	Bottom, sides	Side	Center	Bottom, pieces	21	v.w. bowl[a]
SECOND INDUSTRIAL PERIOD							
34	103	Bottom, sides	Bottom, sides	Absent	Bottom, side	14	dish
16	103	Bottom, sides	Bottom	Absent	"Pieces" sides	21	dish
6	95	Bottom, sides	Bottom, sides	Center	In fill/ sides	23	v.w. bowl[a]
THIRD INDUSTRIAL PERIOD							
18	101	Sides	Bottom	Absent/ "side"	Bottom, sides	20	dish
14	107	Bottom	Bottom	Absent	Sides	22	dish
LATER INDUSTRIAL PERIODS							
21	?	Bottom	Absent	Center	Absent	?	v.w. bowl[a]
23	95	Bottom	Bottom	Absent	Sides	12	dish
24	93	Bottom, side	Bottom	Absent	Sides	12	dish
25	86	Bottom, sides	Bottom, side	Center	Bottom	20	v.w. bowl[a]
26	103	Bottom	Bottom	Center/ "side"	Sides complete	20	v.w. bowl[a]
QUESTIONABLE AS FURNACES							
28	(77)	Bottom	Bottom, over feature 29	Absent	Absent	14	v.w. bowl[a]
31	(65)	Bottom, sides	Absent	Absent	Absent	10	dish
32	(85)	Bottom, sides	Bottom	Absent	Absent	11	dish

[a]vertical walled bowl

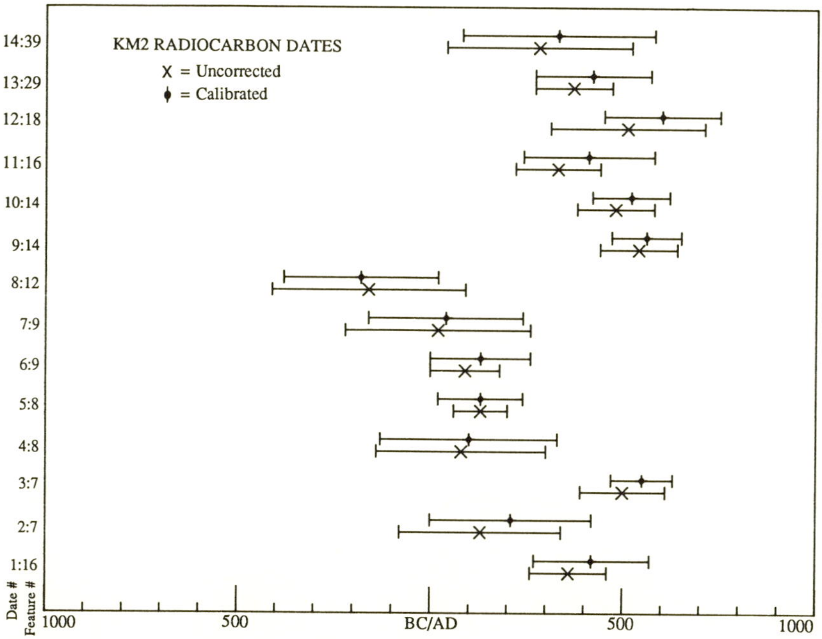

FIGURE 9.6. A plot of all radiocarbon dates from the KM2 site. Each plot line represents a date at a 95 percent level of confidence. The uncorrected dates are designated by an X, while the calendrically corrected dates are designated by a 0.

Mahe/Katuruka dates of similar age (Schmidt 1975, 132; 1978, 179, 195), without taking into account, as does Clist (1987), that the archaeological evidence for a forest fire stratum that dates to the late second millennium and Schmidt's interpretation that those dates associated with the Early Iron Age forge may very well be on old wood (see Soper's similar analysis, Soper 1982, 225; Schmidt 1978, 179, 191).

There is clearly a need for multiple dates for each furnace structure to help eliminate such uncertainties; this is a method that we intend to follow in our future studies. At this time there is no reason to believe that iron smelting was an established technological system in the interlacustrine area during the mid-first millennium B.C. on the basis of a series of single C^{14} dates from solitary furnace pits that constitute the only evidence from specific sites. The furnace 8 anomaly at KM3 poignantly illustrates one dimension of this problem and testifies to the need for a different documentary strategy. Despite the san-

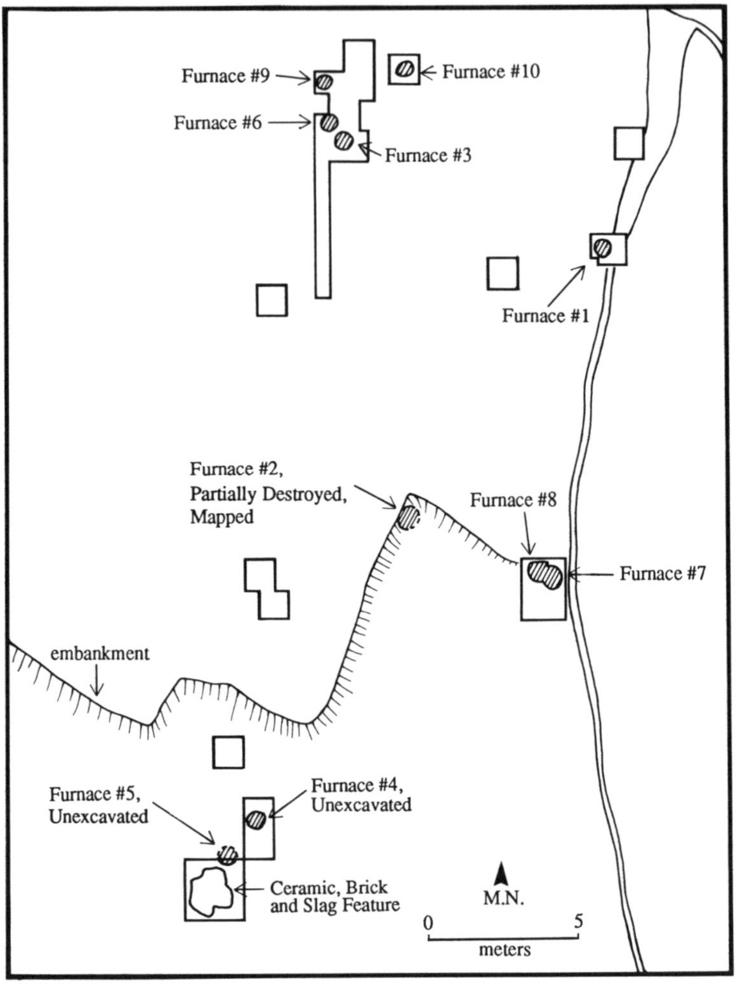

FIGURE 9.7. Furnace locations in the northwest section of the KM3 site.

guine positions of Clist (1987) and van Grunderbeek (1992) that iron industry is established very early in the region, we await more fulsome evidence before endorsing their assessment.

The second temporal cluster at KM3 does not have the tight spatial relationship of the Block B furnaces. Furnace 1 was located in the main path running through the site, while furnace 10 was located just outside of Block A about 5–8 m northeast of a cluster of three later furnaces. Furnace 1 resembled features 7 and 9 at KM2 and furnace 9 at KM3 in that it was packed with furnace bricks and slag (figures 9.8a and 9.8b). It has been dated to 1660 ± 70 B.P. (290 A.D. ± 70) and 1717 ± 37 B.P. (233 A.D. ± 37), which result in a weighted average of 1705 ± 33 B.P. (245 A.D. ± 33, corrected to 260 A.D. ± 85). Furnace 10 has been dated to 1740 ± 70 B.P. (210 A.D. ± 70, corrected to 243 A.D. ± 91). The t-test applied to the corrected dates indicates more than a 90 percent probability for their contemporaneity. Furnace 10 has large quantities of furnace brick, but it had been disturbed by local agriculture, which had stirred up and partially dispersed the fill and had obliterated the original furnace floor. Both furnaces 1 and 10 represent a short-lived episode of industrial activity on the site during the middle of the third century A.D., a period during which iron production is absent at the KM2 site. The differential use of the two sites may be related to a host of different reasons, including patterns of resource availability, the location of social groups that controlled the industry, and ideological notions about sacred space and the practice of iron smelting.

The third cluster of dates is associated with a spatially coherent clustering of three furnaces, numbers 3, 6, and 9. These furnace pits are tightly grouped in close proximity to a set of stone features that we interpret as iron ore processing areas (figure 9.9). Furnace 3 and 6 resembled one another, as their fill lacked furnace bricks larger than 5 x 5 x 5 cm. Furnace 9, more similar to the two furnaces of the earlier period with its brick and slag fill, has been dated twice: 1500 ± 60 B.P. (450 A.D. ± 60) and 1684 ± 55 B.P. (266 A.D. ± 55), which yield a weighted average of 1602 ± 41 (348 A.D. ± 41, corrected to 365 A.D. ± 60). As reported in earlier publications, this furnace contained the definitive tuyere evidence that confirmed beyond doubt that preheating had developed at least by the fourth century A.D. in Africa (Schmidt 1980b, 1981, 1983a).

The first of the two remaining furnaces of this cluster, no. 3, has been dated to 1560 ± 80 B.P. (390 A.D. ± 80, corrected to 434 A.D. ± 86), while no. 6 was dated to 1620 ± 70 B.P. (330 A.D. ± 70, corrected to 408 A.D. ± 86). After calendrical correction, it is clear that these three furnaces form a coherent temporal group with no significant difference in age. The third use period at the KM3 site occurred during the latter part of the fourth century and first half of the fifth century A.D. This era also saw utilization of the KM2 site, though we did not find any evidence there for the placement of furnace bricks in the furnace pit after the dismantling of the furnace superstructure. Moreover, the KM3 site was not used for iron production during the last period at KM2, namely, the sixth and seventh centuries A.D., a further commentary on the differential and

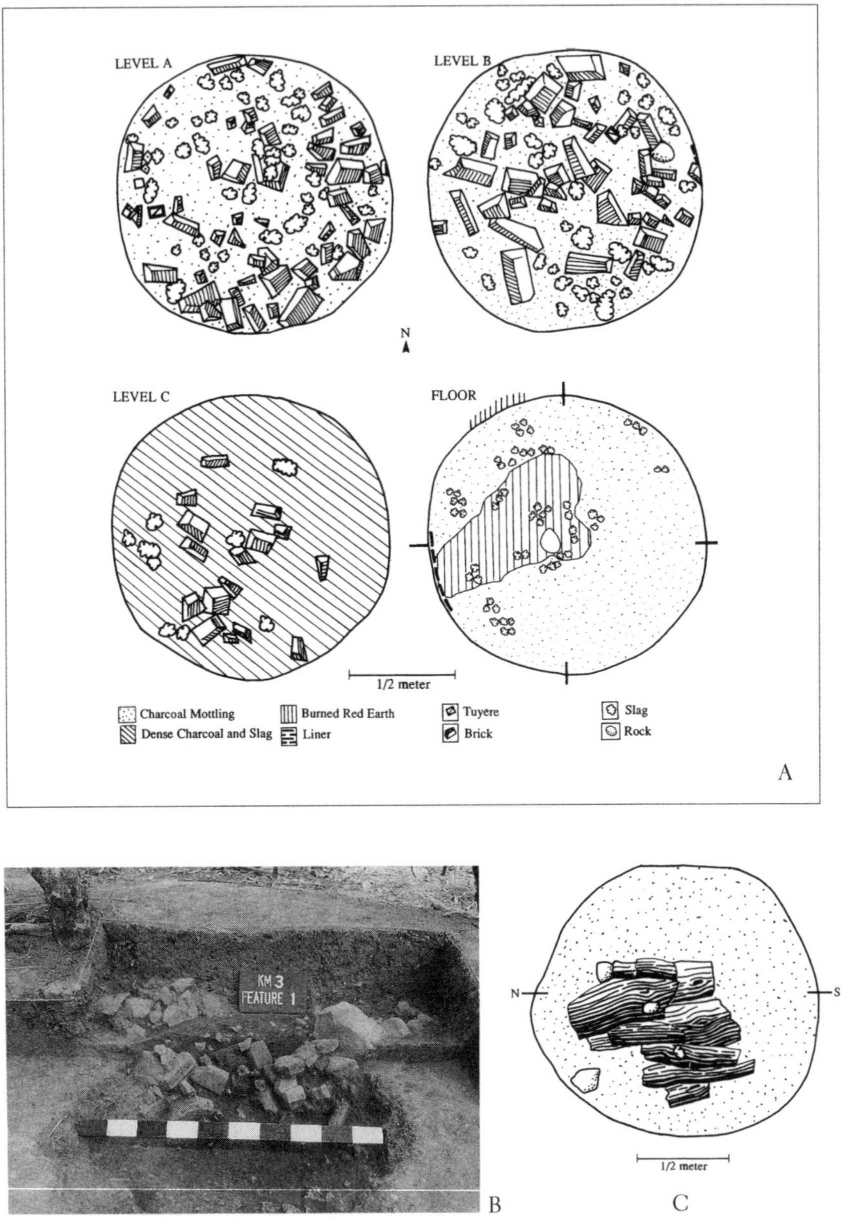

FIGURE 9.8. A, plan views of successive stratigraphic levels excavated in feature 7 at KM2; B, a profile view of the massive concretions of furnace bricks and slag in furnace 1 at KM3; C, large charcoal logs placed in parallel alignment on the floor of furnace 9 at KM3.

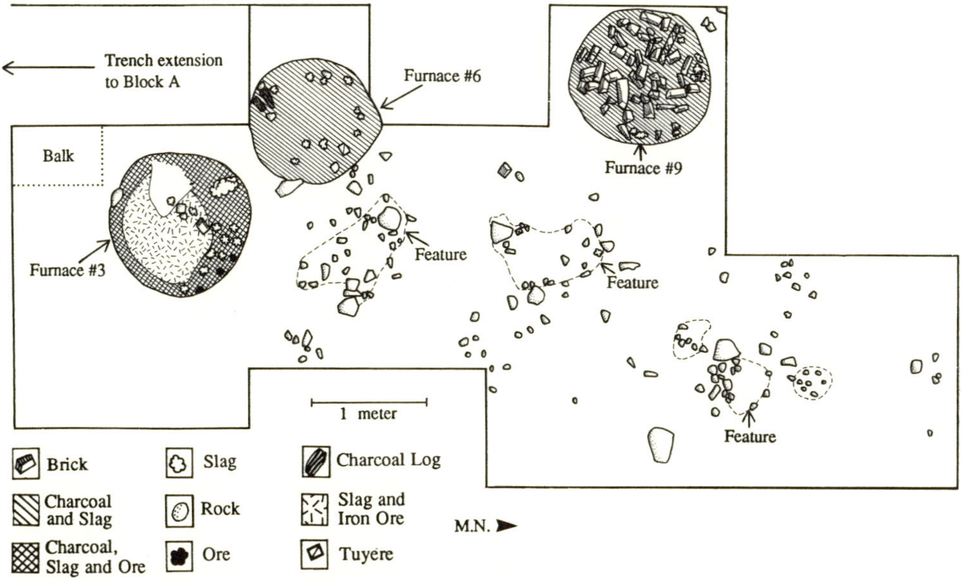

FIGURE 9.9. A plan view of Block A at KM3 with pits of Early Iron Age iron smelting furnaces that date to the period of use at KM3. The small stone features appear to be iron ore processing areas.

periodic utilization of these two areas during the first millennium of iron production in northwest Tanzania. A summary of the KM3 use periods, using the same criteria applied to KM2, allows the following divisions to be seen: (1) the first period, represented by two contiguous furnaces that date to the first two centuries A.D. (or, between c. 50 B.C. and 250 A.D.); (2) a second era during the third century A.D., during which time iron production was restricted to one contemporaneous episode at KM3 and absent at KM2; and (3) activity during the fourth and fifth centuries, also an active time at KM2, but the end of iron smelting at KM3 (figure 9.10). For purposes of simplicity and clarity in our remaining discussion, we will collapse the two later periods at both sites and refer to each as the later period.

Furnace Attributes

KM2

The furnace pits of KM2 share a number of attributes during all time periods on the site although there are also some clear associations between age and form (table 9.3). Most apparent is the greater depth of the furnaces that have been dated to the early industrial period, the first and second centuries A.D. However, the difference between these and the later furnaces is not significant, and it may simply be the result of agricultural activity that removed the walls of furnaces such as nos. 14, 16, 18, 23, 24, and 34. However, we cannot be certain that this happened in the past, and we must keep open the possibility that there was a change in furnace depth and form between the second and fourth centuries A.D. Two furnaces, nos. 7 and 9, along with the partially destroyed no. 21, date to the earliest industrial period and also have abrupt angles between the floor and their vertical side walls. This is a characteristic conspicuously lacking in all but one of later dated furnaces (furnace no. 6). Several other pits in the Block B extension have similar profiles (figure 9.11: 25, 26); these pits are undated, but we have assigned them to the later period because they are surrounded by many later period furnaces and they are morphologically similar to no. 6.

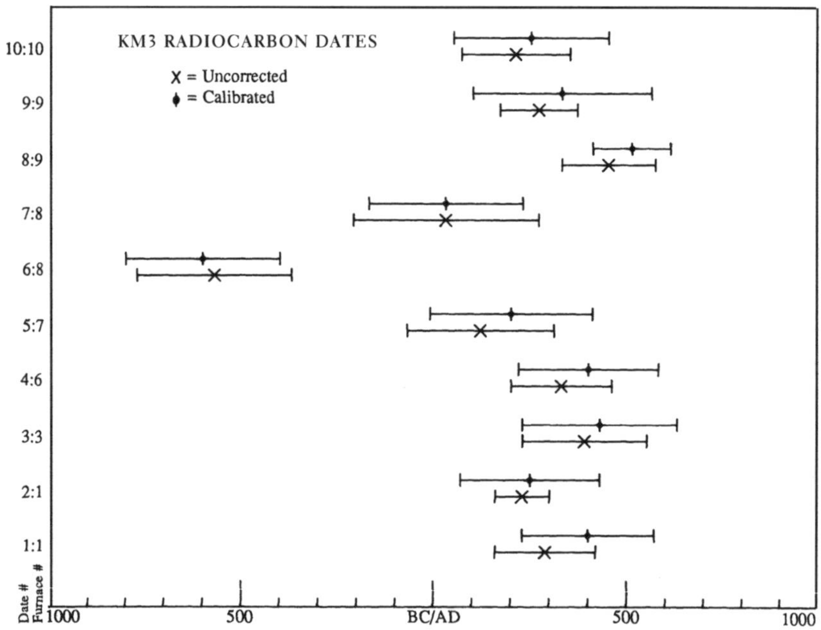

FIGURE 9.10. A plot of all radiocarbon dates from the KM3 site. Each plot line represents a date at a 95 percent level of confidence. The uncorrected dates are designated by an X, while the calendrically corrected dates are designated by a 0.

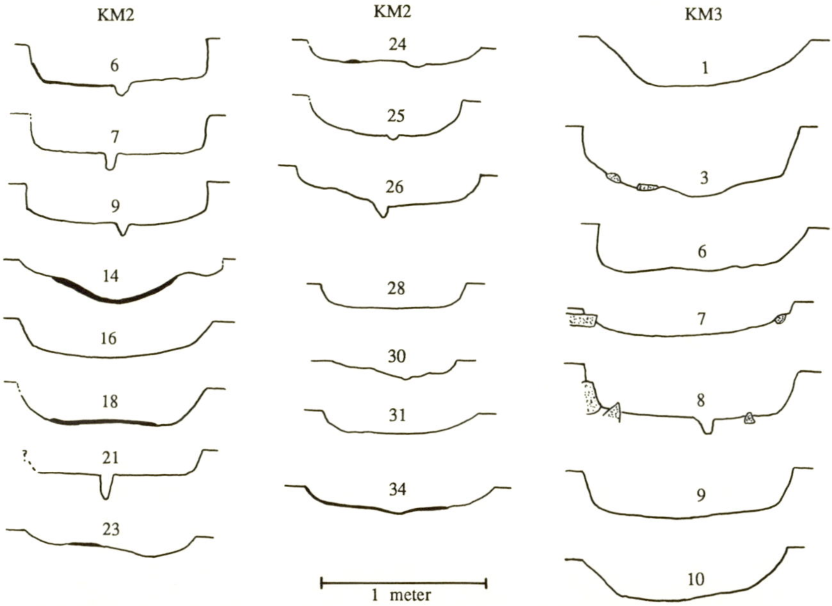

FIGURE 9.11. Profile sections of all furnace pits (including the doubtful features) at the KM2 and KM3 sites. A heavy line on the bottom represents liner that was *in situ*.

It is questionable whether the shallow furnaces of the later era would have been deep enough to drain slag from the blast zone. We feel that most of the furnaces at KM2 have been truncated, a conclusion affirmed by the irregular distribution of a red oxidized line around the furnace boundary. This absence of a continuous and deep oxidation boundary indicates that in most cases the upper furnace walls have been removed; only small and irregular lines of oxidation were left where we would expect to see a continuous and 10–20 cm deep and 1–3 cm thick oxidized zone immediately bordering the upper wall of an undisturbed furnace.

A formal characteristic that shows continuity from the first century A.D. through the sixth century A.D. is the small hole found in the floor of some furnaces. Schmidt has documented the ritual and symbolic meaning of furnace floor holes among the Barongo, a technological culture just to the south of Buhaya in Buzinza (Schmidt, chapter 5). The association between the symbolic/ritual domain and the productivity of the technology leaves a distinct material signature in the furnace floor that is unambiguous and undeniable. The rituals are a powerful mystification that symbolically identifies the ritual specialists as separate and distinctive, a process that helps to ensure their economic domination.

The historical documentation for the excavation of ritual holes in smelting furnaces is extensive in Bantu-speaking Africa. But interviews also tell us that there were differences within the Haya culture area in the use of ritual holes in furnace floors. We feel confident that when found archaeologically such features refer to the ideational system that supports the productive edifice. The ethnological evidence from related and neighboring cultures overwhelmingly supports this view. There is no material explanation for this furnace attribute. More comparative ethnography remains to be done, but there is a reasonable possibility that this attribute marks different social groups with different symbolic/ritual identities. Thus the variability through time seen in the occurrence of the furnace floor hole may reflect different groups that may have utilized the site during the same period of time; this is an interpretation that may be applied to the affinities that we see among features nos. 6, 25, and 26 and their considerable differences from nos. 14, 16, 18, 23, 24, 31, 32, and 34, all of which lack a floor pit.

There is also the possibility that some of the large shallow pits from the later period functioned as roasting pits. This functional alternative causes us to examine each of these pits using several additional criteria to assess possible function of all additional features. Several features in the Block B extension have been excluded from table 9.3. We have assigned these different functions for the following reasons: first, we see that feature 27 had little evidence for burning, no charcoal logs, a very small amount of slag, a low slag to iron ore ratio (table 9.4), and no slag beads on the floor, all of which point to a function not directly associated with the technological process. If we refer to table 9.5, which details the KM3 slag and iron ore densities, then we find a very clear pattern in a set of features that are, with one exception, furnace pits. The regularity of the KM3 evidence provides us with another criterion by which we can judge the function of the KM2 pits. For example, every furnace had a slag wt. to ore wt. ratio of not less than 1.17 and a mean ratio of 4.0. Thus, we must question if pits with small amounts of slag and significantly greater amounts of iron ore were used for iron smelting. This observation and the other criteria suggest that nos. 28, 30, and 31 must be eliminated as smelting pits. They do not display sufficiently strong evidence for burning, the presence of slag beads, and the expected densities of slag and iron ore. Moreover, none had a liner, and no. 28 appears to have had several distinct fill episodes during which slag

TABLE 9.4. Slag and ore distributions at the KM2 site

Feature no.	*Slag fragments*	*Slag weight (kg)*	*Iron ore fragments*	*Iron ore wt (kg)*	*Slag: iron ore wt ratio*
6	1662	27.690	1417	10.470	2.64
7	1872	39.661	1088	7.568	5.24
8	6424	69.731	14,979	94.713	0.74
14	5548	31.888	2989	18.161	1.75
16	321	1.829	1517	10.301	0.18
9	4240	21.627	1207	6.796	3.18
17	367	5.468	1044	6.597	0.83
18	3507	16.234	2662	15.640	1.22
22	1648	10.820	1030	9.739	1.11
23	1689	8.288	1485	6.934	1.19
24	644	4.410	1388	9.208	0.48
25	173	1.158	302	2.125	0.54
26	1186	4.724	2007	9.122	0.52
27	51	0.306	130	0.532	0.58
28	166	1.049	288	1.482	0.71
29	3313	39.404	424	2.417	16.30
30	22	0.140	163	1.723	0.08
31	2	0.013	7	0.042	0.31
32	762	4.693	1002	5.490	0.85
34	1664	8.168	628	4.842	1.69

and domestic globular pots were deposited. It is also important to observe that no. 31 is much too shallow and departs significantly from the mean diameter of the larger furnaces, as do nos. 28 and 30. Finally, there is a reasonable alternative hypothesis that these three smaller pits may have functioned as secondary refining furnaces, in which the bloom was further consolidated and slag removed prior to forging. But the absence of significant amounts of slag and the fact that such refining furnaces are not found elsewhere tend to argue against this position.

There are several further observations to be made about the KM2 features. It appears that features nos. 16 and 18 functioned as roasting pits as well as

TABLE 9.5. Slag and ore distributions at the KM3 site

Feature no.	*Slag fragments*	*Slag weight (kg)*	*Iron ore fragments*	*Iron ore wt (kg)*	*Slag: iron ore wt ratio*
1	2400	30.398	1975	6.145	4.97
3	7717	38.836	7837	33.098	1.17
6	2790	10.039	138	2.222	4.51
7	14,808	29.809	861	2.811	10.60
8	4853	16.187	2836	10.178	1.59
9	4240	21.627	1207	6.796	3.18
10	874	8.551	693	3.778	2.26

smelting furnaces. Both have large charcoal logs on their floors, but in no. 16 we see a massive concretion of partially reduced iron ore packed on the bottom of the pit; no. 18 also had charcoal logs, but the iron ore was mixed in with the charcoal logs that in turn overlaid small slag prills (figure 9.12a). We see this as an example of roasting in a pit that previously had all but the small slag pieces removed from its floor.

The importance of charcoal logs appears to be explained by the actualistic model based on observations of Barongo iron smelting (Schmidt, chapter 5). Large pieces of charcoal allow fluid slag to collect both above and below the logs, a phenomenon we saw in many of the KM2 furnaces. When the charcoal logs are intermixed with slag, smelting is indicated, whereas when they are intermixed with ore, roasting is indicated (figure 9.12a). The latter interpretation derives from our actualistic study of roasting iron ore among the Haya. During smelting such large pieces of charcoal sometimes caught dripping slag so that thin casts of the logs were made (figure 9.13a). But the large pieces of charcoal found on the bottom of many furnaces and mixed into the lower fill were not used in the blast zone or the higher zone of the furnace where most of the slag formation took place. Most of the large slag prills and the overwhelming remainder of porous slag were formed in an environment where 1.5–3 cm cubes of charcoal predominated (figure 9.13b). Small charcoal ensures a high-carbon contact area with the fluid fayalite-wustite slag that forms immediately in front of and below the tuyeres. The stratified remains inside the KM2 furnace pits sometimes resembled the stratified deposits of charcoal

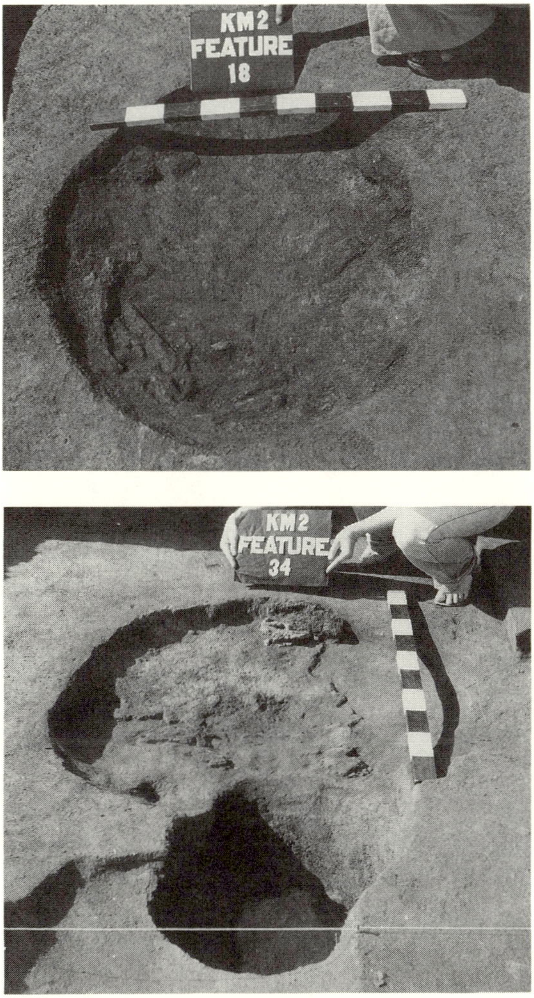

FIGURE 9.12.
Top, charcoal logs on the floor of feature 18 at KM2 were mixed with iron ore, suggesting that the last function of the pit was to roast iron ore; *below*, feature 34 had the most significant preservation of furnace liner, seen in the center of the upper pit. The deep pit in the center was a slag refuse pit that cut through the furnace, a sequence confirmed by the radiocarbon dating.

and slags in contemporary furnaces. The charcoal used in the actual reduction process was not the charcoal logs found on the floors of the KM2 pits; such logs were used to drain slag and, secondarily, to roast iron ore in smelting pits.

We want to caution that the data on furnace liners in table 9.3 do not always refer to large sections of liner (see figure 9.12b for an exception). Often only small pieces were left *in situ*, usually accompanied by large contiguous zones of red burned earth. The red burned earth marks sections where the liner was apparently removed after the last smelt, for we observed that similar

FIGURE 9.13. *Left*, slag from feature 14 at KM2, showing the imprint of a large charcoal log; *right*, slag from furnace 3 at KM3 showing the imprints of a larger charcoal cube (*top*) and a small charcoal cube (*bottom*), indicating the use of charcoal rather than grass in the furnace pit during the Early Iron Age.

earth was located beneath liner that was *in situ*. Had smelting occurred in the pit after removal of the liner, then the reducing conditions should have obliterated these oxidation characteristics. There also remains the possibility that the thick charcoal layers found overlying red zones in some of the KM2 pits were placed there purposefully to protect the exposed areas from further slag contact and that refractory charcoal protected the floor from reduction as well.

The composition of the KM2 furnace liners differed markedly from the soil that filled the pits from the surrounding area. For example, a particle size analysis of the liner from no. 14 shows that it had 53.5 percent coarse sand and 29 percent fine sand compared to 5 percent silt and 12.5 percent clay, whereas the soil from the same feature had 25.2 percent coarse sand, 24.8 percent fine sand, 16.2 percent silt, and 33.8 percent clay. These proportions occurred elsewhere on the site in furnace pits and indicate the purposeful selection of a liner material that differs significantly from the surrounding soil matrix and that resembles contemporary termite mounds. This similarity between recent and ancient lining practices suggests long-term continuity, over scores of generations, in resource selection for a specific refractory function in furnace liners (Childs 1986).

KM3

The KM3 furnaces have much greater consistency in design and contents than the KM2 furnace pits. The most obvious contrast is the regular furnace diameter, ranging only between 112 cm and 116 cm, with a mean of 113.8 cm. The furnace profiles (figure 9.11) show that there is also much more consistency in form through time. We prefer to exclude furnace 10 from consideration, for it is clear that many of the critical formal attributes have been obliterated (table 9.6). The mean depth of 28 cm among KM3 furnace pits is significantly greater than the mean depth of 18 cm at KM2, a difference that we believe is attributable to more complete preservation at KM3. The only departure from the deeper bowl-shaped profiles that prevail on the site is furnace 7, which in profile seems to resemble the large shallow furnaces of KM2. This is likely the result of its close proximity to the surface and the removal of its upper portions. Otherwise, we find that most of the remaining furnace pits, including no. 8 from the early period, have vertical sides in their upper sections, but that the transition to the floor often resembles a bowl profile. This profile differs from the pits with vertical sides at KM2; significantly, the lower half of the profile closely parallels the truncated shallow furnaces of KM2. Furnace 1 displayed a more rounded bowl shape, but this seems wholly consistent with the other furnace pits. Our conclusion is that we do not see any significant shift of furnace pit form through time; what variability we do see is partly the result of our excavation methods, peripheral rodent activity, and excavation techniques of the prehistoric smelters. This conclusion differs from that of van Grunderbeek, who sees a change in the Butare, Rwanda furnaces from a shallow and small type to a deeper and larger type after the third century A.D. (van Grunderbeek et al. 1983a, 16).

It is apparent that we currently lack in African archaeology any standard criteria to determine technological functions of pit features associated with iron production. We are not able to judge from data available, for example, if some pits were smelting furnaces rather than, say, refuse pits. A case in point is the pit at the Nyaruhengiri I site, which is photographed with bricks at the bottom of a deep pit (van Grunderbeek et al. 1983a, 12); it is not clear if other artifacts had been removed or if this was a pit built expressly to receive industrial refuse, as was no. 8 at KM2.

Oral testimony of elderly smelters in Buhaya and observation of the variable sizes of historic smelting pits indicate that the volumetric capacity of a furnace of the shaft-bowl, slag-tapping type varied some to suit the labor requirements and preferences of the recent smelters. The circumference of the furnace pit (and hence diameter and chimney height) increased to accommodate more than the minimum number of eight smelters and eight tuyeres. An

TABLE 9.6. Furnace attributes at the KM3 site

Feature no.	Mean diameter (cm)	Burned earth	Liner	Bottom hole	Charcoal logs	Depth (cm)	Shape
FIRST INDUSTRIAL PERIOD							
8	116	Absent	Sm. sandstone	Center	Sides (m)	27	v.w. bowl[a]
7	114	Bottom	Sandstone	Absent	Bottom (m)	17	dish
SECOND INDUSTRIAL PERIOD							
10	?	Absent	Absent	Absent	Absent	?	?
1	113	Bottom, sides	Sides	Absent	Bottom, sides	30	dish
THIRD INDUSTRIAL PERIOD							
9	112	Bottom, sides	Absent	Absent	Bottom, sides	29	bowl
6	112	Bottom, sides	Sandstone	Center?	Sides (m)	23	v.w. bowl[a]
3	116	Bottom, sides	Bottom	Absent	Bottom, sides	40	bowl

(m) = minimal amount [a]vertical walled bowl

increase from eight to twelve tuyeres (with a corresponding increase in the number of smelters) would not have significantly altered reducing conditions inside the furnace and suggests to us that this recent variability should be taken as a cautionary tale for making too much of variably sized furnaces in the prehistoric past. More important is the observation that the morphology and size of the Early Iron Age furnace pits at KM3 bear very strong affinities to the recent "eight man" smelting pits of the Haya.

A consistent similarity between the KM3 and KM2 furnaces was the use of massive logs of charcoal along the sides and in the bottom of the furnace pits (tables 9.3 and 9.6), such as nos. 1, 3, and 9 (figure 9.8c). Along with the sandstone rubble in the bottom of some of the furnace pits, the charcoal may have functioned as a furnace liner, because charcoal is an excellent refractory material. This property and the parallel alignment of the charcoal logs in furnace no. 9 suggest that this is a reasonable hypothesis, but one that is lacking in the actualistic models. Identifications of wood charcoal from furnace nos. 9 and 3 provide important evidence for the exploitation of forest resources for smelting. The massive logs on the floor of furnace no. 9 are *Microdesmis puberula* (*Euphorbiaceae*), which is commonly found in dense forest clusters, but also in regenerative clusters in secondary regrowth (M. C. van Grunderbeek and H. Doutrelepont, personal communication). The many medicinal uses of parts of this species opens the possibility that its placement on the furnace floor may have had a ritual association. Charcoal from the lowest stratum of furnace 3 has been identified as *Entandrophragma* sp. (*Meliaceae*), which is indicative of dense forest clusters between 1300 and 2000 m in Uganda and Tanzania (H. Doutrelepont and E. Roche, personal communication). These results are consistent with other environmental data that indicate that the region was under forest cover 2500 years B.P., before widespread clearance of forest that may be related to iron production and settled agriculture (Kendall 1969, 159–62; Schmidt 1978, 23–24; Schmidt 1994).

Not all KM3 furnaces had charcoal logs as large or plentiful as no. 9; the earliest furnaces had large pieces of charcoal rather than logs near their edges, while such evidence was only minimally present in furnace 6. Most slags with charcoal imprints indicate that slag formation occurred in a microenvironment with smaller charcoal cubes that rarely exceeded 5 cm along the largest side. We have also inspected slag for grass or reed imprints. The importance of this observation lies in the need to document the history of filamentous grass and ash, especially of high alkali content, as a reduction medium in African iron smelting. This technique confers a number of critical technological and environmental advantages in the African direct steel process, particularly in a distinctive reduction process that occurs within that particular carbonaceous environment (Schmidt and Avery 1978, 1088). The utilization of a rapidly renewed resource such as swamp grass has environmental implications as well. For example, the Haya use of swamp grass (*Miscanthidium* violetti) during the later Iron Age and historic era is an adaptation that is both fuel- and labor-efficient (Schmidt 1981, 29). This was a significant development in the face of severely depleted forest resources. There is a strong possibility that the use of grasses evolved long ago, for the slags at Taruga show signs of having been formed in a grass environment (Tylecote 1975a, 6). Van Grunderbeek and colleagues document a papyrus reed imprint and a leaf imprint on two of the Rwanda slags (van Grunderbeek et al. 1983a, 37). They interpret this to mean that the bottom of the furnace was filled with these materials. If indeed these materials functioned as pit fill, then we should expect to see numerous imprints and slag casts of the reeds. The evidence instead seems to point to a furnace pit liner that resembles the lining of the Barongo smelters, who line their pits with leafy twigs (Schmidt, chapter 5, this volume). It is critical that the differences between these two practices—pit filling and pit lining with grasses, sedges, and twigs—be distinguished on the prehistoric slags if the technological process is to be documented accurately.

It is difficult to know if furnace liners were not as commonly employed at KM3 or if they had simply been removed from furnaces. We know from small patches of liner in furnace 1 and under slag in furnace 3 that furnaces were lined at KM3. Other furnaces in Block A such as numbers 6 and 9 had large, clearly defined areas of oxidized soil on their floors. In this instance we lack inferences based on material observations of recent technologies, but we believe that the reduced liner had been removed after the last smelt, thus exposing the underlying oxidized layer previous to the filling of the pit. The floor of no. 6 was also filled with small sandstone rocks, suggesting that extra sandstone rubble had been added purposefully as a liner and that larger pieces of sandstone, such as in furnace 3, functioned as a natural refractory liner after the pit had been excavated into the sandstone rubble. Sandstone is one of the best natural refractory materials available and its plentiful supply at KM3 was advantageously employed at KM3 from the time of the early furnaces.

The most apparent departure from the KM2 furnace pit attributes is the dearth of small holes in furnace floors at KM3. We searched carefully for such evidence and found unequivocal signs of purposeful modification of the floor in only one instance, furnace 8. In this furnace floor, concealed under a small piece of sandstone, we found a small hole that contained a piece of smelted iron (figure 9.14a) (Schmidt 1981, 26–28; Schmidt 1983a, 425–27). Later metallurgical analysis showed that this piece of iron had been produced by a smelting process that had formed large, highly perfect crystals, a structure that closely resembles that seen in the massive steel pieces produced in traditional, preheated Haya furnaces (figure 9.14b).[3]

The function of the hole is consistent with our interpretation of similar holes at KM2 and elsewhere; its sealed context closely resembles an Early Iron Age pot (sealed with a pot sherd) found under a smelting furnace at Kabuye II in Rwanda (van Noten 1979, 65–67). It is noteworthy that in furnace 8 the iron had been heavily carburized to about 1 percent carbon after prolonged heating between 800–850°C, presumably during repeated smelts in the same furnace in much the same way one episode of ritual medicine functions over many Barongo smelts (Schmidt, chapter 5, this volume). This suggests a ritual function wherein the properties of the ritual material are symbolically identified with the properties of the furnace product (Schmidt 1983a, 424–25; 1983b). Unlike the KM2 site, the group(s) that used the KM3 site during the later period did not employ ritual devices in the furnace floors. However, there may be other ritual markers, such as the use of specific wood species laid on the floor, that distinguish different groups of technicians and their ideologies.

The observation of red burned earth on the upper borders of furnaces was not as obvious at KM3. We failed to observe this attribute around the early furnaces, perhaps as a consequence of poorer preservation. The characteristic isotherm was present at all the later furnaces (except no. 10), though it was only apparent after wetting the very dry and hard packed soil. There is further reason to question if no. 10 was a furnace pit rather than a refuse dump, an alternative that must be considered. All furnaces but no. 10 were marked by the diagnostic slag beads embedded in their floors. While the feature had been disturbed, it is nonetheless significant that no slag beads were recovered from its fill. The fill of furnace no. 7 was also distinctive, for the pit was solidly packed with small pieces of slag resting on larger flow slag on the floor. The absence of furnace bricks in this furnace, as well as no. 3, points to the fact that the chimney walls of some furnaces were apparently dismantled and removed for either reuse or disposal elsewhere. This contrasts with furnace pits such as no. 1 and no. 9, both of which were filled with bricks mixed with slag. The mixing of bricks with slag in the upper strata, plus the inclusion of sections of tuyeres, points to the purposeful disposal of these materials in the furnace pit. Not all the bricks ended up in the pit, thus again suggesting that the remainder of the bricks were removed and possibly reused. The presence of concave and convex sides on these bricks would have provided ideal surfaces that could have been fitted together with a good seal at a later furnace.

Bricks

Much remains to be puzzled out about how bricks were manufactured, the source of the raw materials used and their refractory properties, whether the bricks were reused, and what the significance of the decorative applications was. These are but a few of the many possible questions that arise, and we want to address only several of these questions in a preliminary way. Furnace

FIGURE 9.14. *Left*, a partially corroded iron bloom, with grass imprints, in situ within the furnace hole found in the floor of furnace 8 at KM3. A ritual function is attributed to its placement in the furnace floor; *right*, a photomicrograph of a section of the iron from furnace 8 at KM3. The dark area is the corrosion product; within that area the structure of the prehistoric iron is seen in the form of massive carbide plates that formed along the intergranular boundaries under prolonged reducing conditions, thus providing a skeletal microstructure of the bloom. The spherical slag inclusions show the unworked condition of the iron. Photos by Peter R. Schmidt and Donald H. Avery.

bricks dating to the Early Iron Age were first described by Hiernaux and Maquet (1957, 1960), and sites with these technological artifacts have sometimes been referred to as "brick" sites (van Noten 1979, 69). Any report that mentions furnace bricks invariably emphasizes the so-called decorations that appear on the outer or convex sides to the exclusion of other attributes. We want to suggest here other possible meanings of the putative decorations.[4]

Most bricks are rectangular with variably shaped ends and opposed convex and opposite sides. As might be expected, the interior or concave side is usually black, having been exposed to reducing conditions inside the furnace, while the convex, oxidized side is orange. Bricks that were 110 mm or larger are of particular interest in feature 8 at KM2. There was a significant increase in very large bricks toward the bottom strata of the large refuse pit. These bricks were among the first to be thrown into the pit and, conceivably, the first to be removed from the furnace. They clearly did not come from the top of the furnace; a reasonable hypothesis is that they were removable doors in the base of the furnace and, hence, that they were among the first pieces to be removed from the furnace.

These massive bricks also have "ridges" bordered by supposed decorative grooves (figure 9.15a) that form a series of transverse parallel grooves that often typify bricks in Rwanda and elsewhere in Buhaya (van Noten 1979; van Grunderbeek et al. 1983a; Schmidt 1978) and that Hiernaux and Maquet originally referred to as "finger impressions"; s-shaped and more curvilinear motifs are found outside Buhaya. Most decorated bricks are the massive bricks that exceed 110 mm in width, a characteristic that may indicate that the application of the outer layer had functional value rather then stylistic or symbolic meaning. We posit that the application of this outer, roughened layer may have facilitated handling and removing the putative door. Decorated bricks are absent from the later period at KM2 and the early period at KM3, though they are found at a frequency of 1 percent in furnace nos. 1 and 9 of the later period. The absence or presence of such motifs may be related to the symbolic identities of different craft groups. There are little ethnographic data to support this line of argument among contemporary peoples; but it must be noted that the Barongo smelters used decorative symbols on their tuyeres and that the markings on the putative door closely resemble the parallel grooves applied just above the opening to Shona furnaces to symbolize female genitalia (Childs and Dewey, chapter 7, this volume). Other furnace bricks also show signs of purposeful decoration. There are a few small bricks from the refuse pit, no. 8, at KM2 that were decorated by punctate impressions made by reeds (figure 9.15b). These small bricks appear to have been located on the rim of the furnace. Their decoration resembles the wavy grooves that van Grunderbeek finds applied to rim bricks at the Gisiza site in Rwanda (van Grunderbeek, personal communication with author).

Tuyeres

One of our primary goals in this study has been to obtain more definitive evidence for the attributes of prehistoric tuyeres. We have looked at a variety of attributes with the view to discern how and when a preheated iron technology developed in Africa. A fundamental requirement in preheating that employs clay blowpipes inside the smelting furnace is that the clay be highly refractory. If the heat transfer through the clay is sufficiently high to heat the air passing through the clay chamber, then the result will be a higher temperature in the blast zone; this in turn leads to more heat transfer through the clay walls of the tuyere, but it also requires particularly high refractory clays. Attempts to pre-

cm

cm

FIGURE 9.15.
Above, a massive "decorated" brick with deep grooves and intermediate ridges, excavated from the bottom of the large refuse pit at KM2 and therefore dating to the early (pre–200 A.D.) industrial period. Such bricks are not part of the apparent coils of the midsection of the furnace and appear to have been used in the base of the furnace, possibly as doors; *below,* small furnace bricks from the early (pre–200 A.D.) period at KM2. The bricks were decorated by punctate impressions that embellish the top or rim course of the furnace, similar to treatments of the top course on some EIA furnaces in Rwanda.

heat with clay tuyeres that are not sufficiently refractory inevitably lead to failure of the tuyeres and, consequently, failure of the smelt. Cross sectioning of tuyeres from recent smelts shows that there is often slag penetration into one-third of the tuyere wall or about 3 mm into the tuyere, an isotherm of about 1150°C, showing significant capacity for heat transfer. This and the fact that temperatures of 1500–1700°C prevail immediately in front of the tuyere show that such clays have excellent refractory properties.

Heat experiments on the clays and a sampling program of known clay resources and their mineralogical identification show that the Haya smelters were keenly aware of the clays that would withstand high furnace temperatures. For example, experiments with tuyere clays showed that most samples were refractory and can withstand temperatures up to 1500°C. (in oxidizing conditions) before bloating and collapse occurred (Childs 1986). As we proceed with our technical analyses we hope to determine if regional technologies and economic wealth varied according to access to the most advantageous clay resources. In the case of the KM2 and KM3 sites, we analyzed all tuyeres by looking at critical attributes such as the presence of oxidation and reduction, vitrification, slagging, wall thickness, inclusions, and diameter (all of which are not discussed here). One of our initial assumptions in this analysis, that black and gray colors result on tuyeres inserted in furnaces, was based on the observation of replicated smelting. However, we encountered some KM2 and KM3 tuyere fragments that were very light gray, a color that also results when test tiles are heated to high temperatures in oxidizing conditions (Childs 1986). A whitish-gray zone between areas of slag wetting and reduced areas appears to mark an oxidizing backwash, a feature common to the tuyeres in situ at the NG5 site (Schmidt 1980b, 82–83). We now feel confident that the overwhelming majority of tuyere fragments show convincing evidence in their colors to indicate that they were inside the prehistoric smelting furnaces of KM2 and KM3.

At KM2 for all industrial periods we found that 43 percent of the fragments were totally reduced, while another 21 percent were reduced with an oxidized interior. Twenty-six percent of the KM2 sample were completely oxidized, which proportionately agrees with the Haya examples; the remainder show signs of fluctuating gases. The KM3 data show similarly stronger evidence for preheating, for 58 percent of the tuyere fragments were completely reduced, 7 percent were reduced and oxidized inside, and 23 percent were oxidized. If there is a noteworthy contrast between the two sites, it is in the difference between the frequencies for tuyeres that were reduced outside and oxidized inside. We are currently following out this problem in the field, working with the hypothesis that the higher interior oxidation frequency is linked to the plugging of tuyeres after the smelt: the distal end of an open tuyere will oxidize inside, while a plugged tuyere will remain reduced. This difference between KM2 and KM3 may point to different practices relating to the sealing of the furnace after the smelt, a practice with significant technological implications.

Our actualistic studies show that slagging and vitrification occur on approximately 26–50 percent of the inserted portion of tuyeres used to preheat. We can use the frequencies of slagging and vitrification as an index to assess the probability of preheating as well as to assess the proportion of the tuyeres that were inserted inside the furnaces. A significant proportion of the tuyere fragments at KM2 and KM3 were vitrified and slag wetted. This supports the idea that the furnace tuyeres had been placed inside the furnaces where they had been exposed to high temperatures and a slag wash. During the earliest period of production there is a distinct difference between the KM2 and KM3 sites, with a significantly higher frequency, 38 percent, at KM2, compared to 23 percent at KM3 (figure 9.16, graph a). Related to this difference is the question of whether the prehistoric smelters of this region changed their practice of tuyere insertion inside the furnace. It seems that the smelters at KM3 during the later period realized or learned of the increased advantage of deeper tuyere insertion; the initially lower frequencies at KM3 changed in the later period, so that these tuyere attributes were virtually identical to the KM2 frequencies in the later period (figure 9.16, graph b). This suggests a subregional acceptance for deep tuyere insertion and relative standardization for preheating at least in the Kemondo basin. If we include tuyeres from all archaeological contexts (figure 9.16, graph c), then we see that there are higher frequencies at KM3 over all time periods, a fact that may indicate a higher rate of tuyere fragments with the key attributes discarded outside of the furnace pits.

One of the primary goals during the KM3 research was to find more definitive evidence for the use of preheating in northwest Tanzania. The 36 cm section of tuyere recovered from furnace 9 provides such evidence (figure 9.17). Found at the base of the second stratigraphic unit and among pieces of brick and slag, this tuyere section provides the unequivocal proof for which we were searching. The distal end showed deep slag penetration and vitrification for 9.5 cm, some slagging and oxidation (gray) for another 4–5 cm, and then severe (black) reduction for another 10–16 cm (varying from top to bottom). Thus 39 percent of the preserved section displayed the two attributes discussed here, while about 75 percent of the tuyere had the diagnostic markers for preheating. This archaeological evidence, then, provides some of the most important information to date on the certainty and antiquity of a technique in iron technology that marks the African technological experience as distinctive and independent from that of Europe. And it shows that the assumptions that have guided our analysis and interpretation of tuyere fragments have been germane and appropriate.

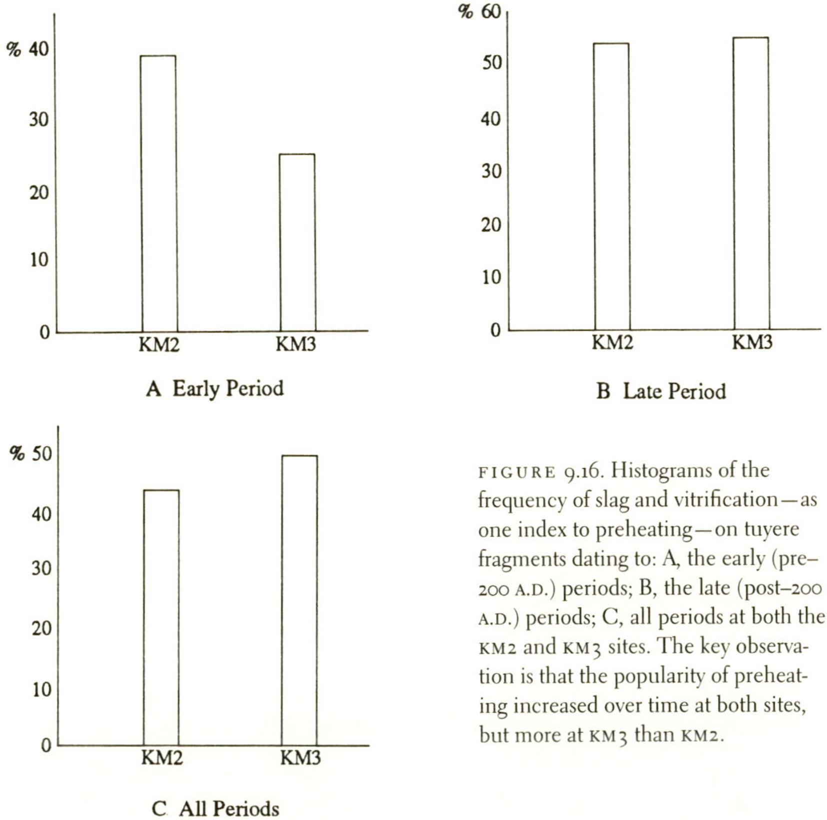

FIGURE 9.16. Histograms of the frequency of slag and vitrification—as one index to preheating—on tuyere fragments dating to: A, the early (pre–200 A.D.) periods; B, the late (post–200 A.D.) periods; C, all periods at both the KM2 and KM3 sites. The key observation is that the popularity of preheating increased over time at both sites, but more at KM3 than KM2.

A key element in the success of preheating is the refractory quality of the clays and the acceptance of standards of production that accounted for such properties. It is clear that the clays vitrified and were deeply penetrated by slag, but it was critical that they withstood these assaults without slumping, melting, or blocking. When clays fail, the tuyeres become completely plugged, break, and collapse. If we can see this in the archaeological record, then we can devise an index based on slag and glass coating of the edges of tuyere fragments that helps to assess the failure rate of the prehistoric tuyeres. Such properties on the broken edges would have developed only under normal smelting temperatures, not after cessation of smelting. We suggest that a high proportion of this attribute in the total sample points to clay failure in a high-temperature environment and that a low proportion points to selection of clay that was sufficiently refractory to withstand the high temperatures.

There was a relatively low proportion of altered fragments at both sites during the early period, suggesting a high rate of success with clays during this era (figure 9.18a). However, there is a distinct shift in rate of clay failure during the later period, when it doubled at KM2 and exceeded that of KM3 by more than 2 to 1 (figure 9.18b). This tentatively points to the utilization of different clay resources for tuyeres at the two sites, especially in light of the observation that the practice of preheating was standardized during that era at both sites. The evidence seems to suggest, then, that preheating was more successfully practiced at the KM3 site during the later period. A significantly lower failure rate in the clays and more success in the practice of the technique appear linked to differential access to good refractory clays, a resource that is central to the success and productivity of the technology.

The extremely high temperatures achieved in an Early Iron Age furnace required not only good clay but also good tempering in order to survive the high thermal shock and the growth of cracks during major temperature fluctuations. The presence of pores in the paste, partly determined by tempering, provides voids at which cracks stop as they begin their development. In a high-temperature environment such as a smelting furnace, it is important that the

FIGURE 9.17. A 36 cm long section of tuyere from KM3 site showing characteristics that indicate preheating in the fourth century A.D.

nonplastic inclusions have a thermal expansion rate similar to the clay. Thus, pieces of broken ceramic, or grog, are particularly good tempering as they reduce the chance for bloating to occur. Firing tests on experimental test tiles made from tuyere clays showed that several clays were unaffected structurally at 1500°C under oxidation, but that all experienced some degradation under reduction because of the fluxing action of FeO and the resulting bloating. The addition of grog significantly retarded bloating at the highest temperatures tested (Childs 1984; 1986). Almost invariably grog (broken sherd) was the type of inclusion added. This practice continued throughout the Early Iron Age, though the amount and size of inclusions varied by site. For example, at KM2 the inclusions are typically over 4 mm, whereas at KM3 the sherd inclusions are generally smaller, with an upper range of 2–4 mm. This selection for smaller tempering material is paralleled on the domestic ceramics from KM2 that are also tempered with grog. Tuyeres are an excellent yardstick by which we can begin to measure the development of complex technology and the process of technological innovation in the African past. The results of these analyses suggest that we have ignored these important sources of information too long in African archaeology.

Conclusion

The KM2 and KM3 sites provide critical evidence for the assessment of the development of innovative forms of iron production in Africa. The density of similar sites along the western shores of Lake Victoria indicates that this area was the center of a highly productive, technologically efficient, and innovative Iron Age culture. The origins of this technology perhaps originated in the dessicated zones of West Africa, such as southeastern Niger. In that region at Agadez, Grébénart has dated iron smelting (with a southern origin) to 490 B.C.± 100 (Grébénart 1983, 14), and iron smelting has been documented to the seventh century B.C. at Azelik, 145 km northeast of Agadez (Calvocoressi and David 1979, 10), dating that has recently been reevaluated (Killick et al. 1988) and assessed to be approximately 500 B.C. It is always possible, though, that the evolutionary progression we have seen in the development of the preheating technique during the first half of the first millennium A.D. may indicate an invention indigenous to the region.

We hope that the KM2 and KM3 investigations will help prepare the way for a more thorough assessment of that idea and the impact of a highly productive

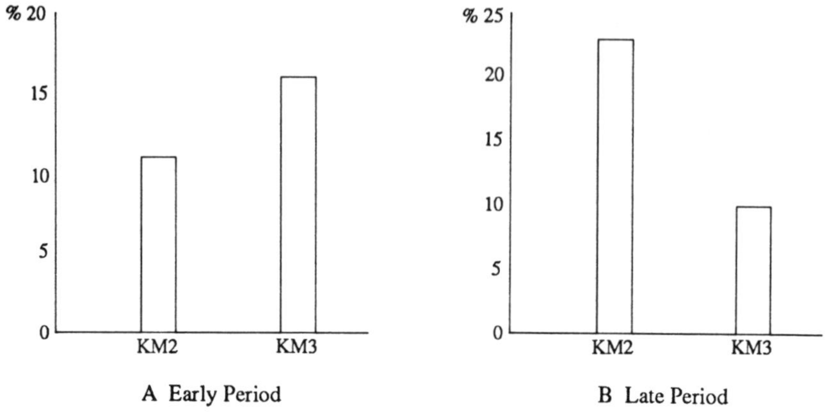

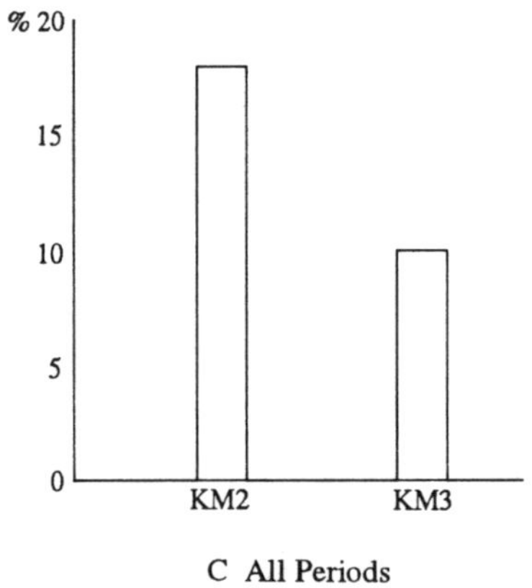

FIGURE 9.18. Histograms of the frequency of slagging and vitrification on the edges of broken tuyeres, as an index to the refractory competence of tuyere clays during: A, the early (pre–200 A.D. periods); B, the late (post–200 A.D.) periods; C, all periods at both the KM2 and KM3 sites. The key observation here is that the smelters of KM3 appear to have developed much better mastery over refractory tuyere clays over time than their KM2 counterparts.

and unique technology on the history of the African peoples. We have suggested several new perspectives, analytical applications, and experimental approaches that may assist in this endeavor. It is apparent that actualistic studies are a basic requirement in any reliable interpretation of prehistoric remains in Africa, particularly if we are to open ourselves to the powerful symbolic realities that so clearly affect this important material domain. Interpretation of the remains of iron-smelting furnaces has been helped significantly by actualistic studies of both Haya and Barongo iron smelters. These empirical models allow us to assess more successfully the significance of the archaeological remains. The application of new analytical approaches along with experimental procedures has provided important insight into the development of technological life during the Early Iron Age in East Africa.

Notes

We want to thank the Antiquities Unit and the then National Scientific Research Council of Tanzania for permission to conduct this research. The cultural officers of Kagera Region and Bukoba District provided every assistance within their means. Research was supported by the National Science Foundation (BNS 75-19611 and BNS 78-07855) and the National Endowment for the Humanities (RO10570-83).

1 The furnaces in the KM3 sites have been numbered by a different scheme in several prior publications (Schmidt 1981; 1983a). For example, furnaces 3, 7, and 8 have been referred to as nos. 2, 3, and 4. The reference system here will be used hereafter.

2 The mean diameters of furnaces published previously (Schmidt 1981; 1983a) have been refined slightly for KM3 by averaging two profile sections for each furnace (a change from 114.5 to 113.8 cm) and have been revised significantly for KM2 by similar averaging plus the reevaluation and exclusion of three (28, 31, 32) smaller pit features (a change from 93.6 to 90.2 cm). This also means a downward revision of furnace pits from fifteen to twelve.

3 A full report of the metal artifacts from KM2 and KM3 has been published in Schmidt 1983a and by Childs in chapter 13 of this volume.

4 See Schmidt and Childs (1985) and Childs (1986) for a comprehensive treatment of furnace bricks. We present only a summary of the decorative applications here. Moreover, we do not discuss here the variation of bricks, the fact that most are not prefired, and their application in what appears to be a coil type of construction.

References

Arkell, A. 1961. "The Valley of the Nile." In *The Dawn of African History*, ed. R. Oliver, 7–21. London: Oxford University Press.

Avery, D., and P. Schmidt. 1979. "A Metallurgical Study of the Iron Bloomery, Particularly as Practiced in Buhaya." *Journal of Metals* 31(9):14–20.

Calvocoressi, D., and N. David. 1979. "A New Survey of Radiocarbon and Thermoluminescence Dates for West Africa." *Journal of African History* 20(1):1–29.

Childs, S. T. 1984. "Clays to Artifacts: Resource Selection in African Early Iron Age Iron Making Technologies." Paper presented at the eighty-third annual meeting of the American Anthropological Association, Denver.

———. 1986. "Early Iron Age Ceramic Technology in East Africa." Ph.D. diss., Boston University.

Childs, S. T., and D. Killick. 1993. "Indigenous African Metallurgy: Nature and Culture." *Annual Review of Anthropology* 22:317–37.

Childs, S. T., and P. Schmidt. 1985. "Experimental Iron Smelting: The Genesis of a Hypothesis with Implications for African Prehistory." In *African Iron Working*,

Ancient and Traditional, ed. P. Shinnie and R. Haaland. Bergen: University of Bergen Press.

Cline, W. 1937. *Mining and Metallurgy in Negro Africa.* Menasha, Wisc.: George Banta Publishing Co.

Clist, B. 1987. "A Critical Reappraisal of the Chronological Framework of the Early Urewe Iron Age Industry." *Muntu* 6:35–62.

David, N. 1982. "Prehistory and Historical Linguistics in Central Africa: Points of Contact." In *The Archaeology and Linguistic Reconstruction of African History*, ed. Ehret and M. Posnansky, 78–93. Berkeley: University of California Press.

Grébénart, D. 1983. "Les metalallurgies du cuivre et du fer autour d'Agadez (Niger), des origines au debut de la période médiévale: Vues generales." In *Metallurgies Africaines: nouvelles contributions*, ed. N. Echard. Mémoires de la Societe a des Africanistes 9:109–25.

Hiernaux, J., and E. Maquet. 1957. "Cultures prehistoriques de l'age de metaux au Ruanda-Urundi." *Bulletin de l'Academie royale des Sciences coloniales* (Bruxelles) n.s. 2(6):1126–49.

——. 1960. "Cultures prehistoriques de l'age des metaux au Ruanda-Urundi et au Kivu (Congo Belge)." *Memoire de l'Academie royal des Sciences d'Outre-Mer, Classe des Sciences naturelles et medicales* (Bruxelles) n.s. 10(2):5–102.

Kendall, R. 1969. "An Ecological History of the Lake Victoria Basin." *Ecological Monographs* 39:121–76.

Killick, D., N. van der Merwe, R. B. Gordon, and D. Grébénart. 1988. "Reassessment of the Evidence for Early Metallurgy in Niger, West Africa." *Journal of Archaeological Sciences* 75:367–94.

Klein, J., J. Lerman, P. Damon, and E. Ralph. 1982. "Calibration of Radiocarbon Dates: Tables Based on the Consensus Data of the Workshop on Calibrating the Radiocarbon Time Scale." *Radiocarbon* 24(2):103–50.

Long, A., and B. Rippeteau. 1974. "Testing Contemporaneity and Averaging Radiocarbon Dates." *American Antiquity* 39(1):205–15.

Mauny, R. 1961. *Tableau geographique de l'Ouest African au Moyen Age, d'apres les sources ecrites, la tradition et l'archeologie.* Memoires de l'IFAN, No. 61. Dakar: IFAN.

Oliver, R., and J. Fage. 1962. *A Short History of Africa.* Baltimore: Penguin.

Phillipson, D. 1985. *African Archaeology.* New York: Cambridge University Press.

Schmidt, P. 1975. "A New Look at Interpretations of the Early Iron Age in East Africa." *History in Africa* 2:127–36.

——. 1978. *Historical Archaeology: A Structural Approach in an African Culture.* Westport, Conn.: Greenwood Press.

——. 1980a. "Steel Production in Prehistoric Africa: Insights from Ethnoarchaeology in West Lake, Tanzania." In *Proceedings of the Eighth Pan African Congress of Prehistory and Quarternary Studies*, ed. R. E. Leakey and B. A. Ogot, 335–40. Nairobi: TILLMIAP.

———. 1980b. "Early Iron Age Settlements and Industrial Locales in West Lake." *Tanzania Notes and Records* 84–85:335–40.

———. 1981. *The Origins of Iron Smelting in Africa: A Complex Technology in Tanzania.* Research Papers in Anthropology, No. 1. Providence: Brown University.

———. 1983a. "Further Evidence for an Advanced Prehistoric Iron Technology in Africa." *Journal of Field Archaeology* 10:421–34.

———. 1983b. "Cultural Meaning and History in African Myth." *International Journal of Oral History* 4(3):165–83.

———. 1994. "Historical Ecology and Landscape Transformation in Eastern Equatorial Africa." In *Historical Ecology*, ed. C. Crumley, 99–125. Santa Fe: SAR Press.

Schmidt, P., and D. Avery. 1978. "Complex Iron Smelting and Prehistoric Culture in Tanzania." *Science* 201:1085–89.

Schmidt, P., and S. T. Childs. 1985. "Innovation and Industry during the Early Iron Age in East Africa: The KM2 and KM3 Sites of Northwest Tanzania." *African Archaeological Review* 3:53–94.

Shaw, T. 1969. "Archaeology in Nigeria." *Antiquity* 23:187–99.

Soper, R. 1982. "Bantu Expansion into Eastern Africa: Linguistic Evidence." In *The Archaeological and Linguistic Reconstruction of African History*, ed. C. Ehret and M. Posnansky, 223–38. Berkeley and Los Angeles: University of California Press.

Stuiver, M. 1982. "A High-Precision Calibration of the A.D. Radiocarbon Time Scale." *Radiocarbon* 24(1):1–26.

Tylecote, R. 1970. "Iron Working at Meroe, Sudan." *Bulletin of the Historical Metallurgy Group* 4(2):67–72.

———. 1975a. "The Origin of Iron Smelting in Africa." *West African Journal of Archaeology* 5:1–9.

———. 1975b. "Iron Smelting at Taruga, Nigeria." *Journal of the Historical Metallurgical Society* 9:49–56.

van der Merwe, N. 1980. "The Advent of Iron in Africa." In *The Coming of the Age of Iron*, ed. Wertime and J. Muhly, 463–506. New Haven: Yale University Press.

van der Merwe, N., and D. Avery. 1982. "Pathways to Steel." *American Scientist* 70:146–55.

van Grunderbeek, M.-C. 1992. "Chronologie de l'Age du Fer Ancien au Burundi, du Rwanda et dans la région des Grand Lacs." *Azania* 27:53–80.

van Grunderbeek, M.-C., E. Roche, and H. Doutrelepont. 1983a. *Premier Age du Fer au Rwanda et au Burundi: Archaeologie et Environment.* Brussels: IFAQ.

———. 1983b. "Metallurgie ancienne au Rwanda et au Burundi." *Journées de Paléométallurgie* (February 22–23, 1983):407–23.

van Noten, F. 1979. "The Early Iron Age in the Interlacustrine Region: The Diffusion of Iron Technology." *Azania* 14:61–80.

Willett, F. 1971. "A Survey of Recent Results in the Radiocarbon Chronology of Western and Northern Africa." *Journal of African History* 12:339–70.

10 Use of Preheated Air in Primitive Furnaces: Comment on Views of Avery and Schmidt

J. E. Rehder

and

Use of Preheated Air in Ancient and Recent African Iron Smelting Furnaces: A Reply to Rehder

Donald H. Avery & Peter R. Schmidt

Use of Preheated Air in Primitive Furnaces (Rehder)

The opinion has been published that extended tuyeres (air-injection tubes) in primitive charcoal-fired iron-smelting furnaces preheat the combustion air to an important degree and that the increased combustion temperatures thereby created gave metallurgical advantages in early African furnaces. Measurements taken of air temperature inside tuyeres in an operating native furnace have been presented in support of this thesis.

It is shown here that the method of temperature measurement was incapable of measuring air temperature with even approximate accuracy, the reported values being seriously high. It is also pointed out that heat transfer to air in a hot tuyere is largely by convection and that under the conditions of operation of the experimental furnace the temperature increase of the air should be 22°C or less, an amount very minor in effect. Furthermore, it is pointed out that maximum furnace temperatures over 1,600°C are readily obtained in charcoal furnaces with combustion air at ambient temperature and so no special explanation is necessary for the observed smelting results.

It is concluded that there is little air-preheat in long tuyeres and therefore an early novel smelting technique in Africa has not been demonstrated.

Introduction

In the remains of primitive furnaces, tuyeres are occasionally found that clearly protruded some distance into the hearth. There has been speculation in the literature that such tuyeres served to preheat combustion air, and recently the thought has been taken seriously by D. H. Avery and Peter Schmidt. In a series of publications over the last few years, they have described the operation of a primitive furnace in Tanzania, reported temperature measurements in tuyeres, and advanced two hypotheses: (1) that substantial preheating of combustion air occurs in tuyeres protruding into a smelting furnace; and (2) that use of such an arrangement in early furnaces in Africa produced a major advance in iron-smelting technology due to the high furnace temperatures supposedly developed thereby. These views seem to be entering the literature as facts, but since they are questionable a review is in order. This will be done by (1) reassessing their experimental evidence; (2) estimating by standard heat transfer methods the extent of air-heating possible in the circumstances; and (3) showing that very high temperatures occur in fuel beds without preheated combustion air. This will demonstrate that air-heating in tuyeres is of very minor extent and that the high furnace temperatures observed are to be expected in charcoal beds blown with ambient air.

Experimental Evidence

Tuyeres that protrude into a furnace must become heated by the burning fuel bed and in fact reach temperatures near their noses sufficiently high to become coated or fluxed with slag. This is commonly observed and requires temperatures of about 1200°C, but this is not unusual in charcoal-fueled bloomery furnaces. The thought that such a hot tuyere would heat the air passing through it is natural, and steps were taken by Avery and Schmidt to measure the air temperature in tuyeres. This was done on a primitive-style iron-smelting furnace which was built and operated by natives on the basis of remembered practices. The building of the furnace, the recovery of ore from an old surface deposit, the making of the charcoal, and the operation of the furnace were described in some detail (Avery and Schmidt 1979).

The method of measuring air temperature in a tuyere was "with a ceramic-insulated bare-bead thermocouple placed in the air flow and held free from the tuyere," and temperature readings were evidently taken at intervals from the exterior end to the nose of the tuyere. As the smelt continued, "hot slag dissolved the tuyere and its length slowly decreased" to a point near the furnace wall at the end of the smelt, and air temperature-distance measurements were made at several time intervals. These temperature measurements were shown in their figure 5.3 (Schmidt and Avery 1978). The melting of several Pt-Pt Rh thermocouples at 1,820°C is mentioned, and the statement is made that "preheating is clearly in the 500–600°C range." There is no indication or description of how this temperature range was selected. In summary, Avery and Schmidt felt they had demonstrated that an extensive degree of combustion air preheating had occurred, and that temperatures in excess of 1820°C existed in the fuel bed which could only have been reached because of such preheating.

TEMPERATURE MEASUREMENTS

Accurate measurement of the temperature of air flowing in a hot tube with one end open to a high temperature source is not easy. There are three pitfalls: (1) the air flow is not uniform across the cross-section of the tube; (2) air picks up very little heat by radiation; and (3) solids absorb heat readily by radiation, increasing as the fourth power of the absolute temperature of the source.

In the present case the temperature of the tuyere wall undoubtedly increased considerably from furnace wall to the inner end, or nose, of the tuyere. As a bare metallic thermocouple bead was traversed toward the nose of the tuyere the bead would absorb increasing heat by radiation from the tuyere wall, increasing the temperature of the bead which of course would register on the external measuring device. This will occur independently of whether air was flowing or even present. In addition to this, as the bare bead approached the nose of the tuyere beyond which temperatures in excess of 1,600°C existed (Hiles and Mott 1944, 154), more heat was added to the bead by radiation onto its end from the contents of the furnace. The net result will be a temperature-distance curve that rises steeply as the tuyere nose is approached, which is what is shown on the published curves. These, however, are only the temperatures of the thermocouple bead and have little or no connection with air temperature.

A further point is that the temperature-distance curve in their figure, which is taken near the end of smelting, when the tuyere is nearly completely fluxed away with little length left for air heating, shows the same maximum temperature as from a full-length tuyere at the start of smelting. It must be concluded that the measurements shown do not represent air temperatures in the tuyeres and are not evidence for air preheating.

HEAT TRANSFER CALCULATION

Preheated combustion air is widely used industrially and calculation for design of air heaters is well developed. Heat transfer from a solid to air is almost entirely by convection and engineering calculations are so based. The tuyeres were described (Avery and Schmidt 1979) as being eight in number, each about 60 cm long, 7 cm outside diameter, 4.5 cm inside diameter, made of tempered clay, and extending into the furnace about 40 cm. The exterior of the tuyere was fluxed by molten slag with a fusion temperature of at least 1,200°C. The tuyere wall would decrease in temperature from nose to furnace wall, so an average exterior temperature of 1,100°C is assumed here for an estimate of heat transfer.

In order to calculate the rate of heat transfer it is necessary to estimate the rate of air flow in a tuyere, and this requires an estimate of the average total air flow to the furnace, which (using standard formulae for nozzles, packed beds, and chimneys) is about 1.0 m^3 per minute for natural draft only. Bellows were also used, but their effect in the present case is very uncertain. There were sixteen small bowl-type bellows with leather covers but no valves, and their nozzles had no direct connection to tuyere entrances, so they acted as periodic injectors. From the number of strokes given per minute, and estimate of bellows size from the photographs in Avery and Schmidt (1979, 14–20), possibly 0.5 m^3 per minute was contributed. This would give a total flow of roughly 1.5 m^3 per minute and an air flow rate per tuyere of about 0.2 m^3 per minute.

The total air flow rate could be checked by the amount of charcoal consumed in a smelt, but the amounts charged and remaining in the hearth were not given. With an internal diameter of 80 cm in the shaft, the specific airflow rate would be 3.0 m^3 per m^2 per minute, which is a little lower than those typical of reported experimental work on bloomery furnaces.

Using methods described in Trinks (1951), a widely used handbook of furnace and recuperator design, it is calculated that air flowing at a rate of 0.2 m^3 per minute through a clay tube of the dimensions of the subject tuyere, whose exterior is at an average temperature of 1,100°C, will increase in temperature by about 22°C. If the air flow rate were to be doubled, the temperature increase would be halved. Tylecote (1965, 342) notes that calculation of air preheat in a two-foot-long protruding tuyere (with air rate and external temperature unspecified) indicated an air temperature increase of 10°C. In the present case the calculated air-temperature rise is a maximum, occurring only at the start of smelting when full tuyere length existed. As the tuyeres were fluxed away, the air preheating would have become very insignificant.

It is therefore clear that only minor preheating of air could have occurred in the tuyeres observed by Avery and Schmidt, and this agrees with the conclusion above that their temperature measurements taken in tuyeres were not those of the air passing through. In this connection it must be realized that if preheating of combustion air in protruding tuyeres were practical, blast furnace and cupola operators would long ago have adopted them instead of purchasing expensive stoves and recuperators. Protruding tuyeres are indeed used, but the protrusion is short and the tuyeres are heavily watercooled.

HIGH TEMPERATURE IN FUEL BEDS

Combustion of lump fuels in beds has been extensively explored. The basic pattern is that of very rapid development of CO_2 content with resulting high flame temperature close to air entry, and then more or less rapid decrease of CO content and temperature due to reduction to CO and to heat losses. In a major review of experimental findings Hiles and Mott (1944) noted that the maximum initial CO_2 content in a bed of charcoal is 10–11 percent, which translates to an adiabatic flame temperature, corrected for dissociation, of 1,950°C. Preheating the combustion air (hot blast) increases flame temperature, a 22°C increase in air temperature in the present case, giving an 11°C increase in flame temperature, which is in practice negligible.

Decrease in initial flame temperature due to heat losses to the walls and top of a small furnace, even when fully preheated, is very rapid and, at the low air rates used in primitive furnaces, the volume of bed near maximum temperature adjacent to the nose of each tuyere will be only a fraction of a cm^3. That, however, is frequently enough to melt a Pt-Pt Rh thermocouple and there is no need to postulate the use of preheated air. Tylecote et al. (1971, 351) have measured 1600°C, maintained for hours on many smelts in a reconstructed charcoal-fueled bloomery furnace. The subject of very high temperatures in coke beds has been discussed by Rehder (1979, 44).

Molten cast iron was inadvertently produced by Tylecote, illustrating the problem a primitive smith had in avoiding too large a volume of high temperature that would produce, then, useless molten cast iron. This dilemma was as real as that of obtaining a wider distribution of lower temperature for maximum amount of solid-state reduction to a bloom.

A secondary and minor point is that protruding tuyeres are steadily fluxed away during a smelt to become part of the furnace slag. In the subject case an estimated 15 kg of tuyere material entered the slag, requiring iron oxide as flux and absorbing heat for melting, and so increasing hearth load.

Conclusion

Tuyeres protruding into a fuel bed cannot preheat to any useful degree the air passing through them, because of basic facts of heat transfer, and the thesis that a useful increase in combustion temperature was developed this way in the furnace as described by Avery and Schmidt is not supported. Similarly, since operating temperatures in the subject furnace must have been normal for any charcoal-fueled shaft furnace of similar dimensions, of charcoal to ore ratio, and of air rate, no novel early technology based on high temperature existed. The observed metallurgy in the furnace of their reports must be explained on other grounds.

Use of Preheated Air in Ancient and Present African Iron Smelting Furnaces (Avery and Schmidt)

The first matter of importance to address in J. E. Rehder's paper is the structure and content of his title: "Use of Preheated Air in Primitive Furnaces." We have purposefully used a different title in our response precisely because a major thrust of our research (Schmidt and Avery 1978; Avery and Schmidt 1979; Schmidt 1981; Schmidt and Avery 1983; Schmidt and Childs 1985) has been that the conceptualization of primitive iron smelting, when applied to Africa, betrays a hierarchical ordering of African technology vis-à-vis the Western world that consigns the African to the category of less sophisticated and technologically backward, or "primitive." The implications of our research challenge and reject the legitimacy of the concept and all of its attendant historical implications. Consistent with our historical and scientific perspectives, we offer a different title in our comments, for we feel the archaeological evidence and experimental/observational evidence point convincingly to a technological process of great antiquity that is hardly "primitive," but relatively and comparatively sophisticated and complex.

As Rehder points out, we do not speculate that preheating took place in African antiquity. We have material and comparative evidence that argues that it was an ancient innovation in Africa (Schmidt and Avery 1983; Schmidt and Childs 1985). With this fundamental conceptual and historical problem addressed, let us examine the underlying assumptions of Rehder's comments.

First, Rehder's comments on the use of preheated air follow a well-traveled path of applying data and operating experience on modern furnaces as a model to predict and characterize early furnaces (both in Europe and Africa) and, in many cases, to build and conduct experiments based on a modern model. While some of these exercises have produced interesting and important explanations, the serious question arises as to whether or not they are applicable and germane for the African technological experience. In the case of Avery's research, he had been involved in experimental work in university laboratories on Early Iron Age furnaces before he had the opportunity to work with traditional African iron smelters (Schmidt and Avery 1978; Avery and Schmidt 1979). These experimental furnaces, it should be noted, were singularly unsuccessful in producing traditional iron blooms.

It was precisely this point—the failure of contemporary Western scientific concepts applied to ancient technology—and the recognition of the historical implications of a fully investigated indigenous model (Schmidt 1981) that led to the series of investigations on traditional African furnaces, first in Tanzania and subsequently in other areas of Africa where direct historical knowledge of iron smelting has survived. To be sure, many of the results are surprising and contradict conventional ideas about the history of metallurgy, as they reveal sophisticated techniques that have, for too long, been obscured by a Western preoccupation with processes of mystification such as ritual, magic, and taboos.

We will now answer Rehder's comments in the engineering context in which they are posed. In his note, Rehder makes three technical contentions. First, he asserts that air temperatures cannot be measured with thermocouples; second, he claims that heat transfer calculations show negligible preheating; finally, he says that high temperatures are normal with charcoal-blown ambient air.

In his first claim, Rehder relies on the greater efficiency of radiant heat transfer to solids as compared to convective transfer. This is a problem, particularly in field investigations. Increasingly improved techniques have been developed over the last fifteen years by Avery. We have discarded (though reported) the rapid rise of temperature observed on thermocouples in the last three diameters of the tuyeres as being primarily attributable to radiant heating. With this in mind the amount of preheating, as depicted in our figure (8.4) must take into account our acknowledgment that the temperatures measured in the last three diameters (that is, 13–14 cm) of the tuyere are dominated by radiant heating. We have tried a variety of heat shields, and we have correlated temperatures measured in the tuyeres with readings from bare wire thermocouples embedded in the tuyeres just below the surface. Cracking in the tuyeres and the breaking of thermocouple wires limited these observations to the early stages of the smelt, but we feel that the measured air temperatures and the wall temperatures of the tuyeres are independent. Optical pyrometer readings can be used at tuyere surface temperatures above 800°C. In recent research in Malawi, Avery attempted to measure the surface temperature with an infrared pyrometer, but the instrument had too wide an optical aperture and was not considered reliable in the field.

On the Haya tuyeres there is a sharp step function in wall temperature where the tuyere enters through the wall into the furnace. This is readily picked up in tuyere wall temperature with contact thermocouples. The best technique for air measurement is to use very fine wires and beads with the sensing element bent up at right angles from the end of the inconel ceramic protection tube. Based on our evidence derived from these methods and our calculations and observations, we believe that there is substantial preheating. We feel confident, given the limits described above, that preheating occurs in the range of 400–600°C.

One observation critical to our position is that the bellows in the Haya process are valveless and operate as aspirators in the bell-shaped tuyere ends. This results in a reverse flow of air through the tuyere, with the initial aspiration drawing additional air into the tuyere. On the reverse stroke, the air flow slows down and reverses, thus pulling hot gasses into the end of the tuyere and carrying air back toward the entrance. This process is easily observed when the tuyere ends melt and sag; the reverse flow can be observed when the drooping bits of melting clay sway back and forth. We believe that this unusual reverse flow and the attendant greater effective residence time and travel distance may explain the slightly higher measured temperatures than those we calculated for straight forced convection.

A final point apparent both in the replicated and ancient furnaces is that there is hard physical evidence for high temperatures in the tuyeres. Recent and ancient tuyeres are coated with slag (temperatures greater than 1150°C). The boundary of slag melting and penetration of the tuyere was near the midpoint of the wall thickness. If the furnace temperature outside the tuyere wall is in the 1300°C range, this would put the inside wall temperature of the tuyere in the 1000°C range, given the absence of substantial inner wall (tuyere) heat transfer. The lower temperatures measured on the inner walls of the tuyeres are attributable to significant heat removal from the inner wall. We furthermore find no justification for the assertion that thermocouples cannot measure air temperature, but we recognize that radiant heating, particularly within three diameters of the end of the tuyere, requires caution and careful judgments of the results.

Rehder's second point addresses the conductive heat transfer based on through flow (not open end) recuperator design data (Trinks 1951). Based on an altogether different set of data and observations, we make the assumptions that: (1) all the air entering the furnace is coming through the tuyeres (although air leaks and cracks did admit some additional combustion air); (2) the final combustion product was CO (because measured CO_2 at the furnace top was negligible); (3) for purposes of calculation we have estimated the thermodynamic properties of air at a mean value of 200°C. Then, we may proceed with the following calculations for heat transfer involved in the Haya preheating process:

The average charcoal consumption per tuyere is 8.2 lb/hr; assuming 10 percent ash this is 7.3 lb/hr carbon.

Since $C + \frac{1}{2}O_2 + 2N_2 = CO + 2N2 = 21.9$ lb. air / hr.

$$\text{Reynolds no.} = \frac{DG}{\mu}$$

D = diameter .148 ft.

$$G = \text{mass flow} = \frac{21.9}{AX3600} = 0.28 \text{ lb./ ft.}^2\text{/ sec.}$$

$$\mu = 1.3 \times 10^{-7}\,(T)^{.8} = 1.79 \times 10^{-5} \text{ at } 200°C$$

$$\text{or RE} = \frac{148(.28)}{1.8 \times 10^{-5}} = 2.3 \times 10^{3}$$

The Prandelt no. is approximately 0.74 for air, and

$$\text{the Nusselt no.} = \frac{hcD}{k} = 0.023(RE)^{0.8}\,(PR)^{0.4}$$

$$\text{now Prandel no. } 0.74 = \frac{Cp\,\mu}{k}$$

where Cp = heatcap = 0.25 BTU / lb. ft.

and $\mu = 1.769 \times 10^{-5} \times 3600$ lb. ft./ hr.

$$k = \frac{(0.25)(1.79 \text{ x } 10^{-5} \times 3600)}{0.74} = 0.022$$

$$\text{and the Nusselt no.} = 9.97 = \frac{hcD}{k}$$

$$\text{or hc} = \frac{9.97(0.022)}{.148} = 1.48 \text{ BTU / hr./ ft.}^2\text{ /°F}$$

We have assumed a mean value of air temperature of 200°C = 392°F. We have assumed an inside tuyere wall temperature of 800°C. (This is based on optical pyrometer readings, slag wetting, surface thermocouple readings, and a balanced judgment of all these data.)

$$\text{Then heat } q = (h_c)(3.14)(D)(T_{wall} - T_{air})$$
$$= (1.48)(3.14)(.148)(1080)$$
$$= 744.2 \text{ BTU / hr./ ft.}$$

$$\text{Temperature} = \frac{q}{\text{CpXlb./hr.}} = \frac{744.2}{(.25)(21.9)} = 136°\text{F}$$

$$= 75.5°\text{C / ft.}$$

The calculated minimum temperature is therefore 141°C. We assume, because of the factors of pulsating and reverse air flow, that the total dwelling time of the air in the tuyere is threefold; this would then result in a temperature rise of approximately 323°C.

There are several other factors that will increase the temperatures *above* the calculated value, namely, radiation heating will not be negligible, particularly with some CO_2 dilution from back flow; secondly, air leakage will reduce the air passing through the tuyere and increase its temperature. It is felt that our estimate is compatible with the calculations and is by no means negligible. The presence of back flow will also add to the sensible heat by mixing. While the effects of flow reversal on the boundary layer are complicated and highly specific to conditions, both aspiration and flow reversal strongly affect the boundary layer and will increase heat transfer. As normal methods of velocity and temperature measurement, thermocouples and hot wire anaerometry are sensitive to both radiation and gas temperature. Rigorous measurement of gas velocity and temperature profiles in the tuyeres would probably have to rely on laser optical techniques such as laser Doppler velocity probes and the laser interferometer for temperature measurements. These instruments are too expensive and delicate to be currently considered for fieldwork.

Rehder's final point regarding fuel bed combustion temperatures totally misses the conundrum of traditional iron smelting, namely, the requirement to have strongly reducing atmospheres of CO/CO_2 greater than five or six. With complete combustion to CO_2, adiabatic temperatures in the 1700–1800°C range are easily reached with most charcoals. The atmosphere is strongly oxidizing, however. One example of this can be seen in high temperature and oxidizing forges where there is almost complete decarbonization of forged artifacts made from high carbon bloom. A high flame temperature can be obtained by vigorous bellows blowing or blowing in the presence of adequate outside air around the open forge. This will not reduce iron; it will also oxidize appreciable amounts of iron to produce forge slag. To reduce iron one must keep temperatures down, requiring long smelting times and the excessive consumption of fuel, or make use of heat recycling methods, as in the preheating Haya furnace. The evidence for preheating in traditional furnaces studied in Buhaya is amply supported by physical measurements, by heat transfer calculations, and by the rapid production of iron at high temperatures that would normally be fully oxidizing.

A further point should be made regarding the uniqueness of many early technologies and the difficulty of laboratory reconstructions, mathematical modeling, and technological comparison with modern furnaces, which operate under very different conditions. In the case of the Haya furnaces it is the supposed "inefficient" valveless bellows and aspiration pump that are probably responsible for the substantial preheating that occurred. With a steady blast of ambient air and a stable boundary layer, relatively little preheating would have occurred. It is also important to recognize that the preheated Haya furnaces were highly productive, both in later and in ancient history. Archaeological evidence is overwhelming for the widespread and intensive practice of the technology in the Early and Late Iron Ages, as highly fused iron slag is a major feature in and around almost all villages. We must also point out that the ancient iron smelters of northwest Tanzania clearly recognized the value of preheating, for a significant proportion of the tuyere fragments at the KM2 and KM3 sites (both Early Iron Age) are vitrified and slag melted, supporting the idea that they were inside the furnace and exposed to high temperatures and slag wash (Schmidt and Childs 1985). We find, in fact, that there was increasing standardization and acceptance of preheating at the KM3 site between the first two centuries A.C. and the post–300 A.C. period (Schmidt and Childs 1985). This strongly suggests, along with careful selection of highly refractory tuyere clays, that the prehistoric and recent smelters were master practitioners of preheating along with its technical requirements.

The concluding observations of the previous two paragraphs highlight the fact that the material and experimental evidence are real and that African iron smelters clearly understood the principle of preheating. Speculative mathematical models and comparative data based on the Western laboratory experience tend to deny ipso facto the reality of the African technological experience and must be viewed, because of their context, as academic exercises.

References

Avery, D. H., and P. R. Schmidt. 1979. "A Metallurgical Study of the Iron Bloomery, Particularly as Practiced in Buhaya." *Journal of Metals* 31:14–20.

Hiles, J., and R. A. Mott. 1944. "Mode of Combustion of Coke." *Fuels* 23:154.

Rehder, J. E. 1979. "Heat Economy in the Cupola." *Foundry Management and Technology* (May):44.

Schmidt, P. R. 1981. *The Origins of Iron Smelting in Africa: A Complex Technology in Tanzania.* Research Papers in Anthropology, No. 1. Providence: Department of Anthropology, Brown University.

Schmidt, P. R., and D. H. Avery. 1978. "Complex Iron Smelting and Prehistoric Culture in Tanzania." *Science* 201:1085–89.

——. 1983. "More Evidence for an Advanced Prehistoric Iron Technology in Africa." *Journal of Field Archaeology* 10:421–34.

Schmidt, P. R., and S. T. Childs. 1985. "Innovation and Industry during the Early Iron Age in East Africa: The KM2 and KM3 Sites of Northwestern Tanzania." *African Archaeological Review* 3:53–94.

Trinks, W. 1951. *Industrial Furnaces.* 4th ed. Vol. 1. New York: Wiley.

Tylecote, R. F. 1965. "Iron Smelting in Pre-industrial Communities." *Journal of the Iron and Steel Institute* (April):342.

Tylecote, R. F., J. N. Austin, and A. E. Wraith. 1971. "The Mechanism of the Bloomery Process in Shaft Furnaces." *Journal of the Iron and Steel Institute* (May):351.

11 On Claims For "Advanced" Ironworking Technology in Precolonial Africa

David Killick

One of the major puzzles in the economic history of sub-Saharan Africa is the apparently low rate of technological innovation in the subcontinent over the last 1500 years, relative to that in Asia, the Near East, Europe, or the Americas (Hopkins 1973; Austen 1987). This issue holds more than academic interest, for public awareness of this disparity shapes the contemporary attitudes of the West toward Africa, and indeed of Africans toward themselves. Africans are, as the Negritude poet Aimé Césaire noted in 1956,

> "ceux qui n'ont inventé ni la poudre ni la boussole
> ceux qui n'ont jamais su dompter la vapeur ni l'électricité
> ceux qui n'ont exploré ni les mers ni le ciel
> mais ceux sans qui la terre ne serait pas la terre" (1968, 100).[1]

Reactions to the evident disparity in technology between Africa and the developed Old World have varied. The most extreme thesis is that of the late Cheikh Anta Diop and his followers, who claim that Western civilization (and all its scientific and technological achievements) has its roots in black Africa (Diop 1955, 1974). Diop's argument may be briefly summarized as follows: (1) Western accomplishments in the sciences and arts are founded in the rationalism of ancient Greece; (2) Greek civilization derived from that of Egypt;

and (3) early Egyptian civilization, contrary to the view of most scholars, was the product of black (Negroid) Africans. In spite of its reliance on discredited notions of race, and its implication that later African societies were technologically degenerate, Diop's thesis has found many supporters in francophone West Africa and among certain Afro-American intellectuals in the United States. This idea has recently received much wider exposure with the ongoing controversy over *Black Athena* (Bernal 1987, 1991), a more sophisticated reworking of arguments for the derivation of Greek civilization from that of Egypt. (Bernal appears to accept without question Diop's contention that the Egyptians were black.)

Mainstream African history and anthropology occupies the middle ground between these extremes. Most scholars accept that the forces of production in precolonial Africa were relatively underdeveloped, but do see some evidence of technological innovation in Africa during the last two millennia. There are, however, several different currents within this mainstream. Some scholars have concentrated on explaining both the generally low level of development of the forces of production in Africa and the interesting exceptions to this rule (Law 1980; Austen and Headrick 1983). Others have set out to counter the view that Africans invented nothing by arguing that African achievements in the field of technology have been underestimated. These claims are of more limited scope than Diop's, and do not seek to link sub-Saharan technology to that any other region. On the contrary, they argue that Africans have in fact developed several "advanced" or "sophisticated" technologies that have gone unrecognized because of the historical isolation of Africa from Europe and the Near East.

Most recent claims of this sort have dealt with supposed African inventions in the field of ferrous metallurgy. This chapter seeks to critically assess some of these. It will evaluate the supposedly unique qualities of African metallurgical technology by reference to the evidence accumulated by scholars of metallurgy in other regions of the Old and New Worlds.

African Inventions and Achievements in Ferrous Metallurgy

Recent claims for African achievements in this area of technology include the following:

1 Preheated blast, an invention of enormous importance to the Industrial Revolution, was first employed in Africa, possibly as early as the late first millennium B.C. (Schmidt and Avery 1978, 1983; Avery and Schmidt 1979, 1986).
2 African ironworkers developed a method of making steel (the "African direct steel process") that was distinct from steel-making developed elsewhere (van der Merwe 1980; van der Merwe and Avery 1982, following Stanley 1931, 133).
3 African iron and steel were often of superior quality to those imported from Europe (Goucher 1981).
4 Natural draft smelting furnaces (those that operate without bellows or equivalent devices) are an African invention (van der Merwe and Avery 1982, 153).
5 Iron smelting may have been independently invented in Africa (Diop 1968; Andah 1979; Schmidt and Avery 1983).

There is some degree of disagreement on each of these claims, but it is fair to say that the core thesis underlying all of them—that African bloomery iron smelting is technologically sophisticated and that it is technically distinct from other Old World bloomery processes—is now accepted as fact by most Africanist archaeologists and historians, and is presented as such in recent textbooks on African history and prehistory (Austen 1987, 14; Hall 1988, 29). This view is also featured in a widely used introductory textbook on world archaeology (Renfrew and Bahn 1991, 305).

This chapter will evaluate claims for African priority in the use of preheated blast and for the existence of an "African direct steel process." I have argued elsewhere that the region (or regions) of origin of the natural draft smelting furnace cannot be known, as it usually impossible to deduce from archaeological remains alone that a given furnace was operated by natural draft (Killick 1991a). I will not review C. L. Goucher's claim that indigenous steel was often of superior quality to imported metal, because her case rests entirely on historical anecdote. The validity of this theory could be evaluated by metallography and mechanical testing of contemporary samples of imported and indigenous metal from a given region, but no such study has been attempted to date.[2] The arguments for and against independent invention of iron smelting in Africa are too complex to be dissected in this chapter but have been examined elsewhere (Childs and Killick 1993, 319–22; Schmidt, chapter 1, this volume).

Did African Furnaces Employ a Preheated Air Blast?

The invention of preheated blast is usually traced to the British patent of James Neilson in 1828 (Percy 1864, 394–99; Hyde 1977).[3] It was one of several major technical advances that dramatically lowered the cost of iron and steel (thereby making a crucial contribution to the Industrial Revolution) and remains central to blast-furnace technology today. The claim that African ironworkers were the first to discover the utility of preheated blast has therefore attracted interest far outside the ranks of African studies, and indeed far outside academia.

The first and most extensive case for the use of preheated blast in Africa was made for the iron-smelting technology of the Haya peoples of the Kagera region of Tanzania (Schmidt and Avery 1978). Because all subsequent claims for the application of preheated blast in Africa refer to the mechanism proposed for Buhaya, the case stands or falls with this example.

While monitoring a reconstruction of traditional Haya smelting technology, Peter Schmidt and Donald Avery were intrigued by the high temperatures measured within the furnace. In Avery's judgment, the measured temperatures were far higher than should have been attainable by the use of unheated blast. He therefore proposed that preheating of the blast must have occurred. In searching for a mechanism by which preheating could have been achieved, Avery and Schmidt's attention focused on the placement of the tuyeres, the ceramic pipes that convey the blast from the bellows into the furnace. In the Haya fumace, these pipes were positioned with much of their length inside the furnace. Avery and Schmidt proposed that this portion of each tuyere acted as a heat exchanger to transfer furnace heat to the incoming air. The initial publication of this research generated unusually wide interest and was featured in publications as diverse as *Science* and *Time*. More recently, however, the claim for preheating of the blast in the Buhaya furnace has been the subject of strong criticism (Rehder 1986; Eggert 1987).

In evaluating the evidence for and against preheating of the blast in Buhaya, it is critical to separate what was observed from what was inferred. The following facts are not disputed:

1 Temperatures in excess of 1820°C (the melting point of the platinum/rhodium thermocouples used) were measured at the nose of the tuyere by a thermocouple inserted along its bore;
2 the clay tuyeres were laid so that some 50 cm of their length was inside the furnace at the beginning of the smelt; and
3 temperature profiles made by progressively inserting the thermocouple along the axis of the tuyeres show a steady rise in temperature from the bellows end towards the nose (Schmidt and Avery 1978, figure 5, reproduced as figure 11.1 in this chapter).

The argument occurs over the interpretation of these facts. Schmidt and Avery assumed that temperatures above 1820°C could only be achieved if the incoming air was preheated. They also argued that the observed rise in temperature along the axis of the tuyere (figure 11.1) provided direct proof that the air was being preheated "in the range of 600°C or more" (Schmidt and Avery 1978, 1087).

Each of these inferences has recently been challenged by Rehder (1986). As a retired industrial metallurgist with a lifetime of experience in the design of cupola furnaces who has also served as president of the Canadian Institute of Metallurgy, Rehder is a knowledgeable critic whose criticisms deserve careful scrutiny. He makes the valid point that it is very difficult to use thermocouples to obtain an accurate measurement of the temperature of air entering a furnace, because air does not readily absorb radiant energy, whereas solid bodies (such as the tip of a thermocouple) do. The thermocouple therefore integrates radiant energy from both the tuyere walls and from the incandescent charcoal at the nose, and will inevitably register a temperature higher than the actual temperature of the surrounding air unless elaborate shielding is installed (see Baker et al. 1975, 271–74, 322–33). The curves of temperatures measured at intervals within the tuyere (figure 11.1; Schmidt and Avery 1978, figure 5) cannot therefore be accepted as proof of preheating (Rehder 1986, 352).

Rehder then attempts to calculate the degree of preheat, using standard formulae from industrial practice, and concludes that the preheat obtained was only 22°C. He does, however, make a serious error that invalidates the calculation, as Avery and Schmidt (1986, 355) note in their response. Rehder assumes that the flow of air through the tuyere is continuous and laminar, and selects his formulae accordingly. This is certainly incorrect. The bellows used in the Haya process are valveless, and each time the bellows diaphragm

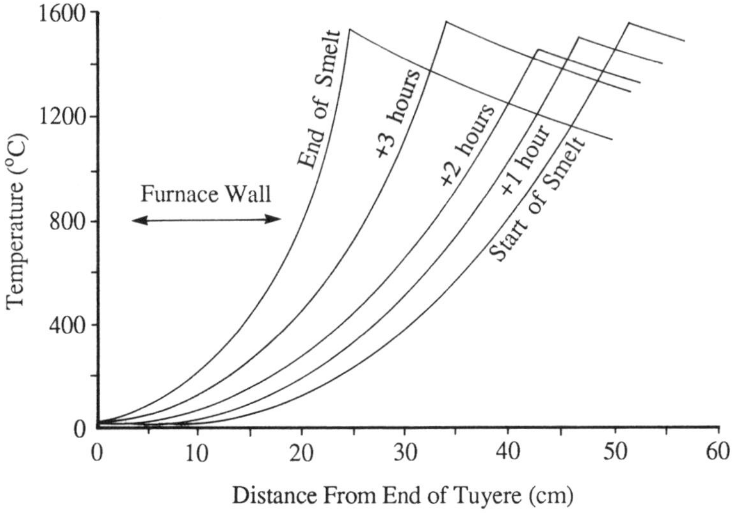

FIGURE 11.1. A schematic profile of temperatures measured along the axis of a tuyere in the Buhaya furnace (redrawn from Schmidt and Avery 1978, figure 5; courtesy of *Science*, copyright by American Association for the Advancement of Science). Note that the tuyere projects beyond the wall into the furnace. The peak in each curve marks the position of the hot oxidizing zone immediately in front of the tuyere nose. The position of the peak moves from the center of the furnace toward the wall as the smelt progresses, because the tuyere is continually eaten away at the nose by reaction with molten slag.

is raised the flow of air is reversed; the passage of air through the tuyere must therefore be not laminar but turbulent. The rapid mixing produced by turbulent flow would promote more efficient transfer of heat to the air than would laminar flow, so Rehder's estimate of preheat must be too low.

Prompted by Rehder's criticism, Avery has tried to calculate the degree of heating using a mathematical model particularly designed for turbulent flow, and he arrives at a figure of at least 323°C—about half the degree of preheat claimed in the original publication. It must, however, be noted that this model assumes *continuous* turbulent flow, and its validity where the flow is rapidly reversed is questionable. Avery's solution to this problem is simply to assume that under these conditions the air remains in the tuyere three times as long; the actual solution to the model (141°C) is therefore multiplied by three to achieve the quoted figure of 323°C (Avery and Schmidt 1986, 356). There is no way to judge whether this factor is correct, but the fact that arbitrary factors are introduced raises questions about the validity of the result.

Rehder also cites experimental evidence that temperatures as high as 1950°C can be obtained in charcoal beds without any preheating at all. Such temperatures are achieved only in the small zone immediately in front of the nose of the tuyere, where the oxygen in the incoming air is rapidly converted to carbon dioxide. This zone is oxidizing, but the carbon dioxide is very rapidly converted to carbon monoxide by reaction with solid charcoal, giving a reducing atmosphere. The latter reaction absorbs heat, so temperatures fall very rapidly with distance away from the oxidizing zone (Ross 1980).

With low blast pressure, as in a hand-blown furnace, the oxidizing zone may be very small indeed (a few cubic centimeters), with temperatures elsewhere in the furnace being very much lower (Rehder 1986, 353). Rehder assumes that the temperatures in excess of 1820°C cited for the Buhaya furnace were measured directly in front of the tuyere and were confined to a minuscule proportion of the total volume of the furnace. Avery and Schmidt have never stated where these high temperatures were measured. Until they can produce isotherm contours (that is, contours of equal temperature) for the Buhaya furnace that prove the existence of a larger volume at very high temperature, Rehder's argument will be the more persuasive.

The only set of time/temperature/location data actually published by Schmidt and Avery (1978, figure 5; figure 11.1 in this chapter) does not support their claim for preheating in the Buhaya furnace. The peak on each of the temperature curves in this graph is the measured temperature in the oxidizing zone in front of the tuyere nose.[4] These peaks are consistently in the range of 1400–1600°C throughout the duration of the smelt. These temperatures do not imply preheating of the air blast; as Rehder notes, similar temperatures have been maintained for hours in bloomery iron-smelting experiments elsewhere. A log of temperatures in one such experiment is reported by Tylecote, Austin, and Wraith (1971, figure 17) and is reproduced here as figure 11.2. Note that the temperatures measured at tuyere level are very similar to the peak temperatures in figure 11.1. Yet in Tylecote's experiment there cannot have been any significant preheating of the blast in the tuyere, as it protruded less than 5 cm into the furnace (Tylecote et al. 1971, figure 5).[5]

Similar findings emerge from a very thorough series of bloomery iron-smelting experiments by the Swedish archaeometallurgist Erik Tholander. The experiments that are relevant to the present discussion are those that he conducted in a low bloomery furnace with a single tuyere. A temperature log for one of these smelts and a series of vertical sections showing isotherms (con-

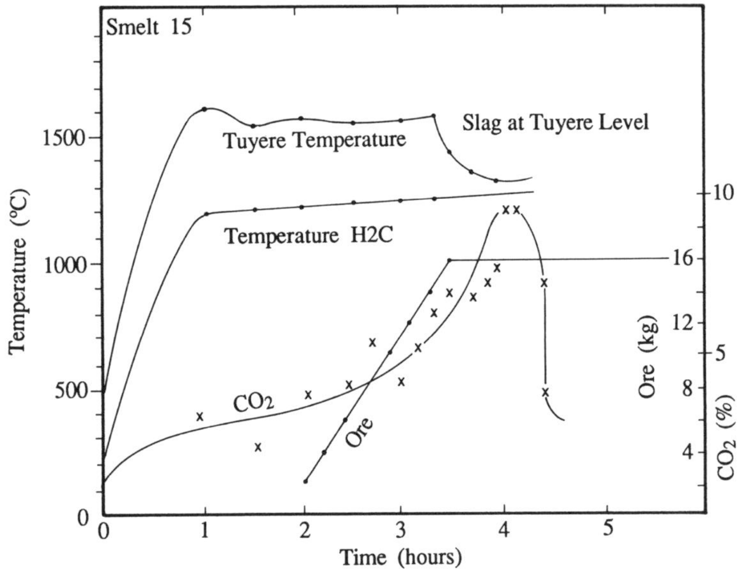

FIGURE 11.2. A graphic summary of the measured furnace parameters for experimental smelt fifteen of Tylecote, Austin and Wraith (reprinted from Tylecote et al. 1971, figure 17; courtesy of the Institute of Materials, London). The line labeled "tuyere temperature" shows the temperature of the oxidizing zone at the nose of the tuyere, and that labeled H2C is the temperature at the center of the furnace fifteen centimeters above the tuyere.

tours of equal temperature) shows that the temperature in the center of the furnace at tuyere level was above 1500°C for much of the smelt (Tholander 1987, figures 6.1 and 7.3). Such temperatures cannot be attributed to preheating in the tuyere, the nose of which was flush with the furnace wall. It is therefore clear from both sets of experiments that the temperature scans provided by Schmidt and Avery (figure 11.1) do not prove that preheating was occurring in the tuyeres of the Buhaya furnace.

There is another anomaly in figure 11.1 that has thus far passed without comment. If the air blown into the Haya furnace were preheated by heat transfer from the tuyere walls, then the temperature rise should have been proportional to the length of the tuyere inside the furnace. Avery's revised estimate is that the rise in the temperature of the air is at least 226.5°C per foot (30.5 cm) of tuyere length (Avery and Schmidt 1986, 356). Note, however, that the tuyeres in the Buhaya furnace shorten steadily during the course of the smelt as they are eaten away by the slag; for the plot shown in figure 11.1 the tuyere shortened approximately 25 cm over the course of the smelt (Schmidt and Avery 1978; figure 8.4, chapter 8, this volume). If one accepts Avery's estimate of 226.5°C per foot,[6] the peak temperature should therefore decrease over the course of the smelt by 185.7°C. But the graph shows no such decrease; peak temperatures at the end of the smelt differed little from those at the start.

There was at least one region in Africa where iron-smelting furnaces should, if the Avery-Schmidt model is correct, show even greater levels of preheat than the Haya model. The extraordinary furnaces of the Mafa highlands in north Cameroon (David et al. 1989) employ a single tuyere placed vertically and blown from above with valveless bellows. Approximately 120 cm of the tuyere is surrounded by hot charcoal, as compared to about 50 cm in the Haya fumace, so the Mafa furnace should attain even greater preheat of the blast than the Haya model. Ronald Tylecote did in fact consider (in 1965!) the possibility of preheating in the Mafa furnace, and he attempted to calculate its magnitude. He concluded that this was "almost negligible" (1965, 345) but his calculation is invalidated by the same error committed by Rehder in his analysis of the Haya furnace—he used a formula for heat transfer that assumes laminar flow of air through the tuyere, whereas the valveless bellows of the Mafa produce turbulent flow and, therefore, more efficient heat transfer. Nevertheless, the degree of preheat in this furnace does not appear to be significant. In 1989 Nicholas David and I monitored a successful smelt conducted by a Mafa ironworker in this type of furnace, and measured temperatures in the blast zone by thermocouple. The maximum temperature measured in the hot zone around the nose of the tuyere throughout the smelt was only 1380°C. Any zone of higher temperature was evidently too small to have been intercepted by the thermocouple.[7] This rather low temperature suggests that no significant degree of preheat was attained.

Finally it should be noted that some examples of bloomery furnaces are known to have used preheated blast, and it is instructive to compare these with the model proposed for Buhaya. Preheated blast was employed between about 1830 and 1900 in the "American bloomery process" of New Jersey and upper New York State. There are several excellent contemporary descriptions of the New York bloomeries, recently supplemented by field survey and laboratory study of the material remains (Allen et al. 1990; Gordon and Killick 1992). The temperature of the preheated air blast is given as 550–600°F (288–315°C) by Hunt (1869, 276) and as 600–800°F (315–427°C) by Egleston (1879, 33). It is not clear how these measurements were made, so they may have been subject to the same sources of systematic error as in the Haya example. Proof that a significant degree of preheating was obtained can, however, be found in the recorded charcoal consumption, which decreased 20 percent with the introduction of hot blast (Hunt 1869, 278).

How does the mechanism for providing hot blast in the Adirondacks compare with that proposed for the Haya bloomery? In the American bloomery, as in all modern blast furnaces, waste heat escaping from the top of the furnace was used to heat the blast. A tall brick chimney was erected over the bloomery furnace and lined with iron pipes, through which the incoming cold air passed en route to the tuyere. In Avery and Schmidt's model of preheating in the Buhaya furnace, by contrast, preheating is claimed to occur within the furnace, as the protruding tuyeres soak up heat from the hot zone of the furnace and transfer it to the incoming air. This would appear to be a zero-sum system, in that the heat absorbed by the incoming air cannot exceed that lost by the furnace to the tuyere. It is therefore difficult to see how there can be any net gain in such a system.

In summary, the case for preheated blast in the Haya furnace is, to employ an old Scots legal verdict, not proven. A definitive answer will require more sophisticated techniques of measuring gas flow and temperature than can at present be employed under field conditions (Avery and Schmidt 1986, 356). There is, however, no reason why the core of this hypothesis—the contention that a reciprocating air blast can be heated to a temperature of several hundred degrees centigrade while passing through a short clay tube immersed in a bed of hot charcoal—cannot be adequately tested in a university laboratory. It is unfortunate that neither the proponents of this thesis nor their critics have attempted this experiment, but have instead resorted to the use of mathematical models of dubious relevance.

Is There an African Direct Steel Process?

The concept of the "African direct steel process" was first formally proposed by van der Merwe (1980) and elaborated by van der Merwe and Avery (1982). The essence of this argument is that African methods of making steel were fundamentally different from those employed elsewhere in the Old World. Outside Africa, they argued, steel was produced by one of two processes. The first was to make carbon-free wrought iron in a bloomery furnace and then convert this to steel by carburization. Carburization could be accomplished either by holding solid wrought iron at a high temperature in a bed of carbon (a technique known as cementation) or by enclosing small batches of iron and carbon in a sealed crucible placed in a furnace (crucible steel making, first developed in India). This, van der Merwe and Avery suggest, was how all steel was made in Europe and the near East prior to the adoption of the blast fumace (1982, 146–49).

The second process was to produce cast iron (2–5% carbon) in a blast furnace, and then to reduce the carbon content by oxidation into the range of composition of steel (0.2–2 percent carbon). Decarburization can be accomplished either by melting cast iron in an oxidizing air blast (in a finery, puddling furnace, or converter) or alternatively by dissolving carbon-free wrought iron in molten cast iron (the cofusion process). Both methods were developed in China before 100 B.C. (Han et al., 1986), and appear in Europe with the earliest blast furnaces in the twelfth or thirteenth century A.D. (Needham 1980; Tylecote 1985).

Both of these methods of steel making involve two separate steps, and are therefore classified by van der Merwe and Avery as indirect pathways to steel. They contrast both of these processes with the direct production of steel in Africa. By this they mean that steel blooms were produced in the bloomery furnace itself and could be forged immediately without any secondary processing (1982, 152–54). This technology is claimed to have been confined to Africa, and is therefore named the "African direct steel process" (van der Merwe 1980; van der Merwe and Avery 1982, 154).

In evaluating this claim, we need to ask two separate questions. Did African iron-smelting processes produce high carbon blooms? And is it true that blooms of direct steel were not produced outside Africa?

STEEL BLOOMS FROM AFRICAN FURNACES

The evidence shows beyond any doubt that many African bloomery furnaces produced steel blooms. Their reported occurrence does not appear to have been restricted to any particular type of furnace nor to any particular region of Africa.[8] Steel blooms have been reported from short shaft furnaces (less than 1.5 m tall) operated by forced draft from bellows in Zimbabwe (Stanley 1931, 132), Tanzania (Avery and Schmidt 1979), Ghana (van der Merwe 1980, 332), and Ethiopia (Todd 1985, 94). Turning to natural draft furnaces (those designed to work without bellows), we find steel blooms recorded from a short furnace (1.2 m) in Nigeria (Bellamy and Harbord 1904, 123), and from tall furnaces (2.5–3.4 m) in the Ivory Coast (Eckert 1986), Burkina Faso (Forbes 1933, 235), Malawi (Killick 1990), and Zambia.[9] The oldest archaeological steel bloom yet reported from Africa is that from the KM3 site in Buhaya, Tanzania (Schmidt 1983; Schmidt and Childs 1985). This is associated with a radiocarbon date of 1920 ± 120 B.P. (RL-1260), the calibrated age range for which is 200 B.C.–A.D. 380 at two standard deviations (Stuiver and Pearson 1986).[10]

Some of the blooms cited above are very high-carbon steel, containing more than 1 percent carbon. African ironworkers have also produced cast iron (greater than 2 percent carbon) on occasion. A sample of cast iron from Namibia was reported by Stanley (1931, 133), while a steel bloom from Ghana analyzed by van der Merwe (1980, figure 2) contains a small mass of gray cast iron. Although these examples were presumably inadvertent, at least one African case of deliberate production of cast iron is known. Some ironworkers among the northern Mafa of north Cameroon smelted magnetite sands to produce a melange of cast iron, steel, and wrought iron. In a successful reconstruction of this process, arranged by Nicholas David in 1986, about 40 percent of the product was cast iron, with the remainder high-carbon steel, low-carbon steel and carbon-free wrought iron (David et al. 1989). The production of cast iron was clearly not a mistake, as the smelter went on to decarburize the cast iron to steel in a strongly oxidizing forge fire. This technology, the only one of its kind yet confirmed from Africa, is therefore a curious hybrid of the direct and indirect methods of making steel. Similar processes may have been more widespread in Africa in the past; there is, for example, a very credible account of such a process among the Fan in the mid-nineteenth century (Du Chaillu 1861, 90).

DIRECT STEEL OUTSIDE AFRICA

We may therefore accept as true the claim that many African iron smelters were able to produce high-carbon steel directly in the bloomery furnace. But the second half of the argument—that steel blooms were not made outside Africa—is clearly false. This error arose because the only two authorities cited by van der Merwe and Avery on this point—Coughlan (1956) and Tylecote (1962)—are both thoroughly obsolete. At the time that these texts were written most historians of metallurgy did in fact believe that steel could not be produced directly in a bloomery. But in the early 1960s researchers in England, France, Czechoslovakia, and Poland began to make systematic use of metallography as a research tool. They soon noted many microstructures that could only be explained by positing the direct product of steel in the furnace itself. The concept of a "direct steel process" was in fact invoked by the great Czech archaeometallurgist Radomir Pleiner in 1968 to account for this evidence (Pleiner 1968).

Steel blooms similar to those from Africa were produced in some areas of Europe at least as early as the Late Hallstatt period, 500–100 B.C. (see, for example France-Lanord 1963; Pleiner 1980, 390). The larger furnaces of the first millennium A.D. in Europe were more favorable to the direct production of steel, and so there are more reported instances of it than in earlier times (Pleiner 1980, 393–96; Tylecote 1986, 167–68). As in Africa, the carbon content of the metal occasionally rose high enough to produce cast iron, which was discarded as waste (Pleiner 1968, 314; Tylecote 1986, 168). From later times, a mixture of wrought iron, high-carbon steel (1.1 percent C) and minor cast iron has been reported in residues from an excavated medieval bloomery in England (Morton and Wingrove 1970).

Little is known at present about the early history of the bloomery process in Asia, but nineteenth-century eyewitness accounts make it clear that some ironworkers of this continent were capable of producing steel in bloomery furnaces if they chose to do so. For example, there is a detailed account of the working of a large hand-powered bloomery furnace at Tendukera, northwest India, in the 1860s. The bloom obtained from this process "usually consists of a mixture of malleable iron and natural steel. . . . When however the object is to produce steel a large proportion of charcoal is employed and a gentle blast applied." The account goes on to note that "sometimes small quantities of cast iron are produced to the great annoyance of the smelters, who have much difficulty in separating it from the rest of the iron. They consider that in this case the iron has been injured by raising the temperature of the furnace too high" (Percy 1864, 260).

Further east, the hand-powered *tatara* furnaces of Japan smelted fine magnetite sand to produce both cast iron and bloomery steel in the same operation; the slag and most of the cast iron were tapped from the furnace, while the steel bloom and the remaining cast iron were removed in the solid state at the end of the run (Rostoker et al. 1989). The tatara process is conceptually quite similar to the extraordinary smelting technology of the Mafa Hills, Cameroon (David et al. 1989). The obvious differences lie in the scale of production (the tatara produced four tons of metal in a run of seventy-two hours, the Mafa process an estimated 20–40 kg per day) and in the fact that neither cast iron nor slag was tapped from the Mafa furnace.

Both of the "high-tech" nineteenth-century bloomery technologies could produce either wrought iron or steel. These are the Catalan process of the Pyrénées (François 1843; Percy 1864, 278–312) and the American bloomery process, which was developed in small rural ironworks in the Adirondack mountains of New York State (Egleston 1879; Gordon and Killick 1992). Both processes used furnaces less than 1 m tall, powered by a pressurized blast supplied by water or steam power; in the American process the blast was preheated. Both the Catalan and the American bloomeries could produce either wrought iron or steel. It is clear from the accounts cited here that the adjustments required to favor the production of either wrought iron or steel were subtle and that a very high degree of skill was needed to obtain a consistent product.

Finally, a great deal of information has been gathered in the past twenty-five years by trials in experimental bloomery furnaces (Pleiner 1968; Guyan et al. 1973; Tylecote and Merkel 1985). The experiments most relevant to this discussion are a series of thirty-one trials conducted by Tylecote et al. (1971). The most significant result of these experiments was their demonstration that the operator of a low-shaft bloomery furnace can determine the carbon content of the bloom at will by selection of an appropriate charcoal-to-ore ratio (other factors being held constant). Low ratios favor the production of wrought iron, intermediate ratios the production of steel, and high ratios the production of cast iron.[11]

In summary, the direct production of steel in bloomery furnaces was not confined to Africa and cannot be presumed to be an African invention. The conditions required for the production of steel fall within the normal range of variation of bloomery processes, and are found in all regions of the world where bloomeries were operated. The concept of a "direct steel process" is therefore unnecessary. There is no sound evidence that African bloomery technology was fundamentally different from that employed elsewhere in the Old World, as the same mechanisms of reduction of the ore and of formation and carburization of the bloom have been observed in both African and non-African bloomery furnaces (Killick and Gordon 1988).

Discussion

Both of the hypotheses reviewed above were important stimuli for the study of African metallurgy, and their authors deserve much of the credit for the surge of research on the topic in the last decade. Their most notable achievement was methodological, in that these studies demonstrated the necessity of collaboration in the field of anthropologists and materials scientists. Studies of African metalworking that lack one or other of these contributions now seem—and are—seriously deficient.

The central argument in both of these hypotheses is that African iron-smelting techniques differed significantly from those employed elsewhere in the Old World. This review has shown that this thesis is mistaken in the case of "direct steel production" in Africa and dubious at best in the claim for preheated blast. The case for an African direct steel process is negated by the fact that steel blooms have been reported from many other areas of the Old World and by historical and experimental evidence that the steel can be produced at will in most bloomery furnaces. The argument for preheated blast is severely weakened by evidence from experimental bloomery smelting in Europe that shows that comparable temperatures are attainable in furnaces that cannot possibly have employed preheated blast.

Even if both of these hypotheses are rejected, ample evidence remains of change in African bloomery iron smelting over the two and a half millennia of the African Iron Age. African ironworkers adapted the bloomery process to a wider variety of ores and invented a greater range of furnace designs than did bloomery ironworkers elsewhere in the Old World (Killick 1991b). There is therefore no lack of invention in African iron smelting; what requires explanation is the lack of growth in productivity of African iron technology over this span of time, during which the productivity of iron smelting in Europe grew by a factor of one thousand. Where specialist iron industries did emerge in later African history, as at Bassar, Togo (de Barros 1986), or on the Ndop Plain of Cameroon (Fowler 1990), increased productivity was obtained by changes in the social organization of production, not by technological innovations that greatly increased the amount of iron produced in a furnace in a given time. One of the tasks that lies ahead is to identify the reasons for the lack of growth in productivity in smelting technology. Were the barriers to increasing productivity social, or environmental, or both?

More generally, a claim that African technology is "advanced" or "sophisticated," or merely different, has meaning only with specific reference to technology in some other time or place. Africanists have tended to use such concepts rather loosely and without much understanding of the history of technology elsewhere. The same criticism can be leveled against most claims for "large-scale" iron production in Africa; only rarely is some external standard of reference supplied for this term, as for example in the fine study of iron production at Bassar, Togo, by de Barros (1986). A more realistic assessment of the history of production in Africa requires much closer acquaintance with the history of technology outside the continent.

Notes

I thank Nikolaas van der Merwe for comments on an earlier draft of this chapter. More generally, I owe him a debt of gratitude for steering me to the study of African metallurgy and for his support and friendship over the past fifteen years. He has always urged his students not to become too attached to their own theories and has actively encouraged critical scrutiny of his own work. Would that all scholars were as open-minded.

1 "those who have invented neither gunpowder nor compass | those who tamed neither steam nor electricity | those who explored neither the sea nor the sky | but those without whom the earth would not be the earth" (translation by Emile Snyders, in Césaire 1968, 99).

2 This presumes that indigenous and imported iron can be distinguished. This distinction can be made with confidence after 1880, when all iron imported to Africa would have been smelted with coke in a blast furnace, then decarburized in a puddling furnace or a Bessemer or Siemens-Martin converter. The sulfur imparted to the iron by the coke was balanced by addition of manganese, giving rise to manganese sulfide inclusions that are not present in indigenous charcoal-smelted bloomery iron. It is much less certain that imported iron produced before 1880 can always be distinguished from the local product, as bloomery furnaces and charcoal-fueled blast furnaces were still used in some areas of Europe and Scandinavia to this date. Their products may well have been exported to Africa.

3 Not to Siemens's patent of 1856, as stated by Avery and Schmidt (1979, 16).

4 *Editor's note:* These temperatures are not inside the blast zone proper. See Avery and Schmidt, chapter 12, this volume, for a clarification of precisely where the peak temperatures were measured.

5 *Editor's note:* See Avery and Schmidt, chapter 12, this volume, for the thermodynamic argument: early in the smelt a tight cluster of eight blast zones, under preheating conditions, leads to elevated temperatures in the internal reduction zones.

6 The calculated value is 75.5°C per foot; Avery multiplies this by three, which results in 226.5°C per foot.

7 The thermocouple was inserted through the furnace door, at right angles to the tuyere, rather than along the bore of the tuyere as in the Haya case. It may therefore not have passed exactly through the center of the hottest zone at the nose of the tuyere.

8 This is not meant to imply that all African bloomery furnaces produced steel. One should also note that iron blooms are rarely homogeneous, and that wrought iron, steel, and even cast iron may be produced simultaneously in a single smelt.

9 The example cited from Zambia is from an unpublished account by G. Quick, entitled "Native Iron Industry, Lunda-Chishinga area" (Zambia National Archives, Lusaka, file NA2/Sec/Nat 147 [Ref. 266/13]). This describes a smelt conducted in 1938 and includes metallographic and chemical analyses of an ingot forged directly from the bloom, the average carbon content of which was 0.75 percent C. I thank Jo Golson for a copy of this document.

10 A second date of 2520 ± 100 (Beta-10493) was rejected by the excavators (Schmidt and Childs 1985, 67–68).

11 G. H. Stanley suggested this more than fifty years ago (1931, 33). It is also implied in the account of the Indian process quoted above, and in the modifications made in the Catalan process to favor the production of steel (Percy 1864, 307).

References

Allen, R. F., J. C. Dawson, M. F. Glenn, R. B. Gordon, D. J. Killick, and R. W. Ward. 1990. "An Archaeological Survey of Bloomery Forges in the Adirondacks." *Journal of the Society for Industrial Archaeology* 16:3–20.

Andah, B. 1979. "Iron Age Beginnings in West Africa: Reflections and Suggestions." *West African Journal of Archaeology* 9:125–41.

Austen, R. A. 1987. *African Economic History*. London: James Currey.

Austen, R. A., and D. Headrick. 1983. "The Role of Technology in the African Past." *African Studies Review* 26:163–84.

Avery, D. H., and P. R. Schmidt. 1979. "A Metallurgical Study of the Iron Bloomery, Particularly as Practiced in Buhaya." *Journal of Metals* 31:14–20.

———. 1986. "The Use of Preheated Air in Ancient and Recent African Iron Smelting Furnaces: A Reply to Rehder." *Journal of Field Archaeology* 13:354–57.

Baker, H. D., E. A. Ryder, and N. H. Baker. 1975. *Temperature Measurement in Engineering*. Vol. 2. Stamford, Conn.: Omega Press.

Bellamy, C. V., and F. Harbord. 1904. "A West African Smelting House." *Journal of the Iron and Steel Institute* 66:99–126.

Bernal, M. 1987. *Black Athena: The Afroasiatic Roots of Classical Civilization.* Vol. 1: *The Fabrication of Ancient Greece, 1785–1985.* New Brunswick, N.J.: Rutgers University Press.

———. 1991. *Black Athena: The Afroasiatic Roots of Classical Civilization.* Vol. 2: *The Archaeological and Documentary Evidence.* New Brunswick, N.J.: Rutgers University Press.

Césaire, A. 1968. *Cahier d 'un Retour au Pays Natal.* Paris: Éditions Presence Africaine.

Childs, S. T., and D. J. Killick. 1993. "Indigenous African Metallurgy: Nature and Culture." *Annual Review of Anthropology* 22:317–37.

Coughlan, H. H. 1956. *Notes on Prehistoric and Early Iron in the Old World.* Occasional Papers on Technology, No. 8. Oxford: Pitt-Rivers Museum.

David, N., R. B. Heimann, D. J. Killick, and M. Wayman. 1989. "Between Bloomery and Blast Furnace: Mafa Iron Smelting Technology in North Cameroon." *African Archaeological Review* 7:185–210.

de Barros, P. 1986. "Bassar: A Quantified, Chronologically Controlled, Regional Approach to a Traditional Iron Production Centre in West Africa." *Africa* 56:148–74.

Diop, C. A. 1955. *Nations Nègres et Culture.* Paris: Éditions Présence Africaine.

———. 1974. *The African Origin of Civilization: Myth or Reality?* Edit. and trans. Mercer Cook. Chicago: Lawrence Hill Books.

Diop, L. M. 1968. "Métallurgie Traditionelle et âge de fer en Afrique." *Bulletin de L'I.F.A.N.*, series B, 30:10–37.

Du Chaillu, G. 1861. *Expeditions and Adventures in Equatorial Africa.* New York.

Eckert, H.-E. 1986. "Bergbau, Verhütting und Eisenschmieden bei den Senufo in Westafrika, Beobachungen in Jahre 1985." *Der Anschnitt* 38:70–79.

Eggert, M. K. H. 1987. "On the Alleged Complexity of Early and Recent Iron Smelting in Africa: Further Comments on the Preheating Hypothesis." *Journal of Field Archaeology* 14:377–81.

Egleston, T. 1879. "The American Bloomary Process for Making Iron Directly from the Ore." *Transactions of the American Institute of Mining Engineers* 8:1–36.

Forbes, R. H. 1933. "The Black Man's Industries." *Geographical Review* 23:230–47.

Fowler, I. 1990. "Babungo: A Study of Iron Production, Trade and Power in a Nineteenth-Century Ndop Plain Chiefdom (Cameroon)." Ph.D. diss., University College, London University.

France-Lanord, A. 1963. "Les Lingots de Fer Protohistoriques." *Revue d'Histoire de la Sidérugie* 4:167–78.

Francois, J. 1843. *Recherches sur le Gisement et le Traitement Direct des Minerais de Fer dans les Pyrénées, et Particuliarement dans l'Ariége.* Paris.

Gordon, R. B., and D. J. Killick. 1992. "The Metallurgy of the American Bloomery Process." *Archeomaterials* 6:141–67.

Goucher, C. L. 1981. "Iron Is Iron 'Til It Rust: Trade and Ecology in the Decline of West African Iron-Smelting." *Journal of African History* 22:179–90.

Guyan, W. U., R. Pleiner, and R. Fabesova, eds. 1973. *Die Versuchschmelzen und Ihre Bedeutung für die Metallurgie des Eisens und Dessen Geschichte.* Schaffhausen: Eisenbibliothek GF Stiftung.

Hall, M. 1988. *The Changing Past: Farmers, Kings, and Traders in Southern Africa, 200–1860.* London: James Currey.

Han, R., G. Watt, and N. Kennon. 1986. "The Development of Iron and Steel Technology in Ancient China." *Materials Australasia* 18:12–16.

Hopkins, A. G. 1973. *An Economic History of West Africa.* London: Longman.

Hunt, T. Sterry. 1869. "Iron and Iron Ores." *Reports of Progress,* Geological Survey of Canada, 245–394.

Hyde, C. K. 1977. *Technological Change and the British Iron Industry, 1700–1870.* Princeton: Princeton University Press.

Killick, D. J. 1990. "Technology in Its Social Setting: Bloomery Iron-Working at Kasungu, Malawi, 1860–1940." Ph.D. diss., Yale University.

———. 1991a. "A Little Known Extractive Process: Iron Smelting in Natural-Draft Furnaces." JOM: *Journal of the Minerals, Metals and Materials Society* 43:62–64.

———. 1991b. "The Relevance of Recent African Iron-Smelting Practice to Reconstructions of Prehistoric Smelting Technology." In *Recent Trends in Archaeometallurgical Research,* ed. P. D. Glumac, 1:1–8. University Museum, MASCA Research Papers in Science and Archaeology. Philadelphia.

Killick, D. J., and R. B. Gordon. 1988. "The Mechanism of the Bloomery Process." In *Proceedings of the 26th International Symposium on Archaeometry,* ed. R. M. Farquhar, R. G. V. Hancock, and L. A. Pavlish, 120–23. Toronto: University of Toronto, Department of Physics.

Law, R. 1980. "Wheeled Transport in Precolonial West Africa." *Africa* 50:249–62.

Morton, G. R., and J. Wingrove. 1970. "Reduction of Iron from Its Ore in the Medieval Bloomery." *Steel Times* 198:237–42.

Needham, J. 1980. "The Evolution of Iron and Steel Technology in East and Southeast Asia." In *The Coming of the Age of Iron,* ed. T. A. Wertime and J. D. Muhly, 507–42 . New Haven: Yale University Press.

Percy, J. 1864. *Metallurgy. Volume 2: Iron, Steel.* London: John Murray.

Pleiner, R. 1968. "Problem of Direct Steel Production in Early Ferrous Metallurgy." *Steel Times* 196:312–18.

Pleiner, R. 1980. "Early Iron Metallurgy in Europe." In *The Coming of the Age of Iron,* ed. T. A. Wertime and J. D. Muhly, 375–416. New Haven: Yale University Press.

Rehder, J. E. 1986. "Use of Preheated Air in Primitive Furnaces: Comment on the Views of Avery and Schmidt." *Journal of Field Archaeology* 13:351–53.

Renfrew, C., and P. Bahn. 1991. *Archaeology: Theories, Methods, and Practice.* New York: Thames and Hudson.

Ross, H. U. 1980. "Physical Chemistry." In *Direct Reduced Iron,* ed. R. L. Stephenson, 9–34. Warrendale, Penn.: Iron and Steel Society of AIME.

Rostoker, W., B. Bronson, and J. R. Dvorak. 1989. "Smelting to Steel by the Japanese Tatara Process." *Archaeomaterials* 3:11–25.

Schmidt, P. R., and D. H. Avery. 1978. "Complex Iron Smelting and Prehistoric Culture in Tanzania." *Science* 201:1085–89.

——. 1983. "More Evidence for an Advanced Prehistoric Iron Technology in Africa." *Journal of Field Archaeology* 10:421–34.

Schmidt, P. R., and S. T. Childs. 1985. "Innovation and Industry during the Early Iron Age in East Africa: The KM2 and KM3 sites of Northwest Tanzania." *African Archaeological Review* 3:53–94.

Stanley, G. H. 1931. "Some Products of Native Iron Smelting." *South African Journal of Science* 28:131–34.

Stuiver, M., and G. W. Pearson. 1986. "High-Precision Calibration of the Radiocarbon Time Scale, A.D. 1950–500 B.C." *Radiocarbon* 28:805–28.

Tholander, E. 1987. *Experimental Studes on Early Iron-Making.* Stockholm: Royal Institute of Technology.

Todd, J. A. 1985. "Iron Production by the Dimi of Ethiopia." In *African Iron Production: Ancient and Traditional,* ed. R. Haaland and P. Shinnie, 88–101. Oslo: Norwegian Universities Press.

Tylecote, R. F. 1962. *Metallurgy in Archaeology.* London: Edward Arnold.

——. 1965. "Iron Smelting in Pre-industrial Communities." *Journal of the Iron and Steel Institute* 203:340–48.

——. 1985. "The Early History of the Iron Blast Furnace in Europe: An Early Case of East-West Contact?" In *Medieval Iron in Society,* 158–173. Stockholm: Jernkontorets Forsknung, Series H, No. 34.

——. 1986. *The Prehistory of Metallurgy in the British Isles.* London: Institute of Metals.

Tylecote, R. F., J. N. Austin, and A. E. Wraith. 1971. "The Mechanism of the Bloomery Process in Shaft Furnaces." *Journal of the Iron and Steel Institute* 209:342–63.

Tylecote, R. F., and J. F. Merkel. 1985. "Experimental Smelting Techniques: Achievements and Future." In *Furnaces and Smelting Technology in Antiquity,* ed. P. T. Craddock and M. J. Hughes, 3–20. Occasional Paper No. 48. London: British Museum.

van der Merwe, N. J. 1980. "Production of High Carbon Steel in the African Iron Age: The Direct Steel Process." In *Proceedings of the 1977 Pan African Congress of Prehistory and Quaternary Studies,* ed. R. E. Leakey and B. A. Ogot, 331–334. Nairobi.

van der Merwe, N. J., and D. H. Avery. 1982. "Pathways to Steel." *American Scientist* 70:146–55.

12 Preheating: Practice or Illusion?

Donald H. Avery & Peter R. Schmidt

The first observation to make in response to Killick's paper is that he has not disproved the preheating hypothesis nor has his critique weakened interpretations of the data used to support the hypothesis. Second, many of the points that Killick raises are not new and have already been answered in our response to Rehder's (1986) critique. It is not clear why Killick wants to revisit the issues that we have thoroughly addressed in previous publications. Because we want to avoid repetitiveness, we do not want to reiterate all of our technical arguments, so in this response we will cross reference key points that have appeared elsewhere but have been ignored in this particular critique.

It is important to correct Killick's construction of how the preheating hypothesis arose and how its subsequent proof was developed. He evidently does not understand the sequence of events in the scientific inquiry when he says, "In searching for a mechanism in which preheating could be achieved, Avery and Schmidt's attention focused upon the placement of the tuyeres, the ceramic pipes that convey the blast from the bellows into the furnace." This characterization inverts the discovery process and seems to suggest that the hypothesis arose spontaneously during the monitoring of Haya smelting. Rather, we were searching for an explanation for refractory slag from very early African smelting at the Rugomora Mahe site, c. 2000 B.P., that evidenced very high flow (melting) temperatures of 1400°C and above. There were also massive amounts of slag in the region fused into large 0.75 m/diam intact refractory skulls from the bottoms of traditional furnaces.

In Avery's judgment, the laboratory-measured flow temperatures were far higher than should have been attainable from the use of an unheated blast under the reducing conditions necessary for iron smelting. He had previously conducted both laboratory and reconstructed furnace smelting experiments and had never seen materials like this. Archaeological evidence also attested to the presence of slag-wetting and penetration, and black (reduced) iron oxides on the outside of prehistoric tuyeres, resulting from their exposure to high temperature reducing furnace conditions (Schmidt and Avery 1978; Schmidt and Childs 1985; Schmidt and Childs, chapter 9, this volume). These facts provided the basis for our hypothesis that the tuyeres were inserted deep into prehistoric furnaces and served to preheat the entering air blast. We were fortunate that there were still traditional smelters alive who were able to reestablish traditional smelting practice. We conducted one unsuccessful "Haya" reconstruction with electric blowers and industrial ore in rural Connecticut in 1975 (Childs and Schmidt 1985). In 1976 we returned to Africa and observed, instrumented, and characterized the reconstruction in the field of bloomery smelting by and under the direction of traditional Haya smelters.

There has been some criticism of the lack of controlled and reproducible laboratory conditions; however, our aim was to characterize an *authentic* traditional Haya process. We scrupulously avoided directing the project and imposing our concepts on it (even when it drove the technical member of the team to distraction). Likewise instrumentation was kept unobtrusive and we accepted the use of traditional methods such as valveless bellows, though at first they seemed inefficient and uncontrollable.

Let us now turn to the most important of Killick's challenges. One of Killick's major contentions is that "Schmidt and Avery assumed that temperatures above 1820°C could only be achieved if the incoming air was preheated." We disagree: *an assumption was not made.* Rather, calculations based on standard and generally accepted thermodynamic data indicated that temperatures over 1742°C could only be achieved by preheating, even under completely oxidizing conditions. A plot of the adiabatic temperature, or maximum temperature (when all the heat of the reaction goes into the gaseous products of combustion with no heat losses to the furnace or the smelting process) vs. $CO:CO^2$ ratio has been addressed in many other publications (see figure 2, Avery and Schmidt 1979; Avery 1981). This represents an upper limit to temperature, as in any real process there will be thermal losses. The specific mention of the 1820°C temperature emphasized by Killick arose from the unfortunate melting of an expensive platinum–10 percent rhodium thermocouple in the oxidizing blast zone of an experimental smelt in Buhaya, and its citation was meant to be one example of high temperatures above the adiabatic curve. Temperatures in the oxidizing zone at the end of the tuyere were also monitored with an optical pyrometer and showed incandescent temperatures of burning charcoal interfaces and fused slag droplets well in excess of 1800°C.

Another of Killick's criticisms is that the thermocouples used are measuring radiant heat from the tuyere mouth and the tuyere walls. In this instance he reiterates Rehder's observation (see Rehder, chapter 10, this volume) that it is difficult to use thermocouples to obtain an accurate measurement of temperature of air entering a furnace. This is indeed true. We were aware from the start that it is difficult, but it is not impossible. We used a variety of heat shields designed to block radiation but allow unimpeded air flow across the thermocouple. These consisted of polished metal tubes of various lengths. The thermocouple was supported in the center of the tube. This was done most effectively by turning the tip of the couple up 90°. The measured results with shields differed little from those without shields, except within 2–3 diameters of the opening of the tuyere. The reflectivity of the shields decreases with exposure to furnace temperatures.

Temperature profiles in the tuyeres were done with both a platinum-rhodium thermocouple that has a reflectivity of 0.90–0.92 (mirrorlike) and a chromel-alumel thermocouple that has a dark oxide coating and a reflectivity of 0.21. We would expect that the platinum-rhodium thermocouple would reflect most radiant heat, while the chrome-alumel thermocouple would theoretically absorb more. However, the similarity of the results between the two types of thermocouples, with and without heat shields, shows a strong consistency, giving us confidence that the temperature increase measured along the tuyere extending into the furnace did in fact reflect preheating of the air. We have never argued that the temperatures were exact to the degree, and have acknowledged that near the inside end radiation will heat the thermocouple bead significantly above the surrounding air. We therefore disagree with Killick's assertion that our temperature curves for preheating within the tuyeres do not constitute proof of preheating.

It is encouraging that Killick, unlike Rehder, sees the importance of our argument about the significance of valveless bellows and reversed flow. Rehder's use of a continuous flow model does not fit the reality of the Haya bellows operation, where the very rapid oscillation of the flow of air into the bellows, caused by reverse flow, is a key factor in the success of the Haya preheating process. To recapitulate, during conventional laminar flow a stagnant boundary layer attaches itself to the interior surface of the tuyere. This provides an insulating layer, just as does the air in a double-paned window, and reduces heat transfer to diffusion across the boundary layer, a very slow process. In turbulent flow, eddies and vortices develop which mechanically mix heated air from the boundary layer into the stream, just as stirring mixes cream into coffee. With reverse flow the hot boundary layer is peeled back like strips of skin and mixed into the air flowing in the furnace. Not only does the reverse flow increase the heat transfer from the wall of the tuyere to the air flowing through it, but it also increases the heat transfer between the air and the thermocouple.

However, Killick challenges the idea of turbulent flow, which we used to make a *lower limit* estimate of temperature. We know that reverse flow will disrupt a laminar boundary layer, transferring additional heat into the air. This will be substantially greater than steady-state turbulent flow. This principle is commonly seen in nature, which is often pretty smart. Common examples include a dog panting to remove the whole heat load from his body through reversed air flow over his tongue, an elephant flapping his ears, and a runner panting to increase his oxygen exchange. The cooling effect of a mild oscillating breeze caused by waving a fan on a hot day also illustrates the principle.

Because there is no adequate mathematical model or a detailed characterization of the reverse flow and its effect on the (time-dependent) makeup of the boundary layer, we calculated a *lower limit* model based on *continuous* turbulence (though the Reynolds number for the low net flow through the furnace would indicate a laminar model). A model that fully accounts for reverse turbulence would indicate much higher temperature. Therefore, the turbulent model calculates a *minimum limit* for preheating, knowing that the actual values will be higher.

Killick also finds that Avery's calculations employ an "arbitrary" quality by suggesting that the travel time of air in the tuyeres under valveless bellows conditions is at least three times as long as compared to continuous flow. Suffice it to say that the dwell time of the air in the tube is well beyond a factor of one, or, continuous passage through the tube. If there are oscillating pulses of air inside the tuyere, then the path is proportionately longer. Figure 12.1 illustrates what would happen with a pulse which is one-half forward and one-third back (3.5 pulses), for example, a path that is four times the length of the tuyere. Under these illustrated conditions the dwell time is longer than we have previously suggested, which is a conservative estimate. In fact the Haya bellows were pumped at a remarkable rate of six to eight strokes per second.

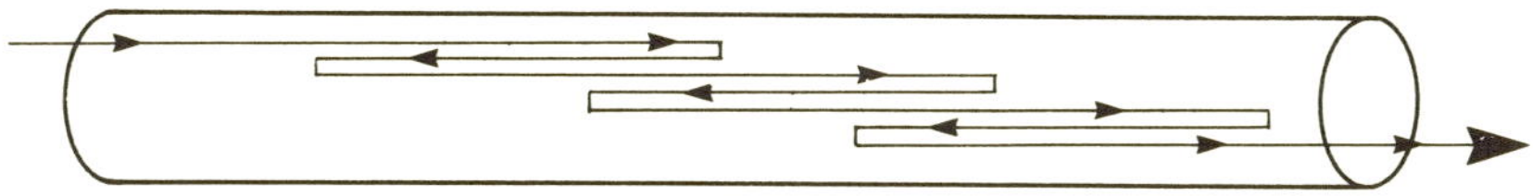

FIGURE 12.1. A schematic drawing—not to scale—showing the increased dwell time of air in the tuyere as a consequence of the oscillating air blast.

The bellows consist of a twin skin drumheads over bowls and are driven at a resonant frequency. This is clearly subject to head tension and the length, resistance, and gap between the tuyere and bellows. We also want to emphasize, however, that there is considerable variability in bellowing activity, with some smelters working at eight strokes per second, others at six strokes per second, and yet others who work at one stroke per second while resting.

Killick says that there is no way to judge if our reckoning is correct. It is our best scientific judgment, based on many observations in the field over many seasons. Our calculations published in the response to Rehder (chapter 10, this volume) represent a floor or lower limit based on turbulent flow; with reverse flow, as is the case, temperatures will be higher. If anything we err on the side of being too cautious. We believe that this is a more responsible position than Killick's; he assumes that because the conditions in the bellows and tuyeres are not precisely characterized, little of significance happens.

The next major misapprehension is Killick's assertion that repeats Rehder's suggestion that we are measuring a minuscule (1 cc) portion of the furnace within the blast zone when we record temperatures of greater than 1800°C. This is not true, for we have consistently measured temperatures, using an optical pyrometer, down in front of the blast zone where there is a molten pool of slag that varies between 1600–1800°C. There is also Killick's failure to take note of the schematic graph, figure 5, published on page 15 of our 1979 article (Avery and Schmidt 1979) in which we illustrate that the highest temperatures are found in the oxidizing zone immediately in front of the tuyere opening (figure 12.3). Our measurements within the oxidizing zone (obtained both by optical pyrometer and by platinum–10 percent rhodium thermocouple) *are not the important issue here.* Killick's focus on the oxidizing zone temperatures (where no reduction to iron occurs) masks *the much more critical observation that the temperatures beyond it in the reducing zone are higher than expected, about* 1375–1500°C. In order that this essential observation will not be lost, we have replotted the data in figure 5 of the 1978 article so that an entire furnace cross-section is represented (figure 12.2).

It should be noted that the furnace has eight tuyeres and is symmetrical. Killick laments the lack of isotherms for the furnace but any point on the gradient when rotated about the axis is a circular isotherm. The important temperatures to note in figure 12.2 are those in the reducing zone, points A to B, where temperatures range generally from 1375–1500°C. The advantage of this high temperature zone is that the slag can be reduced in iron concentration to relatively low values, for example, more iron can be removed from the slag. Usually, in a nonpreheating, iron-smelting furnace, reducing conditions can be reached only at temperatures below 1300°C, with the loss of considerable amounts of iron to the slag, which requires that the ores be high quality or upgraded by special processing. A striking attribute of the Haya process is that it utilizes low quality iron ore and still produces substantial blooms.

Killick then again deflects attention from the important features of the original graph (fig. 5, Schmidt and Avery 1978), with an emphasis on the "peak" temperatures (the pointed top of the curve), maintaining that they do not indicate preheating and that Tylecote sustained similar combustion (oxidizing) zone temperatures for long periods in a small Roman shaftlike furnace. We want to point that the "peaks" on the graph *are not* directly inside the oxidizing zone, but just behind it, that is, slightly further inside the furnace. We could not keep a thermocouple in the combustion zone without its melting. The key observation here is that the most important temperatures are those *between* the "peaks," that is, *in the reduction zone in which elevated temperatures are maintained*—an event that cannot occur without preheating. You simply cannot obtain such temperatures *and have them reducing* without preheating.

Killick's reference to the Tylecote paper is unfortunate, for it seriously obfuscates both Tylecote's and our observations. Tylecote did a remarkable piece of work in a simulated shaft bloomery furnace, but it was *very* different from the Haya furnace (Tylecote et al. 1971). The shaft diameter of 1 foot was fitted with a single tuyere 1 in/diam and blown with an industrial blower at 1.3 bar or 18.9 psi at a rate of 350 liters/min, or 38'/sec, 26 mph, 22.5 knots. Furthermore, to economize on charcoal the furnace was preheated with an air gas burner to 1000°C prior to charging and blowing through the tuyere. The ores were commercial ones preroasted in a gas furnace at 800–900°C.

Tylecote also never reached temperatures much above 1200°C in his furnace, and he indicates isotherms of the whole furnace taken at six levels, showing a maximum of 1200°C restricted to a small volume immediately sur-

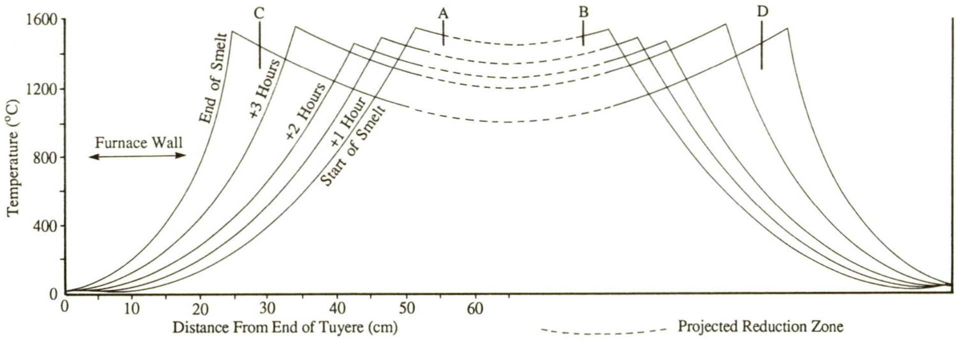

FIGURE 12.2. A schematic drawing of the Haya furnace temperatures over time. This is the extension of one half of a furnace profile taken through one tuyere that was published in 1978 (figure 5). The mirror imaging of the profile, when combined with precise measurements of the furnace dimensions, allows us to show the isotherms with the oxidizing and reducing zones. The most important observation is that the temperatures that prevail in the interior of the furnace are reducing and do not fall below 1200°C until the end of the smelt. This the one of the most important features of the Haya process, a critical innovation that results from preheating and that is overlooked when the higher oxidizing temperatures are the point of focus in critiques such as Killick's.

rounding the tuyere. The only reference to temperatures above 1200°C is the oxidizing zone immediately beyond the tuyere. There is no way one can construe the Tylecote paper as support for furnace temperatures of more than 1500°C. To do so is meretricious.

Similarly, Killick then points to the experimental work of Tholander (1987). This, too, is an argument that emphasizes the oxidizing or blast zone temperatures at the expense of the reduction zone temperatures. For example, in figures 6.1 and 7.3 which Killick cites, it is apparent that the object of his concern is a very restricted, somewhat schematic hot, oxidizing zone of approximately 1500°C in front of the tuyere in a small shaft furnace. Tholander's data cited by Killick are like those of Tylecote. There is again a large, continuous blower pumping 300 liters of air/min. Temperatures of 1500–1600°C are to be expected in such circumstances, but what is more important is that higher in the furnace the reducing zone operates at temperatures considerably lower, approximately 1200°C, exactly what we would expect in conventional furnace without preheating. What, then, is the point that Killick is trying to make by directing our attention to only the oxidizing blast zone?

Killick also notes that the Haya tuyeres are eaten away during the smelt (and the preheating capacity is gradually erased), and asks why the "peak" tempera-

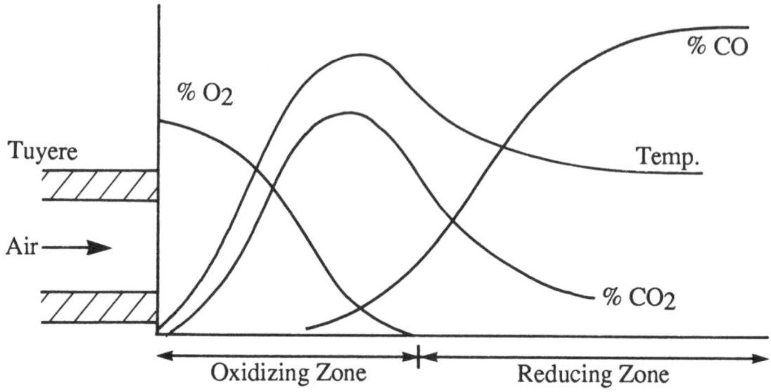

FIGURE 12.3. The schematic drawing published originally in Avery and Schmidt (1979) shows the peaking temperatures immediately in front of the tuyere with a downward curve deeper into the furnace where CO or reducing conditions prevail. It is this zone beyond the temperature peak that is of greater interest in regards to the preheating argument.

tures have not declined correspondingly, a query also made by Rheder. First, the temperature quotients in the *reducing zone decline* over time. This is the important observation. Because the "peak" on each scan is consistently about the same distance from the (oxidizing) combustion zone, this naturally leads to a steady reading over time for the so-called "peaks." We also note that the tuyeres were not neatly shortened by nibbling from the end but were dissolved into the slag, an occurrence that thinned the walls and impregnated the pores in the tuyere, both of which would increase the conductivity in partial compensation for the loss in length.

As to Killick's observation of the Mafa smelting furnace in the Mandara highlands of northern Cameroon, where a large and long tuyere is inserted into the furnace from its top, it is not clear if the one tuyere is the only source of air to the furnace; it appears likely that the one vertical tuyere may be vulnerable to convection processes, which would tend to force air up the tube. There is a peephole at the bottom of the furnace and it may be possible that sufficient air is entering through this aperture to create a convection path up the tuyere. We understand from the published report of this smelt (David et al. 1989) that there were no monitoring instruments used to measure temperatures.

In a later 1989 smelt there was again no attempt to measure temperature in the tuyere, just a horizontal probe that was not located with respect to the tuyere, which melts some 15 in during the course of the smelt. It is curious that someone who has been so concerned with the subtleties of thermocouple errors and the necessity of complete furnace isotherms provides such scanty evidence and commentary on either concern in the Mafa study. It is also not clear if Killick's failure to find an oxidizing zone in the Mafa furnace may be related to the absence of a sufficient detection device or to some observational impediment, such as the bellows attachment. In any event, the tuyere does not seem to have been effective, regardless of the question of preheating. Consequently, the case study does not provide sufficient evidence to comment on preheating.

Finally, Killick reviews evidence for preheating of nineteenth-century American iron smelting in the Adirondack Mountains. He describes the development of a preheating system that used waste heat from the furnace chimney. He then argues that the Haya system is a zero-sum system, which seems to impute that it does not use waste heat. This assertion ignores our clear explanation of how heat is used in the Haya furnace (Avery and Schmidt 1979). For purposes of clarity we refer once again to the figure 5 from that paper, which illustrates how "waste" heat (combustion products) flows back along the tuyere, heating the tuyere walls, a process that is not dissimilar from that in the American system.

Killick's concluding position is that the preheating hypothesis has not been proven. It is curious that he ignores repeated observations that have garnered similar evidence and discounts extensive archaeological evidence (Schmidt and Childs, chapter 9, this volume) pointing to production of iron from within highly refractory slags. As we have demonstrated, there is substantial evidence that indicates preheating, while there has been *no evidence marshaled to disprove the idea* as it is applied to recent Haya furnaces or to those of antiquity in the same region.

It is also necessary to address Killick's sanguine proposition that it is possible to test this hypothesis in a laboratory setting, which is also a point that has been stressed by Rehder. Our answer is essentially the same. An experiment would be a welcome addition and might provide insights, but like Tylecote's furnace it cannot provide a definitive proof. We have been involved with experimental approaches at a variety of universities and in field reconstructions in four different African cultures. Avery and Killick, interestingly, have participated in some of the same smelts and experiments. Field conditions and actual practicing smelters have never failed to challenge preconceived models and laboratory experiments, and to astonish researchers who would have conducted such smelts by very different procedures.

Often the accumulated lore and practice of established smelters have demonstrated competence and utility that greatly exceeded the expectations of academic models. A laboratory approach will never provide a definitive proof, as laboratory conditions can never hope to replicate those of the field. How can you factor in such variables as tuyere failure, wet charcoal/dry charcoal charges, sporadic bellowing as a consequence of fatigue, bellows failure, inebriation, etc.? In sum, how can you account for the fact that there are no controllable variables?

Haya iron smelting can only be studied in the field. This has already been demonstrated as manageable and possible. If Killick does not agree with our interpretations, which are based on sound empirical evidence, and reasonable modeling, then the challenge is his to disprove them. This is unlikely to be done in a university laboratory. To argue that this is a preferred way to continue to test the hypothesis simply perpetuates models that are unrelated to Africa and further contributes to the erasure of an authentic technological history in Africa.

References

Avery, D. H. 1981. "The Iron Bloomery." In *A History of Pyrometallurgy*, ed. T. W. Wertime, 205–15. Washington, D.C.: Smithsonian Institution Press.

Avery, D. H., and P. R. Schmidt. 1979. "A Metallurgical Study of the Iron Bloomery, Particularly as practiced in Buhaya." *Journal of Metals* (October):14–20.

Childs, S. T., and P. R. Schmidt. 1985. "Experimental Iron Smelting: The Genesis of a Hypothesis with Implications for African Prehistory and History." In *African Iron working: Ancient and Traditional*, ed. R. Haaland and P. Shinnie, 121–41. Oslo: Norwegian University Press.

David, N., J. R. Hiemann, D. Killick, and M. Wayman. 1989. "Between Bloomery and Blast Furnace: Mafa Iron Smelting Technology in North Cameroon." *African Archaeological Review* 7:183–208.

Rehder, J. E. 1986. "Use of Preheated Air in Primitive Furnaces: Comment on Views of Avery and Schmidt." *Journal of Field Archaeology* 13:351–53.

Schmidt, P. R., and D. H. Avery. 1978. "Complex Iron Smelting and Prehistoric Culture in Tanzania." *Science* 201:1085–89.

Schmidt, P. R., and S. T. Childs 1985. "Innovation and Industry during the Early Iron Age in East Africa: The KM2 and KM3 Sites of Northwest Tanzania." *African Archaeological Review* 3:53–94.

Tholander, E. 1987. *Experimental Studies on Early Iron-Making*. Stockholm: Royal Institute of Technology.

Tylecote, R. F., J. N. Austin, and A. E. Wraith. 1971. "The Mechanism of the Bloomery Process in Shaft Furnaces." *Journal of the Iron and Steel Institute* 209:342–63.

13 Technological History and Culture in Western Tanzania

S. Terry Childs

The history and prehistory of the peoples occupying the Kagera Region in northwestern Tanzania (map 13.1) attests to long-term and dynamic interrelations among technology, increasing cultural complexity, and changing social relationships. The manufacture and use of iron,[1] in particular, profoundly affected and was affected by local cultural and environmental factors for over two millennia, from Early Iron Age times to the rise of the Haya states in the past several centuries.

Exploring the nature, complexity, and evolution of these relationships in the Kagera Region has been the focus of a major interdisciplinary research effort over the last two decades. The compass of study includes archaeological survey and excavations of ancient iron smelting sites (Schmidt 1975, 1978, 1980, 1981, 1983c, 1993; Schmidt and Childs 1985); collection of Haya oral histories and their structural and symbolic analysis (Schmidt 1978, 1983a, 1983b, 1990); palynological investigations of changing environmental conditions (Schmidt 1981, 1994); ethnoarchaeological, actualistic, experimental, and laboratory studies to understand the complex process of iron production in the region over time and to provide technological models from ethnography to evaluate the prehistoric record (Schmidt and Avery 1978; Schmidt 1983c; Avery and Schmidt 1979; Avery 1982; Childs and Schmidt 1985; Childs 1986, 1988, 1989a, 1989b, 1990). Integrating the findings of such interdisciplinary research is a long process that requires cross-checking, revising, and filling in unanticipated gaps.

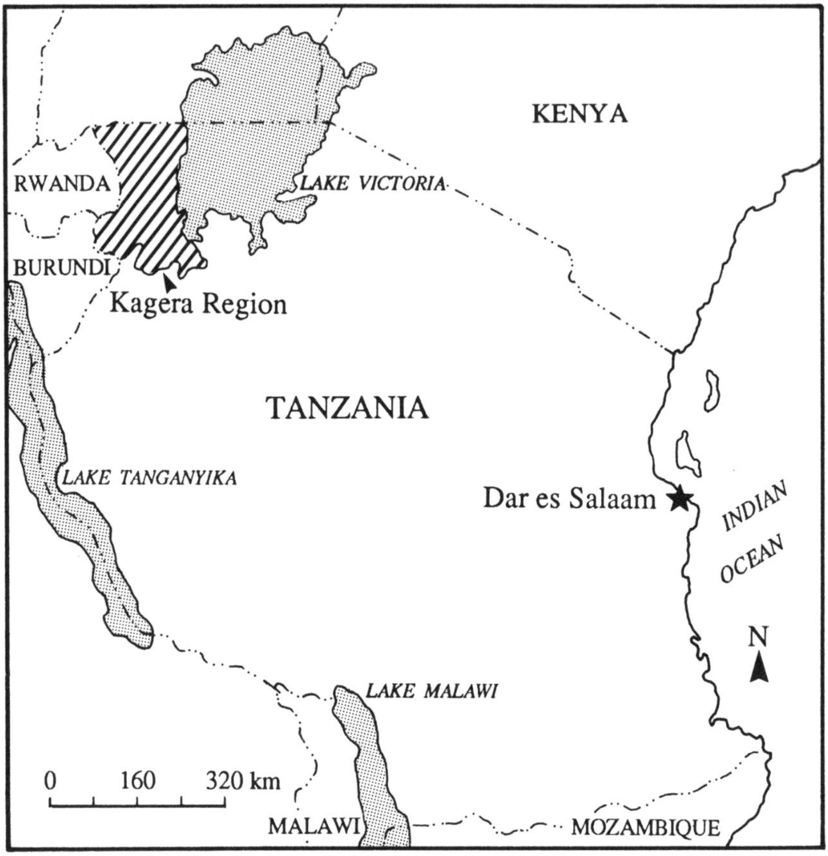

MAP 13.1. Map of the Kagera Region in northwestern Tanzania.

With the ultimate intent of contributing to this integrative process, this chapter has two goals. First, it revises and substantially expands on the existing knowledge of the physical and chemical dynamics of producing iron in the Kagera Region over time by considering some complicated features that have not previously been recognized. This chapter presents a complex smelting process that does not typify either a bloomery or a blast furnace operation, but has some features of both.[2] The process yields a high-phosphorus iron product that presented challenges to the smith during the forging process, both in the Early Iron Age and in more recent times (Schmidt and Childs 1995).

The second purpose of this chapter is to consider whether the chemical and physical reactions produced during the iron-making process and frozen in the smelting and forging products reflect some of the sociocultural and environmental influences on them. This issue relates to the premise that technical and labor decisions and choices of raw materials are intimately linked to the cultural and environmental context in which they are made. Changes in the organization of labor and access to specialized resources over time are indicated through analysis of numerous metallurgical samples from the Kagera Region.

The ability to provide an updated reconstruction of the ancient smelting process, to document the forging techniques, and to examine the evolution of the iron technology in a changing cultural context results from extensive laboratory work that builds upon and incorporates earlier phases of interdisciplinary research. The historical, cultural, and environmental background of the Kagera Region is outlined in the next section, followed, after a short methods section, by a systematic reconstruction of the iron-smelting and smithing processes based on metallographic and chemical analyses. These were conducted on selected slags, blooms,[3] and forged objects excavated from several Early Iron Age furnaces and from a similar range of materials produced during studies of ironworking among the Haya, the current occupants of the Kagera Region. Several traditional Haya blooms, kept as heirlooms, and forged objects were also examined. Comparisons between both sets of data provide a unique view into the regional history of an African technology in its cultural context.

Background: The Development of Iron Making in the Kagera Region

Ironworking was well established along the western shores of Lake Victoria by 300–500 B.C. (Schmidt 1975, 1978). From its inception to the present day, the production and use of iron were interwoven into the economic, sociopolitical, ideological, and ecological fabric of the local culture. Archaeological survey in the Kagera Region indicates that the local population of Early Iron Age farmers, hunters, and fishermen prospered and increased over time (Schmidt 1980). Iron production sites were sometimes located in or very near villages and grew in numbers as settlement expanded and dispersed, suggesting that iron constituted an important component of the productive economy and that the production of food was linked to the manufacture of iron goods (Schmidt 1983b). Efforts to reconstruct the technical operations over time, along with the cultural constraints on iron making, have resulted in some important advances to date.

Despite the probability that iron objects were relatively common in the Early Iron Age, some evidence exists that the technology itself was not practiced by large numbers of people. There is some evidence that ironworking was controlled by part-time specialists who took steps early on to establish themselves as important and distinct in their communities, possibly in ways that had economic and political benefits. One strategy may have involved minimizing the access of the general public and competing work groups to essential craft knowledge and expertise. This might have involved creating long apprenticeships and limiting involvement in iron production to members of particular clans. Although these practices are known for many groups in the ethnographic literature, including the Haya (Cline 1937; Schmidt 1978), they are very difficult to detect archaeologically.

Some of the more important technical information that might have been controlled included knowledge of and access to specialized raw materials. Certain resources could provide immediate benefits to ironworkers through increased iron yields. For example, the tuyeres (pipes through which air was sent to stimulate furnace combustion and heat generation) were made of clay. Highly refractory or heat resistant clays were beneficial, because they helped to generate effective smelting conditions, including high temperatures, yet did not rapidly break down in the heat. Evidence that sources of high-quality tuyere clay were restricted knowledge is that the clays were collected in selected swamps away from the smelting sites, differed from the clays used for other functional needs, such as pottery, and were not used in or available to all the iron-smelting operations studied (Childs 1988, 1989a; Schmidt and Childs 1985).

Other knowledge and practice involved in the iron technology existed outside the technical realm. Good evidence exists that ritual and ideology played significant roles in Early Iron Age iron smelting. Ritual and symbolic behavior may have been invented to complicate and mystify procedure to limit participation in the technology and to limit access to the economic benefits. The furnace superstructures were made of clay bricks, many of them decorated. The effort expended in decorating suggests that a concerted attempt was made to make furnaces meaningful, distinctive, and accessible, but only to those who knew how to understand and use them. The floors of some of the early furnaces also contained a small hole. Ethnographic analogy to several ironworking groups suggests that the holes may have once contained ritual medicines (Schmidt 1981; Schmidt in press; Schmidt and Childs 1985). Such elaboration of the technical process certainly made it more difficult to learn and helped ensure that its practice was confined to and controlled by a relative few who differed from other members of society (Schmidt and Childs 1985).

There is a long period of time, approximately A.D. 600–1300, for which there is little data on population and settlement fluctuations and the relative scale and importance of iron in the Kagera Region. The existing information suggests that ironworking was discontinued in a context of population decline, more restricted settlement near Lake Victoria, and changes in ceramic production (Schmidt 1994).

More changes occurred as sociopolitical complexity evolved and the Haya peoples came under centralized authority. Historical work involving the collection and analysis of oral traditions (Schmidt 1978, 1983b, 1990) provides a unique look into the relations among iron production, the political economy, and the religion of the Haya in the seventeenth through twentieth centuries. Prior to and during the early rise of a centralized state system, Haya ironworkers belonged to indigenous clans, which, independent of royal domination, restricted membership and controlled access to important raw materials (ore, fuel, clay). With an assured resource base, they supplied essential iron for local subsistence and other needs as well as for trade (Schmidt 1983a). The iron-producing clans also controlled the specialized knowledge, including a significant amount of ritual and symbolism, necessary to produce metal from those materials. These ironworkers acquired and enjoyed high social status, wealth, and an independent power base by providing, manipulating, and controlling a critical means of production (iron) for the basic economy (agriculture).

Technical Developments

Early Iron Age iron smelting has been characterized as sophisticated, efficient, and complex through ethnographic analogy and material correlates to reconstructed Haya smelts (Avery and Schmidt 1979; Schmidt 1981; Avery 1982; Schmidt and Avery 1983; Muhly and Schmidt 1988). Preheating is an important technical component that may have contributed to this characterization.[4] Preheating through specialized, refractory clay tuyeres may have offered a means to achieve high smelting temperatures in the brick furnaces *while maintaining critical reducing gases.* These conditions are believed to have led to the production of abundant, relatively slag-free blooms while economizing on fuel consumption. Development of this smelting feature had important ramifications for the local environment and human adaptations to it as discussed below.

The metallurgical accomplishments of the ancient ironworkers were initially revealed by metallographic analysis of a bloom sample excavated from an Early Iron Age furnace and identified as a high-carbon steel (Schmidt 1983c). Utilitarian objects made of such steel would have been durable, strong, and hard. Analysis of an Early Iron Age forged object, however, showed it to be wrought iron (Schmidt 1983c), a soft material that contains less than 0.1 percent carbon, and that wears down with use. These two somewhat conflicting samples have played key roles in interpreting the prehistoric technology. New data will be presented that explain this apparent contradiction and show that the smelting and smithing operations were even more complicated than previously thought.

The smelting technology of the past century or so has been reconstructed by conducting experimental smelts under the supervision and participation of Haya elders. These men were involved in smelting as young men and were smiths as adults (Schmidt and Avery 1978; Avery and Schmidt 1979; O'Neill et al. 1988). When the Early Iron Age and Haya processes are compared, several similarities emerge: furnace size was approximately the same; both ran on forced draft from bellows; both involved some degree of preheating to reach the high requisite temperatures that yielded carbon steel with minimal slag; and both involved a ritual component. Several changes are also clear. The Haya furnace walls were made of large slag lumps held together by termite earth instead of bricks, the slag and earth walls were not decorated, the furnace floor usually lacked a central hole, smelting was conducted in secrecy away from Haya villages, and swamp grasses provided an additional source of fuel along with wood charcoal (Schmidt 1994). The latter change was probably a response to increasing deforestation and labor costs associated with making and transporting charcoal over increasing distances (Schmidt 1981; Schmidt 1994).

The interdisciplinary evidence provided in this overview attests to the iron's importance to economic prosperity and population growth in the Kagera Region for more than two millennia. It further suggests that significant social, political, and religious changes occurred that were strongly linked to a productive yet changing iron technology (Schmidt 1983b). The following discussion considers new laboratory findings that provide a more complete picture of the complex iron-smelting and smithing practices in the region concurrent with the cultural developments. This analysis permits a critical examination and comparison of the productivity and fuel efficiency of the smelting processes, the technical challenges involved, and how they were resolved over time. It also provides some evidence to speculate on whether changes in social, political, and religious relationships significantly affected the mechanics of iron production that are manifested in the smelting and smithing processes.

METHODOLOGY

Iron smelting in bloomery furnaces is a very heterogeneous process. A number of microenvironments involving different chemical and physical reactions evolve during a smelt. This results from localized variation in atmosphere and temperature as a result of constant fluctuations in air flow and pressure from bellowing, the sizes and quantities of fuel and ore, and the rate at which these are put in the furnace and consumed (Childs 1986; Tholander 1987; Avery et al. 1988; Rostoker and Bronson 1990). These localized zones of reaction vary both horizontally and vertically within the furnace and their effects become frozen in the microstructure of the reaction products. It is imperative, then, that a study of iron smelting uses a strategy of rigorous artifact sampling within a furnace to reflect this heterogeneity and to document all the processes that occur during a smelt (Killick et al. 1988). Analysis of one or two slags, often from surface collections, is an inadequate method to reconstruct a complex smelting technology (e.g., Bachmann 1982).

The subjects of the following study are a range of smelting products that include partially reduced pieces of ore, slags, and blooms. Examination of some objects provides insights into the forging technology. The primary laboratory method employed is metallography, by which the microstructures of specially prepared samples are examined under a specialized microscope. This technique, in conjunction with thin section petrography and chemical analysis,[5] provides the means to reconstruct the physical and chemical events that occurred during iron smelting and smithing.

Every effort was made to select similar types and numbers of samples from well-documented, dated contexts for direct comparisons between the Early Iron Age and Haya evidence. Groups of five or six slags, including metal-enriched ones and possible blooms, and pieces of ore that had only begun to react during smelting were chosen from each of three excavated Early Iron Age furnace pits. Furnaces 6 and 34 at the KM2 site were dated (calibrated and reported at a 95 percent confidence level) in the range of A.D. 245–590 and A.D. 65–585, respectively (see Schmidt and Childs 1985 for details on the furnace structures and on the KM2 site in general). Furnace 1 at the NG-5 site is presently undated. Object selection was based on surface morphology and its vertical position in a furnace pit. The same procedure and criteria were used to sample the remains of two Haya iron smelts conducted in 1976 and 1979. Several Haya blooms from early twentieth-century smelts were also examined. The ancient forging process, on the other hand, was reconstructed by analyzing objects (often unrecognizable) excavated from identified contexts. These were compared with several forged objects found in the Haya village of Katuruka and materials produced during a forging demonstration in 1977.

Careful examination of the Haya furnace and the procedure to dismantle it after a smelt provides a warning that it may be impossible to fully compare the ancient and Haya iron-smelting processes, because the complete range of slags were probably not recovered in the ancient furnaces. The Haya furnaces were sampled before any major disturbances of their contents occurred, whereas the ancient furnaces were excavated and sampled after they had been dismantled to remove the blooms. The original positions of many slags were certainly disturbed over time. In addition, Haya furnace pits are approximately the same diameter as those of the Early Iron Age, yet their depths often differ significantly (Schmidt and Childs 1985). Many of the ancient furnaces were also truncated, perhaps by centuries of deep hoeing (Schmidt and Childs 1985).

Given these provisos, three sections follow. The first presents the mechanisms and processes of ore reduction, slag formation, metal formation, and bloom consolidation and carburization during iron smelting in the Kagera Region. The second section documents the forging technology. The physical and chemical mechanisms that operated during Early Iron Age and recent Haya ironworking are compared at both stages. The technological similarities and differences that existed over time are examined in the last section in light of the sociocultural developments outlined above.

The Mechanisms of Iron Smelting in the Kagera Region

The basic components of the iron-smelting operations in the Kagera Region included a furnace structure, clay tuyeres, bellows, iron ore, and fuel. Options existed in the manufacture or selection of these components that could vary according to their physical attributes, location in or around the furnace, and chemical composition. The choices affected the type and quality of the iron and slag products and could change over time. The particular combination of materials and procedures selected by the ancient and Haya iron smelters resulted in operations that can be best characterized as somewhere between those found in a bloomery and a blast furnace. The metal products were dense masses of predominantly carbon steel (0.2–2.0 percent C), along with some cast iron (2.1–4.5% C) and wrought iron (less than 0.1 percent C). All of these types of iron and iron alloy contained significant amounts of phosphorus, an element that affected the entire iron production process from smelting to forging.

The following evidence suggests that many of the same procedural decisions were made over time. First, the furnaces were built so that the combination of the stack and pit created a fairly long reaction zone. Second, the furnace pit was filled with a bed of charcoal logs and cubes by the Early Iron Age workers and a bed of charred grass by the Haya. This bed initially helped to keep the ore and charcoal charge situated high in the stack, where early reduction occurred (Schmidt and Avery 1978). It later provided a medium in which the forming metal underwent critical chemical reactions. Third, both the ancient and Haya ironworkers used ore that was approximately 2 cm^3 in size. Fourth, they used refractory (heat-resistant) clays to make their tuyeres, which they then positioned radially around the furnace.[6] The majority of each pipe length was situated well inside the furnace. Fifth, the ore was roasted prior to smelting (Schmidt and Avery 1978; Schmidt and Childs 1985). The significance of these decisions and evidence of their long-term use will be developed below.

Other procedural decisions, however, changed in fundamental ways between the Early Iron Age and the recent past, particularly the materials used to build the furnace structures (clay bricks to slabs of slag and termite earth daub) and the fuel sources (charcoal lumps and logs to swamp grass and charcoal pieces) (Schmidt in press).

ORE PROCESSING

The evidence to date indicates that similar ores, principally limonite ($FeO\bullet OH\bullet H_2O$) containing goethite ($FeO[OH]$), were used over time in the Kagera Region. These hydrated iron oxides are the accumulated precipitates of iron-bearing minerals weathered by water action. Conditions of long-term, high water levels in Lake Victoria were probably a major factor in the widespread formation of these ores.

Limonite ores are generally regarded as inferior to hematite or magnetite ores because of their lower iron oxide content and the potential problems resulting from associated constituents such as phosphorus (Rostoker and Bronson 1990, 46). They do, however, have several advantages in certain contexts. First, limonites naturally contain enough gangue materials to be self-fluxing. This means that no other materials need to be added to stimulate slag production, a critical stage of the smelting process. Second, limonites are relatively porous. They can be easily broken up into desired sizes prior to smelting and they respond quickly to circulating heat and gases during smelting. Third, their iron content can be increased by a refining or hand-sorting procedure prior to smelting.

From early times onward, the ironworkers of the Kagera Region refined their ore and roasted it prior to smelting although some of the methods changed over time. Ironworkers first broke up the ore into approximately 2 cm^3 pieces and sorted out the undesirable constituents. This process yielded an increase in the overall iron content of the ore, which is evident in the following range of elemental constituents for two ore samples, one prehistoric and one modern: 70–75 percent FeO, 10–15 percent Si_2O_3, 8 percent Al_2O_3, 2–3 percent TiO_2, 0.2–.3 percent MnO, and 0.1–.2 percent P_2O_5.

Several pits filled with reacted ore were excavated at the Early Iron Age smelting sites (Schmidt and Childs 1985, 75). The presence of these pits suggests that the ore was roasted but evidently as an independent procedure prior to smelting (Schmidt and Childs 1985, 75). The Haya, in contrast, combined roasting with smelting (Avery and Schmidt 1979). They laid freshly prepared ore, also approximately 2 cm^3 in size, along the edge of the furnace pit, where it was subjected to a fairly low, steady heat (600–800°C) during smelting. They collected this ore at the end of the smelt and used it in the next operation.

Both refining and roasting caused the ore to dehydrate and crack (figure 13.1). As cracks penetrated the ore, the amount of surface area and iron oxides left exposed increased greatly (Avery and Schmidt 1979). This event was significant because in the presence of sufficient carbon monoxide (CO) gas the exposed iron oxides began to lose their oxygen content and, in some areas, reduced to metal. This began to occur at temperatures only just above 700°C (Blomgren and Tholander 1986, 15–16).

The reduction process and the variety of phases produced during roasting are visible in a metallographic section of Early Iron Age ore. The piece contains fossilized organics that were replaced by iron oxide minerals during ore formation. These structures were reduced by CO gas to wustite (FeO) (figure 13.2) during roasting, while tiny dots of iron metal (Fe) formed along cracks where the reducing gases penetrated effectively (Avery and Schmidt 1979). Other mineral phases also began to develop in the solid state at this time, including some that were rich in titanium and aluminum. The presence of these latter minerals in the partially reduced ores of both the Early Iron Age and recent smelts are testimonies that similar ores were used over time.

FIGURE 13.1. Cracks in limonite ore from Karuyenje, Kagera Region, used in the 1976 experimental smelt. Magnified 50 times.

Roasting potentially served two important functions. First, metal reduction commenced during this process. Although some of the metal formed may have reoxidized at the end of the roast as the gases became oxygen-rich (Avery and Schmidt 1979), the evidence shows that some metal remained and was available for new reactions as smelting began. Second, the Haya roasting process saved a significant amount of charcoal fuel, perhaps as wood resources dwindled (Schmidt 1981), because both roasting and smelting were conducted in one operation.

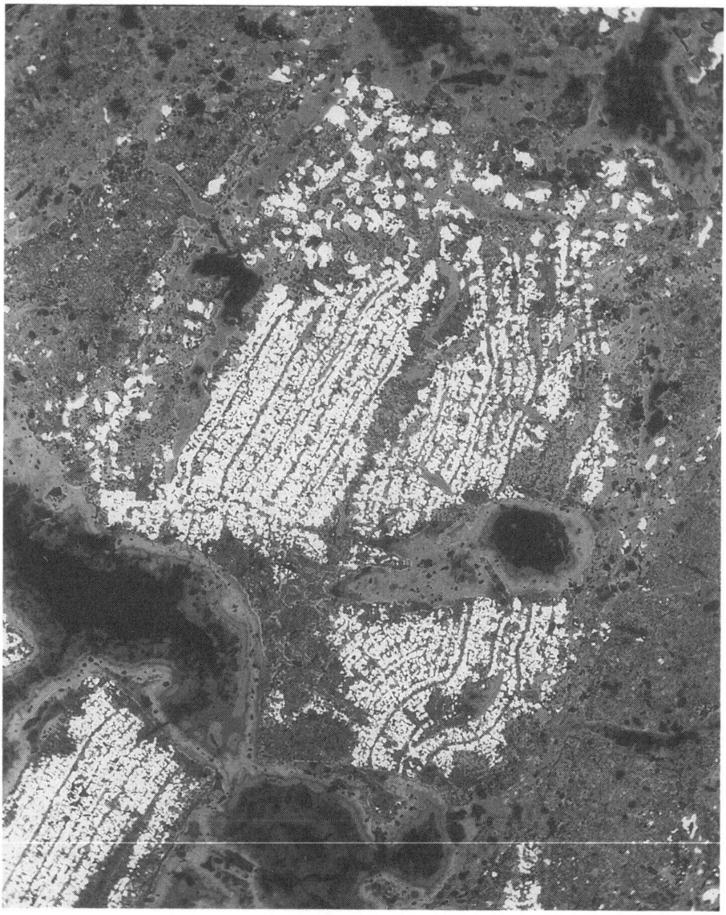

FIGURE 13.2. Reduction to wustite of fossilized organics in a piece of ore from the KM2 site, furnace 34. Magnified 200 times.

Smelting began in earnest after several preparatory steps were taken. First, the ironworkers filled a 50 cm deep furnace pit with a bed of carbonaceous materials, either logs and cubes[7] in antiquity or charred swamp grasses in recent times. The workers placed the tuyeres on this bed in a radial pattern so that only the trumpeted ends were exposed to the air. The smelters then built walls, somewhat over a meter high, over these materials. Once the walls were dry, the workers filled the furnace stack with the roasted, partly reduced ore and charcoal fuel. Finally, the ironworkers lit the charge and started pumping air from bellows into the furnace via the tuyeres to initiate fuel combustion and heat generation. Carbon dioxide (CO_2) was generated as air (O_2) interacted with carbon (C) fuel: $C + O_2 \rightarrow CO_2$. In the presence of excess carbon, reducing CO gas was created: $C + CO_2 \rightarrow 2CO$. Once temperatures reached above 700°C and reducing CO gases were attained in the furnace stack, the process toward metal production recommenced in the roasted ore. Solid-state reduction occurred as the oxygen in the iron oxides was removed by the CO gas to produce wustite (FeO), and with further loss of oxygen, dots of uncarburized iron metal, called ferrite, resulted.

Tholander's (1987) laboratory experiments with four different furnace types from Sweden provide some important comparative data with which to evaluate the initial dynamics of this smelting process. He found that ore no bigger than 3 mm in diam could be smelted in a 1-m high, low-bloomery furnace. Larger ore pieces up to 2 cm^3, however, could be smelted in a high (2.2 m) bloomery furnace and a charcoal blast furnace. Tholander found further that pieces of ore situated in the stack of a blast furnace reacted individually, depending on their size and the local atmospheric and thermal conditions. Larger ore pieces (about 2 cm^3) tended to stay up in the stack longer, where they reduced more fully into a structure with concentric layers.

Analyses of several ancient and modern Haya pieces of partially reduced ore yield two-and three-layered structures similar to the Swedish products as well as to samples from Catalan forges in France (Percy 1864, 303) and isolated pieces from natural draft furnaces in Malawi (Killick 1990, 253). The inner cores consist of wustite and other solid-state components that are surrounded by a middle layer of iron and wustite (figure 13.3). These layers are encased by an outer ring of metal where reduction first occurred. The central core of wustite did not reduce to metal because it was sealed off to CO gas by the outermost layer of metal. Another Haya sample with a thick outer layer of metal, a middle layer of wustite and metal, and a central gap reveals a further step that occurred when pieces fell into hotter areas. Gas pressure built up sufficiently in the core to cause the outer layer to rupture and liquid slag to drain out. A substantial amount of iron was clearly formed in these ore pieces and constituted a significant contribution to the formation of bloom in the furnace pit below.

The burning fuel that produced the reducing gases started to generate higher temperatures in the stack. At approximately 1175°C, the wustite (FeO)

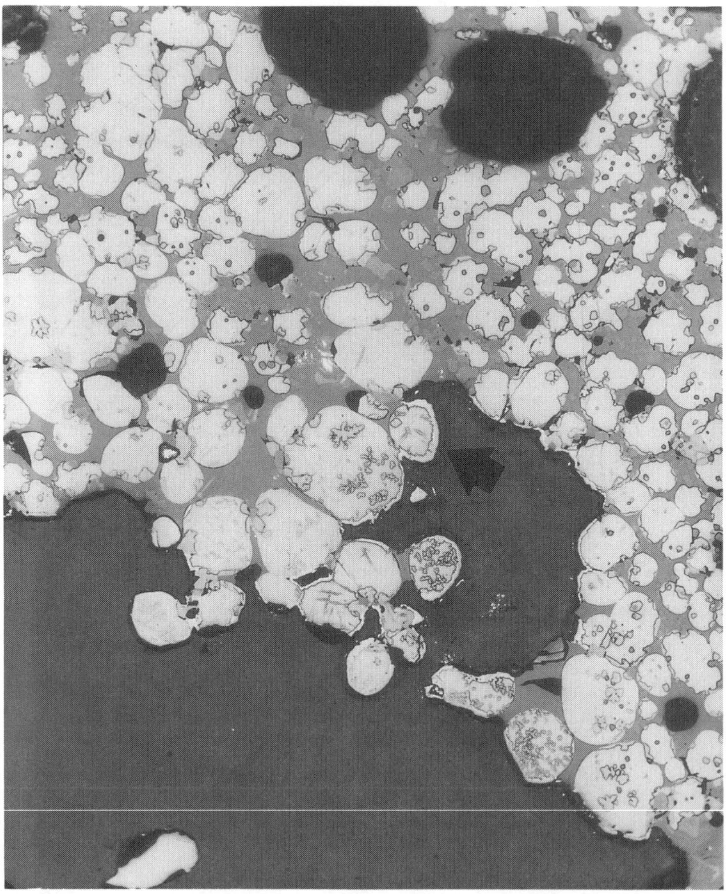

FIGURE 13.3. Wustite reducing to metallic iron around a gas hole (see arrow) in a heavily reduced piece of ore from the 1979 Haya smelt. Magnified 200 times, 2 percent nital etch.

that had not yet reduced to iron and the silicates (SiO_2) in the ore began to react to create slag of the approximate composition of fayalite (Fe_2SiO_4). Reacting within the individual lumps of ore, this slag dissolved the remaining iron oxides and other oxides of calcium, phosphorus, and titanium and enveloped the remaining wustite and bits of metal in the core. It then broke through the layer of metal at the exterior and began its descent through the furnace stack. Some of the iron that formed at the ore exterior may have collapsed into larger masses before it was coated with dripping slag and started to fall through the stack.

The small amount of captured wustite found in the slags from the bottom of a furnace pit testifies to the efficiency with which metal was reduced higher up in the stack (Morton and Wingrove 1972; Tholander 1987). Bloomery slags typically contain large amounts of wustite. Examination of some Early Iron Age (figure 13.4) and Haya (figure 13.5) bottom slags reveals that they contain relatively little wustite, although somewhat more exists in the ancient slags. This occurs, along with the production of more metal, when the ore has a longer time to react to the hot reducing gases before slagging commences. It suggests that the smelting procedure practiced in the Kagera Region over time and particularly in more recent times was more efficient than many bloomery operations in the ancient world (e.g., Morton and Wingrove 1969; Bachmann 1982; Tylecote 1986; Rostoker and Bronson 1990).

The ancient and Haya smelters did several things that contributed to the efficient reduction of metal in the stack and to low wustite[8] production. First, the smelters constructed a furnace with a combined stack and pit height that provided a fairly long reduction zone, about 2 m for the Haya furnaces. Although it is impossible to know the height of the Early Iron Age furnace structures, the intact furnace-pit depths were about the same as the .5 m deep Haya pits. Second, the ironworkers filled the furnace pit with charred fuel that provided a structure to help keep the charge elevated in the stack. Third, they used large pieces of ore that could not fall rapidly through the stack, allowing more time for more metal to reduce in the ore (as opposed to wustite).

The amount of wustite in the slags also helps predict the ease with which the amassing iron might carburize to a carbon steel, the common product of the ancient and Haya smelts. The result of intimate contact between a wustite-rich slag and a carbon-rich iron is the removal of the carbon by the oxygen in the wustite (Rostoker et al. 1989). The low wustite content of the ancient and particularly the Haya slags, therefore, provided an environment conducive to the formation of carbon steels.

Further Metal Formation

Despite the importance of slag formation in the stack, it stopped the further reduction of wustite and metal and halted possible carburization of the metal that had already formed.[9] As the fayalitic slag enveloped the metal and any remaining wustite, it cut off their access to reducing CO gases (Blomgren and Tholander 1986, 153). It also cut off oxygen that might cause a reversal of iron to iron oxide (Avery and Schmidt 1979). New mechanisms[10] that continued metal reduction and consolidation as well as promoted carburization of the formed

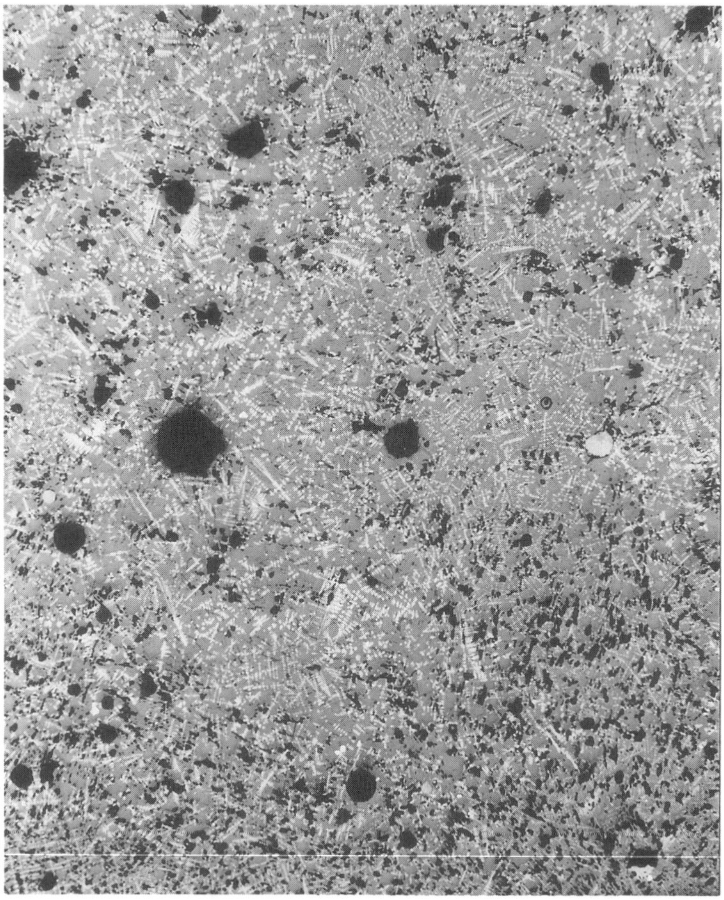

FIGURE 13.4. Typical bottom slag from the KM2 site, furnace 34, with predominantly fayalite (gray matrix), wustite (light gray dendrites), and occasional hercynite and metal droplets. The round black areas are gas holes. Magnified 50 times.

iron commenced as the remaining ore, metal, and slag descended into and through the blast zone, the hottest and most oxidizing area of the furnace.

One important mechanism for the further recovery of iron occurred as the iron oxides dissolved in the slag had intimate contact with the carbon fuel in the furnace stack and below in the furnace pit (Schmidt and Avery 1978; Avery and Schmidt 1979). There is clear metallographic evidence from both the prehistoric and Haya smelts that the slag traveled through and wet the charcoal fuel (both charred wood and grass), thereby exposing the dissolved iron oxides to a new source of carbon. Metal reduced out and formed metallic ribbons next to and aligned with the orientation of the fuel (figure 13.6). Some metal collected where it fell against fixed walls of solid fuel and, on occasion, may have begun to absorb carbon with prolonged exposure to the fuel. Crisp slag casts of the fuel (Schmidt and Childs 1985, figure 15) attest to the degree to which the slag wet the charcoal.

A second critical furnace mechanism was the production of intense heat in the oxidizing blast zone, observed to be 1600–1800°C in Haya smelts (Avery 1982; Schmidt and Avery 1983; Avery and Schmidt, chapter 8, this volume), by vigorous bellowing and preheating in the clay tuyeres. In this environment the refractory tuyere tips slowly began to melt at around 1500°C (Childs 1988, 1989b). Because the prehistoric tuyeres and some local tuyere clays used by Haya ironworkers contained 25–35% Al_2O_3 (Childs 1986), this process introduced considerable amounts of aluminum oxides into the slag. These oxides reacted with iron oxides to yield unusually high amounts of hercynite (FeO Al_2O_3), which formed at relatively high temperatures (figure 13.7). The presence of hercynite increased slag viscosity, thereby slowing down the ability of the slag to quickly fall away from any accumulating metal.

The oxygen-rich heat produced at the ends of the tuyeres also indirectly contributed to rapid and efficient metal reduction and carburization, because the somewhat lower temperature gradients located just above and behind the tuyere mouths, where reducing conditions would prevail, remained very high (Tholander 1987; Rostoker and Bronson 1990). The frequent formation of cubic hercynites around drops of ancient iron and in Haya iron in these areas attests to high furnace temperatures under reducing conditions outside of the blast zone (see Avery and Schmidt, chapter 12, this volume). Solid or liquid metal grains were the only extant bodies in the slag solution and, when temperatures fell to the crystallization point of hercynite (1470°C),[11] the minerals solidified around nuclei of the metal (figure 13.8).

A third mechanism of metal formation, called a carbon boil, involved the recovery of iron from the slag in the zone directly below and beginning some 10–15 cm in front of the tuyeres, which was above the area of bloom formation (Schmidt and Avery 1978; Avery and Schmidt 1979). The mixture of excess carbon from the charcoal and the oxygen expelled from the tuyeres ($C + O_2 \rightarrow CO_2$), along with the reactions between the dissolved iron oxides in the slag and the carbon fuel ($FeO + C \rightarrow Fe + CO$) created abundant CO and CO_2. As these gases bubbled and ruptured through the slag, remaining iron oxides

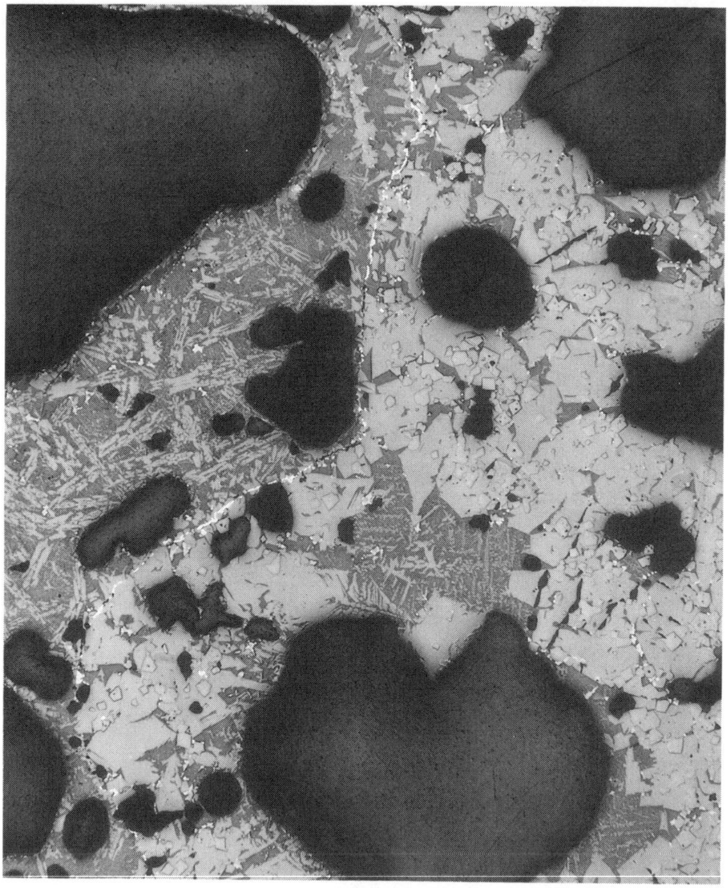

FIGURE 13.5. Typical bottom slag from the 1976 experimental Haya smelt. Predominately fayalite (medium gray) with hercynite (small cubes in relief) and occasional metal droplets (white). The round black areas are gas holes. Magnified 50 times.

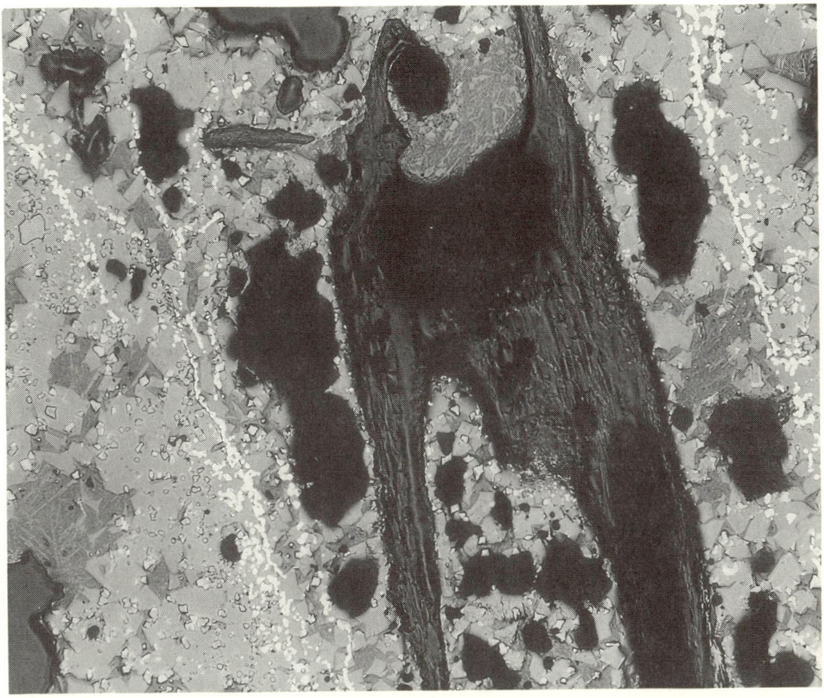

FIGURE 13.6. The reduction of metal (strings of white dots) in the presence of charred grass during the 1976 Haya smelt. Magnified 50 times.

were exposed to the CO gas. This caused iron to reduce out in the slag (figure 13.9) and fall a short distance into the zone of bloom accumulation (Schmidt and Avery 1978; Avery and Schmidt 1979). The slag became increasingly iron-poor, glassy (figure 13.10), and viscous as the iron was efficiently removed from the slag by these three mechanisms.

BLOOM CONSOLIDATION AND CARBURIZATION

The uncarburized, ferritic iron that reduced in the solid state high in the stack or later in the slag fell beneath the central blast zone, where it collected as a bloom. The prehistoric and Haya blooms studied to date, however, were generally well carburized to steel, although heterogeneous and varying a good deal in carbon content. This variation existed because localized furnace conditions allowed different mechanisms to convert the largely ferritic iron to carbon steel and, sometimes, to cast iron.

A critical factor in the carburizing process was the ratio of fuel to ore (Tylecote et al. 1971). Experiments show that a low fuel to ore ratio, such as one to four, yields a reducing atmosphere that can, at best, produce very low carbon steel. A high fuel to ore ratio of two to one results in such a hot, heavily reducing atmosphere that cast iron (2–4.5 percent C) may rapidly develop. The prehistoric and Haya evidence suggests that high fuel to ore ratios were used, certainly greater than one to one. A significant contribution to this ratio was

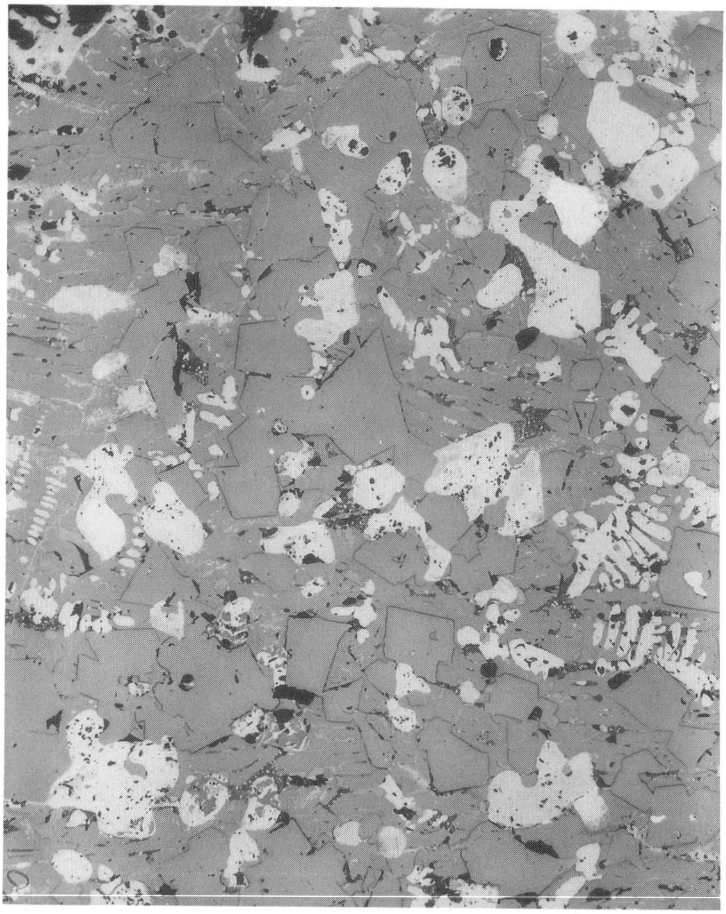

FIGURE 13.7. Early Iron Age slag from the Rugomora Mahe site with abundant hercynite (gray cubes) plus metal (now oxidized and whitish), wustite (whitish and dendritic in shape), fayalite (gray background), and glass (dark gray). Magnified 100 times.

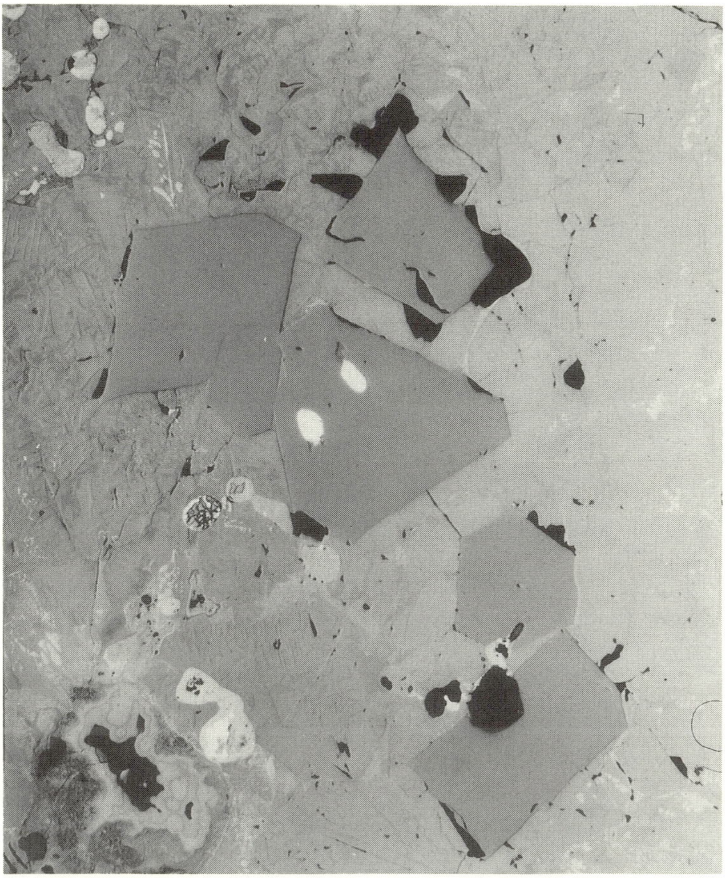

FIGURE 13.8. Prills of metal (white dots) encased in hercynite grains in an Early Iron Age slag from the Rugomora Mahe site. Magnified 100 times.

the abundant charcoal placed in the furnace pit prior to smelting. Although Early Iron Age smelters used charred wood chips or log chunks in the furnace pit and the Haya used charred grasses, both yielded the same results (Schmidt and Childs 1985; Schmidt 1994).

Primary carburization took place in the charcoal bed below and beyond the blast zone. There, direct contact between the metal surface and fuel was possible as the bloom grew into the solid charcoal, carbon slowly diffused into the amassing solid iron, and a low-carbon steel resulted. Sometimes charcoal was trapped in the metal, which provided the essential contact for carbon diffusion (Schmidt and Avery 1978).

Iron smelting in the Kagera Region did not just result in adequately reducing or hot conditions but in quite elevated ones. A mechanism, previously unrecorded for the Haya process, played a dynamic role in bloom consolidation and carburization, and this was the transformation of solid iron grains to liquid. This process occurred in two furnace locations with some different results. In the hot reducing zone just above and behind the tuyere mouths (Tholander 1987; Rostoker and Bronson 1990, 98), two reactions occurred. First, high temperatures reached above the melting temperature of ferrite (1540°C) and

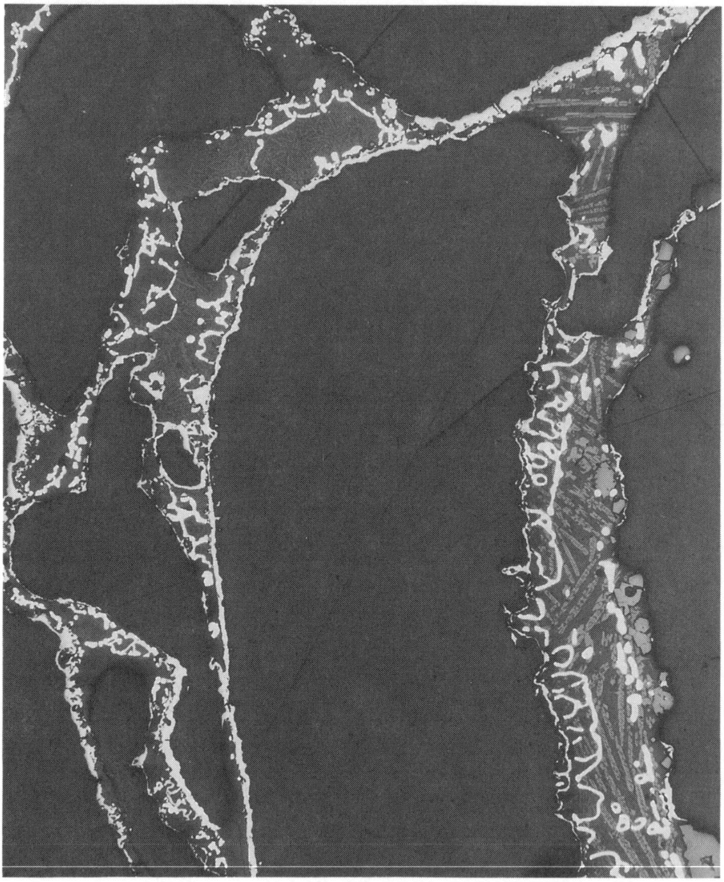

FIGURE 13.9. Shells of fayalitic (gray laths) slag formed by bubbles of CO gas percolating through the slag. Iron (white) has reduced out along the slag edges from contact with CO gas. Some hercynite cubes are also present. Magnified 50 times.

quickly melted the small grains of uncarburized iron caught in the slag. Second, some of the small flakes of iron absorbed large amounts of carbon from the hot CO gases. This interaction lowered the melting temperatures of the now high-carbon alloys and drops of cast iron (figure 13.11) resulted.

The bed of charred fuel in the furnace pit below the blast zone provided the second context in which liquid iron could develop. As the solid iron fell below the fayalite slag[12] and began to coalesce in this bed, the metal absorbed considerable carbon. A high carbon (4.3 percent C)–iron alloy melts at a much lower temperature than pure iron (1147°C versus 1534°C), resulting in a layer or pockets of liquid cast iron at this iron-carbon interface in the furnace pit. As more solid iron accumulated on top of the cast iron, it absorbed carbon from beneath. A layered bloom resulted during the experimental Haya iron smelt in 1979, such as those produced in the highly efficient Japanese tatara process (Rostoker et al. 1989). These blooms consist of cast iron at the bottom, carbon steel in the center, and relatively pure ferrite above, topped with a layer of glassy slag (Schmidt and Childs 1995).

The production of liquid ferrite and cast iron is significant for several reasons. First, the metal coalesces more easily and with less slag inclusions than in the solid state. This promotes the formation of dense, consolidated blooms such as were common to the Haya process. Second, uncarburized liquid iron readily absorbs carbon. Evidence from the ancient and Haya slags and blooms indicates that some drops of liquid iron produced in the high-temperature zone directly behind the tuyere mouths rapidly absorbed carbon from the fuel and the CO gases in the carbon boil below. Drops of cast iron resulted that fell on the amassing bloom and carburized the ferritic iron they encountered (Schmidt and Childs 1995). The heterogeneous distribution of carbon seen in the Haya blooms, often decreasing in concentration from the exterior surface inward, likely was influenced by this process.

Uncarburized liquid iron also absorbs nitrogen and phosphorus. Nitrogen naturally occurs in the furnace gases and is soluble in pure iron at temperatures above 1400°C. Upon cooling, iron nitride needles (Fe_3N) precipitate out in the iron grains (Tholander 1987). Iron nitrides were occasionally found in the Haya blooms in areas where the semiliquid or liquid metal had not absorbed carbon. Although nitrogen could have helped harden iron (Rostoker and Bronson 1990, 23), its contribution was usurped in the ancient and Haya blooms by another element, phosphorus (P).

Phosphorus hardens and strengthens iron when absorbed in small quantities and in somewhat greater amounts significantly affects metal formation.[13] Metallographic and chemical analyses revealed that phosphorus strongly influenced ancient and Haya processes of iron smelting and forging as well as the physical properties of the forged objects discussed in the next section.

The same heavily reducing conditions and high temperatures that promoted carburization in the furnace also encouraged the absorption of available phosphorus. Two sources of phosphorus have been found in the ancient and Haya smelting operations. First, small amounts (0.1–0.2 percent) of the

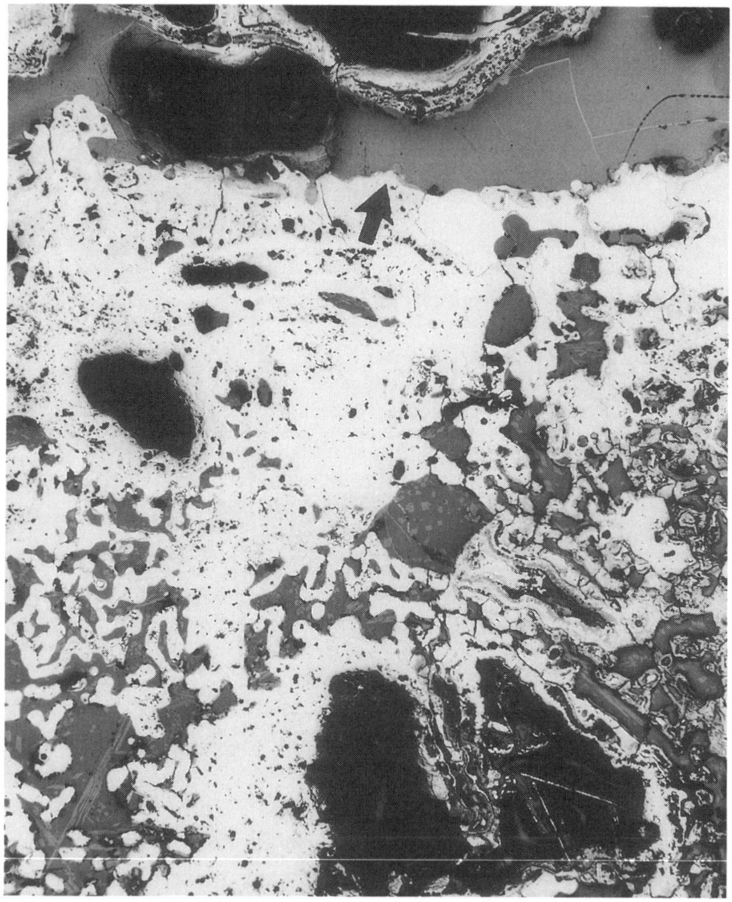

FIGURE 13.10. Piece of slag with abundant metal (white, now corroded) from the Early Iron Age site at Kabale. All the slag (see arrow) is glass, with occasional hercynite and fayalite. Magnified 50 times.

oxide P_2O_5 occurred in the ores. The second, more significant source was the furnace fuels. The Haya used at least two types of fuels during smelting.[14] The wood that comprised the charcoal fuel for roasting the ore and reducing the ore in the furnace stack came from the *Muchewzi* (*Syzygium* guineenis) tree. Its ash contained approximately 0.15 percent P. The ash of the charred grass (*Miscanthidium* violetti) that was packed in the furnace pit and in which the amassing bloom sat averaged 0.34 percent P. Although the charcoals from the ancient smelts have not been analyzed, they probably contained comparable amounts of phosphorus, based on similar characteristics between the ancient and Haya blooms.

Much of the phosphorus present in the charcoal ash or ore probably vaporized by 1000°C (Rostoker and Bronson 1990) and was retained in the strongly reducing atmosphere of the smelt. Other phosphorus was picked up by the slag, especially as it traveled through the grass reeds in the Haya furnace and over the charcoal bed in the Early Iron Age furnaces.[15] Whether the phosphorus existed in the gases, the slag, or the charcoal, it readily combined with ferritic iron and, to some extent, carbon steel as the metal traveled in the slag and was subjected to bubbles of gas in the slag, and as the bloom amassed in the charcoal bed.

As iron absorbed phosphorus, the melting temperature of the iron-phosphorus alloy decreased significantly from that of pure iron or ferrite (1540°C). For example, the iron-iron phosphide eutectic,[16] called steadite, contains 10.5 percent P and melts at 1050°C (Hansen 1958; Stead 1915, 1918). A phosphorus-sensitive etchant applied to several fayalitic slag samples, particularly those from the 1979 Haya smelt, reveals that some drops of liquid iron were phosphorus-rich, forming below the melting temperature of ferrite. These droplets might have caused a heterogeneous distribution of phosphorus in the final bloom product.

A liquid eutectic of iron-carbon-iron phosphide (approximately 2 percent C and 6.9 percent P) forms at the even lower temperature of 953°C (Tholander 1987, 190) and is often found with cast iron. The production of cast iron with this eutectic is clearly demonstrated by the microstructure of the bloom produced in the 1979 experimental Haya smelt (figure 13.12) and of the piece of ancient bloom (figure 13.13) found at the KM3 site (Schmidt 1981). The formation of this cast iron at relatively low temperatures results in a dense, relatively slag-free bloom. It also can significantly alter the overall phosphorus and carbon content of other parts of a bloom. Direct contact between the cast iron and any solid ferrite that accumulates will carburize the ferrite as described above. Phosphoric cast iron is found primarily at the peripheries of several blooms, and its effects penetrate inside (Schmidt and Childs 1995).

The important effects of phosphorus on metal formation are evident in all the blooms studied to date from the Early Iron Age smelts (see, for example, figure 13.13) and the recent experimental smelts performed by Haya elders (figure 13.12), as well as in the traditional Haya blooms (figure 13.14). Areas of fluid cast iron, hypereutectoid steels (greater than 0.8 percent C) interspersed with the liquid iron-iron phosphide phase, and early stages of phosphorus concentration (c. 0.7 percent P) in solid ferrite (figure 13.15) were produced over

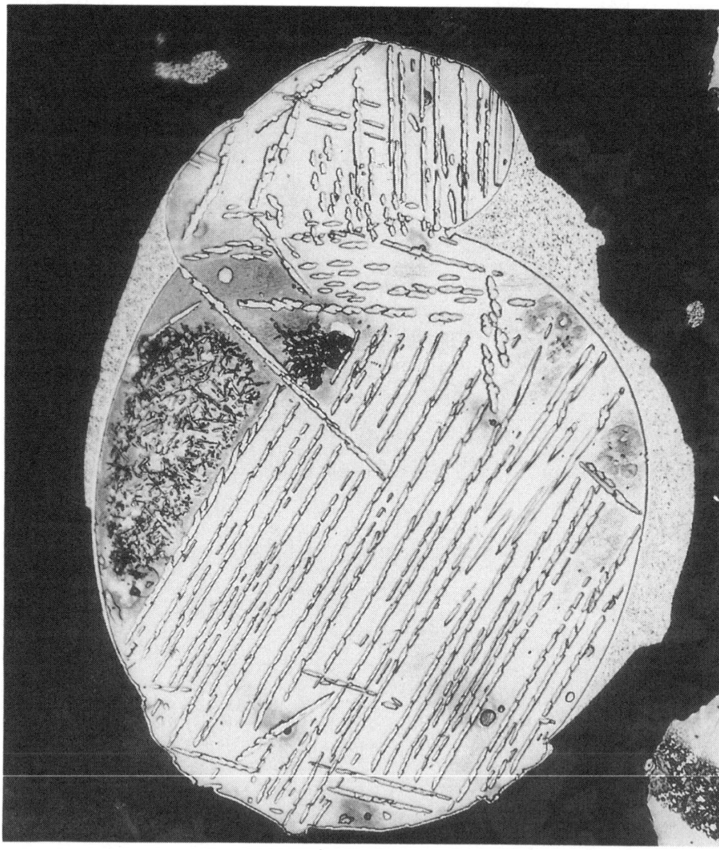

FIGURE 13.11. Prill of cast iron with carbide laths and possible phosphide from the 1979 experimental smelt. Magnified 500 times, 2 percent nital etch.

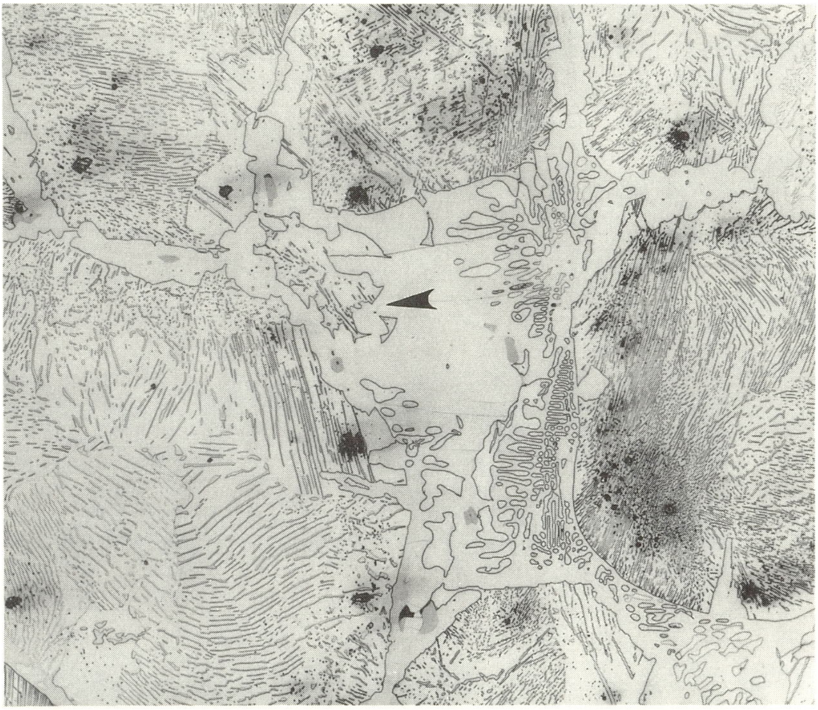

FIGURE 13.12. Eutectic phase of iron-iron phosphide in cast iron from the 1979 smelt. Gray sulfides (see arrow) occur in the phosphorus-rich phase. Magnified 200 times, 2 percent nital etch.

time. The fluid nature of these metals contributed to the efficient formation of large, dense blooms with few slag inclusions, particularly noticeable among the traditional Haya blooms. Phosphorus, in fact, is purposely added during modern cast iron operations to increase fluidity (Rostoker and Bronson 1990, 22). The negative side of these products was their hardness and brittleness. These properties made hammer forging impossible under certain conditions without some adaptations to render the metal workable at the forge.

Forging

Success at forging a usable object from any bloom depends on removing much of the trapped slag and overcoming any physical or chemical problems inherent in the metal. Slag inclusions in an object constitute potential points of weakness where failure might occur, and physical properties such as brittleness can hamper successful forging and use. Because the ancient and recent blooms produced in the Kagera Region were heterogeneous, the smiths had to have the skill to determine their principal constituents (wrought iron, steel, or cast iron) and know how best to forge them.

A few partially corroded objects have been recovered to date from several Early Iron Age sites. Metallographic examination of these provides further information about the skills and techniques of the early ironworkers. First, the objects contain considerable amounts of slag, suggesting that some portions of the original bloom consolidated as a solid, spongy mass of metal and slag that

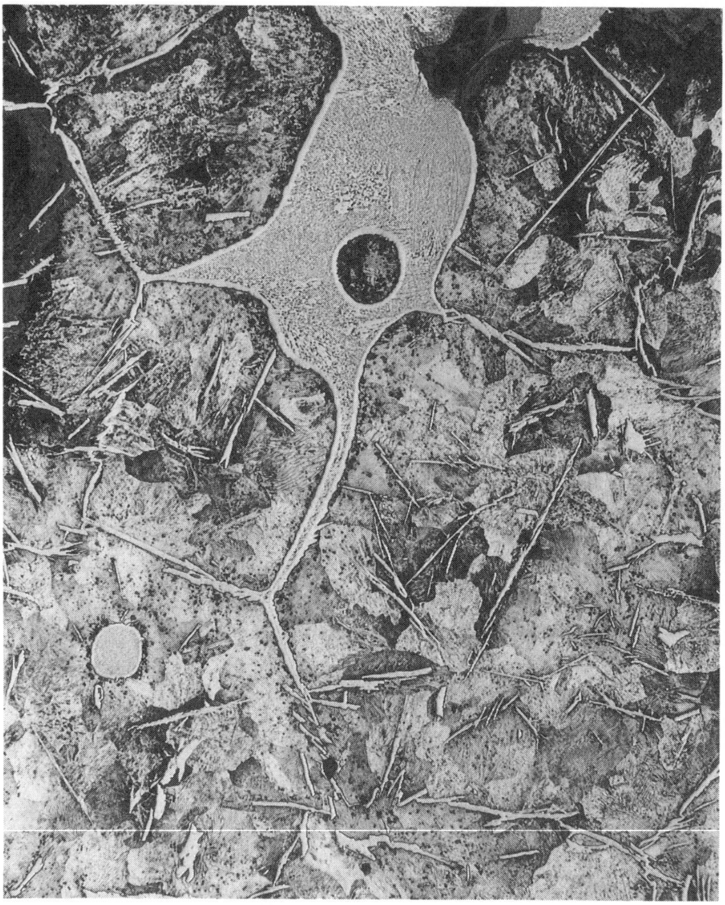

FIGURE 13.13. Iron-carbon-iron phosphide eutectic in a piece of Early Iron Age bloom excavated from furnace 8 at the KM3 site. Magnified 100 times. Etched with Oberhoffer's solution, which shows high phosphorus areas in white.

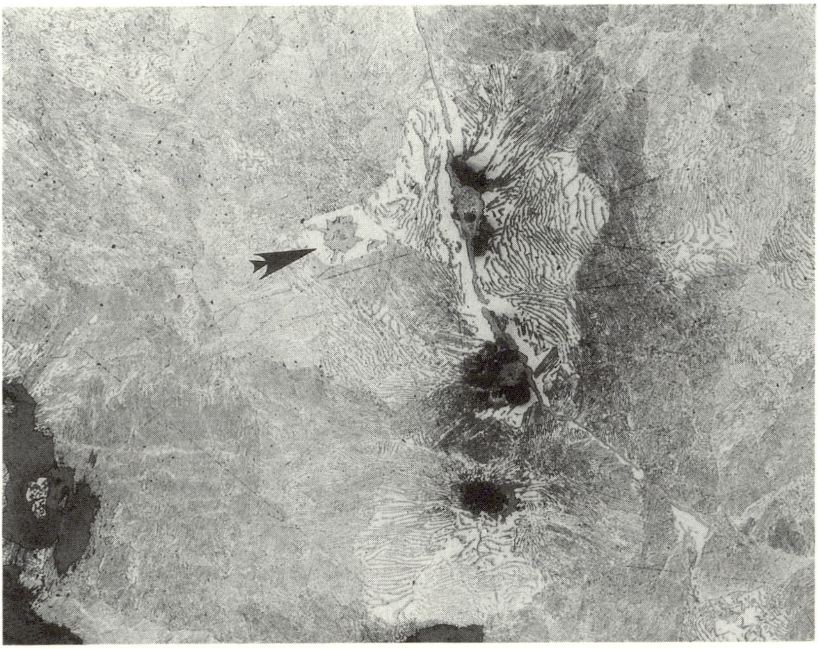

FIGURE 13.14. Areas of iron-iron phosphide eutectic (see arrow) in a traditional bloom from Nyungwe village. Magnified 500 times, 2 percent nital etch.

is typical of standard bloomery products. This evidence stands in sharp contrast with the relatively slag-free cast iron and steels containing liquid phosphorus-rich phases found in the ancient bloom samples. A range of metals clearly was produced in the Early Iron Age furnaces. Many more samples need to be recovered and examined, particularly to determine whether some bloom pieces were smelting discards.

Second, the quantity of slag suggests that either the smiths did not take the time to remove slag inclusions from the objects studied or that they were not skilled at heating the metal to the melting temperature of the slag (c. 1170°C) and removing it by hammering. Careful study of the slag constituents reveals that in two of the three objects examined, slag removal was complicated by hercynite grains trapped in the original bloom (figure 13.16). This mineral could not be expelled from the metal because it is angular and has a high melting point (ca. 1470°C). One object, however, shows that at least some ancient smiths were skilled enough to pressure-weld pieces of metal together by hammering (Schmidt 1983c, figure 12), an operation that generally involves working at and controlling relatively high temperatures (1150–1370°C).

A third characteristic of the objects examined is that they consist of very low-carbon steel or ferrite in stark contrast to the high-carbon steels and cast iron found in the blooms. Two objects were probably made, at least in part, from ferritic bloom that formed in a semiliquid state when it readily absorbed nitrogen. Upon cooling, numerous nitrides (Fe_3N) precipitated out of the metal (figure 13.17). Two of the three objects were also strongly affected by phosphorus, a constant challenge to the smiths. The production of phosphoric cast iron yields an extremely brittle material that cannot be hammer forged. It also embrittles carbon steel so that the metal will crack on impact if

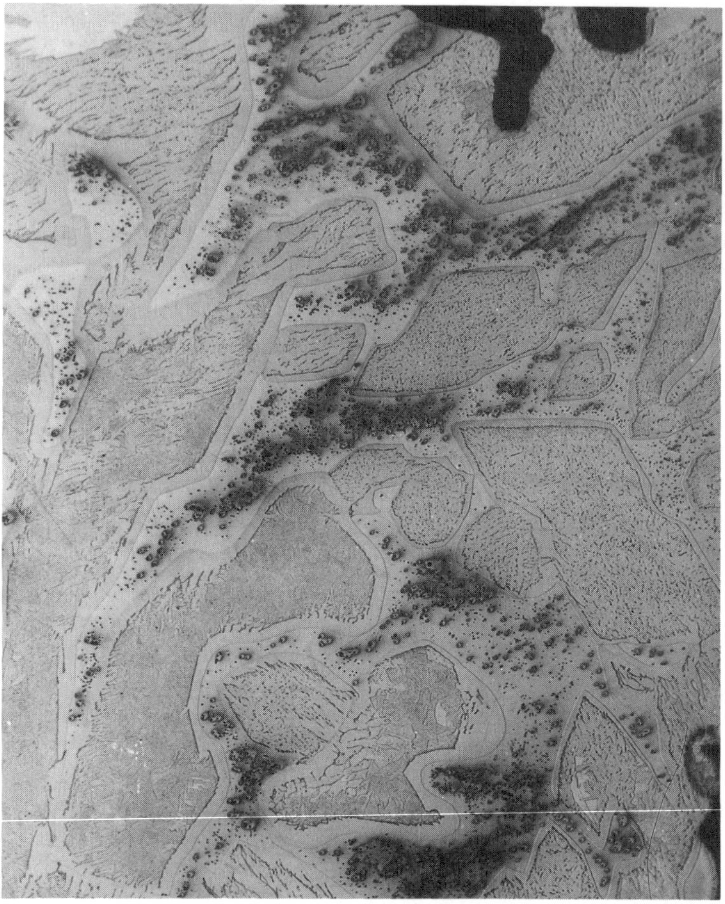

FIGURE 13.15. Areas of elevated phosphorus content in an Early Iron Age bloom piece from the NG-5 site. Magnified 200 times. Etched with Oberhoffer's solution, which shows high phosphorus areas in white and in relief.

not forged at elevated temperatures with care and skill (Goodway 1987; Rostoker and Bronson 1990). In contrast, phosphorus benefits wrought iron or ferrite by strengthening and hardening it almost twice as effectively as equivalent amounts of carbon (Hopkins and Tipler 1958; Tylecote and Gilmore 1986).

The evidence suggests that the smiths from the Early Iron Age onward decarburized the phosphorus-rich steel and cast iron produced in the furnace by exposing it to oxidizing heat in the forge. The result was a malleable yet relatively hard and strong wrought iron. The retention of the phosphorus in the decarburized metal is evident in two ways. First, many of the ferrite grains are veined, showing an internal structure that is sometimes associated with the presence of phosphorus (Curran and Sanford 1933; Tholander 1987; McDonnell 1989, plate 1). Second, areas high in phosphorus are whitened after application of Oberhoffer's etchant to the polished surface of an object. Both Early Iron Age objects have the typical banding pattern of phosphorus segregation (figure 13.17) that occurs as a bloom with a heterogeneous distribution of the element is hammered out (Todd and Charles 1978).

The smithing skills and traditions of the Middle Iron Age can only be assessed from one rusted iron object dated to 1200–1300 A.D. that had a sufficient metal content. The thirteenth- or fourteenth-century blade has the same characteristics as the Early Iron Age objects, including a ferritic structure that consists of strips of welded metal. Many ferrite grains are veined from phosphorus alloying, and the Oberhoffer's etchant reveals clear phosphorus banding.

Similar traditions of forging, including working with phosphorus-rich blooms, continued from the earlier periods to recent times. There is some evidence, however, that the Haya smiths developed some greater sophistication than their predecessors. Their expertise at decarburizing and working phosphoric iron alloys (one averaged 0.5 percent P) and welding is evident from the microscopic examination of six objects. Hercynite, however, remained trapped in the metal on occasion and could not be expelled during forging.

The most notable differences between the earlier and more recent forging skills are apparent in the metallography of a Haya spear holder (figure 13.18) (Schmidt and Childs 1995). The object held the end of a spear handle in one end and was stuck in the ground at its pointed end. Two sections reveal a microstructure that varies from ferrite to high-carbon steel and contains relatively little slag. Evenly distributed through the transitional areas between higher- and lower-carbon steel are numerous spheroids of a high phosphorus phase (figure 13.19). Chemical analysis reveals that these contain 7 percent P by weight, a value that is extremely close to the composition of the iron-carbon-iron phosphide eutectic. The evidence suggests that with some partial decarburization and careful hammering at the forge, this phase was broken up into relic bits and began to be distributed throughout the metal.

Furthermore, the microstructure indicates that the smith knew how to forge phosphoric steels and cast iron without fully decarburizing the bloom and how to prevent the brittle metal from cracking or failing as it cooled. After forging, however, it is likely that the still-brittle phosphoric steel of this spear holder could have cracked if it received hard use, particularly heavy impact. Several of the other objects studied were broken, perhaps for this reason. It is

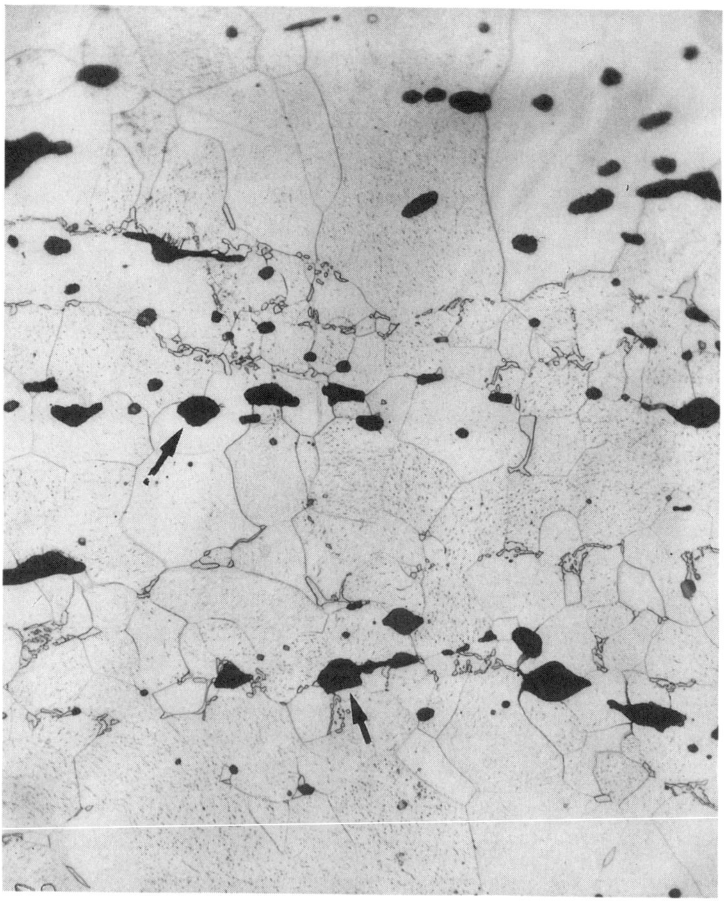

FIGURE 13.16. Ferrite grains, trapped hercynite grains in slag (see arrows) and numerous nitrides in an what is apparently an iron rod from the Rugomora Mahe site, feature 13. Magnified 50 times, 2 percent nital etch.

likely that the spear holder was used for ritual or ceremony in contexts involving minimal impact (Schmidt 1978, 258), so its problematic physical properties were of little consequence. This observation leads to a testable hypothesis that some Haya smiths selected particular types of metal (with or without high phosphorus and carbon) and decided on the degree to which they needed to decarburize a bloom based on the intended function (utilitarian or nonutilitarian) of an object. Metallography of many more objects is necessary to test this hypothesis.

To emphasize the difficulties involved in forging phosphorus-rich iron, evidence exists to suggest that Haya smiths have lost some of this expertise since they began forging scrap steel. An accomplished smith forged a sicklelike blade from a bloom produced during the 1976 experimental smelt. Metallographic analysis of a portion of the bloom reveals that it consists of variable ferrite and low-carbon steel thinly coated by a glassy, hercynite-rich slag. Pockets of the liquid iron-iron phosphide phase and areas of initial phosphorus infusion, in the form of tiny dots, are also present. Interestingly, the smith placed the bloom under the oxidizing blast of the forge tuyere for some forty minutes before he began to forge (Avery and Schmidt 1979). This action may have been stimulated by tradition and memory, because it could not affect the chemistry of the already very low carbon iron.

Examination of a sample from the blade reveals (1) a primarily ferritic structure with one area of low carbon steel; (2) an uneven distribution of phosphorus that reflects its occurrence in the bloom; (3) evidence of pressure welding; and (4) several cracks. When an object is hot-forged, such as is required during pressure welding, phosphorus does not have the same embrittling effect on iron as it does at lower temperatures. Cracks in the blades did occur in several places. It is possible that the smith continued to hammer the metal as it cooled, thereby cold-working it, at which time the phosphorus present did have a negative effect. The white band in figure 13.20 shows where the phosphorus is concentrated,[17] and how its embrittling effect may have contributed to the crack above it (Schmidt and Childs 1995).

Discussion

The laboratory analysis of the ancient and Haya ores, slags, and blooms reveal long-standing continuities in the mechanisms of iron smelting in the Kagera Region. These continuities include the use of large pieces of ore approximately 2 cm^3 in size, the practice of packing the furnace pit with fuel, placement of refractory tuyeres well inside the furnace resulting in high temperatures and hercynite-rich slags, and the important contribution of carbon and phosphorus to bloom consolidation and carburization, primarily from the charred fuel in the furnace pit. The smiths' responses to the phosphoric irons, steels, and cast iron droplets produced in the furnace were also similar, although some increasing sophistication is suggested over time. Given these observations, it is now appropriate to consider if this metallurgical record reflects the changing social relations that supported and helped structure the iron technology.

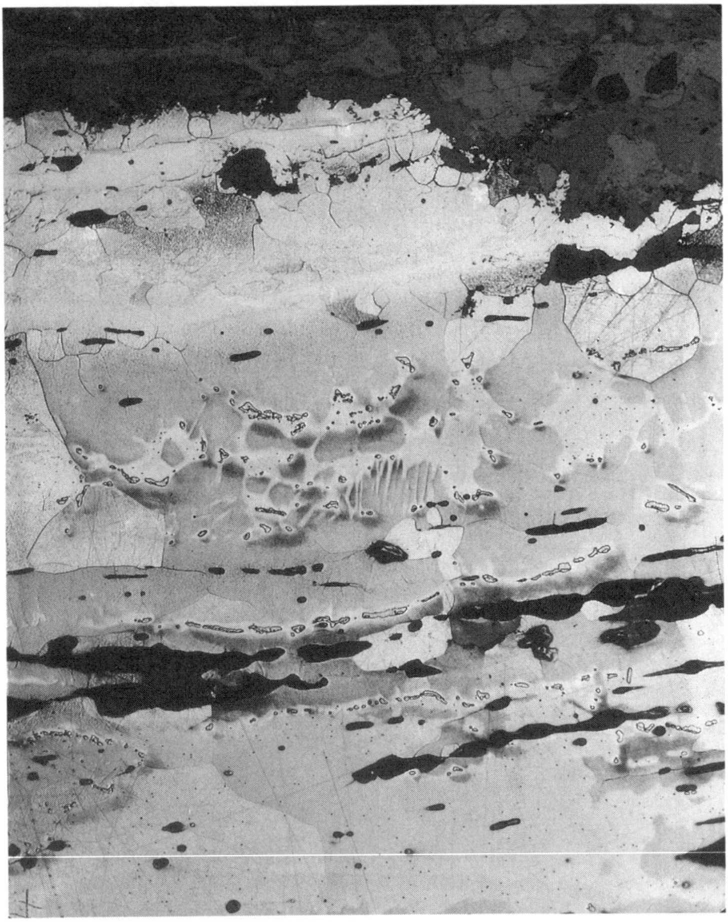

FIGURE 13.17. Phosphorus banding in an Early Iron Age object from the KM2 site, feature 8. Magnified 200 times, Oberhoffer's etch.

FIGURE 13.18. A spear shaft holder from Katuruka village, Kagera Region.

The evidence, admittedly somewhat speculative, indicates that two social variables may have had a considerable and, possibly, changing impact on the technical success of iron making in the Kagera Region. These variables, both related to sociopolitical organization, are the ability to mobilize labor and the ability to control access to specialized resources and knowledge. Some metallurgical data supports the historical record that the organization of ironworking changed from a limited pool of individuals who shared membership in a clan during the Early Iron Age to expanded, less clan-aligned participation among the Haya.

Several pieces of evidence suggest that Early Iron Age iron smelting was organized around competitive family or clan groups, a format that likely encouraged a degree of secrecy and limited the amount of labor available for the arduous operations. First, the metallography of several forged objects and smelted blooms indicates that the original blooms varied considerably from a spongy mass of ferrite, low-carbon steel, and slag to a dense, phosphoric cast iron. Although this variability created significant challenges for the smiths (to which they successfully responded), it also reflects the unpredictable nature of the smelting process. One way to achieve greater consistency might have been to mobilize a large work force that, by continually rotating turns at the bellows, could maintain more uniform thermal and atmospheric conditions in the furnace. Instead, responses to the lack of predictability seem to have been to decorate the furnace exteriors and place substances in a central hole in the furnace floor. The fact that considerable variability existed between sites and over time in the use of these ideologically based furnace features (Schmidt and Childs 1985) suggests that different work groups, perhaps clan-based, applied their own combination of strategies toward this problem.

Another plausible indication that iron smelting was practiced by independent clan-based groups involves resource selection. The refractory clays used to make tuyeres, in particular, were important for an efficient, productive smelt because a clay with a low melting point would have diminished successful preheating and created considerably more slag, to the detriment of metal formation. Petrographic and chemical evidence shows that specialized clays were consistently used for tuyere-making and that some groups, even within a 1 km area, used and had access to much higher quality tuyere clays than others. Differences in the quantity and quality of the resulting metal product would have been significant and, quite likely, conferred some economic advantage to the group that had access to the best clay (Childs 1988, 1989a).

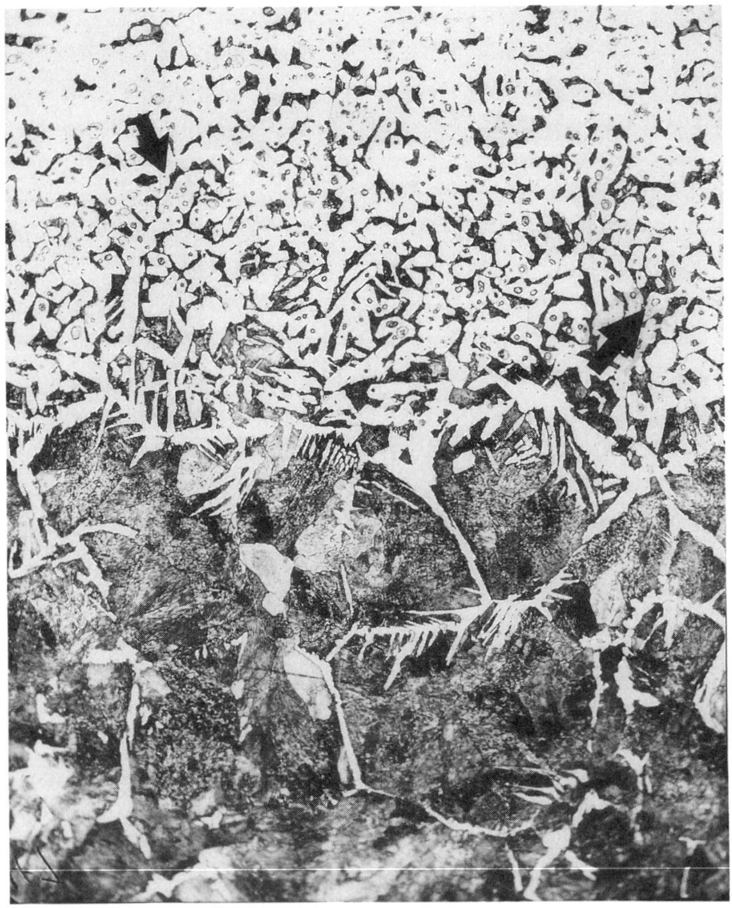

FIGURE 13.19. Tiny spheroids of a high phosphorus phase (see arrow) in an intermediate zone between medium- and low-carbon steels of the spear shaft holder. Magnified 200 times, 2 percent nital etch.

Between the Early Iron Age and the historic period of the Haya, several obvious changes occurred in the process of iron smelting. First, different materials were used to construct the furnace walls. This change did not affect the internal dynamics of smelting; nor did the decisions to abandon furnace wall decorations or the use of a central hole in the furnace floor. Another change involved the source of the charred fuel placed in the furnace pit. Wood charcoal was used in ancient times, whereas the Haya used charred swamp grasses (Schmidt 1994). Palynological evidence suggests that this resulted from the pressures of increasing deforestation as the iron technology grew (Schmidt 1984), yet the change did not affect the important reactions within the furnace involving phosphorus and carbon.

Some metallographic evidence, however, may reflect the effect on the technology of sociopolitical and economic changes under a centralized authority in recent centuries. These changes relate to the availability of an increased labor pool among the Haya. Given remarkable similarities in most of the slags and blooms found in the ancient and Haya furnaces, the fact that less wustite was found trapped in the bottom slags of the Haya furnaces may be significant. A critical factor in efficient metal production, given that a similar size and type of ore was used over time, was the length of time that the ore reduced in the furnace stack before slagging commenced (Tholander 1987). The Haya may have built taller furnaces than their predecessors, which, combined with the high temperatures that were generated, resulted in minimal loss of wustite to the slag.

Furthermore, construction of a higher stack meant that more ore and fuel could be accommodated and more tuyeres might have been used to process this load. Greater numbers of tuyeres, when properly operated, could have generated higher yields of metal with a more predictable carbon and phosphorus content. This is because, as more uniform and high temperatures were produced in front of the tuyeres, increased areas of heavily reducing conditions and only somewhat lower temperatures were created behind them. In the latter areas, some liquid iron and cast iron formed and fell to coalesce in the charcoal bed below. The large, dense traditional Haya blooms with virtually no entrapped slag are testimonies to this process.

The operation of every tuyere required a bellows operator and others to provide periodic relief from pumping. It is impossible at this time to estimate the number of tuyeres and people involved in Early Iron Age smelting in the Kagera Region, although van Grunderbeek et al. (1983) estimate that three to four tuyeres were used in similar prehistoric furnaces in Rwanda. The reconstructed Haya smelts involved eight bellowers (Schmidt and Avery 1978), possibly twice the Early Iron Age number if the Rwanda case provides a viable model for the Kagera Region. It is plausible that eight Haya bellow operators reflects greater availability of labor under a centralized authority, in contrast to greater control over ironworking by clans in earlier times. The associated increase in productivity could have had important economic ramifications, particularly for trade and for stocking the royal coffers with iron for redistribution.

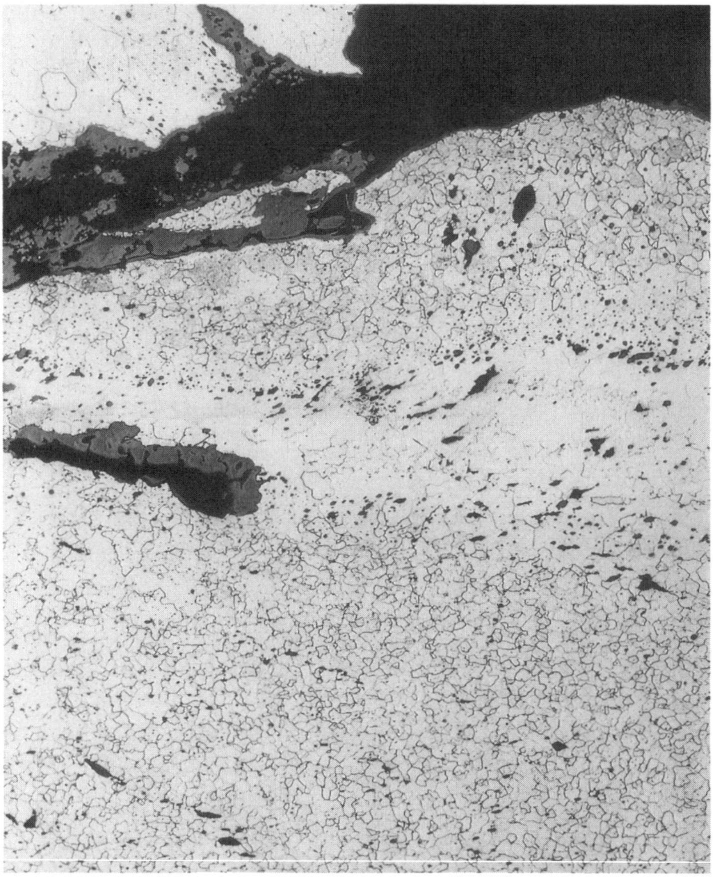

FIGURE 13.20. An area of ferrite in a sickle forged from a 1976 bloom smelted by Haya elders. The central area of white has a higher phosphorus content and may have conferred some brittleness to the metal, leading to the crack above it, if cold worked. Magnified 100 times, 2 percent nital etch.

The evidence from the forging technology is less telling about the social relationships that helped structure it. Although evidence of increasing sophistication and skill in forging exists, even the earliest smiths knew how to cope with difficult phosphorus-rich cast irons and steels to make strong, durable wrought iron objects. Perhaps as the inventory of iron objects increased to include numerous utilitarian objects and items associated with status, prestige, and ritual, the skills of the smiths adapted accordingly. A smith might have learned to form carefully medium- and high-carbon steels containing phosphorus into special objects that he knew would not be subjected to significant physical impact. This would have allowed him to increase his productivity at the forge by spending less time decarburizing blooms.

The iron-smelting and forging processes developed and practiced in the Kagera Region were complex and presented a number of challenges to their practitioners. Because many of the same challenges occurred over time, the basic knowledge was there to meet them. When clans controlled the labor pool, access to resources, and access to knowledge necessary for iron smelting, considerable variability existed in specific smelting procedures and the final bloom products. With the development of centralized authority and the at least partial liberation of iron production from clan control, more people could participate, resources and information were more accessible, and the yield of metal was greater and more consistent (Schmidt 1983b). Only recently, under pressures of colonialism and local competition, have the smiths begun to lose their knowledge of the most effective ways to forge high-phosphorus iron, steel, and cast iron as scrap metal has replaced bloomery iron at the forge.

Conclusions

Many critical choices and decisions made during ironworking are reflected in the microstructure and chemistry of the products. Laboratory studies involving metallography and chemistry provide unique glimpses into the traditions and innovations that ironworkers have developed within specific cultural and environmental contexts. These research techniques also provide a corpus of data that can be compared and used to study and explain the variation that once existed in ironworking techniques around Africa and the world.

This study of the ores, slags, and blooms used and produced by ironworkers for more than 2000 years in the Kagera Region of Tanzania contributes new data to an interdisciplinary project that has documented the importance of ironworking within a changing sociocultural context. This chapter also constitutes an attempt to further the understanding of metallurgical innovation and variation in Africa by elucidating the processes by which phosphorus-rich iron, steels, and cast irons were produced over time, the challenges they presented, and the responses developed.

Notes

This laboratory project would not have succeeded without the support and encouragement of Peter Schmidt, the principal investigator of the Kagera project, and to him I give my greatest thanks. The work was conducted in the laboratory of the Center for Materials Research in Archaeology and Ethnology at MIT, directed by Heather Lechtman, and at the Laboratory of African Metallurgy at the University of Florida. Special thanks also go to the late Edward Barry, for assistance with the chemical analyses, and to David Killick for his comments on an earlier draft of this paper. I take full responsibility for the contents, however. Permission to conduct research in Kagera Region was granted to Peter Schmidt by the Division of Antiquities and the TNSRC, Tanzania.

1 Iron is used in the text as a generic term for all iron-based metals, including wrought iron (less than 0.1 percent carbon), the various grades of steels, and cast iron. When particular types of iron and iron alloys are discussed in specific contexts, their appropriate terms will be used.
2 In a typical bloomery operation, iron metal is smelted from ore at temperatures below the melting point of iron, whereas a blast furnace operation occurs under thermal and atmospheric conditions that yield a liquid cast iron. This characterization of in-betweenness has been made for the Mafa iron smelting operation in northern Cameroon (David et al. 1989). The particulars of the Tanzania and Cameroon processes are different.
3 A bloom is a mixture of metal and slag waste products that results from the reduction of iron ore in a smelting furnace. A bloom must then be forged by a smith to render recognizable objects.
4 Preheating involves placing the tuyeres or blowpipes inside the furnace so that ambient air is heated as it passes into and through them. When this air enters the furnace, it contributes extra heat to the furnace operation but does not jeopardize the requisite reducing conditions or require more fuel. The phenomenon is controversial, however, and the discussion and details are considerable (Schmidt and Avery 1978, 1983; Avery and Schmidt 1979, 1986, this volume; Schmidt 1981; Avery 1982; Rehder 1986; Eggert 1987; Killick 1988, this volume).

5 Most of the chemical analysis was performed by a KEVEX energy dispersive x-ray spectrometer and a KEVEX Quantex-Ray Micro x/7000 data acquisition and reduction system attached to an AMRAY 1200B scanning electron microscope, although some work was conducted by electron probe microscopy.

6 The exact number of tuyeres used by Early Iron Age iron smelters is presently unknown due to lack of evidence.

7 It is impossible to know if these were charred before each smelt began.

8 Some of the FeO may have combined with the aluminum oxides (Al_2O_3) in the ore to produce hercynite ($FeO{\bullet}Al_2O_3$). Because neither wustite nor hercynite occurs frequently in the bottom slags, this reaction was probably of little significance in the early stages of the smelt.

9 Although some of this discussion has been presented previously (Avery and Schmidt 1979; Avery 1982; Schmidt 1981; Schmidt and Avery 1983; Avery et al. 1988), it is reviewed here for two reasons. First, metallographic evidence now exists for these reactions in both the Early Iron Age and Haya smelts. Second and more important, several previously unidentified processes were critical to the long-term tradition of iron smelting in the Kagera Region.

10 It is impossible to give primacy to any of the mechanisms to be discussed because they occur in different parts of the furnace and under varying conditions.

11 This temperature applies only to pure hercynite, because any addition of Fe^{2+} will lower the melting temperature. Only microprobe analysis can provide the data to estimate correctly the crystallization temperature of the mineral.

12 Very little wustite, which would have indicated elevated oxygen and a minimal chance of carburization, has been found in the slags adjacent to the blooms studied.

13 There are a number of excellent discussions about the effects of phosphorus on iron and iron information, including Stead 1915, Lorig and Krause 1936, Todd and Charles 1978, Ehrenreich 1985, Goodway 1987, Tholander 1987, Piaskowski 1989, and Rostoker and Bronson 1990.

14 Four samples of each fuel type were chemically analyzed. It is possible that more significant amounts of phosphorus than the analyses indicate came from the tree bark, where phosphorus is known to concentrate (Tylecote 1986, 223).

15 The slags from the 1976 smelt average 0.29 percent P_2O_5, while those from the 1979 smelt average 0.49 percent P_2O_5. The Early Iron Age slags average 0.25 percent P_2O_5.

16 A eutectic is an alloy whose proportioned constituents solidify at the lowest possible temperature, usually significantly lower than the melting points of each individual element.

17 The presence of phosphorus in this area was clearly delineated by using the phosphorus-sensitive Oberhoffer's etchant.

References

Avery, D. 1982. "The Iron Bloomery." In *Early Pyrotechnology*, ed. T. Wertime and S. Wertime, 205–14. Washington, D.C.: Smithsonian Institution Press.

Avery, D., and P. Schmidt. 1979. "A Metallurgical Study of the Iron Bloomery, Particularly as Practiced in Buhaya." *Journal of Metals* 31:14–20.

———. 1986. "The Use of Preheated Air in Ancient and Recent African Iron Smelting Furnaces: A Reply to Rehder." *Journal of Field Archaeology* 13:354–57.

Avery, D., N. van der Merwe, and S. Saitowitz. 1988. "The Metallurgy of the Iron Bloomery in Africa." In *The Beginning of the Use of Metals and Alloys*, ed. R. Maddin, 261–82. Cambridge, Mass.: MIT Press.

Bachmann, H-G. 1982. *The Identification of Slags from Archaeological Sites*. University of London Institute of Archaeology, Occasional Paper No. 6. London: Institute of Archaeology.

Blomgren, S., and E. Tholander. 1986. "Influence of the Ore Smelting Course on the Slag Microstructures at Early Ironmaking, Usable as Identification Basis for the Furnace Process Employed." *Scandinavian Journal of Metallurgy* 15:151–60.

Childs, S. T. 1986. "Style in Technology: A View of African Early Iron Age Iron Smelting through its Refractory Ceramics." Ph.D. diss., Boston University.

———. 1988. "Clay Resource Specialization in Ancient Tanzania: Implications for Cultural Process." In *Ceramic Ecology Revisited: The Technology and Socioeconomics of Pottery*, ed. C. Kolb, 1–31. Oxford: British Archaeological Reports International Series 436(ii).

———. 1989a. "Clays to Artifacts: Resource Selection in African Early Iron Age Iron Smelting Technologies." In *Pottery Technology: Ideas and Approaches*, ed. G. Bronitsky, 139–64. Boulder, Colo.: Westview Press.

———. 1989b. "Petrographic Analysis of Archaeological Ceramics." *Materials Research Society Bulletin* 14:24–29.

———. 1990. "Refractory Ceramics and Iron Smelting in East Africa." *Journal of Metals* 42:36–38.

Childs, S. T., and P. Schmidt. 1985. "Experimental Iron Smelting: The Genesis of a Hypothesis with Implications for African Prehistory and History." In *African Iron Working: Ancient and Traditional*, ed. R. Haaland and P. Shinnie, 121–41. Oslo: Norwegian University Press.

Cline, W. 1937. *Mining and Metallurgy in Negro Africa*. General Series of Anthropology, No. 5. Menasha, Wisc.: George Banta Publishing Co.

Curran, J., and E. Sanford. 1933. "Wrought Iron—Fact and Opinion." *Metals and Alloys* 4:1–5.

David, N., R. Heinmann, D. Killick, and M. Wayman. 1989. "Between Bloomery and Blast Furnace: Mafa Iron-Smelting Technology in North Cameroon." *African Archaeological Review* 7:183–208.

Eggert, M. 1987. "On the Alleged Complexity of Early and Recent Iron Smelting in

Africa: Further Comments on the Preheating Hypothesis." *Journal of Field Archaeology* 14:377–81.

Ehrenreich, R. 1985. *Trade, Technology and the Ironworking Community in the Iron Age of Southern Britain.* Oxford: British Archaeological Reports, British Series No. 144.

Goodway, M. 1987. "Phosphorus in Antique Music Wire." *Science* 236:927–32.

Hansen, M. 1958. *Constitution of Binary Alloys.* New York: McGraw-Hill.

Hopkins, B., and H. Tipler. 1958. "The Effect of Phosphorus on the Notch-Impact Properties of High Purity Iron and Iron-carbon Alloys." *Journal of the Iron and Steel Institute* 188:218–37.

Killick, D. 1988. "A Comparative Perspective on African Ironworking Technologies." Paper presented at the 1988 African Studies Association meetings, Chicago.

——. 1990. "Technology in its Social Setting: Bloomery Iron-Smelting at Kasungu, Malawi, 1860–1940." Ph.D. diss., Yale University.

Killick, D., N. van der Merwe, R. Gordon, and D. Grébénart. 1988. "Reassessment of the Evidence for Early Metallurgy in Niger, West Africa." *Journal of Archaeological Science* 15:367–94.

Lorig, C., and D. Krause. 1936. "Phosphorus as an Alloying Element in Low Carbon, Low Alloy Steels." *Metals and Alloys* 7:9–13.

McDonnell, G. 1989. "Iron and Its Alloys in the Fifth to Eleventh Centuries A.D. in England." *World Archaeology* 20:373–81.

Morton, G., and J. Wingrove. 1969. "Constitution of Bloomery Slags—Part I: Roman." *Journal of the Iron and Steel Institute* 207:1556–64.

——. 1972. "Constitution of Bloomery Slags—Part II: Medieval." *Journal of the Iron and Steel Institute* 210:478–88.

Muhly, F. and P. Schmidt. 1988. Filmscript for *The Tree of Iron.* Gainesville, Fla.: Foundation for African Prehistory and Archaeology.

O'Neill, P., P. Schmidt, F. Mulhy, and W. Lambrecht. 1988. *The Tree of Iron.* Film. Gainesville, Fla.: Foundation for African Prehistory and Archaeology.

Percy, J. 1864. *Metallurgy.* Vol. 2: *Iron and Steel.* London: John Murray.

Piaskowski, Jerzy. 1989. "Phosphorus in Iron Ore and Slag, and in Bloomery Iron." *Archeomaterials* 3:47–59.

Rehder, J. 1986. "Use of Preheated Air in Primitive Furnaces: Comment on the Views of Avery and Schmidt." *Journal of Field Archaeology* 13:351–53.

Rostoker, W., and B. Bronson. 1990. *Pre-industrial Iron—Its Technology and Ethnology.* Archeomaterials Monograph, No. 1. Philadelphia.

Rostoker, W., B. Bronson, and J. Dvorak. 1989. "Smelting to Steel by the Japanese Tatara Process." *Archeomaterials* 3:11–25.

Schmidt, P. 1975. "A New Look at Interpretations of the Early Iron Age in East Africa." *History in Africa* 2:127–36.

——. 1978. *Historical Archaeology: A Structural Approach in an African Culture.* Westfield, Conn.: Greenwood Press.

——. 1980. "Early Iron Age Settlements and Industrial Locales in West Lake." *Tanzania Notes and Records* 84–85:335–40.

——. 1981. *The Origins of Iron Smelting in Africa: A Complex Technology in Tanzania.* Research Papers in Anthropology, No. 1. Providence: Department of Anthropology, Brown University.

——. 1983a. "An Alternative to a Strictly Materialistic Perspective: A Review of Historical Archaeology, Ethnoarchaeology, and Symbolic Approaches in African Archaeology." *American Antiquity* 48:62–79.

——. 1983b. "Cultural Meaning and History in African Myth." *International Journal of Oral History* 4:165–83.

——. 1983c. "More Evidence for an Advanced Prehistoric Iron Technology in Africa." *Journal of Field Archaeology* 10:421–34.

——. 1990. "Oral Traditions, Archaeology, and History: A Short Reflective History." In *A History of African Archaeology,* ed. P. Robertshaw, 252–70. London: James Currey.

——. 1994. "Historical Ecology and Landscape Transformation in Eastern Equatorial Africa." In *Historical Ecology: Cultural Knowledge and Changing Landscapes,* ed. C. L. Crumley, 99–125. Santa Fe: SAR Press.

——. In press. *Symbolic and Material Views of Iron Technology in East Africa.* Bloomington: Indiana University Press.

Schmidt, P., and D. Avery. 1978. "Complex Iron Smelting and Prehistoric Culture in Tanzania." *Science* 201:1085–89.

Schmidt, P., and S. T. Childs. 1985. "Innovation and Industry during the Early Iron Age in East Africa: The KM2 and KM3 Sites of Northwestern Tanzania." *African Archaeological Review* 3:53–94.

——. 1995. "Ancient Iron Production." *American Scientist* 83(6):524–33.

Stead, J. 1915. "Iron, Carbon, and Phosphorus." *Journal of the Iron and Steel Institute* 91:140–81.

——. 1918. "Iron, Carbon, and Phosphorus." *Journal of the Iron and Steel Institute* 97:389–412.

Tholander, E. 1987. "Experimental Studies on Early Iron-making." Ph.D. diss., Royal Institute of Technology, Stockholm.

Todd, J., and J. Charles. 1978. "Ethiopian Bloomery Iron and the Significance of Inclusion Analysis in Iron Studies." *Journal of the Historical Metallurgy Society* 12:63–87.

Tylecote, R. 1986. *The Prehistory of Metallurgy in the British Isles.* London: Institute of Metals.

Tylecote, R., J. Austin, and A. Wraith. 1971. "The Mechanism of the Bloomery Process in Shaft Furnaces." *Journal of the Iron and Steel Institute* 209:342–63.

Tylecote, R., and B. Gilmore. 1986. *The Metallography of Early Ferrous Edge Tools and Edged Weapons.* Oxford: British Archaeological Reports, British Series No. 155.

van Grunderbeek, M.-C., E. Roche, and H. Doutrelepont. 1983. *Le Premier Age du Fer au Rwanda et au Burundi: Archeologie et Environnement.* Brussels: IFAQ.

Glossary

Air blast. In many kinds of African smelting furnaces the air required for combustion is forced into the combustion zone under pressure. Usually this pressurized air is supplied by bellows and is referred to as the air blast.

Anneal. In metallurgy this means to take a piece of metal through a cycle of heating, holding it at temperature, and then cooling it in order to modify its properties. It is done for a variety of purposes, but most often to soften a piece after cold working, during which the grains are compressed and under internal stress.

Austenite. This is the high temperature crystal form of iron and its iron alloys. It has the capacity to dissolve more carbon than the low temperature crystal form known as **ferrite,** which allows for the heat treatment of steels.

Bloom. This is a sometimes spongy (due to the interstitial spaces that are either voids or filled with slag) aggregate of metal formed in a smelting or refining furnace. Iron blooms contain varying amounts of slag. There are relatively few slag inclusions in the Haya process, for example, while the Barongo iron bloom contains much slag. Most of the slag is expelled under pressure of the hammer on the forge and also drains away within the forge. Finished wrought or bloom iron often contains slag inclusions.

Bloom iron. This is the solid iron product from the reduction of iron ore in a smelting furnace, often including iron, charcoal, unreduced ore, and slag. While some iron bloom, especially what was once thought to be "typical" African iron bloom, has a low carbon content, it is now clear that smelting processes such as that of the Haya produce a partially carburized metal similar to carbon steels. **Wrought iron** is another term used to refer to bloom iron when the emphasis is on the properties—such as degree of carburization—instead of the way it was produced.

Carburization. This occurs when carbon is deposited on the surface of an iron object during forging at a temperature high enough so that the carbon diffuses inward and combines or alloys with the iron to form a steel. It also occurs in smelting under reducing conditions when the carbon content of a steel is increased by diffusion of carbon into the alloy.

Cast iron. This is iron that contains between 2–4.5% carbon. Iron with this amount of carbon can become liquid at temperatures well below the melting points of wrought iron and steel and thus is easy to cast.

Cementite. This is an iron carbide in the microstructure of steel and cast iron. It appears in the grain boundaries of wrought iron (containing about .02 percent carbon) and is one constituent of pearlite in steel. The hardness of a steel or cast iron is proportional to the number and close spacing of the cementite crystals present.

Charge. The placement of ore, fuel, and flux (if any) into a smelting furnace.

Decarburizing. This is the removal of a proportion of carbon in a high carbon steel or cast iron. It occurs when cast iron in the liquid state has carbon removed (for example, in puddling). In Africa it can occur in the forge when high carbon iron blooms are submitted to long-term high temperatures and oxidizing conditions as well as in the annealing (heat treatment) of iron blooms that have an excessive carbon content.

Draft. This is usually the pressurized air that feeds combustion in a furnace. In the African context it commonly refers to furnaces in which an induced or convection draft is created by the density difference between the hot gases in a stack or chimney and the cold air on the outside, instead of the use of bellows.

Eutectic. This is a term that refers to an alloy that solidifies through the simultaneous crystallization of two or more metals or alloy components. A eutectic alloy solidifies at a lower temperature than any of its separate elements. Eutectic alloys are formed in iron with carbon (4.3 percent C), phosphorus (10.2 percent P) and many other elements. The creation of a eutectic alloy decreases the incipient melting temperatures.

Eutectoid. This is a term that denotes the carbon content of a steel at which the temperature of solid state transformation of austenite to ferrite and cementite is minimized. This transformation occurs at about 910°C. In pure iron, at the eutectoid composition (0.8 percent C), it occurs at only 723°C. In the heat treatment of steels the eutectoid temperature is important, for if they are not heated beyond the eutectoid temperature, then no structural changes will occur.

Fayalite slag. This is an iron silicate slag ($2FeO.SiO_2$). While characterized by silica and iron content, fayalite slags often contain more than 60 percent FeO. They are an essential component in most African iron smelting processes as they are fluid at temperatures ranging between 1200–1300°C. Modern blast furnace slags have little fayalite content.

Ferrite. This is the low temperature crystal form of iron. Ferrite transforms to austenite when heated above 910°C. It has the capacity to dissolve phosphorus but little carbon; this is why some steels have distinct zones of cementite—iron carbide—and pure ferrite.

Flux. This is a substance such as lime that is used in the smelting furnace to lower the flow temperature of slag. Materials such as the oxides of iron, manganese, sodium, potassium, and calcium can be used as fluxes.

Forge. This is both the shaping of iron or other metal by hammering as well as the facility—an enclosed or open-air stone anvil in Africa—where forging is done.

Goethite. This is a hydrated ferrous oxide, $HFeO_2$, a mineral sometimes used as an iron ore.

Hematite. Iron oxide (Fe_2O_3) commonly used as an iron ore.

Hercynite. A mineral that results from the high temperature reaction of aluminum oxides with iron oxides to yield hercynite ($FeO\ Al_2O_3$). The presence of large hercynite crystals, which form at 1470°C upon cooling, increases slag viscosity; hercynite is difficult to purge from the iron during the forging process, creating special problems in forging the iron/steel bloom.

Hot blast. In modern smelting processes this is the preheating of air before it is forced through the tuyere into the furnace. In Africa a hot blast is obtained by air passing though a hot tuyere inserted inside the smelting furnace. The increased temperatures that result in and around the combustion zone—including the reduction zones—cause smelting to proceed at a faster rate as well as obtain significant economies in fuel use.

Hypereutectoid. This specialized term refers to steels that are higher in carbon than the **eutectoid** composition. Hypereutectoid steels are harder and more brittle.

Inclusions. These are nonmetallic areas, often particles of slag, and sometimes infusible substances. A piece of worked bloom will often incorporate (see the Barongo example in chapter 5, this volume) many slag inclusions—despite heavy hammering and long forging.

Kaolin. This is the most refractory of the clay minerals, a hydroxysilicate of alumina. A clay with a high proportion of kaolin is necessary for heat resistant tuyeres that penetrate deep inside a smelting furnace to preheat the air blast.

Laterite. A residual, often concretized soil formed in the tropics, usually under conditions of heavy leaching. Laterites contain concentrations of iron and aluminum hydroxides; sometimes there is sufficient iron that some laterites can be used as ores.

Limonite. This mineral is a hydrated ferrous oxide used as an iron ore. It is often of yellowish color and mined from bogs.

Magnetite. This mineral (Fe_3O_4) is widely used as an ore in iron smelting. In its pure form it must be mixed with silica in order to form a slag and to be smelted successfully in African iron smelting furnaces.

Martensite. This is the hardest and most brittle crystal form of steel. When steel cools slowly past the transformation temperature, its austenite structure is changed to ferrite and pearlite. But when hot steel cools quickly, then the product becomes martensite. An increased carbon content increases the hardness of martensite, but it also increases its brittleness.

Nucleation. The initiation of crystallization at discrete sites in a molten mass.

Ore. This is a metal-bearing mineral that is used to produce metals after smelting and refining. Types of iron ore vary widely in Africa. One culture may have magnetite (Fe_3O_4) deposits, while a neighboring culture may have only a limonite or bog ore available for smelting.

Pearlite. This is one of the products of slowly cooled **austenite.** It has a structure of crystals of cementite and ferrite interleaved together. In modern production most steel products are shaped while in a pearlitic state. Pearlite can be reheated and quenched (thrust into water, to cool rapidly) to produce a martensite structure that is harder and stronger.

Prill. A droplet of slag or globule of metal.

Quenching. This process is used when iron is heated to **austenite** temperatures and is then thrust into water, cooling it rapidly. This causes much of the austenite to take the form of hard and brittle martensite.

Reduction. The chemical removal of oxygen from a compound.

Refractory. This is resistance to softening or melting at high temperatures, a property that is especially important in clay products used in iron smelting. The clays used in tuyeres inserted into furnaces for preheating, for example, must be highly refractory in order to stand up to the high temperatures inside the furnace as well as to resist the corrosive action of slags.

Roasting. The heating of ore in open fires. It induces cracking, expels water and sulfides, and deposits carbon in preparation for the smelting.

Sinter. This is a process in which microwelded bonds develop at contact points between solid particles. It occurs at elevated temperatures, forming homogeneous masses without melting. It occurs in many places in the bloomery process, starting in the ore within the furnace stack and continuing throughout the formation of the bloom.

Slag. This is a major by-product of iron smelting that occurs as a ceramic or glass composed of minerals and metal oxides. In Africa it is an important and active part of the smelting process. The slag that is left after smelting is derived from ore, fuel ash, fluxes (if added), and often portions of tuyeres and clay from the furnace superstructure and lining. Slag may contain pockets of sintered iron and therefore be selected for reuse in subsequent smelts (see chapter 5, this volume).

Smelting. Smelting uses thermochemical reactions between fuel and ore to reduce ore minerals to metal. Most smelting processes are carried out above the melting point of the metal, except in the case of iron. In traditional iron smelting, the key reaction involves the reduction of iron oxides by carbon monoxide at high temperatures; one of the keys to success in the African iron-smelting process is to maintain high temperatures at a reducing (CO) atmosphere.

Tempering. This is the final stage in the heat treatment of steel. The brittleness of a quickly cooled steel (Martensite) can be altered through reheating to a temperature between 200–400°C; the desired temperature is judged by observing the temper colors. This process decreases brittleness and can lead to various combinations of hardness and toughness.

Tuyere. In Africa this is usually a clay tube used to deliver air to the combustion zone of a smelting furnace or forge. In most furnaces using forced and induced draft the tuyeres project only a few centimeters into the furnace. However, in smelting furnaces in which tuyeres project up to 50 cm the tuyeres must be made of highly refractory clay capable of withstanding very high furnace temperatures.

Welding. This is the bonding of iron pieces at their surfaces by melting them together, which is fusion welding. Welding also occurs on the forge when microwelds form between adjacent surfaces while the pieces remain in a solid state. Welding on the forge is often aided by the natural fluxing of extruded FeO (wustite) and other artificially applied fluxes which prevent further reoxidation of the surface of the iron pieces and allowing them to be welded.

Widmanstatten structure. This structure occurs in steels that have been rapidly cooled from high temperatures around 1000°C. Ferrite or cementite grains precipitate along certain crystal planes to form a meshlike arrangement to which this term is applied.

Wustite. This is a mineral (FeO) that is an important transition oxide in the reduction of iron ore. In Africa it is found to be a significant component in some slags. As a component in bloomery iron it can also play a role as a welding flux on the forge—oozing out of the bloom to form a film that prevents reoxidation.

Contributors

Donald H. Avery is professor of engineering at Brown University, where he began his collaboration with Peter Schmidt in 1974.

Randi Barndon is completing her Ph.D. studies in archaeology at the University of Bergen, Norway.

S. Terry Childs is an archaeologist for the National Park Service in Washington, D.C., and is a Research Collaborator at the Conservation Analytical Laboratory, Smithsonian Institution.

Nicholas David is professor of archaeology at the University of Calgary.

Pierre de Maret is dean of the Faculty of Letters and Human Sciences at the Free University of Brussels.

William J. Dewey is an assistant professor of Art History at the University of Iowa.

Candice L. Goucher is an associate professor of Black Studies at Portland State University.

Eugenia W. Herbert is E. Novis Rodman Professor of History at Mount Holyoke College.

David Killick is assistant professor of anthropology and materials sciences and engineering at the University of Arizona.

J. E. Rehder is a retired professional metallurgist and past president of the Canadian Institute of Metallurgy.

Ian Robertson holds a master's degree from the University of Calgary and is now a Ph.D. candidate at Arizona State University.

Peter R. Schmidt is former director of the Center for African Studies and associate professor of Anthropology at the University of Florida.

Genvieve Thiry is a recent Ph.D. student at the Free University of Brussels.

Index

Page numbers in italics refer to illustrations.